Coastal Zone Management

HANDBOOK

Coastal Zone Management HANDBOOK

JOHN R. CLARK

Library of Congress Cataloging-in-Publication Data

Clark, John R., 1927-
Coastal zone management handbook / John R. Clark.
p. cm.
Includes bibliographical references and index.
ISBN 1-56670-092-2
1. Coastal zone management. 2. Coastal zone management--Case studies. I. Title.
HT391.C494 1995
333.78'4--dc20

95-10219
CIP

Direct all inquiries to CRC Press LLC, 2000 N.W. Corporate Blvd., Boca Raton, Florida 33431.

Visit the CRC Press Web site at www.crcpress.com

Lewis Publishers is an imprint of CRC Press LLC

International Standard Book Number 1-56670-092-2
Library of Congress Card Number 95-10219
Printed in the United States of America 3 4 5 6 7 8 9 0
Printed on acid-free paper

Preface

This volume is intended as a practitioner's guidebook. Although its viewpoint is clearly environmental, this is not an advocacy book. Nor is it a textbook. It is a manual and a reference source for those who practice the arts of coastal resources planning and management. It describes a strategy for comprehensive management and provides a set of working tools along with a detailed information base for the practitioner. This basic material is followed by 47 case histories on coastal management, ranging around the globe from Australia to the West Indies. How the book should be used is described in some detail in the "Reader's Guide" which follows this preface.

The guidebook is about environmental conservation from a practical point of view. It does not deal with nature abstractly but rather pragmatically. It was written from a human perspective and is highly influenced by the author's international consulting activities in developing countries over many years. Accordingly, it responds to an immediate global need for resource planning and management technology.

In a world of rapid population growth and diminishing natural resources, each country must plan for economic growth in balance with resource conservation and environmental management if it is to make progress in health, food, housing, energy, and other critical national needs. Such basic resources as fuel, water, fertile land, and fish stocks are already in short supply in many countries, and their future prospects are in grave doubt.

In the coastal areas of the world, high population densities, linked with urban growth, expanding tourism, and industrialization, pose major threats to natural resources and biological diversity. The effects of uncontrolled development are destabilizing ecosystems, changing land use patterns, making communities vulnerable to seastorms, and creating demands on the ecological resources of the world that are not sustainable.

This situation is expected to worsen. The current coastal population is likely to more than double in a few decades while resource uses accelerate—for example, international tourism is expected to reach 637 million travelers by the year 2000. Urbanization and tourism expansion will unavoidably lead to further pressure on water supplies and to irreversible changes in coastal environments. Growing amounts of gaseous, liquid, and solid waste also jeopardize the future of marine, coastal, and wetland ecosystems, as well as threaten species survival.

The major resource systems of the coast have no equal on land. Coastal ecosystems and key coastal habitats—such as coral reefs and mangrove forests—are not only distinctive, but also extremely productive of renewable resources—protein food, tourist income, mangrove forest products, and other economic goods and services. These resource systems must be conserved to continue. Sustainable use is the alternative to resource depletion that accompanies excessive exploitation for short-term profit.

Unfortunately, the water's edge is also a place where competition and conflict among users is great and where governments have failed to develop special policies and programs. The water influence not only establishes special conditions, but also dictates unusual and complex institutional arrangements. In most countries a great variety of agencies have interests in coastal waters, interests that sometimes are complementary but more often are competitive. An integrated coastal zone management (ICZM) program is needed to coordinate all the varied interests in coastal resource uses.

Where use of coastal resources is concerned, governments have the mandate to manage public commons wisely. In most countries coastal waters are considered "commons"; that is, they are not owned by any person or agency but are common property available equally to all citizens, with the government as "trustee." This is an age-old public right going back to the Institutes of Justinian: "*Et quidem naturali jure communia sunt omnium haec: aer, aqua profluens, et mare per hoc litora maris*"—"By the law of nature these things are common to mankind: the air, running water, the sea and consequently the shores of the sea" (158). Further, this influential doctrine states that "no one, therefore, is forbidden to approach the seashore, provided that he respects habitations, monuments, and buildings, which are not, like the sea, subject only to the law of nations."

To cope with the complexity of managing the coastal commons, many countries are now working out special ICZM strategies for compatible development and resource conservation management that are for the good of the nation as a whole. ICZM is committed to advancing sustainable multiple use of coastal resources and maintenance of biodiversity through an integrative, multiple sector approach. It may be initiated in response to a planning mandate but more often because of a crisis—a use conflict, a severe decline in a resource, or a devastating experience with natural hazards.

Environmentally planned development adds to the economic and social prosperity of a coastal community. Fisheries productivity, increased tourism revenues, sustained mangrove forestry, and security from natural hazard devastation are among the practical benefits of coastal zone planning and management. This holistic approval is supported by "Agenda 21", the report of the Earth Summit conference of June, 1992, sponsored by the United Nations.

A major purpose of coastal zone management is to coordinate the initiatives of the various coastal economic sectors toward long-term optimal socio-economic outcomes, including resolution of use conflicts and beneficial tradeoffs. My colleague, Chua Thia-Eng, urged me to remind coastal planners that existing poverty and increasing populations in many developing countries contribute strongly to resource depletion and environmental degradation. Therefore, alleviation of poverty and population increase should be addressed in coastal zone plans.

Because it operates at the water/land interface, every aspect of ICZM relates to water in one way or another, whether making provisions for marine commerce, the ravages of seastorms, resource conservation, or pollution abatement. The water influence not only establishes special conditions, it also dictates unusual and complex institutional arrangements.

Because the special conditions of the coast have not always been understood and given due regard by development organizations, program designers, economic planners, and project engineers, there have been losses of revenue, jobs, food, and foreign exchange earnings potential in many coastal countries that could have been avoided. The key is *unitary management.* The necessity is to comprehend and manage the shorelands and coastal waters together as a single interacting unit.

Fortunately, there is now a shift in political reasoning away from exploitive, non-sustainable uses, which may produce a fast dollar but which are costly in the long run because of resource depletion, rehabilitation, and management costs. The current trend is toward more comprehensive and broadly integrated coastal programs of the ICZM type. An effective comprehensive program can be the major force in any country for maintaining coastal biodiversity, for resolving conflicting demands over the use of coastal resources, and for guaranteeing the long-term economic sustainability of the coastal resource base.

Traditional land-based or marine-based forms of management and planning must be modified to be effective for the coast, at the transition between land and sea. But the place where the land ends is also the place where the knowledge and experience of most administrators ends. For example, seafood is recognized as a critically important resource (30); it provides more protein worldwide per capita (16.1 lb) than do beef and mutton combined (12.2 lb). Yet economic planners take slight notice, perhaps because seafood production is a mysterious pursuit carried out far from cities and incomprehensible to most persons. Also, fisheries are variable and, in large measure, unpredictable and unprogrammable.

From the natural sciences point of view, the coastal area is an extremely complex, highly diverse, and complicated system; a complexity that ensures a continuing high need for help from scientists in the coastal planning sphere. As Joseph Conrad said (*Typhoon,* 1902): "The sea never changes and its works, for all the talk of men, are wrapped in mystery".

From a planner's viewpoint, the coast is a place with an inherently incomplete database, where major influences—e.g., storms, erosion, fish migrations—are usually unpredictable and where land-based planning principles are irrelevant. Nor are managers, engineers, or politicians usually well informed about the sea and the seacoast. Moreover, persons who do have knowledge of the coast may possess quite narrow expertise—for example, in navigation, fisheries science, pollution, biodiversity, or hydrodynamics—and know the sea and seacoast from a specialist's point of view rather than from a systems perspective.

Those few practitioners who have prepared themselves broadly to know the sea and seacoast as a natural system and who understand the interactions of sea and coastal land are exceptionally valuable in coastal management programs. Only recently has formal graduate level training been available to those who would choose to make this interdisciplinary pursuit a career. I pray that this book will help them by presenting the basics on biodiversity, resource conservation, and social needs in a single volume.

Without apology, I wish to explain that this book is written by a generalist who has several decades of experience in coastal management but not of a doctrinal type. While this book is built on the knowledge of a multitude of experts, it is not intended to help any of them *within* their own particular area of expertise. The hundreds of topics addressed are presented for the benefit of planners or other generalists or for cross-disciplinary use.

Although I have tried to compile and present information of the most interest to the widest audience of practitioners, personal selection has admittedly influenced the inclusion of materials. Yet this book should provide the reader with a more comprehensive introduction to the practice of coastal management than any source available thus far.

John R. Clark
Ramrod Key, Florida

Reader's Guide

This is a reference book and is not intended as a reader. As a central information source for the practice of coastal zone resources management, the book is a compilation of information gleaned from numerous disciplines. The information is structured by four major categories, which together make up the substance of the guidebook: strategies, methods, information base, and case histories. A description of each of these four parts follows.

Part 1 is a description of the strategic approach to coastal zone management (ICZM), including concepts, problems, and solutions. Potential administrative/legal approaches are described along with the management framework. Strategy planning, including issues analysis, is held to be the key to the process. The material is presented in progressive order, beginning with goals and ending with program development. In this way, Part 1 becomes the "blueprint" for an integrated, multiple sector, coastal management program. The "tools" for building the program are given in Part 2.

Part 2 is a presentation of specific management approaches. This is the what-to-do and how-to-do-it part of the book. It describes primary techniques, gives planning instructions, and proposes specific management outcomes. Whereas Part 1 provides the blueprint for regional or national ICZM programs, Part 2 contains the tools for conducting the program. Subjects are listed alphabetically. With the ideas and instructions here, you can address most management jobs. And while extent of coverage is limited to what will fit within the covers of a single book, the references cited will lead you to a much wider range of technologies. Further information also appears in Part 3.

Part 3 is a general compilation of information about coastal resources and management, arranged alphabetically. It consists mostly of material intended to support the coastal management processes explained in Part 2, but some of it describes additional management techniques (at the second level of need). This is because the separation of subject matter—techniques in Part 2 and supporting information in Part 3—was only partly successful. Each subject is an abstract of a much larger literature and, in so small a space of words, it can only be exemplary, not prescriptive. If a desired subject is not listed under the title you would use for it, please refer to the index.

Part 4 is a collection of mini-case histories of coastal zone situations along with management techniques and their trials. Subjects include conservation, land use controls, pollution abatement, resource management, public reaction, and environmental restoration. Subjects are arranged by country, alphabetically. Most are contributed by experts whose authorship is noted at the end of each case.

Part 5 contains the references cited in Parts 1 to 3. This book uses a numeric citation system; that is, each reference is identified by a number within parentheses. For example, the citation *(18)* indicates the 18th item in the alphabeticized reference list. When the reference appears past the last sentence of a paragraph it refers to the whole paragraph. Institutional authorship is used when personal authorship is not given in the publication referenced.

The reader will find that the same subject is addressed in different parts of the guidebook. This occurs first in Part 1 where the book follows the process of generating an integrated coastal program—here the same items are considered in various stages of program planning and development. Also, a particular subject may be treated in both Parts 2 and 3 because it has specific management application for Part 2 but also additional information content for Part 3. However, care has been taken to minimize direct duplication, *except* where important material is *intentionally* repeated either for emphasis or to make sure the selective reader will encounter it in one part or the other. ICZM is used throughout this book to mean Integrated Coastal Zone Management.

The Author

JOHN R. CLARK has devoted his career to research and conservation of coastal zone resources since graduating from the University of Washington in 1949 and going to the Woods Hole Fisheries Laboratory. He continued in the 1960s as assistant director of the Sandy Hook Marine Laboratory in New Jersey (1960–1970), as director of the Narragansett Marine Laboratory (1970–1971), and as senior associate with the Conservation Foundation, Washington, D.C. (1972–1981). In Washington, his involvement with coastal zone conservation policy matters intensified and he was influential in the formation of national coastal environmental policy, particularly in the new federal Coastal Zone Management and Clean Waters programs. During the 1960s and 1970s he was active as the environmental witness in major court cases.

Returning to government, he continued his career with the Interior Department (International Affairs Office of the National Park Service), where he dealt with research, planning, and training projects in coastal environments, renewable resource planning and management, and the creation of nature reserves, all for developing countries (1982–1987). Now retired from government, he is an international consultant in coastal resources conservation and is affiliated with several institutions including the Mote Marine Laboratory of Sarasota, Florida, and the School of Marine and Atmospheric Sciences (RSMAS) of the University of Miami. He has worked in more than 30 countries.

He has authored 26 books and major works, 39 reports, and 137 papers. He is currently listed in *Who's Who in America* and has won several professional awards. Areas of current activity include research and technical assistance in international coastal zone management, sustainable development planning, environmental impact assessment and environmental auditing, marine reserves/parks design and management, coastal biodiversity maintenance, fishery habitat conservation, practitioner training, information transfer, and community-based conservation.

Acknowledgments

Writing the acknowledgments for a book such as this is a daunting task. I was helped by hundreds of colleagues, literally, and each deserves high praise. I tried to give credit to each person at the place in the book where his or her contribution occurs. But my debt of gratitude extends beyond these valuable contributions to those many colleagues who also offered guidance and encouragement and provided me with valuable sources of information over the years.

Two colleagues stand out as special partners in this project—Jens Sorensen of the University of Rhode Island and James Maragos of the East/West Center (Hawaii). A third partner, my brother Donald R. Clark, rescued me during the critical stage of manuscript preparation. A fourth partner, Bill Keogh was exceedingly generous with his time and talent; he provided many high quality photographs and processed dozens of others. The fifth partner was my wife, Catherine Clark, who labored mightily on this book for years and whose very presence from the beginning to the end was a godsend.

Others to whom I owe a big debt of gratitude include Carl Kittel who wrote on aquaculture, Russell VernonClark who advised on chemistry, Kay Hale who conducted countless library searches, Alan White who helped so much with writings and references, and Peter Burbridge and Peter Saenger, both of whom supplied essential information and lifted my spirits.

Of the dozens of others who helped, I make special mention of: Alec Dawson Shepherd, Timothy Kana, Marc Hershman, Robert Knecht, Tomas Tomasik, and James Spurgeon. Natalie Forsyth created much of the copy, laboring long hours at the word processor. And I received extraordinary support from Gerry Jaffe and the rest of the Lewis Publishers staff.

Table of Contents

PART 1. MANAGEMENT STRATEGIES

PART 3. MANAGEMENT INFORMATION

PART 4. CASE HISTORIES

PART 1

Management Strategies

1.1 INTRODUCTION

Part 1 presents the process of integrated coastal zone management (ICZM) by describing environmental impacts that need to be controlled and by outlining the variety of available methods. The methods underlie major program aspects such as management solutions, strategy planning, master plan creation, and program development. Part 1 thus provides a framework within which conservation of coastal resources can be accomplished.

Coastal zones are unique. Such things as daily tides, mangrove forests, coral reefs, tidal flats, sea beaches, storm waves, and barrier islands are found only at the coast. Because of these features and because coastal enterprise is also distinctive, most countries recognize the coastal zone as a distinct region with resources that require special attention. In all but the larger coastal settlements, the coastal location creates distinctively maritime cultures.

The transitional strip of land and sea that straddles the coastline contains some of the most productive and valuable habitats of the biosphere, including estuaries, lagoons, coastal wetlands, and fringing coral reefs. It is also a place of natural dynamism where huge amounts of natural energy are released and a great abundance of life is nurtured. It is a place of high priority interest to people, to commerce, to the military, and to a variety of industries. Because it contains dense populations, the coast undergoes great environmental modification and deterioration through landfill, dredging, and pollution caused by urban, industrial, and agricultural development.

The land can strongly affect the sea. Impacts on coastal ecosystems from terrestrial activity include industrial and agricultural pollution; siltation from eroded uplands; filling to provide sites for industry, housing, recreation, airports, and farmland; dredging to create, deepen, and improve harbors; quarrying; and the excessive cutting of mangroves for fuel. The impacts affect community security (from seastorms), tourism revenues, biological diversity, and natural resources abundance.

Where fisheries are important for food and income, the effects of pollution and the physical destruction of habitats can be crucial, particularly for those species depending on coastal wetlands and shallow nearcoast waters for breeding and nurturing of their young. In many parts of the world, the construction of dams has blocked the passage of marine species migrating to inland spawning sites.

The coastal area, or coastal zone, is defined by Sorenson and McCreary (326) as the interface or transition zone, specifically "that part of the land affected by its proximity to the sea and that part of the ocean affected by its proximity to the land . . . an area in which processes depending on the interaction between land and sea are most intense." But it must be noted that the border between land and sea is not fixed—it changes daily with the tides, with the moon stages, seasonally with astronomic forces, and sporadically with seastorms and great river floods.

The coastal zone may be drawn wide or narrow in order to meet program goals. For example, it could embrace a wide band of shorelands or a quite narrow strip. It can include coastal islands and the shallower nearshore coastal waters, but it could in some cases extend to the outer edge of the continental shelf.

But the coastal zone always includes the intertidal and supratidal zones of the water's edge that include coastal floodplains, mangroves, marshes, and tideflats, as well as beaches and dunes and fringing coral reefs. This is the place where agency authority changes abruptly, where storms hit, where waterfront development locates, where boats make their landfalls, and where some of the richest aquatic habitat is found. It is also the place where terrestrial-type planning and resource management programs are at their weakest. It is the core of the coastal zone.

Seaward from the tidal limit, the coastal waters are a "commons" (in nearly all countries) and have been for millennia, as per the Institutes of Justinian 2.1.1, (158): "By the law of nature these things are common to mankind—the air, running water, the sea and consequently the shores of the sea." This commons, which is always under central governmental authority, is often the main focus of attention in coastal resources conservation.

Most governments have some variety of environmental, resource management, and development control programs. These may include pollution control, natural hazards management, biodiversity maintenance, environmental assessment, wetlands protection, and so forth. But these programs are typically operated by a variety of agencies and are uncoordinated, with the result that each agency goes its own way, disregarding the others. Nor is there much coordination with various private sector enterprises or with the recognized non-government organizations (NGOs). This non-coordinated and non-integrated situation is inefficient at solving coastal zone problems.

For these reasons many tropical countries are now working out special integrated coastal zone management (ICZM) strategies, and some have already begun to adopt such programs. ICZM establishes a process whereby government intervention can be organized, informed, and effective through programs that are integrated with the various economic sectors and resource conservation programs. The advantage of the ICZM (multiple use) approach over the traditional sectoral (single use) approach is that it provides a framework for broad participation and for resolution of conflicts between a variety economic development and resource conservation needs (28).

Economic development planners, particularly, must recognize that modification of the land area (e.g., land clearing and grading) has a high potential for adverse effects on lagoon, estuarine, and littoral resource systems. It follows that ICZM must address land modification activities, principally those associated with site preparation for development (84).

The distinctive feature of an ICZM program is the fact that it is multi-sectoral and that it seeks to integrate or coordinate activities of existing users. Such efforts have significant political and managerial dimensions. They also have significant legal and managerial dimensions in that an ICZM program must have jurisdictional scope and a clear legal track if it is not to run immediately into conflict with existing jurisdictional powers.

Indeed, it could be argued that the failure or refusal of many countries to adopt an integrated approach to coastal zone management is precisely because of the difficulties of disturbing the plethora of institutions that already regulate activities in coastal areas. But an integrated approach is always preferable, whether accomplished through ICZM or by other means (134).

In narrow sectoral planning it is easy to irreversibly destroy a resource and foreclose future options for use of that resource. ICZM attempts to avoid this for coastal area resources by broad multiple-sector planning and project development, by future-oriented resource analysis, and by applying the test of sustainability to each development initiative. In no other part of the earth is integrated, multi-sectoral, resource planning and management more needed than at the coast.

In the future, more government intervention in development and resource conservation will be needed because of the complex management issues that are emerging in the coastal commons: the growth in coastal populations and the social demands that will arise; industrial development; increasing pressure on the coast to supply seafood, tourism revenue, and other needs; and the intense competition and conflicts that come with coastal area crowding. A balanced approach to government intervention in coastal management is needed.

In summary, ICZM is a *unitary program*—it has to both manage development and conserve natural resources and, while so doing, it has to integrate the concerns of all relevant sectors of society and of the economy. Also it is most important that coastal economic development be generated for the people of a country, not just for those who are already rich and powerful.

1.2 MANAGEMENT GOALS AND PURPOSES

As planners sort out the problems, opportunities, and issues of the coastal zone, those which most naturally fall under an ICZM-type program become obvious. The tableau will vary from country to country according to conservation needs, traditions, norms, and governmental systems. But compatible multiple use objectives should always be encouraged with the goal of strengthening the program, improving its efficiency, and guaranteeing the greatest benefit to coastal communities through equitable sharing of resources.

To provide an example, the following problems were identified by the Philippines government as those to be addressed in an ICZM-type program (234):

- Natural resources degradation: (a) beach erosion; (b) conversion of mangrove swamplands into other land uses; (c) landfill or reclamation of foreshore areas; (d) dynamite fishing; (e) overfishing; and (f) overexploitation of mangrove forests.
- Pollution: (a) industrial sources (industrial waste); (b) domestic sources (household wastes and solid wastes); (c) agricultural sources (pesticide and fertilizer runoff); and (d) other sources (dredging activities).
- Land use conflicts: (a) absence of access to foreshore lands due to human settlements encroachment; (b) unusable beach areas due to excessive pollution; and (c) conservation and preservation of mangrove areas versus conversion of the same into fishponds, or reclamation of the same for human settlements and commercial purposes.
- Destruction of life and property by natural hazards: (a) flooding due to storms; (b) earthquakes; (c) tropical cyclones, and (d) tsunamis.

While it should be efficient to organize a comprehensive ICZM-type program to solve this array of problems, many countries are now at the single-purpose level of management where different agencies have jurisdiction over different resources or development activities. This often results in a lack of intersectoral coordination, key information, intergovernmental liaison, support or political will from the top, and professional resources (326). Such difficulties can be eased by an ICZM coordinating and integrating approach.

The linkages between the "dryside" and "wetside" of the coast—that is, between the terrestrial and the marine realms—precludes the effective management of a marine or estuarine resource system without concurrent management of adjacent land habitats. Therefore, the proposition that coastal ecosystems include both dryside (land) and wetside (water) components and that they should be managed together is considered fundamental. Therefore, the *planning* boundaries for coastal zone planning and management are set to encompass dryside problem areas as well as wetside ones.

There are several driving forces and areas of misunderstanding that can be identified among ICZM-related issues. For example, Chua and White (55) list the following:

- high rates of population growth;
- poverty exacerbated by dwindling resources, degraded fisheries habitats and lack of alternative livelihoods;
- large-scale, quick-profit, commercial enterprises that degrade resources and conflict with interests of the local people;
- lack of awareness about management for sustainability among local people and policy makers;
- lack of understanding of the economic contribution of coastal resources to society; and
- lack of serious government follow-up in support and enforcement of conservation programs.

The technology by which to counter these forces and guarantee sustainability of coastal resources, as well as the method for its application, is available and is the subject of this book.

1.2.1 SUSTAINABLE USE OF RESOURCES

Ecologically, impacts from any uncontrolled development activity located anywhere near coastal areas—in watersheds, floodplains, wetlands, tidelands, or water basins—have the potential to deplete fisheries and other coastal resources. Coastal resources conservation policy should emphasize that it is in the best interests of the country to achieve sustainability of its resources and long-term protection of its natural assets.

The criterion for sustainable use is that the resource not be harvested, extracted, or utilized in excess of the amount that can be produced or regenerated over the same time period. It is important to learn the acceptable limits of coastal environmental degradation and the limits of sustainability of coastal resources. One way is to determine the carrying capacity of an ecosystem and manage it always to remain above the minimal limit (see "Carrying Capacity" in Part 3).

To cite an example problem, sustainability of coastal fisheries is often threatened because the lagoons, estuaries, and other enclosed water bodies that nuture the young stages of so many species have been rapidly degraded by coastal development. Fishery catches are declining worldwide and, in fact, the present levels of seafood production will be difficult to maintain in the absence of effective environmental management. This applies to all levels of fishing from artisanal to industrial (Figure 1.1).

1.2.2 BIOLOGICAL DIVERSITY

Biological diversity conservation is an urgent coastal matter. Thousands of species and subspecies of wild plants and animals are threatened with extinction. The most serious threat is habitat destruction. This destruction takes many forms: (a) the replacement of entire habitats by settlements, harbors, and other human constructions, by cropland, grazing land, and plantations, and by mines and quarries; (b) the effects of dams (blocking spawning migrations, drowning habitats, and altering chemical or thermal conditions); (c) drainage, channelization, and flood control; (d) pollution and solid waste disposal (from domestic, agricultural, industrial, and mining sources); (e) overuse of groundwater aquifers (for domestic, agricultural, and industrial purposes); (f) removal of materials (such as vegetation, gravel, and stones) for timber, fuel, construction, etc.; (g) dredging and dumping; and (h) erosion and siltation. (308)

The loss of species has many dimensions. There are ethical considerations regarding loss of life and the prerogative of the human species to eliminate other species from this planet. There are economic speculations about the potential use of species. Organisms whose properties have not yet been investigated may be important as sources of drugs, or as food, or as raw materials for the emerging field of

Figure 1.1. Artisanal fishing provides the main support for numerous coastal communities such as this village in Thailand. (Photo by author.)

biotechnology. Many not yet described species may possess novel biological properties that may help us understand how nature works. They may also play unique roles in many ecosystems. (320)

A major need for biodiversity maintenance is protection of special, or critical, littoral habitats, including mangrove forest, coral reef, seagrass meadows, shallow water bodies like brackish lagoons, and beaches (308). While it is useful and practical to focus on individual habitat types or species, one must not forget that they exist only as components of wider coastal systems. The complexity of biotic systems and the interrelatedness of their components require that each coastal water ecosystem be managed as a system. Neither piecemeal management nor treatment of single components or single species will fully succeed.

1.2.3 PROTECTION AGAINST NATURAL HAZARDS

The coastlines of many countries face distinctly high risks of damage from certain types of natural disasters. The leading concern is death and property loss caused by the winds and waters of hurricanes, or cyclones. These violent storms born at sea strike the coast with winds up to 200 mph and storm surges to 20 feet or higher. Many coastal nations have developed mechanisms to cope with such threats to coastal communities through controls on type, density, and location of settlement (Fig. 1.2).

A combined approach to hazards protection and resource conservation simplifies the process of management and leads to more balanced decisions on what constitutes acceptable development. Note that the same setback requirement that protects beachfront structures from erosion and storm waves can also preserve turtle nesting sites on the back beach. Similarly, a restriction on clearing of mangrove swamps will not only conserve an economically valuable resource, but also maintain a physical defense against storm waves. (84)

Along many densely populated coastlines, the risks of natural disasters are being increased by population growth and unmanaged development projects. Coastal people become more susceptible to natural hazards such as floods, typhoons, or tsunamis when projects encourage settlement in dangerously low-lying areas or when land clearing and construction removes protective vegetation, reefs, or sand

Figure 1.2. Coastal countries threatened with cyclones (hurricanes) must take special precautions to protect life and property. Arrow in above satelite image shows Hurricane Allen located between Mexico and Florida on August 7, 1980. (Satellite image courtesy of U.S. National Oceanic and Atmospheric Administration.)

dunes (326). It is particularly important to maintain these natural landforms that take the brunt of storms and protect lives and property.

An example is Bangladesh, where more than 130,000 people were lost in one major storm/flood event in 1991. Now Bangladesh has committed large sums of money to maintaining a wide "greenbelt" of mangroves and other trees along its entire coastline (1,760 km), including the planting or replanting of more than 30,000 hectares. (See "Green Belt" in Part 3.)

1.2.4 POLLUTION CONTROL

In most countries, a pollution control authority has been established, is operating well enough, and should usually retain continuing general responsibility for pollution. In this situation, ICZM would most likely coordinate its main activities with the pollution authority to ensure conformance of coastal development projects with national pollution guidelines and standards.

But ICZM in its project review and environmental assessment role would discourage or modify projects with unacceptable potentials for pollution. These would be challenged and only permitted when they conform to water quality standards. ICZM programs should focus on special coastal pollution sources, such as those caused by new development and, especially, those that affect ecologically critical natural areas.

1.2.5 ECONOMIC DEVELOPMENT MANAGEMENT AND PLANNING

An important test of an economic development program is whether its impacts cause a long-term loss rather than a gain in social welfare and economic conditions. Another test is whether the program is planned to accomplish broad multiple use opportunity—that is, whether the program ensures that fisheries, manufacturing, tourism, habitat protection, housing, shipping, and so forth, can all prosper in an optimum mix of economic activity.

In most parts of the world, renewable coastal resource uses tend to be economically limited (316). Over time, the economic demand for a given resource will commonly exceed the supply, be it arable land, fresh water, wood, or fish. Sustainable use management, achieved by conservation practices, ensures that renewable resources are not jeopardized but remain available to us and future generations.

Therefore, ICZM should relate to the broader national economic planning process through components on resources planning, land use planning, and economic planning. Advice should be available from ICZM for government agencies, developers, and other stakeholders and to the general public regarding economic development options. The ICZM entity should also transmit to the national planners any coastal needs for legislation, new programs, improved policy base, and interagency coordination.

Regarding effects on the landed class, Lowry (222) notes that in Sri Lanka private owners of coastal land argue that ICZM is depriving them of ownership, particularly by the imposition of coastal setbacks. The government has been accused of not only depriving landowners of their rights, but also impeding economic gains that could be realized under a more "unfettered" policy of coastal tourism development.

The argument has also been made that coast protection should be left to individual landowners. They take the risks, it is argued; therefore, they should be prepared to pay for the consequences. But only if coastal landowners are prepared to assume the full public and environmental costs of their coastal development actions could a strong case be made for a privatized approach. Thus far that has not been the result. (222)

The Sri Lanka approach to coastal management generally, and to setbacks in particular, is based on a justification for government intervention in private markets drawn from economic theory. The rationale is that private shoreline protection structures can create negative externalities and that one basis for modern management is externality control. Specifically, a groin or revetment may indeed reduce or prevent coastal erosion at the site where constructed but do harm to other sites. Because Sri Lanka's shoreline is so dynamic, the construction of a coastal protection structure may interfere with coastal dynamics and increase erosion someplace else.

Developers must be prepared to pay the full social cost for erosion protection structures at these sites. Until they are, the private approach to coastal protection is fundamentally flawed. Developers profit by imposing costs on others and these frequently end up being borne by the government.

1.2.6 ENHANCEMENT OF SOCIAL WELFARE OF COASTAL COMMUNITIES

How the coast is managed and who benefits is important from a socio-economic standpoint. For example, in rural communities, many people do not receive wages or salaries, but rather depend on activities like subsistence fishing or small scale farming or trade for their survival. An increase in the nation's per capita income is unlikely to affect these people much—improvement for them would mean the ability to continue the activities on which they rely and modest improvements in health and education. Actions that close future options by concentrating on large scale economic developments may create more problems than they solve. When, for example, industrial fishing is developed to the point of squeezing out the artisanal fisherman, the net result to the community can be negative, despite officially claimed economic growth.

Communities react to development in several ways: new allocations of human resources may occur, and the pre-existing utilization of natural resources may be modified. Cultural patterns may change as a result. Increased contact with people from outside the community—their behavior, ideas, and commercial goods—as part of planning, funding, and execution of a development project frequently sets many more cultural changes in motion (94). These potential effects should be considered in ICZM planning.

1.2.7 OPTIMUM MIX OF USES

Integrated planning for coastal areas—including land use, resources, and pollution management—is needed to resolve competition and conflict that occur frequently among residential, tourist, commercial, industrial, transportation, recreational, and agricultural activities competing within limited space (66). Certain sectorally oriented uses might coexist in a multiple use approach while others might not or would have to be severely restricted. ICZM's role is to sort out the uses and recommend the optimal mix.

In marine areas, jet skis, fishermen, dive boats, ports, marinas, yacht anchorings, water sports, factories (Figure 1.3), oil terminals, waste outfalls, and transportation systems are examples of multiple

Figure 1.3. Coastal zones are a favorite location for many kinds of factories because of the ease of shipping and of liquid waste disposal. (Photo by author.)

uses. Tourism can intensify both resource and land use competition and conflict of these uses. The closer the physical relationship between marine resources and coastal tourism, the greater the likelihood that resources will be depleted (78) and the more important is the adoption of some form of ICZM (66).

Burbridge (31) makes a compelling argument for multiple use strategies for coastal resources in demonstrating that a unique feature of the ecological systems of the coastal zone is the wide variety of functions they serve and the broad array of economic and environmental goods and services they provide. In the case of mangroves, over 70 direct and indirect uses arise from mangrove forests. In addition, the mangrove forest acts as a nurturing, feeding, or spawning area for commercially valuable fish and shrimp species. These and other services cannot be replaced at a reasonable cost. (31)

For example, there is little problem in combining fisheries, tourism, and water supply, but mixing port development and tourism with critical area protection (e.g., mangrove forests) may be very difficult. The ICZM planning unit's job is to evaluate the effects of uses in certain combinations in order to advise the decision-makers and managers.

1.3 DEVELOPMENT IMPACTS

The history of coastal occupancy and coastal development around the globe shows a pattern of depletion of coastal resources and loss of biodiversity. Critical habitats have been destroyed, ecosystem processes disrupted, and waters heavily polluted. But with the ecological knowledge that we have now accumulated and the techniques that are available for coastal zone management, this trend can be reversed. Appropriate actions are available by which coastal communities can both conserve their natural resources and invigorate their economies.

A problem in detecting resource damage and lowered biodiversity is that many critical marine habitats are often not visible or evident to most observers. To take one example, submerged seagrass meadows are a major marine habitat and ecological component of shallow tropical coastal waters, but most people are not conscious of their existence, much less of their important role. Consequently, they are being depleted by widespread dredge and fill activities and by water pollution, including brine disposal from desalination plants and oil production facilities, waste disposal around industrial facilities, accidental spill of petroleum and petroleum products, and thermal discharges from power plants. The loss of seagrasses—an important habitat and source of nutrition—can cause a significant loss in marine life and fisheries production. Because so much of the damage is unseen or unrealized, it is often overlooked. The same may be said for coral reefs (usually submerged) and other undersea features.

In this section, examples are given of some of the more prevalent sources of environmental impact to the coastal zone.

1.3.1 GENERAL ISSUES

Shorelands of the coastal zone are used for human settlement, agriculture, trade, industry, and amenity and as shore bases for maritime activities such as shipping, fishing, and sea mining. These various uses of the coast are not always compatible and may result in a wide array of conflicts and problems for resource users and decision makers.

Coastal resources are also affected by activities far distant from the coast, such as the discharge to coastal seas of pesticides, heavy metals, coliform bacteria, and other harmful substances; changes in salinity regimes as a result of the damming of rivers; and siltation due to deforestation and cultivation. Coral reefs have been destroyed and shellfish beds degraded by siltation, increased nutrients from nearby sources, and sewage and agricultural runoff.

Coastal waters are the "sink" for the continents; they receive and concentrate all kinds of pollutants—sewage, pesticides, factory wastes, garbage, street runoff, waste oil, and so forth. Pollution issues arise in all coastal nations, irrespective of the degree of development or variation in environmental and socio-economic conditions. But the greater the rate of economic development, the greater the threat to environmental resources.

Lagoons, estuaries, wetlands, and shallow nearshore waters are particularly vulnerable (Figure 1.4). Every coastal nation with a major metropolitan area bordering an estuary seems to have a pollution problem, usually as a function of municipal sewage, industrial toxins, and/or polluted land runoff entering a lagoon or estuary (326).

There may also be pollution from dumping at sea and an array of conflicts due to the interaction among user groups in congested marine areas (66). Then, too, nearly every coastal nation that actively harvests its coastal fishery stocks seems to have an overfishing problem. Coastal nations with intertidal forest almost always experience mangrove forest depletion from aquaculture, pollution, filling, and the overharvesting of timber for fuel.

1.3.2 AGRICULTURE

The industrialization of agriculture with increased use of chemical inputs poses a threat to coastal environments. Unless there are effective controls, excessive amounts of chemicals run off the land into coastal water bodies. Here they contribute to eutrophication and toxification of river, coastal, and marine areas. Because of the seepage of excess nitrates into surface and groundwaters, the increased use of fertilizers plays a determining role in the eutrophication of rivers and coastal marine areas. Industrial products (e.g., pesticides, herbicides, and fertilizers) are being used more frequently in agriculture, while agricultural produce is undergoing various processing operations involving chemicals. Pesticides can kill or debilitate marine life when they occur in the water in very small concentrations.

Conversion of coastal wetlands and lowlands to rice farming (Figure 1.5) and other agricultural uses should be accompanied by suitable controls to prevent resource conflicts. Uncontrolled coastal agriculture has potentially harmful side effects on coastal environments and natural resources. Examples are runoff pollution by agricultural chemicals and elimination of mangrove forests or other critical wetland habitats. Also, the potential of reclaimed wetland soils may be low because of acid-sulfate soils and waterlogging.

In addition, flooding of coastal lowlands from rainstorms and cyclonic seastorms will be exacerbated by the clearing and draining of coastal areas and uplands. Such floods destroy property, risk lives, drown crops, and carry huge amounts of sediment, fertilizing substances, and organic and chemical pollutants into coastal waters.

Because these impacts can originate in farmlands a long distance from the coast, they may be difficult to identify and control for the benefit of coastal environmental resources. Yet they must be addressed, or lagoons, coral reefs, shellfish and seagrass beds, tourism, and sea fisheries may suffer.

1.3.3 AQUACULTURE

Shrimp farming and other marine aquaculture straddles the land-sea boundary, utilizing land (shrimp ponds), wetlands (*tambaks*), and the sea (salmon cages or artificial reefs). Aquaculture is one of the fastest growing sectors of the coastal zone economy. Socially, its product is seen as a supplement to

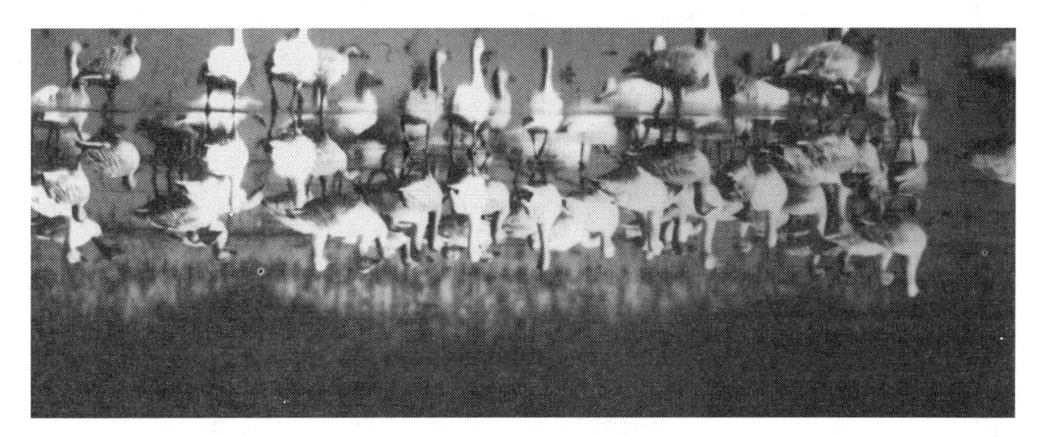

Figure 1.4. The life of coastal lagoons and wetlands is particularly threatened by pollution and inappropriate land use practices. (Photo courtesy of U.S. National Park Service.)

Figure 1.5. Coastal agriculture, such as this rice field in Bali (Indonesia), should be controlled to prevent pollution of coastal waters and loss of biological diversity. (Photo by author.)

local diets and as a valuable means of earning foreign currency through export. Financially, it is seen as a fast profit earner (Figure 1.6).

However, if aquaculture expansion is not controlled, the long-term consequences may be to reduce resource values and to pollute shallow coastal waters. Aquaculture may be either a polluter or a petitioner for a clean environment. On the one hand, success in aquaculture requires a clean water supply from the environment and, on the other hand, intensive aquaculture practices themselves may pollute coastal waters and reduce biodiversity. Also, uncontrolled aquaculture expansion may preempt mangrove and other critical habitats, which would then reduce the reproduction of marine species, including those used in aquaculture.

Aquaculture usually requires the conversion of land and water areas. "Grow-out" pond construction, for example, usually requires conversion of natural coastal lowlands or mangrove forests to open water impoundments. The presently favored management response to requests for additional construction of ponds is to recommend against expansion if intensification of use and increased production from existing ponds is feasible (315). (For details see "Aquaculture Management" in Part 2 and "Aquaculture" in Part 3.)

Many of the same problems arise in solar salt production in open coastal ponds.

1.3.4 FOREST INDUSTRIES

The removal of forest cover in watersheds increases sediment loadings of rivers (see "Watershed and Upland Effects" in Part 3). Water quality is reduced and siltation increased (Figure 1.7). The enhanced runoff may bring chemical pollutants of many types to the coast. A most serious effect is "smothering" of coral reefs and submerged vegetation (seagrasses and "seaweed" beds) of key importance as fish habitat. Sediment runoff also leads to smothering of shellfish such as oysters and clams and reduces the vigor of life in coastal waters.

Maritime forests—those that grow in salt water—are especially important. It is well known that coastal mangrove forests are a major ecolological resource. They nurture an abundance of plants and animals (e.g., crustaceans, molluscs, fishes, and birds), which form a significant part of the mangrove

Figure 1.6. Sorting clam seed stock in the hatchery of a coastal shellfish aquaculture facility. (Photo by author.)

resource, indicating their important contribution to marine biodiversity (see "Mangrove Forest Resources" in Part 3). In addition, mangrove forests stabilize the silt deposited in major deltas and attenuate cyclone (hurricane) surges, saving lives and property. They can be harvested for fuel wood or construction materials if done so sustainably. Also, they can supply many side products, such as tannin, honey, and thatch.

Mangrove ecosystem deterioration, caused by intensive human activities aimed at maximizing short-term gains, and the negative economic consequences on the coastal communities, in both the short and long term, are causing increased concern in many countries and international agencies, resulting in urgent calls for conservation. The other lowland and wetland forest types, such as nypa palm, hardwood hammocks, and freshwater tidal swamps, also need management attention.

1.3.5 HEAVY INDUSTRY

Coastal locations for heavy industry pose a variety of threats that extend considerably beyond the plant sites. For example, a new factory may require dredging of a deep-water channel (which raises a number of ecological problems; see "Dredging Management" in Part 2) or put increased pressure on infrastructure—waste disposal, water and electricity supply, land and air transportation links=each with its own potential for ecological disruption. Also, coastal factory sites often preempt mangrove forests or other critical habitats, which are obliterated by seawalls, docks, landfills, car parks, and buildings (see "Industrial Pollution" in Part 3).

The wastewaters from coastal heavy industries can seriously jeopardize coastal ecosystems. These impacts range from relatively minor disturbances (such as temporary, localized turbidity increase) to major disruptions (e.g., water pollution caused by discharge of toxic chemicals). But much of the industrial pollution can be eliminated by the application of existing, affordable, waste treatment technology. Also, many estuaries, lagoons, and bays that are now polluted can be successfully reclaimed if the wastes entering them are disposed offshore or are adequately treated before discharge.

"Spin-off" industries may often be attracted to the coastal zone by the development of large factories and may be environmentally damaging, imposing higher costs on the community for streets, police,

Figure 1.7. Clear-cut stripping of coastal watershed slopes should be controlled to prevent pollution and hydrological disruption of coastal ecosystems during forest harvest operations. (Photo courtesy of Henry J. Gratkowski.)

fire protection, schools, and other essential services. Therefore, planning decisions relating to industrial siting must consider the secondary development such industry will induce.

1.3.6 INFRASTRUCTURE

Development infrastructure—electricity, water, gas, sewers, roads, etc.—is designed to serve multiple users and is usually planned and provided by government. Developing infrastructure systems in coastal zones can have a wide range of environmental impacts (Figure 1.8). Infrastructure systems are, by their nature, typically linear. Also, several systems may be combined—roads, wires, pipes—to supply linear development corridors or commercial nuclei.

Roadways, bridges, airports, and other transportation infrastructure can create special problems along the coast unless properly designed (Figure 1.9). These structures often pollute the sea, preempt critical intertidal habitats, and obstruct natural water flows. If they are carelessly planned, coastal environmental resources can be seriously jeopardized, sometimes more by the extra development they engender than by the wires, pipes, or causeways that are installed. Therefore, their locations and engineering designs must be carefully planned in accordance with conservation guidelines.

Electric power plants cause major impacts to coastal waters via both intake and discharge of massive amounts of cooling water (required by all open-system, steam-operated, plants). As much as half of the water of an estuary can be sucked into a plant's cooling system, where fragile life forms are killed (e.g., planktonic shrimp larvae). Additionally, the hot water discharge causes thermal pollution, which can unbalance the ecosystem. Then, too, chemical additives (e.g., chlorine) pollute coastal waters. The solution is not to allow "once-through" (open) cooling systems in lagoons, estuaries, or other confined or stagnant coastal water bodies—both intakes and outlets should be placed offshore.

Infrastructure planning is one of the strongest environmental management tools in the hands of government. There are exceptions, but usually there can be no major coastal development without major infrastructure provision by government. All key parameters of development projects—location, type of operation, magnitude, engineering design—are responsive to infrastructure layout strategy. Therefore, ICZM programs should emphasize infrastructure control.

Sewage disposal is also a difficult problem. Because secondary treatment does not remove all bacteria and is of very uncertain value in terms of virus removal, outfalls in bays and sounds can endanger recreational and shellfishing waters. Also, these plants serve as "sewage refineries," releasing large amounts of mineral fertilizers, which may cause harmful algae blooms. In addition, the toxic substances and nutrients released into such water bodies may reach unacceptable concentrations and cause extensive ecological damage (see "Sewage Management" in Part 2). Therefore, sewage should be discharged as far offshore as possible after minimal treatment.

Coastal infrastructure should therefore be controlled and construction initiatives always scrutinized via formal environmental assessment, or EA (see "Environmental Assessment" in Part 2). Planners and system designers can predict the type and extent of development engendered by roadway routings and terminal locations and modify the plans to conserve key environmental resources.

1.3.7 MINING

Mining for beach sand has been an important industry in many countries. Sand for construction is a valuable commodity, but it must be remembered that the key to the natural protection provided to waterfront properties by the beachfront is sand. Quantities of sand are held in storage by the beach and then temporarily sacrificed to storm waves, thereby dissipating the force of their attack. This sand is later restored to the beach by waves, current, and wind. Consequently, taking sand from any part of the beach or the nearshore submerged zone can lead to erosion and recession of the beachfront. Therefore, beach conservation should start with the presumption that any removal of sand is adverse and should be prohibited, unless it can be shown to be naturally replaceable (i.e., a renewable resource).

Coral mining also requires firm controls. It is extensively undertaken in some island countries in the Indian Ocean and Southeast Asia and leaves the shoreline exposed to erosion and storm surges, causing serious loss of beach and shoreland and damage to coastal human and marine resources habitats (see "Coral Reef Management" in Part 2). In addition, loss of coral reef habitats reduces biodiversity, depletes fisheries, and discourages tourism.

"Dryside" quarrying for rock or mining for other construction materials for coastal projects may have "secondary impacts," including loss of environmental quality (aesthetic), reduction in downstream freshwater fisheries production, endangerment of protected species, or loss of other wildlife-related values. The stone and aggregate quarry sites that are in stream beds often flow through valleys that are quite scenic or are important to wildlife. Coastal projects can reduce both species and habitats at quarry sites; therefore, when a quarry site is selected it should be surveyed ecologically to ensure that important species and habitats will not be depleted.

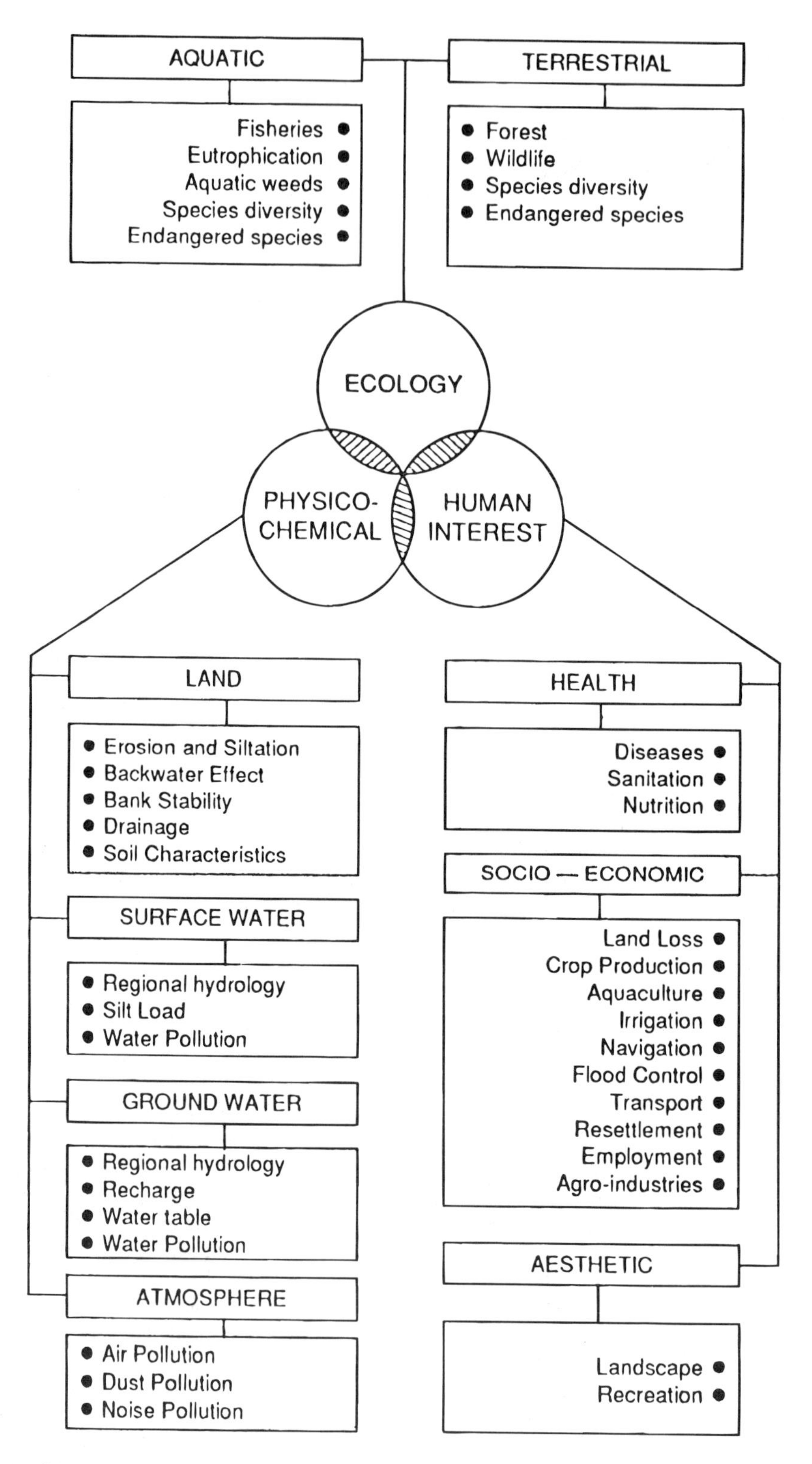

Figure 1.8. "Wetside" (coastal waters) and "dryside" (coastal lands) impacts should be considered in combination when assessing the effects of infrastructure on coastal environments. (*Source:* Reference 218.)

Figure 1.9. Piling structures in coastal waters (e.g., bridges) are ecologically superior to solid fill structures (e.g., causeways) because they do not block water flow. (Photo by author.)

Mitigation potentials for quarries (which should be specified after a particular site is chosen) include restoring the site back to good condition, using checkdams and stilling ponds if needed to slow soil erosion, work scheduling, careful choice of quarry site, and using an existing rather than a new quarry site. Additional mitigation may be necessary for transport of the materials.

1.3.8 NATIONAL SECURITY

Extensive security interests exist in the coastal zone and coastal seas because they are at the frontier, the border zone, where invasion or other negative activities (smuggling) might occur. Naval ports and harbors, coastal airfields, and special bases of all kinds are sited in the coastal zone, usually with high priority and intense security. However, with the correct approach, the military can be expected to cooperate in coastal zone conservation so long as ICZM is not in conflict with national security needs. Therefore, the defense and civil security establishments should be included as a party to any ICZM program.

1.3.9 PETROLEUM INDUSTRY

Oil pollution is an ever present danger to coastal ecosystems (265). But few countries have adequate contingency plans and emergency response procedures (Figure 1.10). Current international agreements have greatly reduced oil pollution from shipping, but they need stricter enforcement, along with provision of guidelines. The International Maritime Organization (IMO) organizes international emergency preparedness and response.

The effects of oil spills in the oceanic environment are more visible, but perhaps less serious in the long run, than are the effects of discharges of nutrients and toxic chemicals. An exception is the big spill in a semi-enclosed sea—such as the huge, recent, war-related Persian Gulf "spill"—where effects of oil on coastal ecosystems and habitats essential to fisheries can be particularly severe.

Sea pollution does not usually come directly under ICZM but rather a national pollution control authority. However, when the expansion and infrastructure needs of shore facilities are addressed, they

Figure 1.10. Oil released from the wreck of the *Ocean Eagle* demonstrates why coastal communities must have oil spill strategies, especially to protect vulnerable beaches, straits, estuaries, bay inlets, and wetlands. (Photo courtesy of U.S. National Oceanic and Atmospheric Administration.)

should be refered to the ICZM agency so that plans and designs can be modified for environmental protection, as is done for any industry. This can be difficult to accomplish because of the political power wielded by big oil enterprises (private or government).

1.3.10 PORTS AND MARINAS

Port development and shipping support facilities are not usually thought of as infrastructure, but rather as *clients* of infrastructure. Ports usually serve one or more of four major types of activities: (a) offshore oil and gas development, which requires general port facilities, oil storage capacity, refineries, land transport, and other infrastructure support; (b) the shipping industry, which requires channels, port facilities, shipyards, and extensive land areas for container storage; (c) fisheries development, which requires breakwaters, channels, ports, processing plants, and other facilities for the fishing fleet; and (d) military operations, which require special port facilities, infrastructure, and support services. The facilities can often be used by more than one general maritime activity, but military operations will usually require exclusive use of certain port and harbor sites.

Major threats from port development are preemption of fringing mangrove wetlands for development sites, dredging of channels with destructive disposal of dredge spoils, dredging and filling operations and obliteration of seagrass meadows, and spills and persistent pollution. In some countries, harbors and marinas built primarily for recreational use by small boats may disturb more of the coastal zone than commercial and industrial uses.

1.3.11 TOURISM

While environmental change is an unavoidable consequence of the growth of coastal tourism, it is necessary to keep the change within acceptable bounds (Figure 1.11). Negative effects can be minimized if priority is given to the identification and evaluation of resources and potential impacts and a planning and control system is established.

Figure 1.11. Tourism—a major source of income for coastal communities in many tropical countries—can have major effects on coastal environments unless scale and type of activities are controlled. Small groups guided by local boatsmen, as shown here for Palau, have very low impact compared to mass tourism. (Photo by author.)

Environmental deterioration threatens the tourism industry in many ways. Biodiversity reduction, resource depletion, and human health problems may result from the accumulated environmental effects of tourism. Beaches are the focal point for most coastal tourism and a major source of revenue for many countries. Good beaches are worth billions of tourist dollars. Degraded beaches are worth little.

One serious impact from tourism development worldwide is that of a decline in local water quality, mostly from sewage (301). This is a factor in the high incidence of waterborne diseases in many regions; already some beaches have high levels of coliform counts, which on occasion may exceed recognized water quality standards. Saenger (302) notes that bacterial levels frequently exceed international standards in tourist areas and also that sewage discharges, particularly if poorly sited and inadequately treated, are the most common source of adverse effects on the biota, destroying valuable fishing and nurture habitats. He notes that, in the Caribbean region, less than 10 percent of the sewage that is generated receives treatment.

1.3.12 SETTLEMENTS

Human settlements may crowd the seas with boats, overexploit resources, generate polluting industries, produce heaps of garbage, and create large amounts of chemical wastes. Shoreline real estate is in strong demand for human settlements, agriculture, trade, industry, amenity, and marine support activities for shipping and fishing and recreation.

While waterfront expansion may be necessary for coastal cities, it can jeopardize coastal resources. This demand, which responds to essential needs for economic growth, leads to a linear approach to coastal development focusing on the land along the water's edge (e.g., the line of mean high water), wetlands that straddle the shoreline, or land with access to recreational beaches or that has an outstanding view of the sea. Planned and managed coastal zone development is the corrective approach.

Mangrove wetlands, tideflats, seagrass beds, and other critical habitats are often obliterated when settlements preempt these areas by land filling to create expensive real estate. Infrastructure impacts—from roads, airports, sewage treatment, water supply, electricity, and other utilities and services—may strain available resources and degrade natural systems.

The ocean beachfront is often a hazardous place to build. Structures are threatened by erosion and seastorms, and protective works such as bulkheads, seawalls, and groins are often a Faustian bargain whereby the structure may hold but the beach is lost. Normally, if nothing is built either on the beach or next to the beach, it will remain intact so long as the processes of natural replenishment continue. The best solution is to keep development back to a safety point, i.e., behind a "setback line" that can be delineated at a safe distance inland from the beach. All permanent construction is kept behind this line.

1.3.13 SHORE PROTECTION WORKS

Severe beach erosion is a problem shared by many countries. Structural solutions to beach erosion and protection of shoreline property from the hazards of seastorms may be expensive and often temporary or counterproductive. The groins, seawalls, breakwaters, and other popular protection structures often have complex and unanticipated secondary effects resulting in major "downstream" erosion and, quite often, total loss of beach. The natural forces working against the beachfront are immense (Figure 1.12).

Mobile and responsive, the beach will usually remain over the years, even if rising sea level or other natural factors cause erosion. It may shift with the seasons, yield sand temporarily to storm erosion, slowly recede landward with rising sea level, or accrete seaward with natural shifts in the flow of ocean currents which bring more sand. Major problems do not arise until hotels, houses, or other structures are placed on or near the beach and the engineering process of "stabilization of the shoreline" commences.

If the natural movements of the sand are restrained with bulkheads or groins so as to fix the position of the beach, a complex chain reaction of problems may be started which can be solved only by the very expensive process of continuously transporting sand onto the beach (e.g., by pumping it from the ocean bottom). This remedy is so costly it is not available to most communities. Keeping development far enough inland to avoid the high risk zone is a better solution (e.g., by establishing construction setback lines).

Coastal habitats, such as mangroves and other wetlands, coral reefs, and coastal barrier islands and lagoons, are often recognized by natural hazards experts as the best defenses against seastorms and

Figure 1.12. The beachfront poses difficult management problems because of the huge amounts of energy released by storm waves. (Photo by author.)

erosion, deflecting and absorbing much of the energy of seastorms. Therefore, it is important to maintain these natural habitats for shore protection as well as for environmental conservation.

1.3.14 WASTE DISPOSAL

Coastal communities create large amounts of waste. Some have numerous septic tanks or cesspools that leach nutrient matter into shallow waters. Others collect sewage and dispose of it in the sea after some level of treatment. Harbors of large coastal cities and industrial ports have the most problems with pollutant discharges. Two major types of pollution—oxygen-depleting organic wastes (e.g., sewage) and toxic industrial wastes—damage coastal environments and resources and pose risks to human health.

The best solution is to discharge wastewater sufficiently far at sea to avoid lagoons, estuaries, and nearshore waters of the open coast. It is better to spend money on extending the discharge pipe to create an "ocean outfall" than on a complicated sewage treatment plant of doubtful reliability and efficiency.

Lagoon and estuary degradation is often caused by pollution. Aside from outright fish kills and other dramatic effects, pollution causes pervasive and continuous degradation that is evidenced by the gradual disappearance of fish or shellfish or a general decline in the natural carrying capacity of the system. Coastal seas are particularly susceptible to pollution conveyed by streams and rivers, including agricultural runoff.

Another problem is the dumping of large amounts of solid wastes along the shoreline, which disfigures the coastal landscape and leaches pollutants into coastal seas.

1.3.15 WATER SUPPLY PROJECTS

Many fisheries (shrimp, striped bass, mullet, oysters) are strongly dependent upon natural river flows that enter the sea. These rivers transport valuable nutrients to coastal ecosystems and establish beneficial brackish conditions for mangrove forests and juvenile fish–nurturing areas in the estuaries, as well as for nesting of colonial water birds and other species like alligator. Dams and water diversion or withdrawal schemes can seriously unbalance such river-dependent resource ecosystems and reduce their productivity and biodiversity by diverting water from the ecosystem or changing the beneficial hydroperiod through use of store-and-release tactics created for irrigation, flood control, water supply, and so forth.

1.3.16 MARINE EXCAVATION

Marine excavation, or dredging, physically disturbs or removes sea bottom life, suspends sediment in the water column, deposits it elsewhere on the bottom, reduces light penetration, increases turbidity, changes circulation, reduces dissolved oxygen, and increases nutrient levels in the water column.

While direct elimination of benthic habitat may not be avoidable in dredging operations, damage to adjacent non-dredged areas, such as coral reefs, is avoidable. Many coastal ecosystems (e.g., coral reef communities) are sensitive to both suspended and accumulating sediments and require long time periods for recolonization. Reduction of dredging impacts has, in fact, been the focus of considerable attention from environmental and engineering interests.

The most widespread and visible consequence of dredging and excavation is the generation of suspended sediments and turbidity. Dredged materials high in organics can generate biological oxygen demand (BOD) and depress oxygen levels. The magnitude of the above effects varies considerably depending upon the type of dredging and excavation method in use. (231)

1.4 SOLUTIONS THROUGH MANAGEMENT

Integrated, multi-sectoral resource planning and management is especially needed in the coastal zone. Yet achieving this need for the coast is more difficult than for most geo-economic zones. Coastal areas and coastal resource systems are governmentally complex because of unclear jurisdiction, dispersal of authority, and the amount of common property resources involved. Therefore, resource management programs must involve all levels from national to village governments. The approach recommended in this book—integrated coastal zone management—incorporates management of natural resources, conservation of biodiversity, maximization of socioeconomic benefits, and protection of life and property from natural hazards (such as cyclonic seastorms).

It is generally understood that there is no one "correct" way to organize, plan, and implement an ICZM program; the plan must be tailored to fit into the institutional and organizational environments of the countries or regions involved, including political and administrative structures, economic conditions, cultural patterns, and social traditions (54).

In one example, large-scale coral mining enterprises in southern Sri Lanka in the 1960s and 1970s left the shoreline exposed to erosion and storm surges, causing serious loss of beach and shoreland and exposure of the coast to storm surges. A local fishery collapsed; mangroves, lagoons, and coconut groves were lost to shore erosion; and local wells became contaminated with saltwater. Sri Lanka reacted by enacting an ICZM program in 1982, with first emphasis on controlling coral mining to protect the shoreline and a later emphasis on natural resources conservation (see "Sri Lanka" case history in Part 4). Even though the original problem was localized to part of the country and specific to one issue, a national ICZM format was set up as the best way to deal with it. This format proved to be ideal for broadening of the program to include the coastline as a whole and, later, to incorporate environmental issues (6, 7, 43, 358).

1.4.1 THE INTEGRATED APPROACH

The ICZM approach can be expressed in a variety of forms of comprehensive, integrated coastal management, whether the focus is on management of privately owned shorelands or of the waters of the coastal "commons" (the areas held for use of all citizens). In purpose, ICZM programs are intended to address all resources in a defined coastal zone and to integrate the interests of a variety of economic sectors in conservation. ICZM-type programs, while distinctive in purpose, are rooted in such well-known and standard approaches as regional development planning, resource conservation and watershed management.

While the main purpose of ICZM is to manage development so as not to harm environmental resources, it does *not* directly manage use of the resources. The resources themselves would usually continue to be managed by sectorally oriented resource agencies, such as forestry, fisheries, wildlife, and water pollution. But the development process would be managed by ICZM, which in this way would assist in conserving environmental resources by serving a multi-sectoral and multi-agency coordinative purpose, a function that would not otherwise be filled.

Nor is ocean management considered a major focus of ICZM. Yet in some countries ocean management and ICZM may be brought closely together (e.g., Brazil and India). Ocean management issues of greatest concern include shipping, offshore fisheries, mineral exploration and development, ocean dumping control, and ocean research. Recent agreement on the "Law of the Sea" regime has stimulated many initiatives for national ocean management. These may in turn enhance ICZM-type programs as ocean concerns are felt along the land/ocean interface. Sweden, Sri Lanka, and Brazil seem to exemplify this trend (194).

Horizontal integration is a term that describes efforts to coordinate the separate economic and governmental sectors and thereby reduce fragmentation and duplication (326). Getting the coordinating mechanism working right is clearly the most difficult part of the creation of an ICZM-type program because it is the major problem to be solved by the process. Numerous private economic sectors and correspondingly numerous government bureaucracies directly or indirectly affect coastal uses, resources, and environments (Figure 1.13).

In its *planning* mode, ICZM examines the consequences of various development actions and proposes necessary safeguards, constraints, and development alternatives that will guarantee sustainable development and the sustainable use of coastal natural resources, at the most productive levels possible.

In its *management* (or "implementation") mode, ICZM assesses the environmental and socioeconomic impacts of specific development projects and recommends changes necessary to conserve resources and protect biodiversity. ICZM coordinates actions of various economic sectors to ensure that

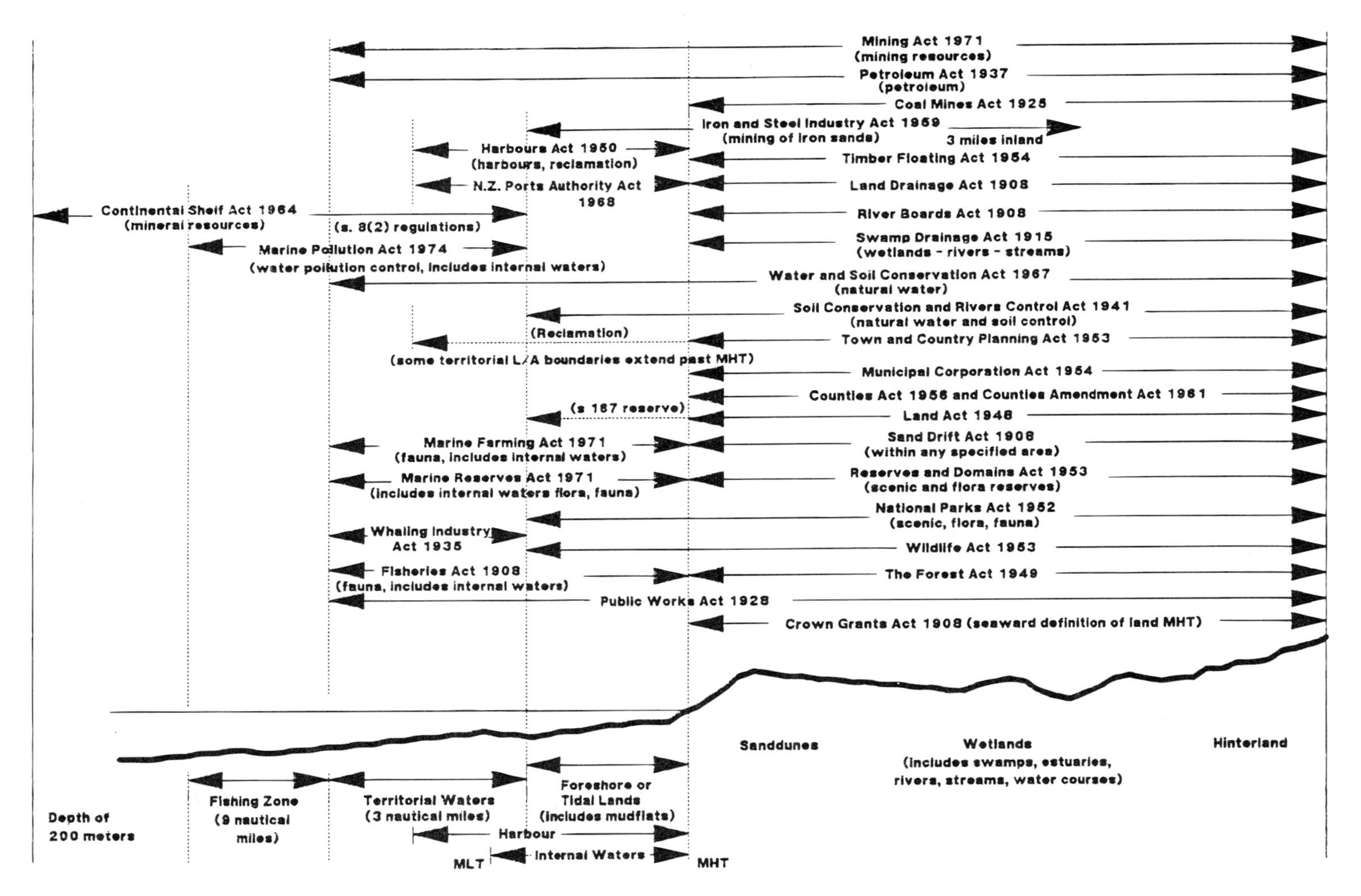

Figure 1.13. New Zealand provides an example of the great variety of interests and agencies involved in the coastal zone. (*Source:* Reference 326.)

advances in one sector do not bring reverses in another—for example, that port development does not unnecessarily diminish local fisheries or tourism.

The ICZM process allows great flexibility. It can concentrate on the hazards of coastal erosion, as in Sri Lanka's ICZM program (6); on fisheries, as in the emerging Philippines ICZM program (234); on coastal and marine protected areas, as in the proposed Saudi Arabia ICZM strategy (83); on shrimp aquaculture, as in Ecuador (315, 237); on a "networking" approach (management dispersed among many agencies), as in Oman (306, 307); or on land use, as in the United States ICZM program (which has 30 separate state subprograms).

According to Kenchington and Crawford (205), the following elements are required:

- A dynamic goal or vision of the desired condition of the oceanic or coastal area and the integration of human use and impact for a period significantly longer than conventional economic planning horizons, say 25 or 50 years
- National objectives are broad, commonly agreed aims or common purpose to which policies and management are directed. For regional and local plans, progressively more detailed objectives consistent with the national objectives are usually required.
- Guiding principles for managers or statutory decision makers exercising discretionary powers for planning, granting approvals, or making changes to the purpose or extent of use and access
- A strategy, commitment, and resources for the objectives to be met through detailed day-to-day management, which may involve several agencies and the community
- Clear, legally based identification of authority, precedence, and accountability for achievement of the strategy in relation to any other legislation applying to the area in question
- Performance indicators and monitoring to enable objective assessment of the extent to which goals and objectives have been met
- Above all, political, administrative, and stakeholder will and commitment to implement the strategy

1.4.2 THE SEARCH FOR SUSTAINABLE YIELD

The Brundtland Commission of the United Nations formalized the concept of *sustainable development* and defined it as as "development which does not compromise the future." However, little practical guidance was offered by the commission on how the concept should be translated into specific plans and programs. Yet common sense tells us that sustainable use should require adjusting rates of use of renewable resources so that we do not degrade or exhaust them.

For sustainability, resources should be maintained so that the ability of a resource to renew itself is never jeopardized. Such management maintains biological potential and enhances the long-term economic potential of renewable natural resources. Pursuing development on a sustainable use basis must be recognized as an absolute necessity to sustain progress in health, food security, housing, energy, and other critical human needs.

Sustainable exploitation implies the wise use (development) and careful management (conservation) of individual species and ecosystems on which they depend so that their current or potential usefulness to people is not impaired. Resources should not be harvested, extracted, or utilized in excess of the amount that can be regenerated over the same time period. In essence, the resource is seen as a capital investment with an annual yield—it is therefore the yield that should be utilized and not the capital.

We must recognize that it is becoming more and more difficult to sustain the output from any one particular resource in the absence of a framework for broad, integrated coastal planning and management. For example, in Ecuador, shrimp aquaculture became so lucrative that up to half the mangroves in certain regions were replaced with shrimp ponds (61, 237, 315), with the result that by 1986 the remaining natural mangrove stands were unable to produce sufficient larval "seed shrimp" to stock the aquaculture ponds and 60 percent of them were deactivated (268).

No conservation policy or program was in effect to guide the industry, which produced up to 44 percent of Ecuador's foreign exchange earnings (in 1982) and provided more than 100,000 jobs. No mechanism existed for cooperation among forest, aquaculture, fisheries, economic planning, and other key sectors. In such a situation, short-term enterprise was free to flourish to the detriment of the long-term economy of the country. Heavy borrowing of foreign capital for ponds and facilities contributed to a serious foreign debt dilemma. (See "Ecuador" case history in Part 4).

While the presence of an ICZM-type integrated planning and comprehensive management program alone may not assure a sustained yield from the coastal natural resources of any country, its absence will lead to their depletion. Very rarely will long-term economic benefits result from development based on excessive exploitation of coastal resources. Economic stability will come from development closely linked to resource conservation, planning, and management elements of ICZM.

1.4.3 BIODIVERSITY CONSERVATION

The need to preserve "biological diversity" and the methods for doing so were terrestrially derived. Therefore, they require modification to fit to coastal habitats. The wetter the landscape, the less well the basic concepts fit. For example, few oceanic species are in danger of *extinction* because of habitat damage. But along the coast and the beaches, many species (turtles, terns) are jeopardized by habitat deegradation and loss (Figure 1.14). According to Ray and McCormick-Ray (292), five aspects of marine biological diversity are paramount for conservation:

1. The diversity of marine fauna is much greater than for terrestrial fauna at higher taxonomic levels.
2. The marine fauna is also much less well known.
3. Most marine species are widely dispersed.
4. Most marine communities are highly patchy and variable in species composition.
5. The response times to environmental perturbations are relatively short.

A major strategic objective for ICZM is to preserve the habitats of species that have been designated as especially valuable or in danger of extinction. Therefore, an important motivation for designating "ecologically critical areas" (ECAs) for special conservation is species protection; other purposes might be protection of especially productive or scenic natural resources. Sometimes these designations are done in response to an international program, such as UNESCO's Biosphere Reserves or the Ramsar Treaty for major wetlands, but more often they are made by independent national action in establishing nature preserves or national parks.

Figure 1.14. A common tern guards an egg in the seclusion of a coastal dune. (Photo by author.)

1.4.4 NATURAL HAZARDS

The measures best suited to conserving ecological resources are often the same as those needed to preserve the natural barriers to storms and flooding. Human activities that remove or degrade protective landforms—removing beach sand, weakening coral reefs, bulldozing dunes, or destroying mangrove swamps—diminish the degree of natural protection the coast affords (57). For example, if dunes are removed by sand mining or because of obstruction to ocean views, the risk to coastal development behind the former dunes is greatly increased. Similarly, mangroves serve to dissipate wave energy and to stabilize the land areas behind them from the erosive forces of storms. The value that these natural resources have for hazard prevention reinforces the need to identify them as critical areas and to give them strong measures of protection.

In fact, a "hazards loss reduction program" should begin with preservation of coastal habitats that provide natural resistance to wave attack, flooding, and erosion (Figure 1.15). Many communities have found that a combined approach to hazards and resource management simplifies the process of coastal management and leads to more predictable decisions on what constitutes sustainable development. For example, the same setback requirement that protects beachfront settlements from erosion and storm waves could also preserve turtle nesting sites there. Similarly, a zoning restriction on development of mangrove swamps could not only conserve an economically valuable resource, but also help maintain a physical defense against storm waves. In a final example, a seashore or coral reef park can protect these natural landforms from both natural hazards and resource depletion (308).

A simple and effective approach to natural hazards protection is for hazards interests to join forces with environmental and natural resource management interests. Some countries have already begun to experiment with this combined approach through comprehensive, multi-sectoral ICZM programs that simultaneously accomplish both objectives.

1.4.5 HINTERLANDS

It would be in the purview of the ICZM planning unit to create strategies for reducing negative effects from upland activities in the "Area of Influence" that lies outside the ICZM boundary. In their

Figure 1.15. Bulldozers make a futile effort to rebuild the beach at Ocean City, Maryland (USA), after a spring storm. (Photo courtesy of United Press International.)

natural state, uplands terrains (forests and grasslands) and hydrological systems (streams, ponds, and wetlands) can hold and detain large amounts of storm water, acting in effect as a natural sponge that holds water during heavy rains or snows for later, more gradual release. This provides an ecologically compatible rate of runoff flow as well as some protection against flooding for downstream communities. Uplands are also important in protecting coastal waters from storm runoff pollution because their vegetation and soils cleanse the water (84).

Obviously, coastal resource systems are jeopardized when the terrain is cleared of vegetation for agriculture; when it is altered to accelerate drainage; when surface water bodies and watercourses are filled, detoured, or channelized; or when the natural flow pattern is significantly disrupted, causing freshwater flow to the coast to occur in sediment-laden surges. Therefore, ICZM should, to the extent possible, try to influence upland development activities. This is difficult because it is uncommon for those who plan dams or land-clearing enterprises in the hinterlands to consult with coastal interests. However, it is worth trying to get ICZM in a position to review and comment on the Area of Influence that includes major upland projects.

1.4.6 MANAGEMENT FUNCTIONS

The overall purpose of the ICZM-type program is to provide for the best long-term sustainable use of coastal natural resources and for perpetual maintenance of the most beneficial natural environment. In practical terms, the ICZM program is intended to support the management goals and purposes identified in Section 1.2 by providing the basis for sustainable use of resources, biodiversity preservation, protection against natural hazards, pollution control, enhancement of welfare, development of a sustainable economy, and optimum multiple use.

Specific purposes may be identified as supporting fisheries, attracting tourists, improving public health, raising public awareness, or maintaining yields from mangrove forests. All of these require coordinated community action for their accomplishment, a need that ICZM fulfills. Specific objectives might include any of the following:

1. Maintain a high quality coastal environment.
2. Identify and protect valuable species (and their intra-specific variations).
3. Identify and conserve critical coastal habitats and identify lands that are particularly suitable for development.
4. Resolve conflicts among incompatible activities affecting coastal and ocean resources and the use of space.
5. Identify and control activities that have an adverse effect upon the coastal and marine environment.
6. Control pollution from "point sources" and from land runoff, as well as accidental spills of pollutants.
7. Restore damaged ecosystems.
8. Coordinate governmental efforts to promote the sustainable development of coastal and ocean resources.
9. Balance economic and environmental pressures as they affect the development and conservation of coastal and ocean resources.
10. Provide guidance for coastal development planning to reduce inadvertent side effects.
11. Create and offer safer options for coastal development.
12. Raise public awareness.

1.4.7 DESIGNING AN ICZM PROGRAM

Every country evaluating the potential of an ICZM-type program will have its own special approach to conservation of resources and will be facing its own distinct array of coastal issues. The first priority has to be getting ICZM on the local and national political agenda and getting favorable action on a mandate for resource conservation.

The particular form of any ICZM program will depend upon the national and regional issues it is meant to address. As the tableau of issues and options varies from country to country, so will the form of the program. No two countries would be expected to have identical programs. But the essential purpose remains the same—to create an interagency coordination mechanism and regulatory program to promote sustainable multiple use of renewable resources within the defined coastal area.

While each country's program will be unique, there are several basic stages in the generation of an ICZM program that will be found to be common to all, in one form or another. These stages are as follows:

1. **Policy Formulation:** Creation of a policy framework to establish goals and to authorize and guide the ICZM program; accomplished by executive and/or legislative action.
2. **Strategy Planning:** Sometimes called *preliminary planning,* this is the stage where the potential impacts of the ICZM policy action are explored (on resources and resource users, on income and jobs, on social and cultural well-being), where benefits are evaluated, where a wide array of data is accumulated, and where a general strategy is created and recommendations are made for organization and administration of the ICZM program.
3. **Program Development:** Once the Strategy Plan is accepted by policy makers, development of the ICZM program can commence and a detailed Master Plan for its implementation can be created.
4. **Implementation:** Once the Master Plan is approved and a budget and staff are authorized, the Implementation Stage can commence.

In practice, the above stages are not so discrete and linear as theory suggests. Instead, there will be feedback and revisions of earlier stages as new facts and opportunities come to light in later stages. For example, there will certainly be the need for policy revision and strengthening as a result of findings and recommendations from Stages 2 and 3. Therefore, the whole program must be flexible and adaptable.

Many tools are available to ICZM planners and managers, most of which are familiar to planners and administrators. It is the mix of tools and their particular applications that are distinctive to ICZM. Some of the main tools are

- Permits for building, land clearing, etc.
- Environmental assessment (EA)
- Natural area protection (marine reserves)
- Zoning, use allocation
- Setbacks, buffer areas
- Environmental standards and guidelines
- Infrastructure control
- Land use regulations
- Land use planning
- Watershed management (Figure 1.16)
- Pollution regulations
- Community participation
- Conflict resolution
- Rural assessment
- Remote sensing and geographic information systems

1.4.8 JOINT MANAGEMENT OF LAND AND SEA

A key factor that distinguishes ICZM-type programs from others is that coastal waters and coastal lands are addressed *together in a single unified management program.* The special role of ICZM-type programs is that they are centrally organized and apply integrated, area-wide resources planning and management to the distinct landforms and waters of the coast. The coast is a place where special knowledge, techniques, and governmental interactions are involved. ICZM focuses on shoreline development, natural habitat protection, and environmental conservation.

ICZM addresses national concerns for the coastal waters and natural habitats (e.g., coral reefs and seagrass meadows), the adjacent shorelands, and the transitional areas such as floodlands and intertidal areas (beaches, mangrove forests, etc.) that lie between the permanent waters and the shorelands. It provides a method to resolve problems of interaction between coastal lands and coastal waters in much

Figure 1.16. There should be controls to protect critical habitats and water quality in the coastal zone from the effects of land stripping, such as this peat mining operation in North Carolina (USA). (Photo by author.)

the same way as river basin planning provides a method for resolving problems of interaction of valley lands and river waters.

An ICZM-type integrated coastal resources management program has the following six attributes, according to Sorensen and McCreary (326):

1. It is initiated by government in response to very evident resource degradation and multiple use conflicts.
2. It is distinct from a one-time project; it has continuity and is usually a response to a legislative or executive mandate.
3. Its geographical jurisdiction is specified. It has an inland and an ocean boundary—with the exception of small, unpopulated islands, which often have only an oceanward boundary; because it is *not* an ocean management program, it must have both shore and landward components.
4. A specific set of objectives or issues is to be addressed or resolved by the program.
5. It has an institutional identity—it is identifiable as either an independent organization or a coordinated network of organizations linked together by functions and management strategies.
6. It is characterized by the integration of two or more sectors, based on the recognition of the natural and public service systems that interconnect coastal uses and environments.

1.4.9 NON-INTEGRATED OPTION

Not all countries want a formally constituted ICZM program. With the development of ICZM as a recognized field, it has become a common perception that countries without a formalized ICZM program cannot, or do not, manage their coastal resources in an integrated fashion. In some cases, however, it may be possible to achieve the same goals through an existing framework of governance if the purposes of ICZM are incorporated in the existing institutional structure and process.

The realities of political, economic, and social life of many countries may require a less ambitious beginning than full-scale ICZM. Limited programs may be targeted on shore erosion, for example, or perhaps on designating critical habitats. Such targeted programs might later evolve into larger scale ICZM-type programs, as occurred in Sri Lanka (358), where ICZM started with beach erosion control and ended with a much broader resources management program (see the "Sri Lanka" case history in Part 4).

Where government departments and agencies responsible for different sectors of the coastal zone have the potential to collaborate and effectively coordinate their management activities, the result could be a program that is just as well integrated as a single ICZM program where one agency takes the lead role. If the existing system works—as it does in the Maldives (see "Maldives" case history in Part 4)—it is unnecessary to develop a new, independent program that could stretch limited financial and manpower resources, introduce a sense of competition, and be seen as a threat to existing institutions.

While many countries may not, in the end, choose to go to a full-scale, integrated, ICZM approach, but rather adopt a less sweeping program, such countries are well advised to strive for the ICZM goals. It is particularly important to address these goals early in the planning phase—in effect, to take a regional planner's approach. For example, the organization of a coastal program to manage mangrove forests for the sole benefit of the fisheries sector might succeed, but more likely it would fail if it lacks mechanisms to incorporate the interests of local villagers, forest industries, upland agriculture, public health, tourism, port development, and so forth. Therefore, if an ICZM "Situation Management" or regional program can be organized, at least in the planning phase, a more satisfactory outcome can be expected. (See "Incremental Approach," Section 1.5.17.)

In some countries the result will be to modify existing planning and resource management mechanisms to accommodate ICZM-type needs. For example, Trinidad and Tobago experimented with ICZM (starting in 1978) and conducted preliminary investigations of the potential for a regionalized approach (see "Trinidad and Tobago" case history in Part 4). The country finally decided in 1984 that the established Town and County Planning Division could handle coastal development management with technical assistance from the Coastal Area Planning and Management Division of the Institute of Marine Affairs. According to McShine (244), "It was decided that a separate Coastal Area Plan as distinct from an integrated National Physical Development Plan involving coastal development should not be formulated." Chua Thia Eng (52) cites Singapore as a good example of workable coastal planning "without the benefits of ICZM."

Another example of using existing authorities rather than passing new coastal laws is the Australian state of Western Australia, where the ICZM strategy is to keep the land adjacent to the edge of the sea in a state reserve and not "alienate" it for private purposes. Thus, the state has, in effect, created a generic construction setback. (See "Setbacks" in Part 2.) The effect of this linear coastal reserve is to provide a buffer strip (of undeveloped public land) between the high water line and the developed part of the shorelands (264). In this manner, development does not crowd into the edge zone, the most valuable of coastal critical areas. This coastal "linear" or "buffer strip" reserve program, as a non-statutory approach to ICZM, does require extensive coordination of agencies and optimal use of existing governmental powers to acquire and manage land and resources.

1.5 STRATEGY PLANNING

Coastal planning refers to a process of comprehensively studying resources, economic activities, and societal needs, including problems and opportunities in the designated coastal planning area, or Coastal Zone, and proposing future actions. The important purpose of planning is to examine the past and the present so as to choose the best outcome for the future.

There are two important stages of ICZM planning—*strategy planning* and *master planning*. Strategy planning is discussed in this section and master planning in the following section. Strategy planning is the process that explores options and develops an *optimum strategy* for a management program. It examines the facts, considers the issues, suggests possible solutions, and proposes specific legal and institutional arrangements. Strategy planning is the key step in the process of organizing an integrated coastal zone management (ICZM) program because the whole strategy of the process is being worked out. In this step the hows, whys, wheres, whens, and whos are pretty much decided. Policies and objectives are chosen and methods determined.

The importance of the strategy planning function cannot be overemphasized—it is the key to all that follows in master planning and implementation of any ICZM program. Therefore, the strategic planning stage is organized to identify major issues and anticipate the questions that decision makers will pose and to provide the data to answer these questions. However, many complexities face strategy planners—numerous sectors are involved and strong political and economic pressures come into play (Figure 1.17).

1.5.1 THE PROCESS

Strategy planning involves all the preliminary investigation, data collection, analysis, dialogue, negotiation, and draft writing that is necessary to enable those responsible to define problems, to identify options, and to proceed to authorize an ICZM program. This is the stage where needs and solutions are examined and recommendations advanced. This is not the detailed planning stage for the management program; that comes in the Master Plan creation phase (see Section 1.6 below).

The special characteristics of the coastal area require special planning and management approaches, such as those found in ICZM-type programs, approaches that anticipate and deal with unusual problems and the unique conflicts that arise along the coast. Those who work with ICZM programs will need special training.

An example planning sequence (linearized) for an ICZM strategy might include any of the seven following steps:

1. Investigation of issues and needs
2. Review of policies
3. Formulation of goals and objectives
4. Pre-planning review activities and preliminary strategy report (resources, legal/institutional, socio-economic, plan boundaries, etc.)
5. Organization of planning program (funding, staff, facilities, equipment, operational strategy)
6. Implementation of master planning program (data collection, analysis, mapping, public hearings, etc.)
7. Drafting, redrafting, and production of final plan

In the end, the Strategic Plan should recommend a Master Plan that includes review, analysis, and coordination functions leading to environmental assessment and issuing of special coastal development permits, a function that could be handled by an appropriate "lead agency" (see "Institutional Mechanism" below).

It is accepted by ICZM experts that only a truly integrated program (i.e., one that includes all major economic sectors affected) can succeed fully. If important stakeholders are left out (e.g., tribal chiefs, port authorities, housing departments, tourist industries, fishermen, economic development planners) ICZM will probably fail. In fact, a major function of ICZM is to provide a framework for coordination of a wide array of interests.

The planner's role is to deal with great complexity and reduce it to simple concepts and program means that are politically and administratively viable. This is particularly relevant to the program implementation stage because *the most difficult aspect of ICZM is getting authorization to move the program from the planning stage into the management mode.* If the program appears too complex, too controversial, too disrupting, or too expensive, the government may stall the program at the planning stage rather than move it to implementation. Therefore, ICZM advocates must avoid unrealistic goals and excessive complexity.

The strategies to be implemented for ICZM must remain flexible within the broad limits of the overall management objectives if they are to meet the changing needs of the population or changing conditions of the resource. Flexibility is best maintained through periodic reviews of the strategies. (156)

Whatever the approach, the essential purpose of strategy planning remains the same—to create a viable mechanism for conservation, that is, sustainable use of renewable resources within the defined coastal area. Accordingly, some of the most destructive "development" practices—massive coral reef

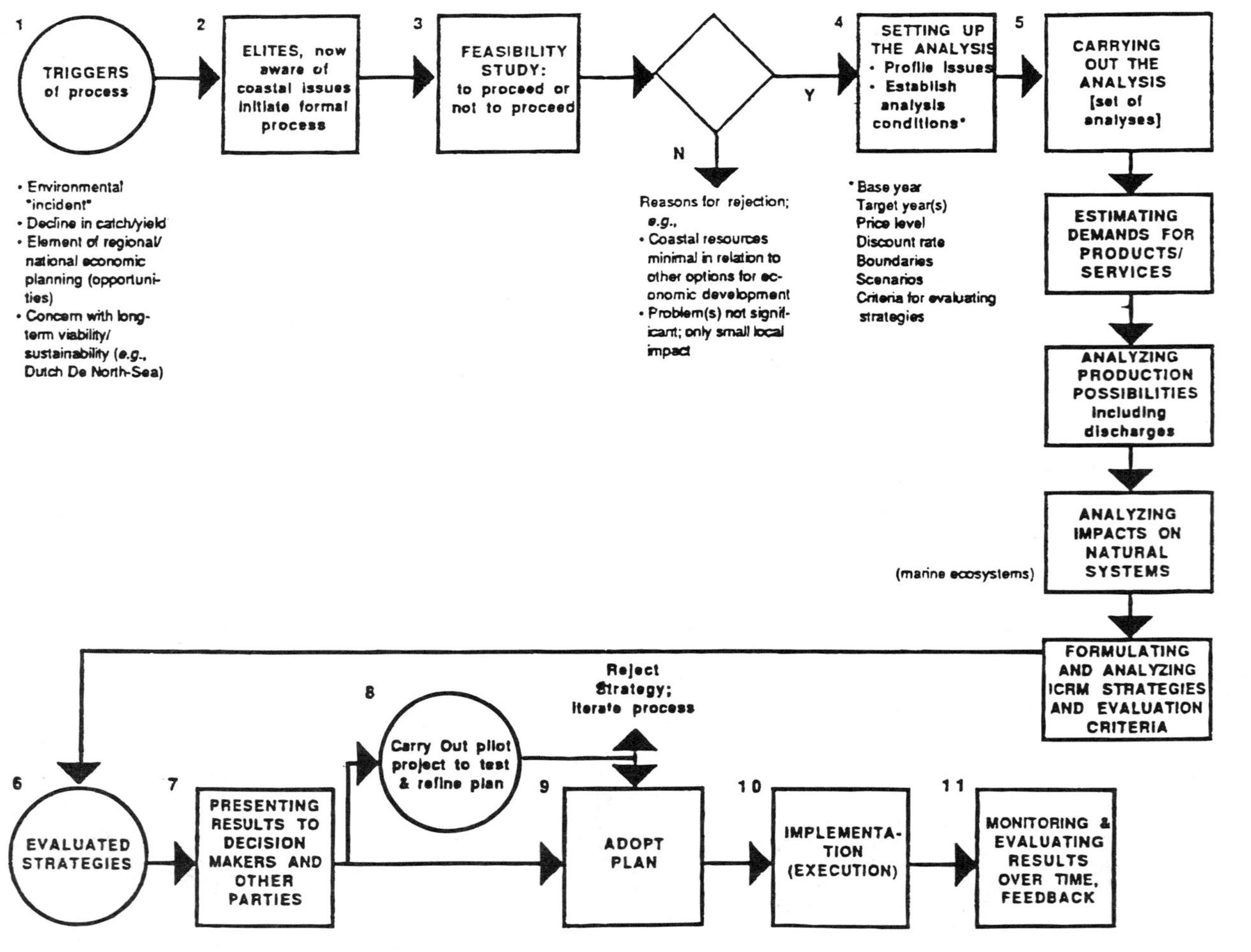

Figure 1.17 Example approach to organizing an integrated coastal zone management (ICZM) plan. (*Source:* Reference 65.)

mining or mangrove forest cutting—will usually have to be curtailed because they are not sustainable and conflict strongly with other economic uses such as fishing and tourism. Alternatively, development activities with a lesser potential for damage can be adjusted in location, design, or scale to meet coastal management requirements.

1.5.2 OBJECTIVES

ICZM has as its central purpose the organization of an integrative, holistic system for ensuring sustainability of coastal resources, the perpetuation of biodiversity, socio-economic improvement and security from natural hazards. A major objective is to overcome the consequences of an uncoordinated sequence of coastal development projects, the result of which is resource degradation and foreclosure of future options for resource use. ICZM attempts to avoid these by broad multiple-sector planning and integrated project development, by future-oriented resource analysis, and by applying the test of sustainability to each development initiative. These central purposes can be articulated as a set of typical ICZM objectives for the Strategy Plan, as shown in the example objectives listed below:

- **Maintain a high quality coastal environment.** The coast is a major resource, providing commerce, food, recreation, spiritual refreshment, and security. But this productive state is not guaranteed to last forever; the coast could easily become polluted, ugly, and unproductive along much of its length if development safeguards are not applied.
- **Protect species diversity.** Maintenance of a natural coastal environment, as provided by regulatory means, is favorable to most species. Yet, many species need the special protection and extra safeguards that can be provided by designating their habitats as nature reserves or national parks.
- **Conserve critical habitats.** Certain ecosystems are of such outstanding diversity and ecological value that they should be set aside and protected from alteration by development; these include many of the productive estuaries and islands along the coast and their adjacent shorelands. Specific habitats of importance include mangrove forests, seagrass meadows, sandy beaches and dunes, and certain tideflat habitats. Wherever these occur on the coast, they are presumed to be "critical habitats," the loss of which would reduce productivity, species well-being, and ecological balance.
- **Enhance critical ecological processes.** Certain ecological processes are critical to the productivity of coastal ecosystems, for example, light penetration through the water (which can be blocked by excessive turbidity) and water circulation (which can be blocked by infills and causeways). These critical processes need protection through regulation of development and appropriate planning guidelines.
- **Control pollution.** Pollution from specific "point sources" as well as that from land runoff can foul coastal habitats and waters causing human health problems as well as ecological disruption and reduced productivity. Accidental spills of oil and chemicals can be most damaging. Strong pollution control regulations are needed to keep the coast clean and productive.
- **Identify critical lands.** Certain areas of the coast have a special potential for recreation, housing, nature protection, economic development, and so forth. The ICZM program would identify lands optimum for development and for nature conservation.
- **Identify lands for development.** Certain areas of the coast have a special potential for recreation, housing, economic development, and so forth. The ICZM program should advise development entities of areas that would be optimum for development.
- **Protect against natural hazards.** Hazards protection is an essential part of an ICZM program. As experienced planners and managers already know, the measures best suited to conserving ecological resources are often the same as those needed to preserve the natural landforms that serve as barriers to storms and flooding. Accordingly, a combined approach to hazards and resource management could simplify the process of zoning and permit reviews and lead to more predictable decisions on what constitutes acceptable development (84).
- **Restore damaged ecosystems.** Many otherwise productive coastal ecosystems have been damaged but are restorable. Wetlands should be restored, species habitats rebuilt, and pollution damaged reversed.
- **Encourage participation.** An important objective is to create public awareness of the coast and of needs for coastal conservation. At the same time, the local communities will expect to benefit

economically and/or socially from the sites. They will want to guide the action and not just listen to talk from government agencies. The public must be consulted and not just placated and propagandized.

- **Provide planning guidance.** To avoid initiatives that would be damaging to the coast, guidelines and advice could be provided to planning entities of all kinds—physical planners, economic planners, development planners, and so forth. Of particular importance is the control of infrastructure so that highways are properly routed and water and power are not provided to places where development should be discouraged, like undeveloped barrier islands.
- **Provide development guidance.** Much of the ecological and scenic disruption of the coast is from inadvertent side effects of coastal development. An effective program could provide specific guidelines and advisory services to development entities, enabling them to design and construct projects that avoid conflict with coastal conservation to the extent possible.

Regardless of the complexities, a comprehensive and fully integrated ICZM program should be visualized, *at least in the planning phase.* This comprehensive approach will guarantee the most satisfactory outcome for strategy planning.

1.5.3 POLICY FORMULATION

The foundation for the larger changes in laws and programs needed to accomplish coastal management should be built upon a realistic assessment of where each nation currently finds itself in terms of coastal policy development and implementation, in the broader context of sustainable natural resources development. These stages or steps, for a given issue, are as follows (270):

- Problem identification
- Analysis of causes and significance
- Placement of the issue on the political agenda
- Consideration of policy options
- Choice of a policy
- The design or modification of government programs
- Organization to implement the policy
- Implementation
- Evaluation of progress
- Reexamination, modification, or termination of the policy based upon implementation experience or new information

The appropriate political authority—executive or legislative—must first authorize a specific government entity to investigate the need for and the potential benefits of ICZM. It should declare in the strongest terms possible that it is the intention of the nation to review and exercise control over development activities affecting the sustainability of coastal renewable resources (141). One specific policy goal should be maintenance of the optimum sustainable use of coastal renewable natural resources, in both the economic and the social context.

The policy should list specific national coastal concerns and issues to be addressed and state the priorities of the nation toward coastal resources conservation. The policy statement should also have a directive component to state the actions that the executive authority or legislative body expects various agencies of government to take. In addition to specific assignments to agencies for the strategy planning stage, there should be assurance that *funds are authorized and available* to pay for the planning work that must be done.

In the policy formulation stage, it is not appropriate to identify specific ICZM program details because the "policy statement" is concerned with general goals and directives for the most part. Because the strategy phase is largely exploratory, diagnostic and formulative, the specifics of management are usually left to the Master Plan formulation stage. The reasoning is that strategy planning comes too early in the process to permit informed decisions to be made about specific program components or objectives. There is danger in getting locked into specifics that may prove to be misdirected and that may seriously hamper master planning and program development.

Because the support and cooperation of numerous economic sectors are essential to the success of integrated, multi-sectoral programs such as ICZM, it will be necessary in the policy formulation stage to consult with representatives of the sectors (public and private) that are likely to be most affected. Many of these sectors will have little to gain from coastal resources conservation. Such public agencies as transportation, housing, military, agriculture, and industry may see the ICZM program as a problem rather than as an opportunity. Also, powerful private sector interests such as manufacturing and aquaculture may object initially. To ensure a minimum of continued opposition from these sectors, persuasion may be needed along with assurance that the policies to be formulated will treat them fairly. (See also "Political Motivation" in Part 3.)

Regardless of how it is accomplished, there must evolve specific ICZM policy to give direction to ministries and to inform the private sector and international donors of the national intention to control coastal development and to conserve coastal resources and biodiversity.

Many strategy issues may need policy direction, including such examples as the following (66):

- Should the ICZM emphasis be on national standards or regional programs?
- What are the expected net benefits of an integrated national ICZM-type approach, in economic and social terms; how can an ICZM program be funded?
- Who are the major proponents and opponents of the proposed ICZM program?
- What can an ICZM-type program do to prevent loss of life and property from coastal natural hazards such as seastorms and beach erosion and what are the benefits of combining hazards protection and resource conservation in a single coastal program?

Listed below are ten policy guidelines proposed by an ASEAN working group for ICZM and reported by Chua and White (55) (abridged by the author):

1. **Industrial development and environmental quality.** Coastal industries should be located in sites that have minimal impact on critical habitats and that do not pollute the waters. Planning for industrial development should include zonation for industries, ports, and shipping facilities. Incorporate environmental impact statements and sustainable use criteria.
2. **Mangrove conversion.** Zonation schemes should prescribe clear guidelines designating areas for conservation, protection, and development. Evaluation of the resources, including replacement costs, must be made for decision making. Public education is needed to reverse the "wasteland" image. Mangrove habitats should be included in management plans for sustainable use and/or protection.
3. **Shrimp farming and other coastal aquaculture.** The use of mangrove habitat for aquaculture must be reconsidered. Land use zonation and water quality for and the environmental impact of aquaculture should be included in local and national development plans.
4. **Coral reef protection.** Enforcement of laws on coral reef fishing is needed. Education and community participation programs to establish local resistance and alternatives to destructive activities must be initiated. Marine parks and reserves at the municipal and community levels should be established. Bans on coral trade are required.
5. **Water quality.** Setting of quality standards is needed. Integration of river and watershed management should be made, if possible, with upland pollution control. Sampling and monitoring must be standardized. Industries and sectors that violate standards should be controlled.
6. **Coastal erosion and sedimentation.** Control measures should be implemented in areas where valuable productive ecosystems, such as coral reefs, seagrasses, mangroves, estuaries, and beaches, are affected by erosion and sedimentation. Maritime construction projects should have environmental assessments. All mining and dredging along inshore coastal areas or on coral reefs should be stopped or regulated.
7. **Tourism development.** Develop guidelines for environmental management on sewage discharge; shoreline erosion; maintenance of beaches, coral reefs, and other ecosystems; marine uses; and general zones appropriate for tourism. Social displacement should be minimized.
8. **Institutional arrangements and capabilities.** Agencies with jurisdiction over coastal areas and trained personnel are needed to analyze management issues and develop plans. All levels

of government should be involved in ICZM. Participation of communities and NGOs will improve institutional building.

9. **Media interest.** Coastal ecology and conservation should be promoted in the mass media and included in the educational curriculum to increase public awareness.
10. **Upgrading legislation.** Existing laws on coastal area management need to be carefully reviewed and improved to be more practical and enforceable, and unworkable laws should be replaced.

1.5.4 ISSUES ANALYSIS

One expects that, after policy formulation, government will next need to know a great deal more about the issues, conflicts, economic tradeoffs and benefits, and working mechanisms before it can authorize development of a program with specific forms of implementation. Exploring this political field is an important purpose of strategy planning. Issues analysis is quite similar to needs analysis, and the two are not separated here.

The knowledge of issues and the answers to policy questions will lead to decisions to authorize or not to authorize the next stages—Master Plan creation and Program Development—or to request more fact finding. Therefore, issues analysis should be organized to answer questions in the minds of decision makers in government. Methods are listed in "Issues Analysis," Part 2.

Strategy planning identifies and examines major issues and conflicts facing coastal resources and coastal development. ICZM is an issue-driven process, and the nature of the particular issues will dictate the type of program to be created. Also, the reality is that the effort and expense required to set up an integrated program would not be justified for countries that do not face serious resource conflicts and multiple use issues.

Sorensen and McCreary (326) make the point that there should be a good fit between the set of issues that the integrated program is attempting to resolve and the institutional mechanisms set up in response to these issues. These authors also remark that the same issues that motivated a nation to create an ICZM program are those that are likely to reappear as the criteria for program evaluation.

1.5.5 DIMENSIONS OF COASTAL ZONE

By virtually any set of criteria, the coastal zone is a linear band of land and water paralleling the coast—a "corridor" in planning parlance—which has a one-dimensional aspect. The second dimension (width from onshore to offshore) tends to be overshadowed by the linearity: thus, people talk about being *at* the coast or *on* the coast, but never *in* the coast.

There is not one standard prescribed set of boundaries. The actual boundaries to be used for defining the coastal zone depend upon the specific functions of the program. *All* ICZM programs are expected to include both land and water within their boundaries.

It is desirable to include in the coastal zone all land areas affected by the sea (the *dryside*) and all coastal water areas influenced by the land (the *wetside*). However, such a definition could encompass all coastal plains and the watersheds of all streams and rivers that drain into the sea, which may extend hundreds of kilometers inland into the hinterlands. At the other extreme, it might be appropriate to set the seaward ICZM boundary at the offshore edge of the continental shelf (depth of 200 meters). Within these extremes falls the littoral area, usually delineated as the ICZM planning zone. This zone should include as a minimum: the coastal floodplain, the intertidal areas, and the lagoons, estuaries, and shallow coastal waters within the usual range of artisanal fishermen (66).

It is important to recognize that ICZM programs evolve through stages from initial planning to final management program. The first set of boundaries are for the planning phase and do not imply that the entire coastal zone delineated for planning will be included in the management program that evolves. In fact, the zone of management that emerges will, in most cases, be considerably narrower than the initial zone of planning. Therefore, for the Strategy Plan, a relatively broad zone should be delineated. Subsequently, a narrower zone or zones might be delineated for the Master Plan.

It is conservation of the coastal *common property resources* of the wetside that is usually emphasized in ICZM. But to do this it is often necessary to control use of public and private property of the dryside.

These resources can be most efficiently managed by a program that focuses attention on the coastal edge—floodplain, tidelands, and immediate shorelands. (66)

To give one example, the Sri Lanka ICZM program has authority over a zone that extends from 300 meters landward of the mean high water line to 1 kilometer seaward of the mean low waterline. However, the zone may be extended to a maximum of 2 kilometers inland where rivers, lagoons, or estuaries occur (43). (See "Sri Lanka" case history in Part 4.)

Where different types, or levels, of management are anticipated in different parts of the designated coastal zone, it may be expeditious to delineate two or more *management tiers* or strip zones. In a simple four-tier system, the coastal waters would be the first zone; the transition areas (flood plain, tidelands) would be the second; the shorelands would be the third, and the hinterlands would be the fourth (see also "Tiers for Management" in Part 2).

Small island countries have a particularly difficult time determining ICZM boundaries. Some ICZM authorities would call entire islands *coastal zones* because most island commerce and societal affairs have a coastal connection. But primary authority over certain aspects of the *whole* of an island seems unrealistic for a *coastal* program. Therefore, for islands, the author prefers to identify a coastal zone in the same manner as for any other landform because management needs can be focused by using tiers. One of the tiers can include all the inner or higher parts of the island (perhaps an upland tier).

1.5.6 SEASTORMS AND OTHER HAZARDS

Coastal natural disasters cut across all economic sectors. Wind or water damage from a cyclone (hurricane), inundation by a tsunami, wreckage from an earthquake, or coastal erosion from seastorms can affect tourism, the fishing industry, port operations, public works, and transportation. Other sectors such as housing and industry are also vulnerable.

Sinking shorelands must be given special attention. Land subsidence may be caused by natural processes (e.g., soil compaction) or by human activities such as excessive pumping of water or oil from underground. The effects of coastal subsidence are multiplied by global sea level rise, causing even greater potential for the shoreline to move inland and settlements to be submerged during flood periods (Figure 1.18).

The consequences of development in coastal hazard-prone areas are so potentially devastating as to require serious governmental attention. A good way to direct this attention is via an ICZM-type

Figure 1.18. More coastal zone flooding events may occur in the future because of sea level rise caused by global warming. (Photo courtesy of Alvin Samet.)

program (326). It is usually neither economically feasible to neutralize hazards totally through engineering (e.g., by building giant seawalls) nor socially acceptable to exclude people and structures totally from the hazard zones. Therefore, a balanced management approach is needed. An ICZM-type program is a good vehicle for this because it can control development patterns and combine coastal natural hazards prevention with resource conservation. (84)

Embracing the dual goals of conserving coastal resources and maintaining nature's hazard protection systems can save money. Because of the link between uncontrolled development and disasters, an important aim of ICZM is to integrate the knowledge of coastal hazards and risks into development standards and planning guidelines. Specific guidelines for estimating the hazard risk level of a project or program should be applied to every development proposal.

Recommendations should be made about inclusion of natural hazards prevention in the ICZM program. As before, there are benefits and disbenefits to be expected from inclusion. But if no other governmental agency is dealing with the subject of maintaining natural storm defenses, ICZM should.

1.5.7 HINTERLANDS

One job of the ICZM planning unit is to create strategies for reducing negative impacts to coastal resources that occur when the hinterlands (shoreland or inland terrain) are cleared of vegetation, paved, or altered to accelerate drainage; when surface water bodies and watercourses are filled, detoured, or channelized; or when the natural flow pattern is significantly disrupted so that freshwater flow to the coast occurs in unnatural pulses (Figure 1.19).

Most nations have less control over the hinterlands generally than over their coastal areas. Exceptions may be found among nations with strong programs for land use planning or town and country planning (e.g., British Commonwealth), those with centralized economic development programs (326), or those with a high proportion of land in government ownership.

Because the hinterlands (and offshore waters) that may lie outside of a defined coastal zone are linked to the coastal resource system, an institutional method must be found to coordinate the management of this "Zone of Influence" with that of the ICZM program. The best approach may be to coordinate with a regional land use entity such as integrated rural development, watershed management, or regional planning.

1.5.8 POLLUTION

While ICZM's main business is management of coastal developments and conservation of environmental resources, this management should be construed to include pollution control. The Strategy Plan should recommend the level of involvement of the ICZM program with pollution control, for both liquid and solid wastes—coastal areas have some unique pollution problems (Figure 1.20). Most countries have an existing agency to handle pollution and other environmental quality matters.

Whether any of the coastal pollution control functions should be transferred to an ICZM program may become a major question. The answer depends upon the individual circumstances because there are both advantages (better integration into the full program) and disadvantages (increases complexity of program) of incorporating pollution control in ICZM programs.

In any event, ICZM cannot leave pollution problems entirely to another agency because many aspects of permit review and environmental assessment relate to the potential problems of pollution. Also, the engineers and architects of any project have to provide waste control facilities in their design studies and need the type of advice they can receive from an ICZM office. (See "Pollution" and "Solid Wastes," Part 3.)

It would be reasonable to assume that "non-point" pollution (that which is not collected, concentrated, or conveyed for disposal like sewage or factory waste) would be addressed in ICZM. A particular source of non-point pollution that might come under the primary responsibility of ICZM is storm water runoff from shorelands. Such runoff, including sediment from construction sites, farmlands, and forest cutting and land clearing operations can be seriously damaging to the productivity of coastal areas and can result in the fouling and filling of waterways. Also, pollutants flushed from the land by storm runoff into coastal waters can create toxicity (biocides, oil wastes, etc.) and bring excessive nutrients (fertilizers, animal wastes).

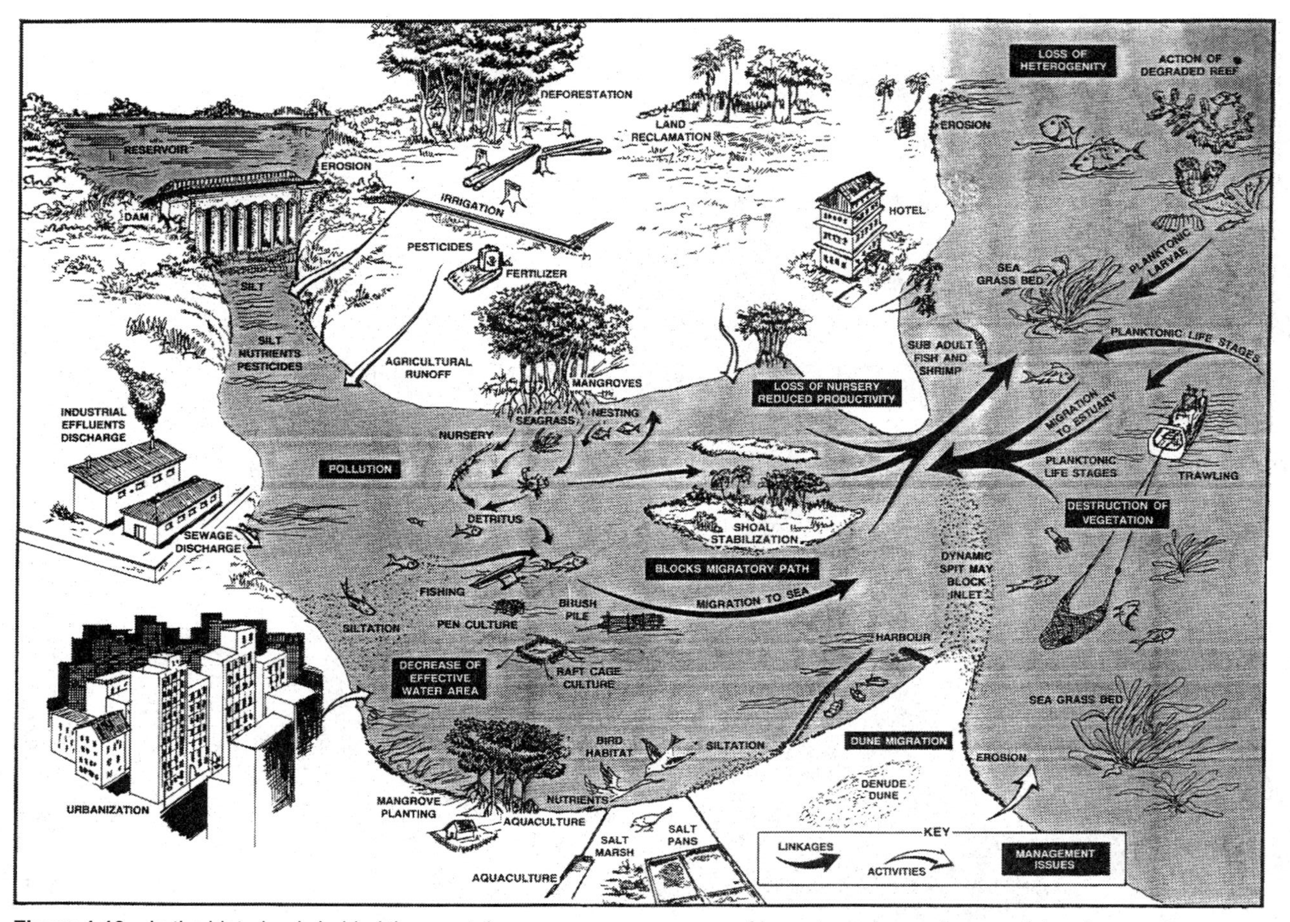

Figure 1.19. In the hinterlands behind the coastal zone, numerous sources of impact may jeopardize coastal environments. (*Source:* References 43.)

Figure 1.20. The coastal zone has unique environmental problems, such as this burning oil platform. (Photo courtesy of U.S. National Oceanic and Atmospheric Administration.)

1.5.9 BIOLOGICAL DIVERSITY

Threats to the productivity of the unique biological systems of the coastal zone—species and their habitats—arise from development activities and their side effects, including reef and beach mining, shoreline filling, lagoon pollution, sedimentation, and marine construction activities that are quite distinct from those on land.

The Strategy Plan must recognize that species and the habitats of coastal zones are so different from their terrestrial counterparts as to require different and special forms of conservation. For example, coral reefs, beaches, coastal lagoons, submerged seagrass meadows, and intertidal mangrove forests have no counterparts in terrestrial systems. In addition to habitat management, there may be other appropriate actions to be taken under the ICZM program, for example, banning exploitation of endangered species.

The species at risk are quite different from terrestrial ones—e.g., oysters, octopus, porpoises, whales, sea fishes, sea turtles, dugongs. The occurrence of endangered and threatened species is less in the sea because it is an open system with few boundaries to migration. While several of the sea mammals and sea turtles are endangered, the fishes and shellfishes are usually not.

Species protection by designating protected nature reserves is relatively inexpensive and simple to administer. This strategy can be implemented on a site-specific basis, commensurate with available information, staffing, or expertise. It can be reinforced with regulatory measures that combine "wetside" (estuarine or marine area) protection with "dryside" (shorelands) management strategies, offering the possibility of managing whole coastal ecosystems. But, first, it is necessary for the ICZM process to identify, during Strategy Planning, the critical coastal habitats that merit a high degree of protection so they can be addressed specifically in the Master Plan.

1.5.10 MULTIPLE USE

The exclusive use of a particular coastal resource unit for a single economic purpose is discouraged by ICZM in favor of a balance of multiple uses whereby economic and social benefits are maximized and conservation and development become compatible goals. The simultaneous achievement of development goals and resource conservation goals may require that communities modify some development patterns. However, with innovative management based upon sustainable use, communities may be able to achieve a desirable balance without serious sacrifice to either short-term development progress or longer term conservation needs.

Because achievement of balanced multiple use is a goal of ICZM, decision makers need to know which economic and social uses best coexist in an area and which conflict strongly. The following have been listed by Sorenson and McCreary (326) as major types of coastal economic sectors whose interests should be integrated into the ICZM program:

- Fisheries
- Water supply
- Port development
- Recreation development
- Agricultural/forestry development
- Aquaculture development
- Tourism development
- Energy development
- Biodiversity
- Industry

Many of these uses could coexist in a multiple use approach while others might not or would have to be severely restricted. The ICZM role is to sort them out and recommend an optimal mix. For example, there might be little problem in combining fisheries, tourism, and water supply, but mixing port development with critical area protection (e.g., mangrove forests) might be difficult. (See "Multiple Use" in Part 3).

The integrated approach of ICZM is particularly helpful in multiple use programs. The concept of greatest yield from the best multiple-use plan takes into account that specific resource systems are always components of larger ecological systems that may contain many other resources with economic and social values (316). Also taken into account is the fact that component resource systems naturally tend to be highly integrated and dependent upon one another. For example, in many parts of the world, freshwater is considered to be a limited resource for agricultural and domestic use (316). In the recent past, water development specialists tended to allocate water usage based on the most obvious economic and domestic demands but failed to take into account the key role of freshwater discharge in the maintenance of coastal estuaries and their fisheries (e.g., the serious depletion of fisheries after damming the Nile in Egypt and the Indus in Pakistan).

Contrary to some current impressions, conservation and economic development are not conflicting ideas. In fact, well-planned, conservation-oriented development will add to the general economic and social prosperity of a coastal community, while bad development will sooner or later have a negative effect.

In short, development and conservation can co-exist. Indeed, they must co-exist if resource-dependent societies are to prosper in coming years. First, *conservation* is necessary to adjust the pressure on the resources so that they are not overexploited and to preserve biodiversity. Second, control of *development* is necessary to protect resources from gross pollution and flagrant destruction of habitats. Both needs must be met if we are to have resources for the future.

While multiple use should be advocated for ICZM overall, there are specific situations where exclusive use would be recommended as the best approach. In fact, a general multiple use program will often be made up of limited use, or exclusive use, *subzones*. For example, some coral reef areas might be set aside exclusively for recreational diving. Or other areas might be reserved for marinas or factories. Yet this would all be done within the scope of a larger multiple use program.

The ICZM strategy planning unit should be prepared to investigate compatibility of uses and make recommendations for multiple or exclusive use. The planning unit's job is to evaluate the effects of uses singly and in their various combinations in order to advise the decision makers and managers as to the optimum management approach. The final details are resolved in the Master Plan process.

1.5.11 INTEGRATION

A major purpose of ICZM is to coordinate the initiatives of the various coastal economic sectors toward long-term optimal socio-economic outcomes, including resolution of use conflicts and beneficial tradeoffs. This integrated, multiple-sector approach is designed to coordinate and jointly guide the activities of two or more economic sectors in planning and management (Figure 1.21). This supports a programmatic goal to optimize resource conservation, public use, and economic development.

Integration is the *sine qua non* of ICZM. The ICZM management strategy requires "horizontal integration" among various economic sectors and the national agencies (including their private sector clientele). As an example, integration may be needed among fisheries, tourism, oil and gas development, and public works where these sectors are all attempting to use the coastal zone simultaneously (326). Both fisheries and tourism depend to a large extent on a high level of environmental quality, particularly coastal water quality (Figure 1.22). Both sectors may receive "spillover" impacts such as pollution, loss of wildlife habitat, and aesthetic degradation from uncontrolled oil and gas development (66).

International assistance agencies commonly use the term *sectoral* planning, or management, to describe these activities within a particular economic sector or development area. Some of the more common sectors are agriculture, forestry, fisheries, energy, transportation, manufacturing, tourism, housing, military, and public health. Any of these sectoral areas can be further divided into more specialized coastal components; for example, transportation may be divided into shipping, ports, and surface transport (326). A further "sector" to be considered in ICZM is the public, or social, sector, including the various publics who are affected by private and government decisions.

Almost every sector has a strong stake in the coastal area. It would be virtually impossible to allocate the coast to a single one of these economic sectors for development or even to give one or two sectors a priority for coastal development. In fact, it is the intense conflicts over use of the coast that arise among the various sectors, or *stakeholders* that make the ICZM process so necessary.

As an example, fisheries may require port services similar to those that tourism depends on—an infrastructure system that supplies water, sanitation, transportation, and telecommunications. Therefore, planning for both should be integrated with that for transportation and public works sectors.

It is clear that the 30 or so developing nations in the humid tropics with extensive mangrove forests should have a strong incentive for integrated coastal resources management. Most developing nations with extensive mangrove forests are confronted with similar stresses, which threaten the sustainability of this renewable resource. Conversion of mangrove forest to aquaculture ponds or croplands often presents a particularly difficult conflict of uses (326).

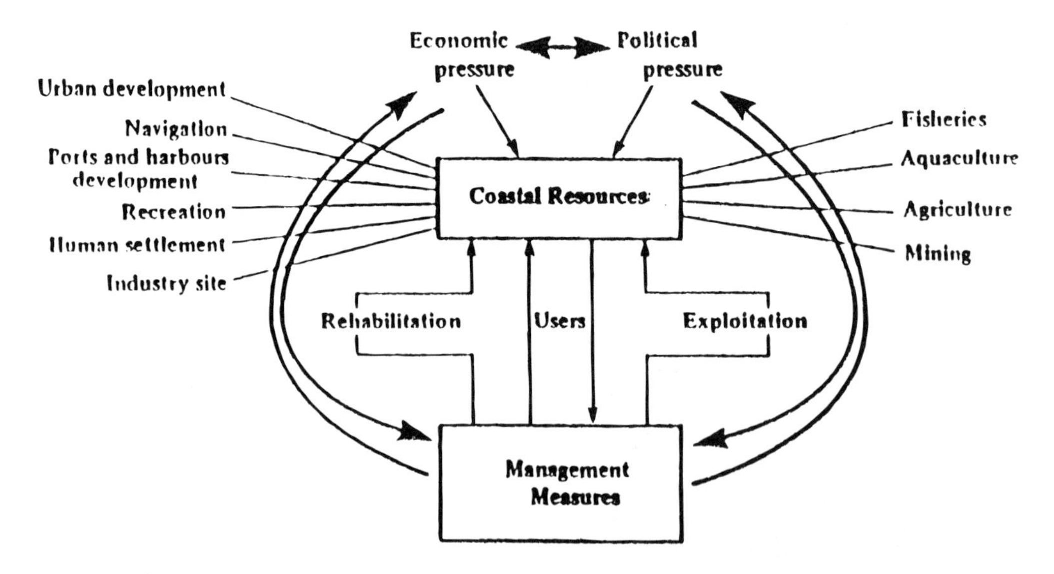

Figure 1.21. The stakeholders in coastal resources include a variety of economic sectors and political powers, each with a need to influence an integrated coastal zone management (ICZM) program. (*Source:* Reference 25.1.)

Figure 1.22. Fishermen have an important stake in protection of coastal environments and natural resources, but they must have the cooperation of many sectors and agencies. (Photo courtesy of U.S. National Oceanic and Atmospheric Administration.)

A major goal of ICZM is to identify conflicts over coastal land and coastal renewable resources and to find ways to allocate and manage uses for the optimum long-term benefit of the nation in a multiple use format. In narrow sectoral planning it is easy to irreversibly destroy a resource and foreclose all future options for use of that resource. ICZM attempts to avoid this for coastal area resources by broad multiple-sector planning and project development, by future-oriented resource analysis, and by applying the test of sustainability to each development initiative.

Because the ICZM process operates at the interface between land and water, there is often intense conflict between private (or quasi-private) property-based operations in shorelands and public (common) property-based activities in the tidelands and coastal waters. Thus, the ICZM process may have an important mediating role between conflicting wetside and dryside interests. This essential role should not be played out competitively, sector against sector. Rather, the ICZM mediating/coordinating entity must look at all sectors with legitimate interests to find the most broadly compatible solutions.

The integrated approach of ICZM is particularly essential for effective multiple use approaches. The concept of greatest yield from the best multiple-use plan takes into view that specific resource systems are always components of a larger ecological system that contains many other resources with economic and social values (316). Also taken into account is the fact that component resource systems naturally tend to be highly integrated and dependent upon one another. In summary, in no other part of the earth is integrated, multi-sectoral resource planning and management more needed than at the coast.

1.5.12 COORDINATION

ICZM is a complex undertaking in terms of the number of stakeholders. In comparison, most social policy areas (e.g., criminal justice, drug enforcement, public health) involve relatively fewer government sectors than does coastal management. The greater the number of sectoral divisions within a policy area (such as ICZM), the greater the potential for fragmentation of governmental responsibility and duplication of effort. The ICZM authority must have influence on a wide range of ministries and agencies—e.g., finance, agriculture, economic planning, commerce, tourism, forestry, and transportation—and must often take a position on coastal development or conservation that may be viewed as adverse by one or more of these agencies. To enhance strategy planning it is important to create a strong interagency coordinating mechanism. This will ensure the widest and most effective participation of government agencies.

Because coastal areas are governmentally complex, they require an especially high level of intergovernmental coordination. Some of the causes are given below (66):

- The amount and complexity of public interests in coastal areas is high.
- The effects of conflicts and impacts of one sector on another that require government intervention is exceptionally high.
- There is considerable involvement with public (common property) resources and their conservation.
- Water is a fluid resource that is not containable or ownable in the usual sense and that simultaneously affects all coastal interests.
- A greater variety of coordinated multi-governmental policy decisions is required in coastal areas.
- There tends to be a high level of international interest in coastal matters.

1.5.13 INSTITUTIONAL MECHANISMS

Coastal resource management requires involvement by all levels of government. The local governments are involved because they govern where development takes place, where resources are found, and where the benefits or disbenefits are mainly to be felt. The central government has to be involved because responsibility and authority for marine affairs inevitably rests there (navigation, national security, migratory fish, international relations, etc.). Intermediate levels of government—e.g., state (provincial) or regional—are involved because all entities that have responsibility in the coastal area have a role in the ICZM process. It should be made clear that ICZM is an overlay program; it does not replace existing institutional arrangements in most cases but rather is meant to strengthen them. (66)

But the integration of multiple agency interests into a single program is *very difficult.* Without exception, institutions will defend their turf and only yield authority and prerogative grudgingly. Getting institutions to cooperate in multi-sectoral activities toward ICZM goals that no single institution can accomplish singly is certainly one of the toughest jobs for the ICZM authority.

The foundation for the larger changes in laws and programs to accomplish coastal management should be built upon a realistic assessment of where each nation currently finds itself in terms of coastal policy development and implementation in the broader context of sustainable natural resources development. For example, many countries formulate policy through a continuous series of five-year economic development plans.

For full-scale, comprehensive programs, it may be desirable to create a new agency, such as a joint authority, provided it will have the governmental support, power, and resources necessary to perform its function. In other cases, an existing agency may be designated as the lead agency, provided it can be motivated to carry out conservation management, has clearly stated objectives consistent with the objectives of the legislation, and is given the necessary responsibilities, powers, and administrative and technical resources. Without primacy of authority—i.e., above the level of the individual ministries or departments of government—ICZM success may be minimal.

There is no single answer to the question Where should the ICZM authority be lodged within the government institutional structure? The correct answer will be different for each country depending upon answers to other questions, such as Would a coordinating office be sufficient? If so, within what ministry should the office be lodged? Or would an agency with power to act independently be needed? What kind of staff skills are necessary? How would such an agency integrate the roles of the several sectorally oriented agencies with strong interests in the coast? These are crucial points that must be addressed.

Institutionally, most countries will find that it is appropriate to fit their ICZM program into the current governmental structure in a manner that causes the least disruption of present institutional alignments. The political priorities of most countries are such that a new agency with strong powers that would preempt the authority of existing agencies would not usually be formed for ICZM. Therefore, it may be desirable to locate the ICZM office within an agency that already has appropriate regulatory powers, such as a natural resources, fishery, or environmental ministry, as shown by Alternative A of

Figure 1.23. Or the authority can be spread over many agencies if there is strong central coordination, as in Alternative B of Figure 1.23, exemplified by Oman (306, 307). Specifically, the ICZM regional unit plan for the Greater Capital Area of Oman recommends a lead agency only for coordination; management responses are allocated to the separate "line agencies," a process that is sometimes termed *networking* (see "Oman" case history in Part 4).

However it may be implemented, ICZM requires effective integration of appropriate sectors into an umbrella program. This can occur only if there is strong interagency coordination. Therefore, it will be desirable, if not esential, to establish an "interagency coordinating committee" to participate in preparation of the Strategy Plan. The same group will later participate in Master Plan formulation with a mandate to review progress, consider program changes, discuss proposed new rules, and provide technical information and advice. Later, after implementation, the committee would review specific development applications and resource management proposals (as the Sri Lanka national coordinating committee is now doing).

The ICZM office within the lead agency should be mandated, staffed, and budgeted to accomplish at least the following three tasks: (a) inter-institutional coordination on coastal development and resource conservation matters, (b) environmental assessment and permit issuance for all major coastal developments, and (c) gain compliance of all sectors with ICZM rules and decisions. Other tasks should be added, as possible, to build a full-service ICZM operation. It may be useful to designate a particular ministry for the strategic planning phase and another for implementation, including program development and management.

1.5.14 LEGISLATION

Strategy planning lays the foundation for the legislation or executive order needed to authorize the ICZM program and should specifically recommend the legislative approach by which to (a) authorize

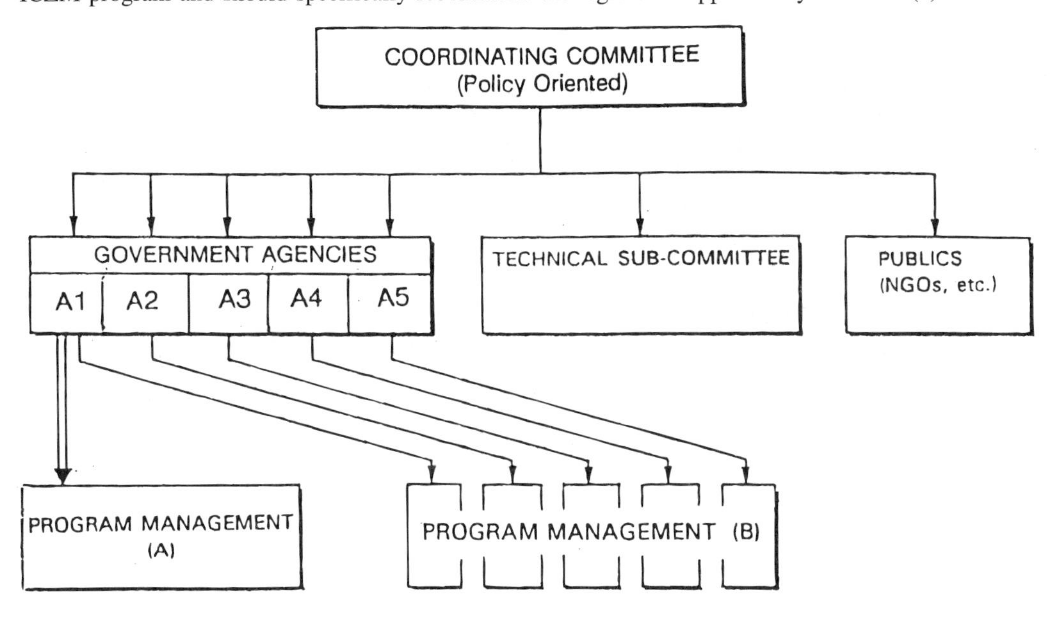

Figure 1.23. Alternative institutional arrangements for administering an integrated coastal zone management (ICZM) program. (*Source:* Reference 77.)

the Program Development phase and the necessary funding; (b) assign responsibility for Master Plan preparation to a particular agency; (c) pronounce the goals and purposes of the ICZM program; (d) prescribe a method for collaboration among the various stakeholders (sectoral agencies, private sector, and the affected public); (e) limit the time to complete the various stages of program development and master planning; and (f) require a specific step-by-step program development and organizing process. (66)

In the ICZM process, legislation should enable the administrative arrangements to be as flexible and cost-effective as possible and should adhere to simple guidelines, such as the following (308):

- New agencies should be created only where existing agencies cannot be adapted, motivated, and empowered to carry out the task adequately.
- Existing agencies with jurisdiction over coastal/marine activities should be involved by interagency agreement as necessary and appropriate to meet the conservation objectives.
- Existing regulations and regulatory mechanisms should be continued when they are consistent with conservation objectives.
- Regulations and management plans should be as simple as possible.

In either case, the relationship between the lead agency and other concerned agencies must be clearly defined in legislation, particularly with regard to potential conflict or overlap in different pieces of legislation. Processes for resolving conflicts and for consultation between relevant agencies should be defined in the legislation, which should additionally specify that the lead agency has ultimate authority over marine conservation and area protection. (308)

1.5.15 PROJECT REVIEW, PERMITS, AND ENVIRONMENTAL ASSESSMENT

The Strategy Plan should define a requirement for a special review of each major development project and state some particulars about the environmental and socio-economic assessment procedures to be used. Most developing countries have some experience with Environmental Assessment (EA) systems for project review, often in conjunction with the international aid community of doners and loanors. With such experience as a background, the ICZM program should be able easily to create its particular project review and EA system.

The main mechanism for environmental control in ICZM is a requirement that a permit be obtained for each coastal zone development project. A project review—including EA—is conducted for each permit application. The EA process includes the prediction of a proposed project's effects on renewable coastal resources, biodiversity, and the quality of the human environment.

When EA (including socio-economic assessment) is added to the traditional engineering and financial studies of a project, the package of impacts is complete and sound decisions can be made. Experience shows that requiring an EA is not the death knell of a project, but it could necessitate design changes. Said another way, it would be just as rare for a project to be stopped because of an EA as it would be for a project to emerge from an EA unmodified. Assessors are typically more focused on finding practical changes to conserve resources than on frustrating development initiatives. (See also "Project Review and Permits," "Environmental Assessment," "Economic Impact Assessment," and "Social Impact Assessment" in Part 2).

1.5.16 SETBACKS

Shoreline setbacks (or *exclusion zones*) are very useful in coastal management. The purpose is to exclude certain uses from areas close to the shoreline. The objectives of such setback zones include avoidance of damage from flooding and erosion, protection of ecological functions, and protection of the viewshed and of public access to the shore. Shoreline setbacks may be set at a uniform distance (such as 100 feet), or may vary as dictated by natural features, such as the 10-foot topographic contour. (65)

Setbacks have many uses in an ICZM program. For example, along eroding and receding beaches, a predicted "recession line" of say 50 years into the future can be estimated by beach geologists, with no structures allowed seaward of the line (Figure 1.24). In this way, the recession line becomes a setback line that anticipates the horizontal effect of sea level rise (predicted to be as much as 1 meter in the next 100 years). A setback policy of 100 meters is recommended by Cambers (40) for Caribbean beaches suffering severe erosion.

In addition, setbacks are a good way to protect certain ecologically critical areas (ECAs), such as mangrove wetland habitats. For this, it is necessary to use a setback provision whereby coastal development is prohibited in a buffer zone between the edge of the critical habitat and any structure or major land clearing or conversion. Setbacks have other uses in an ICZM program, as discussed in "Setbacks" in Part 2.

1.5.17 INCREMENTAL APPROACH

While the ICZM program is to be planned and organized in a comprehensive and nationwide format, it can be initiated incrementally. The increments could be divided into any of the following: year by year, function by function, resource by resource, issue by issue, or region by region. The idea is that the incremental approach works with *generic* categories that can be introduced incrementally. On the other hand, *situation management* handles the *specific,* or one of a kind, type of situation (see Section 1.6.8 below for details).

As an example of the function-by-function approach, the program might start with just a review-and-permit function and without land use planning or technical services, which would follow at later phases of implementation. In the resource-by-resource approach, it may be decided to focus on a particular resource, as Costa Rica did in concentrating its ICZM-type program to protect mangrove forests. In the issue-by-issue approach a country may start with one main generic issue, as Sri Lanka did by starting with beach protection (see "Sri Lanka" case history in Part 4). Alternatively, it may be expeditious to focus initially on loss of coral reefs or degradation of estuarine fish spawning areas, using the strategies of critical area designation or shoreline exclusion setbacks (326).

Region-by-region implementation refers to applying a comprehensive ICZM program to one or more regions within a country. This region may be designated for special attention because it is politically advanced or is identified as one unit in a decentralized ICZM program, as the United States has done, Sri Lanka is attempting, and Yucatan, Mexico, is studying.

It is recognized that ICZM programs involve a strong central government role for many reasons, not the least of which is that central government typically retains most jurisdiction over coastal and ocean waters. Nevertheless, ICZM programs can operate at a subnational level. In fact, many countries will find that the most effective program is one that recognizes the distinctiveness of resource programs in various coastal regions of the country and focuses on regional, rather than national, solutions.

This approach should be feasible in countries with a tradition of regional development planning, by means of which coastal resource management can be considered concurrently with economic and social development needs. Then a balance can be struck between short-term payouts and the long-term benefits of management for sustainable use of natural resources. Canada's experience with a regional approach did not work out well. In fact, Hildebrand (167) states that "the momentum for a nationally coordinated CZM program was lost" in disagreements over administration of a federal/provincial agreement for the Fraser River Estuary (British Columbia, Canada) regional program.

Hildebrand lists the following reasons to explain why the Canadian initiative for a nationwide ICZM program failed in the early 1980s:

- Lack of agreement on a satisfactory definition of the coastal zone
- Political boundaries versus ecological boundaries
- Coastal zone treated as a common property resource
- Lack of awareness of coastal zone problems
- No clear motivation for coastal zone management (CZM)
- Administrative fragmentation
- Lack of clearly stated goals
- Dominance of short-term management over long-range planning
- Inadequate information on which to base decisions
- Attitudinal problems
- Antagonism to the political and economic grain of the time

Figure 1.24. Houses fall into the sea at Fire Island, New York (USA), because in the absense of a designated "setback" they were built too close to the sea. (Photo by Robert Perron.)

The incremental approach to ICZM implementation provides the opportunity to test concepts and approaches as a pilot effort before committing energies and political capital to a full-scale nationwide effort. The risks and consequences of failure should be considerably less with incremental program implementation. Also, the experience gained during the incremental effort could increase the likelihood of success of the later, expanded effort. Nevertheless, one should review Canada's failed experiment with ICZM whereby a political backfire in a test province, British Columbia (in the 1980s), caused the central government to dismiss the idea of a national program.

1.5.18 PARTICIPATION

ICZM programs require high levels of participation. Most likely, people who live along the coast and have traditionally used coastal resources will be greatly affected by any new rules and procedures. Therefore, if they are to support the program they must have a voice in the formation of coastal policies and rules on resource use. Broad participation will also provide an opportunity to resolve conflicting points of view among various stakeholders so that later the ICZM program will meet lesser political resistance (Figure 1.25).

The stakeholders should be encouraged to go beyond dialogue and to participate to the maximum possible in goal setting, data collection, conflict identification, and so forth. The decision-making and implementation processes must be shared with the people and their organizers, requiring efficient communication and effective dialogue through "participatory planning."

The responsibility for communication with the variety of general public stakeholders is complex and important. Information sharing is especially important. Public consultations should be held in advance of important decisions. Many people are reluctant to attend public meetings and particularly to speak out. Therefore, a participation preparation stage of training and encouragement may be necessary. (See also "Public Participation in Planning" in Part 2.)

According to Yves Renard (293), public participation is a tool available to the entire management community (resource users, public agencies, non-governmental organizations, etc.) to ensure the quality and the effectiveness of the management solutions that will be implemented. Participation is also a duty because the issue remains, above all, one of human development and because "people are not the *object* of that development but the *subject* of development and the makers of their own history."

The ICZM dialog should include private sector and military spokesmen as well as government and NGO groups. Some typical economic sectors of coastal countries include the following (25.2).

Sectors Often Coastal or Ocean Specific

1. Navy and other national defense operations
2. Port and harbor development (including shipping channels)
3. Shipping and navigation
4. Recreational boating and harbors
5. Commercial and recreational fishing
6. Mariculture
7. Tourism
8. Marine and coastal research
9. Shoreline erosion control

Sectors Rarely Coastal Specific but with Impacts

1. Agriculture
2. Forestry
3. Fish and wildlife management
4. Parks and recreation
5. Education
6. Public health—mosquito

7. Housing
8. Water pollution control
9. Water supply
10. Transportation
11. Flood control
12. Oil and gas development
13. Mining
14. Industrial development
15. Energy generation

1.5.19 MOTIVATION

For most countries, the motivation for implementing a coastal mangement program will have to be very practical. Sorenson and McCreary (326) list some major incentives for coastal programs: (a) to benefit fisheries; (b) to enhance tourism; (c) to improve management of mangrove forests; and (d) to protect life and property from natural hazards (e.g., hurricanes, shore erosion, land subsidence, sea level rise, landslides).

Advocates of conservation will usually have to demonstrate that definite economic benefits will accrue from ICZM. Explicit and persuasive social benefits will also be needed. The values that developed nations put on biological diversity, saving endangered species, and protecting environmental quality will not so readily be embraced by developing countries.

In trying to convince a supervisor, decision maker, or legislator that conservation is essential, nothing is more important than economic evidence. National income, foreign exchange earnings, employment, and local self-sufficiency are most important factors. Because of the importance of economic justification, it may be necessary to employ a professional resource economist to assist with economic valuation of resources. (See also "Political Motivation" in Part 3.)

1.5.20 ADDRESSING SOCIO-ECONOMIC CONCERNS

Environmental impacts and their social and economic ramifications are especially important aspects of planning for economic development. While environmental impact assessment and socio-economic impact assessment are usually done for a specific project, they can be done in a general or "programmatic" sense for regional-type planning; that is, a generic assessment can be made based upon possible development proposals and the environmental vulnerabilities of coastal ecosystems.

Also relevant in addressing rural communities is the human ecology perspective, which takes into account traditional uses, rights, and special needs of tribal minorities and how environmental change might affect them. Migration and population expansion also must be considered (263). This can often be efficiently accomplished by use of rapid rural appraisal methods (see "Rapid Rural Appraisal" in Part 3).

Because economic development should be premised on the concept of social benefit, the equity aspects of social impact assessment (SIA) of development are an important part of the planning and evaluation process (94). For one reason, SIA is a vital part of environmental impact assessment (EIA) because people live in the environment right along with the natural flora and fauna. Environmental changes affect people. People in turn affect their environments, and these changes affect social well-being.

Since people are the subject of development, not its object (293), opportunity and social equity should be the important parameters. In addition, it should be within the context of human action and reaction that all impacts are assessed. Environmental impacts also have economic consequences, which should be evaluated in terms of their effects on the people involved.

Burbridge (32) recommends that, in practice, ecological and economic impacts should be jointly evaluated. With increased perception of ecological functions, it is possible to improve the economic expression of their value to society. For example, mangroves, once thought of as worthless swamps unless developed for real estate or converted to make shrimp ponds, are now seen as extremely valuable resources capable of supporting a variety of activities.

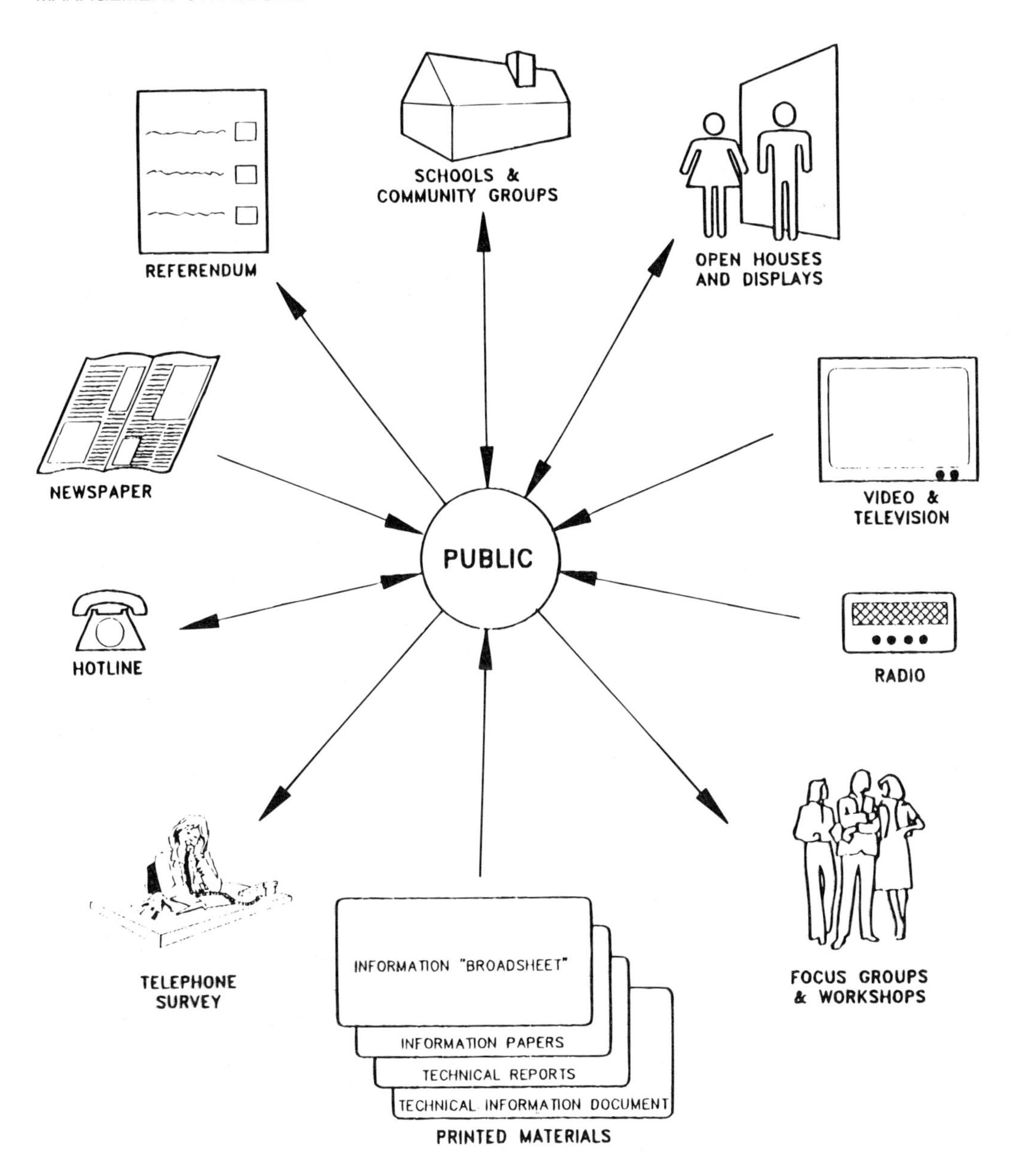

Figure 1.25. The participation "wheel" shows a variety of methods for creating public awareness and gaining the opinions of stakeholders. (Courtesy of the Capital Regional District, Victoria, B.C., Canada.)

If economic analysis is done carefully, it is feasible to incorporate the full range of products and services and to demonstrate their value either in explicit monetary terms or in a qualitative manner that is fully defensible. One goal of an economic analysis is to help decision makers to identify all relevant factors and to assess the benefits versus costs to society from different management alternatives (32).

1.5.21 ALTERNATIVE LIVELIHOODS

In coastal areas many rural communities are dependent on resources that are rapidly declining and yet they have few other means to survive. Such marginalized people do not have the cushion to risk management actions that might deny their means to a livelihood even for a short period. In this situation, according to DuBois (115), politicians may be willing to put a program in place but not enforce it if it means forcing people beyond the margin. As Dubois puts it: "One might say management does not translate into reduced livelihood. However, a management regime that prevents dynamite fishing . . . directly translates into reduced catch for the short term." Moreover, there exists the question of *perceived* threat to livelihood, which is just as serious as the reality of the situation with respect to political will for controls, enforceability, and effectiveness of management.

Therefore a desirable ICZM approach in those rural areas undergoing new conservation regulation is the development of alternative livelihood systems designed to provide relief during a waiting period while resources are regenerated. One assumes that the rehabilitation will, in turn, provide a more productive long-term natural resources base and support a truly sustainable system of resource utilization. (For examples see "Sri Lanka" and "Indonesia, Bali" case histories in Part 4.)

1.5.22 INFORMATION

Data collection and assimilation is an important part of the Strategy Plan formulation phase. It is during this phase that the most important decisions will be made about the future of the ICZM program or even about whether it will have a future. Clearly, these decisions should be made in the most data-rich circumstances, so that the consequences of taking or not taking specific actions are knowable.

The traditional planning model may assume that decision makers have (a) a well-defined problem, (b) a full array of alternatives to consider, (c) extensive baseline information, (d) complete information about the consequences of each alternative, (e) full information about the values and preferences of citizens, and (f) adequate time, skill, and resources. Yet, in reality, planners often face the opposite of these conditions (327).

Because the information needs for a Strategy Plan depend upon the issues to be addressed and because these vary considerably from country to country, it is not possible to set forth a standard list of information requirements for an ICZM program from which to prescribe a data compilation program (but see "Database Development" in Part 2 and "Information Needs" in Part 3).

It will be expeditious to organize the Strategy Plan so that essential information can be mapped and to display as many categories of data as possible on maps. For example, a Caribbean Conservation Association report states that "it has been found time and again that perhaps the most useful way for the environmental planner to discover trends, conflicts, and problem areas that can otherwise be easily overlooked, is by mapping information" (140).

In collecting data and synthesizing information for the Strategy Plan, it should be remembered that the purpose of this stage is to bridge between the policy formulation and the program development stages. The information is to be used in getting a management program approved or disapproved if it should be discovered that the country does not need or want ICZM.

Therefore, the kinds of information needed for the strategic plan are those that will enhance the decision-making process, that clearly depict the tradeoffs between the present situation and an integrated approach, and that lead to the clearest and least ambiguous set of objectives and mandates to the governmental agencies who are to manage the ICZM program. Examples include the following:

- **Users of coastal areas and resources:** Tourism, manufacturing, maritime trade, mining, urban, and oil industries; jobs, revenue, investment, and tax yields
- **Coastal renewable resources:** Fisheries and aquaculture activity and yields, by species and seasons; mangrove forest exploitation, activities, and products
- **Environmental impacts:** Impairment of coastal resources and ecosystems: pollution, habitat losses, species depletion, sedimentation, and visual degradation
- **Upland effects:** Impairment of coastal resources from river dams and diversions, accelerated sediment transport, reduction of freshwater inflow, disruption of natural hydroperiod, and reduction of beach nourishment with erosion and pollution
- **Socio-economic conditions:** Economic statistics for coastal communities; social organization of coastal communities and dependencies on coastal resources
- **Critical habitats:** Habitats of critical importance (ECAs) such as mangroves and other wetlands, beaches, dune fields, seagrass meadows, coral reefs, tideflats, estuaries, lagoons, shellfish beds, and special breeding and feeding areas for coastal species; restoration needs

- **Critical species:** Identification of the coastal species of particular significance or economic value and their habitats; restoration needs
- **Resource problems, issues, conflicts:** Information on special problem situations, such as highly polluted estuaries, extensive mangrove clearing for aquaculture ponds, destruction of coral reefs, and so forth
- **Natural hazards:** Identification of areas of high risk to natural hazards such as badly eroding beaches, floodable lowlands, and landslide prone slopes
- **Nature reserves:** Description of areas that should be designated as reserves or other protected areas

For ICZM, it is often useful to start with the regional context. At the regional scale, information is usually readily available and fairly easy to assemble. To be relevant to management, natural units such as ecological systems and morphological units should be related to political jurisdictions, which represent the loci of decision making. The boundaries of political and natural systems rarely coincide.

1.5.23 COMPARABLE PLANNING

The ICZM strategy planning process for coastal conservation may seem innovative, and it certainly is distinct because of its subject. However, the closely related "strategic planning" for economic purposes is commonplace, particularly in countries with centrally planned economies and large corporations with diversified activities and holdings. In regard to private sector strategic planning, Michael Porter (287) comments that:

> Strategic thinking rarely occurs spontaneously. Without formal planning systems, day-to-day concerns tend to prevail. The future is forgotten. Formal planning provides the discipline to pause occasionally to think about strategic issues. It also offers a mechanism for communicating strategy to those who have to carry it out, something that seldom happens when the formulation of strategy remains the private province of the chief executive.

Strategic planning has long been recommended for countries considering national systems of protected natural areas, including marine and coastal parks and reserves (308). Kenton Miller (246) discusses the advantages:

> Strategic planning for the maintenance of living resources examines the major overall problems facing species, communities, people and life support systems. It focuses upon the requirements for ensuring the long-term maintenance of species and habitats while assuring the flow of short and immediate-term benefits. Strategic planning reviews major trends, analyzes the relevant factors, synthesizes the relationships among key factors, and identifies priorities for action. In addition, and to keep the planning process up-to-date and on target, strategic planning periodically reviews all field activities which have been derived from the strategy in order to glean learning from experience, and to constantly revise the information upon which planning is based.
>
> The strategic planning process provides the planner with means to rise above the details and grasp entire ecosystems. Trees are envisioned as elements of forests; fish are elements of complex water columns. Through the use of these means, the areas of land or sea which merit special forms of management to sustain human development and environmental stability can be identified. The ways, places, and time periods when certain development alternatives will involve negative or positive impacts for the environment can be noted.

1.6 PROGRAM DEVELOPMENT

The next activity after completion of an ICZM Strategy Plan is to develop the management program, should the country (or province) decide to establish one. The key item in the Program Development phase is an integrated comprehensive management plan, or coastal "Master Plan."

The Master Plan will dictate the content, form, and scope of the program. An integrated, centralized coastal management operation will usually be a special function placed in a newly created ICZM office, which could be either a coordinating entity or a semi-autonomous management authority. At the center of the ICZM program is a project review and permit system that includes environmental assessment (EA). The information and analyses generated in the EA provide the basis for the decision on whether a coastal project is disapproved, is approved as it is, or is approved with conditions attached. Lately it has become common to include economic and social considerations as part of the assessment.

1.6.1 ORIENTATION OF PROGRAM

In creating the ICZM program, it is important to recognize that the primary emphasis is on management of coastal development, which would be a new and a special function to be addressed by ICZM. Lesser emphasis would be on resources management—fisheries, forestry, and so forth—because this is usually an ongoing function of government.

An example of resource management is fisheries management. In most countries there is a department of fisheries, which has responsibility for the management of coastal and marine fish stocks (as common property resources). This department implements conservation by prescribing closed areas for fishing, closed seasons for certain species, prohibitions on certain methods of fishing, and certain other controls. ICZM does not preempt this resource management role. Instead, it supplements and enhances the work of the fishery managers by controlling coastal development and habitat destruction (such as coral mining), which conflict with fisheries and which could have an adverse impact on fish stocks and particularly on the critical habitats that provide life support for fish species.

An example of a critical habitat for fishes and shellfishes to be protected by ICZM control of development activity is the mangrove forest (Figure 1.26). ICZM attempts to prevent obliteration of mangrove forests by constraining excessive mangrove clearing, as for aquaculture. It also offers control of other destructive development activities that are unreasonable and have a long-term net negative effect on biodiversity, the coastal economy, and the welfare of coastal people. (See "Mangrove Forest Management" in Part 2.)

Figure 1.26. Mangrove forests—such as this one in an Ivory Coast lagoon—are key resources providing a wide range of benefits for fisheries, storm surge protection, and shore stability as well as dozens of wood products. (Photo by author.)

1.6.2 MASTER PLAN

The Master Plan is the framework of the ICZM program. It identifies options for human progress in the coastal area and recommends governmental and private actions to accomplish beneficial and sustainable change, change that is "economically sound and socially just and that maintains the natural resource base" (263). Depending upon the degree of comprehensiveness and multi-sectoral integration involved in the program, the planning function of the program may be either limited or extensive in subject and geographic coverage.

Each country's (or province's) Master Plan will reflect its own special set of conservation issues. For example, the Sri Lanka ICZM program is focused on erosion control and habitat conservation, as shown in the structure of its 1990 Master Plan (43) abstracted below:

Introduction: Need for ICZM, scope of plan, basis of plan, definition of coastal zone (cz), management framework

The Regulatory System: Permits, criteria, variances, prohibited activities, setbacks, environmental impact, monitoring

Coastal Erosion: Problem, causes, mangement needs, objectives, and policies

Coastal Habitats: Problems; evaluation of habitats; management needs, objectives, and policies; listing of habitat categories (reefs, mangroves, seagrass, beaches, etc.); habitat-specific management program

Historic Sites: Problem, listing of categories/types, management objectives, policies, methods

Institutional: Needed revisions to law, needed improvements in administration, financing of the program, information and awareness needs/improvements, land acquisition

1.6.3 JURISDICTION

It is a major objective of ICZM to facilitate improved coordination among levels of government and their various bureaucracies toward defining specific resource conservation goals. This may require realignment of legal authorities at various levels of government. The very fact that an ICZM program is being considered for a country suggests that more attention needs to be given to conservation of coastal renewable resources and that some different alignment of responsibilities and jurisdictions may be needed.

The matter of jurisdiction will arise in the master planning activity. As mentioned above, there is *real* danger in expanding the ICZM program into areas of jurisdiction that are already occupied by other governmental authorities: for example, while ICZM programs do address fish habitat, nursery areas, and the water quality of fishery areas, they do not usually address fisheries operations, such as controls on harvesting (e.g., quotas, sizes, closed seasons, gear type). Because these functions typically involve only the fishery sector, they may be handled separately from ICZM, which emphasizes multiple-sector concerns and control of development. However, for smaller countries that must consolidate bureaucratic function, the ICZM format could be expanded to include fisheries harvest management and such controls as closed areas and maintenance of traditional fishing practices.

Few countries will have an interagency, or interministerial, entity already in existence that is positioned jurisdictionally to take on an ICZM program. Therefore, a lead agency with an interagency mandate will usually have to be organized to accomplish the coordinative management and planning functions of ICZM. As Joliffe (44) states the case, "what is quite clear is that existing planning and management agencies are essentially land or sea-based and that the littoral zone normally marks a jurisdictional boundary rather than the vital focus for coastal planning and management."

It would be relatively easy to identify the coastal zone *grosso modo,* according to Freestone (31.1), but for an effective ICZM program to work the implementing legislation would need to identify clearly the area over which the management authority would be able to exercise direct administrative and regulatory powers (for the granting of planning permission, permits, concessions, etc.). This area should reflect and include the main areas in which problems are envisaged (predominantly those that triggered the whole ICZM process) but otherwise may well be essentially arbitrary in that there is no single boundary that will be useful or feasible in all situations (Figure 1.27).

1.6.4 LAND USE

Because coastal land is so scarce and so often ecologically sensitive, it may have to be used in a different manner than other land in order to protect the environment and keep options open for future priority development and resource use. For example, in the United States, "non-waterfront dependent" uses of waterfront land are typically discouraged and usually denied if negative environmental effects are predicted (by exercise of state CZM or federal permit authority).

The very notion of resource limits engenders planning and land use control as a framework for making coastal development decisions. To a large extent the strategy for such developments has been undertaken by the private sector alone, in response to market demand and largely in the absence of government policy or guidelines and without the essential inputs of socio-economic and environmental assessments. This default is to some degree caused by the non-availability of planning guidelines developed specifically for ICZM-type programs.

Blommestein (27) notes that there is a concentration of often-conflicting human activities in the coastal zone. Activities like housing, agriculture, tourism, manufacture, and infrastructure often compete for the same scarce resources; therefore, conflicts can and do frequently arise. The continuance of such conflicts may reflect a weak commitment to land use planning.

Waterfront development is as vital to coastal cities as it is threatening to coastal resources. But high quality shorefront land parcels are in high demand, particularly those that extend to and/or across the intertidal zone and carry values for ship berthing, access to fishing, access to the beach for tourists, or unobstructed view of the sea from home or hotel. With such demand for sites at the water's edge, the development planning context must give special attention to properties that lie along the shoreline.

Land use zoning is quite common in many countries, whereby specific parcels of land are identified for particular classes of use—ports, warehouses, condominiums, houses, shops, open space, and so forth. This could be a quite useful adjunct to an ICZM program because it would make clear to public or private landholders that certain types of land are reserved for particular uses. Provision for land use zoning should be incorporated in the Master Plan.

Because of the extreme scarcity of good shorefront land and its importance to coastal communities, both centrally planned and free market countries have reason to allocate coastal land uses within a broader social perspective than ordinary land and to use a longer term, more comprehensive and integrated approach, one that includes conservation and social equity needs as well as economic growth objectives.

At the interface between land and water—the coastal zone—there is often intense conflict between private (or quasi-private) property-based operations in shorelands and public (common) property-based activities in the tidelands and coastal waters. Thus, the ICZM process may have an important mediating role between conflicting wetside and dryside interests. This essential role should not be played out sector against sector—the mediating/coordinating entity must look at all sectors with legitimate interests to find the most broadly compatible solutions. It is by preparation of a Master Plan that the various development sectors are assessed for their effects on the various resources in a given geographic area.

Also, in ICZM planning it might be desirable to identify areas that are particularly well suited for certain types of development anticipated for the future, for example, tourist facilities, port expansions, or aquaculture ponds. Environmental evaluation and public dialog *in advance* of specific project applications could be most beneficial to both development and environment interests.

One of easiest sectors to analyze is resort development, where density parameters can translate into direct standards with respect to rooms per square kilometer, plot size, site coverage, and so forth, which may be imposed by town and country planning and development control authorities. Such standards can be used to reflect directly the type of tourism desired. But other, more coastal-specific standards are needed for coastal resorts because of issues of water quality, beaches and dunes (Figure 1.28), hurricanes, coral reefs, and so forth.

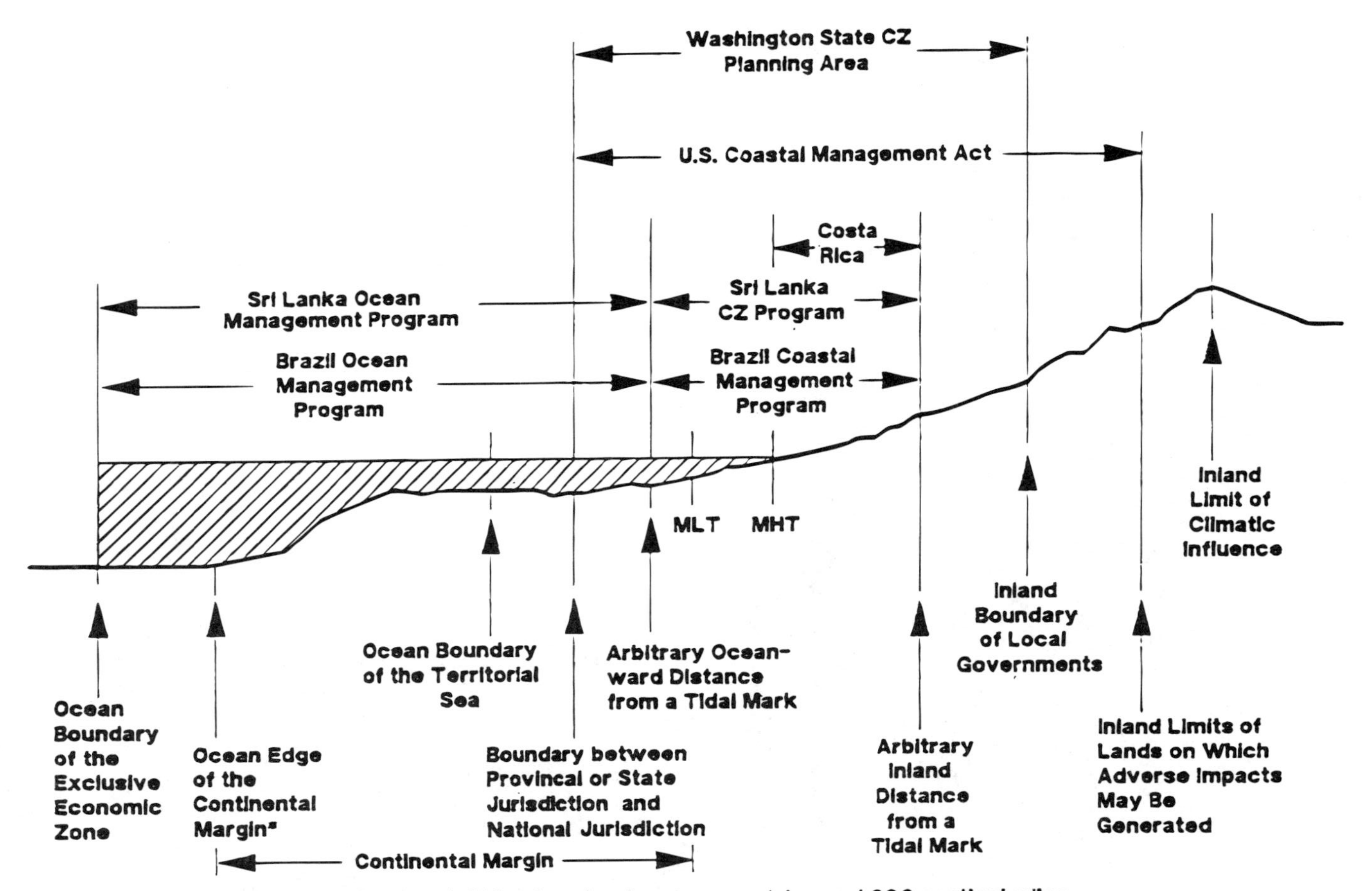

Figure 1.27. Boundaries of the coastal zone—biophysical and jurisdictional. (*Source:* Reference 75.)

Figure 1.28. Beaches are important coastal resources that must be protected in a variety of ways. (Photo by author.)

1.6.5 REGULATORY PROGRAM—PERMITS AND REVIEWS

A major purpose of ICZM-type programs is to review development projects to determine the impacts they will have on coastal resources and biodiversity. The result of the review is to recommend changes in project design or its location to eliminate or reduce any negative environmental impacts before a permit is issued.

Environmental considerations and analyses should be an integral part of project and program planning and of urban planning and land use planning. This regulatory function will require some formal type of "environmental assessment" (EA) procedure as a principal fact-finding procedure. EA, a complex subject in its own right, is discussed in detail below and in "Environmental Assessment," Part 2.

Environmental review should not wait until plans are finalized or project design has been completed. Most development involves several phases: preliminary feasibility studies; selection of a plan and detailed engineering and design; construction; operation; and decommissioning of facilities (232). Environmental assessment would apply to each of these phases.

While the review process may cause some confusion at first, over time, experience gained from reviewing different kinds of projects and different resources will raise the level of knowledge and confidence of the ICZM program staff. This will make it possible actually to speed up development by providing useful strategic guidance to coastal communities and to development interests.

1.6.6 ENVIRONMENTAL ASSESSMENT

At the center of the development evaluation process is the formal prediction of impacts for proposed projects, termed "environmental assessment" (EA) or—at its most complex level—"environmental *impact* assessment" (EIA). EA must be linked to some kind of permit system whereby approval of the development is dependent upon EA (Figure 1.29).

According to Maragos *et al.* (232), the EA process is the mechanism by which the ecological and other environmental consequences of proposed development are estimated and recommendations are provided to decision makers to reduce or avoid impacts. Significant impacts are identified and, if they

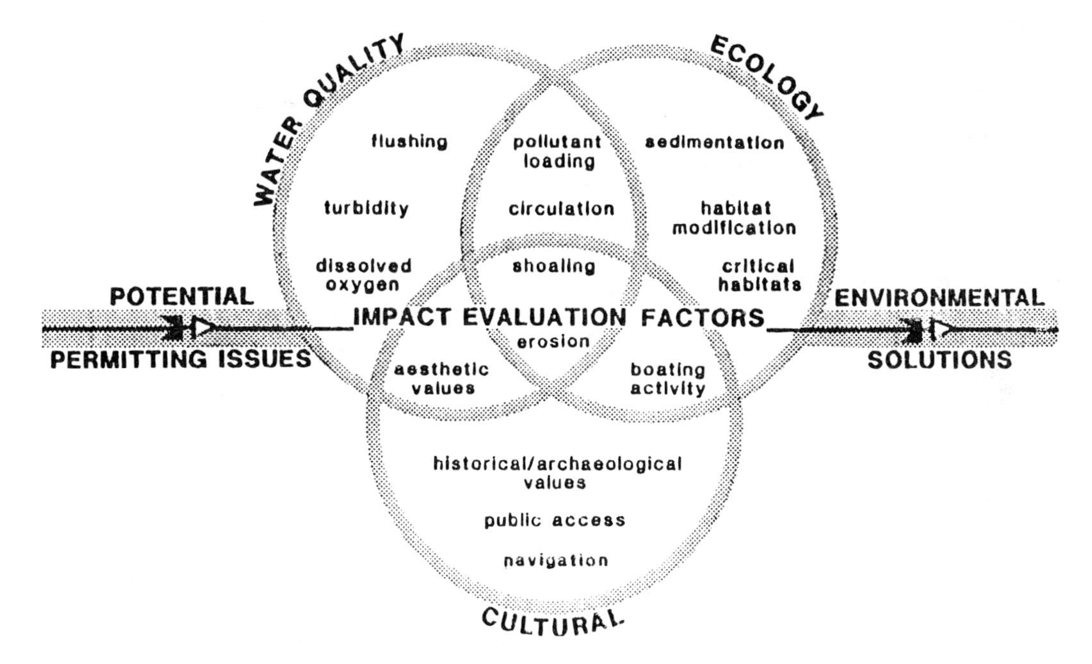

Figure 1.29. Impact assessment (EA), a key part of the permitting process for projects proposed for the coastal zone, incorporates a broad range of physical, ecological, and social factors. (*Source:* Reference 41.)

are eliminated or reduced to a minor level or if they are appropriately "mitigated" (i.e., offset by compensatory activity such as environmental enhancement elsewhere on or off the development site) in the development proposal, the project will be granted development permission.

The information and analysis generated in the EA provide the basis for the decision on whether a coastal project is disapproved, approved as is, or approved with conditions attached. In the case of specific development projects, environmental assessment should commence in the pre-feasibility stage and continue through the various stages of project design. Lately, it has become common to include economic and social considerations as part of the assessment (see "Social Impact Assessment" and "Economic Impact Assessment," Part 2). Nevertheless, it is still usually termed an *environmental assessment* (EA).

The assessment process includes prediction of the effects of a proposed project or economic plan on renewable coastal resources as well as on biodiversity and the quality of the human environment. A series of standard steps for EA analysis is suggested in "Environmental Assessment" in Part 2. The process, when mandated by law or executive decree, generally involves a procedure that requires the following minimum information: 1) the characteristics of the project site; 2) description of the project, and 3) description of the environmental impacts of the project. Usually it is required that alternatives to the project be identified and comparatively assessed, and measures to avoid or mitigate impacts be spelled out in an environmental management plan (326).

1.6.7 PROTECTED AREAS

Protection of species and their habitats is a necessary part of ICZM and is the most important aspect of biodiversity maintenance. Effective programs will include *both* regulatory and custodial (protected areas) components. On one hand, the regulatory component provides a broad framework for controlling uses of coastal resources, including regulations, permits, environmental assessment, and development planning, operating through administrative process and police power. On the other hand the custodial component provides special protection for natural areas (reserves, national parks, etc.) of special resource value, operating through the owner's (government) exercise of proprietary rights. By combining the two (83), there is created both an "umbrella" regulatory scheme for resource conservation and orderly development and a specific custodial scheme for high level protection of ecologically critical areas (ECAs).

Marine and coastal nature reserves, national parks, and other types of protected areas are integrated into an ICZM program. Managing a nature reserve or marine park in isolation from surrounding land uses and peoples and without interagency cooperation may fail. Protected areas alienated from a wider program of coastal resources management exist as islands of protection surrounded by "zones of influence" where resource exploitation is not well controlled. ICZM provides an appropriate framework for incorporation of protected areas into a larger system of protection and a method of consensus building for their support.

ICZM can be organized to control uses in the zones of influence adjacent to protected area boundaries and thereby prevent encroachment into these areas as well as reduce pollution from external sources, limit destruction of special "nurturing" habitats, and minimize other types of external impact that could be damaging to the protected area.

1.6.8 SITUATION MANAGEMENT

Situation Management is a specific integrated approach to the management of particular coastal situations that are, by definition, issue (or problem) driven. Recent examples are Sri Lanka ICZM's two programs for ecosystem rehabilitation—one at Hikkaduwa (degraded coral reef) and one at Rekawa (degraded lagoon). Situation Management has often been termed *Special Area Management.*

For example, major port and lagoon (or estuary) complexes are the focus of the greatest intensity and number of coastal resource conflicts and the greatest environmental degradation (Figure 1.30). As a result, national interest in ICZM may focus on one specific port complex or on an urban center and its associated bay, lagoon, or estuary ecosystem. Situation Management is an appropriate tool to use for this type of problem.

Other examples of problem areas could include a major area of coastal lowland agriculture, a coastal aquaculture mega-complex, an extended tourist resort strip, or the location of natural spectacles like

Figure 1.30. Commercial ports are a potential major source of pollution and critical habitat loss in the coastal zone and must be planned carefully. (Photo by author.)

flamingo gathering areas or giant sea turtle "arribadas" (mass nesting on beaches). Situation Management may be invoked wherever the regulatory (rather than custodial) approach is used to reduce conflict and conserve coastal resources (both for land use and sea use).

Situation Management (SM) would most likely be found as part of a national ICZM program, although it could be international, as is the situation management program for the Waddenzee of northwest Europe. In any event, SM requires the same approach as a country-wide ICZM program—Strategy Plan and Master Plan with all relevant aspects, such as issue analysis, broad participation, cooperation, permits, and environmental assessment.

1.6.9 INFORMATION SERVICES

Coastal and marine resource management agencies have to resolve conflicting user needs and provide sound environmental controls. Such resource management provisions depend on timely and action-oriented information in a form that is meaningful to administrators. The more quality information an agency has on the resources of a region, the better it is able to assess the impact of present and potential activities on those resources and to inform the public via educational programs (Figure 1.31). (89).

Up-to-date mapping and evaluation of resources can be a very important component of data management for an ICZM program. It can identify in advance the most valuable, or critical, coastal habitats, ecological functions, tourist attractions, and places most subject to natural hazards damage. Such information can be useful in the review and assessment of coastal development projects and in identifying candidate sites for coastal parks and protected areas. A survey would also locate highly polluted waters and degraded resources that need rehabilitation. Another benefit would be the location of good sites for future development activities.

The ICZM staff unit will find that its management activities may require the production of considerable technical information, including surveys and delineations of special areas. Much of the output would be maps and overlays. Planning and use zoning of coastal zones involves making provision for

Figure 1.31. Environmental education programs should be sponsored by integrated coastal zone management (ICZM) programs. (Photo by J. Phillips.)

human activities—for example, transportation corridors (shipping lanes, roads, pipelines, etc.), settlements, form of resource use (fishing, collecting, tourist activities), and public utilities (waste and sewage disposal, industrial activity)—which may have some positive or negative impact on available resources. Good information is necessary to make these assessments (89).

Current mapping and resource characterization technologies and information systems available should be assessed. Geographic information systems (GIS) can play a useful role for data inventories and collection, promotion of data analysis, elaboration of synthesis and historical perspectives, monitoring of coastal development, and development of coastal degradation indexes (see "Information Needs" and "Geographic Information Systems" in Part 3).

To take the example of ICZM guidelines and standards, management actions for which these should be prepared are listed below:

- Facilitation of sustainable economic growth (aquaculture, tourism, fishing, non-polluting industry)
- Conservation of natural areas, resources, and critical habitats (wetlands, coral reefs, endangered species habitats, etc.)
- Control of pollution
- Control of site alteration in adjacent shorelands (landfill, land clearing)
- Control of beachfront alteration (removal of dunes, excavation of beach sand)
- Control of watershed activities (clear cutting, dams, cropland clearing, urban construction, etc.)
- Control of excavation/mining and other alteration of coral reefs, water basins, and sea floors
- Rehabilitation of degraded habitats (coral reefs, mangrove forests, estuaries, seagrass beds, etc.)

1.6.10 TECHNICAL SERVICES

The ICZM planning unit may find an important role in providing technical information to various coastal interests. The users of this information might be NGOs, developers and their consultants, design engineers, various government agencies and their planners, environmental interests, and the media. Also, a need for technical descriptions and guidelines will emerge.

With an ICZM management unit reviewing development proposals and often requesting changes in project specifications, developers may find that they can benefit particularly from prior consultation with ICZM staff.

1.6.11 RESTORATION AND REHABILITATION

Regarding environmental restoration, critical habitats that have been degraded should be rehabilitated to the highest possible level of productivity and biodiversity. Although all coastal resources that have been lost cannot, in a practical sense, be returned to productivity, many of them can. Marshlands can be replanted (Figure 1.32), dikes can be dismantled, normal freshwater or tidal flows to mangrove forests can be restored, and coral reefs can be started toward gradual renewal. The ICZM planning unit should identify, through survey, those critical areas that have been degraded and can be repaired at reasonable cost and effort. These can be mapped, priorities assigned, and a strategy for rehabilitation created. (See "Rehabilitation and Restoration" in Part 3.)

1.6.12 OPERATIONAL FORMAT

An ICZM program can be designed to fit into any governmental structure. The variety of institutional arrangements is sufficiently flexible to match whatever administrative system a country may use (see "Institutional Mechanisms," section 1.5.13). In concept, ICZM can be as simple as a program of impact assessment for development projects. Or it can be as complex as a comprehensive, full-service program of economic development, conservation, education, and social well-being. In its most rudimentary form, the program can be initiated with not more than three components:

1. A central government ICZM coordination office
2. A project review/permit system for major coastal developments (including environmental and socio-economic assessment)
3. Empowerment to ensure compliance with the program, its requirements, guidelines, and standards

All ICZM programs require these three components as a minimum. In this minimal mode, it is anticipated that the program would mainly review developments and issue permits for acceptable projects.

Elaborations to the managment program can be added as appropriate. In an expanded mode, the ICZM program would broaden out to include additional functions needed for a more complete program. Examples of some of the more useful additional functions are listed below:

- Technical services: planning, information management, survey and mapping, monitoring, research, economic analysis, and technical advice to developers, government agencies, and interest groups
- Education services: information dissemination, public education, staff training, extension services
- Guidelines: formulation of standards and guidelines for coastal development and publication and update of the standards and guidelines
- Restoration activities: assessing the condition of coastal renewable resources, determining restoration needs, designing restoration projects
- Protected areas: determining the need for the establishment of a system of parks, reserves, or other types of protected areas, priorities for site selection, creation of management plans

It should be recognized that two quite distinct fields of endeavor may be involved in coastal resources management: (a) the broad, ICZM-type of development management and (b) the traditional type of single purpose resource management. The ICZM-type people who look after development control and

Figure 1.32. Defunct coastal wetlands can be rehabilitated by planting with mangrove seedlings or with marsh grass. (Photo by author.)

land use planning may be quite differently trained and mandated than are those who look after wildlife, fisheries, or forest resources.

1.6.13 PROGRAM EVALUATION

Two conditions are required before implementation of the ICZM program should be evaluated: (a) an adequate post-implementation time period to allow a program to reach maturity and (b) the creation of a set of indicators for measuring performance. All evaluation studies seek to assess program performance, although they differ markedly in the evaluative criteria used. Two basic types of evaluation can be distinguished. One type of evaluation focuses on the policy-making process (such as the number of permits issued), and the other type focuses on the eventual outcomes (such as improvement in water quality). Of course, the evaluation may measure both processes and outcomes. (326)

Process evaluation examines the means by which goals are achieved. Process indicators include the clarity of goal statements and legislative mandates, measures of the rationality of organizational structures and the process and information flow, the adequacy of yearly budget allocations, the number of permits issued, and the number of agreements executed to promote interagency cooperation.

Outcome evaluation measures the extent to which the program's goals or objectives are achieved. Outcome indicators can be subdivided into instrumental factors and environmental/socio-economic conditions. Instrumental indicators measure goals whose achievement is thought necessary to the achievement of environmental and socio-economic goals. These may include the extent of the information base, the efficiency of permit review, and the extent of public participation. Environmental or socio-economic conditions measure such things as the extent of protected wildlife habitat or the number of jobs created. (326)

Specific environmental and socio-economic outcome indicators can be identified. Some examples are given below (326):

Process Indicators

- Budget allocation per year
- Number of permits issued, denied, conditional
- Consistency of law dealing with coastal management
- Number of agreements relating to interagency cooperation
- Availability of appropriately trained and educated staff
- Number of subnational programs initiated or approved
- Quality of information used in program development

Outcome Indicators

Instrumental factors:

- Cost and length of time for permit review
- Number of procedures and steps eliminated ("streamlining")
- Public participation—number of individuals and groups
- Geographic scope and issue coverage of information base

Environmental or socio-economic conditions:

- Water quality (toxics, oxygen, nutrient levels, etc.)
- Fishery yields
- Food, income, provided by coastal fisheries
- Amount (linear distance) of coastal access
- Kilometers of coast in public ownership
- Number of tourist user days
- Number of coastal species under special protection
- Acreage of wetlands protected or restricted
- Amount of special housing provided within the coastal zone
- Tonnage and value of commodities handled in ports
- Employment derived from fisheries, ports, and tourism sectors
- Natural hazard effects; amount of lives and property saved

PART 2

Management Methods

2.1 AQUACULTURE MANAGEMENT

Demands for seafood are high because of increasing prosperity, growing human populations, rapid worldwide transport, and shrinking supplies. For high priced items like prawns and tuna, there is virtually a single worldwide market, strongly influenced by European and Japanese buyers.

One solution to meeting the soaring demand is to build "fish farms"—or aquaculture facilities—to supplement the natural supply. While aquaculture has contributed significantly to world seafood supplies and to tropical economies, conflicts with other coastal uses must be adequately addressed.

Coastal aquaculture projects are usually built near the shoreline to gain access to saltwater supplies and seed stock. They often run into direct confrontation with environmental conservation objectives, requiring government intervention to balance the benefits of aquacultural development against the losses to the environment and to other economic sectors dependent on the resources of the coastal zone (see also "Aquaculture" in Part 3).

Many of the environmental implications are well known: preemption of critical fishery habitat (e.g., mangrove); pollution of lagoon waters; excessive exploitation of natural stocks of larvae and juveniles (Figure 2.1); importation of inferior, sick or non-compatible seedstock; and jeopardy from the introduction of exotic species. Also, conflict over space and water supplies can be expected with coastal rice culture, nature reserves (mangrove forests), solar salt production activities, and many other enterprises. Therefore, governments should consider both positive and negative aspects through an environmental assessment (EA) or equivalent procedure to reach balanced multiple-use decisions.

From a worldwide viewpoint, the major coastal environmental impact of aquaculture seems to be from expansion of shrimp farms in the tropics. Shrimp culture projects often involve the conversion of massive amounts of mangrove forests to shrimp "growout" ponds, with all the biodiversity and ecological implications that loss of such wetland implies (Figure 2.2).

Although aquaculture ponds have also been located in tide flats, marshes, lagoons, beaches (Japan), and other critical natural habitats, the most controversial location is coastal mangrove forests. For example, during the peak aquaculture expansion period in the Philippines (1967 to 1977), aquaculture accounted for 80 percent of the loss of mangrove (84). As stated by Snedaker *et al.* (315): "To the extent that mangroves are being destroyed . . . there is loss of marine seafood production."

Major problems are associated with pond siting, including the direct removal of mangrove wetlands, the development of acid-sulfate sediment conditions, inadequate to poor flushing of ponds, and local depletion of larval and juvenile organisms for pond stocking. Acid-sulfate soils develop when previously waterlogged sediments that are high in organic material are exposed to the air or oxygen-rich water. The problem is aggravated in situations where there is inadequate pond-water exchange or flushing, which otherwise would reduce the acid water and replace it with new water. In severe cases, even the highest rates of flushing will not maintain a healthy pH (level of acidity) in the pond water.

In the Ecuadorean case, lack of evaluation, planning, and control led to severe problems in one of the world's leading shrimp-producing areas (involving over 200,000 workers). Large scale pond construction started in the 1970s and continued rapidly until 150,000 hectares of ponds had been built by 1983. Production increased continually until several problems began to be encountered. A most serious problem was a scarcity of wild seedstock (postlarvae), of which over two billion were stocked annually. The scarcity may have been associated with conversion of 25 percent of Ecuador's mangrove forests to

Figure 2.1. When excessive numbers of young stages (seed stock) are taken from natural waters for fish or shrimp ponds, the supply can rapidly diminish, curtailing production (Egypt). (Photo by author.)

Figure 2.2. In an effort to restore the original intertidal ecosystem, leases are being canceled by the Bali government (Indonesia) for shrimp ponds built in mangrove areas. (Photo by author.)

shrimp growout ponds and to diminishing water quality caused by pond effluent. Conclusions are confused by large natural swings in natural seedstock abundance.

A partial remedy for the seedstock problem is the construction of hatcheries. Unfortunately, hatcheries increase costs to producers and, with current technology limitations, they may give unreliable results. Now that the Ecuador coast is under an ICZM-type program, the management and planning approach best suited to control the expansion of the industry and, at the same time, accomplish conservation and other needed management functions is being implemented. (See Ecuador case history in Part 4.)

Kapetsky *et al.* (201), who studied aquaculture siting in Costa Rica, advise that "semi-intensive" shrimp farming should be done outside of mangrove areas. There is a high rate of failure for shrimp pond systems located in converted mangrove areas because of the acidic conditions which, in turn, lower shrimp growth and survival. Also, shrimp culture in mangrove areas can be self-defeating from the point of view of replacing natural shrimp nursery areas (as well as the nurture, feeding, and breeding areas of other valuable species) with shrimp ponds.

It has proved quite practicable to use "salinas" and other less productive lowlands that lie behind the mangrove wetlands and to preserve a substantial fringe of mangrove along the shoreline. In the Caribbean country of the Turks and Caicos, potential aquaculture uses for salinas (abandoned for salt culture) are listed as culture of fish, shrimp, crab, conch, or brine shrimp (374). The problem with these somewhat higher areas is that they would receive replenishment water only on the highest tides unless supplied with expensive diesel pumps. Another important factor is that the soil may be more appropriate for building of dikes inland from the mangroves. The special soil considerations involved are shown in Figure 2.3. (201)

The main requirement for all uses is efficient water management, that is, allowing for the optimal flow of water, the appropriate depth, and the most suitable quality. Other management requirements depend upon the particular conditions of the salina in question.

Studies have shown that pumping water into aquaculture ponds can increase yields to economic advantage in comparison with tide-driven circulation (201). With pumped circulation there is no need for ponds to be at intertidal level (Figure 2.4). Furthermore, pond construction costs are less outside of the mangrove habitat than inside, and management is facilitated. Finally, there is considerable pressure from environmentalists to conserve mangrove areas in most countries (see "Mangrove Forest Resources" in Part 3).

A serious problem in such places as East Java and Bali (Indonesia) and Khulna (Bangladesh) is the conversion of coastal rice fields to shrimp ponds. These conversions require bringing saline water (for the ponds) into paddy and other agricultural areas (Figure 2.5), where fields adjacent to the ponds can be heavily damaged by salination (63, 86). The salination has been so bad that dozens of rice farmers have been forced off their fields.

The most common recommendation to discourage further conversion of mangrove forests is to increase yields of existing ponds instead. This intensification would be done by raising the yield of shrimp from 100–250 kilograms/ha/year to 400, 1,000, or even several thousand kilograms per hectare per year. This can be done by increasing the number of small shrimp used in stocking the ponds, by fertilizing the ponds to increase the natural food supply, or by other techniques such as providing prepared feeds to the shrimp. One problem of the "intensification solution," or any other production increase, is that seedstock requirements will increase, which puts further pressure on the sources (natural or hatchery). Another problem is that high-tech approaches would be disadvantageous to countries like Bangladesh, whose low labor costs and extensive lowlands allow them to be competitive on the world market (210).

Another method of increasing production without opening up new ponds is "time sharing" between salt production and aquaculture in existing ponds now used for salt production only. It is feasible to use these ponds for shrimp in the monsoon period when high rainfall makes solar salt production uneconomical. Such time sharing is successful in some countries, such as Bangladesh and India.

"Cage culture"—growing finfish in cages in fjords, canals, and bays—can also cause some environmental problems, such as local eutrophication and depletion of benthic fauna through the discharge of food residues and excrement and also toxic pollution through the escape of pharmaceutical and antifouling products. Figure 2.6 shows that less than 30 percent of the nitrogen and phosphorus in food materials is utilized directly by the fish being fed; the rest is discharged to the environment. But such problems

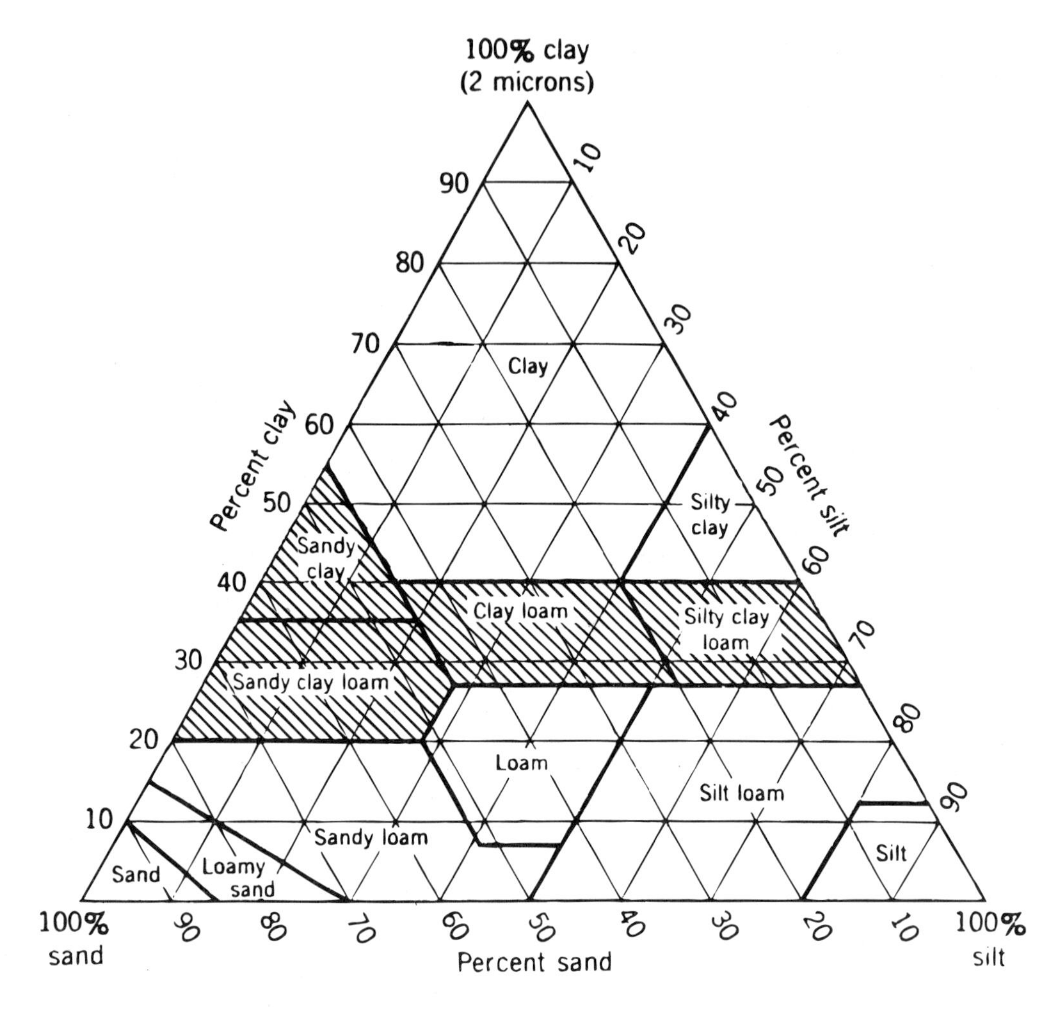

Figure 2.3. The triangle diagram of basic soil texture classes, with soils favorable for building fish ponds indicated by shading. Particle sizes are clay, <0.002 mm; silt, 0.002–0.05 mm; sand, 0.05–2.0 mm. (*Source:* Reference 91.)

can often be mitigated by locating cages in places of vigorous water circulation and by regulating the use of additives.

In recent decades, considerable initiative by entrepreneurs and governments resulted in a wave of introductions of "exotic" marine species, involving many tropical species. This wave has been met with skepticism when the objective was release into natural waters because of uncertainty as to the full ecological ramifications (see "Exotics" in Part 3). Whatever precautions are taken, it must be presumed that the introduced species may sooner or later invade natural waters. A major concern about the introduction of exotic species is that they may carry disease that will be uncontrollable in the new environment, where resistance to the disease is absent. In this regard, quarantine may be helpful to assess the animals for the presence of disease before they are bred or widely dispersed.

Tomascik (345) recommends the following pollution and general controls for tropical aquaculture installations:

1. Discharge all effluents from aquaculture projects downstream of sensitive habitats like coral reefs.
2. Site aquaculture ponds at rear of the mangrove forest.
3. Include performance requirements for the protection of the natural habitat surrounding the pond.
4. Maintain natural patterns of surface-water flow.

Figure 2.4. Aquaculture ponds supplied by saltwater pumps can be situated inland from the edge of the sea so as to spare the wetlands, as here in Ecuador. (Photo by R. Murray.)

Figure 2.5. In Bangladesh (Khulna area) and elsewhere, saltwater brought in to convert paddy fields to shrimp ponds has polluted adjacent fields and made them unsuitable for rice culture. (Photo by author.)

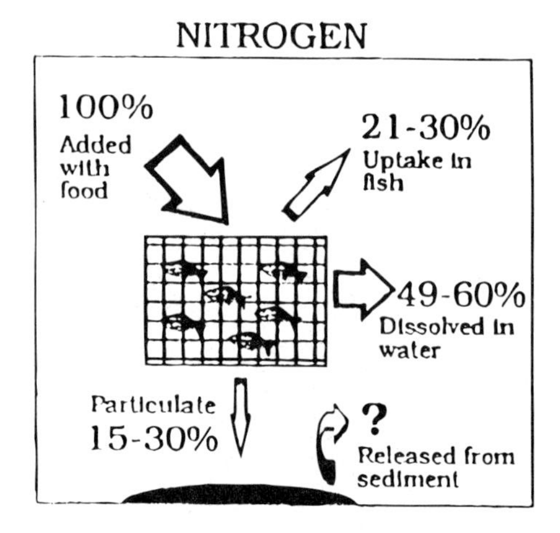

Figure 2.6. Less than one-quarter of the phosphates and nitrates in cage feedings may be taken up by the fish, with the remainder entering the water as pollutants. (*Source:* Reference 15.)

5. Ensure that an appropriate disposal of excavated material is in the planning process.
6. Use biological treatment methods if adequate dilution and dispersion of aquaculture effluents is not possible.
7. Avoid the use of toxic substances, such as the organochlorine pesticide Thiodan, for cleaning shrimp and fish ponds; for cleaning of aquaculture ponds the use of Chemfish and other rotenone-based compounds is advisable because they cause less environmental damage.
8. Restrict reef pen or cage culture to reef areas where the wastes produced will have minimal impact on the adjacent habitats.
9. Avoid the use and discharge of harmful substances such as pesticides, antibiotics, and fertilizers in all coral reef areas.

The following performance requirements, or standards, are recommended by Kittel (210) to protect the natural habitat surrounding aquaculture sites, as a condition for obtaining construction permits:

1. Water flow: A specific requirement is that the natural patterns of surface-water flow (i.e., surface-water runoff, tidal ingress and egress) should not be interrupted. It is suggested that baseline measurements be made of these flow patterns and other specific data such as organic loading of the adjacent waters, nitrogen levels in the water, total bacterial counts in adjacent water, and species present in the adjacent water or sediment as a precondition.
2. Wetlands buffer: In mangrove areas, the project site should be located behind the mangrove forests. When available, areas defined as salt flats or salt pans are preferred sites in terms of production as well as environmental protection. For pond production, yields are frequently greater in salt pans than in mangrove areas, where acid sulfate soil conditions prevail. A wide buffer (+100 m if possible) of natural mangrove forest should be left between the pond sites and open waters. This buffer will allow the mangrove to continue to function as a habitat which will serve as a nursery for many aquatic organisms. In addition, the buffer can serve as a corridor for animals living in remaining inland mangrove forest (monkeys, squirrels, etc.). For example, in Malaysia, troops of monkeys were frequently noticed traveling along a 40-meter-wide mangrove buffer to gain access to river mouths where the monkeys foraged for food.
3. Seedstock: Evaluation of aquaculture ventures should consider whether natural stocks will be harvested for seedstock or as food for the species to be cultured. Large-scale aquaculture demands for seedstock or feed can overtax natural stocks, reducing recruitment to the capture fisheries or creating other problems. Where such concerns exist, hatcheries may be required to supply the seedstock (Figure 2.7).
4. Monitoring: Post-construction (operations) monitoring can then be used to assure that, if the

Figure 2.7. The bay bottom is seeded with small hatchery-grown oysters in a shellfish culture operation (Oyster Bay, New York, USA). (Photo by author.)

parameters change beyond agreed limits, the permit-issuing agency will take action against the enterprise.

For administrative evaluation of aquaculture projects, Kittel (210) suggests the following guidelines:

1. Compliance with ICZM Master Plan: The approval and development of coastal aquaculture projects should be based on ICZM-type plans that identify resources and conflicts over their use. ICZM plans should include the protection of coastal habitats for sustaining capture fisheries, supporting tourism, and maintaining a high level of ecological functions and diversity. Projects that are in conflict with such plans should not be approved.
2. Efficiency: In well-managed aquaculture systems, a relatively high yield can be obtained by various methods—increased stocking and improved water exchange through tidal fluctuations or large-volume pumps. In ponds where feeds are not used, organic and/or inorganic fertilizers can be used to increase production. Where direct feed is applied, an appropriate food should be selected. Proper monitoring of feeding will yield improved food conversion ratios and reduce pollution of the pond and surrounding water. Project proposals should indicate a clear statement of management practices to be followed.
3. Intensification: Preference should be given to improving the efficiency of existing aquaculture enterprises as opposed to converting new coastal areas to ponds or aquaculture enclosures. Approval requirements should include a demonstration that the proposed site is, in fact, suitable for sustained-yield production (a planning horizon of 20 to 50 years should be used).
4. Disease: Outbreaks of disease are a common problem for aquaculture. A management system for monitoring and controlling disease outbreaks should be in place before approval is given along with prevention of disease outbreaks caused by species being imported for culture. A quarantine facility can be required.
5. Water quality: Small-scale aquaculture (yields below 500 kg/ha/year) with low stocking densities and minimal use of chemicals and artificial foods will have a low potential for

pollution. Alternatively, large-scale, high-tech operations (yields over 2,000 kg/ha/year) are a pollution threat, particularly those where pond discharge goes into lagoons or bays with poor flushing. The major pollutant is organic wastes and their nitrogen and phosphorus breakdown products; it is unlikely that toxic substances would be introduced into culture facilities. However, fertilizers and cleaning chemicals are often used (e.g., bleach) and antibiotics may be introduced in culture water or through feeds (especially for hatchery operations). Therefore, proponents of intensive aquaculture facilities in the coastal zone should provide a detailed analysis of amounts of polluting substances to be discharged and probable effects on water quality. (Some governments require treatment of effluent water from facilities that hold more than a specified mass of fish, shrimp, shellfish, or other organism per volume of water.)

The above information relates to culture of marine animals, but there is also a growing industry of algae culture utilizing such algae as *Gracilaria* and *Eucuma,* which can readily be grown on thin twisted ropes or netting stretched across shallow waters on a rigid frame. The starter plants are merely threaded into the twists of the rope. At harvest 2–3 months later, there is a weight increase of 5 to 14 times (yields of 120 tons wet weight/ha are expected in the Gulf of Mannar, India). Algae have more than 60 trace elements and vitamins; the main commercial products are agar-type thickening agents (196).

How expanded seaweed culture might affect littoral ecosystems is little known. However, it seems to have a good, and ecologically safe, potential for community-based "low-tech" industry (Figure 2.8).

The following are some general policy guidelines for environmental management of aquaculture development, suggested by Barg (15) (abstracted by the author):

1. Regarding traditional fisheries, special emphasis should be given to ensure that aquaculture development efforts are compatible with existing and projected conditions of capture fisheries, both coastal and inland. Effects of coastal aquaculture development on coastal fishery resources and on consumer acceptance and marketability, especially distribution and marketing of fishery products, must be considered and forecasted. Expansion of coastal aquaculture must be limited to avoid detriment to coastal fishery resource users, including traditional small-scale fishery

Figure 2.8. Seaweed from which agar will be extracted is sun dried in Bali (Indonesia). (Photo by author.)

communities. Selection of aquaculture sites, methods, and species should take account of both employment opportunities for fishermen and local demands for species that are not provided by capture fisheries.
2. Re rural development, planners of coastal aquaculture development should actively participate in the formulation and implementation of integrated, cross-sectoral management plans aimed at coordinated land use and resource development in coastal areas. The potential sites and allocation of resources to aquaculture and the selection of forms of aquaculture practice should be preceded by adequate on-site surveys and evaluations. Planners of coastal aquaculture development with experience in aquaculture should participate in coastal surveys leading to designation of zones with the resource uses specified. Economic viability and social acceptability of existing aquaculture practices may be further promoted through increased horizontal and vertical integration within local economics. Aquaculture may stimulate acceptance and support by stakeholders of other local activities if the aquaculture planning has taken sufficient account of changes in patterns of land and water use and waste disposal, use of agricultural by-products, promotion of the local marketing systems, and promotion of locally based processing facilities.
3. Re economic policies, success of coastal aquaculture development will depend considerably on the degree of prioritization and integration in national plans for economic development. Likewise, the success of coastal aquaculture development will be determined by the degree of compatibility with development plans for other sectors.
4. Re environment, aquaculture planning and management should take account of established national policies aiming at environmental protection. In aquaculture development planning, it should be stated how the general management framework adopted for the protection of coastal environment has been adapted to meet the specific needs and characteristics of the proposed aquaculture practices. Environmental protection requirements for aquaculture should be integrated into planning for other development sectors. Requirements and specifications on environmental assessment and monitoring of aquaculture practices should form an integral part of an organized development plan, which should also contain options for mitigation/offsetting of adverse environmental impacts. Environmentally acceptable aquaculture practices should be prioritized in development plans. Coastal aquaculture development planners should participate in the formulation of environmental legislation governing aquaculture practices.
5. Re compatibility, government development strategies in support of sustainable aquaculture should give increased emphasis to environmental considerations when promoting the following aspects:

- Demonstration of technical and economic feasibility of aquaculture systems with guidelines of appropriate species and culture systems and appropriate farm operating procedures;
- Infrastructure in the aquaculture zone in terms of water supply, electricity, roads, post-harvest and marketing facilities;
- Availability of credit facilities and insurance schemes for aquaculture investment;
- Technical and management training for small farmers, technicians, and managers;
- Technology development and transfer, supported by research.

2.2 AWARENESS

Awareness plays a major role in accomplishing coastal management objectives. If conservation is to become effective, conservation awareness must be high among the public, the managers, *and* the developers concerning the cross-sectoral scope of the problem, the consequences of environment and natural resource mismanagement, and the importance of the maritime dimension in national planning and development strategies.

The most important goal is to convince people of the value of protecting resources to gain the long-term, sustainable benefits that conservation can provide. Awareness programs aim to provide the community with both information and an ethic so that its members can make informed decisions about the use of their resources. But self-serving agency propaganda should not be part of an awareness campaign. (65)

Awareness programs should be carefully designed to meet specific objectives for each "target group" or audience and a specific message or messages should be defined. Artisanal fishermen, dive operators, farmers, herdsmen, and tourists are examples of target groups found in marine and coastal settings. For example, Hudson (177) recommends the following for Australia's Great Barrier Reef:

Target group	Message
General public:	Nature of coral reef environment
	Need to protect reef areas
Local fishermen:	Economic benefit of proper management
	Provisions of plan regarding fishing
Tourist operator:	Suggested tourist activity on reefs
	Provisions of plan regarding tourism
Govt agencies:	How plan interacts with their mandates

Next, specific objectives must be established in terms of knowledge, attitudes, and behavior to be changed or influenced within each target group. For example, in the Central Visayas project of the Philippines, fishermen who were educated about artificial reef construction and use were able to increase their catches (see "Artificial Reefs" above) and abandon destructive dynamite fishing at the same time.

For a broad public awareness program, a multimedia approach is recommended, one that combines printed materials, audiovisual presentations, and face-to-face interaction. Depending on the target audience and budget, a variety of additional options can be employed: mass media (press, television, radio), fixed exhibits, tours, training workshops, the use of promotional items such as T-shirts, and informal recreational activities with an educational focus. (177)

There are many techniques to be used to get a message across. Each one on Hudson's list below has a relevance to a particular type of situation (177):

1. Television: Has general audience, raises general awareness of situation, and can motivate people to do something about an issue that they may not have known about before. It is a passive medium for the receiver but, handled well, can be of great benefit in general public education.
2. Video: This has many benefits to the environmental educator. You can make your own specific television programs using your environment and people to get your message across to your target groups. The equipment needed is relatively cheap and easy to use. With a little experience one can make short programs quickly (but it is advisable to secure professional advice).
3. Radio: An excellent medium, it is available in most countries and is extensively used by schools and other teaching institutions. Radio series on environmental issues, especially if there is a strong story line, are often appealing to children; not only children listen to school radio, often their families do too.
4. Print media: Either general or specific target groups can be reached by newspapers and magazines. Careful analysis must be made of the type of material needed for each, how it fits into your education program and goals, and the gains you hope to make from it. Journalists must be dealt with carefully but can be most helpful in environmental education work.
5. Books and pamphlets: Each type must be evaluated for suitability. Books are expensive to produce but can be useful in schools and for sale to a selected adult audience. Pamphlets and leaflets can help in specific cases (e.g., rules for management).
6. Posters: They are attractive, easy to make, sellable, and good as a general educational tool. They can be useful in schools and other institutions. Not so good for specific target groups.
7. Printed clothing: T-shirts may be a very good educational tool with specific target groups, especially those in remote areas and where people interchange belongings within the groups regularly. They can become prestige items, showing that the wearer is *au courant* and part of a "new movement." They are cheap to produce and can be sold to support conservation.
8. Badges: Many of the things said about T-shirts are applicable. Badges are cheap to produce and make in quantity. Young people like wearing them, and they can be used as rewards with

school groups. Teenagers and even older people often like badges. They are highly visible and a good talking point.

9. Entertainment: Locally acceptable forms of drama are one of the best ways of getting your message across to your audience. People like to be entertained and, if they can be made aware of issues and motivated at the same time, all the better.
10. Open meetings: They are most often held to discuss specific issues concerning developments in government plans and may be aimed at soliciting ideas from the public. Such meetings should encourage interactive participation. Meetings held just to propagandize the people should be avoided.

Changes of basic public attitudes are rare and not often affected by short-term awareness campaigns, according to Hudson (177). She advises that it may be possible to sell soap powder in this way, but to cause fishermen to use less destructive techniques for fishing, for example, would take much longer. But such changes of public opinion are possible in the long run, especially where the coastal management staff have a good relationship with the people affected and where the need to change attitude can be backed by facts and is reinforced by the experience of the people themselves. (177)

An important part of awareness work is feedback to determine program effectiveness. Feedback provides a conduit that can tell planners and other management staff what is really happening in the field. Field staff interact with the people constantly and therefore are in the best position to gauge the actual effects of the program (177).

Main source of this section: B.E.T. Hudson (consultant) and R. A. Kenchington, Great Barrier Reef Marine Park Authority, Townsville, Australia. (177, 206).

2.3 BASELINE AND MONITORING

When the decision is made to approve a coastal development project with possible environmental impacts, an oversight process should be set up to verify the predictions. The environmental assessment (EA) process, particularly, should recommend a monitoring step both to ensure that mitigation and other countermeasures are carried out (according to the environmental conditions of the development permit) and to determine the actual impacts of the action as implemented (see "Environmental Impact Assessment" below). Sorensen and West (327) list the following six purposes that may be accomplished by monitoring:

- To ascertain that mitigation measures are being implemented as agreed.
- To warn agencies of unanticipated adverse impacts or sudden changes in impact trends.
- To provide immediate warning whenever a pre-selected impact indicator approaches a pre-selected critical level—this will also be the means to ensure that the legal standards for effluents are not exceeded.
- To provide information which could be used to verify estimated impacts and validate impact prediction techniques—based on these findings the techniques (e.g., statistical models) could be modified or adjusted as appropriate.
- To provide information which could be used for evaluating the effectiveness of implemented mitigation measures.
- To provide information which could be used by agencies to improve meaures for future control of timing, location, and level of impacts of projects—control measures could involve planning as well as regulation and enforcement measures.

From the above listing, the advantage of setting up two major types of monitoring—strategic and tactical—is evident. This is explained below (67, 70):

1. The **strategic monitoring level** is the "retrospective," or "hindsight," monitoring that is done as the main activity of the monitoring program. The object is to compare measurements of certain key characteristics of the environment both before and after project work is done so

that a determination can be made of the project's effects. This, of course, requires that "benchmark" information be collected before the project starts to provide a statistically sound before-the-project "baseline" of information to compare to the after-the-project information parameters.
2. The **tactical monitoring level** is the "oversight," or "real-time," monitoring that is done in conjunction with the construction phase. It should be specified in the environmental management plan for the project; the object is to monitor the construction operation day by day so as to detect any major negative impacts that may be occurring. If problems are detected, construction is halted until the situation is remedied.

In choosing the parameters to be included in the baseline study, it is important to consider the space and time limits for the assessment. The space limits, for example, may cover a hillside watershed, a lake and its wetlands, an urban area and surroundings, or an island. A general rule to follow is to start with the physical processes involved. For example, you may establish boundaries based on the area expected to be covered by the silt plume from a dredge, the sediment plume from a river where it enters the ocean, a drainage basin supplying wetlands, the area determined from groundwater movements, or a section of coastline determined by a study of coastal currents. (18)

In selecting parameters for study in baseline projects for pollution, for example, consideration should be given to measuring the effects of pollutants on plants or animals, rather than on measuring the amount of the pollutant itself. Plants accumulate pollutants because they are constantly exposed to them; using them as "indicators," the researcher may be required to take only a few samples and these will relate directly to the condition of the plant. Similarly, because they concentrate pollutants from the water and do not move around much, shellfish can be tested for the accumulation of toxic materials, such as heavy metals, relatively easily as compared to intensively measuring the amounts of metals in the water. (18)

Consequently, it is very useful in environmental monitoring programs to record in the baseline data certain "index," or "indicator," organisms, those that signal a significant change in the ecosystem through changes in their own abundance, distribution, or habits. As stated by Gomez and Yap (145), "The usefulness of species as 'indicators' is gauged by their ability to provide the observer with measurable signs of environmental stress at the earliest possible time." It is assumed that, if the organisms selected for monitoring are integral parts of the reef system, their responses should adequately reflect the processes that threaten that system as a whole and, consequently, its other various components. (145)

The most obvious time over which to consider future impact is the period over which the project will operate. However, this may be too long or too short a period. Beanlands (18) believes that it is more important to consider the natural variation in the parameters you are measuring. He states that, ideally, baseline studies should continue for at least one year before the start of the project and that monitoring should continue for at least three years after the project, in most cases.

To take an example, in setting up a monitoring project for a beachfill project in Bali, Indonesia, we found it necessary to establish a turbidity baseline to represent the pre-project condition (see "Indonesia, Bali" case history in Part 4). Our initial, multi-parameter survey (March 1992) included turbidity measurements and showed readings of 0.10 to 0.86 NTU (turbidity units) for three project areas, Sanur, Nusa Dua, and Kuta (25.2). This covers a range of clarity from almost unlimited visibility to a minimum transparency of around 10 m, indicating rather good conditions throughout the area in the more or less open areas of lagoons and offshore waters. To refine the baseline, it was decided to conduct a special turbidity survey for purposes of defining a baseline for monitoring of actual construction operations. This would specifically represent the areas in jeopardy from turbidity and siltation.

Consequently, during a coral reef descriptive survey (September), turbidity readings were taken from all the transects, at the surface and at the sea bottom. In addition, readings were taken inside the lagoon and along the beaches (108) in total). The following average turbidities (in NTU) were calculated (70):

Location	Range of NTU	Mean
Beach edge	1.8–13.5	7.0
Lagoon	0.1– 3.3	1.4
Reef gap	0.0– 1.0	0.7
Reef slope, surf	0.0– 3.4	0.6
Reef slope, bottom	0.0– 5.5	0.9

There is a gradation from beach to reef slope, that is, a decrease in NTUs. This is reflected in the communities of life, too, because there are quite different species living in the lagoon than living at the reef slope (see reef and lagoon survey results), and turbidity (i.e, water clarity in reverse) is clearly one of the controlling factors. But there is also a great deal of variation in the ranges; therefore, averages have to be used to interpret the results.

In baseline studies every effort should be made to learn from the results of previous projects of a similar nature. For example, if an impact assessment was being done for a project to build a large coastal pier, it would be helpful to look at other piers that have been in place for some time and determine what impacts they have had and also to learn what level of disruption was caused during construction. (18)

2.4 BEACH MANAGEMENT

Special risks are attached to development at the seafront, where houses, hotels, and other structures are directly in the path of storm-driven waves. Beaches shift with changes in the balance between the erosive forces of storm winds and waves, on the one hand, and the restorative powers of tides and currents, on the other. Consequently, the coast can be a risky place to maintain habitation. Costs in property losses and human lives are high, and enormous sums of private and public money are spent to stabilize inhabited beaches. These efforts are too rarely rewarded with success.

Beaches serve as the main protective bulwark for property along the shores of oceans and large sounds. Most beaches can absorb heavy surface use, including vehicle traffic. But in other ways beaches are easily damaged. Removal of sand, improper building, or blocking of sources of sand replenishment may severely damage, or obliterate the beach.

The natural forces at work are immense, making conservation of beaches a most difficult and often elusive endeavor. The general management goal is to maintain the beach profile by protecting both the natural processes that supply the beach with sand and the sand-storage capacity of the beach elements themselves. Meeting this goal will require careful examination of conservation needs, natural processes, building practices, and corrective engineering proposals that affect whole beach systems.

Problems appear only when people try to put structures too close to the beach (Figure 2.9). If nothing is built on or next to the beach, it will remain as long as the process of natural replenishment continues. It may shift with the seasons, yield sand temporarily to storm erosion, slowly recede landward with rising sea levels, or accrete seaward with natural shifts in the flow of ocean currents, which bring more sand. Mobile and responsive, the beach will remain over the years. But when structures are built and occupants try to restrain these natural movements with bulkheads or groin fields so as to hold the beach, you may start a chain reaction of problems that can be solved only by the expensive process of "replenishment," that is, continuously pumping sand from the ocean bottom onto the beach. This remedy is so costly it is not available to most communities.

It must be remembered that the key to the natural protection provided by the beachfront is the sand, which is held in storage and yielded to storm waves, thereby dissipating the force of their attack. Consequently, taking sand from any part of the beach—dry beach, wet beach, bar, or the nearshore submerged zone—can lead to erosion and recession of the beachfront. Therefore, beach conservation should start with the premise that any removal of sand is adverse, whether for construction fill, concrete aggregate, or any other purpose, and should be prohibited or tightly controlled.

In addition to excavation of beach sand for construction, the worst impacts may come from mining of protective coral reefs and the building of seawalls and groins, which deplete or eliminate the beach (see also "Shoreline Construction Management" below and "Beach Resources" and "Beach Erosion" in Part 3). These impacts are multiplied by the recent global tendency toward a rising sea level, caused by global warming, which is threatening to erode beaches all over the world (see "Sea Level Rise" in Part 3).

When sandy shores are occupied, roads are built, and investment capital is committed, it may seem desirable to retard the natural recession of the shore with seawalls and groins. These structures typically serve only a temporary purpose (often for only 5 or 10 years), however; through a false sense of security, they set the stage for larger scale disasters than would have occurred without structural intervention.

Figure 2.9. These Florida (USA) barrier island condominiums provide an example of unwise land use—a hurricane (cyclone) strike could destroy them all in a few minutes. (Photo by author.)

The most troublesome erosion of beaches occurs in developed areas where buildings and roadways have been placed too close to the water's edge and are being undermined or threatened by storm-induced erosion. In such cases, the beach is often "armored," that is, seawalls or groins are built to protect the threatened properties, or jetties are built to keep inlets open. But these structures are very expensive and may even worsen the situation. Therefore, they should be closely reviewed to see if less expensive and more successful "soft-engineering" alternatives exist (see "Beach Erosion" in Part 3 and "United States, South Carolina" case history in Part 4).

Three common types of shore defenses are shown in Figure 2.10: the seawall, the stone revetment, and the bulkhead (of steel or timber). These structures, intended to stabilize the beach may actually deflect and reduce supplies of sand to a level no longer capable of replenishing losses caused by storms—the beach may be lost, even while the structure remains.

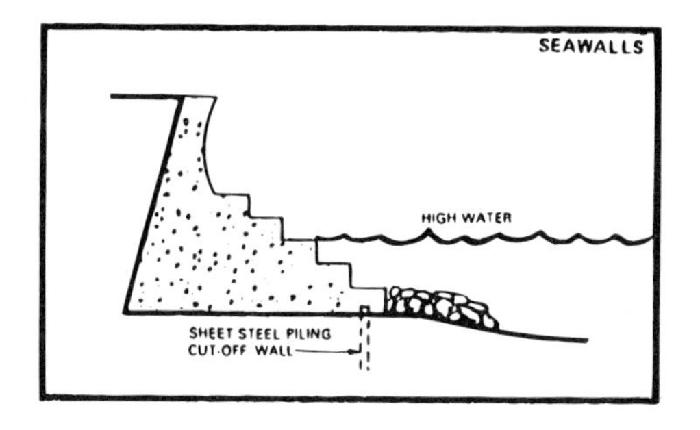

ADVANTAGES

Provides protection both from wave action and stabilizes the backshore.

Low maintenance cost.

Readily lends itself to concrete steps to beach.

Stabilizes the backshore.

DISADVANTAGES

Extremely high first cost.

Subject to full wave forces, fail from scour, flanking of foundation.

Not easily repaired.

Complex design and construction problem. Qualified engineer is essential.

Slope design is most important.

More subject to catastrophic failure unless positive toe protection is provided.

BULKHEADS (STEEL OR TIMBER)

EXISTING SLOPE
SPLASH APRON
FILTER MATERIAL REQUIRED FOR TIMBER PILES BACKFILLED WITH SAND
GRANULAR FILL
HIGH WATER
CABLE TIEBACK
WOVEN FILTER CLOTH
STONE TOE PROTECTION

ADVANTAGES

Provides positive protection.

Maintains shoreline in fixed position.

Low maintenance cost.

Materials are available locally.

DISADVANTAGES

Vertical walls induce severe beach scouring. Adequate toe protection required.

High first cost.

Subject to flanking; bulkheads must be tied back securely.

Pile driving requires special skill and heavy construction equipment.

Complex engineering design problem.

Limits access to beach.

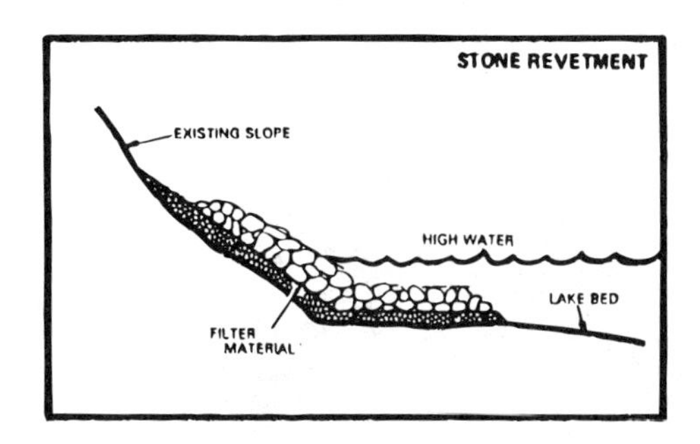

ADVANTAGES

Most effective structure for absorbing wave energy.

Flexible — not weakened by slight movements.

Natural rough surface reduces wave runup.

Lends itself to stage construction.

Easily repaired — low maintenance cost.

The preferred method of protection when rock is readily available at a low cost.

DISADVANTAGES

Heavy equipment required for construction.

Subject to flanking and moderate scour.

Limits access to beach.

Moderately high first cost.

Difficult construction where access is limited.

Figure 2.10. Of three common types of shore protection structures, the revetment is preferable from both engineering and ecological standpoints. (*Source:* Reference 362.)

Bulkheads tend to accelerate beach loss because they reflect the force of waves downward and back into the sand, which causes the beach to be scoured away (see "Shoreline Construction Management" below). A row of parallel groins may force sand to move further offshore with the littoral drift, from one groin tip to another, instead of moving along close to the beach. Thus, improperly designed structures intended to stabilize the beach may actually reduce the reserve of sand to a level no longer capable of replenishing losses caused by severe storms. In such cases, storm waves may remove enough beach to erode under and around the structures, causing the beach line to move inland as the berm regains its equilibrium slope. (57)

A beach disturbed by improperly designed bulkheads and groins may have only a small remaining area available to store sand. If sand is shunted to sea because of groins or bulkheads, for example, the reserve sand in storage may be reduced to a level no longer capable of replacing sand losses from severe storms (Figure 2.11). The beach system becomes unstable, slumps in places, and attempts to re-establish its old equilibrium profile, or "angle of repose." But with less sand, the equilibrium angle of repose can be established only at a position inland of the previous beach profile. When this occurs, erosion cuts away the land (57).

Figure 2.11. Structures built to protect beachfront property often eliminate the sandy beach entirely, leaving direct water contact with the structure. (Photo by author.)

As an example of mismanagement, Waikiki Beach (Hawaii, USA) has now been reduced to a series of seawalls and groin fields to protect what was at one time Hawaii's most important tourist attraction (231). In addition to the obvious aesthetic impacts of the structures, beach sand must now be periodically trucked in to replace sand lost to erosion. Placement of resort structures too close to the shoreline has compounded the problem by requiring more groins to be built at adjacent stretches to the beach to protect other property.

Sand replenishment, or artificial beach nourishment, may be the main hope for restoration of badly eroded beaches, with structures playing a secondary role. Rebuilding beaches artificially by replacing lost sand permits the natural process to continue. In addition, it may be necessary to remove improperly designed structures, such as bulkheads and groins, that deplete the sand supplies. This remedy is so costly that it is not available to most communities.

Although beach nourishment (i.e., pumping sand onto beaches to stabilize them) is normally preferable to building fixed shore protection structures, close scrutiny must be given to nourishment projects because they require a large, continuing expenditure that must be justified. Nourishment projects should not become substitutes for balanced long-term beachfront management (see "Beach Fill" in Part 3).

The general goal for the management program should be to maintain the beach profile by protecting both the natural processes that supply the beach with sand and the sand-storage capacity of the beach elements themselves. Meeting this goal will require careful examination of conservation needs, natural processes, building practices, and shore protection proposals that affect whole beach systems.

Since the main threat to the beach is usually from development on the land next to it, beach protection requires coordinated management to include both the beach and the land behind it, including actions to limit construction, require setbacks, prevent excavation, and control beach, harbor, and inlet structures (57). The general goal for a management program should be to maintain the beach by protecting both the natural processes that supply the sand and the sand-storage capacity of the beach (Figure 2.12). Meeting this goal will require careful examination of conservation needs, natural processes, building practices, and corrective engineering proposals that affect whole beach systems. This process can best be developed through an ICZM program (as it is in Barbados).

Environmental management experts have been able to help shore protection engineers most effectively when their advice is solicited early enough in the design stages so that options are open. These options can include sources and types of rock and unconsolidated materials, type of structures, excavation methods, placement methods, transport methods, seasonal work schedules, and so forth. Big changes can be made at the front end, if necessary, without big increases in cost.

ICZM should treat the shoreline as a *dynamic* system. Beaches erode and accrete in response to varying sediment supplies, sea level changes, storms, and a range of small-scale processes. If a balance between shoreline development and conservation of beaches is the goal, the ICZM program must provide for some reasonable range of shoreline fluctuations (198). But the flexibility cannot be *ad hoc;* it must have a degree of predictability about it. This can be done by means of setback lines that are tied to the *local* erosion rate, as explained below.

For preservation of the natural beach profile, roads, buildings, utilities, and other permanent structures should be prohibited in the frontal dune area. The mining of dunes for sand should also be completely banned. A most important rule is that permanent buildings should be placed well back of the predicted beach recession line. It is unwise to allow development of property that will certainly be lost to the sea or to transfer the risk to the public, as when erosion creates owner demands for protective works. These are not only costly but may further imperil the whole beach system. Therefore, structures should be located behind a setback line that accommodates the predicted long-term erosion of the beach (say, 50-years).

A setback line should be entirely landward of any shifting frontal dune system. Moreover, the line should be far enough landward to allow for predictable recession of the beach. We suggest that the beach and dune system be surveyed to establish a "50-year recession line"—the limit of expected recession and consequent landward movement of the frontal dune in a period of 50 years—to be designated as a setback line. This approach assumes a "useful length of life" of a structure of 50 years. For significantly receding beaches, new recession lines may have to be established periodically. Buildings and other structures should be placed behind the 50-year recession line. Existing structures seaward of the recession line should be designated as non-conforming uses. (See "Setbacks" below.) Also, every effort should be made to prevent disturbance of dunes.

Sometimes, beachfront development has already progressed so far that it is too late for nonstructural remedies alone. Some commendable engineering structures have been devised to supplement natural processes and to reduce further damage where development is quite intense. But there are many examples of failure of the structural approach. Miami's beach was all but eliminated by extensive seawalls and groin fields (57). The lost sand had to be replaced in a beach replenishment project that cost over 80 million dollars (U.S.). The sand will be washed away sooner or later by big storms, and Miami will have to start over.

It is often necessary to determine the degree of erosion taking place by comparing beach "profiles" over time. When standard topographic methods are not available for this, beach profiles can be measured by means of one or the other of the following two simple methods (181):

Figure 2.12. Protection of beaches involves protection of the sand dunes that lie behind them. (Photo courtesy of the U.S. National Park Service.)

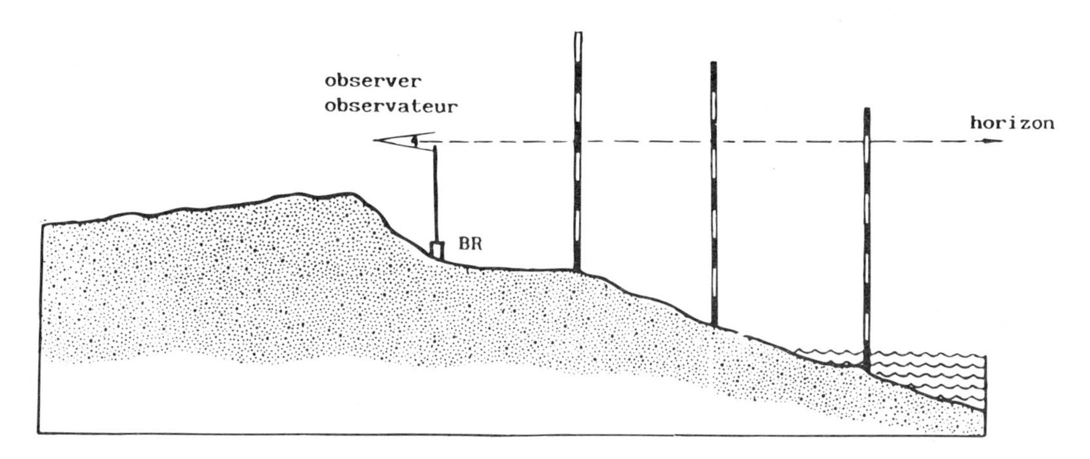

Figure 2.13. Beach profiles can be estimated by sighting the horizon from a fixed point; see text for details. (*Source:* Reference 181.)

1. If no leveling instrument is available, one can use a 1.5-m surveyor's staff placed vertically on the bench mark BR, with the observer sighting the horizon from behind the staff (Figure 2.13). The line of sight intercepts a height H on a graduated staff, which is displaced along the profile of the beach, as in classic leveling. The distance from the stake to the observer is measured by means of a graduated tape. This simple leveling method is relatively accurate (±10 cm), when the measured distances and the slope of the beach do not force the observer to change position several times along a single profile in order to estimate the height of interception on a 4- to 6-m-long stake, with graduations every 20 or 50 cm.
2. In an alternative method, a row of marked galvanized steel pipes is driven vertically into the sand to a depth of 2 to 3 m, at intervals of 5 to 10 m, and protruding from the sand to a height of 2 m. The pipes can be driven into the sand with a pneumatic hammer on the exposed beach or with a pneumatic water gun on the underwater beach. Topographic variations are then measured by direct reading of the distance between the sand level and the top of the pipe. Such readings can be taken regularly by any qualified observer on the exposed beach, but they require an equipped diver for the underwater readings. The major drawback to this simple observation system is the risk of accidents, caused by the protruding metal pipes, to users of the beach, fishermen, and tourists. Therefore, it can only be used with the permission of the public authorities and with posting of very conspicuous warnings.

The ultimate objective of beach profiling is to enable the calculation of volumes of beach materials lost or gained over a period of time. The end areas method can be used in computing the volume of sand between adjacent profiles and progressively for the whole beach, as follows (181):

$$V + D(A_1 + A_2)/2$$

where

V = volume,
D = distance between adjacent profile, and
A_1 & A_2 = cross-sectional areas of the two adjacent profiles 1 and 2.

The following is a simple set of beach protection rules that can be incorporated in any ICZM program—these are "the seven golden rules" of Hayes (163):

1. Understand the natural beach system before it is altered. Site-specific studies may be required at many localities to ensure wise planning decisions.

2. Develop a setback line before construction begins.
3. Where a major obstruction to longshore sand transport is built, such as a large harbor, allow for an adequate sand-bypassing system.
4. Where possible, use *soft* solutions, such as sand nourishment or diversion of channels, rather than *hard* solutions, such as revetments or seawalls, to solve beach erosion problems (e.g., see "united States, South Carolina" case history in Part 4).
5. Maintain a prominent foredune ridge.
6. If a beach is valuable for tourism, recreation, or wildlife habitat, do not mine the sand from dune, beach, or nearshore.
7. Do not panic after a storm and drastically alter the beach. Whenever possible, let the normal cycle return the sand.

An added, *eighth rule,* is to use the best professional expertise available (i.e., engineers or geologists with extensive experience in solving beach problems). Beach erosion is a subject which, for some reason, attracts opinions from self-styled experts. Beach management is not a job for inexperienced or amateur practitioners.

Various quick fixes have been tried to trap and funnel sand for eroding beaches according to Cruikshank (98). Engineers and inventors have made claims for artificial seaweed, fuzzy nets that lie on the ocean floor, specially designed underwater catchment systems made of metal or concrete, rows of used auto tires, and all sorts of other devices. The problem with all of these devices is that (a) they seldom work as well as advertised and (b) they are, for the most part, simply removing sand from some other part of the beach "downstream." (98)

2.5 BOUNDARIES

The land-ocean interface has two principal axes (326)—one axis along shore and the other perpendicular to the shore (or cross shore). For the longshore axis (length of the coast), relatively little controversy arises as to the definition. By contrast, there is considerable discussion about the cross shore axis (the second dimension). While it may be agreed that this axis profiles a zone of transition between the ocean environment and the terrestrial environment, the width of that zone is subject to interpretation.

Boundaries for ICZM programs should be located to capture, and enable resolution of, all major coastal issues, following one of the few maxims of ICZM—boundary lines should be determined by the issues that led to creation of the program (326). Because there is a broad array of possible coastal issues, there is a broad array of possible ICZM *management* boundaries (Figure 2.14).

Narrow boundaries are best suited to deal with use conflicts occurring along the immediate shoreline. But where impacts generated in the hinterlands are a serious threat, then a coastal management boundary extending inland to the ridge line of the watersheds that drain into coastal waters would be justified.

In practical terms, the designated coastal zone should include at a minimum: (a) all coastal lands that are subject to storms and flooding by the sea; (b) all intertidal areas of mangrove, marsh, deltas, salt flats, tideflats, and beaches; (c) all permanent shallow coastal water areas such as bays, lagoons, estuaries, deltaic waterways, and near coast water that include seagrass meadows, coral reefs, shellfish beds, or submerged bars; and (d) all small coastal islands and other important nearshore features. Each of these can be set up as an administrative "tier" with a somewhat different management approach (See "Tiers for Management" below).

Where it is convenient to subdivide the coastal zone into tiers for varying types of management, the "marine and coastal waters" tier (wet side) should be adjusted to reflect fisheries and port sectors, the importance of nearshore spawning and nursery habitats, and regional and international matters. The "edge zone" or "transitional area" tier (wet side) should address shallow water and intertidal areas, while the "shoreland tier" (dry side) should address coastal development and storm-flooded areas. The "upland tier" (dry side) should address watershed problems.

It is in the shorelands (third management tier) where the most difficulty will arise. Here on the "dry side" of the coastal zone all the problems of governance are met—crime, housing, transportation, taxes, waste disposal, etc.—not just coastal resource problems. But this is not to say that the dryside

Figure 2.14. In some cases coastal management boundaries will include urban waterfronts as well as rural areas (Los Angeles). (Photo by author.)

should never be included; indeed, the focus of the U.S. Coastal Zone Management Program—control of waterfront development—is on the dry side. The wet side is mostly controlled by pre-existing governmental programs.

Use of the term *coastal area* or *coastal zone* indicates that the governmental unit administering the management program has distinguished a coastal area or zone as a geographic area apart from, yet between, the oceanic domain and the hinterlands (the terrestrial or uplands domain).

From a strategic and political perspective, success in creating a national ICZM-type program may come more easily if the zone for management is narrow than if it is wide. A mandate to manage a narrow strip of transitional area, comprising mostly tidally influenced habitats, may win approval when a broader zone including deeper parts of the sea and higher and dryer parts of the valuable inland real estate might not.

Small island countries have a particularly difficult time determining ICZM boundaries. Some ICZM authorities would call entire islands "coastal zones" because most island commerce and societal affairs have a coastal connection. But *primary* authority over the *whole* of an island seems unrealistic for an ICZM program. Therefore, for islands, the author prefers to identify a specific coastal zone in the same manner as for any other landform.

Calling a whole island a coastal regime and abandoning the concept of a coastal zone as a definable and separable entity may jeopardize chances to achieve any special integrated resource management for the coast. To abandon the idea of governing a particular defined coastal zone or segment—an area—and instead to try to use ICZM to govern the use of the resources of the country at large could result in a program that lacks focus and distinctiveness and, particularly, cohesiveness. The strong political support needed to initiate an effective program may dissolve. The narrower the coastal zone or coastal area, the more authority ICZM can expect to gain. The broader it is, particularly a whole country, the less authority it can gain because of the appearance of duplication and competition with existing authorities as well as the vagueness of function.

For **planning,** it is advisable to delineate a broad area, including all shorelands (e.g., fringing islands of the coastal plain, the various coastal settlements, industrial areas and agricultural lands, as well as the transitional areas of the edge zone and the open waters). A broad planning area has the advantage that planners can look synoptically at all resource uses and economic and social factors that relate to coastal conservation.

Conversely, for **management** the boundaries should be as narrow as feasible. The narrower the management area, the more clout the management authority can expect to exercise. The broader it is, the less clout it will have because of the increased conflict with other government authorities and the appearance of generality and vagueness of function.

In fact, the two major program development phases of ICZM may well utilize different boundaries; that is, a wider area will quite likely be identified for the *planning* phase than for the subsequent *management* phase. In fact, one purpose of the planning phase is to refine the boundaries and reduce the width of the designated coastal zone to the minimum needed for the management phase.

Official coastal area boundaries are often defined by use of proxy boundaries, at least for the planning phase. For example, prominent physical landmarks or other physical criteria, political boundaries, administrative boundaries, arbitrary distances, or selected environmental units are often used (54).

Locating and mapping the management boundary would be fairly simple if an existing mapped boundary were used (e.g., the "25-year flood isometric line," a coastal highway, or a topographic contour line such as 5 meters elevation). Where a special boundary line has to be delineated for ICZM, the task could be complex, particularly where a high degree of precision is required. Also, landowner resistance and controversy is expected whenever land use zoning of any kind is done.

As an example, for the inland boundary of the *planning* area, it may be convenient to use a major highway paralleling the coast, the foot of a coastal mountain range, the inland political boundary lines of the coastal counties or municipalities, or other recognized political or physical features. This will often be more practical than an arbitrary distance of say, 100 meters or 1 kilometer. This practical type of planning area boundary is most convenient for regional planning and social demographic and economic analysis, as well as for defining interest groups.

The zone defined for *planning* should also include the areas most threatened by seastorms and other natural hazards. Thus, to the extent possible, ICZM planning boundaries should fit to natural function, encompassing natural ecosystems and natural forces. But at the same time they should reflect the political boundaries of towns and industrial centers; that is, where possible the *planning* boundaries should be modified to include an entire community and not divide it into two parts, a coastal zone part and a non–coastal zone part.

Typically, boundaries in the coastal zone have been drawn on the basis of habitat type, flood exposure, littoral/upland vegetation, or elevation, according to Kana (198), and the delineation in the field has often been based on the present situation without regard for changes in habitat area that may occur in the future. Kana stresses that it is important to recognize these boundaries as *dynamic* demarcation lines and that boundaries will change naturally in response to erosion, sedimentation, subsidence, or evolution of habitats or artificially in response to development, new land uses, or altered political mandates and leadership (198).

To summarize, the geographic jurisdiction of ICZM is variable, but programs will be most effective when their geographic scope is limited. Ideally, a coastal nation or subnational unit should set the boundaries of the zone only as far inland and seaward as necessary to achieve the objectives of the planning and management program. Sometimes the boundaries must be adaptable to new circumstances.

Given the environmental, resource, and governmental differences among coastal nations and subnational units (provinces, towns), there is considerable variety in the selection of boundaries to delineate both seaward and inland extent of the coastal zone. Small offshore islands without major populations would be included and protected as natural areas so long as they were within the coastal zone oceanward boundary (Figure 2.15).

International law will provide a limit to the maximum possible seaward extent of the administrative area—namely, the width of the territorial sea (12 nautical miles from the low water mark or coastal baseline)—for comprehensive control. Coastal states may exercise jurisdictional powers over pollution, fisheries (Figure 2.16), research, and related matters to the 200-nautical-mile limit.

Figure 2.15. Small islands, such as this one in the Red Sea, are often included within coastal zone boundaries and conserved as natural areas because they provide critical life support for valuable species. (Photo by author.)

Figure 2.16. Coastal countries usually take jurisdiction over fisheries and environmental matters in a defined Exclusive Economic Zone. (Photo courtesy of U.S. Fish and Wildlife Service.)

Example boundaries for eight countries follow (326):

Country	Inland boundary	Ocean boundary
Brazil	2 km from MHW	12 km from MHW
Costa Rica	200 m from MHW	MLW
China	10 km from MHW	15-m isobath (depth)
Israel	1–2 km variable	500 m from MLW
South Australia	100 m from MHW	3 NM from the CB
Queensland	400 m from MHW	3 NM from the CB
Spain	500 m from highest storm or tide line	12 NM (limit of territorial sea)
Sri Lanka	300 m from MHW	2 km from MLW

Abbreviations: MHW = mean high water; MLW = mean low water; NM = nautical miles; CB = coastal baseline.

2.6 CONSTRUCTION MANAGEMENT

Many types of projects are constructed along the shorelines and beaches and in the shallow waters of the coastal zone—bridges, causeways, channels, airports, piers, jetties, refineries, hotels, factories, swimming pools, power plants, warehouses, and so forth. Specifics are discussed in various sections of this book. In this section we present some management guidelines that apply generally to coastal construction operations.

Threats to the productivity of unique coastal natural resource systems arise from *uncontrolled* development activities and their side effects, such as reef and beach mining, shoreline filling, marine construction, lagoon pollution and sedimentation, and other development activities that are distinct from those on land. Projects that excavate large amounts of materials from coastal waters, even sandy materials, usually pollute the water with fine silt. Therefore, marine excavation should be conducted with extreme care. Also, the construction of large concrete and rock groins and seawalls can cause problems, including the operation of storage and casting yards and heavy equipment on the beach and the local roads. (See "Shoreline Construction Management" below.)

Among the most common coastal zone construction practices that jeopardize marine ecosystems are the various types of excavation. *Excavation* is a general term encompassing activities that include digging, dredging, quarrying, and mining. Common to all is the use of mechanical techniques (and sometimes explosives) to remove bottom materials from submerged habitat. *Dredging* is usually associated with the removal of loosened submerged deposits of sediments or rocky materials for the purpose of deepening a channel (or basin) or obtaining fill or aggregate for construction use. *Mining* is usually associated with the removal of minerals for their chemical and industrial properties. *Quarrying* refers to the fracturing and removal of medium to large pieces of consolidated (hard) rock for masonry and armor stone placement. (231)

Wetlands are particularly at risk from shoreline development because wetlands simply do not have the carrying capacity that land areas have for supporting commercial and industrial activities or urban occupancy. Excavation, draining, grading, and filling should be prohibited in wetlands for any purpose other than the installation of supports for permitted elevated structures (Figure 2.17). Engineers are beginning to learn that innovative designs can solve many development-versus-environment conflicts (Figure 2.18).

Regarding biodiversity, significant disruption of wildlife species and habitats could have a serious impact. Critical habitats like sea turtle nesting sites on the ocean beach are particularly vulnerable and could be in need of protection, as are such high value resource units as mangrove forests, seagrass beds (which can be rapidly degraded by sedimentation and turbidity), and, of course, coral reefs. Any reduction in biodiversity (habitats and quantity and type of fauna/flora) could apply to either the short term (construction period disturbance) or to the long term (permanent loss of habitat), but the latter is less likely.

Construction effects on the human environment arise from noise, air pollution, nuisance factors, and uglification, as is the case for all heavy construction work. During construction operations there are pollution threats including smoke, oil releases, and dust. The trucks and heavy construction equipment can crowd the highways and cause traffic disturbances. Many projects require excavating and moving materials (e.g., sand, rock, concrete) and the use of much heavy machinery operating close to inhabited

Figure 2.17. Drainage channels can dry out wetlands and greatly reduce their biological diversity. (Photo by author).

areas over a more or less prolonged period and also moving of large equipment and materials through the streets. This can create general disturbance, smoke, noise, and traffic. Also many projects need stock yards, casting yards, truck parks, and so forth which can be a source of disturbance.

Regarding animal species, significant problems can arise from fright disturbance of animals by machinery and workers (if their work patterns change often), by annihilation or nest destruction, or by physical disturbance of habitat (feeding, resting, and nesting sites). Shorebirds that occupy construction sites will be displaced temporarily, but care should be taken to ensure that the impact is not permanent. Potential reduction of natural areas and wild species of commercial, ecological, or intangible value should be prevented.

Any construction work that causes soil loss can be a major source of impact to coastal biodiversity. Quarries that supply rock or aggregate can give rise to water quality and supply problems from release of fines (clays) or release of other pollutants from excavated materials (Figure 2.19). But these can be controlled with simple countermeasures. Dikes and stilling ponds can be employed. Other problems could arise from discharges of oil or chemicals from storage tanks or from machinery at yards, on transport routes, or at work sites.

Management implications of construction activities in lagoons and estuaries include the following examples (308):

1. Dredging of channels, marinas, and ports
   - Creates the problem of dredge spoil disposal.
   - Should be discouraged in small estuaries and through or near the critical habitats of valuable species.
   - Should be preceded by impact assessment studies.

Figure 2.18. The materials dredged in constructing the Dinner Key Marina (Miami, Florida, USA) were used to build natural breakwaters to provide shelter for the boats. (Photo by author.)

2. Channelization and diversion of freshwater inputs
    - Generally increases channel bed and bank erosion, resulting in siltation of estuaries, and should be discouraged.
    - Diversion of waters out of estuaries may result in increased salinity, stimulating widespread change in biotic structure and composition, and should be discouraged.
    - Diversion of water into estuaries alters the water balance, decreasing salinity and often increasing siltation, and should be discouraged.
3. Channelization and opening of lagoon mouths
    - In most instances should be discouraged to allow natural cycle of closure and opening.
    - May enable intrusion of seawater and marine sands, which increases salinity and decreases depth.
    - May necessitate periodic dredging of inlets to remove marine sands accumulated there by longshore drift and tidal currents.
4. Reclamation for industrial, urban, aquacultural or agricultural, and port development
    - Should be preceded by national survey and classification of wetlands by both natural values and best-use alternatives.
    - Should avoid interference with freshwater inputs from streams, rivers, and sheet flow.
    - Should avoid contamination of freshwater inputs.
    - Should be preceded by impact assessment studies.
    - Should be sited in areas with suitable soil chemistry to avoid expensive remedial measures, crop failures, and low yields.
    - Should alternate with natural areas managed to provide nurseries (in many cases for seed stock) and the range of traditional uses maintained at sustainable levels.
    - Should be components of multiple use management and should be confined to zones in wetlands where it interferes least with critical habitats and the coastal protection function.
5. Diking and construction of retaining walls, groins, docks, piers, causeways, and roads
    - Should be preceded by studies of water and faunal movements.
    - Should be sited and constructed to avoid interference with the freshwater inputs from streams,

Figure 2.19. Quarrying rock from stream beds may be ecologically disruptive and may cause sedimentation of coastal basins (Bali, Indonesia). (Photo by author.)

rivers, and sheet flow (e.g., roads should follow direction of stream flow and include adequate culverts).

- Should be sited and constructed to avoid interfering with the tidal flushing of wetlands.
- Should be sited and constructed to avoid interfering with movements of detritus and fauna (larvae, juveniles, and adults).
- In many instances, retaining walls may be inferior substitutes for barriers of natural vegetation.

2.7 CORAL REEF MANAGEMENT

Coral reefs, are important resources for a great many tropical countries—they are usually designated for special protection. Reefs are the core of marine diversity and productivity where they occur. They provide essential supplies of seafood for many tropical countries. They are natural breakwaters that protect beaches and shores from erosion and storm damage. They manufacture carbonate sand, which is supplied to beaches by waves and currents. Reefs and their associated sandy beaches are important attractions to tourists, a major source of income and foreign exchange to dozens of tropical countries. Therefore, the productivity, beauty, and health of coral reefs are of great importance, ecologically and economically (see also "Coral Reef Resources" in Part 3).

But coral reef resources have been seriously degraded in many parts of the world. Coral reef degradation has serious consequences for tourism, fishing, beach stability, and biodiversity. For example, most of the 21 countries and 49 marine parks (or reserves) in the Caribbean with coral resources have noted significant damage (166). Some reefs are virtually beyond repair (those closest to settlements), but many that are degraded could still be returned to functioning condition. Extreme loss of corals can ruin a park and cut severely into tourism.

The sources of degradation of coral reefs are varied and extensive. Many sources are known in detail, others more generally, and still others are awaiting discovery. Unfortunately, most coral reefs that receive human use or are located near settlements suffer damage (combined human and natural causes), lowering their value significantly and even, sometimes, reducing them to rubble. Tomascik (345) identifies the following as leading causes of environmental impacts to coral reefs: tourism, sewage and runoff, forest cutting, agricultural practices, aquaculture, ports, mining and dredging, power plant heavy industry, and petroleum industry. We have tabulated the following list of impacts to be managed in the interest of rehabilitating coral reefs (390):

- General urban encroachment
- Pollution from agriculture, sewage, and industrial wastes that inhibit feeding, growth, and reproduction and that are toxic to the corals, fish, and invertebrate life
- Siltation and sedimentation created by dredging, filling, and related construction activities and increased erosion
- Discharge of large volumes of freshwater as may result from diversions and stormwater outfalls
- Oil pollution from drilling, extracting, and transport (tankers and pipelines)
- Physical damage from anchors, diver activities, and coral collecting (Figure 2.20)
- Great storms that smash the coral and "sandblast" their tissues
- Diseases like whiteband and blackband disease that kill corals
- Excessive spear, trap, and cage fishing that removes too many species important for keeping the reef in balance (e.g., parrot fishes)
- Fishing with dynamite, which disintegrates coral
- Destructive fishing practices, including collection of juveniles for aquariums (Figure 2.21) and use of chemicals for same
- Ship groundings, which smash and/or scrape away corals
- Marine construction activities
- Massive loss by disease, depletion, or migration of essential symbionts (e.g., sea urchins that keep the reef healthy, when in appropriate abundance, by cleaning it of algae).

A major concern in coral reef management is the protection of coral reefs from such impacts. For a thorough review of the subject, the reader is referred to *Human Impacts on Coral Reefs: Facts and Recommendations,* edited by B. Salvat (309).

Marine excavation (dredging) near coral reefs suspends a mass of particulate matter, ranging from very coarse (>0.5 mm) to very fine (<25 μm). Because turbid waters are damaging to the coral reef environment, dredging methods should try to reduce or avoid the suspension of fine particles by using suction dredges, pipelines to transport slurry to shore, or protective curtains. Concerning the settlement of particles, a large quantity of either coarse or fine particles will bury the corals, which are unable to withstand cover for periods longer than one or two days. These particles will fill in all crevices and

Figure 2.20. Diver moving anchor to a sandy bottom area to avoid damaging coral. (Photo by author.)

Figure 2.21. Diver collecting fish for a public aquarium is careful to avoid harming reef life in Palau. (Photo by author.)

cavities together with the numerous species and organisms that live there and that are indispensable to the structure and the functioning of the reef ecosystem. (78)

The impact of silty, turbid waters is one of the main causes of destruction of coral reefs. Siltation is a consequence of deforestation and bad land management as well as of urban development such as housing and highway construction. For example, in Kaneohe Bay, Hawaii, in 1969, because of land clearing and a 24-hour storm, a river poured into the bay the equivalent of 1 kg of suspended matter per m^2 of estuary bottom. Between 1888 and 1969, the mean depth of the estuary bottom decreased by 1.5 m, thus burying a lot of corals. Similar examples are found in Okinawa, Puerto Rico, and the Indian Ocean. (78)

A major problem for corals is to rid themselves of any covering by fine silt. For this, they must spend energy that weakens them in their other functions. Considerable literature exists on corals and fine sedimentation. The capacity of corals to clean off fine particles of sediment by the action of their polyps and ciliary movements depends on the corals themselves. Those living close to the coastline, on fringing reefs, are generally more efficient cleaners than are species that live on the outer reef slope.

Coral mining—the excavation and removal of coral materials from reef ecosystems—requires control. Such mining is extensively undertaken in some island countries in the Indian Ocean and Southeast Asia, where it degrades the ecosystem and leaves the shoreline exposed to erosion and storm surges, causing serious loss of beach and shoreland and damage to coastal human and marine resources habitats.

Mining is the most explicit of all physical impacts to coral reefs. The demolishing of coral reefs to produce commercial construction materials (blocks, quicklime, roadbed materials, etc.) not only destroys habitats and disrupts fishing, it also jeopardizes coastal settlements. Mining of coastal reef barriers leaves the shoreline exposed to waves and storm surges that erode beaches, undermine shorelines, and destroy property (see "Indonesia, Bali" case history in Part 4).

Because of its destructiveness, many countries (or provinces) have banned coral mining. Among the examples of restrictions on coral mining enacted by various countries are the following: Panama passed an outright ban on coral mining in 1992; Sri Lanka established an ICZM program with a primary purpose of eliminating coral mining (see "Sri Lanka" case study, Part 4); the governor of Bali, Indonesia, successfully banned coral mining in 1982; the Maldives have taken action to control mining; the United

States does not allow coral mining; and the Great Barrier Reef Authority (Australia) allows some mining but under close control.

In another example, the coral formations in the Gulf of Mannar (Tamil Nadu, India) were fast deteriorating, especially in the four islands of the Chidambaranar District coast due to human interference. Before the Forest Department of Tamil Nadu took charge of the islands within the now Gulf of Mannar Marine National Park, illicit removal of corals was rampant. The Forest Department made earnest efforts to enforce laws against any removal of corals, with good effect, as shown below (256):

Period	Cases booked		Number accused	
	Tuticorin Range	Mandapam Range	Tuticorin Range	Mandapam Range
Dec. 91 to Mar. 92	44	2	111	2
Apr. 92 to Mar. 93	49	7	109	8
Apr. 93 to Sep. 93	40	—	56	—

Note: From the above violations, fines of 183,000 rupees (US$6,100) were collected.

Harvesting coral reef organisms to sell as decorative items is another very damaging activity (Figure 2.22). Because corals grow slowly, the replacement of a removed coral by a new one is a process that can take several decades. In the meantime, the coral reef habitat remains degraded and, if the removal rate is high, depauperate. According to Wood and Wells (389), the potential for over-exploitation and damage to the reefs as a result of collecting activities gives cause for concern. The only collecting that should be permitted is that which can be proved to be sustainable.

Sometimes management of fisheries harvest is relevant to the coral reef management entity. For most reef ecosystems, there is no question that, because of overfishing, yields of food species such as lobsters, snappers, and groupers are much lower than they need be. For any given reef system, the maximum level of sustainable catch may be difficult to estimate, but it can be said that coral reef fisheries in general can yield maximum sustainable catches of 4 to 25 tons/km^2/yr. While fish catches can be regulated according to modern "yield formulas" and "stock assesments," there are strong natural limits. (105)

Figure 2.22. Collection and sale of marine curios must be closely controlled, and perhaps prohibited in some areas, to prevent overharvest. (Photo by author.)

Excessive use by recreational (tourist) divers can deplete a reef in a few years by such impacts as hand and foot damage (Figure 2.23). To document the effect, a researcher systematically observed recreational divers in the Florida Keys (USA) between May 10 and August 13, 1989. A total of 206 divers were studied during 66.6 hours of diving. The scuba divers averaged 7 incidents of reef contact while snorkelers averaged 1 incident for each 30 minutes of time in the water. Of a total of 1,164 contacts witnessed, 65% were interactions with scleractinian corals and 34% were with octocorals. Snorkelers in shallow water often tread water, stirring up large clouds of sediments, and are more apt to stand on corals than are scuba divers. Non-parametric tests show that divers with gloves have significantly higher numbers of contact with corals than do divers without gloves, that men have more contact than women, and that scuba divers have more contact than snorkelers. (340)

Natural stressors also present reef managers with problems, sometimes exacerbated by human disturbances. For instance, the prevalence of disease may be higher in corals that are stressed by mechanical damage or pollution. Corals respond to chemical and sediment stress by the secretion of mucus; if this becomes excessive, the mucus could harbor bacteria, with fatal consequences to the coral.

The ability of a coral ecosystem to recover from natural disturbance may be impaired by human impacts. For example, the fast-growing coral *Acropora cervicornis* has been known to recover rapidly from hurricane (cyclone) damage, but recovery may be delayed by predation (snails) and overgrowth of algae. The abundance of both marine snails and macroalgae may be accelerated by the removal of their predators by fishermen. (78)

The first step in coral reef conservation should be *protection,* the elimination or reduction of the stressors responsible for the damage. The second step is *rehabilitation.* Special programs have been shown to be effective in expediting the natural recovery of degraded coral reefs. As explained below, major types of conservation action include the management of watersheds and wetlands, the abatement of pollution, controls on extractive industries, and the prevention of destructive activities such as anchoring in coral. To put it succinctly, one should control what goes in, what goes out, and what goes on. (78)

Figure 2.23. To prevent serious damage, persons visiting coral reefs should be reminded not to step on the corals. (Photo courtesy of the U.S. National Oceanic and Atmospheric Administration.)

For management of a specific coral reef resource, it is first necessary to identify and delimit the reef to be managed. An initial visual (SCUBA) survey may be sufficient to set preliminary boundaries. Afterward, more detailed surveys can be made to support particular management initiatives (see "Coral Reef Survey Methods" below). Salm and Clark (308) recommend that a coral reef designated for biodiversity protection should be large enough to encompass 95 percent of the coral genera known to be present in the region. This will usually require a minimum "core" area (designated for strict protection) of 4 to 5 km^2. The size of coral reef to be managed for fisheries or shore protection will depend wholely on local conditions and management objectives.

Tomascik (345) provides useful and practical advice on selecting the parameters for coral reef management. His guidelines for coral reef "ecosystem maintenance" are abstracted below in Numbers 1–6 (with permission); Numbers 7 and 8 were added by the author:

1. **Salinity.** The ambient salinity should be maintained at 32 to p.p.t. Impacts that should be controlled include freshwater runoff from various sources and hypersaline affluents from desolination plants and aquaculture.
2. **Temperature.** Thermal additions should be minimized and in no case should increase ambient temperature by more than 1°C. This requires that heat producers such as power plants and industrial processing plants.
3. **Dissolved oxygen.** Discharges that reduce dissolved oxygen (DO) should be discouraged and in no case should DO be allowed to fall below 5.5 mg/l. This means stopping agricultural and urban runoff from impacting coral reef areas and limiting industrial sources of biological oxygen demand (BOD), as well as sewage from settlements that carry high BOD loads.
4. **Water clarity.** It is most important to organize counter-measures to turbidity, induced by sediments in the water or by plankton blooms triggered by nutrients in sewage discharge. Turbidity cuts off light penetration needed for growth and survival of coral communities; also the sediment fallout chokes the coral. This means controlling sediment from agricultural, industrial, and urban sources, using a variety of known techniques, as well as control of dredging activities.
5. **Nutrients.** Great damage can be done to coral communities by nutrient pollution (eutrophication). The worst effects come from release of phosphorus (P) and nitrogen (N), most often occurring as dissolved phosphate or nitrate from land runoff or from mineralization of biologically occurring P and N (e.g., by sewage treatment plants).
6. **Water circulation.** Reductions or changes in water circulation can degrade coral communities by disabling their abiity to remove sediments and biological waste products. Therefore, it is necessary to control engineering works that would block or alter current flows.
7. **Physical damage and removal.** When reefs are extensively used for recreation or artisanal fishing, they can be badly damaged by fishing implements, anchors, explosives (used for fishing or construction), chemicals, hand-and-foot contact, and deliberate breaking off and removal of corals. Reef fish catches can be excessive, and methods can be inimical (e.g., spearfishing is not compatible with recreational use of reefs).
8. **Mining of coral.** The extraction from the reef area of coral materials for construction (blocks, quicklime, bedding) needs close control because the all-valuable reef functions can be disabled when its structure is demolished. Mining could only be justified if can be proved that is *sustainable* (i.e., material removed will be replaced simultaneously by new reef growth) or that the reef produces no socio-economic benefits. Nevertheless, mining activity must be closely controlled to prevent damage to adjacent areas.

In general, coral reef management is expedited by either an ICZM approach (regulatory control) or a protected area strategy (proprietary control) and should include both protection and rehabilitation of coral reef resources. Rehabilitation is an important topic because major coral reef degradation has been experienced by dozens of countries. For these countries, it is not enough just to protect what exists; the goals have to include repair of the damage of the past.

Three management goals for coral reefs are listed by Kenchington and Hudson (206), as follows:

1. Cultural and social
   - Preservation of undisturbed environment
   - Protection of aesthetic, historical, biological, or geological national heritage
   - Provision for traditional use
2. Economic and social
   - Self-sufficiency
   - Regional development
   - Economic growth
   - Regional, national, or international trade
   - Foreign currency earnings
   - Fisheries and tourism
   - Mining
   - Leisure and recreation
3. Biological
   - Species diversity
   - Protection of critical habitat for endangered species or commercially important species
   - Maintenance of biological productivity

Management actions should, of course, be preceded by preliminary studies and planning. These will include at least the following activities (78):

- Inventory and assessment of the reef resources
- Diagnosis of the causes of degradation
- Definition of values of the resources and determination of the reef functions to be encouraged
- Assessment of the costs and benefits of various management actions
- Initiation of a baseline and monitoring program to detect natural or human effects

As mentioned above, reefs that receive heavy recreational visitation need special types of protection. For example, Looe Key coral preserve in South Florida, USA, uses the following guidelines:

- Just touching coral causes damage to the fragile polyps; therefore, hands, knees, or equipment should not contact coral.
- Anchor, anchor chain, or line cannot contact coral; mooring buoys are provided.
- Spearfishing is not allowed. This is one reason the fish are so friendly that you can almost touch them.
- Hand feeding of the fish is discouraged. Besides the risk of bodily injury, such activity changes the natural behavior of the fish.
- Hook and line fishing is allowed. Applicable size, catch limits, and seasons must be observed.
- Spiny lobster may be captured during the season except in the Core Area. Quotas and size regulations must be followed.
- Corals, shells, starfish, and other animals cannot be removed.
- Regulations prohibiting littering and discharge of any substances are strictly enforced.

Zoning is the most widely accepted method of protecting sensitive, valuable, or recuperating areas of coral reefs and limiting the impact of users. The simplest form of zoning is that utilized for establishing "replenishment areas," whereby certain refuges for fish breeding, feeding, and resting are zoned as "off-limits" for any consumptive use; that is, set aside as strict protection zones (see "Australia. Great Barrier Reef" in Part 4). Other functional advantages of coral reef protection and/or use allocation zones include the following:

- They permit selective control of activities at different sites, including both strict protection and various levels of use.
- They can establish core conservation areas (sites of high diversity, critical habitats of threatened species, and special research areas) as sanctuaries where people are prohibited.
- They can be used to separate incompatible recreational activities (bird watching and hunting or waterskiing and snorkeling) to increase the enjoyment and safety of the different pursuits.

- They enable damaged areas to be set aside to recuperate.
- They can protect breeding populations of fishes and other organisms for the natural replenishment of devastated or overfished areas nearby.
- They are extremely cost-effective means of managing different uses, since their manpower and maintenance needs are minimal; they are the accepted method to keep people out of the most sensitive or valuable or recuperating areas of coral reefs.

Corals and fishes can be increased in abundance, and damaging algae can be reduced. Active rehabilitation measures have been successfully demonstrated for corals by enhancing recruitment of young corals, transplanting reef corals, repairing injured or sick colonies, and removing predators. However, these operations have mostly been done on a small scale. (78)

Active rehabilitation usually means the manipulation of reef organisms to facilitate recovery of values (e.g., increase in coral cover, decrease of free-living algae, and increase of reef fishes). There are some successful examples of active reef rehabilitation measures, but their cost effectiveness is debatable when large areas are considered. There are many degraded coral units of high value and small size where headstart programs or active rehabilitation would be justifiable with improved technology. But these techniques may be used only on reefs of high value, such as in marine parks. (78)

The coral reef benefits from its associated fauna. For example, herbivorous parrot fishes and tangs graze algae from the reef, as does the long-spined sea urchin, keeping the coral surface clean and functional. The result is beneficial unless their populations are out of balance and these species begin to consume excessive amounts of coral. Because their depletion could be seriously adverse, augmenting their populations might be advantageous for certain reef units of modest scale and high value (e.g., for tourism). The most sophisticated approach would be to set a goal for the appropriate species mix and try to balance the reef toward that goal.

Another option is to transplant to the reef community certain target species of commercial or recreational value (whether they are symbiotic with the coral or not)—e.g., conch, which have been seriously depleted throughout the Caribbean, can be cultured and planted within the reef community (planting breeding-age adults to enhance natural reproduction is considered a better approach than planting juveniles). The variety of options available means that rehabilitation goals have to be carefully defined. At present, the whole subject is open to experiment because little important work has been reported on the subject. (78)

At Key Largo National Marine Sanctuary in the Florida Keys, H. Hudson did succeed in righting and repairing (with cement) large heads of *Montastraea annularis* that had been overturned or broken by storms and ship groundings. Hudson and colleagues have also treated corals infected with "black band" disease (caused by a blue-green bacterium). They used an aspirator powered by air from a scuba tank to suck off infected tissues (five coral heads were treated with one diver tank of air) and covered the affected area with modeling clay or cement. A simpler approach (without use of aspiration) was to cover a newly infected area with hydraulic cement (178).

Where corals are suffering invertebrate predation, it may be possible to remove or kill the predators. An important example is the campaign to destroy the Indo-Pacific coral-eating starfish *Acanthaster planci,* which are so dense that, at times, a diver can kill hundreds on a single dive (mechanically or chemically). Caribbean examples are the snail, *Coralliophila,* and the fireworm, *Hermodice.*

For restoration of particularly valuable sections of reef, the following guidelines (for the Caribbean region) should be considered, according to Woodley and Clark (390):

1. To increase coral populations
   - **Increase recruitment.** The larvae of corals settle and survive best on hard, clean surfaces, with minute pores or crevices to act as refuges (from grazing). In areas with much sedimentation or algal overgrowth, recruitment could be enhanced by cleaning fouled surfaces. This might be done with water jets, air jets, or air lifts or by stripping algae by hand. Recruitment would be greater if surfaces were cleaned at times when coral spat is known to be abundant (August and September in the Caribbean). Second, one could provide artificial surfaces/structures for settlement.
   - **Increase immigration.** Rather than waiting for recruitment from the plankton, one might achieve a quicker increase in coral cover by transplantation. Clearly, a suitable donor area must be

available, not too far away. If transplanted corals have to resist wave surge, re-attachment mechanisms may be needed, such as pegging or cementing to the substratum or to artificial surfaces (start in areas favorable to settlement and then move to host reef). Staghorn corals (in the Caribbean, *Acropora cervicornis*) would be especially suitable for transplantation because stick-like fragments can be wedged into crevices or, if large enough, allowed to sprawl; moreover, they grow very quickly. Many other species can be transplanted successfully; for example, Guzman (151) transplanted live fragments of *Pocillopora* with 80 percent survival.

- **Decrease mortality.** Action can be taken to reduce mortality due to physical damage, disease, predation, and competition.

2. To reduce macroalgae populations
   - **Removal.** Algal overgrowth is often quite damaging to coral reefs. Macro-algae can be stripped off coral heads by hand. Filamentous algae (e.g., *Cladophora* or *Chaetomorpha*), which blooms in reef lagoons subject to nutrient enrichment (as from sewage) can be raked or pitch-forked into a boat. Such algae can be fed to herbivorous fishes in cage culture (e.g., acanthurids)
   - **Increase predation.** Manipulations to increase algal predators are possible. For example, in the Caribbean herbivores were greatly reduced by a combination of overfishing (removing herbivorous fishes) and natural mass mortality of the grazing sea urchin *Diadema antillarum.* Artificially increasing the abundance of *Diadema* can cause a dramatic decline in the prevalence of algae—an abundance of 1–10 urchins/m^2 would be beneficial (Figure 2.24). However, the prospects of collecting and transporting sufficient numbers of these prickly animals may be somewhat daunting. Sammarco's data suggest that, ideally, the sea urchins should be raised in culture and large numbers of small juveniles released on the overgrown reef. The gastropod *Cerithium eburneum* is another potential algal predator.

Figure 2.24. Sea urchins are beneficial in the Caribbean Sea area, where they clean algae from coral surfaces. (Photo courtesy of Edwin Towle. Island Resources Foundation.)

3. To increase fish populations
   - **Increase recruitment.** To recruit more herbivorous reef fishes, one should consider increasing both the supply of larvae and their survival. Unfortunately, most reef fishes are difficult to propagate artificially. Reducing planktivorous fishes such as (in the Caribbean) creole wrasse and blue chromis might greatly increase the survival of larval recruits, but accomplishing this selective fishing would be difficult. Rehabilitating adjacent seagrass beds, which provide sanctuary for tiny fishes (against predation), could be helpful in increasing the recruitment of herbivorous fishes to coral reef communities.
   - **Increase immigration or re-stock fishes.** The loss of reef structure may reduce its carrying capacity for fishes. The addition of artificial habitat on the reef might be useful, such as the provision of hollow concrete blocks. There also might be a role for floating "fish attraction devices" (FADs) moored above the damaged reef. Importation of desirable species to a coral reef is a possibility, but the logistic problems of transferring live specimens are daunting. But restocking reefs could be beneficial if it could be done economically. For example, importing parrotfish might increase the amount of grazing on algae and lead to faster restoration of the damaged coral colony but also enhance touristic interest in the reef, parrotfishes being among the most colorful and interesting members of the reef community. Therefore, considerable research might be profitably targeted on this area.
   - **Decrease mortality of favored species.** It is conceivable that the abundance of, for example, herbivorous fishes might be increased by reducing the number of their predators. Other ways might be found to increase the carrying capacity of a damaged reef for desirable species, such as increasing the area of adjacent seagrass beds, creating more nocturnal feeding grounds for fish that reside on the reef during the day, like parrotfishes.

4. To increase other beneficial species
   - **Mariculture.** Culture systems are now producing young conch (*Strombus*) in the Caribbean and giant clam (*Tridacna*) in the Pacific. Since these have commercial and recreational value and since there are no reported incompatibilities with healthy reef communities, their introduction to a depleted reef area can be considered. The edible green snail and many other species might also be considered. But it is essential to give the utmost attention to sanitation so that introduced animals do not bring pathogens to infect the reef. (See "Exotics" in Part 3.)

Along lightly populated coral coasts, there are many opportunities for community management of coral reefs. Community management requires (384) (a) the creation of a "core group" of citizens to generate and "sell" the program; (b) monitoring of results with feedback to the community; (c) effective and empathetic outside help; and (d) a realistic view of obstacles. These issues are illustrated in Part 4 in the case history "Philippines: Community Management of Coral Reef Resources."

2.8 CORAL REEF SURVEY METHODS

Coral reef survey can be done following several methods (see "Australia, Coral Reef Survey Method" case history in Part 4). One method commonly used is some version of the reliable and simple method of quadrats on transects originated by the Marine Sciences Center of the University of the Philippines (145). This method effectively combines linear and spatial techniques. One advantage of this technique is that comparison is facilitated because much of the work in different countries is now being done in this manner. The following explanation was adapted from Gomez and Yap (145).

> In brief, the technique involves laying out a transect line perpendicular to shore starting from the reef crest and going seaward to a selected depth contour. This depth may be all the way out to where the reef ends (often at the seaward "drop-off"). Or a shorter transect can be chosen, depending on the objective of the survey—perhaps to a depth of 15 m, or maybe 10 m will suffice.
>
> The quadrats cover representative areas, while being set at a manageable size—from 1 × 1 m to as much as 5 × 5 m (145). On fairly horizontal surfaces, a frame of 2 m (or arbitrarily, 1 or 3 m laid along the transect at about 20-m intervals—on relatively steep gradients, such as slopes greater

than 45°, the frame could be laid down every 10 m. The quadrat defined by the frame may be divided into sixteen 25- × 25-cm squares (Figure 2.25).

It is recognized that the regions where reef monitoring is needed the most are often those having a scarcity of resources. Thus, in the development of reef monitoring techniques, simplicity of design and equipment has been given primary consideration by Gomez and Yap (145). Such techniques usually call for a minimum of equipment such as lines, metal frames, some collecting material, and simple measuring and underwater writing implements.

Most techniques require the use of SCUBA. This is expensive and cumbersome, logistically, but is usually a better alternative than such other means of underwater observation as video, unless the depths involved exceed the safe limits for a diver. However, in some instances, snorkeling on the surface may suffice. (145)

At each sampling point, squares covering the following reef components are counted: live hard coral, live soft coral, dead coral, rocks, rubble, sand, and other major living components (e.g., sponges, coralline algae). These are then divided over the total number of squares to obtain relative percentages in terms of cover of the above components.

Condition of the reef is assessed using the values obtained for combined live hard and soft coral reef. Arbitrary categories have been designated as follows: excellent (75–100 percent live coral); good (50–74.9 percent live coral reef); fair (25–49.9 percent live coral cover); poor (0–24.9 percent live coral cover). Results using this method may be found in the various reports and publications of the Marine Sciences Center (145). See also Sukarno (336), DeSilva (106), and Loya (224).

Counts of individual organisms or colonies and/or of the area covered by each would be a useful addition to the survey results. Also, photographs taken of the reef area enclosed by the frame at each

Figure 2.25. A diver uses a standard sampling frame to survey a sloping coral patch reef. (Photo by William Keogh.)

Figure 2.26. Underwater photography is used extensively as a backup to visual survey of coral reefs and sometimes as the primary survey tool. (Photo by author.)

quadrat will be most useful in making a permanent record and in confirming your field identifications of the coral species or genera present (Figure 2.26).

Dahl (100) recommends a specific "practical approach" in which one starts with the most difficult part of the reef to reach, since the surveyor may be tired later and the wave and tide conditions may get worse:

> 1) Fill out the top of the data sheet (locality, date, your name, circle number, reef zone, water depth). As your actions may frighten the fish away, do the fish count first along a 100 m line. Then attach a short rope to the marker in the middle of the survey plot, so that the free and is four metres long and ends in a loop, as this is easier to hold on to—if you start with a four metre rope, it will be too short by the time you have tied the knots, so use a five metre rope to allow some extra.
>
> 2) Swim or walk around the circle holding the end of the rope in your outer outstretched arm. Choose an obvious feature for a starting point so that you can remember when you have made a complete circle. If the reef is very rough and it is hard to see all of the circle in one turn, it may be necessary to go around again looking at the middle part of the circle.
>
> 3) As you go around holding the rope in front of you, it will help you to measure or count as it crosses things of interest. do not count anything outside the area covered by the rope. On your first time around, look at the amount of bottom (percent cover of the different kinds of sediment (mud, sand, rubble, blocks) and write the code numbers on the data sheet. Do a second circle to estimate the percent cover of live hard coral. Then make circles for the percent cover of soft corals and sponges, dead standing coral, crustose corallines and marine plants.
>
> 4) Mark off the forms present (hard corals, soft corals and sponges, and plants), making as many circles as necessary to look for all the forms on the list. When you have checked all the forms that are inside the circle, go back and estimate which is the dominant form in each group, and write in the code for the largest size of that form.

5) For the counts of animals, make circles to count each kind of animal on the list. You may need to make one or more circles for each kind of animal, unless they are very rare. You can either make the count in your head and write it down when you have finished circling, or make a mark each time you see an animal in the circle, and then add up the marks later.

6) If there are signs of pollution or things made by man (garbage, oil, etc.), write down what they are, and how many are in the circle if it is something you can count. Finish the survey by making notes of any other things that seem unusual or important. Write everything down immediately, and make sure you can read all your notes by going over them again when you get out of the water.

7) All of the survey information for three survey plots can be written on a data sheet like those included with this handbook. Ideally, the data sheets should be printed or photocopied on waterproof paper. They can then be clipped to a clipboard or attached to a board, so that they are easy to write on under water with a pencil. The pencil should be tied to the board or it will float away or get lost. The data sheet format can also be copied on to or scratched into sheets of plastic with a roughened surface on which a pencil can be used. Almost any flat, hard plastic (not polyethylene or vinyl) can be used if the surface is roughened with sandpaper. The data can then be carefully recopied on to plain paper forms after the survey, and the plastic sheet prepared for re-use by erasing, for example with scouring powder or cleanser.

Numerous variations of the above are possible, and numerous other methods exist as well (355): photo quadrats, chain transects, plotless transects, point-quarter, circular stations, and so forth (Figure 2.27). A particularly effective method is described in the case history mentioned above ("Australia, Coral Reef Survey Method" in Part 4). But what works for coral may not work for other purposes; for example, a different approach would certainly have to be used to sample fishes and other mobile species.

A successful approach for expeditiously determining the total coral genera of a reef sector was presented by Salm and Clark (308). They recommend running two to four perpendicular transects down the reef slope (marked only by buoy) from 0 to 18 m, collecting (by hammer) small pieces of problem genera for later identification. Kenchington (207) recommends underwater stereo camera survey because of the permanent record obtainable and the convenience of identifying the corals later under ideal shoreside conditions.

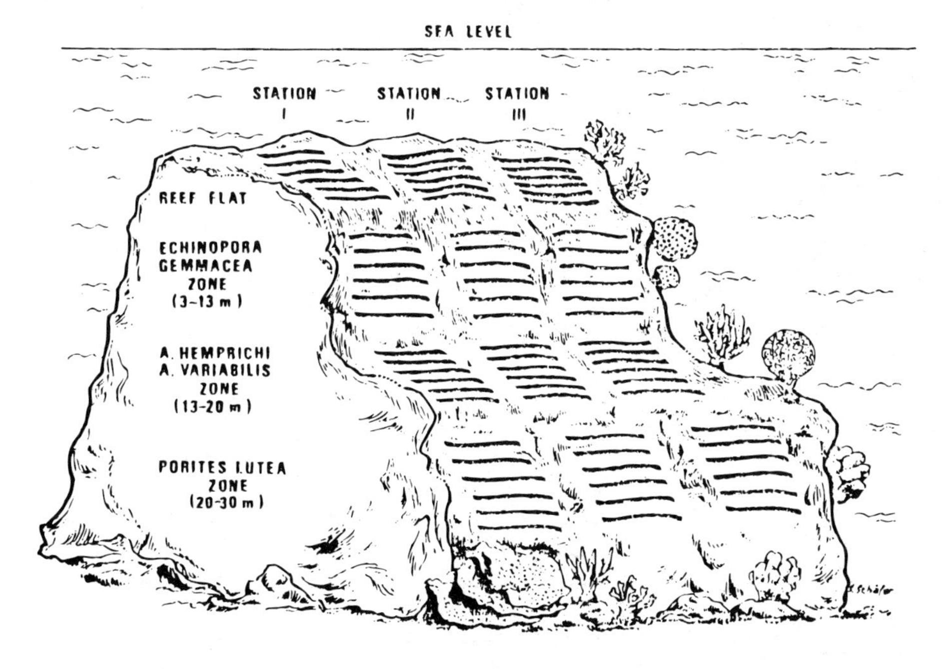

Figure 2.27. The line transect method as used at Eilat in the Red Sea (Israel) to survey coral reef species; note that the transects run along parallel depth contours. (*Source:* Reference 224.)

There are advantages and disadvantages of simplified photo survey. In this approach the survey information is recorded by photography and the identifications and measurements are done in the laboratory. The author used photo survey for reef survey in Bali with some success (70, 393).

DeSilva (106) recommends the linear approach—a transect from reef crest to seaward edge, marked with a nylon rope calibrated at 1-meter intervals to facilitate recording—which is done in the manner of the Bali method discussed above. Time to complete a 100-m transect is $\frac{1}{2}$ hour for an experienced person and 1 $\frac{1}{2}$ hours for a person who has to collect specimens for verification (Figure 2.28).

2.9 ECOLOGICALLY CRITICAL AREAS IDENTIFICATION

In tropical countries—to which this book is oriented—resource conservation programs may be centered on one or more of the following major resource or critical habitat types: mangrove forests, salt marshes, coral reefs, submerged seagrass meadows, beach-dune systems, tidal flats, shellfish beds, and lagoons/estuaries (including embayments). Small, offshore islands are sometimes listed as a critical habitat category, but they are only aggregations of the various habitats mentioned above. All types are in jeopardy wherever they co-exist with human society.

Types of impacts affecting certain critical habitats include the following, as listed by the Sri Lanka ICZM program (43):

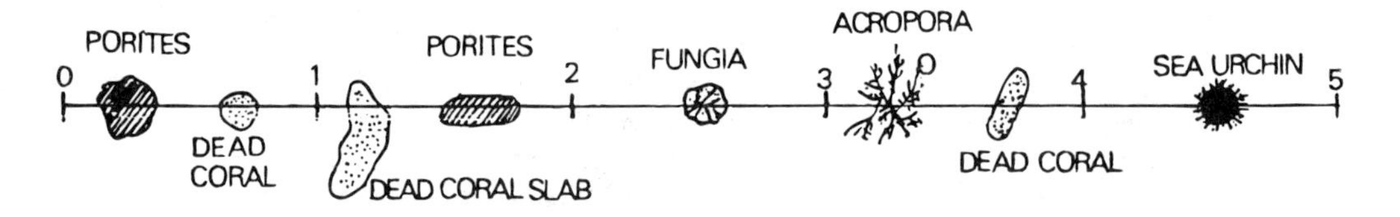

Place: Date: Time: Location:

TRANSECT NO. VISIBILITY: TIDE:

← 1 →

1 (Po) 11

2 (D, Po) 12

3 13

4 (D) 14

5 (SU) DEPTH: RELIEF: 15 DEPTH: RELIEF:

6 16

7 17

8 18

9 19

10 DEPTH: RELIEF: 20 DEPTH: RELIEF:

Figure 2.28. Sketch of a transect line marked at 1-m intervals passing over a coral reef with live and dead corals. The "scoring" based on this passage is shown in the box, where Po = *Porites,* D = dead coral, F = *Fungia,* AC = *Acropora,* and SU = sea urchin. 1_1 1_2 = length 1 and length 2 of intersected coral colonies. (*Source:* Reference 106.)

1. Coral reefs
   - Physical damage to coral reefs and collection of reef organisms beyond sustainable limits
   - Increases in freshwater runoff and sediments
   - Introduction of waterborne pollutants
2. Estuaries/lagoons
   - Encroachment
   - Changes in sedimentation patterns
   - Changes to the salinity regime
   - Introduction of waterborne pollutants
   - Destruction of submerged and fringing vegetation
   - Inlet modifications
   - Loss of fishery habitat
3. Mangroves
   - Changes in freshwater runoff, salinity regime, and tidal flow patterns
   - Excessive siltation
   - Introduction of pollutants
   - Conversion of mangrove habitat and overharvesting of resources
4. Seagrass beds
   - Physical alternations
   - Excessive sedimentation of siltation
   - Introduction of excessive nutrients or pesticides
5. Salt marshes (tidal flats)
   - Degradation of bird habitat or seed fish collection sites
   - Obstruction of storm water runoff
6. Barrier beaches, sand dunes, and spit:
   - Sand mining
   - Erosion
   - Dune migration

In addition to mangroves, intertidal areas of the coast include salt marshes and open tide flats. Salt marshes, where they exist, serve many of the same ecological purposes as mangrove forests. They assimilate nutrients and convert them to plant tissue, which is broken into fine particles and swept into the coastal waters. In addition, the marsh provides a special habitat for many valuable species.

Extensive areas of tide flat (mudflats, sand flats, etc.) are often found in estuaries and lagoons. Such flats are important in processing nutrients for the ecosystem and providing feeding areas for fish at high tide or birds at low tide. Mud flats are often important energy storage elements of the estuarine lagoon ecosystem. The mud flat serves to catch the departing nutrients and hold them until the returning tide can sweep them back into the wetlands. In many estuaries and lagoons, tideflats also produce a high yield of shellfish. At the higher latitudes there are extensive beds of kelp in certain areas (e.g., Southern California and Western South Africa), which provide food and shelter for marine species.

It should be recognized that, in the administrative context, three categories of ecologically critical areas (ECAs) might be recognized for coastal zones:

1. **Generic types of habitats** that are widely recognized as highly valuable and that should be given a high degree of protection through *regulatory* mechanisms—wetlands, seagrass meadows, coral reefs, species nesting sites. In the process of project review and EIA, developers would be told they must not disturb these types of habitats. Therefore, developers should be informed ahead of time, before they design projects, that restrictions exist. In addition to ecologically critical areas, other generic types of areas should be identified, such as flood-prone lowlands (those that are regularly flooded), which would be designated under the natural hazards prevention category.
2. **Geo-specific habitats** that would be identified as *specific* areas needing *regulatory protection.* These would include certain named and specifically delineated lagoons, estuaries, islands, mangrove forests, river deltas, coral reefs, etc. Each would be described, mapped, and announced for the knowledge of all interested parties. Such geographic specific habitats would

be given special consideration for protection by the reviewing authority as "red flag" areas in the development review process, regardless of ownership. Parts or all of such areas might be government owned or might be locally controlled "commons" or might be privately owned—but the controls should apply to all categories.

3. **Nature reserves,** a third administrative category, which exercises control by *right of ownership,* includes nature parks, reserves, and other protected natural areas. Their formation and care is an existing, traditional, and well-recognized governmental responsibility in most countries. In this category, the areas have boundaries and are designated for particular types of nature protection, special protection rules are created, and government ownership is maintained.

An area would be categorized as "ecologically (or environmentally) critical" and listed as a critical area if (a) it contained an outstanding example of one of the generic habitats, (b) it contained two or more of the critical habitats that individually were not outstanding but in combination created a major coastal or marine ecosystem, or (c) it had other outstanding characteristics.

In practice, the identification process might proceed in the following order: (a) identification of generic habitats to be given protection nationally (regionally, if there are regional ICZM programs), (b) delineation—listing and evaluation of sites that qualify as environmentally critical and that would receive special regulatory attention, and (c) selection from the list of critical areas those that should receive the highest level of protection through government proprietorship.

The result would be a single integrated set of "environmentally critical" sites for the entire coast, which then would be sorted out for two different kinds of protection: (a) *regulatory* protection under an ICZM approach and (b) *proprietary* site protection under a traditional protected areas approach.

In the selection of habitats for special forms of environmental protection—either regulatory or proprietary—it is useful to employ a specific method. Various schemes are used for determining whether a coastal/marine area is "critical" (or "ecologically sensitive," "special," "vital," "of concern," etc.). Most of these schemes depend on professional judgment rather than determinative analysis to evaluate importance and establish priorities, that is, to distinguish the *more valuable* from the *not as valuable.* Such judgments must be exercised against the background of social purpose underlying the mandate for identification and designation of critical areas.

These terms of reference vary greatly from program to program—e.g., the terms of reference that apply to designating a UNESCO/MAB Biosphere Reserve site would be quite different from those that apply to designating the local nesting place of a particular bird species. Nevertheless, there are generic types of coastal and marine areas that are widely accepted as being especially valuable ecological resources. (See "Protected Natural Areas" below.)

It is often better for an ICZM program to coordinate with the protected areas, or nature reserves, function rather than try to take it over. One exception would be where protection of environmentally critical areas is not being attempted in a systematic way by the agency in charge; then, it might be appropriate for an ICZM-type program to take over their care.

A summary list of generally recognized types of coastal and marine critical areas might include:

- Coral reefs (Figure 2.29)
- Giant kelp beds
- Dunefields
- Saltmarsh wetlands
- Mangrove wetlands
- Estuaries/lagoons
- Beaches
- Tideflats
- Seagrass beds
- Shellfish beds/reefs and ridges
- Raised banks, dropoffs, and other "high profile" features

Factors that lead to identification of an area within any of the above generic types as specific habitats include, *inter alia:* size, ecosystem context, structural characteristics, species present, and geographic location. For example, a coral reef of several thousand hectares surrounded by seagrass

Figure 2.29. Coral reefs are among the most productive and beautiful habitats of the earth and are ecologically critical, as exemplified by this one in Indonesia's Pulau Seribu National Park. (Photo by Rod Salm.)

beds with a good mixture of *Acrophora, Colpophyllia, Porites,* and *Oculina* corals and a variety of food species of fish and invertebrates along with decorative species might be rated as extremely valuable, particularly if is near to a population center.

In designating special areas for coastal zone programs, it may be desirable to use additional criteria (e.g., historic, scientific, scenic, recreational, and natural hazards protection). For example, the State of North Carolina (USA) coastal management program recognizes the following types of "areas of particular public concern" (254):

- Marshlands and estuarine waters
- Areas with significant impact on environmental, historical, or natural resources of regional or statewide importance
- Areas containing unique or fragile ecosystems that are not capable of withstanding uncontrolled development
- Areas such as waterways and lands under or flowed by tidal waters of navigable waters, which the state may be authorized to preserve, conserve, or protect
- Areas such as floodplains, beaches, and dunelands wherein uncontrolled alteration or development increases the likelihood of flood damage and erosion and may necessitate large expenditures of public funds
- Areas significantly affected by, or having a significant effect on, existing or proposed major public facilities or other areas of major public investment

A survey of the nesting and breeding habitats of local species that use the beach or dune is particularly needed for many coastlines to identify specific nesting sites. Once identified, these critical habitat areas should be protected during the particular breeding and nesting seasons and regulations promulgated for keeping people out of these areas, as is practiced in England for tern.

2.10 DATABASE DEVELOPMENT

Coastal management programs are information driven and, because information is derived from data, a database always will be required for coastal management programs. The database, or data bank, is a system for collection and storing of relevant data, defined according to the main components of the coastal management strategy. It follows that planning a database is largely a matter of how much data of what type is needed and how to store and retrieve it.

The program need will define the data system. In most cases a first priority is diagnosis of the present status of coastal zone resources. This may require that parameters be defined for mangrove obliteration, fish abundance, eutrophication, incidence of ciguatera, coliform bacteria, coral abundance (percent live cover), salination of ground water, endangered species, and heavy metal pollution, to name but a few examples. Additionally it may be necessary to identify "dryside" (land-based) impacting activities (agriculture and others). The database, with maps, then is first used as a tool for issue identification, and for directing attention to coastal situations that need immediate attention.

Maragos *et al.* (232) stress the importance of concentrating on the primary goal of the assessment process which is to avoid or reduce significant adverse impacts of development. (See also "Information Needs" in Part 3.) By identifying sources of environmental impact and resources degradation, databases can play a central role in facilitating a more integrated and better informed approach to coastal resource management, according to Sorensen and McCreary (326). By drawing together data from different aspects of the environment—on mangrove location, shrimp production, and land use designations, for example—databases emphasize the interaction of specific components of the environment. Note that a coastal database should also include economic, technological, ecological, and institutional aspects.

A coastal database can be divided into four basic components (350): social, economic, spatial, and environmental, each of which can be further elaborated. Thus, for example, population characteristics, as part of social development, can be defined by demographic indicators, distribution within settlement, and migrational characteristics. Individual data sets (e.g., total population in a particular area) may be less useful for planning purposes than when combined with the size of the planning area to indicate density of settlement or, with a time component, to indicate a trend.

For coastal zone management purposes, the database is most often organized by geographic units. For example, a database may be keyed to parcels or townships of land, an offshore tract, or a particular length of coastline. Often the database is conceptually organized as a table with information on a set of natural resource parameters (geological material, soil type, vegetation cover, prevailing land use, agricultural suitability) keyed to each geographic unit. Alternatively, a coastal pollution database might be organized as a network of points reflecting the location of monitoring stations for water quality. According to Geoghegan (140), it has been found time and again that a most useful way for the environmental planner to discover trends, conflicts, and problem areas that can otherwise be easily overlooked, is by mapping information (see "Mapping" below). With the advent of reliable, low-cost computer automation, there is a pronounced trend toward computer storage of such geographically oriented databases. The computers can be programmed for direct production of mapped information from a variety of sources including satellites (see "Remote Sensing" in Part 3). These computerized databases are often called "Geographic Information Systems" (GIS) and are explained in Part 3. With the advent of more "user friendly" GIS systems and the increased availability of satellite and low altitude imagery, building a complete coastal atlas need not be prohibitively expensive.

The value of a coastal database is critically dependent on the quality and quantity of raw information. In many countries, the available data is often uneven with regard to accuracy and consistency of coverage. Also, the methods by which data are compiled, scaled and aggregated have an equal impact on the utility of the data base or atlas. This is especially evident in considering the map scales at which data are obtained and reproduced. For instance, maps compiled at 1:250,000 are useful for large-scale regional planning, but finer grain is needed (perhaps 1:50,000 or 1:24,000) for overview of the coast, and a scale of 1:10,000 would be preferred for in-depth physical planning of a region and for preparation of land use plans. Even more detailed maps are needed for site plans of particular projects. (326)

The question of the database needed should be tackled from two perspectives: (a) What basic and applied types of data are needed? (b) How can the information derived from these data be presented in its most useful form?

The listing below includes suggested priority data items that coastal planners often need to answer specific program questions:

- Coastal renewable resources: Fisheries and aquaculture activity and yields, by species and seasons; mangrove forest exploitation, activities and products

- Users of coastal areas and resources: Tourism and recreation, manufacturing, maritime trade, mining, urban, oil and gas industry, jobs, revenue, investment and tax yields
- Impacts: Impairment of coastal resources and ecosystems; pollution, habitat losses, species depletion, sedimentation, visual degradation
- Upland effects: Impairment of coastal resources from river dams and diversions, accelerated sediment transport, reduction of freshwater inflow, disruption of natural hydroperiod, reduction of beach nourishment with erosion and pollution
- Socio-economic status: Economic statistics for coastal communities and information on social organization of coastal communities and dependencies on coastal resources
- Critical habitats: Habitats of critical importance such as mangroves and other wetlands, beaches, dunefields, seagrass meadows, coral reefs, tideflats, estuaries, lagoons, shellfish beds, and special breeding and feeding areas for coastal species; restoration needs
- Important species: Identification of the coastal species of particular significance of economic value and their habitats and trends of their populations; restoration needs
- Resource problems, issues: Information on special problem situations, such as highly polluted estuaries, extensive mangrove clearing for aquaculture ponds, destruction of coral reefs, and so forth
- Natural hazards: Identification of situations that lead to increasing risk of natural hazards, such as badly eroding beaches, floodable lowlands and islands, destabilizing landslide prone slopes, removing protective mangroves, and degrading coral reefs
- Protection areas: Description and evaluation of areas that should be designated as parks, reserves, or other types of protected areas

Following are some general conditions for a database (350):

- It should be capable of continuous updating.
- The data should be reliable and sufficiently disaggregated.
- Data should be readily accessible and easily manipulated.
- It should be possible to use data for different purposes.
- Combining data from different sources should be enabled.
- Use of similar data for different levels of disaggregation should be possible.

A coastal database usually has political decision makers as important end users and therefore it should generate information that policy makers think they need and that is readily understood by them. It helps to have a person involved with database design and operation who has an interdisciplinary background and who can distill data into action options for the policy makers. (65).

In a typical ICZM-type multiple-use management program, data analysis is often driven by the need to allocate coastal resources among competing uses—i.e., to answer the question: what mix of natural goods and services should be produced from the given area over time (65). Examples are water-based recreation, fisheries, marine transport, and waste disposal. This recognizes that ICZM is a means for regional and/or national economic development as well as system for protection of biodiversity.

It is recognized that, while information goals may seem difficult to reach, the existing data base is often adequate to *commence* a coastal management program. Rarely does the lack of technical information need to limit progress on ICZM programs, but it is often used as an excuse for inaction. The management process should be scheduled to commence once the *minimal* information base is available.

2.11 DREDGING MANAGEMENT

Dredging and the disposal of "dredge spoil"—the term commonly used for sediments and other material excavated by dredges—have caused serious losses of estuarine resources. Spoil has often been deposited on wetlands or vital bottom habitats such as grass beds or shellfish beds. Large spoil banks or landfill deposits in water basins have caused high turbidity, restricted water flow, and stagnation. Large portions of lagoons and estuaries area have been degraded, and sometimes eliminated. Spoils from shipping ports and industrial sites often contain high levels of toxic matter. But there are ways to prevent or minimize damage, or even to use the spoil constructively.

According to Maragos (231), dredging has the following adverse impacts: it physically removes and/or disturbs the sea bottom, deposits sediment on the substrate, suspends sediment in the water column, reduces light penetration, increases turbidity, changes circulation, reduces dissolved oxygen, and increases nutrient levels in the water column. The most widespread and visible consequence of dredging and excavation, however, is the suspended sediments and turbidity (Figure 2.30). Dredging or excavating materials high in organics can generate biochemical oxygen demand (BOD) and depress oxygen levels. A summary of environmental impacts from harbor dredging follows (67):

Figure 2.30. The hydraulic dredge, as shown above, usually creates less turbidity than other types and offers the potential for exact placement of the dredge spoil. (Photo by Kenneth Woodburn.)

1. Excavation of sediment: negative
    - Toxic substances
    - Organic nutrients
    - O_2 depletion
    - Turbidity
    - Settlement at bottom
    - Disruption of fisheries
    - Change of bathymetry and tidal currents
2. Disposal of fine sediment: negative
    - Same as above
3. Sand excavation: negative
    - Destroy coral and other critical habitat
    - Undermine coral beds
    - Release fines; smother corals
    - Weaken foundation of seawalls
    - Disrupt fisheries
    - Create pits that become anaerobic, mud filled

    Positive
    - May expose hard bottom for coral attachment
4. Sand deposition: negative
    - Release of fines

    Positive
    - Creation of improved bottom habitat
5. Operations: negative
    - Wastewater discharge; ships and shoreside
    - Spill; oil, chemicals
    - Solid waste; overboard disposal
    - Storm water runoff; contaminated
    - Traffic and general disturbance
6. Onshore support activities: negative
    - Habitat preempted
    - Disturbance, noise, smoke, dust
    - Local water pollution
    - Oil leaks, etc.
    - Traffic congestion
    - Wildlife disturbance
7. Socio-economic: negative
    - Construction disturbance; noise, dust, etc.
    - Operations disturbance; traffic, smoke, etc

    Positive
    - Some improvement of jobs, income, and perhaps of ambience

When dredging occurs in productive coastal water basins, care must be taken not to damage, directly or indirectly, critical habitat areas such as grass beds, shellfish beds, coral reefs, and productive basin-floor habitats. Adequate protection often requires a surrounding buffer strip of several hundred feet (or thousands, in some cases) from which dredging should be excluded, depending upon the method used (see "Dredging Techniques" in Part 3).

Physical barriers such as silt screens and earthen berms can be effective in reducing the area affected by a dredging and filling operation. Silt screens are curtains of plastic, fiberglass, or other fabric vertically suspended from the surface using a system of floats and anchors (Figure 2.31). Normally silt screens are effective where wave action is low and water currents are −0.5 knots (50 cm/sec) or less. Silt screens have been successfully used at many shallow water project sites, including those with shallow reef flats. The screens are less effective in deeper water (Figure 2.32). Placement of armor rock, revetments, rubblemound structures, and sheetpiling prior to filling operations can also confine sedimentation. (231)

Figure 2.31. A silt curtain, used to prevent siltation of a marina during channel dredging, is suspended from its floats. (Photo by author.)

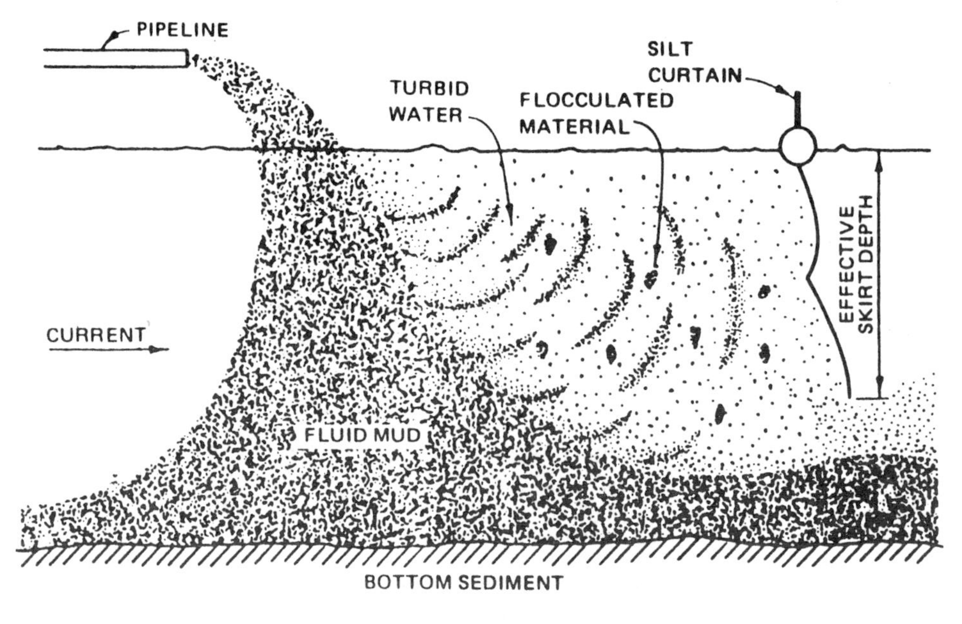

Figure 2.32. Factors affecting the capability of silt curtains to control dispersion of dredge spoil. (*Source:* Reference 364.)

Disposal of dredge spoil in the open sea can potentially cover or modify extensive underwater bottom environments and should be avoided, unless done strategically. In most modern dredging the excavated materials are deposited onshore. The material is usually placed in a contained site, or "retention basin" (diked area), and allowed to de-water. The receiving area is designed as a "settling basin," whereby the still water condition will allow the dredged material ("spoil") to remain in the basin (settle out of the slurry) while the water runs back to the adjacent sea. But care must be taken to assure that sufficient time for settling of particles has occurred and that the effluent is low in suspended solids.

Containment basins range in size from less than 10 acres to over 4 square miles and have life expectancies from less than 1 year to over 100 years. Nearly all containment areas are enclosed by dikes and equipped with height-adjustable spillways to accommodate varying filling rates. The design of these basins is most important. Inadequate sized basins will require reduction of input to prevent slurry waters from overtopping the basin walls and flowing back into the environment prematurely.

Special care should be given to prevention of sediment pollution from marine excavation, reclamation, and runoff from sites. This can be done by standard countermeasures (sediment traps, silt curtains) and by selection of appropriate construction equipment.

The following are general environmental management protocols:

1. The dredging work area should be screened by containment known as a *silt curtain* to prevent excessive release of fine sediment, or *fines* (work to cease if curtain malfunctions).
2. Methods should be chosen which release the least amount of fines (e.g., suction dredge instead of bucket).
3. No dewatering of fines should take place onsite; all silt to be pumped into barge with no overflow and carried to disposal site.
4. Work should cease during major natural events such as storms or excessive currents, and so forth.

Implementation of these environmental protection guidelines needs to be monitored to make sure they are working successfully; if not, corrections should be applied.

Regular containment is not sufficient when the dredge spoils are polluted with toxic matter or ecologically degrading materials because toxic chemicals dissolve or suspend in the water and go out with the discharge. Pollutants released by dredging occur in two categories. The first category includes volatile solids, biological oxygen demand (BOD), oil and grease, and nitrogen, typical of materials such as harbor sites and organic oozes from sewage discharges. The second category consists of the heavy metals, including mercury, lead, and zinc, that are either physically or chemically sorbed or bound within the sediment matrix. Associated PCBs (polychlorinated biphenyls) and pesticides may also have long-term effects (59). Where chemical pollution is a problem, special pre-disposal treatment may be required (231).

To ensure that damaging amounts of toxics will not be released into harbor waters, there needs to be a toxics analysis of sediments in suspected areas (e.g., commercial harbors) before dredging is permitted. Some examples of key substances to be measured are: mercury, lead, zinc, cadmium, copper, chromium, PCBs and hydrocarbons (see "Water Quality Guidelines" below). It is also advisable to determine the organic content of the sediment to see if high BOD problems might arise. With the above information at hand, a strategy for handling and disposing of the dredge spoil can be determined.

The sediment to be sampled should be representative of that to be dredged vertically and horizontally. Therefore, a series of boreholes should be drilled and at each borehole samples should be removed representing the upper, middle, and lower strata. This should be done by contacting an appropriate testing laboratory and following their recommended technique for amount of sample, collecting, storing, and shipping. Each sample might weigh around 0.5 kg (fresh weight), but this should be checked with the laboratory.

Pollution control can be done by standard countermeasures (sediment traps, silt curtains) and by selection of appropriate construction equipment (e.g., suction dredge is much better than bucket or drag line). Only if high concentrations of toxic substances are found in the silt would *pretreatment of silt* be required before disposal; monitoring in the disposal area for toxics might also be necessary.

Water contamination criteria can be used to control dredging. Using an example of a harbor dredging and spoil containment project at New Bedford, Massachusetts (USA), baseline data were recorded and maximum allowable concentrations were set, as follows (133):

Contaminant	μ/l	
	Baseline data	Maximum
PCBs	0.60	1.4
Cadmium	0.23	9.3
Copper	5.3	13.0
Lead	2.6	7.2

While dredging activities may result in potentially adverse ecological effects in terms of habitat loss, fishery potential reduction, and circulation disruption, some positive effects can also result. Dredging can increase circulation by expanding waterways, which may be beneficial for littoral basin water quality and aesthetics. Certain isolated or brackish/freshwater areas could be opened up to estuarine circulation and could become productive nurturing areas. The habitat loss from marina construction is also lessened by the colonization of new surface areas provided by marina structures, particularly rip-rap and piling surface areas. Algae and benthic invertebrates that live on these surface areas may serve as fish forage.

Clean, coarse spoil (coarse sand) from channel dredging can also be deposited as as an estuarine breakwater to protect small boat harbors. Properly designed and stabilized, the breakwaters will allow adequate circulation around the marina area and create, as additional benefits, useful habitats like "spoil islands."

Sometimes the spoil removed in a dredging operation is coarse and clean—that is, it consists of sand or gravel without much clay, mud, or organic matter—it may be used for beneficial purposes such as creation of artificial islands to increase breeding habitats for birds and to expand wetlands along the island fringe. If properly located and designed, these islands may increase ecosystem carrying capacity. However, they must be planned with the utmost care.

The following criteria are suggested for the design of spoil islands: (a) avoid covering existing vital areas, including grass beds, shellfish beds, and wetlands; (b) use coarse sand or other material not susceptible to rapid erosion (fine, organic sediments or polluted spoil should not be used); (c) locate the spoil island in a protected area away from heavily used boat channels to minimize erosion from boat wash; (d) vegetate the island with both upland plants and marsh grasses as soon as possible; and (e) shape the island so as to facilitate water movements (e.g., make it elliptical and parallel to water flows).

Dredging impacts should be assessed holistically, so that all species and all life stages are considered as part of an interrelated community system. Species of particular concern are those that are of commercial, recreational, or aesthetic value; geographically restricted; sensitive; or protected. Their food supply network and habitat vegetation also are factors of concern. Non-motile species/life stages such as sessile (attached) benthic animals, drifting plankton, and the eggs and larvae of fishes and invertebrates are particularly subject to dredging impacts such as siltation since they cannot avoid areas of high turbidity. Loss of seagrass, wetland, reef, and other habitats on which organisms depend may also damage existing populations and hinder or preclude the survival and subsequent recruitment of organisms. (366).

Direct disposal from hydraulic dredging in a particular spot may concentrate the spoil with ecologically damaging results. The largest dredges can discharge up to 38,000 gallons per minute (GPM) of dredged spoil by a pipeline, as shown below (59):

Pipeline diameter (in.)	Discharge rate (for flow velocity of 12 ft/sec*)	
	cu ft/sec	gal/min
8	4.2	1,880
10	6.5	2,910
12	9.4	4,220
14	12.8	5,750
16	16.5	7,400
18	21.2	9,510
20	26.2	11,740
24	37.7	16,890
27	47.6	21,300
28	51.3	23,000
30	58.9	26,400
36	84.9	38,000

To obtain discharge rates for other velocities, multiply the discharge rate above by the velocity (ft/sec) and divide by 12.

Sometimes the preferred solution is to dispose the material so that it dissipates harmlessly into the marine environment. Where chemical pollution is not present, deep-water sea dispersal is an option. The objective of dispersal is to spread the material in as efficient a manner as possible so that the concentration drops rapidly as the material is carried with water currents, rendering it as harmless as possible as quickly as possible. This is best done by choosing a dump site and dump method that allows the natural water current to disperse the material away from areas of value and to spread it down current in a broad plume for gradual dissapation and settlement in a natural manner. Optimally, the natural process of sediment entering the sea from rivers is simulated. (See "Indonesia, Sulawesi" case study in Part 4.) The dumping is best done by barge or hopper dredge that can release the material while moving and thereby spread it over a long track.

This approach requires that the dumping be continually monitored so that undesired effects do not occur. Monitoring can be done most simply by maintaining a given water turbidity—say, not higher than 100 mg/liter or equivalent measured in NTUs (by Hach Turbidity Meter)—outside the boundary of an agreed "mixing zone," say 500 m distant from the dumping point. Work would cease and corrections made if excessive silt was released. Also, the bottom could be monitored outdide the dump area boundary and the dump site moved if siltation of more than 1 cm depth occurred.

The mining of sand for beachfill or construction is often done in coastal waters by dredge. Sand should not be removed from coral reefs or near to coral reefs. Any sand dredging should be done by suction dredge with silt curtains used to keep fine sand ("fines") from being carried onto the reefs. The boundaries of important reef areas should be marked, including a buffer strip, so as to keep dredgers from accidentally moving into coral reef areas. Turbidity monitoring should be conducted and excavation halted when fines and other suspended matter exceed permissable limit, say 100 mg/liter (PPM) or are being carried by currents directly onto reef. Work should be halted in storms or times of extraordinary high currents.

A multinational working group of dredging experts concluded the following in regard to disposal of dredge spoils (284):

- The primary step to minimize effects is the selection of a proper disposal site.
- Careful evaluation is essential as land sites are more likely to have a high potential for damage to the environment.
- Techniques are available for managing disposal at sea between underwater dams, in borrow pits, by capping and by the construction of artificial islands.
- There is no evidence that greater environmental protection can be realized by moving disposal sites further out to sea, except in situations where the discharged material may again accumulate, for instance in an estuary.
- Equipment and techniques currently in use are largely adequate to remove, transport and dispose of dredged material in an efficient and environmentally acceptable manner in coastal regions.
- Special equipment or special techniques may be necessary to ensure accurate removal and minimal dispersion. This would be required when sediments are excessively contaminated as a result of spillage of toxic chemicals or come from uncontrolled point sources.

To minimize effects during and directly after disposal it is generally necessary to attain maximum density of the dredged material to be discharged. The greatest possible density can be obtained by keeping the dilution of the silt as low as possible. Even if artificial or natural dewatering processes are chosen between dredging and disposal the efficiency increases with the density of the dredged material (284).

The following devices and methods were therefore developed, generally serving two of the following purposes: (a) firstly, minimizing effects during the dredging and (b) secondly, aiming at high densities of the dredged materials by use of (284):

- Watertight clamshells
- Silt dragheads with a degasing system
- Turbidity-limiting overflow system
- The IHC siltmaster
- Direct loading of the dredged material into high capacity ocean-going hopper barges

The plan for environmental management of the project should involve realtime monitoring of the suspended solids (SS) load, or other parameter(s) to detect if the construction work is creating unacceptable impacts, according to proposed standards.

2.12 DUNE MANAGEMENT

Sand dunes are the ridges of sand that often lie behind the active part of a beach. There may be several parallel rows of natural dunes, each built in response to forces of waves, which put the sand high up the beach, and to forces of wind, which finally carry the sand grains up onto the dune. Such dunefields are an integral part of the beach system and must be managed as such (see "Beach Management" above).

If a dune is attacked by storm waves, eroded material is carried onto the beach and then to an offshore deposit—often forming a sand bar parallel to the beach. The accumulation of sand absorbs or dissipates, through friction, an increasingly large amount of destructive wave energy that would otherwise focus on the beach. It is this capacity of the berm-and-dune system to store sand and yield it to the adjacent submerged bottom that gives this system its outstanding ability to protect the shorelands.

Dune vegetation promotes large-scale trapping of sand, whereby the sand reserves of the dunes expand. The frontal dune remains fluid, alternately receiving and yielding sand. But the back dunes tend to become stabilized into more permanent features of the landscape. Because dunes aid the forebeach to protect property, they should be preserved. If the dunes are removed or the shore bulkheaded, the reserve sand in storage will be reduced to a level no longer capable of replacing sand losses from severe storms. The beach system then becomes unstable and it erodes and slumps.

The main protection needed for dunes is to prohibit any removal of sand—no taking of sand from dunes should ever be permitted. In addition, the vegetation that binds the dune together needs protection. Vegetation that grows on shifting dunes is adapted to withstanding the rigors of wind, sand, and salt, but not human feet, vehicles, or herds of grazing animals (Figure 2.33). Even slight alterations of dune formations, such as minor erosion or displacement of vegetation, may lead to significant dune loss.

Figure 2.33. Sand dunes should be protected in the coastal management program as an integral part of the beach system. (Photo courtesy of U.S. National Oceanic and Atmospheric Administration.)

Figure 2.34. An efficient method of rehabilitating sand dunes for beachfront protection is to use flexible fencing (snow fence). (Photo by author.)

Once a frontal dune is worn down by vehicles or foot traffic or by consequent loss of vegetation, it may be eroded by wind or wave action and no longer serve its unique protective role.

Community and government projects to restore and stabilize dunes should be encouraged. Two inexpensive and effective methods to protect beachfronts are simple structures, such as snow fences (Figure 2.34) and revegetation programs. Fences have two initial advantages over vegetative planting that often warrant their use before or with planting: (a) sand fences can be installed during any season and (b) the fence is immediately effective as a sand trap once it is installed. There is no waiting for trapping capacity to develop, in comparison with the vegetative method. It is not necessarily "either/or" because it would be a benefit to plant the dunes as an aid to fence stabilization.

Relatively inexpensive, readily available slat-type snow fencing is used almost exclusively in artificial, nonvegetative dune construction. Field tests of dune building with sand fences under a variety of conditions have been conducted in the United States. The following guidelines are based on these tests (362):

1. Fencing with a porosity (ratio of area of open space to total projected area) of about 50 percent should be used. Open and closed areas should be smaller than 5 centimeters in width.
2. Straight fence alignment is recommended. Zigzag alignment does not increase the trapping effectiveness enough to be economical. Lateral spurs may be useful for short fence runs of less than 150 meters (500 feet) where sand may be lost at the ends.
3. Efforts have been most successful when the selected fence line coincided with the natural vegetation or foredune line prevalent in the area. This distance is often greater than 60 meters shoreward of the berm crest.
4. The fence should parallel the shoreline. It need not be perpendicular to the prevailing wind direction; it will function if placed at an angle to sand-transporting winds.
5. A 1.2-m fence with 50 percent porosity will usually fill to capacity within 1 year. The dune will be about as high as the fence. The dune slopes will range from about 1:4 to 1:7, depending on the grain size and wind velocity.

6. Dunes are usually built by installing a single fence and following it with additional single-fence lifts as each fence fills. Succeeding lifts should be parallel to and about 4 times the fence height from the existing fence.
7. The trapping capacity of the 1.2-meter-high fence averages 5 to 8 cubic meters per linear meter (2–3 cubic yard/linear ft).
8. Fence-built dunes must be stabilized with vegetation or the fence will deteriorate and release the sand. The construction of dunes with fence alone is only the first step in a two-step operation.

In choosing species to plant, one should be aware that only a few plants thrive on the dunes, and these are adapted to conditions that include abrasive and accumulating sand, exposure to full sunlight, high surface temperatures, occasional inundation by saltwater, and drought. They are long-lived, rhizomatous or stoloniferous perennials with extensive root systems, stems capable of rapid upward growth through accumulating sand, and tolerance of salt spray.

The most hardy species are native beach grasses and sea oats. In dune planting, plants are often gathered from the wild, trimmed, sorted, bagged, transported, and replanted, as any plant might be. They are planted according to the design strategy for the dune rehabilitation project. (362).

A shore protection plan should include regulations to preserve the frontal dune intact by controlling foot and vehicular traffic. Access to the beach should be limited to elevated steps and boardwalks over the dunes that allow unobstructed movement of sand beneath them, and foot traffic should be limited to these walkways. Fences should be erected to keep grazing animals off dunes. Vehicular traffic anywhere on the frontal dune system should be prohibited. Dune buggies, trail bikes, and other off-road vehicles should be restricted to the beach below the berm and to places where traffic will not interfere with other beach uses. Finally, a coastal construction setback line should be placed on the land side of the dunefield, with due regard for its possible retreat (see "Setbacks" in Part 2).

2.13 ECONOMIC IMPACT ASSESSMENT

The complexities of evaluating coastal renewable resources and measuring the economic impacts of environmental change require that special methods be devised and used in integrated planning. This requirement applies to countries with economic and political systems ranging from laissez-faire to centrally planned economies. Regardless of the type of economic system that is in place, a special mix of economic tradeoff analyses is required for coastal zone management because of the nature of coastal problems and the common property resources that are usually involved.

To assess the merits of alternative forms of development, it is necessary to consider the economic value of the products and services produced by the natural resource systems and how they would be affected by development schemes (a distinction is made between an economic analysis which requires that economic factors external to the physical system be included and a financial analysis which would normally ignore such factors) (32).

In the present context, economic impact assessment is meant to analyze the economic consequences of environmental change—it is not meant to be a free-standing assessment of the economic impacts of a project of the type included in traditional benefit-cost analyses. (See also "Economic Valuation" in Part 3.)

Development investors often attempt to avoid responsibility for their project's externalities; that is, environmental damage occurring outside their property boundaries. For example, a factory operator may wish to avoid financial responsibility for the degradation of fisheries or inhibition of tourism caused by his factory wastes polluting a bay (Figure 2.35). Such effects can be identified, evaluated, and corrected through economic impact assessment. The economic assessment, usually part of an overall full environmental impact assessment (EIA), examines the effects of the "externalities" of a project on the environment and on other sector's welfare and the public interest. It is particularly important to recognize the effects of "dryside" (shoreland) private activities upon the common resources of the "wetside" (intertidal and open water areas) of the coastal area.

For EIA to include major "externalities" generated by a development project, the boundaries for study should be broad enough to incorporate all of the major effects of the project. For example, a mangrove conversion project would have study boundaries broad enough to include the expected

Figure 2.35. Economic impact assessment attempts to predict the full range of community costs of proposed developments, including environmental costs, which can be considerable for the pulp mill shown above (Photo by author.)

environmental costs (or benefits) that occur off-site; examples might be reduction in fish catch from destruction of mangrove-based food chains or salination of groundwater. In the first example, the study boundary would include the fishing areas supported by the mangrove forest, and in the second example, it would encompass the area of the groundwater aquifer that might be salinated. (326)

Extended analysis of the costs and benefits of different development alternatives may be used in environmental assessment. This approach has the advantage of presenting results in the economic language familiar to decision-makers. And while some environmental factors are difficult to express in monetary terms, benefit-cost analysis can often provide a valuable framework for interpreting the biophysical findings of environmental assessment in economic welfare terms. (41)

A benefit foregone is a cost ("opportunity cost"), and a cost avoided is a benefit. For example, the net value of improved wastewater treatment should be analyzed to include not only direct costs of equipment, operation, and maintenance but also the benefits, or damage costs avoided, in further downstream uses of the water. The broader economic analysis seeks to capture these externalities—e.g., through EIA—and establish monetary values for them so they may be included in benefit-cost analysis. Only the additional benefits and costs actually due to the project are considered however.

The typical environmental impact assessor can deal with the arithmetic aspects of analysis but advanced economic analysis is a job for professional economists. While the assessor can competently analyze a wide variety of data and intrude into numerous disciplines successfully, environmental economics is especially tricky. Simple, "cookbook" methods are not available to non-economists for the most part. Unless you are trained as an economist, you are advised to avoid attempting to do detailed economic analysis because it is a very complex subject and many of its techniques require special knowledge and extensive experience.

2.14 ENVIRONMENTAL ASSESSMENT

A major environmental tool for developers, decision makers, technicians, and the public is the process known as environmental assessment (EA) or environmental impact assessment (EIA). EA refers to the general process, while EIA usually refers to an intensive and comprehensive analysis.

The EA procedure is used to predict the environmental effects of a project (or a program) on coastal resources sustainability and biodiversity. Lately, it has become common to include economic and social considerations as part of the full-scale EIA. The final result of EA will often be a project "Environmental Management Plan" (EMP), which describes and schedules the counter-measures necessary to mitigate environmental impacts.

EA is a term used to describe both a governmental process and an analytical method. As a process, EA is imposed by government to require public agencies and private developers to predict environmental impacts of major projects, to coordinate aspects of planning, and to submit development proposals for review. As an analytic method, it is used to predict the effects of a project or a program. With EA at the center of the development review and control process, government decisions on approval of projects and on needed modifications and mitigations can be made on an *informed* basis.

The assessment process includes prediction of a proposed project's effects on natural renewable coastal resources as well as the potential effects on the quality of the human environment. The process, when mandated by law or executive decree, generally involves a procedure that requires the following information: (a) the characteristics of the project sites, (B) description of the project, and (a) description of the environmental impacts of a project for different dimensions of the environment (Figure 2.36). Usually it is required that alternatives to the project be identified and comparatively assessed, and measures to avoid or mitigate impacts be spelled out (326).

Note that EA is *not* just a report, it is a *process*. It is a decision-making process, not just a compilation of data. EA is a process of exploration, analysis, and verification leading to a determination. The EA report is only a record of this accomplishment.

According to Sorensen *et al.* (326), three fundamental benefits of EA are that:

- Cause and effect relationships can be determined with reasonable accuracy and presented in terms understandable by policy makers
- Prediction of impacts will improve planning and decision making
- The government can enforce decisions emanating from the environmental process

But still EA should be seen as enhancing, not diminishing, long-term economic growth. According to McManus *et al.* (242), it is important to emphasize that the purpose of EA is not to hinder development, but rather to enhance long-term, rational development. EA provides access to all of the various aspects of the total development picture, so that optimal decisions may be made.

Of course, some interests may try to avoid EA requirements because they fear delay or negative outcome. For example, in Sri Lanka, hotel developments of over 80 units are subject to EA—as might be expected, some recent projects promoted by hotel developers are planned for 70 units to circumvent

Figure 2.36. The schematic shows the cause-and-effect pathway from development activity, through physical change, to ecological and socio-economic consequences. (*Source:* Reference 110.)

these requirements. Obstacles to effective use of EA also have been reported in Sri Lanka, where project proponents were affiliated with key government members and tried to get politically exempted from EA requirements. (65)

EA has been extensively used during the last 15 years as an analytic method to predict the environmental effects of coastal development projects in many countries, including Indonesia, Sri Lanka, and Malaysia.

The idea of environmental assessment took form in the United States in the late 1960s and has changed little since then. While the idea has spread to numerous countries around the world, the fundamentals of the EA process remain. For instance, one can find a common basis and common principles for EA in Mexico, Sri Lanka, the United States, Indonesia, and Australia.

Economic and social impacts can be incorporated into the EA as evaluators of the effects of environmental change on the local community. Coastal zone management will prosper in most countries to the extent that the economic and social consequences of development can be objectively predicted and modulated. The same may be said for environmental concerns. With increased perception of ecological functions, it is possible to improve the economic expression of their value to society. For example, mangroves, once considered worthless unless harvested are now seen as extremely valuable ecological resources.

Not only government, but also the private sector (recently) is assuming some responsibility for environmental conservation (261). Even though government controls the process and output, the private sector often hires the analysts and pays for the EA activity. Also, the private sector now uses a process termed "environmental audit" to assess their business operations and ensure compliance with environmental regulations, so they can market "green" products and claim an "eco-friendly approach". (See also "Environmental Audit" in Part 3.)

The sources of impacts to the environment are varied and are categorized and defined in different ways by different sources. We make the following distinctions for EA purposes:

1. Naturally caused impacts—those derived from natural systems, such as global temperature changes, hurricanes (typhoons), floods, and diseases, which may result in the gradual or sudden death of organisms and in which humans have had no discernible input
2. Non-natural impacts—always somehow the result of human activities that have direct or indirect effects on ecosystems and include most of the following categories:
   - Direct impacts—coral collecting, trampling, spear fishing, over-fishing, speed boating, trail walking, and tree cutting are examples that have visible immediate effects.
   - Indirect (or external) impacts—inland terrestrial activities such as deforestation, improper soil coverage, and the indiscriminate use of agri-chemicals all contribute to environmental degradation directly or indirectly on other coastal and marine systems.
   - Physical impacts—these can be natural or non-natural in origin, as well as direct or indirect, and result in the physical alteration of a particular ecosystem, such as anchor damage, boat grounding on coral reef, or eutrophication by sewage in a lagoon.
   - Ecological impacts—they may be a result of any of the above categories but pertain to a broader scale of changes with more long-term affects and the possibility of permanent alterations to an ecosystem and to basic biological diversity.
   - Socio-economic impacts—those that reflect a feedback from environmental impact are considered here. They are caused by negative environmental effects of development, which reduce long-term social and/or economic well-being.

Because there are great differences in scale and type of projects to be assessed, it is beneficial to focus the EA process on the expected type and level of impacts, starting *very early* in the design phase. Therefore, it is useful to have two or three levels of assessment, starting simple and then going to increased complexity.

In a three-stage process, the *first* level would be a preliminary review to see if there are potential serious impacts. The *second* would be an initial analysis, or initial environmental examination (IEE), as it is often called, which would be conducted if the preliminary review revealed potential environmental problems. The *third* (if needed) would be a full-scale EIA, invoked when serious environmental problems are forseen in the IEE and it is decided to subject the project to the higher level EA or EIA, a more intensive assessment (usually requiring special data collection).

Typically, only the more threatening of developments are subject to the full-scale EIA, for which the "trigger" is either the type of project or the size of the project (231). To cite an example, this type of three-stage EA process is incorporated into Indonesia's national assessment system, AMDAL. Mexico, Bangladesh, and other countries also use—or propose to use—three levels of EA requirements: preliminary, IEE, and EIA.

EIA reveals conflict and describes the extent and type of project impacts. While EIA does not make decisions—this being a political process—it often comes close. This is particularly true where there are pre-determined criteria and standards against which to compare the predicted effects of a project.

If adverse impacts are predicted, tradeoffs among environmental, social, and economic factors will be necessary. From the economic aspect, this will take into account economic impacts, some of which can be measured using standard evaluators (e.g., internal rate of return, opportunity costs, and benefit-cost ratio) and others that must be estimated with special techniques. From the social standpoint, impact evaluators might include food security, family income, land tenure, credit availability, health, social unity, education, and so forth.

In coastal project EAs, several different impact types may often be identified; some are positive, but most are negative (see "Impact Types" in Part 3). In this situation, the balance of net benefits and losses must be determined in some fashion based on qualitative and quantitative factors. If no predetermined scheme is available to convert the "apples to oranges," the process may have to be more judgmental (i.e., more political).

In the official United States Fish and Wildlife Service mitigation policy it is stated that "the net biological impact of a project proposal is the difference in predicted habitat value between the future with the action and the future without the action". In effect, this encourages the developer to present an actual "balance sheet" in support of his application (including voluntary enhancements) which shows for each of the important functional categories the extent to which the project will benefit or harm the ecosystem in the "without project" and "with project" scenarios (75). The developer's obligations are shown in the environmental management plan (EMP).

EA approaches for coastal zone development projects are mostly modifications of standard terrestrial or marine EA approaches, about which there is an extensive literature. Sources include Thompson (342), Ahmad and Sammy (3), Beanlands (18), and Gammon and McCreary (136). For specific coastal EIA, Sorensen and West (327) have written an excellent guide, as have Carpenter and Maragos (41) also see Marchand (233).

The work of the EA analyst generally consists of the following aspects (59, 326, 327):

- Data gathering on project characteristics (design, location, etc.), alternatives, and coastal resources to be affected, including existing uses of these resources
- Estimates of the negative (and positive, if any) environmental impacts to be caused by the development project
- Identification of alternative designs and locations and operational precautions to prevent *avoidable* negative impacts
- Identification of measures to reduce effects of *unavoidable* negative impacts, including habitat rehabilitation
- Recommendations to decision makers for approval or disapproval of the project and, if approved, the best combination of measures and alternatives to minimize and/or mitigate impacts

Training in the field is easily available, and there is a large literature on EA methods.

The following six-step procedure (which the author has field tested) would be quite satisfactory for ICZM purposes—it is called *STEPS* (Standard Track for Environmental Predictions):

Step 1: Project Identification—This is a simple description of the project including textual (verbal) and descriptive (spatial, maps and plans) components. It should be focused on the aspects that are most likely to cause environmental disturbance. Later in the sequence it may be necessary to obtain more detailed descriptions of aspects of the engineer's plans. If there are alternative project proposals, each one is described.

Step 2: Resources at Risk—This is a description of the particular resources *at risk* to the proposed project, including water, soil, plant, energy, and animal components. Analysts should not waste effort on trivial matters or resources *not* to be significantly affected by the project. Discuss climate *if* relevant, soils *if* involved, hydrology *if* implicated.

Step 3: Screening—This is the reconnaissance step, where the analyst should try to think of all possible environmental problems that could be caused by the project and sort them out into two categories, major and minor. The minor ones are discarded with a brief explanation for each, and the major ones are passed on to the scoping step, *if* it is decided to do a full-scale EIA. Checklists and matrices are used at this stage.

Step 4: Scoping—In this step, you, the assessor, examine all the major impacts predicted and determine levels of significance. You discover all feasible engineering alternatives that result in lesser environmental impact. These initial findings are presented at special meetings, or by other means, to all the "stakeholders" (those affected by the project) for comment. The results of these consultations along with the technical information and any recommendations from the assessor are passed to Step 5.

Step 5: Evaluation—First, several variables have to be agreed upon, including viable engineering alternatives and boundaries for the spread of impact. Mitigation requirements are set. Then an evaluation of the project can be completed from information generated, and a strategy for EIA can be drawn. Finally, a T.O.R. can be prepared to initiate the EIA work.

Step 6: Analysis and Report—Under the T.O.R., the necessary data collection, analysis, and evaluation are completed, along with further stakeholder participation. Then three reports are written: (a) impact assessment, (b) environmental management (including mitigation), and (c) baseline/monitoring (real time and post-audit).

This type of sequence is shown graphically in Figure 2.37.

The initial environmental examination (IEE) will more or less follow the same sequence but more succinctly and will often compress it so there are fewer steps. Sometimes Steps 4 and 5 are eliminated and compressed into Step 3. Step 6 can be reduced, but each of the elements will be presented in the IEE report.

The necessity for the assessor is to ensure that all important impacts have been reviewed and that the conclusions are truly integrative; that is, that the EA weighs and balances all the various aspects (see Figure 2.38).

For EA to be effective, consultation with all significant stakeholders is needed. An efficient way to plan for broad EA consultations is via "Scoping Meetings" (as suggested in Step 4 above). This may be the best approach for meeting with (a) officials responsible for the project and the site, (b) experts in the technology and environmental sciences, (c) affected groups such as local residents and businesses, and (d) representatives of other agencies with expertise or jurisdiction (41). The scoping meeting should be announced well ahead of time and a preliminary assessment should be used to describe the development objective and tentative project plan. Sketch "maps" of the project at a scale of about 1:10,000 can be used to organize the discussion.

All participants are encouraged to add items to the sketch and to propose alternatives and issues to be assessed. Flows of materials, energy, and people are indicated on the sketch map. Impacts are tentatively predicted. Ecologically sensitive areas (e.g., lagoons, coral reefs, flood plains, wetlands) are located. Later, a fresh version of the sketch map may be prepared, but for scoping, neatness is not required—the purpose is to capture all reasonable ideas and comments. Specific sites may be sketched at a larger scale to allow portrayal of more detail.

Simplification is commendable. The EA process may get overly complicated, rigid, and expensive. It also may become too methodologically ambitious, requiring large numbers of highly trained specialists, and also too separated from the planning process. Most experts would agree that emphasis should be put on EA approaches that:

- Are not unnecessarily elaborate
- Do not involve complicated methods
- Are geared toward identifying mitigation and management measures
- Are adaptive to the uncertainty of natural system effects and implementation problems
- Take account of multiple objectives and consequent tradeoffs

In summary, the procedural approach should emphasize flexible, cost-effective, and simple methods for obtaining the appropriate information and conducting the EA analysis. Even with simplification, EA may seem a large order in beginning programs. What helps is a de-mystification of EA and an emphasis on sound policies supported by realistic procedures and suitable technical analysis.

A variety of techniques is available, including checklists, matrices, networks, map overlays, numerical evaluations, and other approaches (18). A comparison of the various techniques is presented in Figure 2.39 (124). The two techniques used most frequently are the checklist and the matrix.

Impact assessors should be wary of mechanistic approaches, those that reduce complex problems to meaningless ratings and rankings or that demand expensive modeling and simulation. Too often the mechanistic approach turns out to be less useful for decision making than a simple common sense approach. Certainly, there are problems that demand sophisticated analysis or judgment. But suitable technical expertise and judgment are more important than elaborate methods.

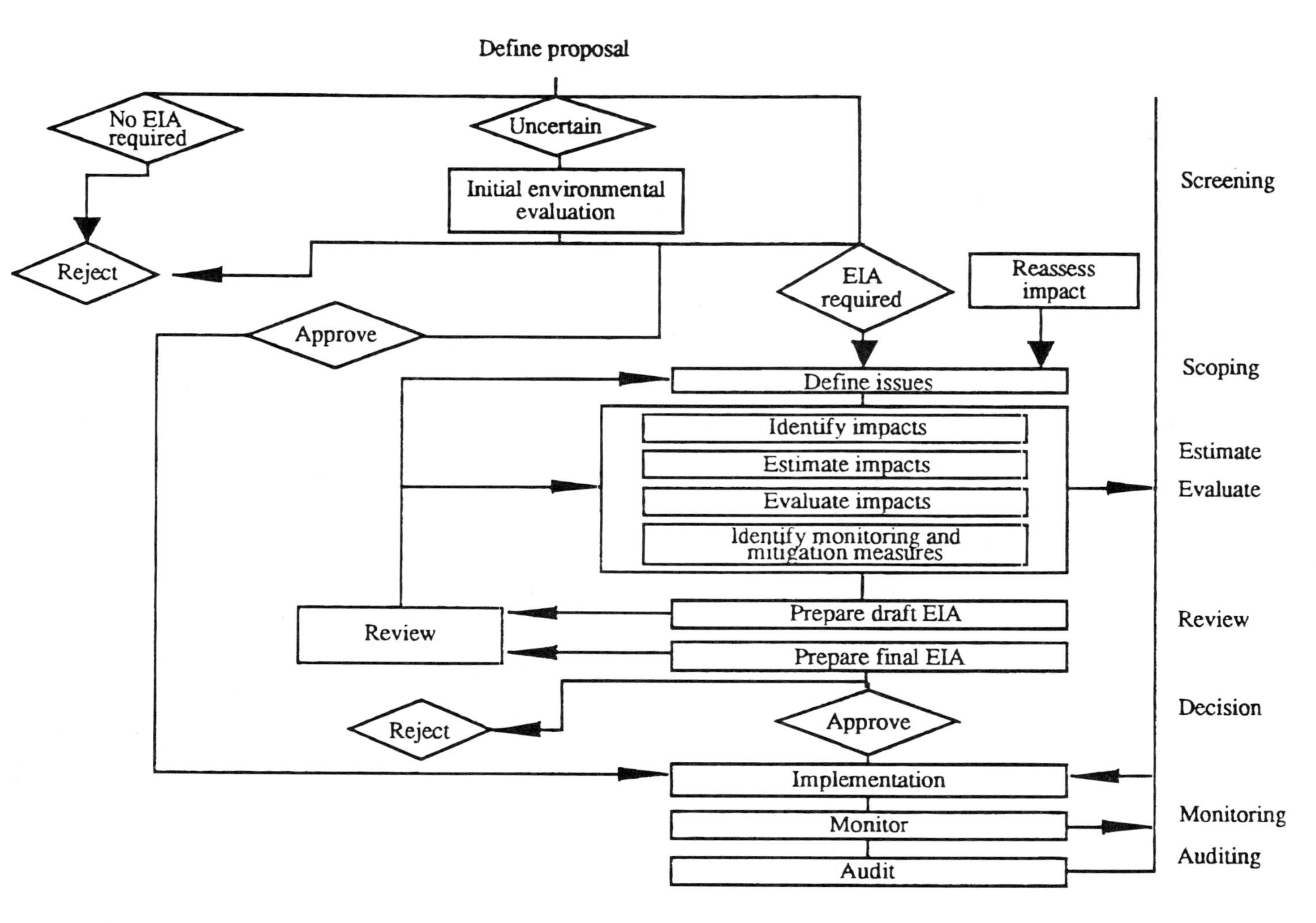

Figure 2.37. Network diagram shows linkages among the main components of an environmental assessment system. (Source: Reference 327.)

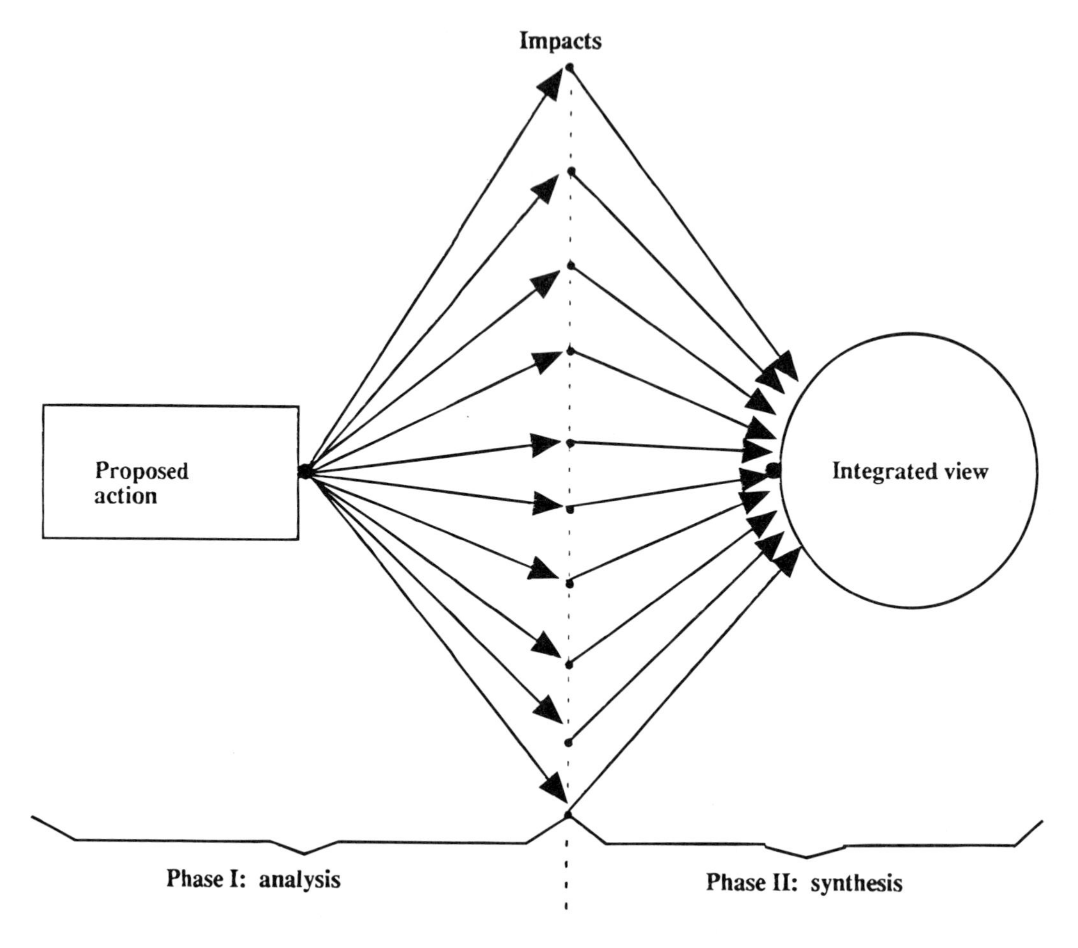

Figure 2.38. An environmental assessment reviews many factors, judges their impacts, and attempts to integrate the results and produce a balanced conclusion. (*Source:* Reference 327.)

A checklist is often used in the preliminary stage of EA and/or in initial environmental examination (IEE). On the basis of the type and significance of impacts identified by the checklist and evaluated in Step 4, recommendation will be made whether to conclude with IEE or proceed to full-scale EIA.

The general checklist below covers some of the common types of impacts encountered in coastal projects and that may require attention in the EA:

1. General Ecology
    - Sewage pollution (including phosphate detergents)
    - Other organic pollution (e.g., fish waste)
    - Land drainage
    - Waste oil pollution
    - Fuel leaks/spills (from tanks, pipes, hoses, couplings)
    - Deterent chemicals (negative behavior modificants)
    - Other toxic wastes (any to be stored on premises)
    - Land clearing, tree cutting
    - Excavation
    - Dredging
    - Filling of wetlands
    - Structures (bridge, causeway, seawall—currents, stagnation, migration block, beach erosion)
    - Marine construction activity (forms, pile driving, concrete pouring, explosives)
    - Shoreline construction activity

- Other sediment producing activities
- Pest control
- Solid waste, including garbage
- Ocean dumping (planned, deliberate, and casual)
- Ship operations (e.g., propellor disturbance)
- Disturbance of species
- Disturbance of species habitats
- Offsite (e.g., quarries)

2. Human Ecology
   - Land/water traffic
   - Disturbance (dust, smoke, odors, machines)
   - Noise
   - Air pollution
   - Aesthetic and viewscape
   - Drinking water
   - Pathological effects
3. Natural Hazards
   - Sea level rise
   - Seastorms, tsunamis
   - Erosion
   - Earthquakes, landslides, liquification
   - Flooding
4. Socio-economic
   - Vocation (fishing, farming, aquaculture)
   - Jobs, income
   - Foreclose future opportunities
   - Safety issues
   - Land use problems
   - Displacement
   - Intrusion (tourists, economic competitors)
   - Effects on culture (life-style, religion, history)
   - Political stability (conflicts with government or neighbors)
   - Equity; winners and losers

Figure 2.40 shows a simple form of the matrix method for identifying impacts—this should be used at the most general level. But a matrix can be developed for any level down to the most specific details.

Some countries establish a pre-determined list of types and sizes of projects that must always have EA (either IEE or EIA). Others may define those that *do not* need EA, whereas others apply guidelines on a case-by-case basis. As an example, Thailand's list of projects that categorically require full-scale EIA follows (327):

Type of project	Magnitude
Dam or reservoir	Storage volume >100,000,000 cubic meters or surface area >15 km^2
Irrigation	Irrigated area >12,800
Commercial airport	All sizes
Hotel or resort adjacent to coast	>80 rooms
Mass transit system/expressway	All sizes
Mining	All sizes
Industrial estate	All sizes
Commercial port and harbor	With >500-ton vessel capacity
Thermal power plant	Capacity >20 MW
Chemical industries	>100 tons/day of raw materials
Oil refinery/natural gas processing	All sizes
Iron and/or steel industry	Raw materials >100 tons/day or >5 tons/batch
Smelting other than iron/steel	Production capacity >50 tons/day
Cement industry	All sizes
Pulp industry	Production capacity >50 tons/day

Criteria	Check-lists	Overlay	Network	Matrix	Environ-mental Index	Cost-benefit Analysis	Simula-tion Modelling Workshop
01. Comprehensiveness	S	N	L	S	S	S	L
02. Communicability	L	L	S	L	S	L	L
03. Flexibility	L	S	L	L	S	S	L
04. Objectivity	N	S	S	L	L	L	S
05. Aggregation	N	S	N	N	S	S	N
06. Replicability	S	L	S	S	S	S	S
07. Multi-functions	N	S	S	S	S	S	L
08. Uncertainity	N	N	N	N	N	N	S
09. Space-dimension	N	L	N	N	S	N	S
10. Time-dimension	S	N	N	N	S	S	L
11. Data Requirement	L	N	S	S	S	N	N
12. Summary Format	L	S	S	L	S	L	L
13. Alternative Comparison	S	L	L	L	L	L	L
14. Time Requirement	L	N	S	S	S	S	N
15. Manpower Requir.	L	S	S	S	S	S	N
16. Economy	L	L	L	L	L	L	N

L = Completely Fulfilled, or Low Resources Need
S = Partially Fulfilled, or Moderate Resource Need
N = Negligibly Fulfilled, or High Resource Need

Figure 2.39. Various assessment techniques evaluated according to their capacity to fulfill their role in environmental assessment and the degree to which other resources would be needed. (*Source:* Reference 218.)

Over time, the following development activities have emerged from various approaches as most usually justifying a full-scale EIA-type environmental assessment (41):

- Large industrial and manufacturing plants
- Large construction projects—deep draft ports, highways, airports
- Water resources structures—dams, irrigation systems
- Electric power plants
- Mining and minerals processing
- Hazardous chemicals manufacture, handling, storage
- Sewerage and sewage treatment plants
- Municipal wastes and hazardous wastes
- New human settlements
- Large-scale intensified forestry, fisheries, or agriculture
- Major tourism facilities

Throughout the environmental assessment process a variety of physical, ecological, and other environmental data will be gathered. This information normally progresses from general, qualitative,

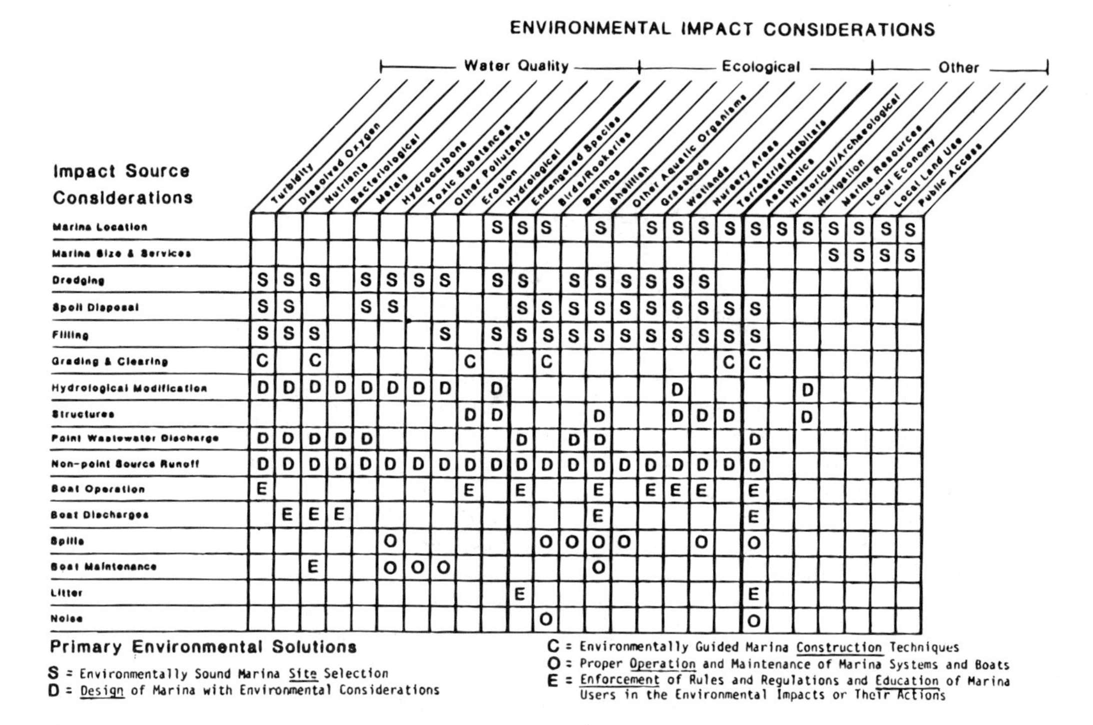

ENVIRONMENTAL IMPACT CONSIDERATIONS

Water Quality — Ecological — Other

Impact Source Considerations	Turbidity	Dissolved Oxygen	Nutrients	Bacteriological	Metals	Hydrocarbons	Toxic Substances	Other Pollutants	Erosion	Hydrological	Endangered Species	Birds/Rookeries	Benthos	Shellfish	Other Aquatic Organisms	Grassbeds	Wetlands	Nursery Areas	Terrestrial Habitats	Aesthetics	Historical/Archaeological	Navigation	Marine Resources	Local Economy	Local Land Use	Public Access
Marina Location										S	S	S		S		S	S	S	S	S	S	S	S	S	S	S
Marina Size & Services																							S	S	S	S
Dredging	S	S	S		S	S	S	S		S	S		S	S	S	S	S	S								
Spoil Disposal	S	S			S	S					S	S	S	S	S	S	S	S	S	S						
Filling	S	S	S					S		S	S	S	S	S	S	S	S	S	S	S						
Grading & Clearing	C		C						C			C							C	C						
Hydrological Modification	D	D	D	D	D	D	D	D		D							D					D				
Structures									D	D				D			D	D	D			D				
Point Wastewater Discharge	D	D	D	D	D						D		D	D						D						
Non-point Source Runoff	D	D	D	D	D	D	D	D	D	D	D	D	D	D	D	D	D	D	D	D						
Boat Operation	E								E		E			E		E	E	E		E						
Boat Discharges		E	E	E										E						E						
Spills						O						O	O	O	O			O		O						
Boat Maintenance			E			O	O	O						O												
Litter											E									E						
Noise												O								O						

Figure 2.40. Impact evaluation matrix using a proposed small boat harbor (marina) development as an example. (*Source:* Reference 165.)

and broad in scope to detailed, specific, and quantitative. Although *qualitative* information is invariably required for all types of development, the choice of alternatives to be implemented dictates to a great extent the need for *quantitative* information. Often the need for expensive quantitative data gathering can be avoided if the project propoponents agree to a direction that minimizes adverse impacts (232).

Timing of the EA process is a key to its effectiveness. Problems arise in cases where EA is done *after* the project is completed, thus providing no opportunity to use the EA to redesign the project. Most project-specific development involves several phases: pre-feasibility study, feasibility study, selection of a plan and detailed engineering and design studies, reconsideration of design, construction, operation, and decommissioning of facilities, if applicable. The EA process should be incorporated in *each* phase of major coastal development projects, as shown in Figure 2.41.

For the EA process to be effective, there must be standards and guidelines to which all interested parties can refer. There must also be a permit system whereby project sponsors request formal permission to proceed with coastal development, after the coastal agency conducts an EA (IEE or EIA), evaluates the project, and recommends approval, rejection, or modification of the project. This procedure should pose no administrative problem to most countries because they will have already had considerable experience with permit processes.

The top ten causes for EA failing to lead to sustainable development are given as (18):

1. EA documents are prepared to justify a decision already made on a project.
2. EA is limited to a single phase or ends too early, missing subsequent project changes and impacts.
3. Lack of interaction of assessor and integration of EIA with remainder of proponent's technical development team.
4. EA fails to consider and compare alternative proposals, including no action.
5. Mitigation is inadequately assessed or selected.

6. Monitoring after project implementation is not conducted.
7. EA documents are not prepared in draft form for full public review, comment, and revision for the better.
8. EA fails to tell the whole story (both the good and bad news) compromising credibility and public support.
9. EA screening and scoping is inadequate, failing to focus on the key environmental issues.
10. Environmental supporting studies are inadequate or mislocated due to changes in project siting or design.

An ICZM program must usually formulate an efficient EA procedure and often a permit letting arrangement. Without the EA process there can be no effective ICZM program. There has to be a systematic way to review project impacts, to modify them to reduce negative impacts on resources, and to reject the worst of them. One reason that the EA methods explained above work best for the review of coastal development projects is that this process has been used for over 20 years and its principles, practices, methods—as well as its limitations—are well known. Also there is a large literature on it and training in the EA field is available.

While impact assessment is usually focused on the *project* level, assessments may also be done for *programs,* such as for a regional economic development initiative. Program level impact assessment—when done for a large geographic area—is conceptually similar to regional planning but does not include a mechanism to compel actual plan making and implementation.

If an appropriate administrative system exists and there are mechanisms for execution, the most urgent need will be for skilled and experienced assessors. Staff training is essential. Every country should have a cadre of professionally trained EA analysts. Such training is regularly provided by certain

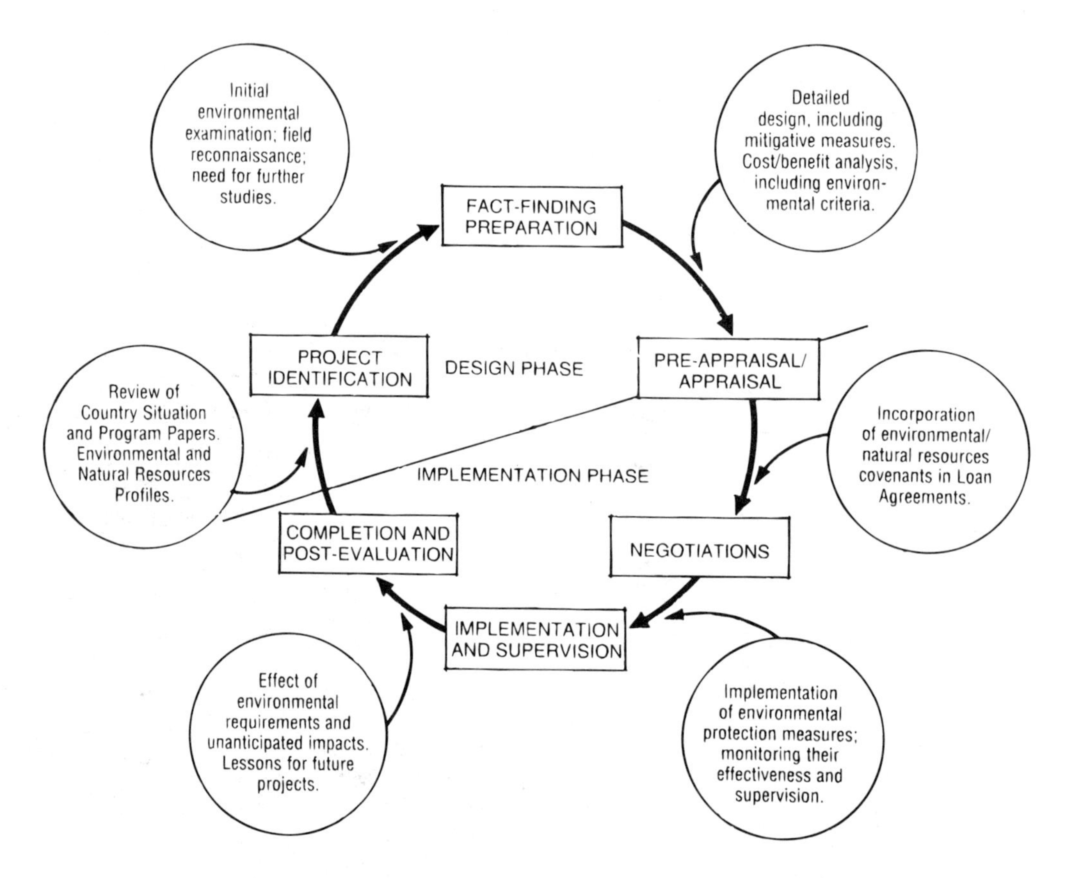

Figure 2.41. Impact assessment should start at the beginning, during the initial project design, and continue through the whole process. (*Source:* Reference 110.)

institutions (such as the University of Aberdeen in Scotland), as well as many international organizations. So far the number of professionals trained may be less than 10 percent of the number needed.

The social and economic parts of the EA process are usually done by professionals in the social sciences and resource economics and not by ecological experts, unless the needs for analysis are simple. The danger is that the social and economic parts can become independent studies not integrated with or balanced against the environmental. Effective inter-disciplinary coordination should solve this problem. (See "Social Impact Assessment" below and "Economic Impact Assessment" above).

Without EA, the narrow financial interests of certain developers are often facilitated at the expense of the environment, which inevitably adversely affects the economically disadvantaged. In the Philippines, this would be evident in places like Bolinao where fishers and farmers, who depend directly on the maintenance of a healthy environment, constitute 80 percent of the human population (242). It seems unarguable that there must be a systematic way to review coastal project impacts, to modify them to reduce negative impacts on biodiversity and natural resources, and to reject the worst of them.

Voluntary Environmental Assessment — While most environmental assessment (EA) is conducted by the public sector, recently the private sector has begun to support or conduct EA for its own reasons. Also the private sector has been conducting reviews of its facilities and operations. It is interesting to reflect on the private sector's reasons for involvement, which are listed by North in an OECD report as follows (261):

- Assurance of adequate procedures for managing environmental risks, and compliance with procedures
- Improved statutory compliance
- Identification of environmental risks and problem areas, early warning and prevention of potential adverse environmental effects (risk identification, assessment and management)
- Improved financial planning, through the identification of future and potential capital, operating and maintenance costs, associated with environmental activities
- Improved preparation for emergency and crisis situation management
- Improved corporate image and positive public relations
- Enhancement of environmental awareness and responsibility throughout the corporate hierarchy
- Improved relations with regulatory authorities
- Facilitation of obtaining insurance coverage for environmental impairment liability

From the private sector point of view, the following questions need to be asked about any major project in order to consider all phases of a project life cycle—construction, operation, extension, liquidation—according to North (261):

Can it operate safely, without serious risk of dangerous accidents or long-term health effects?

- Can the local environment cope with the additional waste and pollution it will produce?
- Will its proposed location conflict with nearby land uses or preclude later developments in the surrounding area?
- How will it affect local fisheries, farms, or industries?
- Is there sufficient infrastructure, such as roads and sewers, to support it?
- How much water, energy, and other resources will it consume, and are these in adequate supply?
- What human resources will it require or replace, and what social effects may this have on the community?
- What damage may it inadvertently cause to national assets such as virgin forests, tourism areas, or historical and cultural sites?

North (261) lists the following private sector principles and guidelines for managing EAs:

Principle 1: Focus on the main issues

- It is important that an environmental impact assessment does not try to cover too many topics in too much detail.

- At an early stage, the scope of the EA should be limited to only the most likely and most serious of the possible environmental impacts, which could be identified by sreening or a preliminary assessment.
- Where mitigation measures are being suggested, it is again important to focus the study only on workable, acceptable solutions to the problem.

Principle 2: Involve the appropriate persons and groups

- Those appointed to manage and undertake the EIA process.
- Those who can contribute facts, ideas or concerns to the study, including scientists, economists, engineers, policy makers, and representatives of interested or affected groups.
- Those who have direct authority to permit, control, or alter the project—that is, the decision-makers—including, for example, the developer, financial aid agency or investors, competent authorities, regulators, and politicians.

Principle 3: Link information to decisions about the project

- An EIA should be organized so that it directly supports the many decisions that need to be taken about the proposed project. It should start early enough to provide information to improve basic designs and should progress through the several stages of project planning and implementation.

Principle 4: Present clear options for the mitigation of impacts and for sound environmental management

- To help decision makers, the EIA must be designed so as to present clear choices on the planning and implementation of the project, and it should make clear the likely results of each option.

For instance, to mitigate adverse impacts, the EA could propose:

- Pollution control technology or design features.
- The reduction, treatment, or disposal of wastes.
- Compensations or concessions to affected groups.

To enhance environmental compatibility, the EIA could suggest:

- Several alternative sites.
- Changes to the project's design and operation.
- Limitations to its initial size or growth.
- Separate programs which contribute in a positive way to local resources or to the quality of the environment.

And to ensure that the implementation of and approval projects is environmentally sound, the EA may prescribe:

- Monitoring programs or periodic impact reviews.
- Contingency plans for regulatory action.
- The involvement of the local community in later decisions.

Principle 5: Provide information in a form useful to the decision makers

- The objective of an EA is to ensure that environmental problems are foreseen and addressed by decision makers. To achieve this, decision makers must fully understand the EIA's conclusions, which should be presented in terms and formats immediately meaningful.

2.15 ENVIRONMENTAL MANAGEMENT PLAN

The main tool for project-by-project environmental management is the Environmental Management Plan (EMP). An efficient program of environmental management is a necessity in a world that is growing fast in both economic and demographic directions. Like Environmental Assessment (see above section), the purpose of EMP is not to slow economic development but to encourage it. The idea is to ensure that development and conservation co-exist in on an enlightened pathway to economic progress.

The EMP is a guidance mechanism for design and construction of economic development projects. It may be invoked as a result of the Environmental Assessment process (as it is in Indonesia's excellent

AMDAL program, the national impact management scheme). It addresses sources of impact as far as they can be known in advance. For example, it addresses environmental management needs for the immediate site, for adjacent areas, and for areas offsite which are subject to project-generated impacts—e.g., staging yards and borrow pits. The EMP also addresses, secondary (linked) impacts, such as those generated in the second order like sewage or roadway traffic that would be induced by more commerce and tourism business. Both environmental and socio-economic impacts are considered.

In the EMP program, a variety of efforts may be recommended for project management, including prevention, reduction, mitigation, enhancement, restoration, monitoring, education, outreach, and coordination. An essential of the EMP is mitigation of impacts; that is, counter-measures to offset significant negative impacts. In marine construction, for example, methods would be prescribed for handling the machinery and materials that would do the least damage to water quality or to critical natural habitats, such as coral reefs. Counter-measures might be proposed for onsite, adjacent, and offsite locations and for both direct and indirect impacts.

For "non-mitigable" impacts, one should attempt to determine whether these impacts are significant or, if trivial, whether they are isolated or if they might accumulate with others to reach significance and therefore require a management response. Coping with these "cumulative impacts" can be a serious problem to coastal resources conservation.

Then, too, the EMP may have to cope with troubling offsite impacts not caused by the particular project under consideration; for example, sediment runoff from adjacent construction sites, paddy fields, forest, land-clearing operations, and so forth, which can be seriously damaging to coral reefs and other coastal habitats. Also, chemical pollutants flushed down rivers into coastal waters can create toxicity (e.g., biocides, oil wastes) and bring excessive nutrients (e.g., commercial fertilizers, animal wastes). However, control of these external impacts is difficult for the designers of a specific coastal project. Control of such widespread effects must usually be referred to the provincial or national government for consideration in an ICZM program.

Many settled coastal areas have exceeded the *pollution threshold*—increased sewage outflows from hotels or residences or increased amounts of fertilizer runoff from paddies could stimulate serious algal attack of the coral reef. When this happens (e.g., at Kaneohe Bay, Hawaii), the algae can overgrow the reef, killing the corals. Then the corals may be worn down by bioerosion (drilling, grazing, and burrowing animals) and dissolution, and the coral reef can begin to disappear.

A major problem is that intermediate (secondary) sewage treatment is known to remove only small amounts of the important nutrients—phosphate and nitrate. Instead, this treatment level refines various organic matters into high powered dissolved mineral nutrients. These nutrients create algae blooms and stimulate reef algae overgrowth when they reach coastal waters. Therefore, the secondary treatment plant acts as a "sewage refinery" and is one of the great threats to coral reef and other ecosystems and to the beaches that they now protect. It is recommended that the effluent be piped offshore to deeper waters where dispersive dilution and transport by currents are effective counter-measures (see "Ocean Outfall Placement" below).

Clearly, it will be important to manage project construction to avoid water pollution. The EMP should call for oversight (real-time) monitoring of pollutants to detect if the construction work is creating unacceptable impacts, according to environmental guidelines and standards.

The beachside work sites in the shorelands adjacent to resort, commercial, and residential areas have the potential to reduce the quality of the human environment because of noise, air pollution, social disturbance, nuisance, and uglification. The use of heavy machinery operating close to inhabited areas over a more or less prolonged period and also moving of large equipment and materials through the streets can be socially disturbing and may cause smoke, noise, and traffic. Counter-measures may help; for example, optional methods for handling machinery and materials in order to cause the least disturbance and do the least damage to ecologically critical areas.

The general approach to preparing an EMP is to "red flag" all *significant* impacts identified in the EA and to recommend specific management guidelines for each. The results of the impact identification, field work, and analysis that were completed for the EA provid the basis to decide for which particular impacts specific counter-measures will be needed during construction, operations, and maintenance. Three major categories can be identified:

- First, investigate all potential sources of environmental impact.
- Second, refer any impacts to the EMP that have the potential to become significant if uncontrolled.

- Third, identify impacts that cannot be remedied by design modifications and must be remedied, or mitigated (eliminated or substantially reduced), through the EMP by operational modifications.

Once the impact sources needing countermeasures have been "red flagged," the environmental control or mitigation plan for each red-flagged item can be presented in the EMP document. The requirements must be stated with sufficient clarity and detail that the engineering management of the project can direct the environmental work and ascertain afterward that it has been satisfactorily completed.

Socio-economic impacts to be addressed in the EMP include the adverse aspects of community change that go with economic development such as cultural breakdown, environmental deterioration, unequal opportunity for economic and life-style improvements, etc. It may seem like using plain common sense to say that the common people should be treated fairly in the management of development projects, but this aspect is often overlooked by the elites who see benefits as mainly accruing to themselves. The common people, who often share the burden of debt for the elites, should get their fair share of benefits.

Problems to be mediated through the EMP include any unexpected diseconomies and inequitable social conditions that could occur as direct or secondary project impacts, such as housing problems, employment dislocation, unfair distribution of profits, non-transfer of benefits to poor people, and loss of jobs for fishermen.

Close surveillance, or monitoring, of many types of projects may be necessary to ensure that environmental damage is not done during the construction period. This "tactical monitoring" (or "oversight monitoring") needed for real-time feedback to EMP is quite different from the usual type of strategic monitoring used for environmental audit of completed, or partially completed, projects (see "Monitoring and Baseline" below).

2.16 FLOODLANDS

Coastal flooding is distinctly different from riverine flooding. When a river floods, the runoff and subsequent damage generally follow the river's course. The real damage of coastal flooding, unlike riverine flooding, does not occur in easily identified runoff channels, but over broad areas that alternately flood and drain during hurricanes and intense winter seastorms. Mounting losses can be expected when residential, commercial, and industrial uses are increasingly located on flood-prone coastal sites. The characteristics of coastal landforms affect the intensity of storm impacts on coastal communities. Three characteristics that have a major effect on the intensity of potential storm hazards are elevation, drainage, and topography.

Elevation — In many coastal areas the land is rising in relation to the sea. In others, it is falling. This is partly from sea level rise and partly from land subsidence augmenting a rise in the sea level. This is a factor of particular importance in managing floodlands (floodplains). Rapid subsidence may result from human actions—for example, excessive pumping of groundwater. Natural subsidence, by contrast, is a slow process that may be caused by the drying and shrinking of geological deposits, the decline of water tables, and movement of large geological deposits. When subsidence is rapid, regardless of the cause, structures built above the floodlands may sink to unsafe elevations (see "Subsidence" below).

Drainage — During a storm, any part of the floodlands not reached by the flood can retain water in its soils and hydrological system, thereby reducing the probable height of the floodwaters. Stream channels and other watercourses can contain floodwaters; so can lakes, ponds, and, particularly, wetlands. The absence of these features leaves only the natural retention capacity of the soils and vegetation of the terrain.

Topography — Topography, or the configuration of the land surface, affects the intensity of storm impacts because the normally dry depressions of floodlands can temporarily retain considerable amounts of floodwaters from both ocean and upland sources. On the other hand, if salt water is held long enough, it can damage soil fertility (by penetration into the earth) or groundwater quality (by penetration into subsurface aquifers).

The management objective for coastal floodlands should be to encourage development that is consistent with conservation of coastal ecosystems and protection of life and property from the threat of periodic flooding (Figure 2.42). Planners need to consider the siting, density, design, and construction of residential, commercial, and industrial facilities and sewage plants. Houses need to be elevated above the forecasted "100-year flood" mark—the highest elevation expected to be reached by a flood having a 1 percent probability of occurrence in any year. Non-residential structures need to be flood-proofed. Within the "high hazard" portion of coastal floodlands, where wave forces are severe, additional constraints on design and location of structures are needed.

There has been a global loss of coastal wetlands because of a general lack of appreciation of this zone's ecological value and its role in resisting flood and erosion hazards. The wetlands (mangrove or marsh areas) that lie at the lower edge of the floodlands have often been preempted by a builder who wants to build and to landscape right to the water's edge or a farmer who wants to open up as much land as possible for planting or grazing. When wetlands are cleared of vegetation, graded, built on, or otherwise obliterated or seriously altered, the result may be the loss of critical wildlife habitat and natural visual screen and an increased potential for water pollution, bank erosion, and damage from storm surges and waves.

Because of variations in landform, floodlands may have edge zones of greater or lesser value; therefore, you have to evaluate each case. The most concentrated ecological values would be expected where the edge zone is wetland backed by a strip of scrub or bush.

The best way to protect the floodlands and wetlands is to define a "setback" from the water's edge creating a "buffer area" or "buffer strip" and to prescribe only non-altering uses of it through special performance standards. In addition to conserving critical wildlife habitat and lowering flood hazard and erosion potential, the unaltered buffer strip of natural vegetation and soil provides a visual screen and a "filter strip" to intercept runoff and helps to purify water by soil infiltration and vegetative "scrubbing." A setback distance of 150 feet will often be sufficient for soil erosion control.

Figure 2.42. Huge areas of Bangladesh are sea-flooded lands "reclaimed" by surrounding them with dikes and controlling water level with the type of structure shown here (at Khulna)—these "poldered" areas are now beginning to fail because of sea level rise, land subsidence, and sedimentation of drainageways. (Photo by author.)

Additional width may be required to provide for removal of nitrate and other agricultural chemicals or for wildlife.

Floodwater retention, which is a key factor in mitigating the severity of flooding, can be significantly influenced by man-made alterations of terrain and watercourses (Figure 2.43). Alteration of terrain also changes the amount of "diffuse source" water pollution—erosion sediment, fertilizers, pesticides, and the like—that reaches coastal waters. Soil conservation programs and controls on land clearing, paving, drainage, and channel alteration are among the measures needed to protect against these problems.

For site preparation in settlements, grades should be designed to direct runoff along natural drainage courses and through natural terrain where the vegetation can cleanse and filter the water. In paving, surfaces should cover a minimal area; to permit water infiltration, permeable surfaces rather than solid paving should be used insofar as possible. These surface-management techniques are also valuable in protecting the recharge potential of groundwater resources. Aquifers are naturally recharged by rain percolating through from the land surface or laterally from a lake or stream. Using impervious surfacing, removing vegetation, and draining land in recharge areas will divert waters that otherwise would filter into groundwater aquifers.

There are places within the floodlands where excluding development is especially important from the twin perspectives of avoiding hazards and protecting ecological values. Regulatory techniques are often sufficient to protect the floodlands with a simple setback or buffer requirement in zoning, subdivision, or building controls. (See "Natural Hazards" in Part 3.)

2.17 HISTORICAL-ARCHAEOLOGICAL SITES IMPACT ASSESSMENT

Areas where valuable physical traces have been left by past inhabitants are known as *archaeological sites.* These may be prehistoric if they date to periods beyond the memory of local communities (where

Figure 2.43. The large-scale alterations of floodlands, as in the type of causeway project illustrated, may hold floodwaters and cause flooding. (Photo by Mike Fahay.)

there are no written records) or historic if knowledge of them is retained in the oral or written records of society. It is through information from archaeological sites that prehistories must be constructed.

Archaeological sites, as the sole remaining manifestation of prehistory should be protected and studied. Some sites have sacred or religious importance (such as a cemetery or sacred reef) or depict the location of an important historical event, such as a major battleground or a site of the first colonization of an island by a group of settlers. But large numbers of sites have disappeared or have been degraded by human impacts, especially construction activities. This includes many shipwrecks exploited by divers and treasure hunters.

It is not unusual in some countries to require an archaeological impact analysis along with the environmental impact assessment if the development site is in an area of known archaelogical interest. The four main steps are: (a) discovery of archaeological resources, (b) identification of type of resources, (c) evaluation of their significance, and (d) conservation strategy.

Preservation of sites and artifacts is often made possible by modest change in location or design of a development project. If damage or destruction is inevitable and the site is deemed to be "significant" but not to be worthy of protection at any cost, some form of mitigation of its destruction may be warranted. This might take the form of a detailed recordation of the site, collection of artifacts exposed on the ground surface, or excavation of structures or stratified archaeological deposits. If the site is deemed "insignificant," it may simply be destroyed without further action.

Detailed archaeological surveys can be expensive, particularly if large areas, excavation, radio-isotope data, pollen analysis, midden analysis, osteological analysis, and so forth are involved. If cost is a major concern, archaeological surveys can be phased over time, beginning with relatively inexpensive reconnaissance level surveys with limited subsurface testing. Decisions can then be made to move or redesign projects to avoid potentially significant adverse effects.

Taking the Pacific region as an example, the following types of sites can be identified:

- **Occupation Sites.** These comprise mainly surface scatters of stone artifacts with or without pottery and food remains, such as burnt bone and shell.
- **Rock Art Sites.** Paintings and engravings occur widely throughout the Pacific region on the walls of caves.
- **Agricultural Sites.** Agricultural systems involving drainage or irrigation commonly left archaeological traces in the form of terraces, stone and earth walls, and ditches.
- **Burial Sites.** Before the influence of Christianity, there were many different ways of disposing of the dead besides burial in the ground, and many of these have left distinct archaeological traces.

The likelihood of archaeological sites being detected is dependent on factors such as the following:

- Surface visibility, which is determined by the nature and extent of the ground cover of grasses and shrubs and of organic litter, such as leaves, bark, or twigs
- Burial of the original land surface on which the site occurred by, for example, slope wash material, flood alluvium, or windblown sand
- Exposure of the original land surface by erosion or by human activities
- Site obtrusiveness—some sites, such as visually spectacular paintings on a rock shelter wall, are easier to detect than others, such as sparce surface scatters of stone artifacts and pottery, especially in well-vegetated terrain

Archaeological sites are normally located by survey techniques that range from cursory inspection to intensive, systematic survey of the area. Oftentimes, visually prominent sites (e.g., agricultural terraces) can be located using air photo interpretation techniques. Buried sites can be detected by excavation techniques ranging from probing the deposit with a steel rod or soil augers through the excavation of test pits (each normally about 1 m^2) into the deposit to full-scale excavations covering large areas.

In discussing the significance of archaelogical sites, it is important to realize that there are several, often interrelated, ways of viewing significance:

- Significance to the local community
- Significance to the wider public

- Educational significance
- Scientific (or archaeological) significance

The various interest groups will normally have different views about how significant a particular site might be. Surface scatters of prehistoric stone artifacts and pottery containing examples of rare types of artifact may be of considerable scientific importance, but because they are not visually spectacular they are unlikely to be considered important by educators and the wider public, nor are they likely to have traditional significance for the community at hand.

On the other hand, a spectacular display of red ochre hand stencils in a rock shelter close to a tourist resort and community schools may be of considerable interest to the wider public (especially tourists) and educators, as it can be used to illustrate aspects of the local culture. But hand stencils normally have little scientific significance.

When significant archaeological sites are identified in an area of proposed development, and where they might be affected by development, several outcomes are possible. These might be summarized as destruction, destruction with mitigation, or preservation (Figure 2.44). Full preservation of sites is possible only by full protection in a custodial institution. But virtual preservation is often possible by modifying the project. If damage or destruction is inevitable and the project is socially valuable and not substitutable, some form of mitigation is warranted. This may take the form of a detailed recordation of the site, collection of artifacts exposed on the ground surface, or excavation of structures or of stratified archaeological deposits.

Detailed archaeological surveys can be expensive, particularly in large areas where excavation, radio-isotope data, pollen analysis, midden analysis, osteological analysis, etc., are involved. If cost is a major concern, archaeological surveys can be phased over time beginning with relatively inexpensive reconnaissance level surveys with limited subsurface testing. Such surveys can differentiate areas that are likely to include or exclude significant archaeological resources. After such a survey is accomplished, decisions can be made early during the planning process to move or redesign projects to avoid potentially significant adverse effects.

Source for this section: P. J. Hughes, Environmental Sciences, University of Papua New Guinea, Papua, New Guinea (179).

2.18 INSTITUTIONAL ANALYSIS

Over the long term, major institutional changes are needed in most countries to improve coastal resource management (326)—new environmental laws, stronger sectoral resource management institutions, deliberate efforts to strengthen the capacity for joint action among resource agencies, better salaries and training for staff, improved research and coastal environmental assessment capability.

The solutions that are required to manage the use of coastal resources are often difficult for citizens to understand and support and for governments to implement. Some examples are (270):

- Multi-agency planning and decision making
- Resolution of multi-party use conflicts
- Planning to solve several issues simultaneously
- Arrangements for networks of agencies and NGOs to take a share of responsibility

It is essential to recognize the importance of carefully analyzing what existing institutional mechanisms are at work in the coastal zone and what specific circumstances are interfering with the preparation, adoption, or implementation of sound resource management policies (270). Management strategies must be tailored to reflect the institutions, laws, and customs now in place, the geographic extent and severity of issues, and the available expertise and staffing (326).

Institutional analysis is most relevant to the ICZM-type of large-scale, comprehensive management programs where cross-sectoral and interagency coordination are vital. According to the experiences of governments that have engaged in the ICZM planning process, the extent to which the existing institutional structure, laws, and management instruments provide an adequate basis for planning and implementation needs to be determined. An analysis is needed of the environmental and socio-economic

Figure 2.44. Protected archaeological site in Saudi Arabia near the Red Sea. (Photo by author.)

interconnections among existing government programs and the management of coastal resources, environments, and hazards. (322)

Analysis of interagency coordinative mechanisms is essential. Many government ministries and associated departments are likely to be involved in the development of an ICZM plan. Two networks need to be analyzed: the horizontal network operating at equivalent levels among government agencies and the vertical network operating within individual economic sectors (and their government agencies). "Horizontal integration" describes efforts to coordinate among the sectors. In a complex policy area such as coastal management, the potential for fragmentation of government responsibility and duplication of effort increases with the number of sectors involved. (322)

Experience in ICZM programs clearly shows that the characteristics of the governance system are important. Too often the mechanisms, the tools themselves (such as zoning to preserve and protect portions of coastal systems, the mechanics of a permit system, or environmental impact assessment processes), come to be the dominant concerns of ICZM.

Institutional analysis should proceed simultaneously at the local, regional, and central levels of government because all three will be involved. A centralized program of coastal zone management involving, for example, control of wastes, regulation of construction in tidal waters, or mangrove forest restoration would be very difficult to implement without devolution of administrative functions to local units. The partnership should also involve local NGOs.

Institutional analysis of the present day framework of resource management is an important first step. This may be best done on an issue-by-issue basis, using an historical perspective. The basic list of questions for performing this analysis includes (270):

- Which laws govern decisions about the resource use action?
- Which agencies and units have responsibility for planning, decision making, enforcement?
- What is the strategy of management?
- What is the overall decision-making process, including how individuals are brought into the system and the decision rules employed?
- What enforcement is carried out, and what sanctions are imposed?

There is no "best" institutional arrangement for managing coastal resources. The "goodness" of an institutional arrangement can best be judged by the effectiveness and efficiency with which coastal use conflicts are resolved. (322)

To look more specifically at the coastal situation, the following questions might be asked (270):

- What are the existing government programs for coastal renewable resource sustainable use (conservation) at national and local levels; how effective are they; what are the shortfalls; what changes in governance have been considered?
- How much of the coastal conservation problem does the existing program effectively address?
- How effective are the existing mechanisms for interagency and intersectoral coordination on coastal matters; what changes have been considered?
- What is the capability of the responsible organizations for performing decision-making tasks?
- What proposals exist for improving the program?
- How much support exists for these proposals, and can existing units successfully implement them?
- What is the status of personnel training for ICZM; is there sufficient expertise; have improvements been considered?

The second step is to evaluate the future prospects. As a precursor to an ICZM program, institutional analysis attempts to compare the known capability of government and private institutions with the future task of organizing and administering an integrated management mechanism. Six examples of specific capability factors that need examination are (326):

1. To have the capacity and resources (personnel, financial, political) to undertake analysis, planning, implementation, and monitoring of management strategies
2. To involve the public, industry, and interest groups in coastal zone management
3. To transcend narrow administrative boundaries and include consideration of the coastal hinterland as well as the territorial seas, perhaps as far as the exclusive economic zone at times
4. To be able to manage resource allocation as well as pollution externalities (spillover effects)
5. To have the authority to raise and distribute funds
6. To have the competence to handle particularly difficult or important areas of integration, such as
    - Coastal developmental/environmental protection
    - Coastal water/inland water
    - Coastal land/coastal water
    - Coastal management/fisheries management
    - National/regional/local government integration

An improved institutional mechanism for coastal zone management will involve improving the linkages among the aforementioned sectors and the development of better regulatory and economic instruments. The necessary bureaucratic/administrative links would involve collaboration among representatives from national, regional, and local governments of the coastal zone areas concerned, designed to develop policies for specific regions or local areas. This will not necessarily entail creating new bodies, but rather the bringing together of existing interests. (324)

Of primary importance for enhanced coastal zone management is a consistent government policy at national or regional level that provides clear direction and support for integration in general and the creation of an institutional (administrative) mechanism in particular. The analysis should evaluate this.

Past experience in ICZM has demonstrated that, if all the major interests, or stakeholders, are not invited to participate in the preparation, adoption, and implementation of the Master Plan, they are more likely to oppose both the plan and the agencies associated with it. The successful adoption and implementation of an ICZM Master Plan usually depends on the degree to which stakeholders form a coalition (or constituency) to support it.

2.19 ISSUES ANALYSIS

In a program of coastal management, the conservation issues and problems must first be identified before counter-measures can be proposed. This identification process—known as "issues analysis"—is mainly done in the "strategy planning" stage (see "Strategy Planning" in Part 1.5). Issues analysis means more than just listing the items; it requires categorizing, weighting, balancing, evaluating, and prioritizing the issues in light of the political, social, and economic background of coastal zone development.

Because ICZM-type programs cannot solve all community problems, the issues to be addressed may be prioritized in a kind of "triage" exercise. The easily resolved issues can be put in a category for modest effort because they can be worked out with minimum effort. The intractable issues can also be given a lower priority on the basis that much time would be wasted on them with little to be gained. Most effort would then go toward the realistically resolvable issues.

A few common themes emerge from issues analysis. Virtually every coastal nation with development bordering lagoons or estuaries appears to have special pollution problems, usually because of municipal sewage and industrial toxins. The lagoon or nearshore pollution issue arises in all coastal nations, irrespective of the degree of development or variation in environmental and socio-economic conditions.

Coastal nations with substantial mangrove acreage almost always experience stresses from watershed practices, pollution, filling, and the overharvesting of mangrove for fuel. As a result, nearly every coastal nation that actively harvests its coastal fishery stocks appears to be suffering from depetion of fish stocks, partly as a consequence of these environmental factors.

The following is a list of generic issues to be explored and questions to be answered (66):

- Which coastal resources are seriously degraded; to what level have yields fallen; what are the economic consequences; what actions are needed to correct this situation?
- What are the causes of the degradation; what type of developments and activities need to be controlled; what are the economic effects of the controls; in consideration of the variety of possible tradeoffs and their effects, what actions are recommended?
- Who are the principal users of coastal renewable resources; how many jobs are at stake; how much income and foreign exchange earnings are involved in tourism, fisheries, and other resource dependent industries; what further losses are expected if ICZM is not implemented?
- What are the *priority* issues; what critical habitats need special protection; what species need protection; what is the best approach, regulation or protected areas?

It is not enough to simply identify and list the issues. Each issue should be evaluated for important aspects, including at least the following: (a) the extent of socio-economic disturbance and resource loss that it causes; (b) the degree to which it could be resolved by an ICZM-type approach; and (c) the consequences of not resolving it (326).

Fisheries conservation and the maintenance of tourism or recreation quality clearly emerge as universal arguments for ICZM. The economic importance of fisheries and tourism will strongly influence the extent to which developing nations will want to initiate coastal resources management programs. Mangrove forestry operated on a sustainable yield basis is a significant matter for many nations, as is coral reef conservation. Other major issues are water quality, beach management, allocation of scarce waterfront land, and natural hazards protection. As an extended example, Table 2.1 shows the issues reported in a literature analysis (mostly for the United States) by Healy and Zinn (164).

2.20 MANGROVE FOREST MANAGEMENT

Mangrove forests are a resource of great value to coastal communities (see "Mangrove Forest Resources" in Part 3). In general, the mangrove ecosystem is fairly resistant to many kinds of environmental perturbations and stresses. However, mangrove species are sensitive to excessive siltation or sedimentation, stagnation, surface-water impoundment, and major oil spills (Figure 2.45). Among other things, these pollutants reduce the uptake of oxygen for respiration, which results in rapid mangrove mortality.

Salinities high enough to kill mangroves (+90 ppt) result from reductions in the freshwater inflow and alterations in flushing patterns from dams, dredging, and bulkheading, which raise salinities. Seawalls and coastal structures often restrict tidal flow, also killing mangroves (316). On the other hand, mangrove forests help maintain coastal water quality by extracting chemical pollutants from the water.

Table 2.1. Issues Reported in an Extensive Analysis of 1,044 Literature Citations (Mostly for the USA).

Development concerns		Environment concerns	
Recreation/tourism	140	Coastal resources	118
Petroleum industry	65	Wetlands	103
Energy facility siting	65	Fish/fisheries	85
Facility siting	62	Public access	82
Ports/waterfronts	58	Critical areas	48
Energy	45	Vegetation	47
Transportation	43	Marine biology	46
Industry/commerce	42	Wildlife	46
Population	29	Water quality	44
Dredging	29	Beaches/dunes	36
Marinas	28	Water resources	35
Boating	25	Estuaries	34
Growth	22	Aesthetics	28
Shipping	18	Resource protection	22
Agriculture	18	Living marine resources	21
Water devel project	14	Cultural/historical	12
Navigation	14	Groundwater	10
Oil and gas	14	Barrier islands	8
Marine mining	9	Open space	7
Coal transport/storage	9	Preservation	5
Energy transport/storage	9	Coral reefs	4
Forestry/logging	7	Air resources	3
Waterfront development	7		
Aquaculture	6		
Total	778	Total	844

Source: Reference 164.

A major problem that affects mangrove habitats is the profitability of converting mangrove areas to residential, commercial, industrial, and agricultural real estate by diking, draining, or land filling. In addition, there is an increasing demand for forest wood products that results in the exploitative clear-felling of the forests. In these situations, the basic habitat and its functions are lost, and that loss is frequently greater than the value of the substituted activity on a long-term basis. In general, destructive development activities result from ignorance of the values of the functioning mangrove system and the absence of integrated planning that takes these functions and values into account (Figure 2.46).

A most difficult conflict facing many countries today is conversion of mangrove areas to aquaculture ponds (shrimps, milkfish, etc.). For example, in the Philippines between 1967 and 1977, aquaculture facilities accounted for 80 percent of the loss of mangrove (391). In Ecuador, mangrove loss for ponds is implicated in the crisis in shrimp production ("the post-larvae crisis") of the early 1980s. As stated by Snedaker *et al.* (315): "To the extent that mangroves are being destroyed . . . there is loss of marine seafood production."

Natural systems, like mangrove forests, should be managed as an investment, where the "interest" earned by the investment is analogous to the sustained productivity of the system. The basic principle of sustained management is to live off the interest, rather than reducing the capital. Some extensive, fast-growing, and highly productive mangrove forests (investments returning high rates of "interest") can be sustainably managed for multiple uses (e.g., charcoal or timber, as well as supplying fish and wildlife to nearby human populations) (Figure 2.47). This management strategy has the advantage that it retains the maximum use options open for the future.

Countries in which commercial use of mangrove timber resources occurs must require that these activities be conducted on a sustainable basis, based on detailed resource assessments and management plans. They must also equip themselves to conduct inspections, control the logging operation, and monitor natural regeneration. It is desirable to provide areas managed for wood, fuelwood, or charcoal extraction with adequate buffer areas to minimize the disturbance to associated animal populations.

It is important to recognize that many of the forces which detrimentally alter mangroves have their origins outside the mangrove ecosystem. Therefore, programs for mangrove conservation and utilization require coordination with entities that control coastal waters and inland areas via integrated planning.

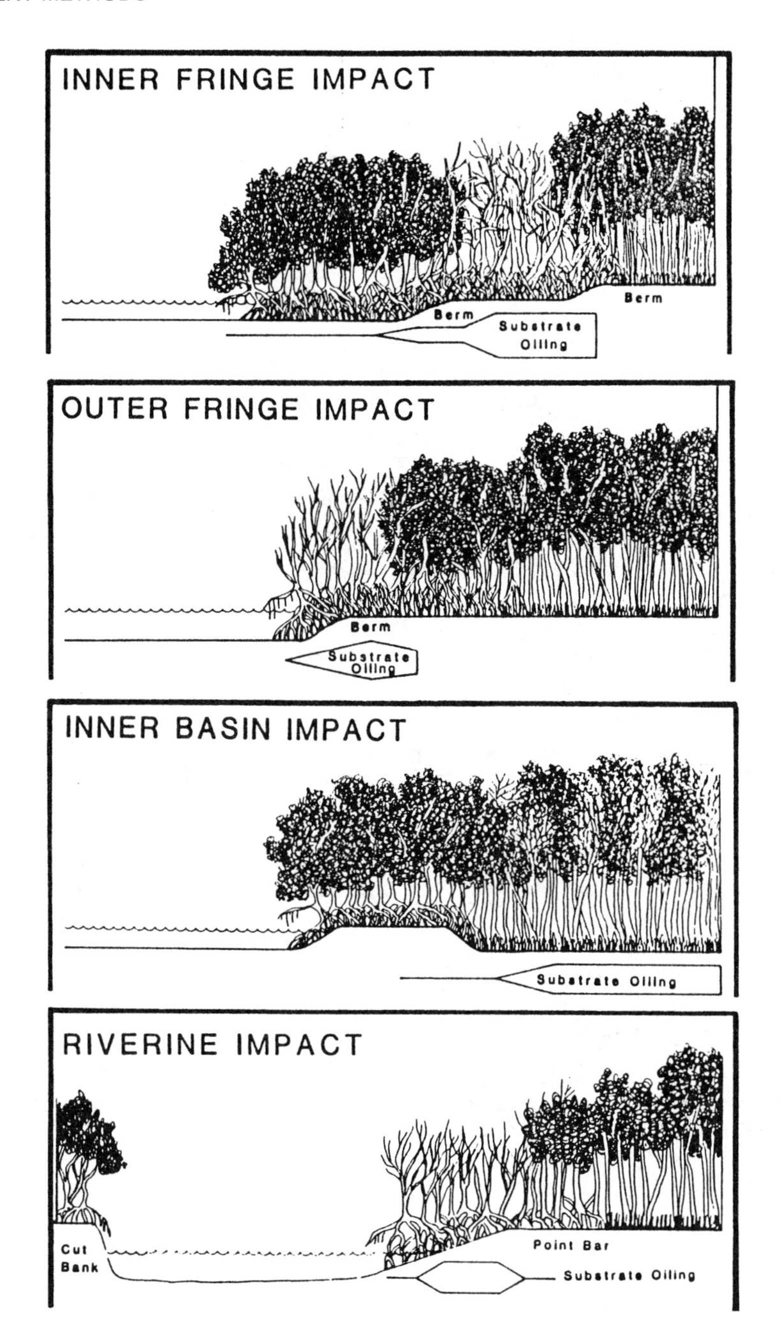

Figure 2.45. Potential effects of oil spills on various mangrove forest configurations. (*Source:* Reference 316.)

One solution is to ensure that proposed developments that could affect the mangrove ecosystem are assessed and approved by the agency operating an ICZM-type program (316).

To develop effective management plans for mangrove resources, it is necessary to relate them to management of the adjoining tidal lands and estuarine waters. Mangroves must be viewed as a part of a complex system of interrelated habitat and dependent biota which, in turn, is maintained by natural drainage patterns and rates of freshwater discharge from the catchment on the one hand and the natural tidal and salinity regimes on the other. It is the natural movement of water that provides the essential linkage of the terrestrial and aquatic elements in these coastal ecosystems (180).

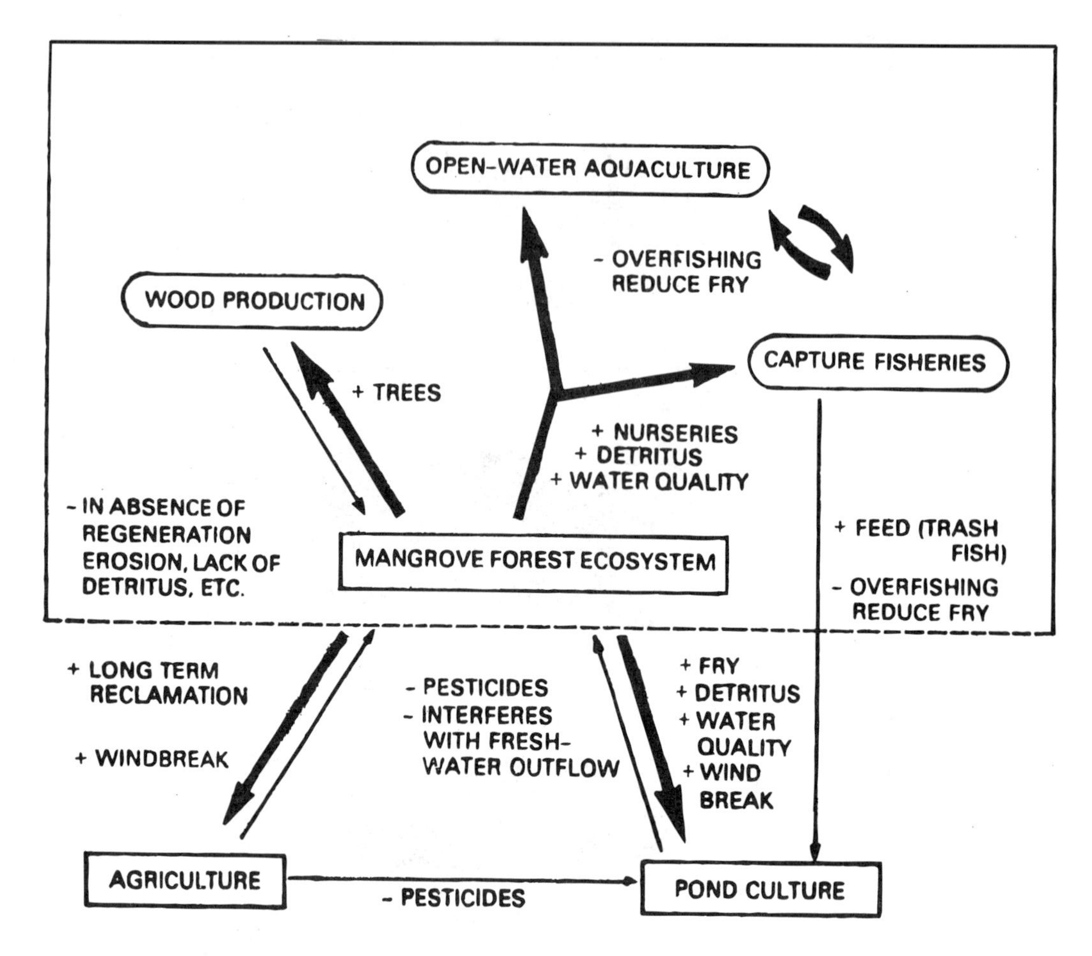

Figure 2.46. Ecological interactions with various sources of adverse environmental impacts that affect the mangrove forest ecosystem. (*Source:* Reference 126.)

Careful planning and management is essential (28). Mangroves are a renewable, sustainable resource, providing, under proper management, a steady supply of foods, timber, and fuel for human use, while supporting a diverse wildlife. Mangroves can provide jobs in rural areas in agriculture, forestry, and fisheries and have the potential to be incorporated into the tourism industry. In addition, they provide many free services that would otherwise require investments in money and technology.

In theory, the annual area available for extraction on a sustainable basis is obtained by dividing the total area suitable for management by the number of years in the rotation. In practice, the management of large-scale logging operations requires a large amount of information and know-how to maintain sustainability. The harvesting system to be used, the techniques to extract the trees from the forest and the felling or thinning programs all require specialized knowledge and considerable investments (28).

Competing interests of different user groups must be reconciled during the planning process. All groups that depend on the mangrove should be involved in this process. The goal should be long-term sustainable management. Comprehensive coastal planning at a national and regional level (ICZM) is important, so that mangroves are managed along with related terrestrial and marine ecosystems, including river estuaries, submerged grass beds, and coral reefs, as stated above.

The management of renewable resources such as mangroves requires a blend of strategies. The designation of mangrove reserves is one such strategy. If well administered, these areas can become important sources of valuable species (shrimp, fish, birds) that can then migrate, become established, and restock depleted areas (Figure 2.48). Another advantage is that they can be integrated into the

Figure 2.47. Charcoal production in the mangroves of St. Lucia, West Indies. (Photo by author.)

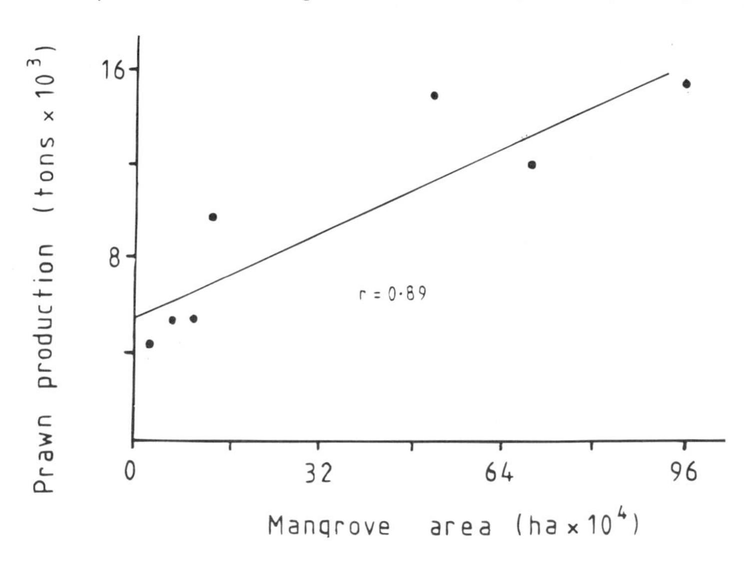

Figure 2.48. Studies of shrimp production, like this one for Indonesia, usually show higher catches of shrimp where mangroves are in greater abundance. (*Source:* Reference 101.)

economy through well-planned tourism as well as instruments for education and the development of a well-informed, concerned citizenry (28).

Community education may be necessary to counter a low regard for mangroves. Lack of awareness about the value of mangrove resources has lead to their rapid destruction. Policy makers, developers, and the public have generally not understood the need for or the urgency of protecting these resources. The attitude that the benefits derived from their protection are minimal when compared to the benefits of the activities that degrade or destroy these ecosystems must be changed (See "Awareness" above).

Loss of a major community mangrove forest resource to an aquaculture pond is an example of an unwise use.

It would be useful for each tropical country to conduct an assessment of the role of mangroves in the national economy, with as much site-specific detail as possible. Data items should include the following (156):

- Direct products (e.g., forest production)
- Indirect products (e.g., fish shellfish, honey)
- Economic activities amenable to sustained yield management (e.g., recreation, flood control, education); does not include conversion uses; may or may not include aquaculture, depending on how it is done
- Numbers of persons involved
- Numbers of persons dependent on benefits
- Mangrove extent and type involved
- Ability to atabilize coastal shorelines
- Buffer against stormtide surges
- Potential substitute areas for any destructive activities
- Economic importance regionally and nationally
- Social importance regionally and nationally
- Market trends for mangrove products
- Current status of the area in present use (e.g., sustainable, presently not sustainable, threatened by conversion activities)
- Need for aforestation or restoration of new areas

The management of mangrove resources to optimize their yield through multiple use and sustainable unilization includes restoration of areas that have been degraded through over-exploitation or lack of proper management. Because mangroves have an exceptional capacity to recover from traumatic events, given the removal of the agent causing the stress, they can often recover naturally. Usually no special knowledge or high technologies are required. What is needed is to protect the areas from the agents causing the stress, for instance by limiting charcoal and firewood concessions, or by controlling pollutants (28).

In some cases, it may be desirable to speed up natural processes by replanting of mangroves. As an example in Bali, Indonesia—which has less than 2,000 ha of mangrove (335)—the Forestry Department began to phase out leases on 297.5 ha of "tambak" (brackish shrimp ponds) in 1990, requiring that they should be replanted with mangrove. The phase-out plan was for 215.7 ha in 1991, 115.7 ha in 1992, and zero in 1993 (338).

At a much larger scale, Bangladesh has already planted more than 120,000 ha of mangrove trees (303) along its coastline in a long-term program. The primary purpose of this "greenbelt" is attenuating storm surge from giant cyclones to protect life and property (see "Bangladesh" case study in Part 4). Other purposes were to stabilize sediments and "build' land as well as gain the products and conservation benefits of mangroves (see "Mangrove Resources" in Part 3).

Nursery, planting, and maintenance routines for mangrove forest replanting are well developed and widely known (Figure 2.49). However, according to Bossi and Cintron (28), a word of caution is in order. Sometimes people try to plant trees where none were ever found. For mangroves this usually is a waste of effort, since the same forces that eliminated nature's plantings may do away with those planted artificially unless the environment can be adjusted or special conditions prevail—for example, in Bangladesh 100,000 ha have been planted in the newly accreted silt banks of the Ganges/Brahmaputra delta.

When nature does its plantings, it disseminates huge numbers of propagules over broad areas and through time. Only those which reach the proper locations develop into mangroves. It is very difficult to mimic this natural process. Usually the best strategy is to prepare the area for restoration by eliminating problems such as pollutant discharges and fill material and letting nature do its own planting; for example, in Puerto Rico, a large urban renewal project cleared a slum along a mangrove channel in the San Juan metropolitan area. Within a short time after the shacks and pilings were removed, mangroves had naturally colonized the banks of this channel. A planting of this scale would have cost an enormous sum of money. It was done free by nature (28).

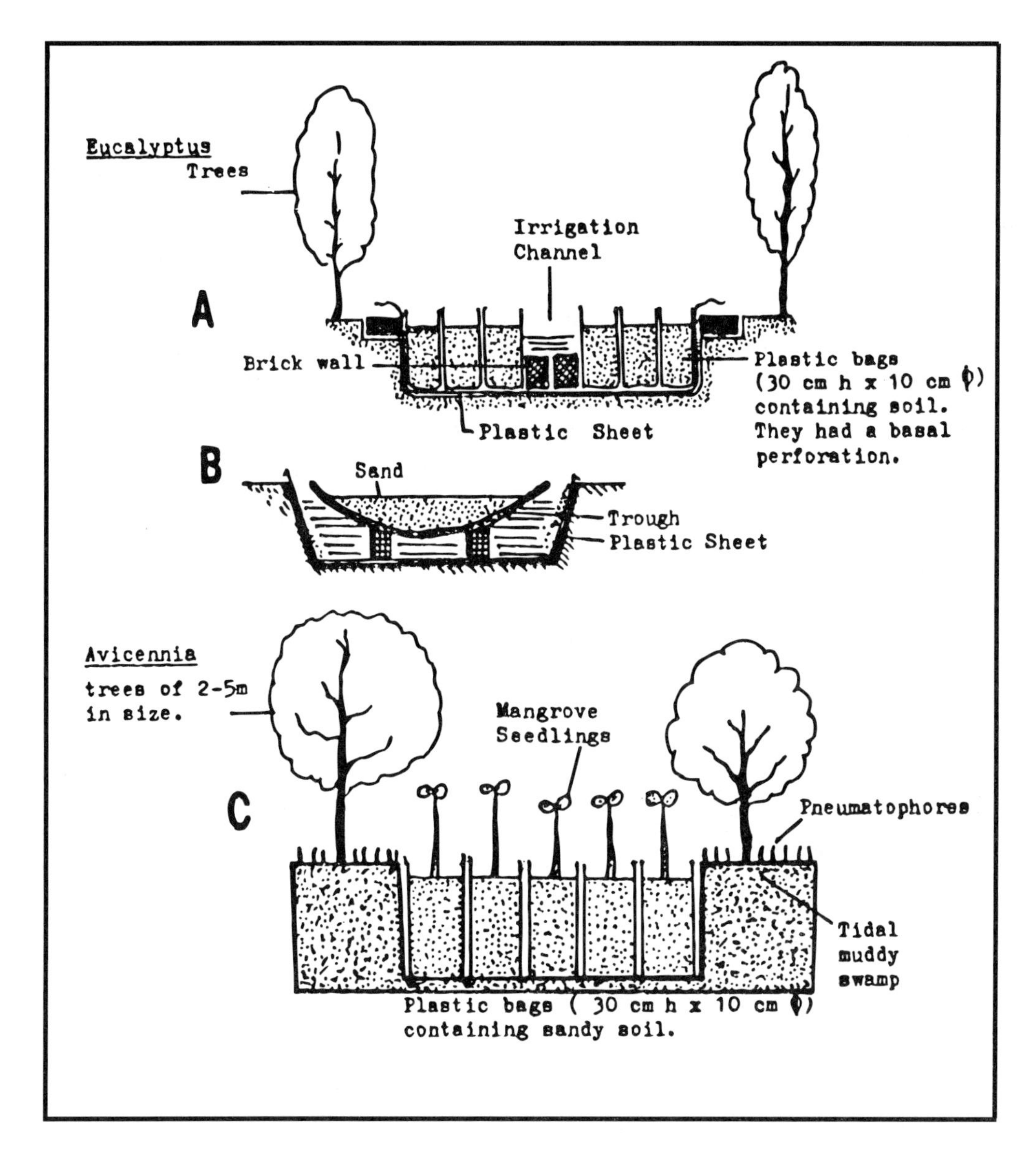

Figure 2.49. Mangrove planting system used in Pakistan. (*Source:* Reference 289.)

Ideally, a plan for managing a country's mangrove resources should be part of an integrated national coastal zone management plan, according to Hamilton and Snedaker (156). It is recognized, however, that coastal zone management planning has not been undertaken in many countries. Such planning is difficult because it involves many interests (often extremely conflicting), resources, government agencies, and political jurisdictions. If mangrove planning embraced the adjacent, linked, living resource systems such as coral reefs, seagrass beds, and lagoons, such a plan would be especially appropriate (156).

Using the Philippines as an example, there are major problems concerning mangroves that might be addressed by an ICZM-type program (234):

- There is conflict between ecological function and economic resource.
- From 1952 through 1975, the area used for aquaculture has increased from 88,681 hectares to 176,032 hectares. The increasing trend of developing the mangrove area into fishponds may have negative impacts on the coastal ecosystem.

- Thousands of hectares of mangrove areas are being declared for such uses as land reclamation, human settlements, and other commercial and industrial sites and purposes.

Under the typical conditions of the tropical intertidal zone, mangroves exhibit abilities to colonize suitable habitats rapidly, to develop complex forest structures, and to be highly productive. However, they are extremely sensitive to factors that alter the prevailing hydroperiod, salinity regime, and physical/chemical properties of the substrate. Conservation of the ecosystem and its resources can be achieved most simply by preventing any significant change from occurring in these factors. Guidelines for resource sustainability are (316):

1. Maintain the topography and character of the forest substrate and water channels. Processes that might lead to excessive sedimentation, erosion, or alterations to the chemical characteristics (such as fertility) should be avoided.
2. Perpetuate the natural patterns and cycles of tidal activity and freshwater runoff. Coastal structures should be designed to ensure that those patterns are maintained.
3. Maintain the natural temporal and spatial patterns of surface and groundwater salinity. Reductions of fresh water by diversion, withdrawal, or groundwater pumping should not be undertaken if they affect the salinity balance of the coastal environment.
4. Maintain the natural equilibrium between accretion, erosion, and sedimentation. Coastal activities, which include construction, have the potential to alter the balance between accretion and erosion. Such activities should be designed for minimum impact on the mangrove environment.
5. Set maximum limits on mangrove harvesting equal to the sustainable annual production. (The current tendency is toward maximum harvesting for maximum short-term profit.) Cutting schedules should be based on rigorous plans that ensure sustained yields and perpetuation of the ecosystem.
6. In areas subjected to spills of polluting materials, develop contingency plans to protect the mangroves from the damaging effects of oil and other hazardous materials.
7. Avoid all activities that would result in the impoundment of mangroves. The cessation of surface-water circulation leads inevitably to the mortality of trees.

2.21 MASTER PLAN

There are several types of plans involved with coastal zone management: three types are discussed below:

1. **Strategy Plan.** Strategy planning is the first step. It explores options and develops an optimum strategy for a management program. It examines facts, considers issues, suggests possible solotions, and proposes specific legal and institutional arrangements.
2. **Master Plan.** This is the complete plan for a coastal management program, either national or fractional, as in Situation Management, and is the type of plan *discussed here.*
3. **Environmental Management Plan.** This is a plan for a specific project intended to minimize or eliminate adverse environmental impacts (see Section 2.14 above).
4. **Protected Area Management Plan.** This is the plan developed for management of a nature reserve, national park, or other type of protected area (see Section 2.29 below).

The strength of an integrated coastal zone management program comes from its Master Plan (see "Program Development," Section 1.6). The Master Plan is the operative basis for the ICZM. Its components reflect the need for the program, the method and legislative basis, rule making, implementation strategy, staffing and financing, public participation, permits and environmental assessment, and other relevant aspects.

The Master Plan has a fixed time frame, of course, to which the strategy of its creation must be adapted. And it must be recognized that each version of the plan is only an interim version, since the plan basis will be continuously changing and therefore in constant need of updating.

A Master Plan should do the following (see also Section 1.6.2):

1. Indicate options for human progress in the coastal area; for example, recommend governmental and private actions needed to accomplish beneficial and sustainable change, change that is economically sound and socially just and that maintains the natural resource base.
2. List a complete set of objectives as its foundation along with updated and refined policies underlying the approved ICZM program.
3. Identify the permissible type of uses of coastal resources and the constraints on these uses.
4. State the procedure for permit approval processes, for environmental assessment (EA), and socio-economic impact analysis (SEIA).
5. Specify the procedures for ensuring public participation, multi-sectoral integration, and inter-agency coordination.
6. Detail the procedures for monitoring of activities and enforcement of compliance.
7. Authorize an institutional arrangement for managing the program.
8. Provide a detailed representation of the coastal zone and an inventory of the resources, with an identification of critical areas in need of special conservation attention.
9. Identify and map areas that offer the best potential for resource compatible development and provide the authorization and technical basis for land use zoning.
10. Provide authority for identifying and protecting natural areas (parks and reserves).
11. Authorize continuing external assistance functions, including information gathering and processing; preparation and updating of development guidelines; and consultative activities (for NGOs, developers, communities, conservation interests, etc.).

Some specific items that may be appropriate for a Master Plan are:

- Major issues
- Goals, objectives
- Resources inventory
- Indicated solutions
- Coastal zone boundaries
- Existing government interventions
- Past experience with integrated approaches
- Authorities: power sharing for an integrated strategy
- Management mechanisms: permits, environmental assessments
- Institutional arrangements: empowerment, linkages
- Guidelines and standards for coastal development
- Land use suitability and project compatability
- Wherewithal, budgets, external support
- Staffing, training, motivation
- International aspects
- Public participation
- Information needs

2.22 MAPPING

Geographic data in the form of aerial photographs, other airborne imagery, and maps are among the most important information for both project-specific or long-range environmental planning. Such information collected at the same place but at different times gives a historic record of environmental change, or stability, according to Carpenter and Maragos (41). Geographic data provide the basis to calculate the area of mappable resources and impacts; to help site settlements, roads, ports, and agricultural areas; to formulate land use and coastal zone plans; and to facilitate impact analysis using overlay techniques.

Photographs and maps are easily read and interpreted and transcend language and cultural barriers to communication and analysis. Aerial photographs can also be used to pre-plan field surveys and sampling strategies to reduce cost, improve efficiency, and ensure adequate sampling of all relevant habitats and environments (41).

Types of information that have been found to be the most useful include basic geographic mapping at appropriate scale—generally 1:50,000 or larger. It is also more effective if the gathering of information is based on objectives set in advance. These objectives may reflect areas of known management concern or suspected resource importance or sensitivity. The single most important map parameter is topography using appropriate elevation contour intervals.

Most modern large-scale maps (those that show considerable detail within small areas) now rely in part on aerial photography. Photographic interpretation and ground-truthing (spot checks in the field to verify photo-interpretations) are important components of mapping. Great strides in both map making and aerial photography were made during the World War II era as part of military intelligence gathering (41). The standard aerial cameras of the time (the Wild RC-10 and 11 with 9″ × 9″ negative format, for example) still provide among the best quality photographs available, and capabilities have been enhanced through decades of improvement in film quality and processing. Film is readily available in black and white, visible color, near infrared color, print negatives, and positive transparency types (41).

Production of detailed accurate maps generally follows the steps listed below (41):

- Establishing ground control points
- Planning and collecting aerial photography
- Correction and calibration of the photographs
- Transfer of imagery data to base maps
- Addition of coordinates and scales to maps
- Addition of thematic (subject matter) information, usually on separate film overlays
- Assembling all overlays to produce the composite print negatives
- Printing and publication of maps

Because of the number of steps and sophisticated procedures and equipment involved, map making is often slow and expensive if conventional procedures are followed. Recent advances in remote sensing (satellite and aerial photographic images) and computer graphics and digital cartography have served to accelerate some forms of map making. Collection of the base information (remote images) is a significant benefit itself because of the many uses of the imagery and can serve as the important first step. Preparation of theme maps can be deferred, as the second step, until funds and capability become available (41).

A special "coastal atlas" for any country would be most useful, one that meets the criteria outlined above and, in addition, includes a reproducible set of maps prepared on a common scale. In some cases, the map may represent the final output of the data base. In other cases, preparation of a series of descriptive and interpretive maps may be part of the analytic effort. For example, an initial round of maps might be prepared to delineate biological, geographic and land use features on a stretch of coast. Next, a second round of maps may be prepared, including all natural hazards and indicating levels of risk for new development from storms, earthquakes, landslides, erosion, etc. (326).

The same approach could be used to combine maps of shellfish beds, wetlands, and endangered species habitats into a single map of sensitive biological resources, according to Sorenson and McCreary (326). The resulting maps would give planners and policy makers tools to guide the type and intensity of new development or to choose priority areas for protection or acquisition.

Like the strategies of impact assessment and acquisition, a coastal atlas can yield valuable educational benefits. The educational benefits are derived not only from the product, but also from the compilation process. This is especially true if an open process is used involving all relevant government agencies and non-governmental organizations. If the product is presented in a clear, attractive format, maps of the coastal zone can also help convey the need for regulation, acquisition, or capital investment. This in turn can help generate support for coastal management policies among citizens, interest grfoups, agency personnel, and elected officials (326).

"Overlay mapping" is especially useful in resource conservation programs, whereby multiple theme maps are used to spatially analyze environmental components, to derive new parameters, or to select "least impact" alternatives. The method was originated by Ian McHarg as portrayed in his classic 1969 book, *Design with Nature* (241).

In this technique, a typical base map is prepared at an appropriate scale (88) and transparent overlay maps are prepared for each of the environmental components or attributes to be compared or analyzed. For example, transparent maps of the terrain, slope, and soil can be overlaid and placed on the base map to identify areas suitable for development under various constraints, for example, development easily done, done with some constraints, done with difficulty, or not feasible. Any other mappable information could also be overlain, such as soil erosion susceptibility (soil type, thickness). (41,88)

One can overlay more environmental components and/or development constraints/attributes as required (e.g., critical habitat, endangered or rare plant communities, historic buildings) until satisfied that all essential aspects have been covered and an optimum, least impact development scheme (or set of alternatives) has been identified (88).

The result is a graphic representation of the type, area, and location of impact. It permits a better than educated guess about the intensity and risk of impact that may occur. It does not make the final decision but provides additional information in a readily understood medium so that the decision can be made with a greater degree of confidence. (88)

2.23 MITIGATION

Impact mitigation is the term used for counter-measures employed to prevent or minimize environmental damage from development projects. These counter-measures are often usually required by environmental laws and regulations enacted by Governments. Mitigation needs are often identified through the environmental assessment (EA) process (see "Environmental Impact Assessment" above).

Mitigation is a mechanism which is, in effect, offered by developers, or required by regulators, as a *quid pro quo* for obtaining waterfront development permits. This approach is only successful when the mitigation process under the permit program is well organized. Long-term future goals should be established, priority types of mitigation decided in advance, and guidelines formulated for developers. In urban settings using mitigation to accomplish restoration may be encouraged because often no other mechanisms are available for repair of damaged habitats. (80).

The term *mitigation* is used in many different and often conflicting or ambiguous ways. The Federal Code of the U.S. Government (44) has defined mitigation as follows (abridged by author):

- Avoiding an adverse impact altogether by not taking certain development actions
- Minimizing impacts by limiting the degree or magnitude of the development project
- Rectifying the impact by repairing, rehabilitating, or restoring the affected environment
- Reducing or eliminating the impact over time by conservation actions during the life of the project
- Compensating for the impact by replacing or providing substitute resources or environments

Mitigation may take any of several forms, such as the following (73):

1. ***Enhancement*** is a form of mitigation that simply implies improvement of an ecosystem; for example, enhancement would be improving or restoring water circulation, plant growth, or a species habitat.
2. ***Minimization*** (or reduction) of impact is a form of mitigation that seeks to reduce adverse impacts to the minimum; for example, minimization would be preventing the spread of silt in dredging or not bulldozing in breeding time to avoid disturbing an adjacent eagle nest.
3. ***Compensation*** is a *quid pro quo* form of mitigation that implies the tradeoff of an unavoidable ecological loss for an ecological improvement; for example, the enhancement and dedication of a piece of upland game habitat as a tradeoff for some riparian habitat lost to a reservoir.
4. ***Replacement*** is a *quid pro quo* exchange of a particular resource for another of the same type; for example, 10 acres of new *Spartina* marsh built on dredge spoil to replace 10 acres lost to marina development.
5. ***Indemnification*** is a *quid pro quo* form of mitigation that implies a monetary recompense for loss of ecological resources; for example, the payment to a public agency of a million dollars in cash for damages to 10 acres of urban wetland converted to housing sites.

Minimization is often the simplest approach. One key to minimizing impacts is the timing of projects (99). As an example, for beachfill projects, warmer months yield higher impacts than winter months because (a) species density and diversity are higher, (b) certain species may be nesting, and (c) warmer waters have less capacity to hold dissolved oxygen (South Carolina, USA). Therefore, if the project construction can be avoided in certain months, impacts on endangered species could be greatly reduced or eliminated (99). For example, where beach nesting of turtles occurs from May to September and least terns from March to June, construction should be planned for October to February.

The preferred *location* of mitigation is usually on the devloment site itself; that is to say, all damage caused by a project whould be mitigated within the development site or project area where damage occurs. The preferred *mode* of mitigation is direct rather than substitute. That is to say, the same functions are to be *directly* restored, replaced, or compensated as those that are lost, on site if possible. (73).

However, two factors—motivation for intensive use of waterfront land and the limited options for onsite mitigation—create strong pressure to find *offsite* locations for mitigation. One solution is the "mitigation bank" whereby developers are "taxed" for impacts and the proceeds "deposited" in a habitat creation/restoration account. A second solution is a different kind of bank whereby areas in need of habitat restoration or suitable for habitat creation are set aside ("banked") so as to be available in the future for mitigation requirements levied against developers. A third solution is a regional cooperative restoration plan, whereby mitigation for various projects is done at pre-designated sites. (80)

2.24 MONITORING AND BASELINE

Two common types of information used in environmental management are (a) baseline information that measures the environmental conditions and status of resources before a project is commenced and (b) monitoring information that measures the changes, if any, that occurred after the project was built and operated. The statistical reliability of parameters used in baseline surveys and monitoring programs is a key factor.

The goal for a quantitative monitoring program is to detect, with statistical reliability, whether a significant environmental change has occurred after intervention. Environmental impact predictions depend on understanding cause-effect relationships and the status and trends of environmental characteristics. Baseline studies establish the current state of ecosystems (41). Ecological baseline studies provide valuable sources of information for planning.

Cognizance should be taken of existing information before embarking on new baseline inventories. Most locations already have at hand some historical data on climate and weather, soils, vegetation, and land use which will be helpful, initially. An inventory of existing inventories should be made first, and judgment should be made of the values of each, their accuracy, scope, applicability to present problems, and how the data can fit into the overall scheme. Appropriate agencies should be consulted, as well as universities and private institutions. Only then should new inventories be started. Areas that are already intensely managed (e.g., agriculture, forestry, or fisheries) merit less study than those that are sensitive to degradation, unique, or as yet undeveloped. (41)

Although the baseline study can be as simple as recording the location and abundance of a species so subsequent monitoring can observe the changes in these parameters, usually it is a more comprehensive effort which entails recording various ecosystem characteristics in the attempt to learn why changes occur. Some of the obvious ecosystem parameters to be measured (for both the "wet side" and the "dry side" of the coastal zone) include weather and climatic factors; vegetation information such as plant species, distributions, abundance, and utilization by animals; soil characteristics such as compaction, fertility loss, or increased erosion potential; and changes in the distribution, quantity, and quality of water.

Environmental assessment (EA) procedures often include a requirement to review completed projects and judge the predictions and recommendations made in the EA against later experience with the project. The purposes are (a) to determine whether consequences were accurately predicted and to identify additional significant effects warranting corrective action and (b) to use the results to refine the impact predictions for future projects of the same type and magnitude. But the findings are often so imprecise and vague that their accuracy cannot really be ascertained. Not surprisingly, physiographic information is usually more complete and precise than is biological information.

All development projects should be managed with the expectation of surprising outcomes and the necessity to adapt and change implementation actions if the goals are to be met. Construction monitoring provides an early warning that adverse impacts are occurring (see "real-time" monitoring in "Environmental Impact Assessment" above). (41).

There are two major types of monitoring—strategic and tactical—as follows:

1. The strategic level is the "retrospective" or "hindsight" monitoring, which is often the main focus of the monitoring program. The object is to compare measurements of certain key characteristics of the environment both before and after project work is done so that an "environmental audit" can be made of the project's effects. This of course requires that "benchmark" information be collected before the project starts, to provide a statistically sound "baseline."
2. The tactical level is the "real-time" or "oversight" monitoring that is done in conjunction with the environmental management program (EMP). The object is to monitor the construction operation day by day so as to detect any major negative impacts that may be occurring. If problems are detected, construction is halted until the situation is remedied.

Detailed research is usually not possible for every area, but transfer of findings from a well-studied site to another one of interest that has not been studied may provide a reasonable beginning. Therefore, it will be important to use standard methods. For example, it would be extremely difficult to compare benthic studies made by different people if a great variety of sampling devices and techniques is used. Species diversity indices are such useful tools for measuring the effects of water quality on aquatic plant and animal communities that some standardization is necessary. (41)

In baseline and monitoring activity, there can be either Type I and Type II errors, according to Carpenter and Maragos (41). A Type I error is to conclude that a change has occurred when nothing has changed, while a Type II error is to conclude that nothing has changed when change has in fact occurred. A Type I error results in considering, or taking, corrective action when none is needed; an error in favor of the environment, but costly. A Type II error leads to failure to act when corrective action may be needed.

Typical measurement errors for physical and chemical parameters are within 25 percent of mean observed values, while measurement errors associated with observations of living organisms are 50 percent or more of mean observed values. Natural variations due to temporal or spatial factors appear to be much greater than collection or analytical errors. Typical values of natural variations range from 100 to 400 percent of observed mean values for both physical and biotic parameters. (41).

Using water quality baseline data as an example, the cost to *obtain* a sample of water may be low, but the cost to perform analyses on all possible parameters can reach several hundreds of dollars. Typical analytical costs (in $US) are (41):

Analyses requiring extensive digestion or wet chemistry such as BOD, COD, nutrients—$10–15/parameter

Trace metals (including preparation of metal for analysis)—$50 each

Suspended solids, volatile solids—$10–15/parameter

Coliforms—$10–15/sample

Oxygen, temperature, conductivity, pH, oxidation/reduction—$1–5/sample or $100/day to rent a probe

Trace organic compounds (gas chromatography/mass spectrometer)—$2–3,000/sample

Suite of common analyses including most of the above except trace metals and organics—$100–150/sample

Using the coral reef as an example, checking on baseline reference points is the best way of detecting changes in a reef, provided such points are situated strategically, according to Gomez and Yap (145).

Where transects or quadrats are to be laid down, they take into account the natural zonation of the reef, given its particular biological and ecological regimes, and known or probable focal points and gradients of an existing environmental stress, if any. This ensures that comparisons of before-and-after project are meaningful (see "Coral Reef Survey Methods" above).

In an example of a dredging project (see "Indonesia, Sulawesi" case history in Part 4), we found it necessary to monitor turbidity at both dredge and disposal sites to prevent excessive release of silt. The following four steps were involved:

1. It was necessary to select a parameter to measure. It was decided that the best parameter would be "light scattering" because it would instantaneously reflect the amount of suspended matter in the water, so that the environmental manager would instantly know and could immediately ask for correction of an excessive silt load.
2. Second, we needed to select an instrument. It was decided that the electronic nephalometric turbidity meter was the simplest and most reliable instrument for this purpose with a direct immediate readout in NTUs (nephalometric turbidity units), which are equivalent to JTUs (Jackson turbidity units).
3. Third, a baseline was established, based upon the normal background turbidity, or reference turbidity, that would be used for the monitoring operation. A Hach turbidity meter was used (the methods and the instrument are described in "Turbidity Measurement" below).
4. Fourth, the following control standards were established:
   - **Numerical standard:** Water turbidity shall not exceed the national standard of 8 NTU (nephalometric turbidity units; ≈ 80 mg/l s.s.) as measured by an electronic turbidity meter, in any area of the sea influenced by the dredging operations, except within the designated non-compliance "work area" and the permitted "mixing zone."
   - **Work area:** The immediate area of discharge of silt has no NTU limit, but this shall include only the area within a radius of 20 m of the point of dredging.
   - **Mixing zone standard:** Water turbidity shall not exceed 20 NTU (nephalometric turbidity units) as measured by an electronic turbidity meter, in a designated mixing zone defined as the area within 100 m from the point of dredging.

In all techniques, it will be readily seen that some amount of skill is indispensable. Meaningful monitoring work could not be carried out without a minimum of knowledge in biology and ecology. In addition, a certain degree of discipline and training is required in making the measurements and observations. Thus, some provision must be made for the training of personnel. (145)

Another approach, an intensive quadrat technique, has been developed for baseline surveys where the object is to measure possible change in the coral community because of an anticipated disturbance of the reef ecosystem. It is an intensive ("saturation") survey of relatively few 10- × 20-m quadrats, with replicates for each one, in order that the statistical reliability of the procedure can be determined. Each quadrat is precisely located so that the exact area can be sampled again in a future monitoring program to see if change is taking place and, if so, what the nature of the change may be.

As an example, at Sanur Beach (Bali, Indonesia), we selected depth contours for survey of 3 meters and 7 meters because (a) preliminary reconnaissance had indicated that 3 meters was the shallowest that a SCUBA diver could work successfully during "east monsoon" wave and current conditions and (b) 7 meters was about the lower limit of the "upper reef slope" coral community and also would be most vulnerable to project-induced sand transfer operations—the project being "beach fill" or "beach nourishment" to replace sand on Bali's tourist beaches that had been lost by careless development actions (70, 388). It is important when using SCUBA to go no deeper than necessary to minimize the rate of compressed air usage and to avoid blood nitrogen problems.

We selected two transects to represent the reef slope, and each was located by triangulation from shore with two theodolites. Along each transect, two 10 × 20 quadrats were placed, one at 3 meters (referenced to local "zero tide" datum) and one at 7 meters, plus there was a replicate for each. So with two transects, each with two sample depths, and with two quadrats at each depth, we had altogether eight quadrats to sample. Each was marked at its four corners by driving steel rod ("re-bar") into the reef (388).

As designed by R. van Woesik (388), each 10-m leg of a quadrat was marked at 0.5-m intervals and half of these were chosen (at random) for transects. In this manner, 10 transect lines of 20 m were selected. A fiberglass measuring tape was laid down over each transect while a diver (with coral identification expertise) proceeded on the transect recording and measuring all objects at the bottom (on waterproof paper on a clipboard)—live coral (hard and soft) by species, sand, dead coral, bare carbonate, and so forth (224). By this means, 200 m of transect were covered for each quadrat, or 1600 m in total. One-half day was required for setting up and surveying each quadrat (with a team of two divers and a boat operator) (388). Where species identification is difficult in this type of sight-recognition survey, identification of coral to genera rather than species might be adequate. (See Case History 4.2 in Part 4).

The basic output of the Bali baseline survey was a linear record, of course—divers recorded so many centimeters of this type and so many of that type. These records are readily converted into such measures as percent live cover (a most useful index) or dominant species of coral. Photographs were taken for verification, and samples of "problem" species were brought ashore for close scrutiny and identification (388). Much incidental information was recorded. The essence of the Bali survey was to produce an objectively defensible and statistically definable baseline record. It was that, but it also was complex and expensive (70).

2.25 NUTRIENTS MANAGEMENT

Nutrients are the dissolved organic and inorganic substances that support all plant life in the sea. There are many types of nutrient of interest to marine ecologists, but to resource managers, two types are of primary concern—compounds of phosphorus (P) and of nitrogen (N). The ratio of P to N to C (carbon) in aquatic plant tissues is normally 1 to 16 to 104, respectively.

In their organic state, both phosphate and nitrate are abundantly present in our wastewaters and find their way via sewer discharge, septic tank leaching, and so forth to coastal waters. In excessive quantities either type can wreak havoc with coastal waters because of *eutrophication,* the overgrowth of marine flora, from microscopic phytoplankton to seagrasses and macroalgae. Excess growth of these plants is due to the transformation of P and N from organic to mineral state (Figure 2.50). In mineral form, P and N trigger rapid blooms of algae and grasses.

Most nutrient discharged to coastal waters is utilized biologically; very little goes to waste. For example, many plants can store excess P within their cells for later use. Phosphates are typically utilized to the point where insignificant concentrations are left in the water. Phosphate amounts are often so low as to be nearly unmeasurable in the open ocean. The main users of organic phosphorus are zooplankton and bacteria (Figure 2.51).

Phosphate concentrations as low as 0.02 to 0.1 μg/l (micrograms/liter or parts per billion) can be expected in oligotrophic (low fertility) ecosystems such as coral reefs. But in estuarine and other terrestrially influenced areas, levels of 10 to over 100 times that amount would be expected. In the

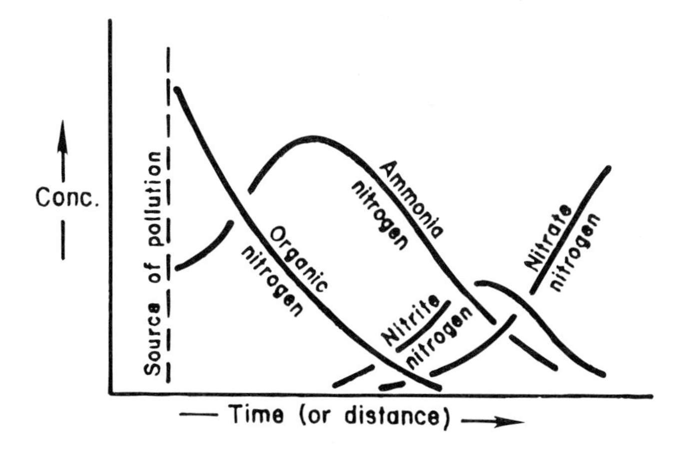

Figure 2.50. In the process of mineralization, organic forms of nitrogen are largely transformed to nitrate, which is often responsible for eutrophication of coastal waters. (*Source:* Reference 371.)

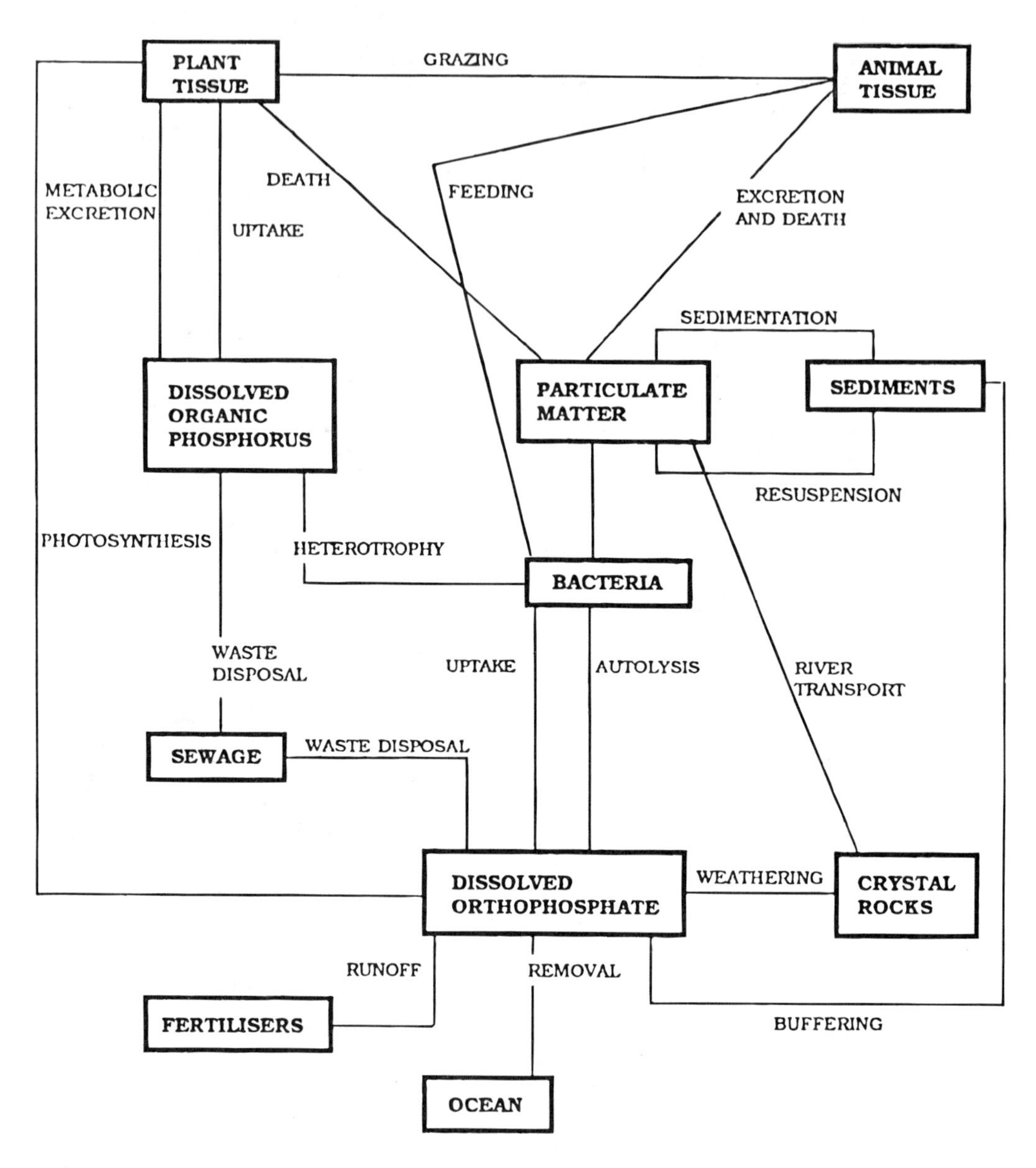

Figure 2.51. The phosphorus cycle in the estuarine environment. (*Source:* Reference 189.)

Mediterranean, for example, phosphate may range from 1 μg/l to 10 μg/l, in the eutrophic northern reaches of the Adriatic Sea (351). Nitrate is always expected in higher concentrations than phosphate, perhaps on the order of 7 times higher.

Oftentimes the result of nutrient imbalance can be seen in the dirty waters of lagoons and harbors, in fish kills, or in "red tide" blooms. But the damage can be more subtle, more insidious, such as algae disabling a coral reef. Coral reefs have naturally evolved life strategies that enable them to exist in oligitrophic waters where nutrients are scarcest but cause them trouble when nutrients become more abundant because (a) benthic algae get the upper hand and begin to overgrow the coral reef; (b) phytoplankton begins to cloud the water, preventing light from reaching many of the corals at good intensity; and (c) coral predators begin to prosper and increase (153).

While N usually triggers eutrophication, in the situation of a coral reef system or a high salinity lagoon, P may be the trigger of eutrophication (because there is usually an excess of N). Therefore, sewage effluents discharged to either of these types of ecosystems should be as free of phosphates as possible. And because sewage treatment does not remove much P, the best hope is to reduce the quantity put into wastewater ("Source Control"). To avoid serious eutrophication that fouls shallow coastal waters, we must seriously reconsider (a) our methods of commercial and residential waste disposal and (b) our enthusiasm for the "secondary" treatment plant. Eutrophication (over-fertilization) causes acceleration of plant growth to the point of nuisance and degradation caused by the presence of excessive plant nutrients in the water (see "Eutrophication" in Part 3).

Official standards for allowable amounts of P and N in coastal waters are most elusive, and many water quality criteria do not list numerical standards, the "appropriate" amount of measurable P and N being quite variable. However, some guidance is offered in "Water Quality Management: Coastal Waters," which appears below. Environmental protection agencies often recommend that effluent for release in coastal waters should contain not more than 1 mg of PO_4 per liter.

From the nutrient management perspective, *source control* is a first line of defense against eutrophication. Obviously, one wants to keep the waste materials flow as low as possible—the storm drains and sewers, particularly. Such tactics as eliminating phosphate detergents and preventing fertilizer runoff could help to reduce the ecological damage. Even if successful, this is only one part of a greater strategy.

The human body excretes about 0.5 kg (1.1 lb) of P per year and, where there are sewers or septic tanks, most goes out with the wastewater. So does other household and commercial waste, such as laundry and cleaning detergent, which contains a surprising amount of P. In countries like the United States, detergents and other phosphorous products contribute about 1.6 kg (3.5 lb) per person per year—more than three times as much P per year as human waste (366). Note that the P in detergents is *already* in the inorganic state as phosphate and represents a clear and present danger to coral reefs and other oligotrophic systems as well as fresh waters (brackish systems are more affected by N, usually. Therefore, many communities control the use of detergent and many manufacturers offer reduced P washing products. The following limits are practicable:

- Laundry detergents: maximum 6 percent phosphorus (by weight)
- Dishwasher detergents: maximum 0.5 percent phosphorus (by weight)

Source control for P in farm and garden fertilizers is more difficult because it is not practicable to reduce the amounts in the product. Consequently, users will have to be constrained against wasteful practices like using excessive amounts of fertilizer or putting fertilizer on a field when a rainstorm is likely to occur. Also detaining rain runoff (in detention basins) or routing it properly (through grassy swales) should help in natural removal and should reduce the loss of fertilizer.

Wastewater treatment takes over after all source controls have been utilized. *Preliminary* treatment is a simple step that removes large solids (those over 5 or 6 mm in size), after which the wastewater can be safely discharged through a properly located ocean outfall. *Primary* treatment plants are more thorough because they screen out the solid materials and remove grit and some organics from the wastewater stream in settling tanks. *Secondary* treatment plants, however, add rather than remove dissolved P and N, acting like *sewage refineries* (biological treatment re-mineralizes the P and N in the organics), and therefore they can actually increase eutrophication depending upon the location of discharge (Figure 2.51). Therefore, if deep flowing water is nearby it is a better plan to use preliminary (or primary) treatment and discharge the effluent at a distance via an ocean outfall where the wastewater will be diluted and carried away by currents (see "Ocean Outfall Placement" and "Sewage Management" below).

2.26 OCEAN OUTFALL PLACEMENT

Very difficult disposal problems have been solved by "ocean outfalls," the pipes that carry polluted liquids offshore for discharge at an ecologically safe distance from shore. The pipe ends as a "diffuser," usually a series of "nozzles" that are designed to disperse the liquid throughout a large volume of ambient water. (See also "Sewage Management" below and "Nutrient Management" above.)

In cases where advanced (tertiary) treatment of sewage or land disposal of the effluent (e.g., golf course or farm) is impractical, the use of *ocean outfalls* offers a sewage treatment option that, if properly designed and located, is relatively compatible with ocean ecosystems (but is not acceptable for estuarine ecosystems). Ocean outfalls must be considered on a case-by-case basis, with a wide range of geological, hydrological, and biological factors being taken into account. The use of ocean outfalls can be supported in circumstances where ocean dispersion is great enough to rapidly dilute toxic and hazardous effluent pollutants below harmful concentrations.

Marszalek (236) notes that nutrient loading and its stimulation of algal growth is a primary cause of the destruction of coastal ecosystems—specifically, coral reefs—in sewage-polluted environments. The most effective means of protecting coral reefs from sewage pollution and eutrophication is to ensure that the affluents are discharged outside of the reef system, according to Marszalek. (See "United States, Florida: Success with Ocean Cutfalls" case history in Part 4.) Diversion to deeper water seaward and downcurrent of the coral reefs is especially desirable because dilution and dispersal by large volumes of seawater neutralizes the pollution potential of the sewage.

This reflects the new approach to sewage management in coastal areas—i.e., provide preliminary (or primary) treatment only and dispose the effluent safely offshore instead of spending large amounts of money on secondary treatment plants, which become sewage "refineries" when they do work. While the intermediate, or "secondary treatment," plant may reduce the organic load of the wastewaters when it is functioning correctly, it also mineralizes organic wastes; that is, the "treatment" converts these wastes from a more benign organic state to an active inorganic state—to dissolved **phosphate** or **nitrate.** This highly refined nutrient is ready for immediate takeup by algal nuisance plants. So, the typical intermediate sewage plant functions as a **nutrient refinery.** And its discharge serves as a **delivery system** for nutrient pollution.

In this way, secondary treatment plants cause eutrophication of coastal waters and present a greater ecological danger than does primary effluent. The nutrients produced by the plants can unbalance a coastal ecosystem by stimulating excessive plant growth with serious effects—overgrowth of plankton or macroalgae, mortality of coral, extreme turbidity of water, depletion of vital dissolved oxygen, and numerous other spinoff effects.

Phosphate and nitrate nutrients in mineral form are not being produced in quantity by preliminary or primary treatment. For all these reasons, the primary (or preliminary) treatment with ocean outfall combination is becoming a popular approach in coastal areas where it is feasible—that is, where the diffuser can be placed an appropriate distance from shore and be located where depth and ocean currents are suitable (Figure 2.52). The technology for ocean outfalls is well developed, and system failure is infrequent. An ocean outfall and primary treatment plant will often be less expensive to build than a

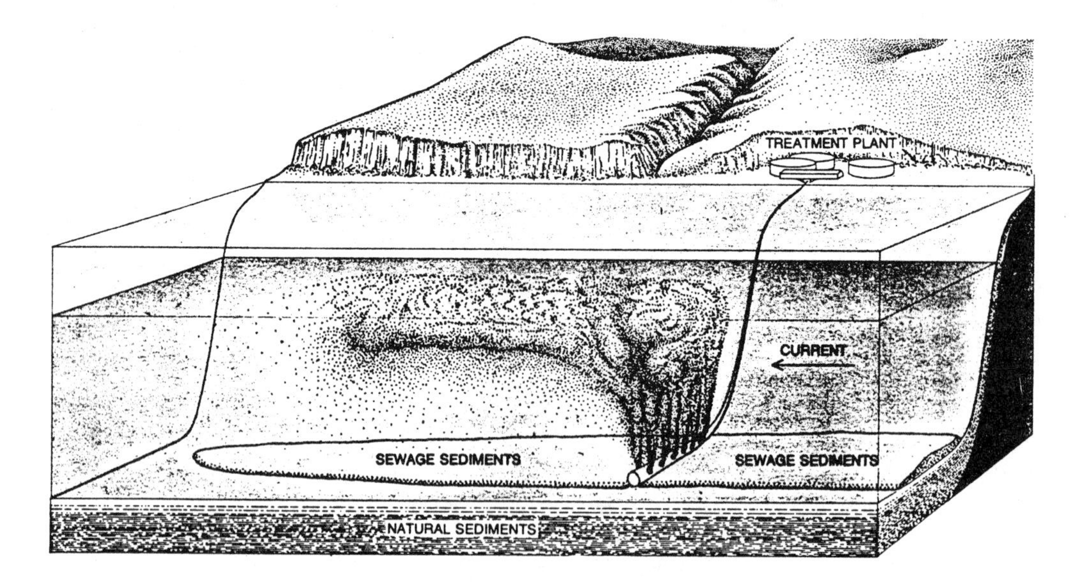

Figure 2.52. Model ocean outfall approach for the California (USA) coast. Effluent flows a distance of 2 to 5 miles through a 6- to 12-ft-diameter pipe. For the last 0.25 mile, the pipe becomes a diffuser with numerous 6-inch ports. The effluent rises to the thermocline and moves away with ocean currents. (*Source:* Reference 59.)

secondary treatment plant, and it should be much less expensive to operate over the years. A preliminary treatment system (removes materials over 6 mm diameter) has an even much lower cost. For Victoria (Canada) estimates were as follows (249):

Type system	Initial cost (millions $US)	Annual operating cost (millions $US)
Preliminary	Existing	0.65
Primary	380	37.6
Secondary	518	54.6

The success of an ocean outfall program is dependent upon swift mixing and dilution of the effluent as well as rapid transport of the water mass away from the diffuser nozzles (usually more or less vertical standpipes).

Because the adverse effects of sewage discharges are minimized by dilution with large volumes of seawater and by natural processes of purification, siting of sea outfalls should be based on the behavior and fate of sewage after discharge. The following are important: the initial dilution that can be expected, buoyancy of the effluent, speed and direction of travel after initial dilution, the rate of dispersion during travel and the rate of self-purification in the receiving waters, relation to recreational boating/fishing areas, and possible storm conditions. The amount discharged should be "tailored" to the assimilative capacity of the receiving system. Fortunately, there happen to be many stretches of coastal waters where currents are strong enough and nearshore water deep enough to provide the needed volume of water and mixing potential.

Experience with some actual projects is helpful to understanding the potential of ocean outfalls for solving coastal pollution problems (particularly those caused by excess dissolved phosphate and nitrate as well as BOD). Ocean outfalls have been utilized successfully by such well-known cities as Miami, Sydney, and Los Angeles. But they have also been designed for smaller communities, such as Larnaca (Cyprus) (352). Other cities have rejected ocean disposal for various reasons—e.g., Port Said feared return of effluent to bathing beaches, but a longer pipe and strategic location of outfall diffusers could have been able to solve that problem (353).

The Sydney Water Board began to consider deepwater submarine discharges in the early 1970s. In 1980, the Water Board adopted the deepwater scheme as its preferred strategy to upgrade the ocean sewerage system. The New South Wales Government approved the project in 1984. The resulting Malabar and North Head deepwater ocean outfalls commenced operation in September and December 1990, respectively. The Bondi outfall followed in 1992. Total cost exceeded $350 million. (175)

The three "ocean tunnels" of the Sydney system extend between 2 and 4 kilometers offshore and lie up to 80 meters below the ocean surface. The outfalls are designed so that the pipelines dispose of treated effluent through diffusers with multiple discharge points. Computer modeling of the performance of the outfalls led the Water Board to predict a high dilution of the effluent discharged into the receiving waters. It was also considered that the saltwater environment would ensure rapid biodegradation and bacterial die-off. Initial reports of the effects of the outfalls are that the water conditions at nearby beaches have improved considerably. Malabar beach is now one of the cleanest beaches along the Sydney coastline. (175)

The Kaneohe Bay (Hawaii) case shows how a coral bay (3 km × 11 km) was badly damaged by receiving secondary sewage effluent from a population of 100,000 and then restored to good condition in six years time by providing an ocean outfall (at a depth of 35 meters). There is an excellent and detailed scientific foundation for this conclusion because Kaneohe Bay was extensively studied by marine scientists. (See "United States, Hawaii: Success with Kaneohe Bay Ocean Outfall" case history in Part 4). Florida's Miami ocean outfall also proved to be an effective and efficient way to dispose of sewage without polluting the coastal environment. (See "United States, Florida: Success with Ocean Outfalls" case history in Part 4.)

All examples clearly illustrate the need for detailed study while sewage disposal systems are still in the design stage. For example, both Kaneohe Bay and Miami outfalls were incorrectly emplaced and later required modification at additional expense. In the case of Kaneohe Bay, an entirely new ocean outfall was constructed and the bay outfalls abandoned; at Miami, the existing outfall was extended to discharge into the Florida Current (Gulf Stream).

An ocean outfall for south Bali (Indonesia) at a distance of 1.75 km offshore was shown to be able to protect a very important coral reef from deterioration by algal overgrowth (see "Indonesia, Bali: Restorative Beach Management" case history in Part 4).

The Larnaca (Greece) outfall design is of particular interest because of its smaller scale and because of a higher level of pre-treatment prescribed (equivalent to secondary treatment). Larnaca is a small coastal city of 52,000 residents (on the eastern coast of Cyprus) whose economy is tourism dependent. The outfall was sited after extensive oceanographic research had enabled the planners to understand the controlling natural factors. The outfall pipe was set at 0.6 m diameter to convey 33,000 m^3/day of effluent in the year 2,000 and the BOD5 level was given as: input, 194 mg/L, output, 20 mg/L (80.5). But no levels of inorganic nutrient were given.

The Larnaca outfall location took advantage of the following natural features (352):

1. A cape (Dates Point) that shortened the length of pipeline because of the land's extension seaward
2. A trough and ridge that extended perpendicular to the coast and provided depth below grade and a "vee-bed" for the pipe
3. Deep water (50 m) less than 3 km from shore
4. A density boundary/thermocline (a horizontal confining layer marked by a sharp gradient in density and temperature) a few kilometers offshore

Because subsurface currents proved to be weak, dilution would be a major dispersal mode with less reliance on mass horizontal transport. In this situation, the arrangement of diffuser "nozzles," or "ports," becomes quite important. Usually, the question is whether to use fewer, larger diameter nozzles or more, smaller ones. For Larnaca, the options for a 0.6-m outfall pipe with an off-bottom current speed of 0.2 knots were:

Number of ports	Diameter of ports	Dilution ratio* achievable
25	100	193
11	150	115
6	200	78

* Within rising "cone of dilution" above diffuser ports.

A vital factor in the success of the project is that a last minute strategic design decision was made to extend the outfall from the initial length of 1.25 or 1.5 km to 2.8 km offshore to ensure that the diffuser would be located under the density boundary, so as to confine the effluent beneath it (Figure 2.53) and reduce potential return of effluent (via upflow and shoreward surface current). Because the density boundary was located between 25 and 45 m under the surface, after considering alternative distances (A–D in Figure 2.54) the designated distance offshore was 2.8 km and the diffusers were placed in a bottom depth of 50 m.

Clearly, numerous tradeoffs are involved and many factors are important, but it must be remembered that **the greatest care should be taken to avoid excessive dissolved nutrients** in coastal waters, and by far the best way to do so is to use an ocean outfall when possible. But planners and environmentalists may find resistance from traditional engineers and sewage system designers. But the coastal planner should not give up easily—you should go on to convince them that BOD_5 is not the only criterion for choice and that different analyses and a different set of parameters (e.g., phosphate and nitrate) may be needed to keep coastal ecosystems healthy. You may also have to convince decision makers that it is worth spending a bit more money today to save a greater amount later.

2.27 OXYGEN: BOD/COD MEASUREMENT

One of the most important measures of water quality is biological oxygen demand (BOD)—the amount of dissolved oxygen (DO) used by organisms to consume biodegradable organic materials. The standard parameter is BOD_5—i.e., a test measuring the amount of oxygen consumed at 20°C in 5 days (in a lightless chamber).

Decomposition of organic wastes consumes oxygen dissolved in water—the same oxygen that is required by fish and shellfish. In general, the higher the BOD, the worse the quality of the water. BOD_5

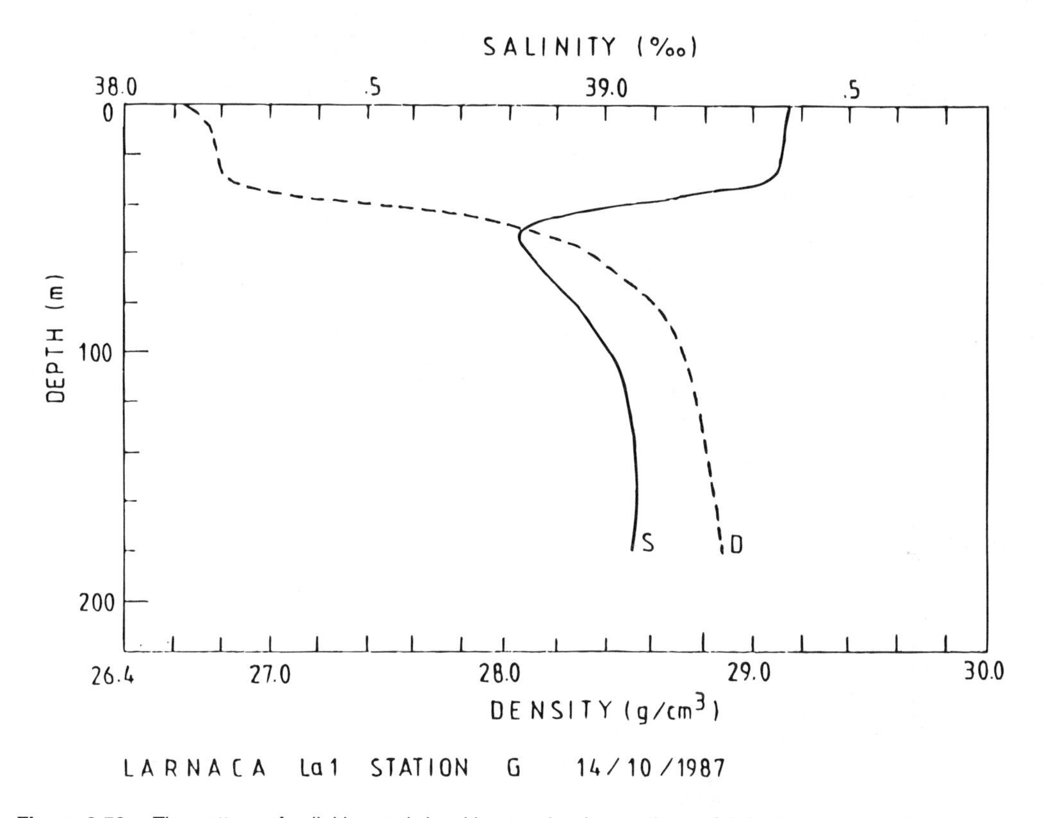

Figure 2.53. The pattern of salinities and densities used to locate the outfall for the Larnaca (Greece) sewer system. (*Source:* Reference 352.)

values greater than 5 mg/l typically indicate polluted water (59). Dissolved oxygen is typically depleted when the BOD exceeds 10 mg/l; then anaerobic bacteria take over and produce hydrogen sulfide, ammonia, and methane from the remaining organic wastes. (41)

BOD is an empirical microbiological measurement of the oxygen consumed (in mg/l) by native bacteria. Success and precision of the measurement depends upon the controlled incubation of the water sample for a period of 5 days (BOD_5) at 20°C and a suitable bacterial "seed" used to inoculate the water to begin the decomposition process.

BOD is a most important water quality parameter and therefore you should understand something of how it is done even though you may never do BOD measurement yourself (371):

1. Two bottles are filled with the water: No. 1 is used to obtain the existing (base) DO reading (say, 7.0 mg/l) and No. 2 (a black, lightless, bottle) is used for the test.
2. No. 2 is kept in the dark (so that photosynthesis cannot operate) for exactly five days at 20°C (hence, BOD_5).
3. After 5 days its remaining DO is measured (say, 3.0 mg/l).
4. Subtracting No. 2 (black test bottle) reading from No. 1 (base DO bottle) reading gives us a BOD_5 of 4.0 mg/l.

The amount of DO that disappeared in five days (4.0 mg/l) was consumed by the bacteria that have been feeding on the organic matter in the water (in the dark) for the five days (and producing CO_2). If only a little DO is consumed, we assume that there was only a little organic matter in the water. Alternatively, if more is consumed, there must have been more organic matter in the water.

But what would have happened above if there were much more suspended organic matter in the sample? What if say, there was enough for 40.0 mg/l BOD_5 in the test sample, instead of only 3.0 and still a base of 7.0 mg/l DO? Clearly the 7.0 mg/l DO wouldn't last five days and, in fact, might go to zero in about one day, abruptly ending the test. So that test would become a failure as a five-day test.

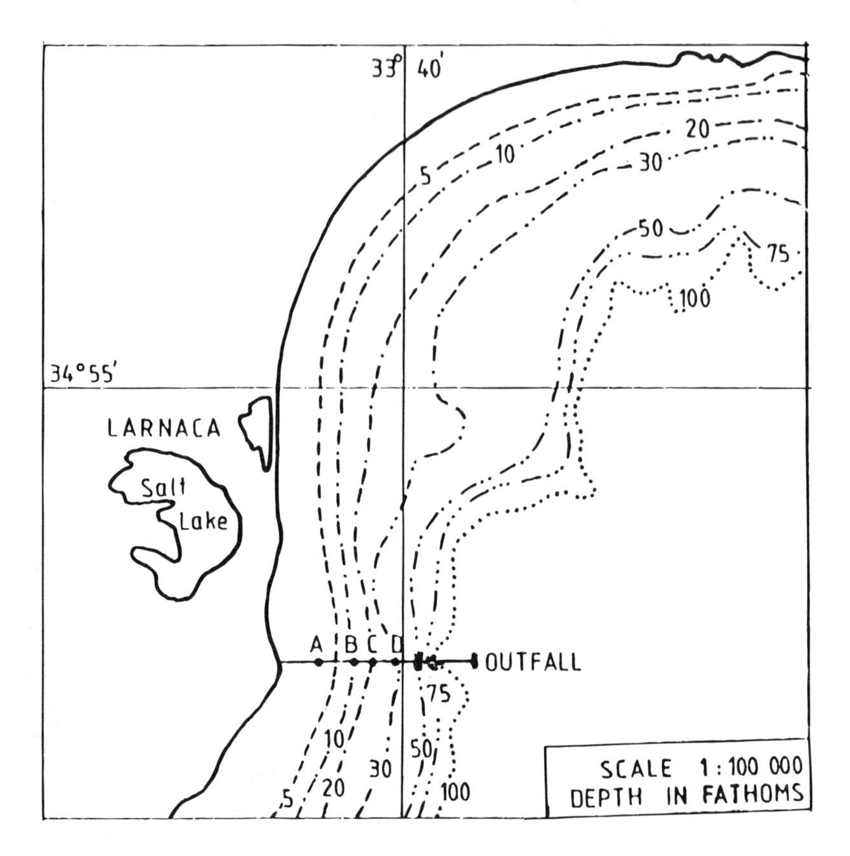

Figure 2.54. The final location of the Larnaca sewage outfall was determined by oceanographic parameters. (*Source:* Reference 352.)

The solution is to simply dilute the original sample. If, in the above case, we dilute the 40.0 mg/l BOD_5 by *10 parts pure water to 1 part sample water,* we will reduce the sample to only 4.0 mg/l BOD_5. Now the sample's DO of 7.0 mg/l will last the whole five days and still leave DO to measure (4.0 mg/l), which is essential to the test. The result will be: 7.0 − 3.0 mg/l = 4.0 mg/l multiplied by 10 to readjust for the dilution factor, yielding a correct BOD_5 value of 40.0 mg/l. Experience will enable the technician to estimate the amount of dilution needed.

Because the BOD_5 test takes so many days to complete, it is often replaced by the Chemical Oxygen Demand (COD) test, which operates on the same concept of measuring oxidation of organic matter but uses chemicals to do it, thereby speeding up the process. A COD test measures the amount of oxygen, from potassium dichromate (an oxidizing agent) in sulfuric acid, that is needed to oxidize all the carbon-containing pollutants (41). And because the COD method also does a more thorough job of oxidation, COD values are always higher than BOD_5 values (371).

2.28 PROJECT REVIEW AND PERMITS

A major purpose of ICZM-type programs is to examine proposed major development projects to determine the impacts they have on coastal resource systems and to recommend design or location changes that can eliminate or reduce any negative impacts. Most probably the appropriate techniques would involve a permit system, whereby the development entity could begin the project only after a permit is issued by the ICZM agency (Figure 2.55). The system is simple: if one wishes to build a structure, reclaim tidelands, clear land, or otherwise engage in economic development activity, a permit must first be issued by government. Many countries have established permit procedures designed to control the environmental effects of construction projects (231).

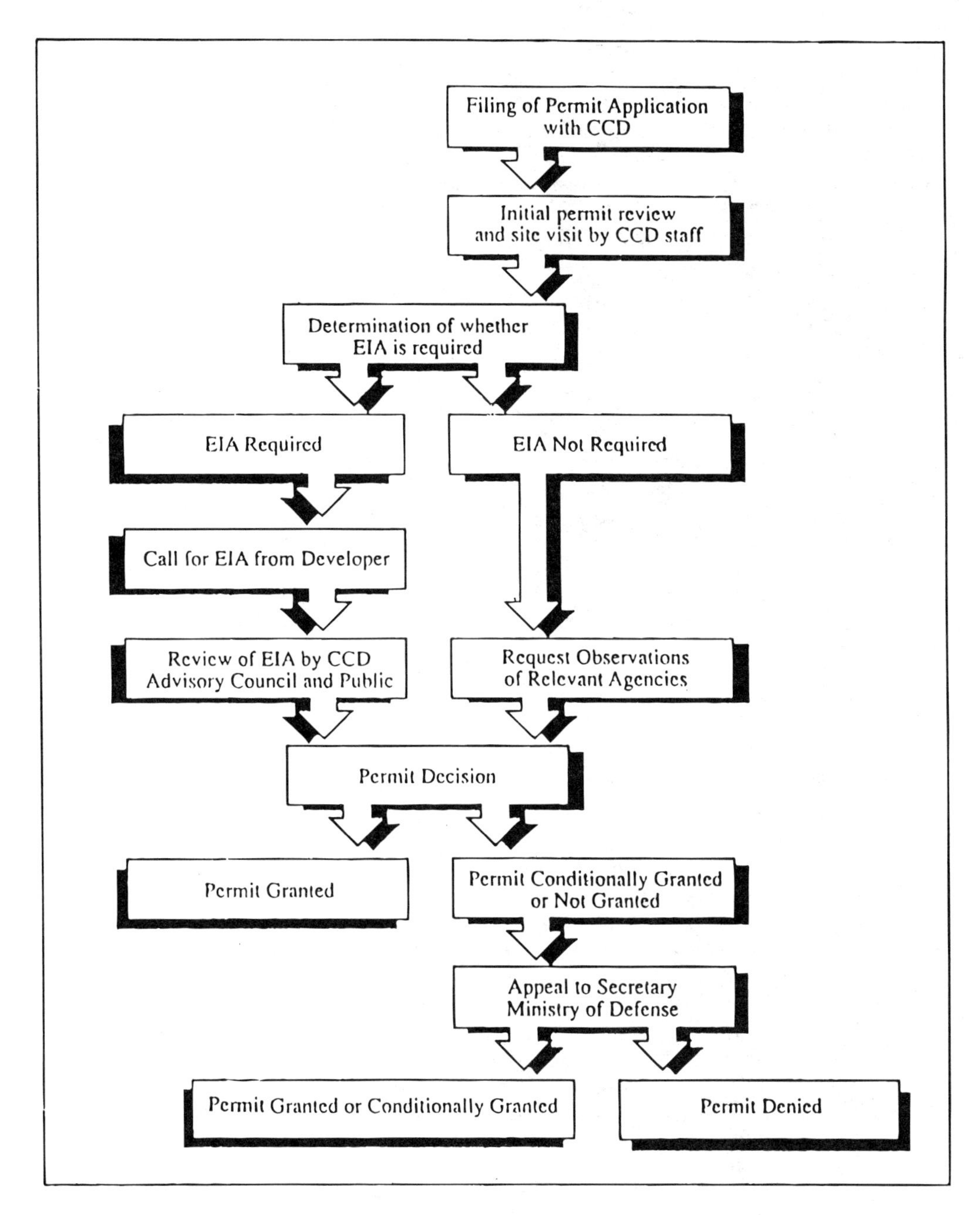

Figure 2.55. The Sri Lanka method for reviewing and issuing (or not issuing) permits for coastal zone projects (maximum jurisdiction 300m from shoreline). (*Source:* Reference 43.)

The permit regulatory process should include (a) environmental assessment (EA) on all major projects, (b) public involvement (including a public meeting or *better* method of broad participation), (c) special conditions to protect the environment as a part of the permit, (d) monitoring of construction in progress to assure permit compliance, and (e) enforcement (confiscation of equipment, penalties) for those projects that avoided permitting or do not conform to permit conditions. The regulatory EA process should include availability of qualified professionals to be consulted in appropriate fields—marine biology, oceanography, public health, environmental engineering, archaeology, vegetation, ornithology, and so forth (231).

If the EA shows the project's impacts to be minimal, if they are eliminated or reduced to a minor level by redesign of the project, or if they are appropriately "mitigated" (i.e., offset by compensatory activity), the proponent of the project will be granted development permission.

It is useful to simplify the permit procedure for smaller projects. To this end two different permits could be used: *major permits* for big projects that require full environmental impact assessment (EIA) and site review (Figure 2.56) and *minor permits,* which require only a preliminary environmental review (IEE). The following is an example of a minor permit checklist that might be used by a provincial permit review official in a decentralized ICZM program (325):

1. Review the application with the applicant, if possible
    - Is all the information required for the application complete?
    - Has the applicant attached a drawing of the structure and or site, in the case of a construction activity?
    - Does the applicant understand the permit process, including the time required for review?
    - Is the application signed by the applicant? For the municipality?
2. Review the proposed activity
    - Order publication of Public Notice.
    - Schedule and inspect the site: Photographs may be desirable to depict any unusual site conditions. During the site inspection, explain the process to any available adjacent/adjoining owners.
    - If other permits are required, or if other agencies have authority or interests related to the activity, circulate copies of the application to those agencies for comment.
    - Review any comments or objections from other agencies or citizens. Review coastal management regulations.
    - Draft any proposed permit conditions that may be necessary.
    - Draft the proposed permit decision.
3. Approve/condition/deny the permit
    - Inform the applicant and any objecting agency/citizen of the proposed action.
    - Inform the applicant and any objecting agency/citizen of their rights to appeal the permit decision.
    - If the permit is approved, issue the permit and mail or deliver it to the applicant.

Using formal permit issuance as a way to control projects is a good way to avoid chaos in coastal development. If the developer must receive a formal permit to build, then environmental and socioeconomic stipulations can be enforced on the development. For coastal management, where a permit program does not exist, a simple but effective permitting authority should be established. This authority must be independent of influence by the entity proposing or sponsoring the construction projects. Close scrutiny must be maintained in some cases to ensure that the power to let permits is not accompanied by corruption and bribery.

Over time, experience gained with different kinds of projects and different resources will raise the level of confidence of the coastal management program and actually enable it to speed up development by providing useful strategic guidance to coastal communities and to development interests.

If the typical permit-by-permit approach of regulatory agencies is ineffective in advancing long-term goals for aquatic ecosystem conservation, it may become a deterrent to strategic system restoration. Improvement requires that individual permit reviews be evaluated wherever and whenever possible through a regional strategy for restoration. Goals and targets should be determined in advance according to a regional strategy and used to guide subsequent permit actions involving restoration and enhancement. For example, Sorensen (321) concludes that "the relative scarcity and abundance of the resource needs to be determined on a region-wide basis in order to set priorities on the types and locations of habitats that should be provided in a restoration site plan."

The permit letting activity of the Sri Lanka ICZM program is an example of a mature system (see "Sri Lanka" case study in Part 4). It is administered by the Coast Conservation Department (CCD). In Sri Lanka a permit is required for all development activities in the coastal zone that are likely to alter the physical nature of the coast, lead to erosion, damage natural habitats, or destroy historical resources (43).

In the coastal area, as elsewhere, some resources are more vulnerable to development impacts than others and some development types and uses are more threatening to resources than others (Figure

Figure 2.56. In conducting the environmental assessment required for coastal project review, an on-site consultation is recomended—including developer, agency staff, and impact assessor—as shown here for the Port Liberté project on the lower Hudson River across from New York City. (Photo by author.)

2.57). Information on these two aspects is useful to ICZM programs, particularly categorizations that can be used in regulatory activities, such as permit review and EIA (useful in development project review and in multiple-use economic development planning). A simple way to do this is by an approach in which:

- Land and water areas are classified as (a) approvable for development, (b) not approvable for development, or (c) approvable for development under specified conditions.
- Development project types are classified as (a) acceptable, (b) unacceptable as designed and located, or (c) acceptable with changes in design and location.

This is a sort of triage approach whereby the acceptable and unacceptable can be dealt with expeditiously and maximum effort can be applied to the in-between categories of either areas or development types. More sophisticated types of land/water classifications might be generated in time (e.g., land suitability analysis).

2.29 PROTECTED NATURAL AREAS

Prudent management of coastal and marine resources—conservation—includes both wise use and conservation of natural resources. Coastal and marine protected areas—nature reserves, national parks, etc.—are an integral and necessary component of that stewardship. Integrated coastal zone management (ICZM) provides a mechanism to achieve a proper balance between use and preservation while increasing the likelihood of long-term survival of protected areas through resolution of potentially divisive conflicts that arise between competing uses of the coastal zone. (65)

A program of protection of ecologically critical areas (ECAs) areas can complement other objectives of ICZM by providing, among other things, juvenile fish nurture areas for fisheries production, generating

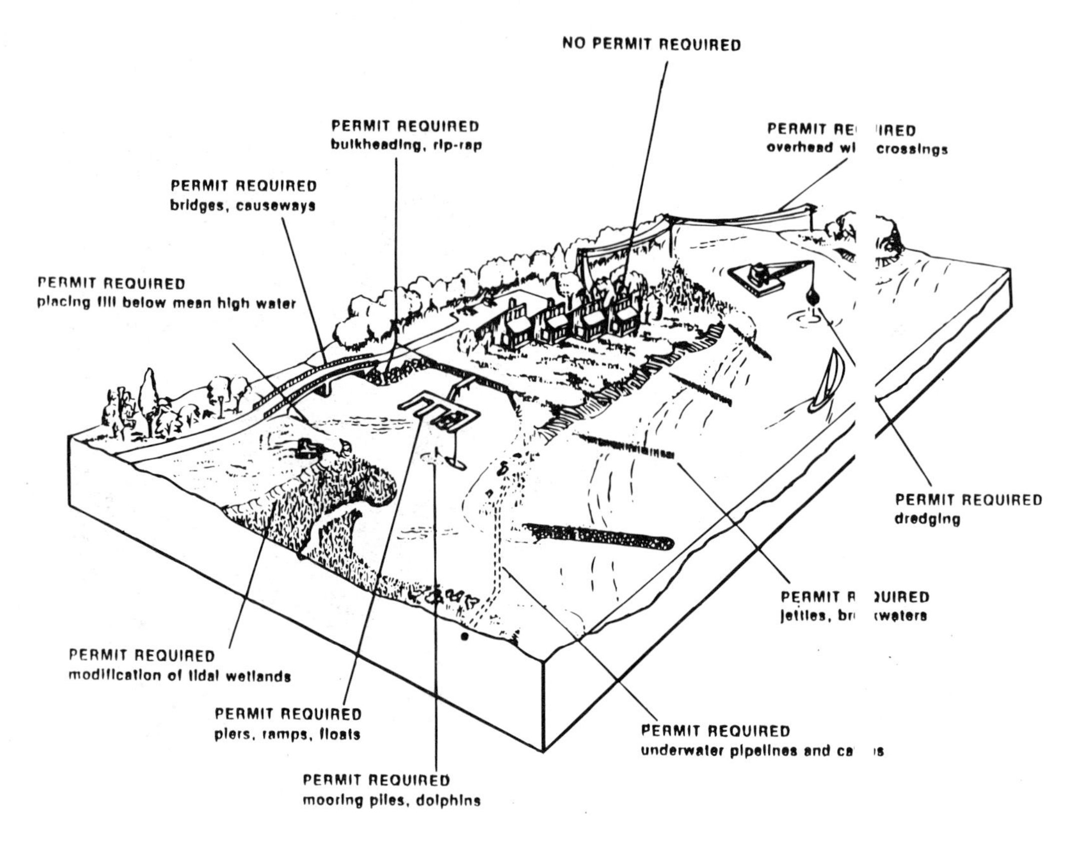

Figure 2.57. Permit requirements for a coastal community in Connecticut (USA). (*Sourc* Reference 42.)

tourism revenues and recreational benefits, maintaining biodiversity, and supplyin ɔaseline scientific and management information for the operation of the ICZM program.

The most certain way to protect ECAs and to conserve biodiversity is to acqu ownership or to assert ownership if the areas are already owned (usually by government). In most c ıntries, those who own the proprietary rights to an area (whether government or private) have the ri t and authority as custodians to determine the way their land or water areas are used, including the ri to control access and use (66).

Government ownership and designation for protection offers the highest assura e that a particular area's resource values will remain intact and is universally the way by which parks ıd other protected areas are formed and made secure. Most simply, if it owns a resource on behalf of the ople, government can dictate, as custodian, the use of that resource. Therefore, government can de nate for resource protection—as parks, reserves, refuges, or sanctuaries—whatever areas it owns. I signation implies that a management program is devised and that appropriate control and surveilla systems are put in place to accomplish specific conservation goals.

Several types of protected areas can be distinguished. In fact, the Internationa Union for Nature (IUCN) has recognized as many as ten different classes (308). But at the Fourth orld Congress on National Parks and Protected Areas (Caracas, 1992) the number of categories v s reduced to six. However, these can be simplified down to two main categories: (a) "national park" t ɔes with a priority on visitation, recreation, and education and (b) "nature (or resource) reserve" type with a priority on biodiversity and/or resource maintenance. Heyman (166) terms these two categ ies *park-like* and *wildlife-reserve-like.* He also offers two other categories: *strict reserve,* which cl es the area to all uses, and *multiple-use reserve,* which opens the area to a variety of uses (which may l regulated or not).

A study by the United Nations Economic and Social Council lists the following r ısons for selecting an area for protected area designation (354):

- It typifies an important ecosystem or habitat type.
- It has high species diversity.
- It is a location of intense biological activity.
- It provides a critical habitat for commercially or ecologically important species or groups of species.
- It has special cultural values (historical, religious).
- It is important for research purposes.
- It is an area of special sensitivity particularly susceptible to damage or disruption.
- It is an area significant for biotic character of species representation (i.e., an area with rare, threatened, endangered, or endemic species).
- It is an area of exceptional human use value, such as a recreational or fishing area.

Marine area protection is usually aimed at practical goals. Except for scenic areas, the habitats, ecosystems, species, and communities that we are trying to conserve have commercial value or potential economic uses. Marine resources may be exploitable, currently exploited, or overexploited. The value of marine conservation can often be readily demonstrated in terms of fish in the diet or cash in the family pockets.

IUCN recommends the following standard ecological criteria for selection of ECAs for protection (308):

1. Diversity: The variety or richness of ecosystems, habitats, communities, and species. Areas having the greatest variety should receive higher ratings. However, this criterion may not apply to simplified ecosystems, such as some pioneer or climax communities, or areas subject to disruptive forces, such as shores exposed to high energy wave action.
2. Naturalness: The lack of disturbance or degradation. Degraded systems will have little value to fisheries or tourism, and make little biological contribution. A high degree of naturalness scores highly. If restoring degraded habitats is a priority, a high degree of degradation may score highly.
3. Dependency: The degree to which a species depends on an area, or the degree to which an ecosystem depends on ecological processes occurring in the area. If an area is critical to more than one species or process, or to a valuable species or ecosystem, it should have a higher rating.
4. Representativeness: The degree to which an area represents a habitat type, ecological process, biological community, physiographic feature or other natural characteristic. If a habitat of a particular type has not been protected, it should have a high rating. (A classification scheme for coastal and marine areas necessary in applying this criterion.)
5. Uniqueness: Whether an area is "one of a kind." Habitats of endangered species occurring only in one area are an example. The interest in uniqueness may extend beyond country borders, assuming regional or international significance. To keep visitor impact low, tourism may be prohibited but limited and research and education permitted. Unique sites should always have a high rating.
6. Integrity: The degree to which the area is a functional unit—an effective, self-sustaining ecological entity. The more ecologically self-contained the area is, the more likely its values can be effectively protected, and so a higher rating should be given to such areas.
7. Productivity: The degree to which productive processes within the area contribute benefits to species or to humans. Productive areas that contribute most to ecosystem sustainment should receive a high rating. Exceptions are eutrophic areas where high productivity may have a deleterious effect.
8. Vulnerability: The area's susceptibility to degradation by natural events or the activities of people. Biotic communities associated with coastal habitats may have a low tolerance to changes in environmental conditions, or they may exist close to the limits of their tolerance (defined by water temperature, salinity, turbidity or depth). They may suffer such natural stresses as storms or prolonged immersion that determine the extent of their development. Additional stress (such as domestic or industrial pollution, excessive reductions in salinity, and increases in turbidity from watershed mismanagement) may determine whether there is total, partial, or no recovery from natural stress, or the area is totally destroyed.

Designating coastal and marine protected areas (under government custody/proprietorship) can help achieve development goals and enhance the benefits of current use. These areas are designed to promote sustainable utilization, whereby resources may be used, but not "used up." Properly designed protected areas provide for a variety of uses and use controls in an integrated resource management scheme (308). In addition, they can assist in the realization of such practical needs as ecotourism development (foreign exchange and jobs), education and awareness, and protection of genetic resources.

In a nature protection program, specific habitat types should be identified as of critical importance to marine resources and to biodiversity—including coral reefs (Figure 2.58), mangrove swamps, marshlands, seagrass beds, and other areas that serve special ecological functions. These functions may include providing sanctuary to (a) valuable species during vulnerable phases of their lives (shrimp, seabirds, fishes), (b) endangered species (whales and turtles), and (c) migratory species (waterfowl and wading birds). These habitats are used for feeding and spawning, as nurture areas and for shelter. Safeguarding these critical marine habitats by designating them protected areas can help conserve species, maintain fisheries, and support tourism.

A national system of protected areas will be more effective if it is coordinated with (or by) an ICZM program. Conversely, protected areas planning is essential to the full development of a country's ICZM program. Planning for a park in isolation from surrounding land uses and peoples, and without interagency cooperation usually will not work because protected areas that are alienated from a wider program of coastal resources management exist as islands of protection threatened by surrounding areas of uncontrolled exploitation.

ICZM can be organized to prevent pollution from external sources along with overfishing, destruction of "nursery" habitats, and other types of external impact that can be damaging to the protected area. ICZM provides an appropriate framework for integration of protected areas into a larger system of protection and a method of consensus building for their support.

Figure 2.58. All critical habitats—such as coral reefs, mangrove forests, seagrass meadows, kelp beds, and beaches—should be scheduled for preservation. (Photo by author.)

A complete program for marine resource conservation includes both regulatory and custodial approaches. The regulatory approach uses the general power of the state to regulate activities of citizens (pollution regulations, fishery regulations). The custodial approach uses the power of a property owner to govern activities of those who enter. On land, the difference is much clearer than on the sea, where the state has both broad policing authority and well tested proprietary rights. Like their land counterparts, marine protected areas are characterized by the allocation of certain use types to certain sub areas by designation of common property by the proprietor (government).

The integrated approach requires categorizing all the most valuable natural habitats, or ECAs, for various of types of management. In the ICZM program it would be relevant to relegate critical areas to the following three management categories:

1. **Generic Type,** where *all* of a habitat type are designated for protection (e.g., mangroves, coral reefs, etc.) through regulatory means
2. **Geo-specific Type,** where areas for regulatory protection are red-flagged
3. **Reserve Type,** where areas are set up as national parks or nature reserves for proprietary protection (for details, see "Critical Natural Areas Identification" above)

Recognizing that coastal and marine protected areas must be considered in the broader context of an integrated program for coastal and marine resources, it is suggested that an "umbrella program" for conservation of renewable resources be considered for ICZM. This umbrella program would encompass both regulatory and proprietary aspects. Its functions would be the following (66):

- Limiting, as necessary, particular exploitative uses of coastal and marine waters and their resources or of linked areas that influence life in these waters (for example, preventing the mining of living coral reefs to maintain their value to fisheries and to protect the coast from natural hazards)
- Protecting particularly vital parts of coastal or ocean ecosystems (for example, critical natural habitats)
- Restoring earlier conditions (for example, closing areas to enable the recuperation of damaged habitats or depleted stocks, or prohibiting activities that are physically damaging or polluting)
- Obtaining and transferring information (for example, through research, education, and interpretive programs)

Each protected area should have a management plan specific to its particular needs. Management plans address legislative background, rules for use, zoning basis, management strategies, interpretation, visitor policies and program, staff training, structures and engineering, fees and finance, public participation, administration, enforcement, O & M, promotion and marketing, and special functions.

Most of the better organized coastal/marine reserves or national parks created in the past few years have used *zoning* because both separation of permitted uses and prevention of certain unacceptable uses require that the overall area of the reserve be subdivided into zones so that "pinpoint" management strategies can be used.

Several types of zoning arrangements have been used with good success in coastal developing countries (see "Zoning" below). The world's largest marine conservation area, the Great Barrier Reef, is managed by an extensive system of zoning, which includes "120 core areas linked by continuous buffer and transition zones" (139). In the Philippines, Sumilon Island presented a case where half of the surrounding water area is designated as a preserve (a "core area"), while the remaining nearshore waters were designated for traditional uses.

In the Mexican state of Quintana Roo, Sian Ka'an Biosphere Reserve includes a core area for preservation and adjacent lagoonal area designated for limited lobster fishing; entry into the fishery is limited and is carried out by a cooperative (65). Numerous other examples of zoning could be cited, including Looe Key Reserve in the United States, Holetown Park in Barbados, and Pulau Seribu Park in Indonesia.

It will not be possible to identify ECAs without having clear guidelines relating to the objectives for which the identifications are to be made. A second necessity for the identification process is full knowledge of the ecosystem setting for each area evaluated. This requires the information base to have boundaries broader than those of the particular special area under evaluation. One needs to know details

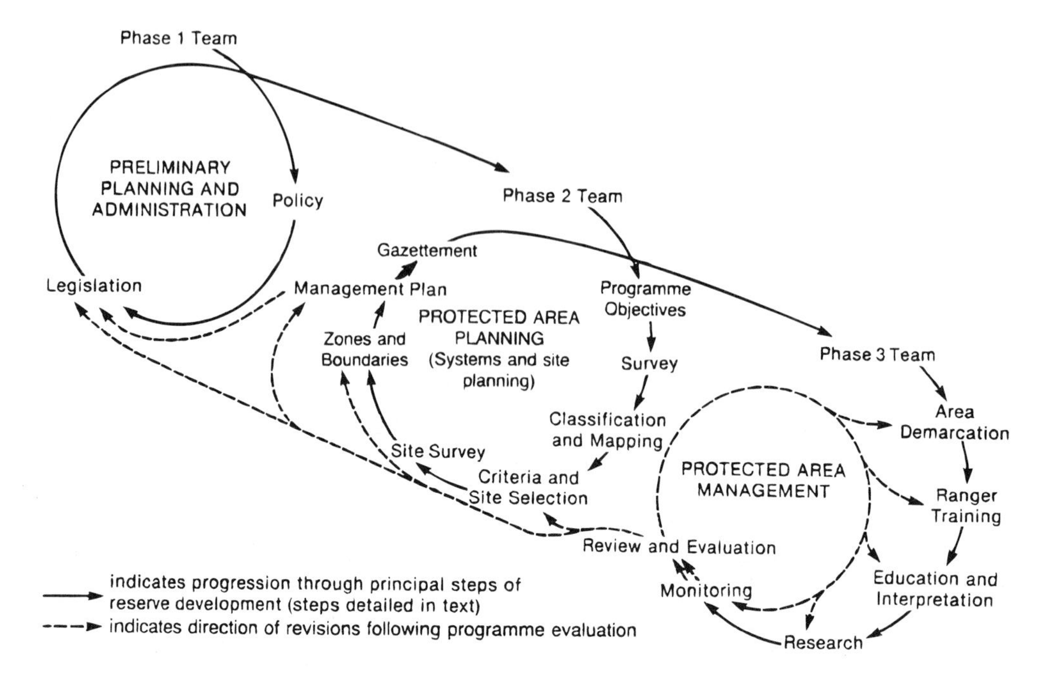

Figure 2.59. The interrelationships among various phases of marine protected area planning and management. (*Source:* Reference 308.)

of the structure and processes of the larger ecosystem of which the subject area is a component, including the life histories of migratory species and the relationship of the area to others within the same system.

A third necessity is to organize the evaluation work according to a recognized classification system such as the one shown below; sorting the areas into biogeographic, ecosystem and habitat type helps in selecting representative areas for protection (if that is a goal) and in tailoring management programs to specific natural characteristics. This needs to be done in a systematic manner (Figure 2.59).

Three major motivations for selection of particular ECAs as marine protected areas are: (a) perceived ecological, social, and economic value; (b) degree of controllable threat to which they are or may be exposed; and (c) potential ability of local, state, regional, national, or international entities to provide protection through management intervention.

Criteria for identifying and selecting protected natural areas depend on the overall objectives of the conservation program. If the objectives are mainly social (recreation, research, and education), the criteria would emphasize safety factors (such as absence of currents and large waves), the presence of cultural or archaeological sites, and accessibility. If economic goals (such as coastal protection, maintenance of fisheries, or development of tourism and appropriate industries) are the primary concern, the criteria would emphasize intensity of resource exploitation, the present and potential economic value of resources, and the degree of threat to them. If ecological goals (such as maintaining genetic diversity, ecological processes, and species replenishment) are stressed, the criteria would primarily concern the uniqueness, diversity, and naturalness of the sites (308).

Such criteria have two functions. They initially serve to assess the eligibility of sites for protected area status in the identification process. But their main role is to rank order eligible sites according to priority in the selection process. Example *economic criteria* for selection of marine protected areas (308):

1. Importance to species: The degree to which certain commercially important species depend on the area. Reefs or wetlands, for example, may be critical habitats for certain species that breed, rest, shelter, or feed there and that form the basis of local fisheries in adjacent areas. Such habitats need management to support these stocks.
2. Importance to fisheries: The number of dependent fishermen and the size of the fishery yield. The greater the dependence of fishermen on an area, and the greater its yield of fishes, the more important it becomes to manage the area correctly and to ensure sustainable harvest.

3. Nature of threats: The extent to which changes in use patterns threaten the overall value to people. Habitats may be threatened directly by destructive practices, such as certain bottom trawls, or by overexploitation. Areas traditionally harvested by local fishermen on these grounds may increase, bringing extra pressure to bear on stocks and habitats.
4. Economic benefits: The degree to which protection will affect the local economy in the long term. Initially, some protected areas may have a short-lived disruptive economic effect. Those that have obvious positive effects should have higher ratings (for example, for protecting feeding areas of commercial fishes or areas of recreational value).
5. Tourism: The existing or potential value of the area to tourism development. Areas that lend themselves to forms of tourism compatible with the aims of conservation should receive a higher rating.

Quite clearly, the major potential trouble for coastal and marine habitats comes from land, not ocean, sources. And the potential diminishes rapidly seaward, at a nearly exponential rate; that is, estuaries are the most impacted, nearshore areas less impacted, and distant ocean sites the least impacted. Serious ocean impacts are typically sporadic events (e.g., oil spills) while coastal impacts are typically continuous or recurrent. Consequently, public and political concern for marine areas is often considerably less than for coastal areas—out of sight, out of mind.

Excessive use (or visitation) of a natural area can lead to its degradation or destruction. Therefore, use should be matched with an estimated (or calculated) *carrying capacity* of the reserve or marine national park (see "Carrying Capacity" in Part 3) as per the following guidelines (64):

- Human use is recognized as a major threat to be managed to protect natural areas.
- Carrying capacity should be used to manage human use of protected areas.
- Carrying capacity should be a major component of any management plan for a protected area.
- The requirement to set a carrying capacity for each natural area should be embodied in legislation.
- Zoning is the most useful framework for application of carrying capacity controls.
- Carrying capacity is seen as a mechanism that should be "pro-active" in protected areas planning, rather than reactive.
- Planning for infrastructure should limit facilities to the physical carrying capacity (i.e., at less than or equal to the carrying capacity).

Following are five recommended stages for the development of a nationwide coastal/marine protected areas system (308):

1. Policy and legislation formalize the governmental decision to initiate and advance the program and to set goals and objectives for its implementation.
2. Preliminary planning interprets policy and legislation, organizes the planning agenda, identifies the planning team, and defines program objectives (though there may also be preliminary planning toward policy and legislation).
3. System planning (strategic planning) looks broadly at program goals and objectives, provides selection criteria, and identifies and selects sites.
4. Site planning provides the initial site design, including use zoning, and the management plan for each protected area in the system as well as for future revisions as needed.
5. Implementation and management develop, administer, manage, and improve the protected area site.

Eight basic guidelines for planning such a system are given below (308):

1. Planning must always work toward attaining specific objectives. Through identifying regional or national resource values, problems, issues and needs, system planning helps to shape national policy and define the conservation objectives of site planning.
2. A plan is not to be viewed as a final product, nor an-end in itself. It is a means to the ends of better program development and site management to attain conservation objectives.
3. A plan should be regarded merely as a "state of progress" document, reflecting the best

information available at the time of its preparation. With time and management experience, new ecological, economic, and social information, and review of the program development and management progress, it may become necessary to review the plan. There must be a balance between the robustness of a plan in the face of possible local or political pressure and its flexibility to respond to altered local conditions or new information.

4. Planning for ecosystem protection requires identifying those areas where important supporting processes originate and those where there are degrading influences (e.g., a coastal river that discharges onto a coral reef).
5. Planning for species protection requires identifying the movements and critical habitats of the species, even if these habitats are in other jurisdictional areas or biogeographic regions, and managing individual populations throughout their range by multinational establishment of protected areas and treaties.
6. It is preferable to select protected areas through system planning rather than by reacting to new threats, such as from a developing oil and gas industry. Nonetheless, degree of threat remains an important criterion both for establishing priorities and for assessing the feasibility of area protection.
7. Planning must have the legal support and consensus of the highest levels possible within all relevant government authorities. It is a tool to coordinate and guide program development among and within agencies toward attaining common objectives.
8. The planning of coastal and marine protected areas must include representatives of all affected parties, including other nations (where appropriate), national agencies, and regional residents. Public involvement is an essential part of planning, especially in site selection and management planning. In addition to aiding public understanding of the need for conservation action, involvement of local user groups (fishermen, divers, etc.) can optimize the benefits of area protection for people and for conservation, while minimizing social and economic disruption.

Probably the most difficult aspect of putting a plan like the above into action is coping with impacts originating externally—pollution and sedimentation from rivers, upcurrent spills and dumps, depletion of distant nurture areas that restock the particular protected area, reduction of wetlands or grassbeds that supply nutrients, and so forth. In an extreme example, the Gulf Stream edge off Cape Hatteras—an especially valuable area both ecologically and for fishing—is dependent on the algal beds of the Caribbean to supply the floating sargassum weed habitat that is so critical to productivity, diversity, and fishing success. (See "United States, Florida: Distant Influence on Coral Reefs" case history in Part 4.)

The above only touches on the complex subject of coastal protected areas creation and management, which provides the content of many publications, including several complete books. As a practitioner's guidebook, aspects of designing "national systems" of coastal/marine parks are not discussed. Persons who expect to engage in creation and management of coastal reserves or parks should refer to the major sources—e.g. the IUCN book *Marine and Coastal Protected Areas* (308) or the Taylor and Francis book *Managing Marine Environments* (204).

2.30 PUBLIC PARTICIPATION IN PLANNING

In most countries a priority goal of coastal management should be to arrange for the most extensive participation possible (starting with the strategy planning stage). Consultations should be held with all relevant agencies of central and local government, with developers, with resource users and other interests that would be affected by ICZM (fishermen, farmers, etc.), environmental advocacy groups, and investment sources (including international donor institutions).

Coastal areas and coastal resource systems are governmentally complex because of the degree of shared jurisdiction and the amount of common property resources involved. Therefore, resource management programs need to involve all levels from national to village governments regardless of the particular institutional arrangement at the central level. People live in the environment right along with other flora and fauna. People affect their environment and in turn the environment affects people, immensely. In addition, it is within the context of human action and reaction that all impacts are assessed. (See also "Participation" in Part 3.)

According to Coetzee (93), a major lesson learned in South Africa is that extensive public involvement should have the first priority in planning and in the review of management actions. The need for public awareness and education is often emphasized in discussion on integrated coastal zone management strategies. Coetzee believes that the recent emphasis on *integrated* coastal zone management is particularly appropriate as it may also help focus the need for integration of the public into the ICZM process—specifically at policy formation, programming of action and implementation stages (93).

An element of South Africa's ICZM program that has proven to be very successful was the "Coastal Liaison Exercise." This program was launched in mid-1988 with the following broad objectives (93):

- To test the public response and susceptibility to the draft policy and objectives for ICZM and to have the various levels of government explain their respective roles in the ICZM process
- To explain in simple terms the physical and ecological process operating in the coastal zone and the interactions between the various ecological elements and management actions
- To demonstrate the use of guidelines for coastal land use by means of case studies with which the specific coastal community is familiar
- To explain the reason for the coastal regulations and emphasize the need for the public to realize that it is in their own interest that they see to the enforcement thereof
- To explain the coastal structure planning process and to emphasize the need for the public to become involved in order to ensure that their needs and aspirations are adequately reflected
- To try to foster the need for effective communication between the various levels of government and the public
- To seek ways in which the public can actively assist with ICZM on an ongoing basis

According to Renard (293), four major objectives of participation are the following:

1. Participation is a way to ensure that popular knowledge and experience is indeed integrated into the planning and management process.
2. Participation gives a better guarantee for the quality of the solution identified and for its adaptation to a particular condition.
3. Participation in planning and problem identification promotes involvement in the actual implementation of decisions.
4. Participation ensures that all needs and priorities are taken into account in the formulation of management decisions.

There are many, many examples of people becoming upset by private and government projects because they have had limited access to the decision-making process. People have marched in protest to stop projects (e.g., polder drainage project in Khulna, Bangladesh), sunk a patrol boat (Tobago Island, Trinidad and Tobago), destroyed signs and structures (Cahuita, Costa Rica), engaged in armed violence (Solomon Islands; see case study in Part 4), and burned down a factory (Thailand, Phuket; see case history in Part 4).

Planners themselves may claim that, since the users have no planning experience, they cannot contribute effectively. Regarding participation of coral reef users in conservation of coral reefs, Mike Gawel (137) says a major stumbling block "is lack of concern for long range impacts and apathy toward planning." Gawel lists the following five groups as interests to be contacted (137):

1. Civil service officials
2. Elected officials

3. Community leaders
4. Commercial resource users
   - Fishery companies and organizations
   - Sand and coral dredgers and extractors
   - Marine oil producers
   - Salvage companies
   - Shipping companies
   - Mariculture developers
   - Businesses supporting tourism
   - Aquarium and aquarium fish sellers
   - Coral and shell harvesters
   - Diving schools
5. Local interest groups
   - Boat owners and their organizations
   - Shell clubs
   - Other natural history or conservation clubs
   - SCUBA associations
   - Scientific researchers
   - Science teachers
   - Subsistence fishing clans
   - Groups supporting traditional rights

Stakeholders who have been involved in the formulation of policies and rules on resource use in coastal areas are more likely to support them (particularly people living along the coast who have traditionally depended on marine resources for their livelihoods). Public participation should be encouraged by the entire management community (resource users, public agencies, non-governmental organizations, social groups, and local communities) to ensure the quality, the effectiveness, and the equity of management proposals. Communities with exclusive rights (traditional or modern) to coastal or marine resources (such as fishermen) could be given active responsibility for their management. The incentive to conserve resources is much stronger where access is limited according to the ability of the resource to provide sustained yields.

Among the various options available for building up capabilities for participatory development, organizing the target members into an effective receiving mechanism has been one that has been favored in many countries. The argument is that individually the rural communities, with low asset and knowledge base, without access to credit and markets, are unable to optimize socio-economic capabilities. Rural institutions for credit, marketing, supply of inputs and technical services, and the like provide an organizational base on which specific programs could be built.

It is difficult to state categorically what kind of rural institutions would provide a clearly supportive structure. Participation of the whole community appears necessary for all programs intended to benefit the community as a whole. A good way to determine the needs of a particular community is to use "rapid rural appraisal" techniques (see "Rapid Rural Appraisal" in Part 3). Effective communication is the key to successful public participation in coastal planning. There are numerous tested methods for communication, as shown in Figure 2.60. (See also "Awareness" above and "Education" in Part 3.)

Implementation of management programs is not and should not be an entirely governmental responsibility, according to Geoghegan (141). Resource management objectives can be pursued by local communities, resource user groups, schools, and other formal or informal collectives (see also "Awareness" above). For example, fishermen's cooperatives have proven to be extremely effective tools for implementation of development objectives in Guyana, Nevis, and Barbados. Government has provided incentive, but the organization has remained non-governmental. Often, local communities and resource users have developed management techniques to sustain their traditional resource use, and these techniques can be exploited. Government regulation, on the other hand, can often make the resource user feel out of control of his destiny and out of touch with his environment (141). It is not always easy for the private sector, government agencies, or even some NGOs to shift into a more interactive mode of operation.

Unless intended beneficiaries are already organized and accustomed to involvement in formal development programs, some procedures for introducing participation may have to be specifically

Characteristics				Corporate planning objectives					
Level of public contact achieved	Ability to handle specific interest	Degree of two-way communication	1 = Low, 2 = Medium, 3 = High; ■ = Capability Public communication techniques	Inform and educate	Identify problems and values	Get ideas and solve problems	Feedback	Evaluate	Resolve conflict and reach consensus
2	1	1	Public hearing		■		■		
2	1	2	Public meetings	■	■		■		
1	2	3	Informal small group meetings	■	■	■	■	■	■
2	1	2	General public information meetings	■					
1	2	2	Presentations to community groups	■	■		■		
1	3	3	Information coordination seminars	■			■		
1	2	1	Operating field offices		■	■	■	■	
1	3	3	Local planning visits		■		■	■	
2	2	1	Community survey research		■	■	■		
2	2	1	Information brochures and pamphlets	■					
1	3	3	Field trips and site visits	■	■				
3	1	2	Public displays	■		■	■		
2	1	2	Model demonstration projects	■			■	■	■
3	1	1	Material for mass media	■					
1	3	2	Response to public inquiries	■					
3	1	1	Press releases inviting comments	■			■		
1	3	1	Letter requests for comments			■	■		
1	3	3	Workshops		■	■	■	■	■
1	3	3	Advisory committees		■	■	■	■	
1	3	3	Task forces		■	■		■	
1	3	3	Employment of community residents		■	■			■
1	3	3	Community interest advocates			■		■	■
1	3	3	Ombudsman or representative		■	■	■	■	■
2	3	1	Public review of impact statement	■		■	■		

Figure 2.60. Techniques developed for the use of the private business sector in communicating with the public. (*Source:* Reference 261.)

worked out. Such procedures should help beneficiaries to acquaint themselves with the project from the beginning, to elicit their ideas and suggestions, to encourage and assist appropriate modes of organization for participation, and to monitor progress so that changes can be made when appropriate. A specific investment in participation and in institutionalizing such participation is suggested for all countries.

Successful organization of rural people in terms of size, sustainability and the realization of goals depends on their receiving tangible short-term benefits in return for their participation. Perceived constraints are that centrally designed programs are seldom responsive to the needs of the poor, and that the agencies through which they are implemented rarely have the capacity to implement the programs as designed. When a governmental agency takes the initiative to organize groups, formal legal structures and procedures are often given preference.

Rural institutions for credit, marketing, supply of inputs and technical services, and the like provide an organizational base on which specific programs could be built. It is difficult to state categorically what kind of rural institutions will provide a clearly supportive structure. Participation of the whole community appears necessary for programs intended to benefit the community as a whole. Effective communication is the key to successful participation in coastal planning and numerous tested methods are available, as shown in Figure 2.60 (see also "Awareness" above and "Education" in Part 3).

2.31 REHABILITATION

While it is an important function of ICZM to constrain further ecological losses, it is also important to plan for the repair of natural systems damaged in the past. Pollution, habitat conversion, and interference with water circulation, among other effects, have seriously reduced the economic and environmental benefits of coastal environments and resource systems (see "Restoration and Rehabilitation" in Part 3 and relevant case studies in Part 4).

If a wetland is covered with fill behind a concrete bulkhead, it is unrealistic to plan to restore it to its original condition. But if a wetland has been diked for rice culture or aquaculture, it would be relatively easy to reconnect the water flow and reconvert it to a nearly natural wetlands condition. Degraded dunes (Figure 2.61) and beaches can be rebuilt. Mangrove forests can be replanted. Polluted estuaries can be cleaned up. Even coral reefs can be rehabilitated under some circumstances. The potential for rehabilitation is discussed for each of many important habitats in relevant sections of this book (for example, see "Coral Reef Management," "Beach Management," or "Mangrove Forest Management" above).

The technology is well developed for most such restoration projects, following the "nature synchronous" approach (see "Nature Synchronous Design" in Part 3). It is usually only a matter of locating a source of funding, which may sometimes be difficult.

In practice, rehabilitation or restoration may be required as part of any development plan through the project review and permitting system. But it is necessary to have a strategy with some advance goals declared for restoration purposes. These can be thought of most simply in two categories: (a) actions to be taken to rehabilitate on-site environmental damage (existing or project caused) and (b) actions to be taken off-site to compensate for damages done on site. Some such actions—those performed as "mitigation" for ecological damage done during development—are discussed in "Mitigation" above.

The first category is simple to implement because the work is done at the project site by the developer. The second category is more complicated because it requires a regional strategy plan and wide cooperation among coastal resource interests, including selection of regional priorities for ecosystem produced goods and services. Most often an environmental professional would be needed to prescribe functional criteria for the rehabilitation and an engineer to design it (75).

In urban settings, coastal resources, particularly wetlands, have been degraded and often are dysfunctional. Therefore, in urban settings a high priority should be given to restoration and enhancement of aquatic ecosystems and to their component wetlands (Figure 2.62). Success in system-wide restoration requires formulation of a regional strategy with goals, objectives, methods, and predesignated restoration sites. All levels of government and private interests must be involved. Moreover, the existing system of site-by-site permit review must be altered to ensure that permit decisions are oriented toward a regional restoration strategy.

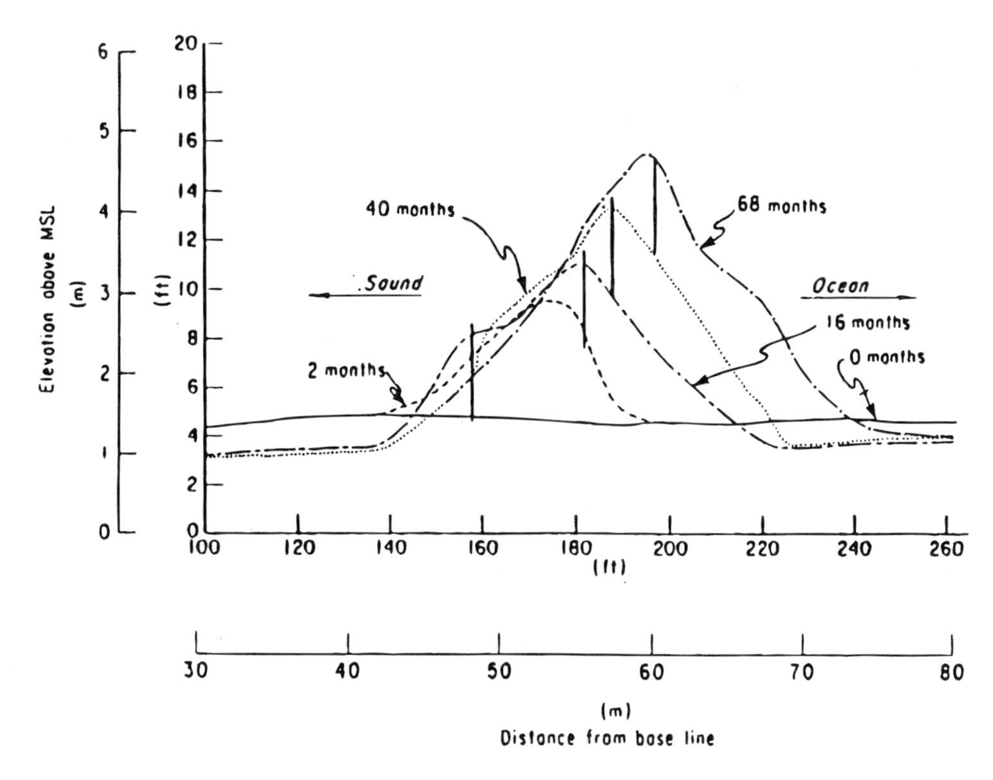

Figure 2.61. Elevation and seaward progression of dune sand accumulated by a series of four "single-fence lifts" (shown as vertical bars) over a period of 68 months (North Carolina, USA). (*Source:* Reference 362.)

Figure 2.62. Community mangrove restoration plantation in Cebu, Philippines, after four years of growth. (Photo by author.)

It is particularly important to recognize that private sector resources may be the main source of restoration project funds through "mitigative restoration" and/or voluntary enhancement by developers (see "Mitigation" above). Therefore, mitigation has to be given a role at the front end of the review process, when projects can be most easily redesigned and the development program is most open to including restoration requirements (80).

2.32 RETREAT

The main threat to stability of the shoreline is often from development on adjacent land. Where the shorefront is composed of erodable materials, such as beach sand, careful planning of development is necessary to avoid disturbing the balance of the beach. Otherwise, the beach may begin to seriously erode during storms, destroying structures along the shorefront (see "Shoreline Construction Management" below).

Coordinated management of the beach itself and the land behind it is needed. This management should include authorization to limit buildings, prevent excavation, and control beach protection and inlet structures. The ocean beachfront is a most hazardous place to build. The sand dune and beach berm system (where these exist) signal high risk locations for buildings (see "Beach Management" above). Keeping development far enough inland to avoid the high risk zone is important in natural hazard prevention efforts. Therefore, a setback line should be delineated at a safe point inland from the beach and all construction kept behind this line (see "Setbacks" below). This is considered the "line of retreat" where the erodable shoreline is already developed.

It is certainly unwise to allow development of property that will probably be lost to the sea, especially when the security of buildings so often creates demands for public money to be spent on groins, bulkheads, and other protective works, which may further imperil the whole beach system. Therefore, management programs should attempt to have structures located behind a setback line that accommodates the predicted long-term recession rate of the beach (26). This contemporary planning solution to anticipated beach recession problems—the concept of "strategic retreat"—follows these steps:

1. Predict how far back the beach will erode in the future (say, 50 years from now).
2. Identify this line on appropriate maps.
3. Prohibit any further building (including rebuilding or expansion) seaward of this line. (Note: the line may need to be recalculated every 5 to 10 years.)

Retreat (or setback) lines also may be established where specific shoreline change data do not exist by prescribing an arbitrary distance considered reasonable and prudent until such time as site-specific data become available.

The "strategic retreat" approach is most effective where no development yet exists. Along already developed shorelines, the strategy for retreat is more complex and requires acknowledgment and accommodation of existing development. Barbados handles this by prohibiting expansion of existing structures or reconstruction of damaged structures in the retreat zone.

Of course, all other anti-erosion counter-measures should be taken in conjunction with the retreat program—prohibiting excavation of sand; maintaining natural defenses like coral reefs, mangrove forests, and greenbelts (Figure 2.63); and controlling shoreline construction.

2.33 SEPTIC TANK PLACEMENT

Septic tank systems—and related systems for underground disposal of household sewage, such as cesspools and pit latrines—often pollute coastal waters, particularly with nitrate and phosphate nutrients. But septic tanks seriously pollute only when placed too near coastal water bodies or sources of drinking water (e.g., wells) or if they are improperly designed or located. A proper septic system discharges effluent into the ground with no harmful side effects and with all of the benefits of public safety, simplicity, and economy.

Figure 2.63. Wherever possible, structures should be set back from the water's edge and the immediate shoreline maintained as a vegetated buffer zone. (Photo by author.)

Ecologically, there are four major problems with septic tanks: (a) wastes leached into coastal waters when septic tanks are located too close to the shore, (b) tidally induced high water tables that provide direct flushing of drain field soils into coastal waters, (c) inadequate drain field components or soil absorption characteristics that cause tanks to overflow and to pollute coastal waters, and (d) each of maintenance. The solution to these problems lies mostly in proper location of septic tanks in relation to coastal water bodies.

The septic system has two main components—a tank and a drainage field (see "Sewage Management" below). The tank is a chamber for partial biological treatment (Figure 2.64). The drainage field functions to provide disposal of the wastewaters within the soil. If the distance between the septic system and a water body is insufficient, the liquid waste leaching through the soil is inadequately treated before it reaches the water. Consequently, it arrives in contaminated condition, polluting the water with a variety of substances, including two troublesome nutrients—phosphate (PO_4) and nitrate (NO_2). Nitrate is particularly mobile in groundwater and is the probable cause of much estuarine eutrophication—average production is 17 g N/person/day for the United States and septic tank effluent averages 62 mg/l of N (202). Phosphate can cause real problems in nitrate-rich water or oligotrophic systems like coral reefs (see "Nutrient Management" above).

Because shoreline soils typically have poor percolation and drainage characteristics and shallow water tables (often less than 1.5 meters), the placement of septic tanks faces severe limitations. However, if the necessary water and soil requirements can be met in the coastal zone and if septic systems are properly maintained and operated within their physical capacities, they should be perfectly adequate, and central sewage systems will not be needed.

Type of soil—gravel, sand, silt, or clay—influences the rate of effluent movement through the ground—movement is faster through sandy and gravely soils than through clayey ones. Soil permeability for septic tanks should be moderate to rapid, allowing a percolation rate of at least 1 inch (2.5 centimeters) per hour (23).

A drainage field setback of 50 feet may be sufficient to allow for removal of such pollutants as coliform bacteria and other pathogens through soil purification. However, it may not be adequate for the removal of dissolved pollutants such as nitrates. Consequently, lagoon and estuarine water bodies

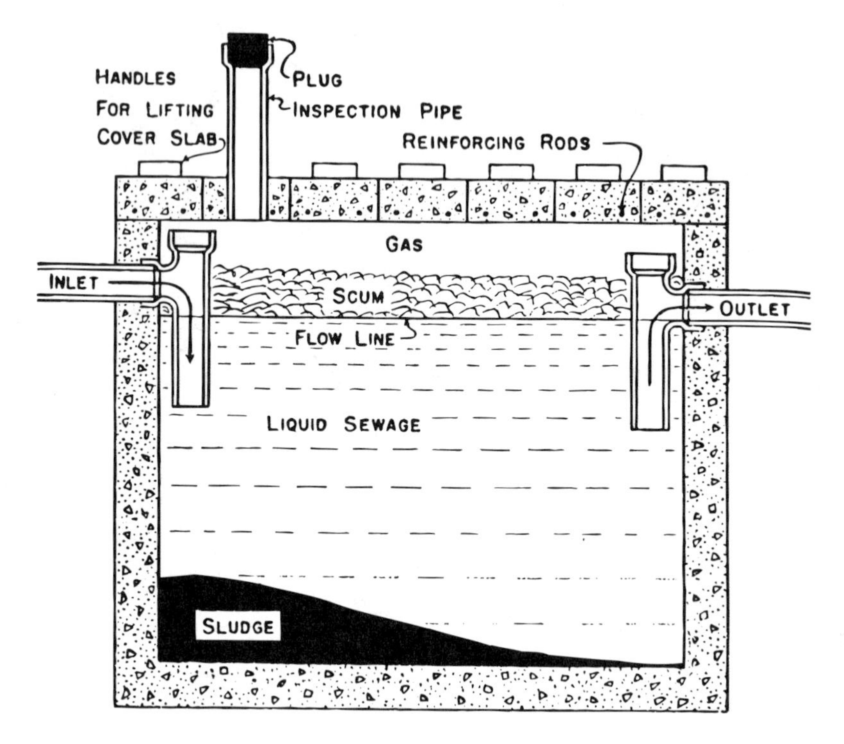

Figure 2.64. Cross-section of a typical septic tank. (*Source:* Reference 252.)

have often become highly eutrophic and degraded from septic tank leaching. Studies have shown that nitrate is commonly found in high concentrations (up to 40 ppm) at distances of 100 feet from septic systems. Other studies have shown unacceptable amounts of nitrate at less than 150 feet. (207)

Strict setback standards should be applied when habitation is denser than one two-bedroom dwelling unit per acre and particularly when an apartment, country club, or other high density structure is involved. There may be pollutants other than nitrate in septic system effluents that would adversely affect coastal waters and these should also be adequately controlled by the setbacks. (59)

After extensive review, L. B. Leopold concluded that "for soil cleansing to be effective, contaminated water must move through unsaturated soil at least 100 feet (30 meters)" and that "it might be advisable to have no source of pollution such as a seepage field closer than 300 feet (90 meters) to a channel or watercourse" (217). Leopold claims that even this setback does not prevent dissolved materials such as nitrates from enriching the water and thus potentially creating eutrophication. Because this much separation is difficult to achieve in practice, other solutions are needed.

The above considerations lead to the recommendation that the drainfield of a septic tank should, when possible, be set back 150 feet (46 meters) from the annual high water line. This is admittedly generic and arbitrary but can be used if there is no site-specific information on soil transmissivity, water table variation, and types of pollutants. Local regulations have typically required the leach fields of septic tanks to be set back a minimum of 50 feet (15 meters) from the edge of the water. Some states require a 100-foot (30-meter) minimum setback.

If the groundwater beneath the septic tank absorption field rises to the level of the drainage field, the now saturated soils cannot absorb the effluent. Consequently, it is recommended that the drainage pipes of the absorption field be placed where they are at least 1.2 meters (4 feet) above the highest expected groundwater level. The worst situation is where groundwater rises to the surface, when it will cause health hazards and can drain directly into drinking water supplies or an adjacent water body.

In unfavorable situations, it may be possible to dispose the effluent at either higher or lower elevations. For example, in the high water table conditions of the islands of the Florida Keys, south of Miami, Florida (USA), wastewater systems near the shoreline are built according to either of two specifications: (a) an elevated (above ground) mound of loose soil with vegetation (to maximize evapotranspiration) or (b) an individual mini-treatment system built of concrete with an injection system to pump sewage into the deeper groundwater (about 75 ft). In the Florida Keys, the domestic water supply comes from the mainland in a pipe (as many as 140 miles distant).

Several other precautions should also be observed. For example, overflow pipes, which convey sewage directly to the water basin when the septic tank fills, should be prohibited. Seepage pits (cesspools) should be discouraged in favor of septic tank and drainage field systems.

2.34 SETBACKS

A key component of a coastal management program is a "setback" provision whereby coastal development is prohibited in a protected zone adjacent to the water's edge. Setback lines provide buffer zones between the ocean and upland property, within which the littoral zone may expand or contract according to nature's dictates without damage to property. Setbacks have many uses in an ICZM program, including designation of a "line of retreat" for eroding beaches (see "Retreat" above).

There are two types of setbacks: (a) those with fixed upland and offshore dimensions and (b) those based on the features of the shoreland. Setbacks—or shoreland exclusion zones—vary greatly in configuration and size. The examples in Figure 2.65 vary from 8 meters to 3 kilometers (326).

In Sri Lanka's ICZM program, a *setback* is defined as an area left free of any physical modification; its width is recommended at 60 meters from mean sea level line, unless particular physical characterics require a wider or narrower setback (43). Such a setback is desirable to allow for shore erosion dynamics—seasonal and long-term fluctuations of the coastline—and to ensure public access to the waterfront and visual access to it.

A major use of setbacks is for adjusting placement of structures along eroding and receding beaches (Figure 2.66). A predicted "recession line" of, say, 50 years into the future can be estimated by beach geologists. No structures would be allowed seaward of the line; in this way, the recession line becomes, in effect, a line of retreat that anticipates persistent erosion, storm effects, and sea level rise (predicted to be as much as 1 meter in the next 100 years). (See "Retreat" above and "Sea Level Rise" in Part 3.)

It is useful to define as a "high hazard zone" the area of the coast that is subject to high risk of damage from wind and storm wave velocity (including erosion and property damage) as well as to the risk of flooding (still water rise) from storm surge. The periodicity of storm and flood events can be calculated as the chance that a hazard event will strike the high hazard zone in any one year. The result, often called the *recurrence rate* or *probability* is given as the percentage chance that an event will occur in a 1-year period. Thus, a recurrence rate of 0.10 means that there is a 10 percent chance that a storm of a particular force will occur in any one year at a particular site.

Having the above information, the ICZM authority can then place the coastal setback at a particular risk point, for example, far enough back from the high water line so that all structures behind it have only, say, a 0.04 or 4 percent probability of being hit by a flood or storm waves. The 4 percent probability level is sometimes called a 25-year event because 4 chances in 100 equals 1 chance in 25. In placing the setback line, the ICZM authority could select the 20- or perhaps the 100-year event (0.05 or 0.01 probability) as the controlling risk factor and relate it to the corresponding distance inland from the high water line. The degree of precision required in delineating and mapping the line depends upon the circumstances of the particular program.

In a U.S. example, South Carolina's 300 km of ocean beaches are extremely variable, with much of it eroding at more than 3 m/yr while others are building seaward. (See "United States, South Carolina" case history in Part 4.) Using site-specific data, the width of the setback is prescribed by state law as a distance 40 times the annual erosion rate measured from the most seaward dune. By tying the setback to a local erosion rate, the law accounts for the natural variation in shoreline trends from one beach to another (197, 198).

Barbados has a firm requirement for a 100-foot setback along its sandy beaches; no new starts nor expansion of structures in the setback zone are permitted. But note that an even wider setback of 100 meters is recommended as best for most Caribbean shorelines by G. Cambers (40).

There are many other applications of setbacks. They can be used to establish a protected buffer zone between the water and the land (Figure 2.67) or between a critical habitat (e.g., beach or mangrove wetland) that needs protection and any encroaching development, such as housing or agriculture (66). In some cases the coastal setback constitutes a "greenbelt," a band of protective vegetation extending

COUNTRIES	DISTANCE INLAND FROM SHORELINE*
Ecuador	8 m.
Hawaii	40 ft.
Philippines (mangrove greenbelt)	20 m.
Mexico	20 m.
Brazil	33 m.
New Zealand	66 ft.
Oregon	Permanent vegetation line (variable)
Colombia	50 m.
Costa Rica (public zone)	50 m.
Indonesia**	50 m.
Venezuela	50 m.
Chile	80 m.
France	100 m.
Norway (no building)	100 m.
Sweden (no building)	100 m. (In some places to 300 m.)
Spain	100 to 200 m.
Costa Rica (restricted zone)	50 m. to 200 m.
Uruguay	250 m.
Indonesia** (mangrove greenbelt)	400 m.
Greece	500 m.
Denmark (no summer homes)	1-3 km.
USSR - Coast of the Black Sea (exclusion of new factories)	3 km.

* Definition of shoreline varies, but it is usually the mean high tide. Most nations and states exempt coastal dependent installations such as harbor developments and marinas.

**Indonesia has both a 50 m setback for forest cutting and a 400 m "greenbelt" for fishery support purposes (see text for explanation).

Figure 2.65. Construction setbacks (exclusion zones) utilized by various countries. (*Source:* Reference 326.)

along the shoreline (Figure 2.68). It can consist of mangroves in the intertidal part and palms or other trees in the supratidal part, as done in Bangladesh.

Another application is to keep septic tanks set back a safe distance from the water's edge so the waste water does not leach through the soil to pollute coastal waters (see "Septic Tank Placement" above). The purpose of the setback is to allow for removal of pollutants—such as nitrate or waterborne pathogenic organisms—from the wastewater through soil purification before it reaches the adjacent water body (18).

In many countries, public ownership of land along the shoreline provides a de facto setback. For example, in Australia and New Zealand, a shoreland zone of Crown Lands constitutes this public area,

Figure 2.66. These condominiums at Kiawah Island, South Carolina (USA), have been set back from the shoreline far enough to be unaffected by major storms for the life of the structures. (Photo by author.)

Figure 2.67. A beach and vegetated dune system of the northern Yucatan coast (Mexico) in near original condition. (Photo by author.)

and in Latin America at least eight countries apply the concept of a "zona publica"—or public zone. Brazil, Chile, Colombia, Costa Rica, Ecuador, Mexico, Uruguay, India, and Venezuela have established shoreland zones based on a specific setback from the shoreline (usually from mean high tide).

As an example, in Costa Rica, the jurisdictional area is a 200-meter-wide marine and terrestrial zone. (See "Costa Rica: Controlling the Zona Publica" case history in Part 4.) The law divides the zone into two components: the *zona publica* and the *zona restringida* (restricted zone). The *zona publica* extends inland 50 meters from mean high tide or the inland limit of the wetlands and the upstream limit of the estuaries as defined by salt or tidal influence. The *zona restringida* covers the remaining 150 meters inland. The *zona publica* is devoted to public use and access, and commercial development is generally prohibited. Exceptions to the prohibition against commercial development are made for enterprises that are coastal dependent, such as sport fishing installations, port installations, and their

Figure 2.68. Each farm in this area of the Chesapeake Bay (Maryland) has a buffer strip of natural vegetation along the water's edge to preserve special wildlife habitats and reduce runoff pollution. (Photo by author.)

infrastructure. In the *zona restringida,* development is controlled by a permit and concession system that is based on a detailed regulation plan formulated at the local level of government (322).

Through its National Coastal Management Program, Greece imposes "strict controls" within a 500-meter band on both sides of the shoreline and thus departs from the general pattern by imposing controls both landward and seaward of mean high tide (326).

2.35 SEWAGE MANAGEMENT

The safe disposal of human wastes is a worldwide problem. In many countries, there are no modern sewage collection systems, even in urban areas. Treatment facilities are often not properly designed, constructed, operated, or maintained, nor are they sited sufficiently far from water bodies. The result is serious contamination of drinking water supplies and pollution of coastal waters important to tourism and fishing. But the irony of it is that modern sewage technology can be ecologically disruptive and therefore counter-productive if the effluent is not properly disposed.

Human waste conveyed in sewage carries pathogens and a high content of decomposing organic matter. This, along with its inorganic content can cause serious ecological problems when it reaches coastal waters. Because most coastal areas are vulnerable to sewage pollution, it is exceptionally important *how* coastal zone sewage is treated and, particularly, *where* it is disposed.

The first step in wastewater management is to control the sources of the waste—the second is to treat the waste. If the sources can be reduced, fewer troublesome pollutants will reach the treatment plants or the environment. This means that industry, commercial operations and households have to cooperate in a widescale program to prevent bad pollutants from getting into sewers or into storm drains. For example, Seattle (Washington, USA) was able to achieve the following reductions of heavy metals in raw sewage over a five-year period (1987–92): cadmium, 50 percent reduction; chromium, 65 percent; copper, 40 percent; mercury, 70 percent; nickel, 50 percent; lead, 25 percent; zinc, 35 percent (97).

Every urban, or semi-urban, community must have a centralized, sewage collection system to gather and transport human wastes—average waste production is 250 g/day/person. All sewage has to be treated to remove solids and reduce organics, nutrients, toxics, and pathogens and then must be disposed in some fashion. The simplest approach is "preliminary treatment," a screening action in which only particles larger than 5 or 6 mm are removed from the waste stream.

The next simplest is "primary treatment," which involves removal of buoyant and non-buoyant coarse solids (40 to 70 percent removal) and initiation of biological decomposition of organic matter

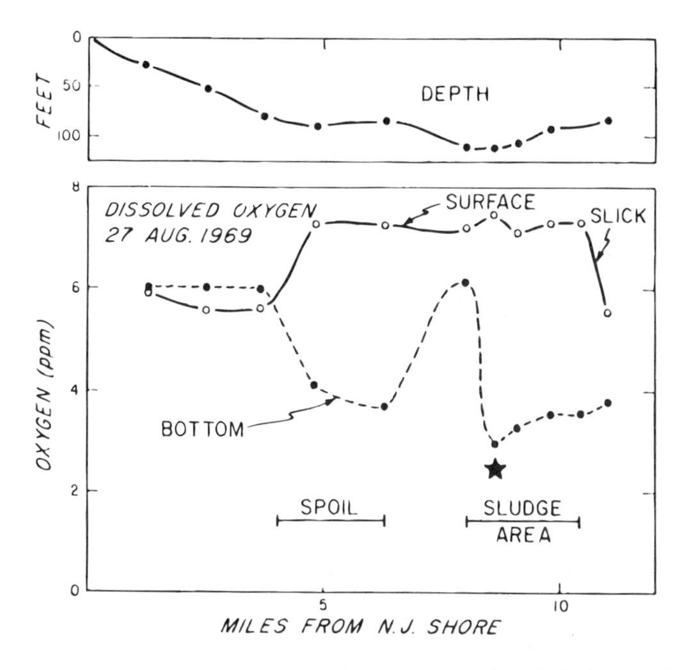

Figure 2.69. Reduction of dissolved oxygen caused by ocean dumping of dredge spoil and sewage sludge from New York City. (*Source:* Reference 361.)

(between 25 and 50 percent removal of biological oxygen demand, or BOD). The waste stream is screened, sometimes chlorinated to kill pathogens, and discharged into a nearby coastal water body (41). This can have undesirable effects unless the outfall is located in the open ocean a goodly distance from shore, so as to release the effluent out of harm's way.

There is some ecological danger from residuals of the chlorine that is used to control pathogens. To prevent problems, Tomascik (345) recommends that free chlorine should not exceed 2.0 mg/l in the plant's waste stream.

Secondary treatment plants specialize in removing most suspended solids including organic matter (about 85 percent) from the waste stream by digestion, or oxidation, in specially designed aeration lagoons (236). The residue, called *sludge,* is disposed of in various ways—dumping in the ocean, burying in landfill, spreading on land, and so forth (Figure 2.69). The liquid remainder, the effluent, is then discharged into a nearby waterway, often after it has been chlorinated.

Sludge is a major byproduct of treatment and contains concentrated organic waste in a decomposing state and must be "thickened and stabilized" before disposal in landfills or used as soil fertilizer. A plant servicing 500,000 population may produce 100 truckloads of sludge per day (353). In New York City the quantity (wet) of sludge produced by sewage plants is nearly 1 quart (about 1 liter) per person per day (or around 1 ton per 1,000 persons per day). Sludge creates a major disposal problem because of its high organic (BOD) content and possible content of toxic substances such as heavy metals.

One of the major constituents of municipal sewage and many industrial wastes is decomposable organic material, which reduces dissolved oxygen in the water through its high BOD. The lower the concentration of dissolved oxygen, the lower the carrying capacity of the system. Marine animals may be killed by a sudden drop in the water's concentration of oxygen, but the usual effect is to reduce their health or, if they are mobile, to drive them away as the waste spreads through the water. Disposal of sludge from sewage plants into coastal waters may create additional oxygen problems.

In addition to the depletion of dissolved oxygen, municipal waste discharges may introduce pathogenic organisms, settleable materials, heavy metals, and inorganic nutrients. A list of the typical contents of secondary treatment effluents is given below (375).

Constituent	mg/l	Constituent	mg/l
Gross organics	55	BOD & COD	25
Sodium	135	Potassium	15
Ammonium	20	Calcium	60
Magnesium	25	Chloride	130
Nitrate	15	Bicarbonate	300
Sulfate	100	Silica	50
Hardness (calcium carbonate)	270	Alkalinity (calcium carbonate)	250
Dissolved solids	73	Phosphate	25

While secondary treatment plants may reduce the organic load of wastewaters—when, and if, they are functioning—in the process they also mineralize organic forms of phosphate and nitrate (such as human wastes). That is, the "treatment" converts these wastes from a more benign organic state to a more active inorganic state (i.e., to dissolved phosphate or nitrate, which is thereby refined and made ready for immediate takeup by nuisance plants). So, the typical secondary sewage plant functions as a *nutrient refinery* and its discharge can serve as a *delivery system* for nutrient pollution (see "Eutrophication" in Part 3).

Typical secondary treatment removes up to about 30 percent of N and 10 percent of P. In one specific example, the sewage plant for Port Said (Egypt) was designed to remove 10 percent of P and 25 percent of N, with the "majority of nitrogen-ammonia" being "converted via nitrification to nitrates" (353). The result is an overload of nitrate and potential N-driven eutrophication (in this case in adjacent Lake Manzala, the receiving water body) (353).

Consequently, something better should be created than the typical sewage plant discharging its effluent along the shoreline. But very high level, or "tertiary," treatment—by which various chemicals are added to the effluent to "strip" the phosphate and nitrate along with metals and some other pollutants—is not usually practicable because of high cost and operational difficulties inherent to maintaining such an advanced system. Therefore, tertiary treatment is rarely an option.

Once rejected as too crude, preliminary and primary treatment plants with ocean outfalls are now coming back into fashion in coastal areas. The main reason is that the more elaborate "secondary treatment" has proved ineffective at nutrient removal or too troublesome or expensive for many communities. One very good reason to choose primary treatment with ocean disposal instead of coastal disposal of secondary effluent is that very often secondary plants malfunction and the effluent is of poor quality. Also secondary plants produce large amounts of sludge, which create a separate disposal problem. (See "Ocean Outfall Placement" above.)

Treatment plants do not operate automatically. They require constant attention, skill, and diligence. Even in the United States, the majority of secondary plants have major problems—in a 1989 survey it was shown that "more than two-thirds of the nation's wastewater treatment plants have known water quality or public health problems" (278). Because of this practical consideration—that secondary plants *rarely* are operated at design efficiency—it is often better to dispose effluent offshore where it will not pollute coastal waters and where it will be sufficiently diluted and dispersed to be ecologically harmless.

Bell and Greenfield (21) make the point that to protect the most sensitive coastal systems—such as coral reefs—very high dilution rates are necessary (1,100 to 10,000 ×) and that these can only be accomplished with appropriate siting of the outfall and correct diffuser design. In eastern Australia diffuser systems may range from 10 to 100 m long and be placed at depths of 10 m or more (21).

It will be seen that most of the options relate to changing the *location of disposal* rather than changing the method of treatment, with ocean outfalls a favored option where feasible. The farther to sea the better, within reason, because there will be deeper water for greater dilution and usually more favorable currents to sweep the effluent away (see "Ocean Outfalls" above). Once dispersed into the sea and diluted, the effluent becomes a nutrient resource for life in the sea.

Thus, it becomes clear that the main ecological strategy would be to dispose effluent offshore, regardless of the level of treatment it has received (see "United States, Hawaii: Success with Kaneohe Bay Ocean Outfall" case history in Part 4). Consequently, it would be a waste of money to build a secondary plant when a preliminary or primary treatment would suffice to prepare sewage for offshore disposal. Also, preliminary and primary plants have the advantage that they are more reliable and much simpler and cheaper to operate. Another advantage is that there will be less sludge to dispose from primary plants.

There are some concerns relative to land use for siting of treatment plants. The major concern is the temptation to site them in ecologically valuable wetlands which are vacant and often available if they are in public ownership. Also the wetland site may be popular because of its low elevation (less need to pump the sewage) and close proximity to a receiving water body. These schemes should be put to a hard test and other options thoroughly investigated before a wetland is preempted for a plant site.

Carpenter and Maragos (41) note that some countries have designed low energy-consuming treatment processes that rely more on human labor, removal of pollutants by aquatic plants and animals, and recycling of treated effluent to agricultural and irrigation lands, as in the 10,000-acre East Calcutta Wetland (W. Bengal, India) where fish and aquatic plants are regularly harvested from the treatment ponds.

The best solution to the problem of effluent is to reuse it, or recycle it. For example, with smaller treatment plants, the effluent can be infiltrated or used to irrigate farm fields, golf courses, or tree nurseries (Figure 2.70). In some cases it is practicable to use greater quantities of effluent. Irrigation use would be most appropriate for areas with low rainfall and nutrient-poor soils, which restrict horticulture and agriculture. The following guidelines to maximum permitted amounts may be useful (374):

Method of re-use	BOD mg/l	Number of fecal coliforms per 100 ml
Irrigation of trees and other non-edible crops	60	50,000
Irrigation of citrus, fruit trees, fodder crops, and nuts	45	10,000
Irrigation of deciduous fruit trees, sugar cane, vegetables, sports fields	35	1,000
Unrestricted crop irrigation, including parks, lawns, and golf courses	25	100

For semi-urban communities, the best sewage disposal option may be to use septic tanks. Wastewater introduced to the earth—when circumstances allow it to disperse harmlessly through the soil—may be a better option than to build a centralized collection system with all the problems created by effluent (see "Septic Tank Placement" above).

With regard to storm water runoff management, in coastal areas it is best to establish a completely separate (independent) system for collection and disposal, that is, *not* a combined sewage and storm water system. Because urban and semi-urban runoff may contain high quantities of BOD, coliform bacteria, and other pollutants, it should be thoughtfully disposed. Storm water can overwhelm the treatment plant, causing a large overflow of untreated pollutants to coastal waters (unless an ocean outfall is used). For example, Reuters (7/19/94) reports that Sydney Harbor (Australia) receives so much overflow that "40 to 70 percent of biological harbor contamination is from overflow." But it is possible to divert the most polluted "first flush" of runoff from rainfall (perhaps 0.25 cm of rain) to the sewage plant for cleanup while the rest is to be directed to the receiving body of water without treatment.

Carpenter and Maragos give the following comments and guidelines for wastewater disposal in coastal zones (41):

- More emphasis should be placed on siting of outfalls to take advantage of areas with greater flushing (potential to dilute sewage rapidly) and containing less sensitive ecosystems. Natural dilution and transport of sewage plumes by currents, tides, and wave action can lead to outfall placement and design to reduce sewage impacts.
- More emphasis should be placed on reducing treatment levels to primary or advanced primary that need less capital investment and fuel.
- Secondary sewage treatment, although removing most solids, results in high levels of dissolved nutrients in a form more usable to marine plants and may stimulate algae "blooms."
- Disinfection, which relies heavily on chlorine, can result in adverse effects on organisms near the outfall.

- Proper maintenance of secondary sewage treatment plants is complex and is often faulty due to inadequate training of personnel or inadequate supplies, repair, and maintenance programs.
- Achieving and maintaining tertiary levels of treatment has been unsuccessful in most cases. For example, some plants in Hawaii, designed for tertiary levels (pure water as the effluents), consistently fail to achieve even secondary treatment goals.
- Recycling of treated effluent to golf courses has met with some success (e.g., in Hawaii) while reducing discharges to coastal water.
- Mangroves and coastal marshes may be suitable sites for limited disposal of treated sewage provided the wetlands are maintained through regular harvesting or removal of biomass.
- A better funding strategy may be to opt for the higher initial costs to extend outfalls and diffusers to deep, open ocean waters far from land. With consequent less costly treatment for sewage required, maintenance and operating costs and complexity can be reduced to help offset the initial investment.
- Water quality standards established for ambient (receiving) waters can be an important control in establishing treatment levels and monitoring programs.

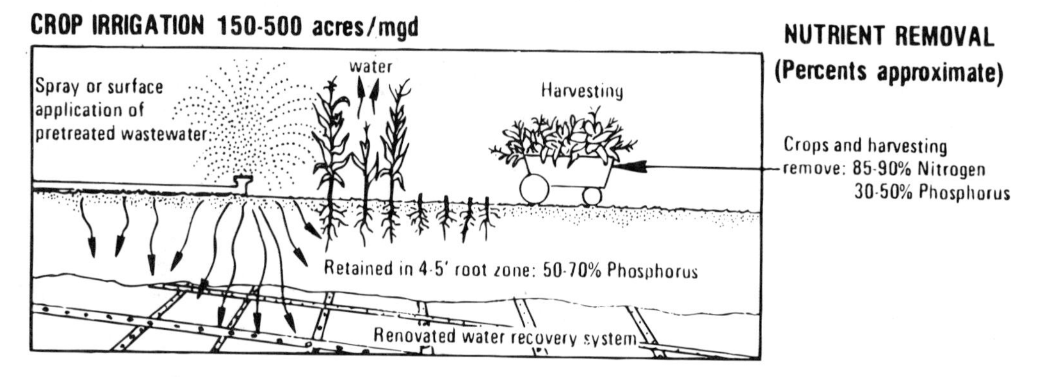

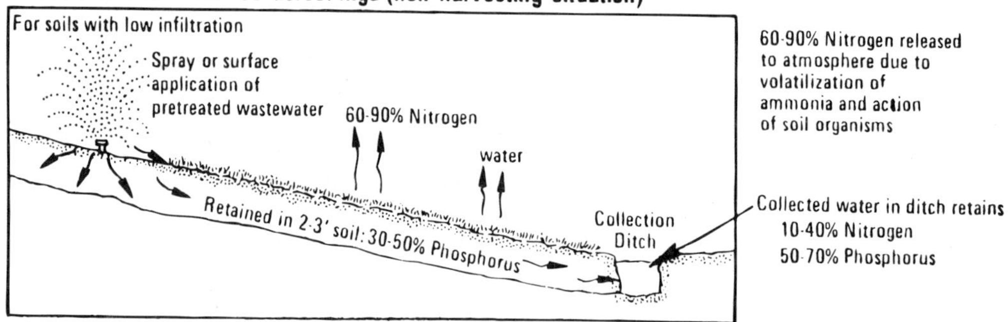

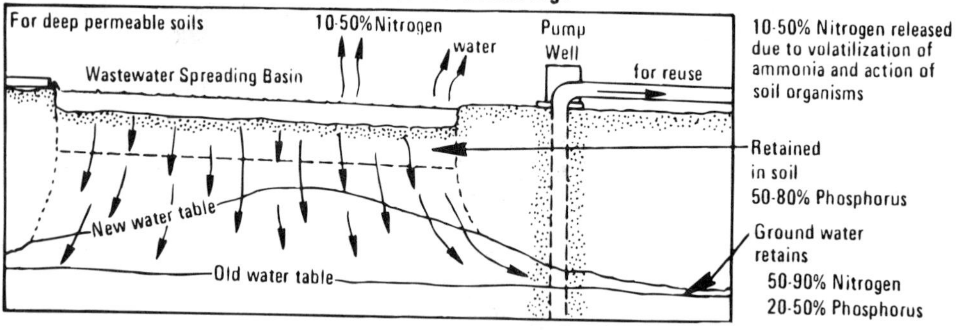

Figure 2.70. Results of various land treatment approaches to disposal of pre-treated sewage effluent (mgd = million gallons/day). (*Source:* Reference 255.)

2.36 SHORELINE CONSTRUCTION MANAGEMENT

Every shoreline construction project affects natural habitats and ecosystem function. Whether it be clearing, grading, landfilling, laying foundations, or erecting structures, there will be effects. Shore construction has complex ecological effects. Often short-term solutions cause serious long-term problems. For example, there are numerous examples of structures causing, or contributing to, destabilization of beaches.

Intertidal areas at the shore edge are especially important parts of the coastal ecosystem, including the mangrove forests, salt marshes, tide-flats, and beaches. They define the ecological boundary between land and sea. But, when shoreline development begins, these habitats are often the first to be obliterated—by bulkheads and other structures.

This section deals with structures that are built primarily to protect coastal developments from seastorms and erosion, although the principles apply to any type of construction (Figure 2.71). The principal shore protection structures include seawalls, revetments, groins, and breakwaters. Each serves a special purpose, and each affects the shoreline in a different manner.

Landfilling is the placement of fill to convert aquatic habitat to dry land. Often dredging and filling sites are located next to each other (such as for harbor projects) so that dredging operations can conveniently generate material required for landfilling. Otherwise, fill material is transported from a stockpile by heavy equipment (usually crane and dump truck) to the landfill site. Bulldozers and graders are normally needed for larger landfills to insure material is distributed, compacted and graded adequately. For projects requiring both dredging and filling, engineers attempt to design projects to balance the quantities of dredging and filling. This minimizes "down" time on site and reduces construction costs. Sides of landfills facing the ocean or exposure to other elements require structural protection to minimize slumping and erosion of fill from currents and wave action (361).

Revetments are protective structures usually consisting of sloping rocky walls. For *rubblemound revetments,* progressively larger sized stones are placed atop smaller stones. The outermost layer, which is directly exposed to the elements, consists of armor stone. The outer basal extension of the rubblemound

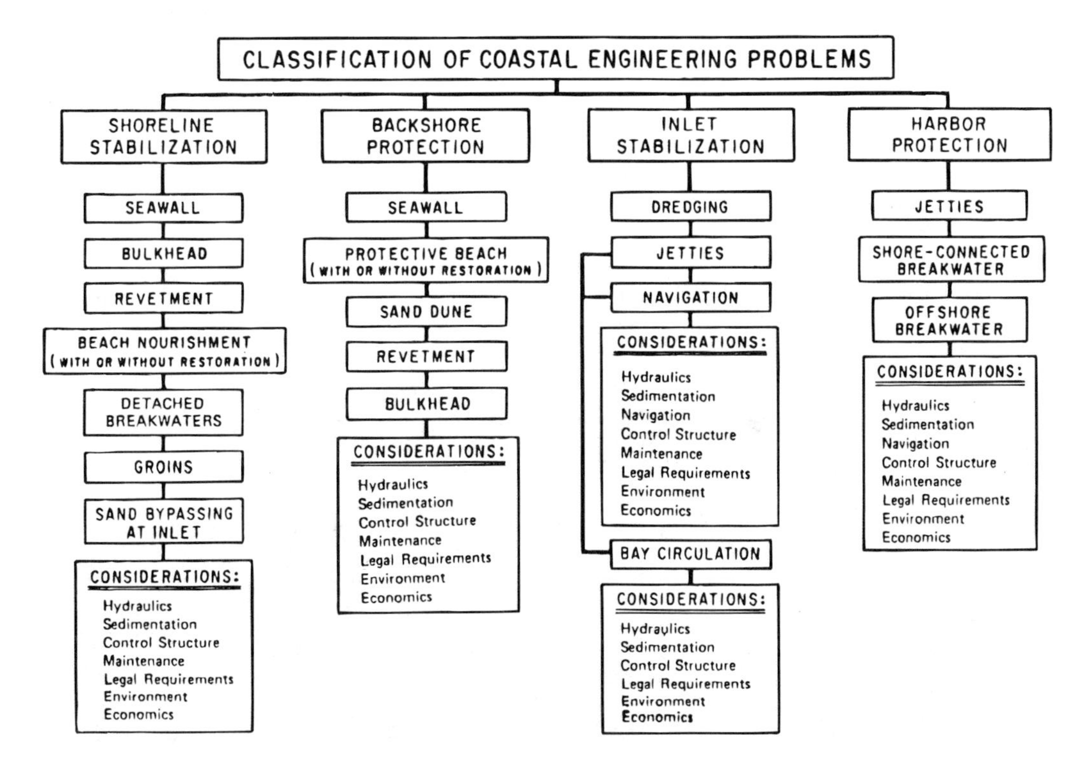

Figure 2.71. A variety of problems and considerations arises with each type of coastal construction. (*Source:* Reference 362.)

revetment (the toe) takes the brunt of wave action striking the slope, and the irregular surfaces and interstices between the stones are effective in dissipating wave energy and minimizing reflected waves. This protection dissipates wave energy with less adverse effect on the beach then occurs with a vertical seawall. But reflected waves can carry sand and other bottom material seaward and away from the shoreline and can undermine shore protection structures unless they are properly designed or placed on a solid rock foundation.

Gabions are shore protection components consisting of smaller rocks and rubble placed in baskets or cages. These can be constructed manually and serve as an alternative where heavy equipment and explosives are unavailable to move larger rock.

Seawalls, bulkheads, and sheet piling are functionally similar and consist of solid vertical walls (either metal, concrete, or masonry) to protect shorelines or fill land. A seawall is a solid barricade built at the water's edge to protect the shore and to prevent inland flooding. Seawalls are expensive and are usually suitable only for special situations, since they often compound shore erosion problems because in the long run they do not hold or protect the beach, the primary asset of shorefront property. The main advantage of vertical walls is that they require less material and take up less space; also, in harbors, ships can more conveniently berth next to docks with vertical walls. However, reflected wave energy is maximized with a vertical facing wall and hence these are only normally constructed along calmer shorelines such as within harbors, protected lagoons or embayments. Sloped or curving seawalls are often used at open sites (Figure 2.72).

Seawalls may be designed for several purposes: to absorb and reflect wave energy, to hold fill in place, and to raise the problem area above flooding elevations. Unfortunately, seawalls (including bulkheads and revetments) commonly accelerate the loss of beach sand as the wall deflects the wave forces downward into the beach deposit. This causes the sand to erode away seaward of the footing and the beach to diminish or disappear. Often the seawall is undermined and collapses.

Groins are rocky protective structures which project or occur seaward of the shoreline. Often these are elongated revetments having two exposed faces. Groins and protective moles are walls constructed perpendicular and connected to the shoreline. A groin is a dam for sand basically; a structure built to interrupt longshore sand movement (littoral drift) and trap sand to stabilize or widen a beach. Groins are constructed of timber, steel, concrete, or rock.

Prevailing currents and wave action normally move bottom sand suspended sand, and other sediment along the shore in a predominant direction (longshore transport). The sand is trapped on the updrift

Figure 2.72. Curved seawalls may be built at the most exposed sites because of their efficiency at reflecting waves. (Photo by author.)

side, causing the shoreline to prograde. However, the downdrift sides of groins can cause shoreline erosion because longshore transport is interrupted (Figure 2.73). Unless groins and similar structures are properly designed and located, they can cause as many shoreline erosion problems than they solve, causing shoreline accretion and shoreline erosion on updrift and downdrift sides of the structure, respectively (Figure 2.74).

Groins are effective only (a) when there is a significant volume of littoral drift, (b) when the drift carries coarse materials (greater than 0.2 mm), and (c) when the beach downstream from the groin can be sacrificed (the sand gained at one place is denied to another). The U.S. Army Corps of Engineers (362) recommends that groins be spaced apart 2 or 3 times their length (from beach berm to groin tip). The Corps also considers it desirable, and frequently necessary, to place sand artificially to fill the area between the groins, thereby ensuring a more-or-less uninterrupted sand supply to downdrift shores.

Figure 2.73. Groins typically build a wide beach on the updrift side but often cause serious erosion on the downdrift side. (Photo by author.)

Figure 2.74. Because the groins are inadequate to cope with storms, bulldozers are called in to rebuild the beach at Ocean City, Maryland (USA). (Photo by James McFarland.)

A row of parallel groins tends to force the littoral drift of sand offshore because some of the sand moves from tip to tip of the groin instead of moving along close to the beach, thereby causing sand starvation of the whole length of the beach. (See also "Beach Management" above).

A **jetty** is a structure that extends into the water to direct and confine river or tidal flow into an inlet or channel and prevent or reduce the shoaling of the channel by littoral material. Jetties located at the entrance to a bay or river also serve to protect the entrance channel from wave action and cross-currents. When located at inlets through barrier beaches, jetties also stabilize the inlet location.

Offshore breakwaters are aligned parallel to the shoreline and separated from it. Offshore breakwaters are constructed to provide safe passage through inlets and to prevent sand blockage. The outer (seaward) face of the breakwater absorbs wave energy and causes currents and wave action shoreward of the structure to be diminished. As a consequence, shorelines opposite properly located offshore breakwaters are less prone to erosion but still allow longshore transport to occur, thus reducing the likelihood of erosion along adjacent shorelines to either side of the breakwaters (Figure 2.75).

Breakwaters placed on the updrift side of a navigation opening may impound sand, prevent it from entering the navigation channel, and afford shelter for a floating dredge that pumps the impounded material across the navigation opening back into the stream of sand moving along the shore. In the absence of wave action to move the sand stream, sand is deposited and allows the shore to build seaward toward the breakwater, creating a groin-like effect.

These various structures that are intended to stabilize the beach (and the land behind it) may actually deflect and reduce supplies of sand to a level no longer capable of replenishing losses caused by storms. Then the beach may be lost, even while the structure remains. A row of parallel groins may force sand to move further offshore with the littoral drift, from one groin tip to another, instead of moving along close to the beach, thus causing a net loss of sand. Bulkheads tend to accelerate beach loss because they reflect the force of waves downward and back into the sand, which causes the beach to be scoured away. (See "Beach Management" above.)

Although it might be simplest to let nature take its course, extensive areas of the coast are already occupied and must somehow be maintained safely until retreat and other protective land use plans can be implemented. Yet even these systems should be allowed to remain as close to their natural dynamic states as possible. Also some structural interference may be necessary to stabilize inlets for navigation purposes.

New "soft engineering" technologies are available to deal with areas susceptible to sea level rise, storm waves, storm surges, coastal flooding, and persistent erosion. They emphasize the non-structural

Figure 2.75. An offshore breakwater is positioned to protect the mouth of this California (USA) harbor and to stabilize the inlet and maintain the adjacent beaches. (Photo courtesy of the U.S. Army Corps of Engineers.)

approach and require much less concrete and rock and are protective of coastal environments (see "United States, South Carolina" and "Indonesia, Bali" case histories in Part 4).

The U.S. Army Corps of Engineers, which has the greatest expertise and capability for large-scale public works projects in the United States, has generally favored an engineering or structural approach to shore protection, but recently has taken more interest in long-term land use controls and other nonstructural solutions (362). The solution is not to go exclusively with either structural or nonstructural techniques but to achieve a balanced plan emphasizing the nonstructural (or nature synchronous approach.

According to the U.S. Army Corps of Engineers (362), shore protection measures by their very nature are planned to result in some modification of the physical environment. However, planning and design enable the environment to be fully considered and understood, including a multidiscipline appraisal of the total impact of the project, which includes environmental quality as well as economic benefits. An accurate assessment of the pre-project environment (baseline) is essential, not only for initial planning and design, but also for later modifications or alternatives in response to monitoring environmental conditions.

The following three interrelated management guidelines are the most important for management of shoreline construction:

- **Conserve natural and protective features:** There is a need to protect to the maximum extent possible all the natural resources and features that protect the coast from storm surge and waves in high hazard areas, such as coral reefs, sand dunes, and mangrove forests (example management responses: setbacks; sand mining prohibition; mangrove cutting controls; coral reef protection).
- **Establish a setback line:** There is a need to delineate no-build zones including "high storm hazard" zones and keep all coastal construction inland of them (i.e., create a coastal development setback line) (see "Setbacks" above).
- **Design with nature:** This guideline stresses the importance of working *with* the tremendous forces of nature not *against* them, of guiding these forces, not countering them (e.g., through soft-engineering); that is, work in synchrony with nature.

While construction can alter the natural environment, the addition of structures can also provide habitat that can be beneficial in less productive environments. Structures can act as artificial reef and supply substrate for animal colonization. Jetties, rip-rap revetments, and rip-rap breakwaters are good examples because they have considerable surface area and irregular, articulated facings that can serve as special habitat ("cover" for sessile, motile, and cryptic species). Resident algal and animal populations also attract grazing and foraging fishes, while above-water rock formations attract waterfowl aggregations (165).

Structures such as bulkheads, pilings, and poured concrete revetments typically provide less valuable habitat because they are vertical, smooth surfaced, chemically treated, and/or have a relatively small surface area. However, most available substrates will be quickly populated in healthy aquatic systems. Encrusting forms such as barnacles are likely marine colonizers of pilings, boat hulls, and bulkheads. Many of these fouling organisms can be used as food by marine organisms (165).

2.37 SITUATION MANAGEMENT

While many coastal resource and environment problems can be solved by central government action, others need to be isolated geographically and solved locally. Centralized programs operate by using generic regulations and nationwide standards. This is clearly necessary for establishing an ICZM base for most countries. But, in addition, it will be beneficial to have a mechanism to focus the program on local *geo-specific, issue-specific situations,* or problem areas.

This situational approach has been attempted with varying degrees of success in many countries through various forms of Situation Management (sometime called *regional program* or *special area management*). The situation might be caused by aquaculture expansion, cyclone damage, resort development, coral reef damage, pollution, lowland agriculture, coastal rural settlements, or jeopardy to natural spectacles like flamingo or ibis migrations. The Situation Management program creates a complete planning and management strategy for one specific situation or problem area.

Situation Management (SM) has emerged as a successful method of managing development in complex settings. The SM process is based on the recognition that existing planning, legislation, and institutional implementation mechanisms are alone insufficient. It accepts the need to integrate the local community at the center of the planning and implementation effort, thereby making them a principal custodian of the resources being managed. (386)

A major SM objective is to allow the local community to accept the role of custodian of the coastal resource with a sense of ownership rather than to view the local community as exploiter of a common heritage. The principle is that coastal resources usually cannot be managed on a sustainable basis unless those who exploit them are committed to this goal and involved in the management process. This, in turn, requires a sound understanding of the social and political structure of the community, the special interest groups and stake holders, and an identification of unbiased local leaders who can play a stewardship role.

Another objective of SM is to resolve competing demands on a resource by selecting optimal sustainable uses. The process is to mediate amongst the competing users and to build a consensus on what use or uses can be harmonious and sustainable. It will incorporate the consensus achieved into a strategy that will lay out the steps for achieving such harmonious and sustainable use and it will put in place the institutional mechanisms needed for ensuring compliance and for implementing the strategy.

Situation Management is said to be a means to achieve resource management within a defined geographic setting. White and Samarkoon (385) state that it can resolve user conflicts and provide predictability for decisions affecting conservation and development interests. The limited geographic area of concern focuses management strategies and makes them effective relative to application in a broader area with more variability. It allows integrated management which includes complex ecological and institutional settings not possible to deal with in a larger context. This type of planning can use and apply criteria for management of resources which are sustainable because the cause and effect factors can be understood within the geographic, ecological, and institutional scope of concern. (385)

Sri Lanka's SM approach makes it possible to organize local communities to manage their natural resources (Figure 2.76). The central ICZM management group plays a catalytic role in organizing the local community, providing technical and financial support for the management effort which is formulated and implemented as a local community effort. Hence, the ICZM agency takes on the role of facilitator rather than that of a superior authority.

Timing is critical to maintain the interest and participation of stake holders and government alike in situation management. If results are not forthcoming within a reasonable time—say, 3 years—all concerned may lose interest. The national and local government are inherently part of any management efforts for coastal resources in Sri Lanka. It is best to realize that community efforts alone may not work without the support and joint participation of government. Sri Lanka's guidelines are as follows (386):

1. Agreement on need for SM and identification of participants (national or local)
2. Compile "Environmental Profile" of area through
   - Secondary information
   - Rapid area assessment
   - Designation of geographic extent
   - Identification of key management issues
   - Identification of key groups/individuals involved
3. Conduct planning/training workshops in the area to
   - Present environmental profile
   - Raise awareness of issues/management possibilities
   - Determine/refine management issues
   - Formulate management objectives
   - Initiate a task force at site level (with outside representatives)
   - Create understanding of resources and environment
   - Develop draft management plan
4. Provide the community with full-time facilitators to
   - Refine management plan
   - Organize work groups to perform planning duties
   - Assist implementation of planned projects
   - Evaluate results

The following are key principles to be kept in focus (386):

- SM planning is an open, participatory process.
- Consensus is the basic principle of SM.
- All decisions must be clear and well documented.
- The Task Force must be of manageable size, and commitment of the members is essential.
- Representation by all key groups must be ensured.
- Tangible results need to be documented and presented within a few months.

Wickremeratne and White list the following common mistakes in SM planning (386):

Allowing the process to take too long

- Drawing boundaries too small from a bio-political perspective
- Establishing too large a task force
- Omitting key group representatives or changing them
- Selecting a consultant or lead agency that is not neutral
- Allowing consultants to dictate task force directions and techniques
- Inadequate records of decisions made
- Too much time before members receive information on meetings, etc.
- Too much time between meetings
- Political intimidation of task force members
- Changing agency policy towards the SM plan
- Inadequate meeting facilities
- Inadequate staff/consultant services
- Inapproprite representatives of agencies or groups or jurisdictions
- Implementation procedures left to end of process

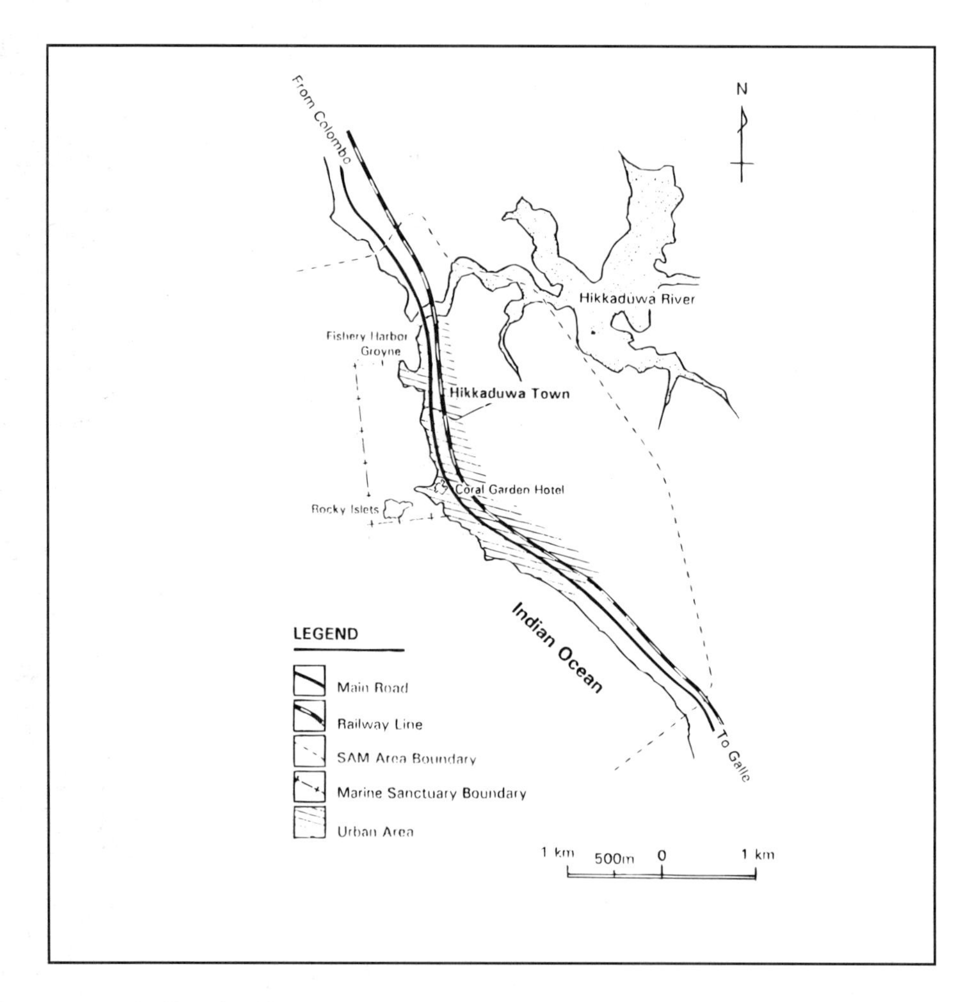

Figure 2.76. The Hikkaduwa "special area management" (or "situation management") site on the south coast of Sri Lanka is enclosed within the dashed line. (*Source:* Reference 385.)

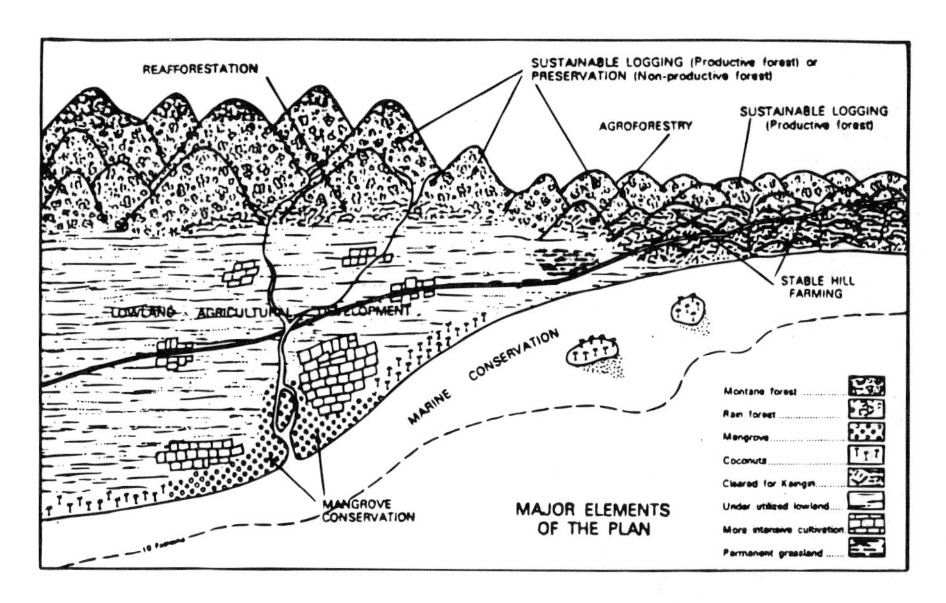

Figure 2.77. A model sustainable development plan for Palawan Island (Philippines). (*Source:* Reference 41.)

By its very nature, participatory SM is time consuming and sometimes stressful. It requires extensive individual "task force" member commitment. Sensitivity and adaptability to emerging concerns of those affected during the SM process will, in most cases, determine how successful the effort will be. In spite of the difficulties, it remains the best way to achieve sustainable resource use by mobilizing the beneficiaries—the local people—in the implementation of planning initiatives (386). But some SM endeavors are aimed more at isolating and solving particular resource problems than at arrangements for participation.

Major port and lagoon (or estuary) complexes are often the focus of great intensity and number of coastal resource conflicts, and great environmental degradation. As a result, the ICZM-type program might focus on one or several of these complexes. Such an SM program might encompass the urban center surrounding a major port along with the port itself and its associated bay, lagoon, or estuary. Or the management unit might be a stretch of coast including beaches, coral reef and lagoon, and surrounding villages and resorts. Or, perhaps, an inhabited island and its aquatic surroundings (Figure 2.77).

Combining "wetside" (coastal waters) protection with "dryside" (shorelands) management strategies offers the possibility of managing entire local ecosystems. Critical area designations and setbacks (exclusion zones) are particularly inexpensive and simple approaches to administer. Either strategy can be implemented on a site specific basis, commensurate with available information, staffing, or expertise. They can be reinforced with regional planning or broad sectoral planning of larger geographic scope.

It would be possible to devote a whole national program to situation management activities; that is, a national ICZM program could operate solely to guide a situation-by-situation management program. In this scenario, central government would not itself execute program components.

In a trial program of interest, the Association of South East Asia Nations (ASEAN) operated a coordinated series of situation management feasibility studies, one in each of the six ASEAN countries. The idea was to start with prototype situational management units in hopes that these will show enough promise to encourage the individual countries to implement nationwide ICZM programs or, at least, to attempt more local or Situation Management programs within each country. The project was funded by USAID and operated by the International Center for Aquatic Living Resources Management (ICLARM) in Manila (51).

2.38 SOCIAL IMPACT ASSESSMENT

Social impact assessment, as an addition to the environmental assessment (EA) process, is meant to explain the social consequences of environmental change. As addressed here, it is not meant to be a free-standing social assessment process but rather an interpretation of the particular chain of impacts that starts with environmental change and leads thence to social and economic change.

Social impact assessment (SIA), as it is used here, is a way of trying to figure out what can and what does happen to people, their organizations, and their communities as a result of a particular environmental change. It involves the use of social science techniques to make predictions and to monitor results and evaluate outcomes. It is aimed at fair dealing with the various people affected.

SIA is an anticipatory endeavor in its predictive role and an evaluative endeavour in its monitoring role. Its primary goal is to facilitate decision making by predicting the full range of social "costs and benefits" of proposed development projects (94). Its secondary goal is to improve the design and administration of projects in order to increase benefits to the common people. Cultural acceptability and social equity are essential components of project planning and hence of SIA. In the SIA modus, social sciences complement the natural sciences in preparation of EAs. The former explains the impacts of environmental changes on people's lives and the latter effects on natural systems, according to Wenzel (379)

Once a development project has been proposed, social science techniques enable planners to identify potential, perhaps inadvertent, impacts of environmental change on an area's social, cultural, and economic fabric, and to determine how the physical impacts of the project will affect opportunities for work, cultural values, settlement patterns, and ways of life. (379).

Environmentally induced social impacts are complex and intertwined. But, because human behavior is not governed by simple laws of cause and effect, social impacts are often difficult to predict and, therefore, scoping—setting the limits of the information needed for SIA—is particularly important. Careful scoping can help identify the best indicators of social change—factors such as crime, unemployment, or family dissolution. (379)

Burbridge and Koesoebiona (34) note that there are two broad considerations concerning the human condition. The first is assessing the negative or positive impacts of development programs and projects on the people who will be affected. The second concerns consultation with these people in the project formulation, planning and implementation processes.

Checklists of potential impacts are commonly used as are flow diagrams and matrices of the impacts of various alternatives. Surveys or questionnaires are often used to discover people's attitudes toward a particular project's impacts. The results provide feedback to policy makers and also give people a sense of participation. This combination is essential if the project is to be successfully integrated into the life of the community. (379)

A sometimes useful approach to the analysis of social impacts is to delineate the types of expected outcomes. People may experience impacts in many areas of their lives including the following (94):

- Economic (e.g., employment)
- Environment-habitat (e.g., relocation)
- Commercial (e.g., price changes)
- Transportation (e.g., accessibility)
- Social (e.g., changing roles)
- Biological (e.g., health risks)
- Psychological (e.g., stress)

Some of the key social impact dimensions are (94):

- Probability: how likely is the outcome?
- Primacy: is the outcome a direct result of the development or is it an indirect one, part of a chain of predictable events flowing from the development?
- Onset: at what point will the outcome occur, immediately or later on?
- Duration: is this a temporary effect or a permanent one?
- Magnitude: how extensive is the outcome?
- Distribution: who will be affected?
- Scope: what will be the geographic limits?

Data gathering and analysis are key elements in the SIA process. A useful procedural framework for this process includes (94):

1. Profiling—Identifying existing conditions, providing a baseline
2. Projecting—Predicting likely changes and their effects (e.g., by using results from similar areas, extrapolation of trends, or creation of scenarios)
3. Assessing—Determining the importance of the effects and ways of avoiding or mitigating them
4. Evaluating—Considering the acceptability of the impact of the project and its alternatives

Once the "right questions" are delineated, the methods for answering them may include the following:

1. Collection and assessment of existing information (e.g., census data, vital statistics, previous studies)
2. Survey methods (e.g., sample surveys, opinion-leader or Delphi panel surveys)
3. Participant observation (e.g., long- or short-term community studies)
4. Unobtrusive techniques (e.g., monitoring media such as newspaper editorials or call-radio shows, observing behavior in public places).

Each method has its own strengths and weaknesses. Secondary data are often available but, having been gathered for other purposes, they may not directly address your particular questions. Survey techniques provide breadth of coverage, but the use of standard questionnaires (in either self-administered or interview format) may not yield sufficient depth. Participant observation, which involves one or more researchers living in an area and studying it intensively, does provide depth, but it is very time consuming, thus limiting coverage. Unobtrusive techniques can yield interesting data, but they all too frequently have built-in biases, which must be taken into account (94). The methods of "rapid rural assessment" are useful.

In designing SIAs the following three criteria should be addressed (94):

1. Appropriateness of methods for the situation, including cultural, ethical, and technical factors. Methods used should "fit" with community patterns of proper behavior and the confidentiality of information provided should be respected; technical aspects should also be appropriate (e.g., phone surveys are inappropriate in areas where few people have phones).
2. Expected quality of resulting data and their likelihood of answering the research questions. Methods used should yield reliable and valid results that directly address the key questions.
3. Time, money, and staffing concerns. Time available to do the assessment before decision making, the available budget, the availability of local expertise, and the advisability of utilizing outside consultants all need to be taken into account.

SIA as a decision-making process is used to inform planners, policy makers, and/or politicians. But SIA can also be viewed as a process of participation involving all stakeholders (affected persons), with emphasis on public involvement in decision-making, in contrast to the more "technical" approach. Both approaches, however, have a common need for information to be utilized by either planners, politicians, the public, or the technical specialists.

As an example, guidelines for the multi-national flood action program (FAP) for Bangladesh state that, when possible, social impacts of environmental change should be identified and quantified. The costs of any mitigation programs are to be calculated and included in the overall economic analysis of the project. The FAP SIA involves the six steps outlined below (condensed by the author). In individual project analyses, full SIAs involving all the steps are to be undertaken; in regional studies, SIAs would exclude Step 4 (131):

Step 1 — Identify the main social groups likely to be affected by the project, both inside and outside the proposed project area, with special attention to the rural poor and women. The numbers of households by attribute (e.g., richer and poorer) should be estimated. Such social groups may include but not necessarily be limited to men and women from the following types of households: agricultural laborers, capture fishing families, aquaculture families, boat operators, rickshaw pullers, rickshaw owners, petty traders, households whose land is acquired, households whose land is erosion-prone, female-headed households, workers in rural industries, farmers (marginal, small, medium, and large farmers). Review the processes of socio-economic change in the area (e.g., in agrarian structure) and attempt to forecast likely changes in social organization between the time of the study and the future "with and without project" situations.

Step 2 — For the main social groups likely to be affected by the project, describe the economics of their livelihoods in terms of subsistence production and cash income. In doing so, any use of common property resources (e.g., fish stocks, grazing lands) will be addressed, as will conflicts with other groups over natural resource use. As far as possible, quantitative estimates should be made of incomes (including the imputed value of subsistence production) and any common property resources used and employment patterns and seasonal variations in these through the year.

Step 3 — Estimate the effects of environmental changes caused by the project (including alternative projects or designs) on the livelihood of these different social groups. Again, where possible, quantitative estimates should be made of changes in income and employment, laborers, fishing households and of other social impacts such as dislocation resulting from resettlement. The SIA analyst should give an overall impact rating for all groups (perhaps on a numerical scale).

Step 4 — Estimate the overall environmental impact of the project on income distribution in the project area and in the proportion of the overall income going to be different classes and make clear any limitations of the analyses.

Step 5 — Assess, likely changes in the general quality of life of men, women and children in the area affected by the project. Indicators of quality of life will include, but not be limited to, security of life and livelihood for different groups, the extent of social conflict, health, nutrition, ease of communications, and safety.

Step 6 — Estimate the initial and recurrent costs of any environmental mitigation neasures, including compensation, needed to offset adverse effects of the subject project (e.g., costs of resettling and retraining fisherman households or those whose land is acquired).

2.39 STRATEGY PLAN

Strategy planning is the key to the whole integrated planning and management process. It involves all the preliminary investigation, data collection, dialogue, negotiation, and draft writing that is necessary to enable the government to define the issues, to understand its options and to proceed to authorize a specific ICZM program (that comes in the Program Development phase; see Part 1.6). The Master Plan that follows contains the detailed plan for management operations (see "Master Plan" above).

The Strategy Plan will provide the justification and lay the foundation for the ICZM program. It will recommend the legislation or executive order needed to authorize program development. It should (a) assign responsibility for the program to a particular agency or council; (b) authorize the funding necessary for program development; (c) state clearly the objectives of the program; (d) recommend a method for collaboration among the various sectoral agencies and private interests involved; (e) state the time limits involved for various stages of program development; and (f) require a specific step-by-step program development and organizing process. (See also "Strategy Planning" in Part 1.5.)

The process by which strategies for the conservation of marine, as well as terrestrial, resources can be planned consists of six interrelated and interdependent planning functions, according to Miller (246). In the planning stage, these functions may be considered as sequential steps. Later, when a strategic system has been set up, they are carried out together:

1. Gather information relevant to the extent and status of marine resources and the problems and issues related to their protection and use. Store the information in a manner useful for, and easily retrieved by, interested users.
2. Verify and gather additional and detailed information in the field. Study areas which are representative of resource types and particular problems to gain relevant field experience.

3. Analyze information to determine concentrations of living resources, the nature of ecological processes important to major species and productive habitats, and locations where present or potential human activities are concentrated. Synthesize the results of analyses to show where potential conflicts and/or compatibilities lie. Present information to guide interested organizations and individuals on strategies for conservation action.
4. Determine priorities for action to address the key problems and issues pertinent to support systems for humans and other species of the Region, and those related to possible conflicts between human activities and natural resources.
5. Identify activities, which through agreements, conventions, and field projects, and in collaboration with other institutions and individuals, can solve the problems.
6. Formulate a process to monitor the program as a whole to learn from real-world experience and to improve the efficiency of future project and program operations. It may be necessary continuously to monitor the status of the coastal environment in areas of concern.

Experience in several countries confirms that, while the policy and planning agenda may cover a wider coastal zone, the operating program itself will usually target the narrower edge zone. In the first ICZM program in the United States—which was for San Francisco Bay—authority was granted to a regional coastal commission (Bay Conservation and Development Commission) for a zone width of only 100 feet, extending shoreward of the water's edge (mean high water), but this was enough to protect the bay. In another example, the ICZM program for Washington State covers only 200 feet of the edge zone. Such narrow zones were possible because in the United States numerous other federal and state laws control actions in the intertidal area and near coast waters seaward of the edge zone and in the shorelands landward of the edge zone.

In fact, the two major program development phases of coastal management may well utilize different boundaries; that is, a wider area will quite likely be identified for the *planning* phase than for the subsequent *management* phase that evolves. The zone of management that emerges will in most cases be considerably narrower than the initial zone of planning for ICZM. In fact, one purpose of the strategy planning phase is to refine the boundaries and reduce the width of the designated coastal zone to the minimum needed for the management phase.

For strategy planning, the designated coastal zone should include, at a minimum: (a) all coastal lands that are subject to storms and flooding by the sea; (b) all intertidal areas of mangrove, marsh, deltas, salt flats, tideflats, and beaches; (c) all permanent shallow coastal water areas such as bays, lagoons, estuaries, deltaic waterways, and near coast water that include seagrass meadows, coral reefs, shellfish beds, submerged bars; and (d) all small coastal islands and other important features (See "Boundaries" above). Each of these can be set up as an administrative "tier" with somewhat different management approaches (see "Tiers for Management" below).

Because strategic planning requires that the planners consult with a wide array of interests, both governmental and private, numerous sectors are involved and strong political and economic pressures come into play. It is the accepted wisdom of coastal management experts that only a truly integrated program (i.e., one that includes all the major economic sectors affected) can succeed fully. If important stakeholders are left out—e.g., tribal chiefs, port authorities, housing departments, tourist industries, fishermen, economic development planners—ICZM will probably fail. In fact, one of the functions of ICZM is to provide a framework for coordination of such a wide array of interests.

Data collection is an important part of the strategic planning phase. It is during this phase that the most important decisions will be made about the future of coastal programs, or even whether it will have a future. Clearly, these decisions should be made in the most data-rich of circumstances so that the consequences of taking or not taking specific actions are predictable. Information needs for the Strategic Plan depend upon the issues to be addressed and these vary considerably from country to country. (See "Information Needs" in Part 3.)

In collecting data and synthesizing and mapping information for the Strategic Plan, the purpose is to bridge between the policy and the program development stages. The information is to be used in getting a program approved (or disapproved, if it should be discovered that the country does not need or want ICZM for whatever reasons). Therefore, the kinds of information needed for the strategic plan are those that will enhance the decision making process, that clearly depict the tradeoffs between the present situation and an integrated management approach, and that lead to the clearest and least ambiguous

Figure 2.78. In Bangladesh, hundreds of thousands of people are dependent upon coastal aquatic resources for sustenance and cash. (Photo by author.)

set of objectives and mandates to the governmental agencies who are to manage the coast, should the country decide to establish one. (See "Database Development" above.)

One way to think of the Strategic Plan is that its purpose is to answer questions in the minds of decision makers in government. The answers will lead to decisions to authorize or not to authorize the next stage (Program Development) or to request more fact finding. To this end, the strategic planning stage should be organized to anticipate the questions that decision makers will ask and to provide the data to answer these questions, whether they are national or regional in scope.

The following are the types of questions that are often addressed in basic strategy plan formulation:

- Which coastal resources are seriously degraded; to what level have yields fallen; what are the economic consequences; what actions are needed to correct this situation?
- What are the causes of the degradation; what type of developments and activities need to be controlled; what are the economic effects of the controls; in consideration of the variety of possible tradeoffs and their effects, what actions are recommended?
- Who are the principal users of coastal renewable resources; how many jobs are at stake; how much income and foreign exchange earnings are involved in tourism, fisheries, and other resource dependent industries (Figure 2.78); what further losses are expected if integrated management is not implemented?
- What are the priority issues; what critical habitats need special protection; what species need protection; what is the best approach, regulation or protected areas?
- What can an integrated management program do to prevent loss of life and property from coastal natural hazards such as seastorms and beach erosion; what are the benefits of combining hazards protection and resource conservation in a single coastal program?
- What are the existing government programs for coastal renewable resource sustainable use (conservation) at national and local levels; how effective are they; what are the shortfalls; what changes in governance are recommended?
- How effective are the existing mechanisms for interagency and inter-sectoral coordination on coastal matters; what can be done to improve the situation; what actions are recommended?
- Should the management emphasis be on nationwide controls or on regional level programs?
- What are the expected net benefits of an integrated national ICZM type approach, in economic and social terms; how can such a program be funded?
- Who are the major proponents and opponents of the proposed management program?

- What is the status of personnel training for integrated coastal management; do we have sufficient expertise; what can be done to improve the situation?

Many countries may not initially go to a full-scale, integrated (ICZM-type) approach, but rather adopt a less sweeping national program or use situation management (see subsection above). Such countries should, nevertheless, strive for the ICZM goals of comprehensive and integrated coastal resource management initially, in the strategy planning phase. For example, managing mangrove forests for the sole benefit of the fisheries sector might fail if it lacks mechanisms to incorporate the interests of local villagers, forest industries, upland agriculture, public health, tourism, port development, and so forth.

2.40 TIERS FOR MANAGEMENT

For planning purposes, it may be useful to subdivide the coastal zone into planning/management subunits, or tiers, that are compatible with existing ecological and political boundaries. Each tier would represent a management unit characterized by different resources, issues, and jurisdictions, as, for example (72):

- Tier 1: **Marine and coastal waters**—The open water part of the coastal zone beyond the transitional area of wetlands, tide flats, etc., to low water mark; permanently submerged resources. High level of central government interest and authority.
- Tier 2: **Transitional area**—The shallow waters of the edge of the sea; the transition from open sea to the land, including intertidal mangroves, tideflats, and beaches, as well as floodable areas. Central government interest; high level of interest regionally and locally.
- Tier 3: **Shorelands**—The lands directly adjacent to the transitional area which generate significant impacts to coastal resources; high value for many purposes; urban waterfront development usually disrupts the edge zone and generates pollution. High level of local interest.
- Tier 4: **Uplands**—A fourth tier should be added to include "uplands"—river valleys, drainage ways, and watersheds—if the ICZM program is designed to address effects of watershed clearing and soil erosion, pesticide or herbicide runoff, and alterations of hydroperiod caused by dams and reservoirs.

The tier system recognizes that management programs can be facilitated by dividing the coastal zone into several distinct subdivisions providing separate management opportunities in each. In the above system, separate management schemes can be devised for each tier as follows: Tier 1, the open sea; Tier 2, shallow and intertidal area; Tier 3, the immediate shorelands; and Tier 4, the hinterlands. Clearly, each of these tiers has very special values, problems, and jurisdictional situations. This approach is most useful for strategy planning even if, in the end, management is homogenized.

In this approach, Tier 1 would be all the marine and coastal waters (wetside) and would be adjusted to reflect fisheries and port sectors, the importance of nearshore spawning and nursery habitats, and regional and international matters.

Tier 2, or "edge zone" (wetside), defined to include all the transitional areas between land and sea, would contain the wetlands, coral reefs, seagrass beds, estuaries/lagoons, beaches, tidal rivers, etc., would be the primary coastal zone for many ICZM programs.

Tier 3 (dryside) would include not only the valuable shorefront that borders the transitional area also coastal lowlands and floodlands. This might be the most complex to manage because of the necessity to control land uses.

Tier 4 would be an area for collateral, not primary, authority for ICZM; however, coordination and cooperation would be expected from those who do address environmental management in watersheds.

An expanded version is a seven-tier arrangement wherein three additional areas are distinguished—coastal flood plains, offshore islands and reefs, and the exclusive economic zone (EEZ). The additional subdivision allows for a more focused planning activitity, leading to a more flexible program format, as shown below:

- Tier 1—Upland areas of major influence;
- Tier 2—All coastal lands subject to storm flooding by the sea;
- Tier 3—All intertidal areas of mangrove, marsh, deltas, flats, tideflats, and beaches;
- Tier 4—All permanent shallow coastal water areas ("wetlands") such as bays, lagoons, estuaries, deltaic waterways, and nearcoast waters that include seagrass meadows, coral reefs, shellfish beds, and submerged bars;
- Tier 5—All the sovereign nearshore coastal waters (territorial waters), often to a distance of 12 miles offshore;
- Tier 6—All small coastal islands and offshore reefs; and
- Tier 7—The country's declared Exclusive Economic Zone (EEZ).

While tiers are most useful for strategy planning for ICZM, the coastal zone must still be viewed as a single interacting system. For example, Chua Tia Eng (pers comm) warns that "one should not lose the holistic aspect" because interactions between tiers may be important in the management strategy.

2.41 TRADITIONAL USE ARRANGEMENTS

There is a striking similarity among the traditional methods of resource management that have been devised by the peoples of very different origins and cultures (299). This commonality of success demands attention to the possible benefits of continuing traditional methods where appropriate, particularly the protection of the outlying community's rights to exclusive use and conservation of its traditional fishing areas. (See "Traditional Uses" in Part 3).

Presently, three approaches can be suggested for integrating traditional uses into coastal management programs, one using a custodial approach, another a regulatory approach, and a third using a combination of the two (see "Protected Natural Areas" above). One result of officially recognizing traditional (or "customary") conservation in a coastal management program will be to confer property rights over an area of "the commons" to local, village, or tribal authority (see "Commons" in Part 3). This can be accomplished by a variety of techniques for transfering property use rights, such as designating the fishing area as a "resource reserve."

Under the custodial approach the subject area would be designated a marine "protected natural area" with joint authority between the local village or tribal governing unit and the national government. With the traditional conservation area set aside as a protected area, rules of access and resource use can be formulated to give priority to the traditional users. In this situation, where the traditional group has exclusive, or priority, access to the resource, more care may be exercised in the quantity of fish or shellfish removed, abstinence may be practiced during key periods (e.g., spawning), and less destructive fishing methods may be used. In some cases, this will provide for better conservation of fishery resources than nationwide, centrally operated management. A protection program of the custodial type can be enhanced by designating or acquiring (if necessary) the appropriate amount of shorelands, transition area, and open water as a protected area, such as a "resource reserve."

A second approach is to designate the subject area for situation management and protect its resources through a special set of regulations and guidelines enforced through the "police power" of government (see "Situation Management" above). This regulatory approach could be combined with the custodial approach through the Situation Management process. The combination would allow a core area under exclusive local jurisdiction and a buffer area under more flexible management. A special management authority could be created.

The combined approach would enable special management authority to delegate rights of use to the area. The central government could keep some control by "leasing" the area to the local authority and by holding memberships therein (the lease could be revocable for cause).

In one example, protection was advocated for the Morovo Lagoon in the Western Province of The Solomon Islands in order to preserve "customary fishing rights." The local culture (population 5,500) is oriented toward the sea and sea fisheries and depends on maintaining fish stocks. This requires not only control over the fish harvest, but protection of habitats against adverse impacts of forest clearing and logging, mining, coastal agriculture, and exploitation by cruising yacht tourists. In this case the management authority is the "Area Council." (12)

To study the potential benefits, a Commonwealth Science Council Study (328) recommends the following guidelines for the assessment of traditional use for the South Pacific (emphasis on fisheries):

- Adaptability of traditional systems in the face of changes in perception, technology, and society
- The place of traditional knowledge in society—its role in social status and differing sex roles
- Responses to commercial development
- Adaptability of traditional fisheries management regimes
- Effective means of obtaining and recording traditional knowledge
- Definition of traditional fisheries
- Clarification of nature of coastal resource use rights
- Social unit(s) on which rights are based
- Principles of boundary delimitation
- Allocation and transfer of coastal resource use rights
- Procedures for sharing of resources with outsiders
- Traditional conservation practices
- Dispute resolution mechanisms

2.42 TURBIDITY MEASUREMENT

Water that contains an excess of "suspended matter" is described as *turbid.* The parameter called *turbidity* measures the **degree** to which water has become turbid and thereby indicates the amount of suspended matter and planktonic life that it contains. The suspended matter may be organic detritus, fine inorganic particles; various colloids, or exotic matter. The planktonic organisms may be any of numerous species of microscopic flora or fauna.

Transparency of the water is controlled by the amount of suspended particulate matter (SPM). While the amounts of SPM (in mg/l) may be the subject parameter, to measure it directly requires filtration, an inconvenience. According to Salvat (309), measuring the weight of particles in suspension is not satisfying because only a few big particles can make up 90 percent of the weight of the particles, these big particles having little or no effect on the turbidity itself. Such a measure would have to be combined with a granulometric study of the particles in suspension. A BOD or COD test will detect how much fine organic matter is dispersed in the water, including plankton (in mg/l of potential oxygen consumption) (see "Oxygen: BOD/COD Measurement" above).

However, the *effect* of SPM can be measured readily by either of the following: (a) determining the transparency of water (as vertical disappearance point) with a Secchi Disk or by (b) determining its turbidity as an "NTU" rating, using an electronic turbidity meter.

The Secchi disk provides a measure of maximum sight distance, which may be called *transparency, visibility,* or *clarity.* The disk is usually about 20 cm, or 8 inches, in diameter; painted with a four-part black-and-white "propeller" design; with a cord fastened to its center. It is lowered into the water until it disappears from sight. The length of cord paid out to the point of disappearance is recorded as the existing water transparency (see also "Transparency of Water" in Part 3).

The Secchi disk is often not appropriate for shallow tropical waters because it may remain visible even when reaching the bottom. Other problems are (a) it can be used only in daylight; (b) the readings may be affected by cloudiness, glare, and water surface disturbances, and (c) the readings will not reflect differences in turbidity of different water strata and horizontal and vertical measurements will often differ. But the Secchi disk can also be used underwater by divers to measure horizontal transparency with good results.

Color can provide a useful index to water transparency, particularly for aerial survey—reconnaissance, satellite readouts, or aerial photography. Figure 2.79 shows the general relation between surface color and Secchi disk readings. An approximation to optical and/or mechanical parameters can also be made using the information below. For example, turbidity readouts of 7.5 and 0.1 are roughly equivalent to M^1 of 1.5 and 0.04, respectively.

Turbidity is usually measured in arbitrary units, originally called Jackson turbidity units (JTUs), which indicated the distance through a sample of water that one could see a candle flame burning. Now electronic devices are used for the purpose and they measure "light scatter" (or "transmittance"). The

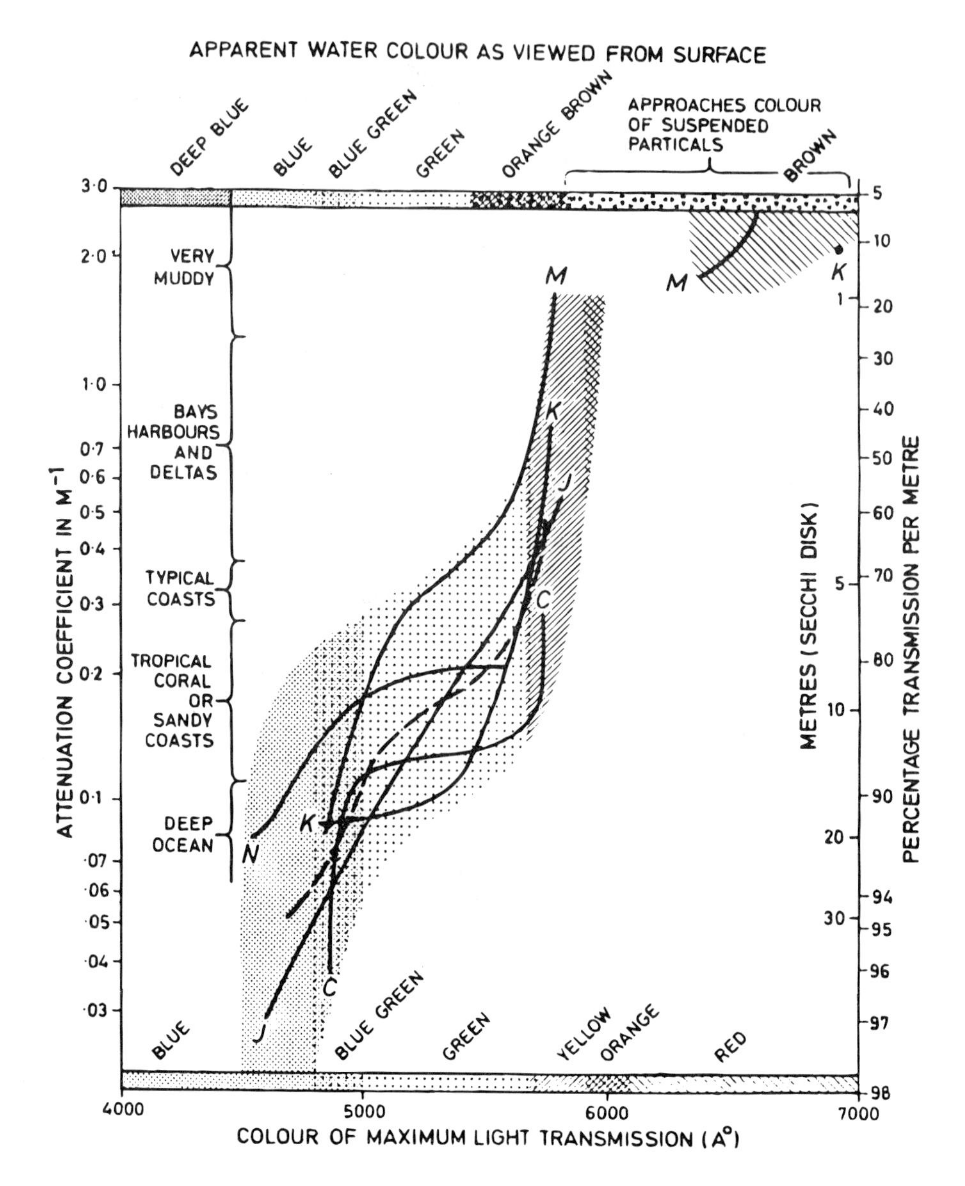

Figure 2.79. Water color can be used as a rough index of water transparency. (*Source:* Reference 174.)

standard is the nephelometric turbidity unit (NTU), which is based upon a reference standard of Formazin. Note that, because all instruments are calibrated to the equivalent of 1 mg/l of SiO_2 = 1 NTU, all their readout units—NTU, JTU, FTU, etc.—are comparable.

Several types of commercial "turbidity meters" are available, for example, the Hach meter made by HF Scientific (3170 Metro Parkway, Ft. Myers, Florida, USA, FAX: (813) 332–7643). The Hach meter is a simple, small, battery-powered device into which water samples are placed in small glass vials ("cuvettes") for direct digital readout—samples can be processed at the rate of one/minute (Figure 2.80). Its drawback is that the size and type of particles found in coastal water samples may cause the digital readout to jump about rather wildly as the lighter particles swirl and the heavier ones quickly settle to the bottom of the 65-mm water column in the cuvette (4 to 10 sec for fine sand of 0.1- to 0.05-mm diam; slower for organics).

Allowing the sample to settle out does not help because non-agitated samples will give artificially low readings since the particles that cause turbidity will have fallen out of suspension—the difference can be great, nearly an order of magnitude. This means the operator has to employ an objective method

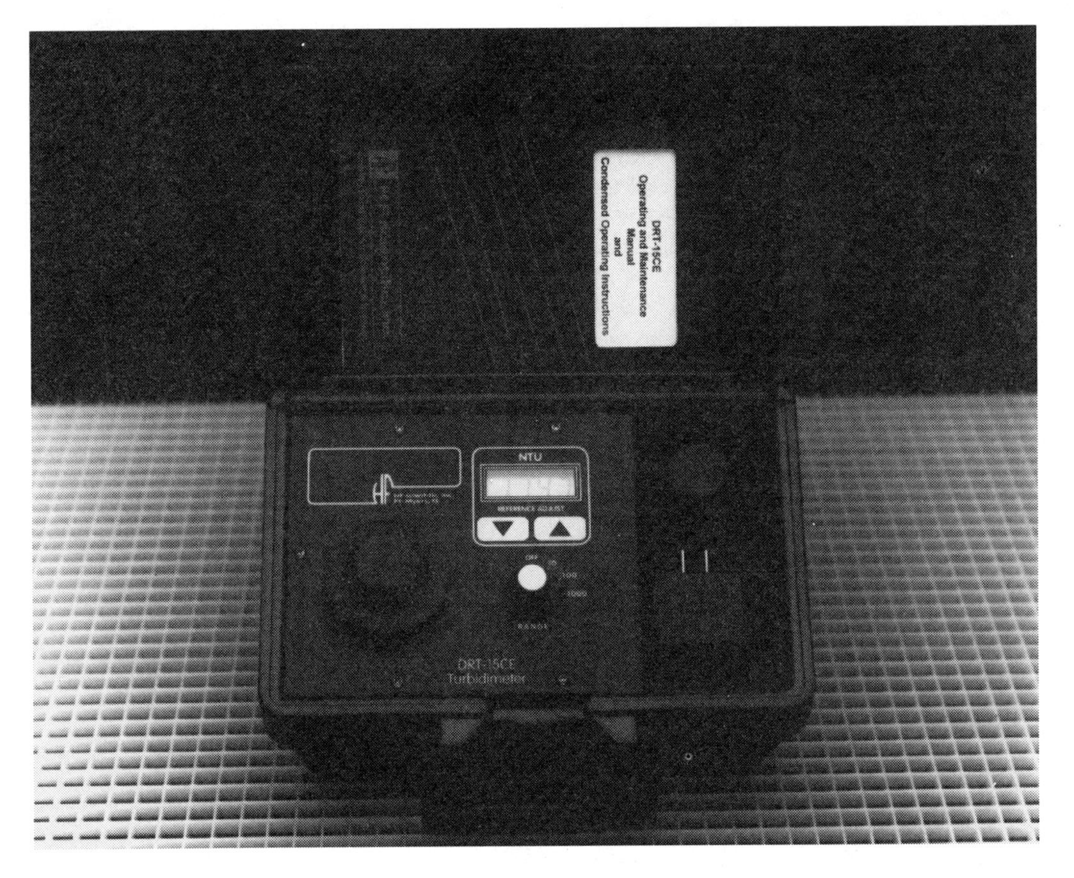

Figure 2.80. A direct reading turbidity meter (in NTU) into which water samples are inserted. (Photo courtesy of H.F. Scientific, Inc.)

for determining (a) when in the rapid settling process to read the meter and (b) which of the jumping numbers to read. The author uses the following approach: (a) I first swirl the cuvette and place it into the optical well of the meter; (b) after waiting just five seconds I immediately record the next to lowest number that flashes on the display as the appropriate NTU—this gives me consistency, if not true comparability.

Turbidity meters are useful for monitoring coastal projects, like dredging, which could put excessive amounts of matter into suspension or activities, like sewage disposal, which could trigger destructive plankton blooms. Of course, the baseline condition has to be known in advance so that the monitoring can reflect a change over the base condition (see "Monitoring and Baseline" above).

A major problem faced by many field workers, including the author, is the lack of available conversions to enable a person to convert from one index of SPM, turbidity, or transparency to another. For example, a Secchi disk lowered from a boat may be inappropriate for shallow oligotrophic waters because "disappearance" depths may not be reached. Therefore, one must collect data by using a horizontal (diver-held) Secchi or a turbidity meter or, perhaps, by weighing the content of suspended solids. Also, pollution regulations, or guidelines, are often given in one index (e.g., mg/l suspended solids) while field measurements are taken in another (e.g., turbidity units).

Also, oversight monitoring has to be done by meter so there is real time, direct, readout of the condition being monitored. While this is best done by an electronic turbidity meter, this instrument does not readout in mg/l of suspended solids as the standard is expressed; therefore, a conversion scale is needed. There is a complexity of variation in SPM data that comes from variation in the type of suspended matter present (plankton, silt, coral "dust") and differences in the size, shape, and "buoyancy" characteristics of the particles present. For this reason we tried to reduce the variables by calibrating our Turbidity meter against the actual silt to be dredged (See Case History 4.18, Ujung Pandang Port Urgent Rehabilitation Project, in Part 4) as shown in Figure 2.81.

The conversions were needed for creating environmental management and monitoring plans for shoreline construction projects (beach nourishment and port expansion) under the Indonesian environmental impact management program. AMDAL (see "Indonesia" case histories in Part 4).

Regarding Figure 2.81, we were not able to find comparisons in the literature. But, for Figure 2.82, there is some (marginally) comparable data, taken from lakes. These measurements (103) gave 20–25 percent lower NTU readings per Secchi reading than our coastal waters sample, but seem to be within the "envelope" of expectation considering the spread of background "noise."

For underwater conversion, a diver-held Secchi disk can be calibrated against diver-collected turbidity samples (for the Hach Turbidity Meter). The "Secchi reading" is often called "visibility", being the limits of human ability to detect an object through water or, in another sense, the extinction point of human visibility under a particular set of conditions.

Figure 2.82 shows the field results in terms of the relationship between Secchi disk and Hach meter (readings at sea bottom) in our survey. While the data are thin (number of data points is small and

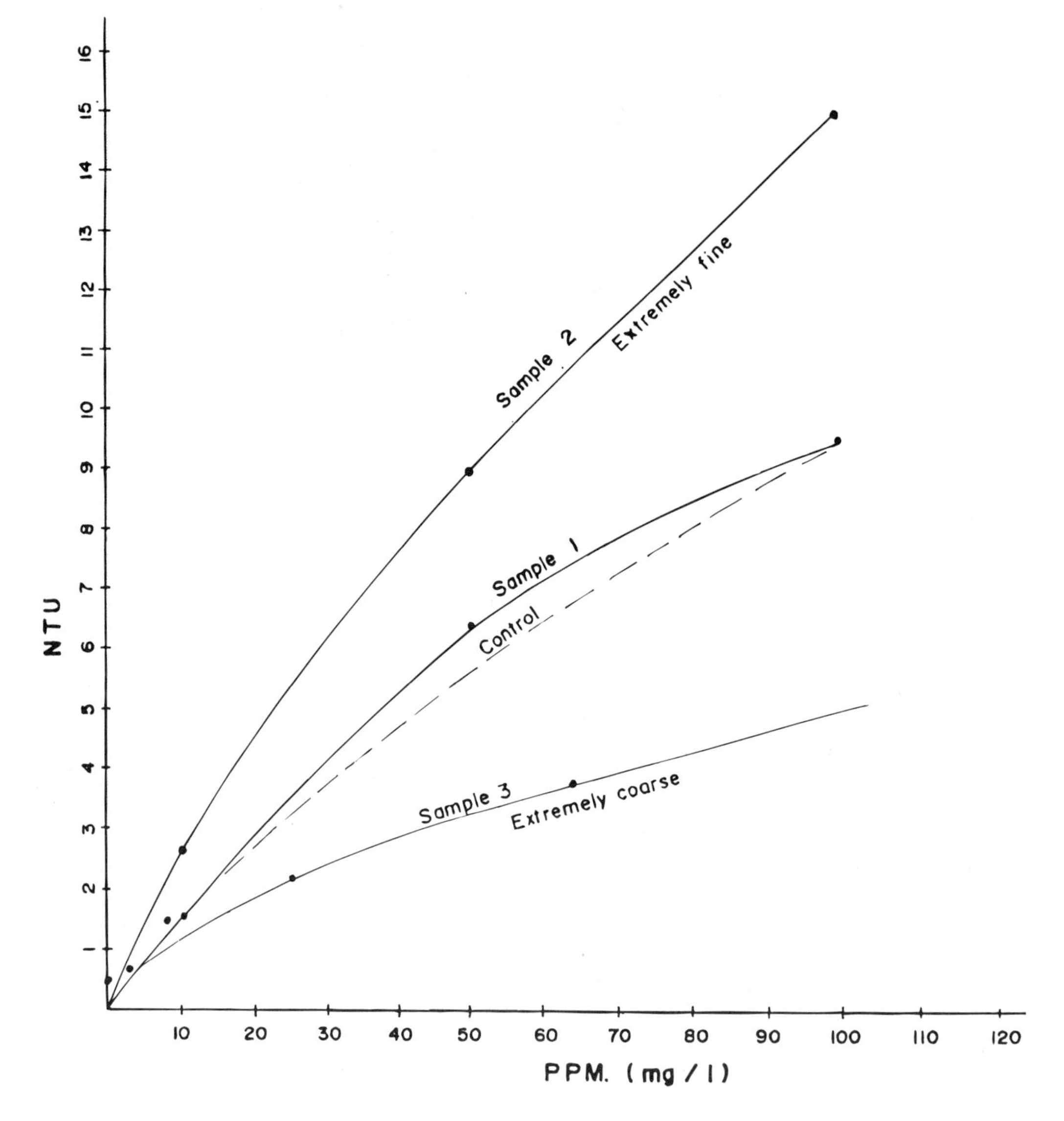

Figure 2.81. Test results for calibration of resuspended solids (in silt from Ujung Pandang, Indonesia) against readout of same samples by a Hach turbidity meter (details in text). (*Source:* author, Reference 67.)

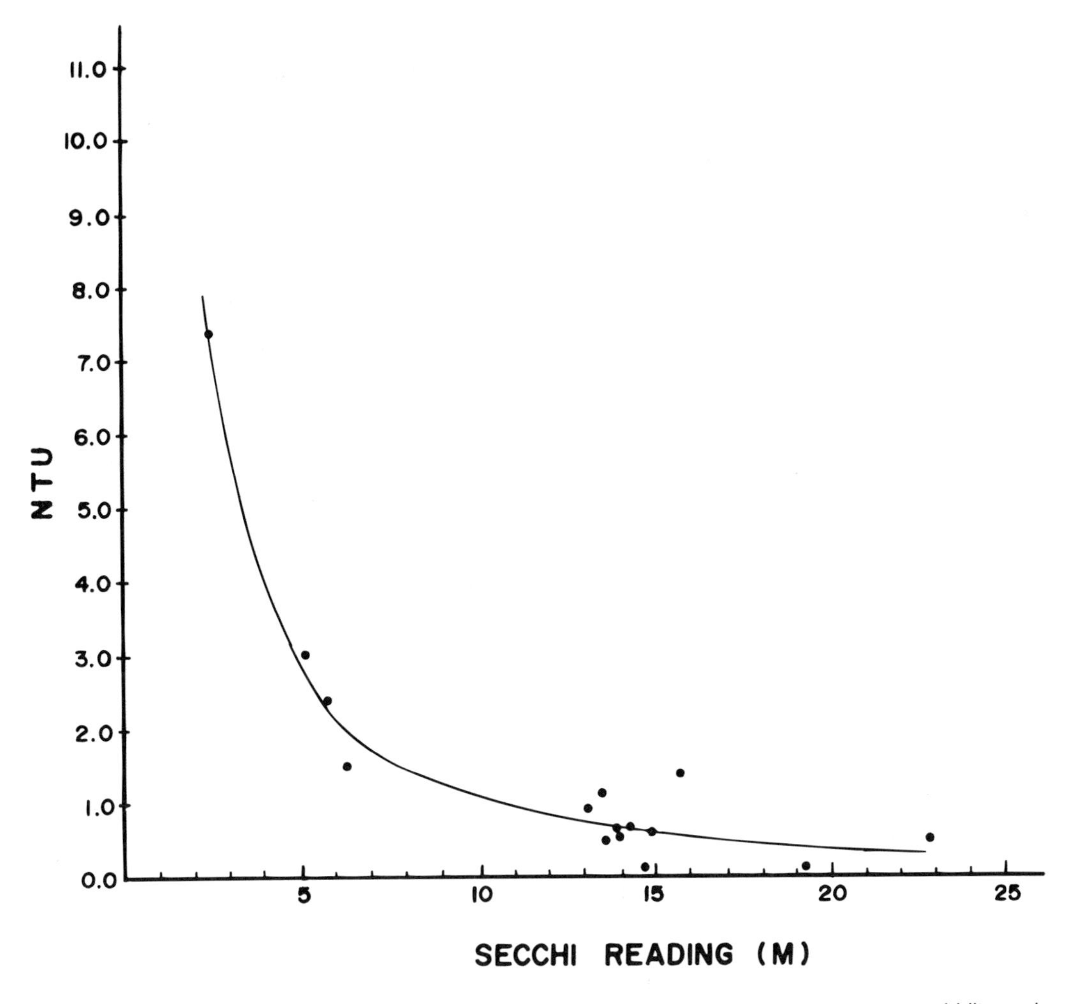

Figure 2.82. Field comparison of horizontal Secchi disk readings (by divers) and simultaneous turbidity read-outs from subsurface samples at same locations (Bali, Indonesia). (*Source:* author, Reference 67.)

concentrated at a center point), one can deduce that, at a Secchi reading of less than 5 m, the water becomes very turbid and is approaching the limits imposed on a project in its Environmental Management Plan. Therefore, turbidity meter readings can be used to check for a possible violation.

It may be useful for environmental management purposes to estimate the time period that SPM remain suspended. For this, one must know the settling rates of particles of different sizes. The following are "still water" rates, which would have to be adjusted for water movement in coastal areas; see also Figure 2.83 for particle size classification (59):

Class of material	Diameter of particles (mm)	Rate of settlement (mm/sec)
Coarse sand	1.0	100
	0.20	21
Fine sand	0.10	8
	0.06	3.8
	0.04	2.1
	0.02	0.6
Silt	0.01	0.15
Coarse clay	0.001	0.0015
Fine clay	0.0001	0.000015

Wentworth Scale (Size Description)	Phi Units ϕ*	Grain Diameter d (mm)	U.S. Standard Sieve Size	Unified Soil Classification (USC)
Boulder				
	−8	256		Cobble
Cobble		76.2	3 in.	
	−6	64.0		Coarse Gravel
Pebble		19.0	¾ in.	
		4.76	No. 4	Fine Gravel
	−2	4.0		Coarse Sand
Granule				
	−1	2.0	No. 10	
Sand: Very Coarse				
	0	1.0		Medium Sand
Sand: Coarse				
	1	0.5		
Sand: Medium		0.42	No. 40	
	2	0.25		Fine Sand
Sand: Fine				
	3	0.125		
Sand: Very Fine		0.074	No. 200	Silt or Clay
	4	0.0625		
Silt				
	8	0.00391		
Clay				
	12	0.00024		
Colloid				

Figure 2.83. Soil grain size scale for soil classification. (*Source:* Reference 362.)

2.43 URBAN RUNOFF MANAGEMENT

Urban storm drainage should be viewed as a source of water pollution contributing undesirable amount of solids, organic material, nutrients, toxic substances, and other constituents. But it confronts us as a pollutant only periodically (i.e., during rainstorms). Nevertheless, the effective management of runoff pollution becomes an exceedingly complex problem.

In general, the pollutant concentrations in urban storm water runoff increase proportionally with the passage of time since the last rainfall. The longer the dry spells between storms, the more contaminants will accumulate. Thus, a storm today will produce higher concentrations than a second or third storm tomorrow. Usually, though, it takes only about one relatively rainless week for a significant amount of materials to accumulate.

The quick washoff or "shock load" effect often complicates our efforts to deal with runoff pollution: (a) shock loads may contain very high contaminant levels, (b) huge pulses of water complicate the design of facilities that can more easily cope with continuous flow conditions, and (c) shock loads in both quantity and quality may overwhelm the absorptive capacities of natural receiving waters, which may likewise be capable of handling more uniform pollution loadings.

The issue of whether to combine storm water with sewage is contentious and has no simple answer. But unless there is strong proof to the contrary, a system that directs only the, "first flush" (first 1–2 cm) into the sewage is recommended—that way the "shock load" is treated without overwhelming the system (59).

Earnest effort at source control can pay off. For example, the swimming beaches of Victoria, B.C, Canada, were very often closed to swimmers because of high fecal coliform bacteria levels (>200

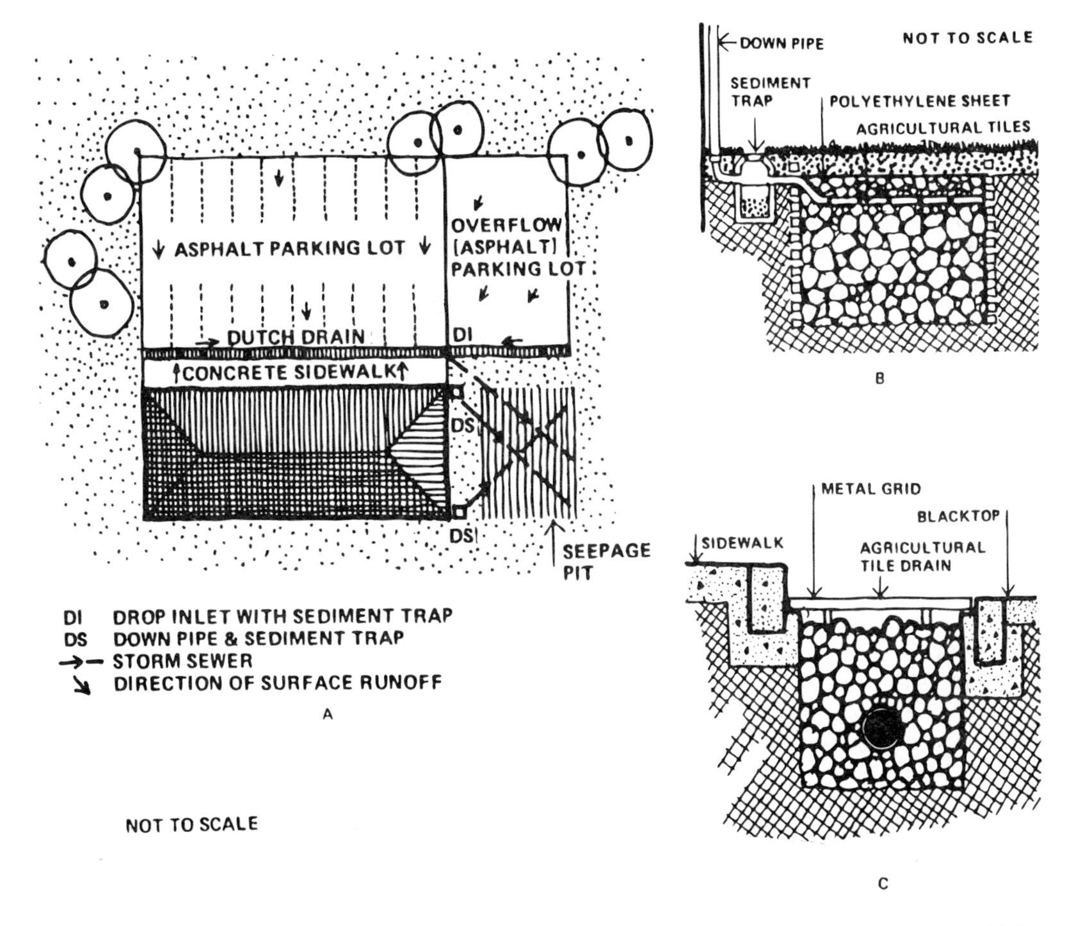

Figure 2.84. Example infiltration systems for runoff: A, plan; B, seepage pit; C, Dutch drain. (*Source:* Reference 348.)

MPN) until 1992, when a "coordinated program to improve the quality of storm sewer discharges" succeeded in reducing these levels to a point where swimming was allowed virtually all the time (249). The lesson here is that urban runoff, *not* sewage, was responsible for microbial pollution of the beaches (Victoria has a successful program of offshore outfall disposal of its sewage; see "Canada, Victoria, B.C." case history in Part 4).

At the micro-management level, each landholder should be required to manage storm runoff in a way that minimizes damage to other interests. Such on-site approaches might include:

1. Temporary ponding of rainwater on rooftops, parking lots, and especially prepared low areas, instead of rapid diversion to downstream receptors
2. A reduction in the area of streets and other paved areas and/or the use of porous materials that would allow water to seep directly into soil and subsurface strata
3. The maintenance and provision of vegetated open spaces and, in particular, vegetated waterways (in contrast to concrete gutters and sewers) that would capitalize on the natural system's ability to retard storm flow and to filter out contaminating substances
4. Provision of infiltration systems such as seepage pits or French drains (Figure 2.84)
5. Implementation of a forthcoming land use program that would provide vegetated open areas where needed and encourage a distribution of land uses more compatible with requirements for proper runoff control
6. Encouragement of aggressive good housekeeping at each site, thereby eliminating runoff pollution at its source

Much more attention has been focused on the advancement of the state of the art in centralized approaches. These include more sophisticated sewage systems, facilities for the temporary storage of large volumes of storm flow, a wide variety of physical, chemical, and biological treatment techniques, and computerized flow regulation systems, to mention just a few.

The point to be emphasized is that the effects of urban runoff can be managed only through a broadly based program of engineering design, land use control, localized good housekeeping, on-site control activities, and public participation to alter our traditional perspectives on storm runoff.

2.44 WATER QUALITY MANAGEMENT: COASTAL WATERS

To know the condition of coastal waters and to provide a system of monitoring for changes in water quality, it is useful to have numerical reference standards (in those cases where the numbers are justifiable). Certain pollution control programs do require numerical guidelines that can signal excessive amounts of contaminants in coastal "receiving waters." Receiving waters would include all coastal waters offshore from the mean high tide level including such littoral areas as estuaries, lagoons, bays, wetlands, and brackish and inland tidal waters or rivers.

A "mixing zone" designation may be appropriate for some "point sources" of pollution. Right at the location of discharge, the concentration of a contaminant may be very high but it may dilute rapidly as it is mixed with the receiving water; therefore, it may be useful to designate a certain size of mixing zone *outside* of which the water quality measurement would be taken.

A somewhat different approach is used by Sri Lanka (30.5), whereby a mixing ratio of 8 is allowed. That is, the pollutant concentration of the discharge is allowed to be 8 times greater than the maximum allowed for the receiving water body. In effect, the mixing zone ends where the pollutant has been diluted by a factor of 1:8. Also, it is often helpful to classify water bodies according to designated uses and assign different water quality guidelines to each. For example, it would be reasonable to ask for higher quality at swimming beaches (e.g., Class "A") than in industrial harbors (e.g, Class "C"), per the classification below (41):

- **Category A:** Waters in this category are primarily for aesthetic enjoyment and recreation, apart from protection of natural aquatic life. These waters should be kept free from pollution attributed to domestic, commercial, and industrial discharges, shipping and intensive boating, construction, and other activities which may impair their intended use.
- **Category B:** Waters in this category are primarily for marine biological activity. Thus, standards must ensure protection for marine organisms, particularly shellfish and coral reefs. Other important uses are mariculture activities, aesthetic enjoyment, and recreational activities (inclusive of "whole body contact").
- **Category C:** Waters in this category are for general use including commercial/industrial uses such as shipping, but should be of sufficient quality to protect aquatic life and permit limited body contact. Aesthetic enjoyment and recreational use should also be maintained.

In setting limits for receiving waters, it is important to recognize that the limits are designed to prevent specific impacts such as the following (41):

- Algal blooms (due to excessive nutrients) and subsequent water quality problems
- Direct toxicity to marine organisms (due to metals, hydrocarbons, or pesticides)
- Significant bio-accumulation in food chains leading to indirect toxicity problems for fish and other life forms
- Health problems to water users (e.g., bacteria, viruses)
- Significant changes in the redox conditions of the waters (BOD, COD, dissolved oxygen level changes)
- Serious damage to marine/coastal ecosystems (due to the influence of sedimentation)
- Damage to engineering facilities (e.g., pipelines, piers, dams) caused by changes in pH, redox conditions

The following water quality guidelines are intended to provide *examples* only and are not appropriate for regulatory use without specific confirmation; that is, they are meant for general world wide guidance and not meant for exact application without special considerations. It is recommended that each country, region, and ecosystem be examined as a separate case so that special guidelines can be selected. The guidelines can also serve as the quantitative basis for the diagnosis of coastal ecosystem condition and for monitoring (see "Monitoring and Baseline" above).

The numbers indicate the maximum amounts of the substances named that should be allowed in coastal waters—these are the generally mesotrophic, terrestrially influenced, waters that lie adjacent to the seacoast. For guidelines that refer to oligotrophic coastal waters, see "Water Quality Management: Coral Reefs" below. The numbers in parentheses refer to sources in the "References" section (Part 5). Abbreviations and symbols are explained below. Where these sources disagree, the more recent ones are used (there has been some "rounding" and averaging). Remember that these are meant to be for general guidance only and are not recommended as the basis for a particular control program.

General Standards

BOD_5 (5-day biological oxygen demand): Outside a prescribed mixing zone, not >10 mg/l (185, 281)

Coliform bacteria: Drinking water, 50 MPN/100 ml; edible shellfish growing, 70 MPN/100 ml; human contact (e.g., swimming), 240 MPN/100 ml (41, 356). Note: Canada uses 14 MPN for shellfish, 200 MPN for swimming (97).

Coliform bacteria, fecal: Not >14 MPN/100 ml (shellfish); not >200 MPN/100 ml (human contact; e.g., swimming) (356)

Detergents: Not >0.2 mg/L

Nitrate: Not >0.5 mg/L (42.1); clear waters, 0.2 mg/L (59)

Oxygen: Not <4.0 mg/L

Hydrocarbon: Not >5 mg/L

PCBs (polychlorinated biphenyls): not >0.001 μg/L (258)

Phosphorus, total: not >0.6 μg/L (185)

Suspended particulate matter (SPM): levels of protection—**high,** not >15mg/L; **medium,** 80 mg/L; **low,** 400 mg/L (366)*

TBT (anti-fouling agent): 0.01 μg/L (162)

Transparency: pollutant not to reduce light penetration by >10 percent (356)

Turbidity: (estuaries) not >5 to 25 NTU, mean reading (59, 257), depending upon ambient level.

*Because coastal waters have a natural content of suspended matter, which varies with season and spontaneously (freshets, winds, plankton blooms), it is not possible to state a single standard.

Heavy Metals (maximum amounts in coastal waters) (59, 258)

Aluminum:	0.1 mg/L	Lead:	1.6 μg/L
Antimony:	0.2 mg/L	*Manganese:*	0.1 mg/L
Arsenic:	0.06 mg/L	*Mercury:*	0.1 μg/L
Barium:	1.0 mg/L	*Nickel:*	7.0 μg/L
Berylium:	1.5 mg/L	*Selenium:*	0.01 mg/L
Cadmium:	7.7 μg/L	*Silver:*	0.5 μg/L
Chromium:*	0.1 μg/L	*Uranium:*	0.5 mg/L
Copper:	2.9 μg/L	*Zinc:*	0.6 mg/L

* Trivalent

Biocides (maximum amounts in coastal waters) (258)

Aldrin:	0.001 μg/L	*Heptachlor:*	0.001 μg/L
DDT/DDE:	0.001 μg/L	*Methoxychlor:*	0.03 μg/L
Dieldrin:	0.001 μg/L	*Toxaphene:*	0.005 μg/L
Chlordane:	0.002 μg/L	*Malathion:*	0.1 μg/L
Endosulfan:	0.001 μg/L	*Parathion:*	0.008 μg/L
Endrin:	0.002 μg/L		

Abbreviations and Symbols

ppt	parts per thousand (o/oo)
ppm	parts per million (mg/L)
ppb	parts per billion
mg/L	milligrams per liter (parts per million-ppm)
μg/L	micrograms per liter (parts per billion-ppb)
μm/L	micromoles per liter
MPN	most probable number (re bacteria counts/ml)
SPM	suspended particulate matter (see also TSS)
TSS	total suspended solids (see also SPM)
TIN	total inorganic nitrogen
<	less than
>	more than

2.45 WATER QUALITY MANAGEMENT: CORAL REEFS

Unfortunately, coral reefs are readily degraded by impacts of human activities. These impacts range from direct physical disturbances to the indirect effects of pollution from ships, coastal settlements, and engineering or agricultural works on adjacent land masses.

The following explains the pollution agents of concern and lists relevant "Quantitative Guidelines" for water quality. These guidelines refer to the special requirements of open reef systems and not to confined coral systems such as coastal lagoons which may be more tolerant of terrestrial influences, including higher background levels of some substances (41). Sources for the listings below are shown by numbers in parentheses, which refer to the numbers in Part 5 ("References").

Pollution Agents

Herbicides: May interfere with basic food chain processes by destroying or damaging zooxanthellae in coral, free living phytoplankton, or algal or sea grass plant communities. Can have serious effects even at very low concentrations.

Pesticides: May selectively destroy or damage elements of zooplankton or reef communities; planktonic larvae are particularly vulnerable. May accumulate in animal tissues and affect physiological processes.

Antifouling Paints and Agents: May selectively destroy or damage elements of zooplankton or reef communities. Not likely to be a major factor except near major harbors, shipping lanes, and industrial plants cooled by seawater.

Sediments: Smother substrate and smother and exceed the clearing capacity of some filter-feeding animals. Reduce light penetration, which may alter vertical distribution of plants and animals on reefs. May absorb and transport other pollutants.

Sewage, Nutrients, and Detergents and Fertilizers: May stimulate phytoplankton and other plant productivity beyond the capacity of control by grazing reef animals and thus modify the community structure of the reef system. May cause eutrophication and consequent death of reef organisms.

Petroleum Hydrocarbons: Have been demonstrated to have a wide range of potential damaging effects at different concentrations.

Heated Water from Power Station and Industrial Plants: Can change local ecological conditions; water temperature is a key factor in distribution and physiological performance of most reef organisms.

Hypersaline Waste Water from Desalinization Plants: Changes local ecological conditions; salinity is a key factor in distribution and physiological performance of many reef organisms.

Heavy Metals: May be accumulated by and have severe physiological effects on filter-feeding animals and reef fish and be accumulated in higher predators.

Radioactive Wastes: May have long-term and largely unpredictable effect on the genetic nature of the biological community.

Herbicides: May interfere with basic food chain processes by destroying or damaging plant communities and their photosynthetic and other processes.

Quantitative Guidelines

Ammonium (NH4): not >0.65 μg/L (37.2); not >+10 percent ambient (345)

BOD_5 (5-day biological oxygen demand): not >0.25–0.78 mg/L (20, 345); not >10 percent above ambient (22)

Chlorine, free: not >0.2 mg/L (345)

Chlorophylla-a: not >0.59 μg/L (162)

Dissolved oxygen: >5.5 mg/L or >80 percent saturation or not < natural level (345)

Hydrocarbons: not >0.25 mg/L (162)

Nitrate/nitrite (NO_3, NO_2): not >1.31 μg/L; not >10 percent above ambient (345)

Nitrogen, total inorganic: not >1.1 μg/L (20)

Nutrients, genl: Not >10 percent above ambient (22); see also phosphate, nitrate

PCB (polychlorinated biphenyls): 0.001 μg/L (258)

Phosphate, inorganic (PO4): not >0.25 μg/L (162); not >10 percent above ambient (345) (note: ambient phosphate in non-polluted coral areas varies globally from about 0.03 to 0.13 μg/L (20)

Salinity: not <30 ppt (20); change not >10 percent of ambient (345)

Sedimentation rate (on coral): not >30 mg/cm^2/day (162)

Surfactants: not >35.0 μg/L (162)

Suspended particulate matter (SPM): not >3.85–4.5 mg/L (162, 345); not >10 percent over ambient (345)

Temperature: not >1.0°C over ambient

Transparency/turbidity: not >10 percent over ambient (345); not >1 NTU turbidity

2.46 ZONING

The management practice of *zoning* serves two main purposes in coastal conservation: (a) *custodial*—for nature reserves, to sub-divide them into particular use allocation zones (e.g, diving, nature study, fishing, protection of breeding areas, water sports) and (b) *regulatory*—for regulatory programs and coastal land use planning, to designate certain areas for particular uses (e.g., hotel, aquaculture, navigation, greenbelt, commercial fishing, nature reserve).

Zones are established in space or time, where certain uses with minimal impact are allowed while others are either prohibited or subjected to special conditions, according to Saenger (301). Means of access may also be included (e.g., vehicular or foot traffic, small craft or larger passenger craft). Some of these limitations are imposed because of impact on the resource, but the majority are usually imposed to minimize the interference with other uses or users (Figure 2.85). Saenger also stresses the need for blanket prohibitions according to which, certain uses would be totally prohibited.

The Great Barrier Reef Marine Park Authority, which has a mandate as a general marine management authority, has used the following zoning scheme (308):

Reef Appreciation Area: An area of a reef, in a zone that normally permits fishing and collecting, in which fishing and collecting are excluded to enable the public to observe reef life relatively undisturbed by human activity.

Seasonal Closure Area: An area, known to be of importance to the breeding of particular animals, that may be closed during the breeding season.

Replenishment Area: Seven areas, two of which may be closed at any time for a period of up to three years. Concept at present experimental, designed to test whether, as has been suggested, periodic closure will increase the productivity of demersal reef fisheries.

General Use "A" Zone: Includes all the shoals on the section as well as Lady Elliott Island; covers more than 80 percent of the area of the section. No restrictions on use other than (a) that provided by Section 38, which prohibits operations for the recovery of minerals except for the purposes of research, and (b) prohibition of commercial spearfishing and spearfishing with SCUBA.

General Use "B" Zone: Includes about 18 percent of the area of the section. Provisions are the same as for General Use "A" Zone, with additional prohibition on trawling and the navigation of vessels greater than 500 tons.

Marine National Park Zone: Conservation management primarily for tourist purposes, with fishing allowed subject to gear restriction (one hand-held line or rod and no more than two hooks).

Scientific Research Zone: Specific provision for scientific research in an area as far as possible unaffected by other uses.

Preservation Zone: Specific provision for management of an island reef and a lagoon reef as far as possible unaffected by human use.

Zoning has recently come into common use in planning of nature reserves. Properly designated areas can provide for a variety of uses allocated through a zoning scheme. This idea is embodied, for example, in the conservation program of the Great Barrier Reef Marine Park Authority in Australia. (66)

Motivations for zoning of coastal/marine parks and reserves have been identified as follows (123):

1. There exist marked and documented conflicts between activities in an area (in protected areas or areas outside the boundaries), for example between heavily used tourism areas and traditional fisheries or between coastal forestry and fisheries.
2. There are current or potential conflicts between the conservation objectives of the designated area (be it park, nature reserve, wilderness area, etc.) and other possible activities (e.g., a sea turtle nesting beach needs to be protected and/or buffered from beach activities or other coastal development).
3. Included in the designated protected area is a unique habitat or cultural feature that will require special management attention (e.g., a shipwreck in a marine park).

4. Included in the designated protected area are severely damaged areas requiring restoration effort (e.g., coral reef restoration after shipgrounding).
5. There is a need to provide specialized or quality services for a group of park users such as SCUBA divers (Figure 2.86).

The following are some general guidelines for developing a workable zoning plan (123):

1. Keep the zoning plan as simple and understandable as possible for users; refrain from complicated schemes.
2. Place the emphasis to minimize interference with customary uses and rights. Make sure that users—e.g., fishermen, villagers, tour boat operators, SCUBA diver operators—are consulted before any implementation of a zoning scheme.
3. Where existing uses are prohibited in one zone, make provisions for these in other parts of the protected area; that is, avoid eliminating any existing uses from the protected area.
4. Zoning of the marine protected area must be consistent with adjacent land area.
5. Avoid sudden transitions in zones, such as having a strict conservation zone adjacent to a general use zone.
6. Resource protection zones must incorporate a range of linked habitats within one unit.
7. Where possible, use discrete physical attributes to delineate zones (e.g., peninsula, a reef, an island, etc.).
8. The zoning should be consistent with existing fisheries closure areas and navigational zones and should complement management regimes in the region.

A cornerstone of coastal conservation programs is the strategic use of strictly protected "core area" zones, each sheltered by adjacent buffer areas and surrounded by multiple use zones. This arrangement, which is embodied in the United Nations' Biosphere Reserves program, is often more practical to administer and less likely to provoke massive resistance than non-zoned reserves because a wider multiplicity of uses can be allowed (65).

Saenger (301) notes that zoning plans must:

- Comply with all statutory and constitutional requirements and any international obligations
- Be technically sound and based on adequate data
- Be capable of administration (including enforcement and surveillance)
- Involve and educate the public
- Develop effective, incremental approaches to use allocation decision making while recognizing the cumulative impact of development
- Provide certainty for future use while remaining flexible to consider individual uses and area
- Balance the interests of competing user groups to achieve political acceptability
- Be supported by adequate levels of expertise for administration and enforcement
- Be cost effective
- Incorporate monitoring and enforcement to ensure implementation and ongoing assessment of policies

and should also:

- Consider local, regional, state, and national interests
- Build upon existing use and resource management programs
- Balance public interests and private expectations and options
- Ensure fair treatment for users, procedural due process, and expeditious review of permits
- Be coordinated with public works construction programs.

Detailed zoning plans need to take a number of practical matters into consideration including accessibility of reefs, shipping lanes, all-weather anchorages, and the provision of a range of zone types near each major population center (Figure 2.87). In addition to the broad zones outlined above, the zoning plan provides for the establishment of smaller areas which can be applied more flexibly. These "designated areas" include the following (301):

- Seasonal closure area: seasonally closed to protect breeding sites from human intrusion
- Replenishment area: temporarily closed for replenishment of living natural resources
- Defense area: notified closure for public safety
- Shipping area: special provisions to facilitate the navigation and operation of ships
- Special management areas: provides for areas to be specially managed for conservation, research, public safety, or public appreciation

Zoning plans must be adapted to specific situations in order to meet the local needs and conditions (301). Obviously, there will be differences in the type of zones and their arrangement depending on the intensity of use of a particular area as well as the overall size of the area to be zoned. An example is shown in Figure 2.88.

On the Great Barrier Reef, where large-scale zoning was required, the various zoning plans have six broad zones. For one of the sections of the marine park, a seventh zone (Marine Park Buffer Zone) was established to deal with the special needs of a large game fishery. In addition, a further zone is currently under consideration which would effectively subdivide the current Marine Park B zone into a "tourism" and a "wilderness" zone (301).

The following is the allocation of uses in Great Barrier Reef Marine Park (1989) for each of seven zones (301):

	NAME OF ZONE						
Activity	**Gen. Use A**	**Gen. Use B**	**Mar. Park A**	**Mar. Park Buffer**	**Mar. Park B**	**Sci-ence**	**Pres-erva-tion**
Boating/diving	A	A	A	A	A	N	N
Pelagic trolling	A	A	A	A	N	N	N
Bait netting	A	A	A	N	N	N	N
Line fishing	A	A	A	N	N	N	N
Traditional fishing	A	A	P	P	N	N	N
Gill netting	N	N	N	N	N		
Spearfishing	N	N	N	N	N		
Crayfishing	N	N	N	N	N		
Cruise ships	A	P	P	P	P	N	N
Trawling	A	N	N	N	N	N	N
General shipping	A	N	N	N	N	N	N
Scientific research	P	P	P	P	P	P	P
Tourist programs	P	P	P	P	P	N	N
Waste discharge	P	P	P	P	P	N	N
Spoil dumping	P	P	P	P	P	N	N
Traditional hunting	P	P	P	P	N	N	N
Collecting	P	P	N	N	N	N	N
Mariculture	P	P	N	N	N	N	N
Tuna fishing	P	P	N	N	N	N	N

Abbreviations: A, allowed; P, by permit only; N, not allowed.

If a zoning plan is to be understood by the public though, a dozen or so zone types would appear to be an upper limit in practice. According to Cocks (92), a zoning plan that is not readily understood is difficult to implement and manage, even though many of the transgressions may be accidental rather than deliberate. To the extent that zoning is used to reduce conflicts between different user groups, the following steps are is suggested (92): (a) identifying interest group demands; (b) identifying conflicts between interest groups; and (c) producing a plan which reduces conflicting demands and develops a "best compromise" plan.

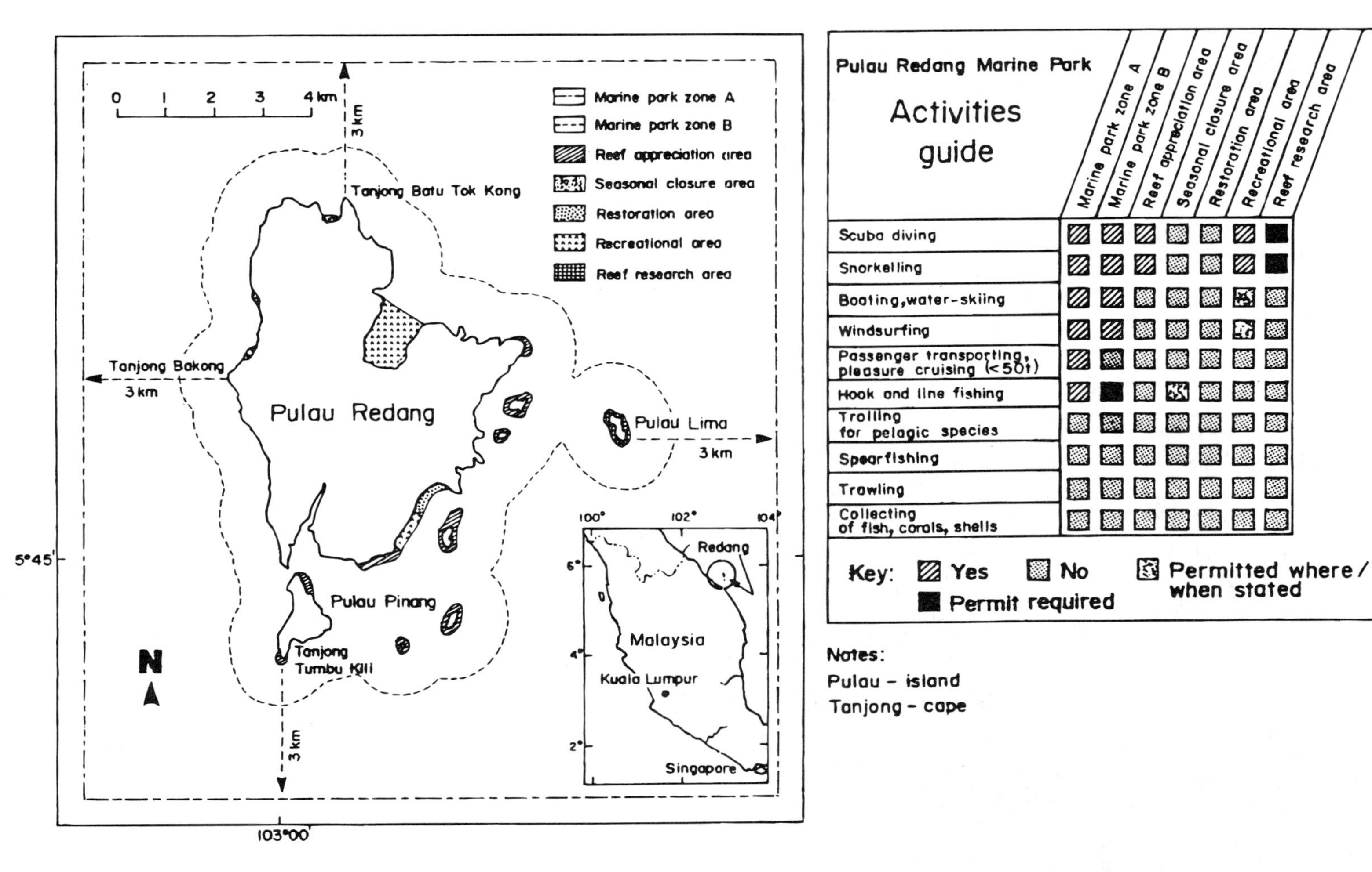

Figure 2.85. Example marine reserve zonation plan (Palau Redang Marine Park, Malaysia). (*Source:* Reference 380.)

Figure 2.86. High impact tourism is allowed on less than 20 percent of the Great Barrier Reef (Australia) where visitors brought by boat are deposited on large, permanently anchored, pontoon rafts, which can withstand a Category 4 cyclone. (Photo courtesy of P. McGinity, Great Barrier Reef Marine Park Authority, Townsville, Australia.)

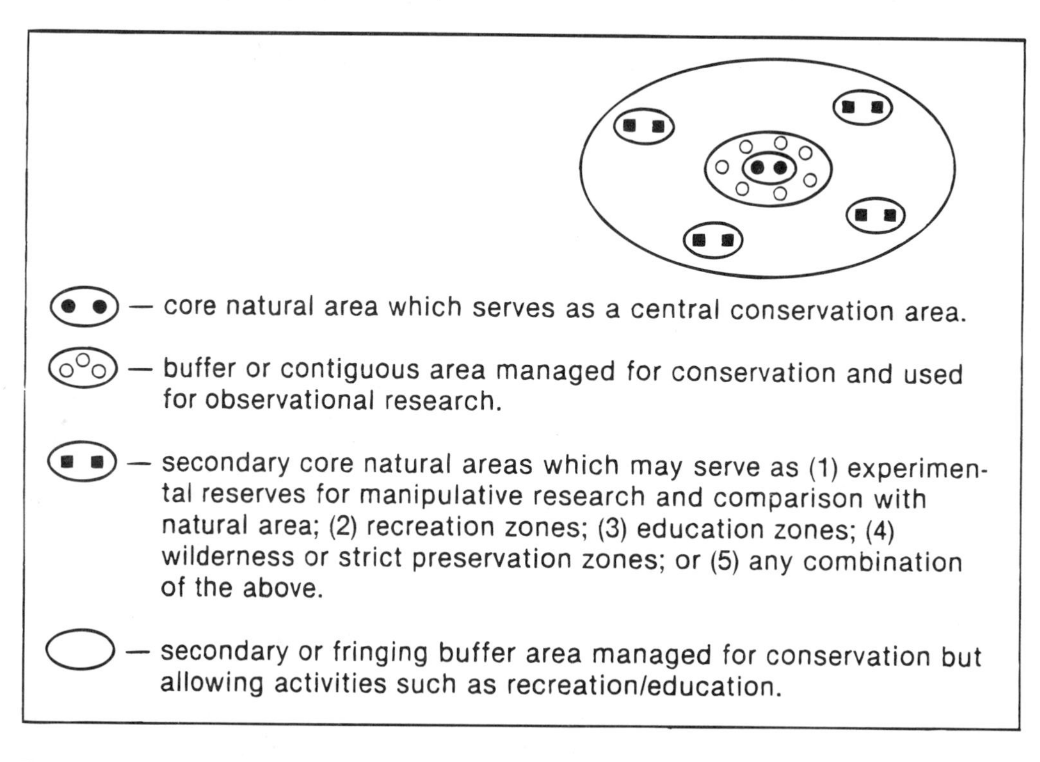

Figure 2.87. Simplified diagram of a marine reserve zonation scheme, with buffers, emphasizing ecological conservation. (*Source:* Reference 308.)

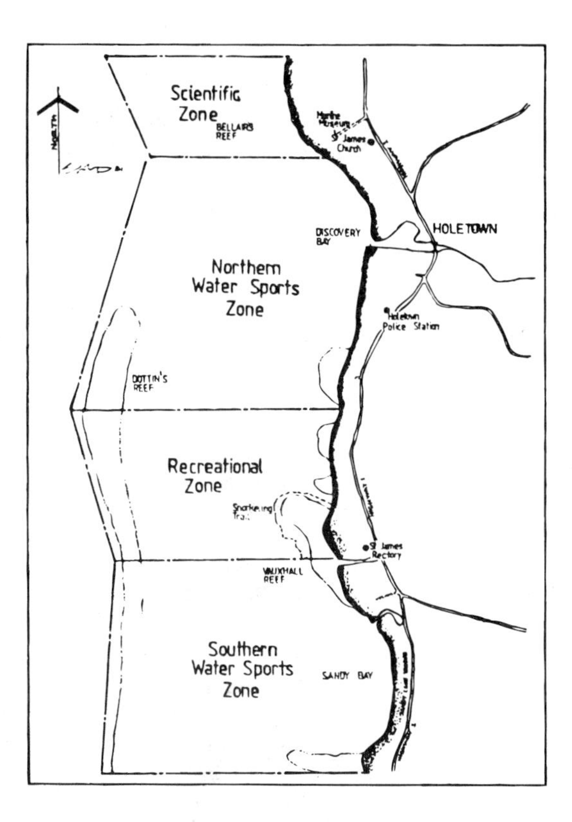

Figure 2.88. Example zoning arrangement—marine reserve at Holetown, Barbados. (*Source:* Reference 96.)

Kelleher and Kenchington (203) suggest the following 15 guidelines for zoning of conservation areas (abstracted by the author):

1. The zoning plan should be as simple as practicable.
2. The plan should minimize the regulation of, and interference in human activities, consistent with meeting the goal of providing for different types and levels of protection for different resources.
3. The zoning plan should maximize consistency with those of other protected areas in terms of zone types and conditions of use.
4. The zones should avoid sudden transitions from highly protected areas to areas of relatively little protection. Buffer zones should be used to surround them.
5. Single zonings should surround areas with a discrete geographic description (e.g., an island or reef).
6. Zoning should be consistent and where possible should be described by geographic features (based on line of sight where possible to aid located them in the field).
7. Critical threatened-species habitats of world, regional, or local significance (for example, dugong, whales, turtles, crocodiles) should be given protective zoning.
8. Significant breeding or nursery sites (particularly for species subjected to harvesting) should be protected by zoning on either a full-time or seasonal basis (e.g., as replenishment zones).
9. Replenishment zones should be located near fishing areas to ensure replenishment of fishing stocks.
10. Areas recognized and/or used for reasonable extractive activities should be given "general use" type zoning.
11. As a general rule, areas of significance or non-extractive activities should be zoned at national park level.
12. When an area is zoned in a way that excludes a particular activity, provision should be made for access to alternative areas.
13. As far as practicable, provision should be made to continue traditional hunting and fishing by indigenous people.
14. Anchorage zones should permit anchoring of vessels to continue in all weather conditions. Ideally,

in sensitive areas such as coral reefs, anchoring should be eliminated and permanent moorings provided where possible.

15. Provision should be made for research. However, reefs should only be zoned exclusively for scientific research where research is frequent and regular.

Five stages in the development of a zoning plan are identified by Kelleher and Kenchington (203):

1. Initial Information Gathering and Preparation: the planning agency, perhaps with the assistance of consultants, assembles and reviews information on the nature and use of the area and develops materials for public participation consideration by the public or appropriate representatives.
2. Public Participation or Consultation: Prior to the Preparation of a Plan the agency seeks public comment on the accuracy and adequacy of review materials and suggestions for content of the proposed zoning plan.
3. Preparation of Draft Plan: Preparation of a draft zoning plan and materials, explaining the plan for the public or appropriate representatives. Specific objectives are defined for each zone.
4. Public Participation or Consultation: Review of Draft Plan: the agency seeks comment on the published draft plan and explanatory materials.
5. Plan Finalization: The government or agency adopts a revised plan, which takes account of comments and information received in response to the published draft plan.

On the "dry side" of the coast, zoning schemes are also useful for ICZM land use planning because of their potential for ecosystem protection and for allocating uses to different user groups; specifically, they are used to legally identify areas for development of various types, areas to be protected for biodiversity reasons, and other management purposes. While regulatory use of zoning is common for general urban land use management, it is still not common for coastal zone management.

In the coastal area, as elsewhere, some resources are more vulnerable to development impacts than others and some development types and uses are more threatening to resources than others. Information on these two aspects is useful to ICZM programs, particularly categorizations which can be used in development project review and in multiple-use economic development planning. The simplest way to do this is by an approach in which:

- Areas are classified as (a) approved for development, (b) not approved for development, or (c) approved for development under specified conditions.
- Development project types are classified as (a) acceptable, (b) unacceptable as designed and located, or (c) acceptable with changes in design and location.

This is a sort of triage approach to zoning, whereby the safe and unsafe can be dealt with expeditiously and maximum effort can be applied to the in-between categories of either areas or development types. This type of simple classification would be useful in the strategic planning phase; later, more sophisticated types of classifications might be generated, for example, through land suitability analysis.

PART 3

Management Information

3.1 AGRICULTURE

Unless there are effective controls on agriculture, an excessive amount of chemicals may run off the land with drainage waters and into coastal water bodies. Here they contribute to eutrophication and toxification of river, coastal, and marine areas. Because of seepage of excess nitrates into surface and groundwaters, the increased use of fertilizers plays a determining role in the eutrophication of tidal rivers, estuaries, and other coastal marine areas. Industrial products (e.g., pesticides, herbicides, and fertilizers) are being used more frequently in agricultural practices, while agricultural produce is undergoing various processing operations involving chemicals. Pesticides in very small concentrations can kill or debilitate marine life.

With the exception of wind erosion and pesticide drift from aerial application, most agricultural pollution is associated with runoff from land being used for agricultural purposes. When the runoff from agriculture areas is intermittent and diffuse, the resulting pollution can be defined as "nonpoint" source pollution, as opposed to pollution from "point sources," which is a discharge from a pipe or something comparable.

Nonpoint source pollutants from crop production systems include sediments, nutrients, and pesticides (Figure 3.1). The pollutants may reach a watercourse by runoff of surface water or through infiltration and percolation to subsurface water.

By volume, sediment is the main agricultural nonpoint source pollutant. The process of erosion has been identified as the single most significant factor to directly affect coastal environments such as coral reefs. Although sediments result principally from the erosion of soils, they may also include crop debris, which may place an oxygen demand on receiving waters during their decomposition.

Nutrients may be transported either in a water-soluble form or absorbed on sediment. Nitrogen (N) and phosphorus (P) are the most important nutrients affecting water quality. These nutrients occur naturally as components of the soil, but in crop production systems are usually supplemented by fertilizer application. N in nitrate form may be leached from the soil and be carried into the groundwater or may be removed by the runoff. N in organic forms may be associated with soil particles and plant debris and be removed by sediment transport processes. Most of the P removed from the crop production system is associated with sediment transport processes, although some movement of soluble phosphorus in runoff does occur. The impact of these nutrients on coastal ecosystems can be enormous. An estuary (e.g., the Hudson in New York) may receive 5 to 10 times more nutrients than it is capable of assimilating.

Pesticides may be water soluble and move with percolating water and runoff or may be adsorbed on soil particles and moved through sediment transport processes. The total amount of pesticide that runs off is usually less than 5 percent of the quantity applied, except when heavy rainfall occurs shortly after application. Although the concentration of pesticide in receiving water may be low, there is little doubt that pesticides are significant pollutants in coastal waters.

Source management is the most practicable approach to reduction of agriculture pollution. It involves integrating water-quality-related management practices into the crop production management practices presently being used. Examples are:

Input Management

1. Quantitative reduction in fertilizer and pesticide application
2. Better timing of fertilizer and pesticide application

3. Use of less persistent pesticides
4. Diversions

Cropping System Management

1. Minimum tillage practices
2. Crop rotation
3. Strip cropping
4. Contouring

Output Management

1. Terraces
2. Debris basins
3. Drop spillways
4. Sod flumes

In Figure 3.2, cost effectiveness—which is assumed to be inversely related to soil loss in tons per acre per year—of five erosion control practices is plotted against the cost of implementing the practices (corn belt of the United States). In analyzing the cost effectiveness of these selected management practices, it is obvious that conservation tillage is the most cost effective, since the farmer actually made more money using this practice. However, if it were determined that a soil-loss limit of 7 tons per year was required to meet water quality goals, then strip cropping might be selected as the most cost-effective management practice, even though terracing and contour farming would also meet these limits. It can be seen from the graph that for only one dollar more per acre, soil loss can be decreased from seven to three by choosing strip cropping over contour farming. It must be emphasized that the cost effectiveness of a particular management practice will vary with local conditions.

To summarize the pollution problem: sediment, nutrients, and pesticides are agricultural pollutants that have significant impacts on coastal ecosystems. When these pollutants enter a watercourse in an

Figure 3.1. Significant amounts of pesticides and other chemicals applied to farm fields become pollutants when carried by rain runoff into coastal waters. (Photo by author.)

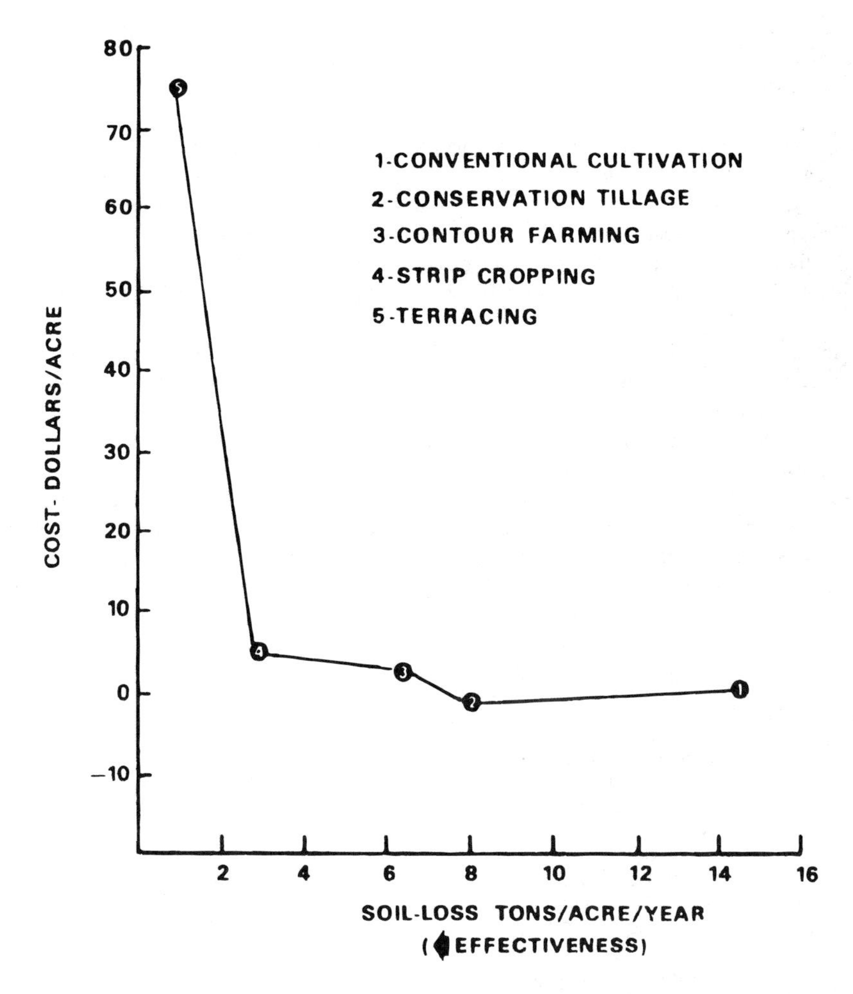

Figure 3.2. Cost effectiveness of selected soil conservation practices (corn fields, USA). (*Source:* Reference 18.)

intermittent and diffuse manner, as in runoff and percolating water from cropland, they are defined as nonpoint source pollutants. The abatement of nonpoint source pollution from crop production systems requires a systems approach. Pollution abatement practices should be selected by considering their effectiveness in controlling pollution and the cost to the owner of the farm of implementing the practices. The industrialization of the agro-food system with increased use of artificial inputs poses a particular threat to coastal environments.

It is the task of the local planning agency to provide the economic incentives and, if necessary, legal sanctions to ensure that water quality and the protection of coastal zone ecosystems are considered by farmers.

Source of this section: Robert C. Hart, U.S. Environmental Protection Agency. Washington, D.C. (Reference 59).

3.2 AIRFIELDS

Airports for coastal cities and resorts have too often been located on sensitive coastal habitats, such as wetland or coral reef areas. Shoreline sites offer easily available, low-priced (or free) sites for landing strips (once they have been filled and graded), as well as unobstructed over-water landing approach paths. The use of such sites for airports has led to pollution of coastal waters and erosion of beaches, as well as to destruction of critical habitats. Discharge of airport stormwater runoff into estuarine waters pollutes them with sediments, oil, organic matter, and toxic substances. The problem is exacerbated in receiving waters that are inadequately flushed or in lagoons or confined bays, where dilution of pollutants is slow.

The impacts of an airport are not confined to the immediate site. Airports and their access roads open new areas and often induce secondary residential, commercial, and industrial growth. As demands for municipal services such as sewer and water hookups, electricity, secondary roads, building lots, and other characteristic urban needs are fulfilled, environmental quality, particularly water quality, will be threatened. Other secondary effects such as air pollution, increased runoff, discharge of improperly or incompletely treated wastes, and disruption of natural drainage patterns must all be anticipated and their effects assessed.

Because so much money is involved, as well as local pride, it is often quite easy for boosters to push through an airport proposal. But approval actions for airport projects should be taken only after careful examination has been made of all potential secondary development, attempts to revise airport plans that would cause major adverse impacts, and a strict procedure to control affected areas have been devised.

To prevent polluted runoff from leaving the site untreated, it is necessary that airports plan for stormwater runoff management to collect and restore these waters before discharge to any coastal water area. As rainwater runs off the runways, parking lots, roadways, and rooftops, it collects oils, jet exhaust particulates, and other airport-related pollutants. Specific water quality problems arise from aircraft washing, aircraft plant stripping, steam cleaning operations, the dumping of oily wastes, and fuel spills. For example, the concentrations of pollutants reported by the Port of Oakland in 1974 for a drainage canal at Oakland International Airport (California) were as follows:

Dissolved solids (mg/l)	3400
Suspended solids (mg/l)	200
Chemical oxygen demand (mg/l)	120
Nitrates (ppm)	4.5
Phosphates (ppm)	6.5
Color (color units)	510
Turbidity (Jackson turbidity units)	125
Lead (mg/l)	0.05
Zinc (mg/l)	0.05
Oil and grease (mg/l)	9.0
Phenols (mg/l)	Trace
pH	8.0

Minimizing the amount of impervious surface (tarmac and concrete) through more efficient airport layout will guarantee a minimum of polluted runoff water. For small airports, the use of porous materials for much of the parking lot surfacing may be practicable (porous surfacing can often virtually eliminate runoff). Problems arising from contamination of runoff can be reduced by keeping work areas clean and free of debris and polluting substances.

Landing strips that project offshore and lie upon or adjacent to sandy beaches have often damaged beaches. The strips function as large groins that cause beach and shoreline erosion on the downdrift sides of the fill structures. At Kosrae Island an air strip destroyed seagrass beds and reef flat ecosystems and disrupted longshore transport of sand (see Figure 4, South Pacific Case Study in Part 4). Similar damage was done at Pala Lagoon, American Samoa. According to Maragos a dramatic example occurred at Kuta Beach on Bali (Indonesia), where the 1967 construction of a large airfield at right angles to the beach and projecting 0.5 km offshore created a giant groin and has caused over 300 m of beach erosion up to 2 km downdrift from the airfield (231). Not only has the beach berm and dune habitat been lost, but millions of dollars of damage and destruction to restaurants, hotels, and residences has occurred. (See "Indonesia, Bali" and "South Pacific: Coastal Construction Impacts" case studies in Part 4.)

The development of John F. Kennedy Airport in New York preempted 4,500 acres of valuable marshlands fringing Jamaica Bay (Figure 3.3). However, additional expansion was denied when a land use study concluded that (128):

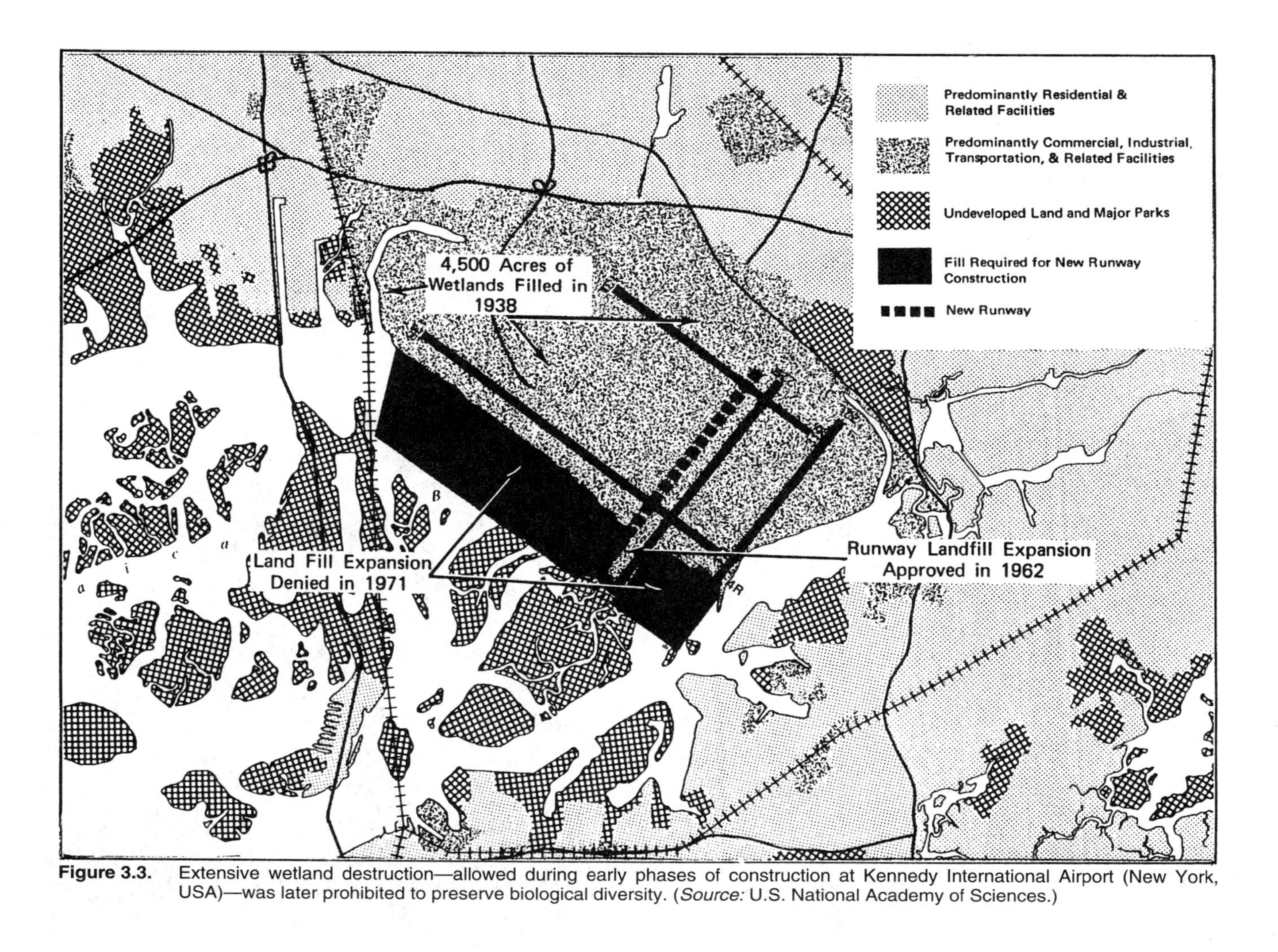

Figure 3.3. Extensive wetland destruction—allowed during early phases of construction at Kennedy International Airport (New York, USA)—was later prohibited to preserve biological diversity. (*Source:* U.S. National Academy of Sciences.)

> Any runway construction will damage the natural environment of [Jamaica] Bay and reduce its potential use for conservation, recreation, and housing. The degree of this impairment will be dependent upon the amount of Bay area taken for this airport extension. . . . A sufficiently large land taking . . . could cause major irreversible ecological damage to the Bay.

3.3 ALTERNATE LIVELIHOODS

Coastal communities often depend on natural resources of the sea for their day-by-day livelihoods. A major restriction on their fishing, or other harvest activities, for conservation or resource rehabilitation purposes could be disastrous for people living so close to the edge, financially as well as geographically. Therefore, the provision for "alternative livelihoods" should be included in the coastal management program if the intervention would result in significant loss of employment and security for traditional users of the coastal zone.

Some coastal management programs have been developed in which resources were reallocated to the detriment of one or more stakeholder groups, for example, coral miners who would be displaced by a tourist development, charcoal makers who would be displaced by a ban on cutting of mangroves, or artisanal fishermen who are not allowed to fish with dynamite. Although their current practices may result in longer term disaster, they still must be assisted to find more resource-harmonius pursuits. (65)

Therefore, it is essential that coastal management planning explicitly includes consideration of such distributional effects. This is not to imply that the coastal agency should become an employment agency. Rather, the agency should have effective linkage with the relevant re-training or re-employment entities, such as a Department of Social Welfare or Economic Affairs or Human Resources (Figure 3.4). Certain NGOs might be very helpful. (65)

The long-term success of integrated coastal management depends on the support of those groups and individuals whose interests will be affected by the implementation of the program. If coastal

Figure 3.4. Coastal management programs should make provision for alternative employment for workers displaced by management programs (e.g., coral miners, factory workers, fishermen)—one possibility is environmental restoration projects such as mangrove forest planting. (Photo by William Keogh.)

management is perceived to have a negative impact on jobs, revenue, or foreign exchange, it is not likely to survive the initial planning stages. An on-going public participation process is needed to involve these interests in program formulation and implementation (65).

The recorded attempts to provide alternative employment for workers displaced by coastal conservation programs have varied outcomes. A useful comparison contrasts the experiences of Sri Lanka and Indonesia (Bali) in creating alternative livelihoods for workers displaced by bans on coral mining (see case histories for each in Part 4). In Bali, the governor banned coral mining (except to provide materials for temples) and directed that the miners be retrained for work in the tourism industry. This was successfully accomplished and no serious problem of enforcement of the ban was experienced. (See "Indonesia, Bali" case history in Part 4.)

However, in Sri Lanka, where coral mining was also made illegal, attempts to retrain the coral miners as agricultural workers met with only partial success. Some miners were not motivated to leave their high paying work to try something else that was unfamiliar to them and perhaps not so profitable; in fact, some were willing to chance a jail term rather than change. Therefore, the Sri Lanka ICZM agency (Coast Conservation Department) is still trying to create new opportunities in fishing, aquaculture, and tourism to replace coral mining. (See "Sri Lanka" case history in Part 4.)

3.4 AQUACULTURE

The coastal aquaculture industry has grown to become a major component of world seafood production with annual sales in the billions of dollars. But this growth has occurred with too few environmental controls in most countries, and the long term net benefit to society may be questionable in many cases.

The Aquaculture Business — Aquaculture can be divided into two categories; a) land-based operations (usually ponds or tanks) and b) open water operations (cages/pens/rafts/stakes, etc.). It is difficult to generalize about marine aquaculture because of the wide variety of species, methods, and intensity of control that are used. On the one hand, a simple aquaculture operation might consist of placement of oyster shell on submerged or tidal mud flats, allowing the collection of oyster seed followed by "growout" to a marketable size. Contrast this with aquaculture operations consisting of a series of pressurized greenhouses built near the seashore to culture large quantities of shrimp which are fed a complicated prepared diet in recirculating water in a completely controlled environment with production of up to 100,000 kilograms per hectare per year (Figure 3.5).

There are two critical limitations on production and yield: excessive predation and a lack of food or primary production. Both problems can be overcome through proper management. Aquatic predators can be removed at the time of harvesting by completely draining the pond or by poisoning water that cannot be removed. Birds can also be serious predators of aquaculture crops, and their control is more problematic. In Hawaii, the largest shrimp farm in the state is located next door to a bird sanctuary. While the farm is under substantial pressure from predatory birds, both operations have survived and generally met their goals through mutual cooperation and understanding for over 15 years.

In ponds, the problem of low primary productivity can be overcome by fertilization and/or supplemental feeding. The feed provides a direct food source, whereas the fertilizer stimulates primary production (phytoplankton and attached-algae production), as well as the formation of detritus. The phytoplankton, algae, and detritus all serve as direct or indirect food sources for the species retained for growout. "Low-tech" unmanaged pond yields are so low (often 30 to 100 kg/ha/yr) as not to justify either the loss of habitat or investment in pond construction.

On the other hand, super-intensive aquaculture systems (production of over 30,000 kilograms per hectare per year) using tanks or ponds do not seem to be economically successful at this time, and they introduce environmental problems of their own. A serious concern is the production of large quantities of waste water high in nitrogen and organics. While it is possible to treat this water before discharge, this is unlikely to be economical. As an example, Figure 3.6 graphically displays some of the critical factors in siting and managing a shrimp growout pond.

Figure 3.5. Aquaculture ranges from crude pond culture to "high tech" intensive operations yielding 100,000 kg/ha. (Photo by Carl Kittel.)

Environmental Problems Related to Land-Based Operations — Problems common to most types of coastal aquaculture include introduction of exotic species (which may be diseased), destruction of coastal habitat and nursery areas for endemic species, disruption of hydraulic flows, pollution of coastal waters, over-harvesting of naturally occurring seedstocks or broodstocks, and aesthetic considerations. Projects should be modified or rejected if serious negative environmental impacts are expected.

Land-based facilities may be located some distance from the seashore (often above the high tide line), and the effects on renewable resources of the marine commons are potentially lower for such facilities than for those located in the coastal waters, such as cages, pens, rafts, stakes, and ponds in mangroves. Projects located within the coastal waters may preempt ecologically critical areas, enter into conflict with mangrove or fishery industries, or conflict with navigation, recreation, and tourism interests. Some of the worst damage has occurred where mangrove forests have been cleared and excavated for shrimp ponds. It is often possible to site land-based operations in areas where such conflicts do not arise.

Aquacultural development has had serious negative environmental impacts in various countries including Ecuador (decreasing quantities of wild fry and deteriorating water quality), France (importation of disease in oyster populations), Thailand (increasing occurrence of disease and deterioration of near-shore water quality), and Taiwan (land subsidence due to draw-down of the water table and increase in occurrence of disease). On the other hand, there has been substantial economic benefit for many developing countries (Ecuador, China, Indonesia, Bangladesh, etc.), which have increased export earnings substantially though aquaculture during the last two decades.

Environmental Problems Related to Open Water Culture — A variety of marine and brackishwater fish, clams, mussels, oysters, and algae can be profitably grown in cages, pens, or other enclosures in open waters. A major consideration is site selection. The effects on navigation of cages and rafts placed in lagoons and harbors are also a matter of public concern. In addition, boat traffic and capture fishing operations can be unfavorably impacted. Also, the cages, rafts, service boats, and other paraphernalia can be aesthetically offensive in some circumstances. For example, in Bermuda, residential complaints

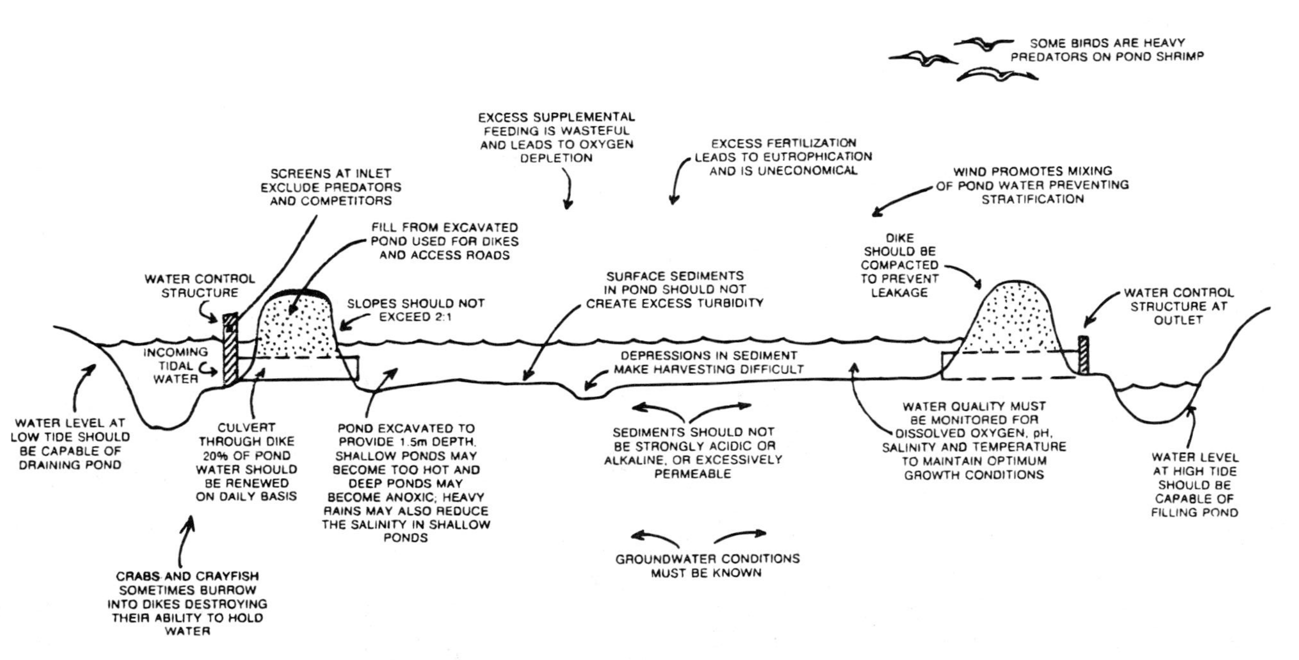

Figure 3.6. Economic success of pond aquaculture depends largely on the initial siting of the ponds and thier subsequent management—some of the main factors are shown in the diagram. (*Source:* Reference 316)

about negative aesthetics have inhibited aquaculture development, and placement of cages in coastal waters in Washington State (USA) is a continuing point of conflict.

In areas where tourism is given a high priority, there will generally be a negative presumption against aquaculture ventures in open waters for aesthetic reasons and because of the strength of feeling that the marine commons should be kept free and open. However, on a small scale some land-based aquaculture enterprises, such as sea turtle and crocodile farms, can be tourist attractions themselves.

In many cases, the least controversial sites for net, cage, stake, or raft-type aquaculture will be those in remote areas where conversion of existing natural habitat is minimal and where the least space competition exists. For example, while Kapetsky *et al.* (201) recommend (for the Gulf of Nicoya, Costa Rica) that the mangrove forest not be converted for aquaculture ponds, they do encourage use of the natural channels in mangrove forests for fish and mollusc culture.

Controls; Planning and Project Review — More government intervention in coastal development will occur in the future because of increased conservation awareness and increasing demands on coastal resources. It is important for planners to be aware of aquaculture as a potential beneficial use of coastal resources, and to be able to intelligently assess and balance the benefits and losses that could result from aquaculture projects.

In their planning, most governments try to accommodate multiple uses in their coastal zones, politically balancing the benefits and impacts of various uses, including aquaculture. Individual aquaculture proposals must be subject to environmental assessment (EA). For commercial operations that intend to produce over ten tons of product per year, or that occupy over one hectare of land, the review should be more formal and complete and should include production of documentation of the project plans, goals, and potential impacts, such as EA (15).

An overall plan for each coastal area is needed to meet broad regional and national needs, in the various sectors such as aquaculture, agriculture, forestry, fisheries, tourism and others. Both regional plans and project EA guidelines should discourage aquaculture projects in designated natural areas or recognized ecologically critical areas such as wetlands, seagrass beds, wetlands, etc. In addition, through coastal zoning, areas can be set aside for development of specific industries such as aquaculture, tourism, shipping, etc.

The development of each region should be planned to accomplish the maximum of multiple use opportunities. Development planning should ensure that aquaculture, fisheries, manufacturing, tourism, nature protection, housing, shipping, and so forth, can all prosper by creating the optimum mix of economic activity (66). Meeting such objectives is also a complex matter for government which needs a special framework for fact-finding, evaluation, negotiation, and resolution of issues as in ICZM.

Proper site selection and appropriate management help to ensure a suitable economic return for the investor/operator, in part through avoiding the conversion of large coastal areas to ponds. It also significantly reduces the risk of failure and abandonment of the facility. In essence, good site selection and management protocols are the bases for both a long-term, profit-generating enterprise and the conservation of the natural sustaining resource base.

Aquaculture proposals pose difficult problems for planners and regulators because the field is so replete with unevaluated successes and failures. Experts vary greatly in their opinions and advice, which seem to take the direction of their own particular advocacies, experiences, and career plans. Unfortunately, critical literature available to planners is typically thin and inconclusive.

As has been demonstrated in many countries, large-scale coastal/marine aquaculture operations have major effects on both human and natural systems. Therefore, in evaluating the benefits of major aquaculture proposals, governments have to look at a wide range of economic and social implications. The major concern for aquaculture, as for any development based upon renewable resources, is whether the use of resources will result in a long term loss or gain in economic benefits and social conditions, a very complex question. Net benefits should be considered over a period of at least 20 to 50 years.

* *Contributed By:* Carl Kittel, Aquatic Farms, Ltd., Ravenna, Ohio (USA).

3.5 ARTIFICIAL REEFS

In the early 18th century, Japanese fishermen discovered that fish catches were more productive in waters around sunken wrecks. This led to the deliberate sinking of wooden structures weighted down with rocks, which enabled fishermen to choose the areas where improved fish catches were desired. This approach, called *artificial reefs* (AR), has since spread and is in extensive use in many coastal countries using a variety of materials, techniques, and configurations.

AR can be either a bane or a blessing, depending upon many circumstances. The use of AR has been extensive in the Philippines, and the results have been well examined and reported. Therefore, the Philippine experience is used as the basis for this section (49, 286, 372).

Deployment — The use of ARs is gaining popularity because of their effective fish-aggregating function. ARs, as a habitat enhancement too, increase the productivity of barren sea floors by providing shelter and, eventually, sources of food. Through time, as the encrusting community develops on the new surfaces, ARs serve as nurseries. In effect, they enhance marine life in the areas where they are deployed. However, it should never be concluded that an AR can serve all the bio-ecological and economic functions of a natural reef.

Site selection is probably the most important consideration for the success of ARs. Placing one in an already rich reef flat may cause more damage than good because the AR would compete with the natural reef as an aggregating habitat and may physically damage the existing reef. Hence, they should be sited in locations that have already been denuded or that support very little marine life. ARs are placed on the bottom, but another innovation—fish aggregating devices (FADs)—are used near the surface of the sea (Figure 3.7).

The choice of materials to be used as ARs and their configurations are further important planning considerations. ARs can be made of inexpensive and readily available materials. Used car tires have become extremely popular because of their availability, durability and low cost. The use of concrete structures, although more expensive, is also widespread. But waste materials which contain toxic chemicals (e.g., old refrigerators and air conditioners containing chlorofluorocarbons) should be forbidden because of their polluting impacts on the marine environment. Materials to be deployed must be effectively fastened and weighted to avoid being swept away by tidal currents. Caution must be exercised in preventing creation of a solid waste dump site for unwanted waste materials.

When setting up an AR program, the agency or country must first critically examine the need for such. Further, those concerned must specify the exact roles ARs are intended to fulfill. The establishment of ARs in Brunei Darussalam and Singapore, which have limited coral reef resources, are good instances. The ARs cater to the needs of recreational fishermen and tourists so that the natural reefs can be completely conserved and relieved of fishing pressure.

But ARs are not panacea and should be viewed as only a part of a larger fishery management program. Definite limits exist on the number of fishermen who can participate if sustainable benefits are to be achieved. What appears to be lacking in many cases is a master plan to ensure that an AR serves as an effective ecological and economic tool. Experience in the developed world has shown that benefits are possible when an AR program is carefully planned, managed and maintained. If not, ARs will be used as excuses for the dumping of waste materials or the over-exploitation of fish and other marine life.

Artificial reef design and construction requires a professional approach. Unskilled persons attempting reef placement can cause trouble, as in Curacao (West Indies) when the local "seaquarium" decided to sink a wreck in the nearby water. The result was that the ship, being improperly secured, slipped into the deep, leaving behind only a much damaged reef and a sea bottom littered with remnants of the ship.

On the other hand, lobster fisheries of Mexico were enriched by the widespread usage of the Cuban "casitas" or shelters; this shelter promotes the aggregation of lobsters facilitating their extraction. Nonetheless, in the same way as happens with conch, most fished animals are either sub-adults or juveniles. The direct effect has been that some important fishery grounds have been depleted.

Effectiveness — AR use can be a cost-effective tool to enhance nearshore reef fish populations, increase fishermen's net income and demonstrate sound fishery management principles. Their abuse can be wastefully expensive, contribute to greater overfishing and perpetuate bad resource management practices. Success or failure, for the immediate user and for the coastal fishery as a whole, depends on how ARs are used in the existing fishery situation.

ARs initially attract fish, particularly "fingerlings", from surrounding waters. Fast-growing tropical species can develop stable populations in 6–12 months' time. The aggregating power of ARs appears to be strong as they have been reported to support up to seven times the fish biomass of natural reefs.

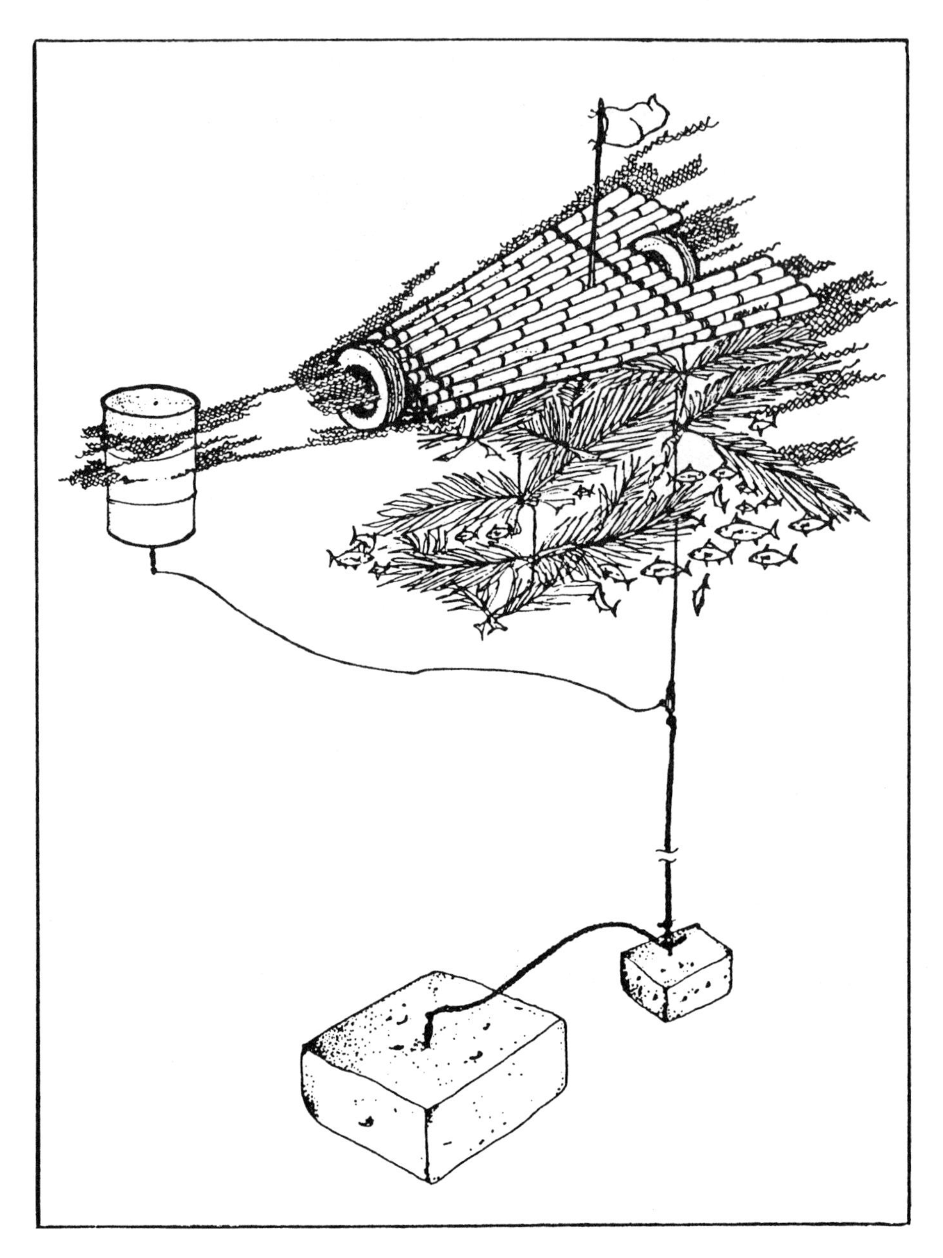

Figure 3.7. The *payao* is a floating, anchored fish-attracting device (FAD) used in the Philippines to attract pelagic fishes. (*Source:* Reference 63.5.)

ARs with relatively high relief (rising 2 meters from the bottom) typically support several fish populations, including (a) residents living within the reef structure, such as groupers and moray eels; (b) mobile bottom feeders (e.g., grunts, goatfish, rabbitfish, and snappers), which range over a larger area as they move from one reef module to another within a cluster and between clusters; and (c) schooling plankton feeders (e.g., fusiliers and surgeonfish), which use the water column above the general reef area. In addition, visiting schools of trevally often linger over reef areas.

The control of both access to the AR and harvest from them is essential to proper AR use. Limited access and harvest management exist only where effective community organization, education and the will of the fishermen, supported by local government officials, provide it. With encouragement, fishermen will volunteer their labor for AR construction (as they did in the Central Visayas Project, Philippines).

Figure 3.8. Bamboo "artificial reef" (AR) unit under construction in a coastal village in Cebu (the Philippines). Bamboo structures have limited underwater life and are being replaced with longer lived concrete units. (Photo by author.)

This reduces project costs while strengthening fishermen's commitment to resource management through an increased sense of participation.

The extent of AR allocation to each fisherman is important. For example, one Philippines program prescribed that each fisherman be given exclusive harvest rights over a 7,850 m^2 circular area (50-meter radius) of sea floor with a cluster of 32 concrete modules placed near the center. The modules have a basal area of 80 m^2 and a volume of 64 m^3. Sustainable harvest was estimated at 512 kg/year (slightly less than 10 kg/week) whereas natural reefs produce about 200 kg. for the same area. Actually, in the Philippines, it is more common to provide such an area to 3–5 families. So the 10 kg per week is at best a supplement to the families' budgets or protein supply.

Good ARs, constructed from bamboo, have lost popularity because of relatively short life span (Figure 3.8). In addition, properly designed concrete modules can provide a semi-permanent reef of similar size at the same cost. Used tires can also be used to construct complex ARs but large numbers are required to build significant reef volumes. Moreover, the tire supply is more limited in developing countries and concentrated in urban centers while the need is largely rural. A few tires may be joined to make a low profile module or six tires formed into a cube which is used to build larger reef modules.

A low cost unit is the concrete tripod—a 2-m-high module that can be constructed by fishermen on a nearby beach and placed from a flatboat or bamboo raft using a block and tackle. Materials and placement cost is about US $10/module or $5/$m^3$. If 25 percent is assumed as an acceptable internal rate of return (IRR) for a ten-year concrete AR project, one finds that the 0.785-ha concrete tripod which costs US$5/$m^3$ is viable with a 4.4 kg/week harvest valued at US$1/kg.

Evaluation — The application of ARs may result in one or more of the following impacts on marine resources:

1. Biomass that is currently exploited is redistributed from natural habitat to AR.
2. Biomass that is currently not being exploited is attracted to AR to increase the total available exploitable biomass.

3. High relief habitat is increased by AR structures, and stocks that are limited by high relief habitat can increase.

ARs may be useful in closing areas to trawling (bottom obstruction), protecting juveniles in shallow nursery grounds, and providing fishing sites for artisanal fishermen using gear that captures older fish. They can substantially reduce travel and research time for artisanal fishermen and improve the catchability of their gear.

ARs have proven particularly effective for artisanal applications in which fishing effort is relatively low. Where fishing effort is uncontrolled, ARs may result in overfishing. Therefore, these structures may not be appropriate except in situations where fishing mortality is controlled.

* *Sources of this section:* "Tropical Coastal Area Management" Newsletter (ICLARM, Manila; Chua Thia-Eng, Ed). Contributors: Sect. 1, L. M. Chou, Natl Univ of Singapore (49); Sect. 2, F. J. van de Vusse, Dept Envir & Natural Res, Quezon City, Philippines (372); and Sect. 3, J. J. Polovina, U.S. Natl Oceanic and Atmospheric Admin, Hawaii (286).

3.6 BARRIER ISLANDS

The seacoast of some countries is edged, in part, by elongated sandy islands or peninsulas. These "barrier islands" and "barrier spits" are mobile, not fixed, geological features. They grow or shrink in response to storms and to fluctuations in sea level, currents, and sediment supply. They also move inland or seaward and up or down the beachfront, according to changing conditions. The changes are the net result of erosion and deposition. The multiple rows of parallel ridges (inactive dunes) that form barrier islands are often visible in the patterns of vegetation. While classified as islands they are functionally the edge of the continental coastline and the frontier where great battles against nature often take place. (81)

The natural properties of barrier islands and beaches provide a strikingly unique combination of values. A typical barrier island, with its ocean beach, sometimes jungle-like interior, and broad expanse of marsh, has scenic qualities unparalleled in the coastal zone (Figure 3.9). They provide habitat and food for hundreds of species of coastal birds, fish, shellfish, reptiles and mammals. They enclose and protect lagoon estuarine resources.

Barrier islands and barrier beaches typically support mangrove swamps or marsh areas on the lagoon/estuarine side. The characteristics and values of these coastal wetlands are described in detail elsewhere as providing essential habitats for many forms of life, supplying basic nutrients to the coastal ecosystem, stabilizing the shore, absorbing floodwaters and removing contaminants from the water. Barrier islands also have a fragile water balance (Figure 3.10).

The perpetuation of barrier islands along high energy coasts depends on perpetuation of beach width/depth and the sand dune system. Dunes are the island's frontal defense against the forces of wind and waves because they store sand in reserve to replace that lost to big storms. They are also the means by which islands move and grow.

Barrier islands cannot be held in one place easily. Typical seawalls, groins, and beach-restoration projects meet with limited success in controlling the powerful oceanic and meteorological forces at work. The great storms may sweep completely over these islands. Sea level rise causes the shoreline to migrate or, retreat, landward.

In the United States, barrier islands are the front line of storm defense for a thousand miles of coastline. When undeveloped, they have scenic qualities—vividness, variety, and unity that are unparalleled elsewhere in the coastal zone. And because they offer broad, sandy beaches and a score of other important recreational features in their natural state, the government is trying to discourage private development of barrier islands.

Also, public costs of occupying barrier islands are very high because they require bridges and extended lines for water (island water is often salty and unpalatable) and utilities and because the costs of storm protection and reconstruction are great (81).

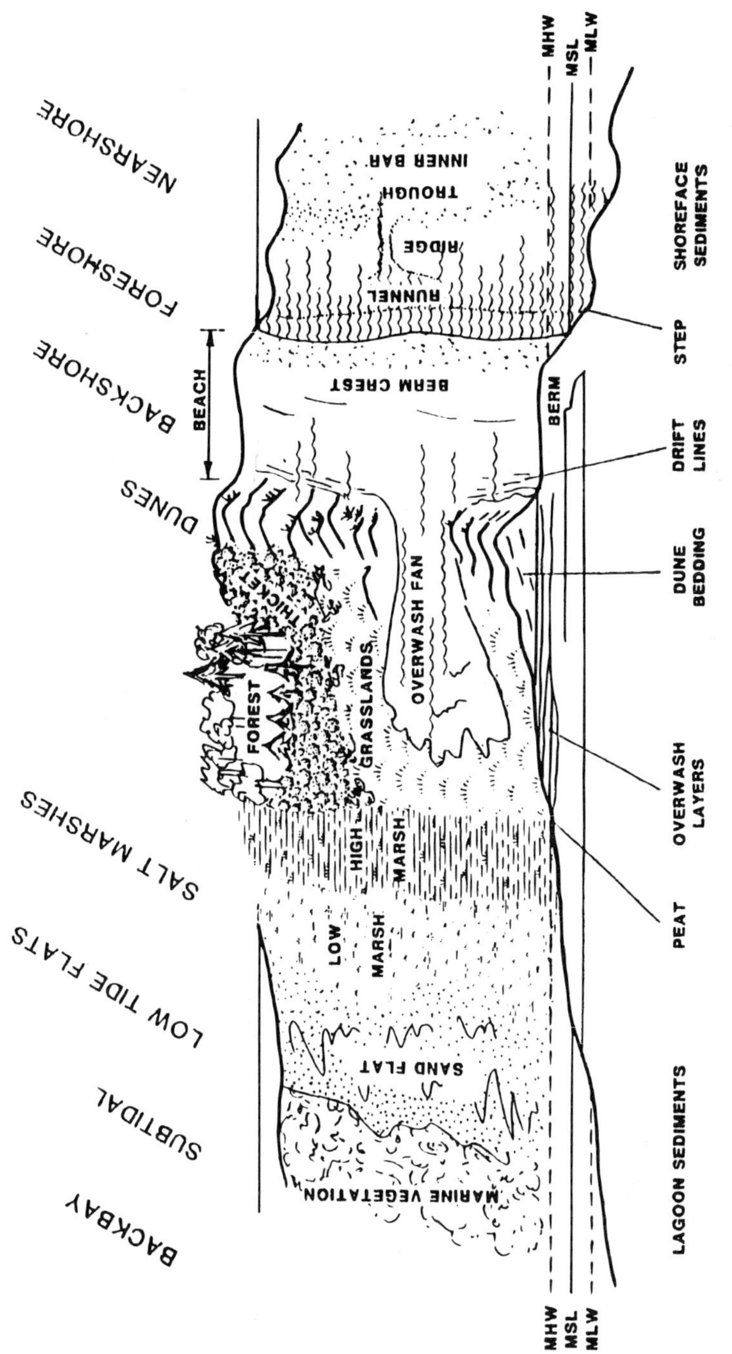

Figure 3.9. Section of a typical barrier island showing hydraulic, geological, and biotic components. (*Source:* Reference 28.5.)

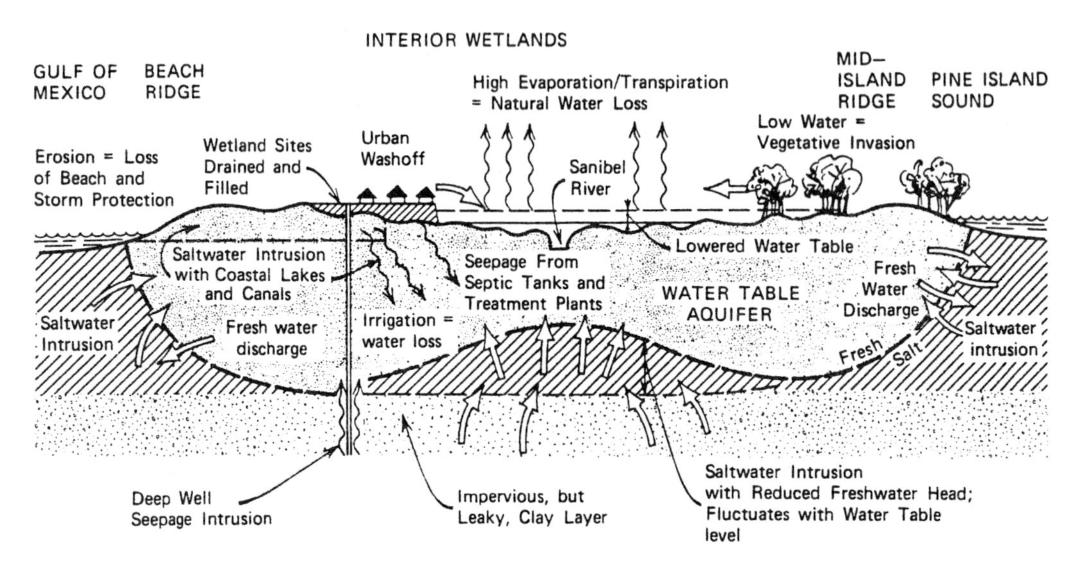

Figure 3.10. Barrier islands and sandy peninsular strips are especially vulnerable to several types of environmental impact (shown is Sanibel Island, Florida, USA). (*Source:* Reference 370.)

3.7 BEACH EROSION

Beaches are being lost at a rapid rate through much of the world. No tropical region is free of the erosion threat. Serious socio-economic problems may arise with erosion and loss of beaches (e.g., risk to property, loss of international tourism revenue). The threat will increase as long as sea level continues to rise—presently estimated at perhaps 1 meter of vertical rise per century (see "Sea Level Rise" below).

The ocean beach in its natural form exists in a state of dynamic tension, continually shifting in response to waves, winds, and tide and continually adjusting back to equilibrium. The beach itself is too hazardous a place to serve as a building site. But people do build very close to the beach, close enough to court danger from erosion and seastorms. If they try to stop the sea forces with structures they can completely disrupt the beach (Figure 3.11). (See "Beach Management" in Part 2).

Alternate erosion and accretion may be seasonal on some beaches; the winter storm waves erode the beach and the summer swell (waves) rebuilds it. Beaches also appear to follow long-term cyclic patterns, where they may erode for several years and then accrete for several years.

Along much of the world's sandy shoreline, sea level is rising at a significant rate—e.g., more than 1 foot (30 centimeters) in the past 100 years in some places. As the sea level rises, the shoreline is forced inland and there is little to anchor it permanently in place. If left natural, beaches and whole sandy islands shift landward while still maintaining their equilibrium slopes (see "Sea Level Rise" below).

The foreshore is the steepest part of the beach profile. The equilibrium slope of the foreshore is a useful design parameter, since this slope, along with the berm elevation, determines minimum beach width. The slope of the foreshore tends to increase as the grain size increases. Slope depends largely on sand size and also significantly on an unspecified measure of wave energy. When the beach erodes, the slope becomes steeper.

Where the beach has a gentle slope, breakers spill gradually toward shore, producing a translational wave that pushes sand up the beach. As sand accumulates further up the beach, the slope increases and this eventually changes the character of waves to a more plunging form. Plunging-type waves have a tendency to erode the profile and shift sand back offshore (99).

Variation in foreshore slope from one region to another appears to be related to the mean nearshore wave heights—the gentler slopes occur on coasts with higher waves. The inverse relation between slope and wave height is partly caused by the relative frequency of the steep or high eroding waves that produce gentle foreshore slopes and the low accretionary post-storm waves that produce steeper beaches.

Figure 3.11. The shoreline is temporarily stabilized, but the beach is totally lost at this once-upon-a-time tourist paradise in Indonesia (Kuta Beach, Bali). (Photo by author.)

In summary (362):

- Slope of the foreshore on open sand beaches depends principally on grain size and (to a lesser extent) on nearshore wave height.
- Slope of the foreshore tends to increase with increasing median grain size.
- Slope of the foreshore tends to decrease with increasing wave height.

Onshore-offshore transport is determined primarily by wave steepness, sediment size, and beach slope. In general, high steep waves move material offshore, and low waves of long period (low steepness waves) move material onshore.

A key feature of beach dynamics is *littoral transport,* defined as the movement of sediments in the nearshore zone by waves and currents. Littoral transport is divided into two general classes: transport parallel to the shore (longshore transport) and transport perpendicular to the shore (onshore-offshore transport). The long-range condition of the beach—whether eroding, stable, or accreting—depends on the rates of supply and loss of littoral material. The shore accretes or progrades when the rate of supply exceeds the rate of loss. The shore is considered stable (even though subject to storm and seasonal changes) when the long-term rates of supply and loss are equal. Thus, conservation of sand is an important aspect of shore protection.

Each part of the beach is capable of receiving, storing, and giving sand, depending on which of several forces is dominant at the moment. This keeps the slope or profile intact through balancing the sand reserves held in various storage components in the beach system—dry beach, wet beach, submerged offshore bar, and so forth—and in the duneland area behind the beach. When storm waves carve away a beach, they take sand out of storage. In the optimum natural state, however, there is enough sand storage capacity in the ocean beach berm (or in the dunelands behind it) to replace the sand lost to storms; consequently, the effects are temporary, with the beach gradually building up again.

Fortunately, nature provides extensive storage of beach sand in bays, lagoons, estuaries, and offshore areas that can be used as a source of beach and dune replenishment where the ecological balance will not be disrupted. The sources are not always located in the proper places for economic utilization nor are they considered to be sustainable without proof of renewability. When these sources are depleted, increasing costs must be faced for the preservation of the beaches. Offshore sand deposits will probably become the most important source in the future (362).

In essence, the dynamic response of a beach under storm attack is a sacrifice of some beach, and often dune, to provide material for an offshore bar. This bar protects the shoreline from further erosion. After a storm or storm season, natural defenses may again be re-formed by normal wave and wind action.

Following a storm there is a return to more normal conditions which are dominated by low, long swells. These waves transport sand from the offshore bar, built during the storm, and place the material on the beach (Figure 3.12). Winds then transport the sand onto the dunes where it is trapped by the vegetation (see "Dunes" below). In this manner the beach begins to recover from the storm attack. The

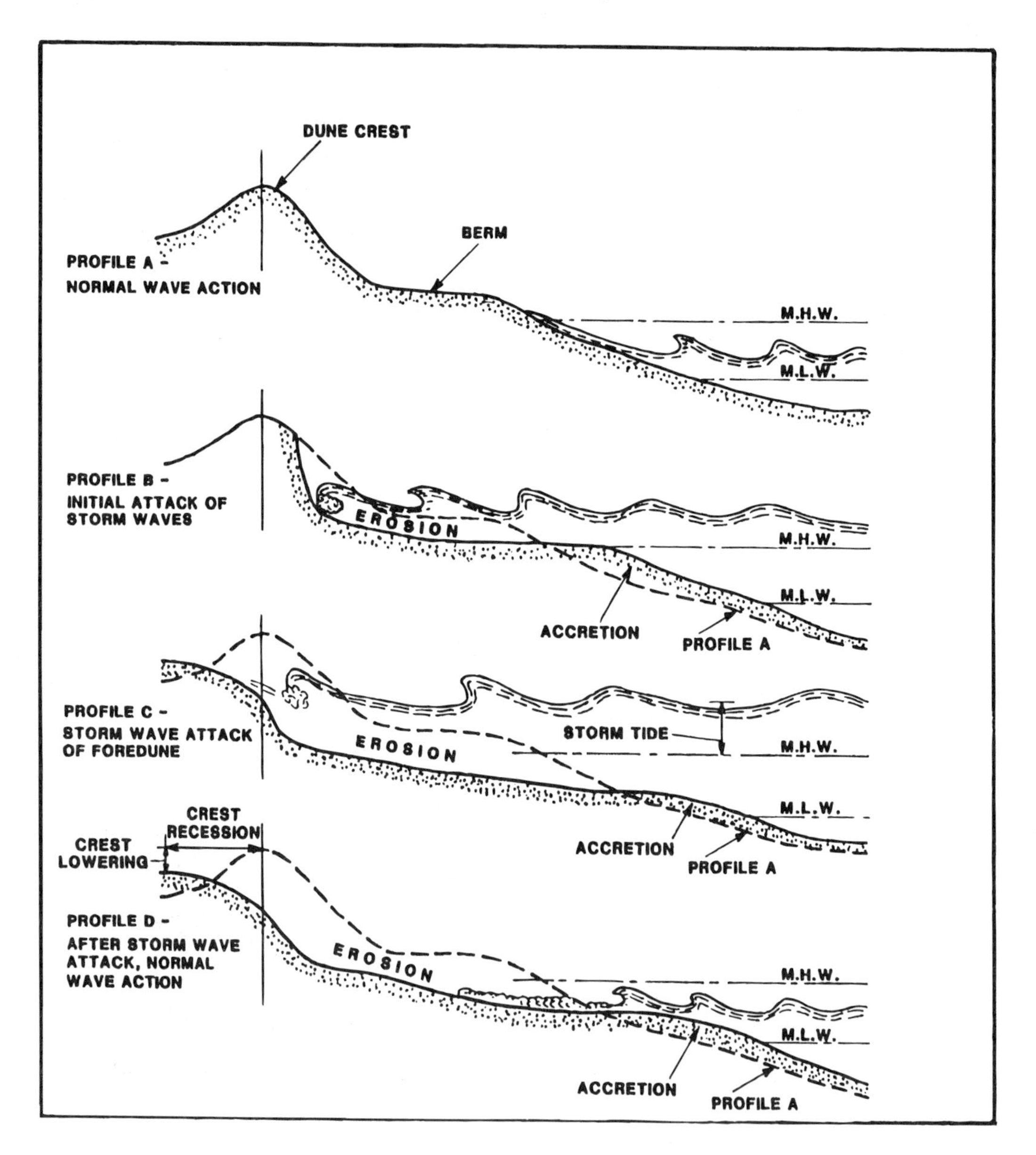

Figure 3.12. In their natural state, beaches yield sand during storms but gain it back afterward; therefore, coastal managers should persuade beachfront property owners to wait for the natural process of restoration and not rush into engineering solutions. (*Source:* Reference 362.)

rebuilding process takes much longer than the short span of erosion which took place. Therefore, a series of violent local storms over a short period of time can result in severe erosion of the shore because natural processes do not have time to rebuild the beach between storms.

It is important to realize that the erosional and depositional cycles of beaches respond to forces acting far from the beach itself. Of special importance are offshore shoals and currents, inland dune systems, and river outflows. It is estimated that rivers of the world bring about 3.4 cubic miles of sediment to the coast each year. Sometimes only a small percentage of this sediment is in the sand size range that is common on beaches; for example, the sediment load of the Mississippi River is 50 percent clay, 48 percent silt, and only 2 percent sand (362).

It should be clearly understood that beach problems usually result from human actions. Beach and dune systems in their natural state provide a buffer against storm-caused erosion and storm breaching. The natural forces at work are immense; therefore, structural solutions to beach erosion and protection of shoreline property from the hazards of sea storms may be expensive and are often temporary or counterproductive. Normally, if nothing is built either on the beach or next to the beach, it will remain as long as the process of natural replenishment continues. Mobile and responsive, the beach will usually remain over the years, but even if sea level or other natural factors cause erosion, problems will not usually emerge unless there are houses or other structures placed near the beach.

3.8 BEACH FILL

Artificial beach renourishment is a desirable, if expensive, method of beach protection and is nearly always preferable to structural methods when affordable. However, beachfill is still only a substitute for proper regulation of land use and beachfront construction (see "Indonesia, Bali" case history in Part 4). Unfortunately, engineers used to consider sand stored in ecologically sensitive bays, lagoons, estuaries, and nearshore areas as an unlimited source of sand for beach nourishment. With the recent awareness of the sensitivity of shallow coastal ecosystems, and the fact that lagoon/estuarine materials are often too fine for good beachfill, however, estuarine and bay sources are now not usually considered available for beach nourishment.

Since dunes and adjacent beaches as well as nearshore areas and lagoons/estuaries are generally considered off limits for sand removal, there are two appropriate sources of supply for beach nourishment: (a) around inlets or other areas of accretion, where the supply is constantly replenished by natural forces, particularly when navigation dredging is being done, or (b) the open ocean or broad non-estuarine bays beyond a depth of about 40 feet (14 meters). Beyond a depth of 40 to 50 feet, the sand reserve is generally independent of the sand reserves of the beach area, and therefore its removal should not upset the beach profile.

However, care must be taken during offshore dredging to avoid any critical habitat areas and to prevent excess siltation of the water. Siltation and induced turbidity may occur as a direct result of dredging, generally at and surrounding the dredging site, or from erosion of beach fill by waves, currents, or rain. Excessive silt may bury portions of reefs, suffocate fish by clogging their gills, or settle out and smother bottom-dwelling life (see "Suspended Matter" below). Potential damage to coral reef communities is minimized when dredging is not done adjacent to reefs, when there is proper selection of equipment and care in dredging activities, and when rehandling of fill material between borrow areas and the beach is avoided. For an example, see the "Indonesia, Bali" case history in Part 4.

The possibility of reef and environmental damage during dredging is a proper concern, but proponents of renourishment in Florida (USA) argue that reefs can be protected in two ways: (a) the judicious selection of a "borrow" site for the sand to be pumped back on the beach and (b) efficient management of the work so that there are no errors on the part of the contractor (98).

Often a supply of sand of suitable quality (type and size) is not readily available; the grain size generally must exceed 0.25 millimeters to remain on the beach and therefore be useful for restoration. A minimum size as large as 0.4 millimeter may be required under unfavorable (high energy) conditions.

Acceptable but qualitative visual estimates of mean size of sand grains are possible with some experience. With more experience, such visual estimates become semi-quantitative. Visual comparison with a prepared standard is a useful tool in reconnaissance and in obtaining interim results pending a more complete laboratory size analysis. A more exact analysis can be made with a series of sieves, graduated in size of opening according to the U.S. standard series (see "Sediments and Soils" below).

It has been argued by beachfill proponents that sand suitable for renourishment is often "between the reef lines requiring tight parameters" for dredging and "takes good design and good control" (98). But the ocean is very difficult place to work and the seafloor is most difficult to excavate with precision. The argument continues that if the project is properly designed and monitored, "one can avoid many ecological problems" (98). To avoid excessive sediment release and turbidity, it is recommended that less than 5 percent, and at the most 10 percent, of the material being deposited on the beach should be silts and clays.

Each replenishment project should be thought of as only the *first* in a series of replenishments. Periodic replenishment will be required at a rate equal to natural losses caused by erosion forces. Replenishment along an eroding beach segment can best be achieved by stockpiling suitable beach material at its updrift end and allowing longshore processes to redistribute the material along the remaining beach—this is called a *feeder beach* and the process is sometimes termed *artificial beach nourishment* (362).

Groins may be included in a beach restoration project to reduce the rate of loss and therefore the renourishment requirements. When groins are considered for use with artificial fill, their benefits should be carefully evaluated to determine their justification, for if a beach is restored by impounding the natural supply of littoral material, a corresponding decrease in supply may occur in downdrift areas. (362)

It is very difficult to design the new beach profile. A good assumption is that the profile after the beachfill eventually will be the same as before, provided the same type of sediment has been used. Directly after the beachfill, the shape of the beach is not optimal. Nature will adapt the shape of the fill (but an overcharge of 40 percent of the design quantity is usually needed to cover the losses). Therefore, it does not matter very much where the sand is placed in the profile. After one or two smaller storms the complete profile is reworked by nature and the natural profile formed. From this, one may conclude that one should dump the sand on that place where dumping is the cheapest and environmental impacts the least. (98).

If the purpose is to prevent storm flooding, the best place to put sand is as high as possible on the beach. If the purpose is to combat chronic erosion, the best place is in the breaker zone (98). The cheapest way may be placing the sand on the higher section of the beach. All discharge pipes can be placed out of the reach of the waves, and after the fill a good wide beach is formed. However, the slope just under the low-water line will be overly steep. But the first storm in autumn will spread out the sand. (98)

However, from a psychological point of view this may not be a good beachfill. Hotel and home owners may see a beautiful beach in the summer, but during the first minor storm, they observe that much of the beach disappears (without knowing that the sand is deposited just below the low water line). So the owner might conclude that the beachfill was not successful—the wide beach has shrunk. But it is only a temporary condition.

Unfortunately, very few communities can afford to engage in large restoration projects on their own. Groins may cost US$500,000 or seawalls, $200 to $500 a foot. The cost of sand used for beach nourishment may range from US$2.00 to $4.00 a cubic yard—for sand pumped by a dredge over a short distance—to as much as US$5.00 to $7.00 a cubic yard if the sand is hauled by truck. (See also "Shoreline Contruction Management" in Part 2).

Beach nourishment provides (a) a beach suitable for recreational purposes, (b) an effective check on erosion in the problem area, (c) a supply of sand to adjacent beaches, and (d) a practicable, if expensive, answer to beach erosion where large quantities of sand are available (Figure 3.13). However, beach nourishment usually does not permanently restore the beach (58).

Sources of sand for beachfill are often scarce. Any removal of sand from the beach system itself will threaten the beach—whether the sand is taken from dunes, the beach per se, or from the longshore bar or nearshore zone. Therefore, you should not try to solve an erosion problem in one part of the beach system by using sand from some other part of the same system. Since dunes, adjacent beaches, nearshore areas, and estuaries are generally considered off limits for sand removal, there are two appropriate sources of supply for beach nourishment: (a) the open ocean or broad non-estuarine bays beyond a depth of 40 feet or (b) areas around inlets or other places of accretion, where the supply is

Figure 3.13. The results of a US$75 million beach restoration project at Miami Beach (Florida, USA)—(A) before beach fill; (B) after beach fill. (Photos courtesy of the U.S. Army Corps of Engineers.)

constantly replenished by natural forces (particularly suitable in conjunction with navigation dredging whereby the sand supply is a by-product).

Figure 3.14 shows a beach that is out of equilibrium with the incident wave field and not aligned with incoming waves which form a broad arc between the inlets. In this situation, beachfront property owners regularly demand publicly financed rehabilitation, which in this case (Hunting Island, South Carolina, USA) needed beachfill ***five times since 1968,*** amounting to over four million cubic yards (99).

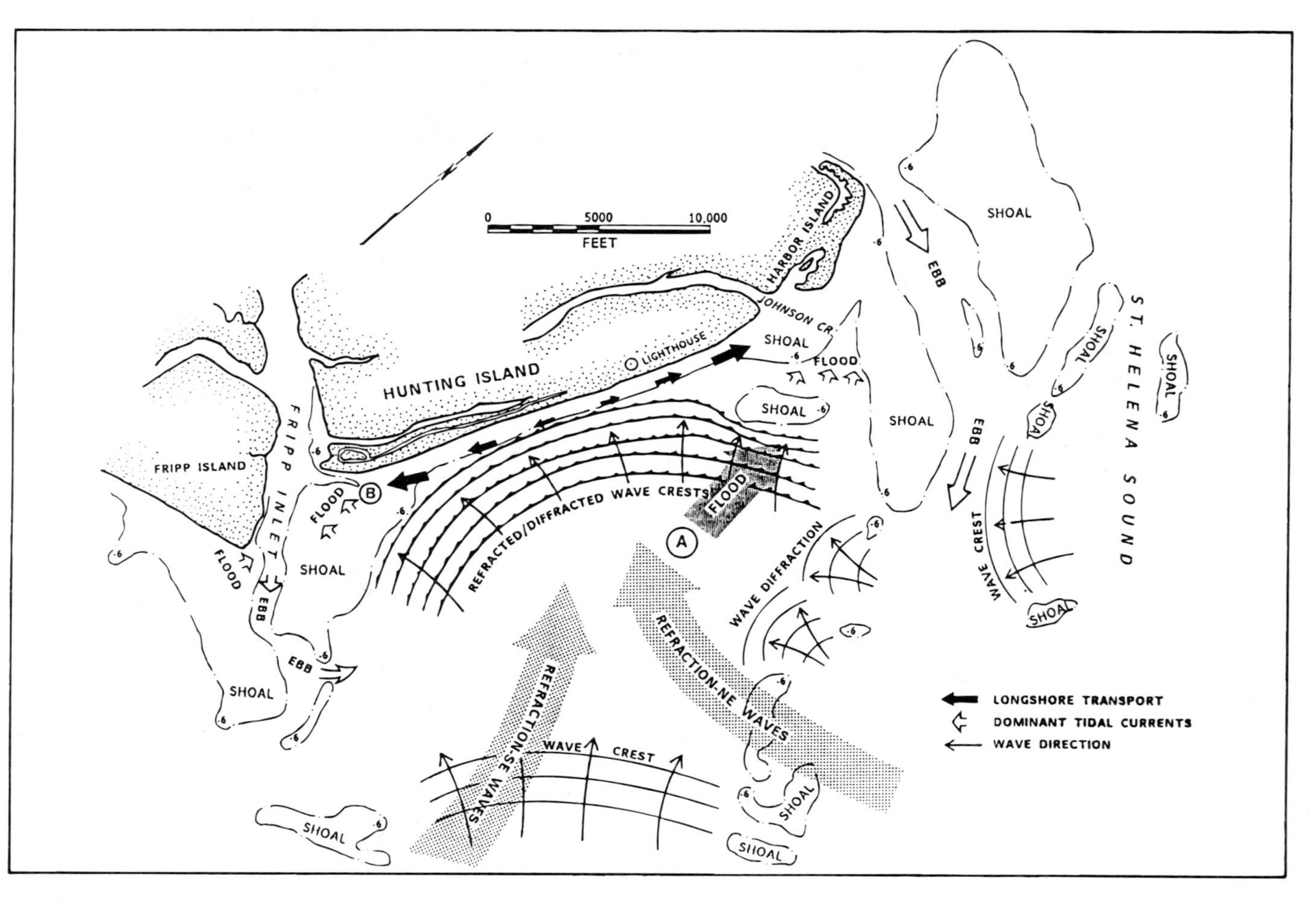

Figure 3.14. The beach shown is losing sand in both directions from the center because of its hydraulic circumstances—note the extensive formation of sand shoals, or sandbars, formed by diurnal tidal movement in and out of the passes. (*Source:* Reference 99.)

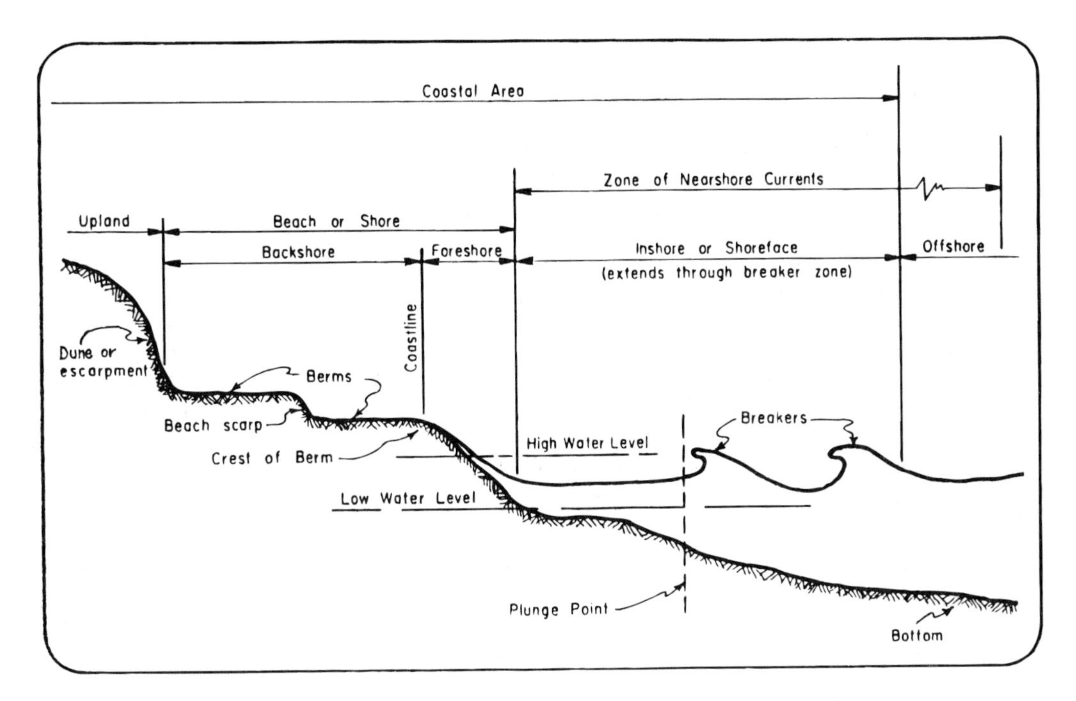

Figure 3.15. Standard beach profile with description and nomenclature. (*Source:* Reference 362.)

While there are ecological problems with renourishment, there also may be some benefits—e.g., one study showed a heavy increase in turtle nesting on a renourished beach (98).

3.9 BEACH RESOURCES

Beaches are used by more people than any other habitat in the coastal zone. Beaches are the focal point for coastal recreation and tourism. People are willing to travel thousands of miles and spend thousands of dollars to lie, sit, or walk on the beach. Consequently, the land immediately adjacent to the beach is far and away the preferred site for coastal tourist hotels in coastal areas around the world.

Beaches are also the first line of defense against storms and erosion. Yet, in spite of all their values, usually beaches are poorly cared after. The means of protection are simple: **first,** permanent development should be placed well inland of the active part of the shore, including erodable shores that would be expected to become active in the future; **second,** positive action should be taken both to prevent the removal of sand from **any** storage element and to prevent blocking the free transport of sand from any one storage element into the active part of the system. (See "Beach Erosion" above and "Beach Mangement" in Part 2.)

Beaches are not stable forms; they are, instead, dynamic landforms, constantly subject to erosion and/or accretion. Differences in beach form and position reflect the local balance or imbalance between deposition (gain) and erosion (loss). On a worldwide basis, erosion (natural and human-induced) dominates over deposition, which is partly due to the global rise in sea level. Consequently, there is serious loss of beach and beachfront in many parts of the world.

A ***beach*** may be defined as the unvegetated part of the shoreline formed of loose material, usually sand, that extends from the upper berm to the low-water mark. The typical beachfront complex is composed of the following seven parts (see Figure 3.15):

1. **Bar.** An offshore ridge that is submerged permanently or at higher tides.
2. **Trough.** A natural channel running between an offshore bar and the beach or between offshore bars.
3. **Foreshore.** The part of the shore lying between the crest of the most seaward berm and the

ordinary low-water mark; it is ordinarily traversed by the uprush and backrush of the waves as the tides rise and fall.

4. **Backshore.** The arm of the beach that is usually, dry, that lies between the foreshore and the duneline and that is acted upon by waves only during storms and exceptionally high water.
5. **Berm.** A ridge or ridges on the backshore of the beach, formed by the deposit of material by wave action, that marks the upper limit of ordinary high tides and wave wash; berms have sharply sloping leading edges.
6. **Beach ridge.** A more or less continuous mound of beach material behind the berm that has been heaped up by wave action during extreme high-water levels (if largely wind built, usually termed *dunes,* often vegetated).
7. **Dunes.** More or less continuous mounds of loose, wind-blown material, usually sand, behind the berm (often vegetated). The first layer dune is termed the *foredune,* or the *frontal,* or *primary,* dune; those behind the frontal dune are called *secondary, rear,* or *backdune.* An active dune is one that is mobile or in the process of visibly gaining or losing sand; such a dune is usually vegetated mostly with grasses rather than woody vegetation (see also "Dune Management" in Part 2).

Many important birds, reptiles, and other animals nest and breed on the berm and open beach, as well as feed and rest there (Figure 3.16). For example, sea turtles (including such endangered species as the loggerhead and green turtle) come ashore during the spring and summer to lay their eggs in the "dry beach" above the high-water line (see "Turtles" below). Also, terns and other seabirds frequently lay their eggs on the upper beach or in the dunes (84).

Beaches also provide a unique habitat for burrowing species such as ghost crabs, coquina clams, razor clams, and others. There may also be a complex intertidal or intertidal community of crustacean organisms that attract shore birds. The shallow waters of the nearshore zone provide habitat for shellfish of many kinds and a wide variety of forage species, which in turn attract fish and birds to feed on them (84).

The plant communities of the beachfront thrive on the continuing stress of natural disturbances, to which the grasses and other plant species living here are especially adapted. The vegetation plays a

Figure 3.16. Beaches are the only habitat for some species and the primary or key habitat for many others; therefore, ecological balance requires that the beach be protected and critical stretches be reserved for biological diversity purposes. (Photo by author.)

significant role in stabilizing the dune front, trapping and holding the sand blown up by the wind, and thereby allowing the dunes to build and stabilize.

3.10 BIOLOGICAL DIVERSITY

Biological diversity, or "biodiversity," can mean many different things. In its general sense it refers to the dazzling variety of life forms that occupy Planet Earth. In its working sense, however, it mainly encompasses efforts to protect species of flora and fauna from extinction (see "Endangered Species" below). Protection of species requires protection of their special habitats as well as preventing the hunting or harassing of them (Figure 3.17).

World concern regarding the loss of biological diversity is not felt equally in all nations nor is it shared equally by all members of society. It stems from the realization that humans have been transforming natural landscapes worldwide, both terrestrial and aquatic. People have been cutting down forests, changing natural grasslands and savannas into agricultural fields, using rivers, lakes, and oceans as dumping grounds for their wastes, and modifying the natural composition of the atmosphere by the addition of millions of tons of carbon dioxide, various oxides of sulfur and nitrogen, and other chemical compounds (320).

According to Solbrig (320), the reason for the divergent perceptions about the real impact of the loss of biodiversity, is that ecosystem change, while incurring costs, also produces benefits. Natural landscapes are being transformed because their conversion fills certain human needs and is of advantage

Figure 3.17. The brown pelican made a dramatic comeback after it was declared endangered and the environmental poison DDT (which weakened its eggs) was banned in the United States. (Photo courtesy of U.S. National Oceanic and Atmospheric Administration.)

to many of those involved in the alteration. For example, the spread and intensification of agriculture is the direct result of the need for food and fiber of a large and growing human population. Likewise the felling of forests answers the needs for timber and fuelwood, the intensification of fishing the need for more protein, etc. Because benefits are typically economic and immediate, while costs are ecological and long term, it is very difficult to objectively balance costs and benefits. (320)

The lesson to be learned is that, to ensure the maximum quantity and quality of renewable natural resources for ourselves and our decendants, we must learn to use resources sustainably. This means we must learn to attach proper value to the benefits and costs of using forests, savannas, grasslands, and all other ecosystems that ultimately provide the oxygen, food, fiber, and recreational opportunities for people worldwide.

Preserving biodiversity is an important reason for protecting natural areas. Endangered species are major beneficiaries of coastal habitat protection; for example, coastal birds, turtles, and even marine mammals. Other protections for species are mostly regulatory, that is, providing legal protection against killing or disturbing endangered species whether inside or outside a designated protected area. While efforts to protect aquatic species are to be encouraged, extinction of marine species is not so probable as for land species. (See "Endangered Species" below.)

Some marine ecosystems—coral reefs, for example—have high species diversity, but despite increasing marine pollution and degradation of coastal habitats, there is little evidence for an imminent major loss of marine biodiversity at the species level. This may be due partly to lack of knowledge, but an important factor is the typical marine life history (390). Marine species live in a wide open system, the sea being continous around the earth, and have greater ranges and fecundity than terrestrial species. This confers on them a greater resilience to exploitation and environmental change. Similarly, endemism is rare. Nevertheless, all aspects of biodiversity protection are referred to in this guidebook.

Many marine species, particularly fish and invertebrates, are so-called *r-strategists,* producing large numbers of offspring, but having short lives. Recent species extinctions are almost unknown among marine organisms with a planktonic larval stage and among the many species of migratory, highly mobile, and widespread fishes. Species that are large-bodied, long-lived, and slow-breeding, producing few offspring with much parental investment (K-strategists) are rarer in the sea. However, these species are most vulnerable to exploitation. The few known cases of marine extinctions in historic times include marine mammals (Steller's sea cow, Caribbean monk seal, sea mink) and flightless birds (great auk), whose demise generally resulted from excessive exploitation (305).

McManus (242) explains the possible effects of exploitation on the diversity of fishes as follows:

1. **Fall-off.** The abundance is reduced at many levels including both abundant and rare species, and some of the rare species are reduced to zero abundance. The would only be expected if:
   - Fishing was uniform regardless of the abundance of a species (false in our case) and if many species are involved in the fishery (true).
   - Some of the abundant species normally act as switching predators which prevents competing species from competitive exclusion—i.e., one decimating the other in competition for food or space. This could be happening here (see "Tilt-off").
2. **Add-on.** Humans remove predators that normally had a fall-off effect. For example, removing most sharks from a reef (essentially true in our case) might cause a general rise in successful recruitment including that of species normally totally incompatible with the predators. This would cause a rise in abundances, species richness, and diversity, and have an unpredictable effect on evenness.
3. **Tilt-on.** Humans become switching predators, causing an increase in evenness and freeing niche-space for other species. If the species pool is large, this could conceivably lead to an increase in species richness, simple diversity, Shannon-Weaver diversity, and of course, evenness. Otherwise, only the latter one or two of these would rise and the rest remain unchanged.
4. **Tilt-off.** Humans remove existing switching predators, causing some species to become dominant relative to other competitors. If the switching predators are responsible for maintaining some of the species richness, then their removal might result in losses of richness, diversity and evenness. It must be noted that a pulse of successful recruitment of a species in the midst of a fall-off decline process could result in a tilt-off pattern.

5. **Terminate.** Humans over-exploit selected species to local extinction. This might be true especially when certain species are very valuable, as with certain aquarium species, or if the docile seahorses of the reef flat seagrass beds were to be collected systematically for sale as folk medicine (a realistic danger). The activity would have to occur on a wide enough scale (hundreds to thousands of kilometers of coastline) to impair recruitment processes. We would expect minor drops in richness and diversity, and conceivably a drop in evenness.
6. **Scramble.** The dominance order of species is merely rearranged, with no substantial net changes in abundance or diversity. This could be the case if recruitment was not strongly limiting and settling space or other resources were.

The reader may also wish to refer to "Genetic Diversity" and "Endangered Species" below. Management responses to the need for biodiversity protection are discussed in Part 1 (Sect. 1.4.3) and Part 2 ("Critical Natural Areas Identification" and "Protected Natural Areas").

3.11 BIOSPHERE RESERVES

Biosphere reserve is a UNESCO-devised type of multiple-use, semi-protected natural area, usually in a rural area. Each reserve embraces a large and varied landscape, which may have farms, livestock, and production forest, but each has a core where nature protection has first priority. The idea is to harmonize nature protection and the work that people must do to survive, following the ecodevelopment approach (see "Ecodevelopment" below).

In this way, biosphere reserves offer a "humanistic" approach to nature conservation such that plants and animals may not be *a priori* considered more nor less important than humans. Humans may be considered as a positive, key factor in the maintenance of a given biosphere reserve; in return, humans learn how to live in harmony between their cultural and natural environment. (81)

The different interests of a biosphere reserve are made compatible by a system of zonation—three more-or-less concentric zones. In this zonation the inner "core area" is strictly protected. The next one, the "buffer zone" can be used for non-destructive sustainable activities. A national park designated as a biosphere reserve often includes a core area together with a buffer zone of this type. The outer zone, or "transition zone" covers other functions of the biosphere reserve, including much lighter controls on traditional uses, serving as a partnership area (between reserve management and the surrounding community) within the biosphere reserve. Each of the three are explained below (81):

The **core area** consists of examples of minimally disturbed ecosystems characteristic of one of the world's terrestrial or coastal/marine regions. A core area has secure legal protection, for example, as a strict nature reserve. Only non-destructive activities that do not adversely affect natural ecosystem processes are allowed. Although natural processes normally operate unimpeded by human intervention, active human intervention, such as prescribed fire or controlled grazing, may be needed in certain subclimax ecosystems to maintain the natural characteristics of the site.

The **buffer zone** adjoins or surrounds the core area; its limits are legally set out and usually correspond with the outer limits of a protected area such as a national park. Here, the activities are diverse and are coordinated in such a fashion that they help to buffer the core from any harmful outside disturbance. These activities serve the multiple objectives of the biosphere reserve and can include basic and applied research, environmental monitoring, traditional land use, recreation and tourism, general environmental education, and specialist training.

The **transition zone,** as the outermost part of a biosphere reserve, is thought of as a dynamic, cooperative zone, where the work of the biosphere reserve is applied directly to the needs of the local communities in the region. This zone—also known as the *zone of influence* or *outer buffer zone*—may contain settlements, fields, pastures, forests, and other economic activities that are in harmony with the natural environment and the biosphere reserve. This zone of cooperation is particularly useful in helping the biosphere reserve to integrate into the planning process of its surrounding region (Figure 3.18).

The alert reader will understand that much of the above is a theoretical construct and that it would be very difficult to find a real world situation to match exactly to this theory. Also, much of it has a ring of "internationalese," the type of language that international assistance folks use when addressing Third World people and their problems. Nonetheless, a Biosphere Reserve program serves coastal

Figure 3.18. The wetlands Biosphere Reserve in Uruguay—Los Baños del Este, which covers over 300,000 ha of land, wetlands, and water—is struggling to balance conflicting uses in a multiple-use agenda. (Photo by author.)

communities by offering a new multiple-use model that can integrate development needs and conservation priorities.

Official biosphere reserves are those sanctioned by UNESCO, but there are reserves in many countries modeled after the UNESCO plan without the UNESCO imprimatur. In fact, the foundation of the biosphere reserve idea—to enclose protected areas within controlled multiple use buffers—is now the main approach for the entire protected areas program for India and many other countries. Success is most certain when the protected area (core) management can offer *real* aid and assistance to the buffer area inhabitants.

The tilt of the biosphere reserve program is more toward achieving sustainable use of resources than it is toward nature protection—traditional nature reserves and national parks do that. The modern view is that biosphere reserves should serve three functions: economic development, science, and conservation. The priority to nature protection holds only for the core zone—which can be a relatively small part of a site. The core is simple to manage because it is usually dedicated exclusively to nature protection. The tough parts are the multiple use buffer and transition zones where controlled use is the mandate and where human presence and activity is accepted.

For any particular proposed site, preparation of a formal Management Plan is enormously important. In addition to the usual content of a site management plan, it must suggest boundaries and a zoning scheme for allocation of uses within those boundaries. It must show full-scale involvement of the people in every planning step. It must deal with economic issues and ascertain that jobs, food, housing, land, and other securities are guaranteed. The specific aid that will be provided to the community must be stated. It must be specific about what public lands are to be included and what effects the reserve would have on private holdings.

At the time of this writing there were 266 official biosphere reserves located in 70 countries, which correspond to 266 separate dots on the world map (Figure 3.19). A challenge to UNESCO is to join these dots into a functional network that can promote the exchange of scientists and management personnel to learn from the experience of other biosphere reserves, particularly in their quest for integrating conservation with local needs and socioeconomic development. Another is to constantly

Figure 3.19. The Sian Ka'an Biosphere Reserve in Yucatan, Mexico, incorporates over 500,000 hectares of forests, wetlands, bays, settlements, and beaches. (Photo by author.)

update and reconceptualize the program to make sure it remains current in terms of world needs and directions. The international biosphere reserve network has an important role to play in creating models for sustainable use of resources.

3.12 BIOTOXINS

Certain marine organisms produce toxins that are poisonous to humans and possibly to sea mammals like dolphins. These biotoxins are defined as "colloidal proteinaceous poisonous substances that are metabolic products of organisms." The causative organism is often a species of dinoflagellate, a microscopic size floating alga, which has the ability to "bloom" in such massive numbers as to change the color of the sea to red or brown, thus gaining the title of *red tides* (see "Red Tide" below).

The following are the most common types of occurrences (351):

- ***PSP, "paralytic shellfish poisoning":*** Caused by *saxotoxin,* and other toxins, accumulated by bivalve shellfish, which consume (by filtering seawater) phytoplanktonic organisms of the type called *dinoflagellates* that contain the toxin. In humans, PSP causes respiratory paralysis, which can lead to death by asphixia. DSP (the "diarrhoeic" version), which comes from the same type of organisms, can cause severe gastrointestinal disturbance but is not fatal. Another source of biotoxin is the unicellular planktonic diatom known as *Nitzschia pungens.* This organism is taken up from the water by certain shellfish and, to those that eat them, can cause "amnesic shellfish poisoning."
- ***"Ciguatera":*** Caused by ciguatoxins, which are found in a variety of tropical and sub-tropical fish that graze on toxic dinoflagellates that live (epiphytically) on macro algae. The grazers are in turn fed on by predatory fish, which accumulate enough toxin, when they reach larger sizes, to poison people, although the fish themselves are little affected. With symptoms that are neurological,

cardiovascular, and intestinal, ciguatera is often extremely disturbing to humans but rarely lethal (fatality—0.1–4.5 percent). (See "Ciguatera" below).

- ***"Plankton blooms":*** These range from benign to hazardous depending upon the particular organism involved. While often they are formed by harmless phytoplankton, the dangerous PSP dinoflagellate often occurs in seasonal blooms. Visible plankton blooms—known as "red tides" or "brown tides"—can also cause fish kills and human eye and nose discomfort. A massive flagellate bloom in Scandinavian waters in 1988 caused fish kills (including salmon farms) and "damaged" seaweeds and invertebrate species. These blooms require great amounts of nutrient (e.g., phosphate and nitrate), which may be furnished by sewage effluent from resorts and cities.

In one example, a massive kill of dolphin (*Tursiops truncatus*) on the east coast of the United States in 1987–88 was diagnosed by a scientist as caused by "brevetoxin" (a type of red tide toxin produced by a planktonic microalga) passed to the dolphins via a schooling prey fish called *menhaden.* But this diagnosis was not confirmed by other scientists, some of whom believed that death was caused by PCBs or an AIDS-like disease that produced an immunosuppressant in the dolphins' systems. Similarly, a pelagic prey fish, the mackerel, is suspected of passing sufficient toxins up the food chain (algae/zooplankton/fish/whale) to kill humpback whales in the North Atlantic. (8)

But society is more concerned usually about human suffering. So the poisoning of seafoods—particularly shellfish, like mussels—caused by toxic dinoflaggelate outbreaks (red tides) is often quite a relevant matter. These outbreaks are often accompanied by massive fish kills (windrows on the beach) and aerosols of toxin carried by the wind can cause such human problems as respiratory disturbance, stinging of the eyes, and so forth.

Many scientists suspect, through circumstantial evidence, that outbreaks of biotoxic organisms are stimulated by "works-of-man," coastal/marine construction or pollution. However, direct cause-and-effect relations that can be used in coastal management programs have not been readily forthcoming. But, it does appear that anything that can be done to reduce either (a) nutrient pollution (eutrophication) from sewage and industrial discharges or (b) construction disturbances of coastal waters would be helpful.

Birkeland (25) speculates that massive land clearing and subsequent enhanced nutrient input (from barren land runoff) could be responsible for the rapid increase in PSP deaths in the Pacific. He notes that casualties from PSP have been accelerating lately and that, prior to 1960, PSP was "virtually never recorded in the tropical Pacific." The countries affected include Philippines, Indonesia, Papua New Guinea, Fiji, Solomon Islands, Brunei, Malaysia (Sabah), and India (both coasts).

Unfortunately, the toxicological studies of red tides, the deaths of marine mammals, and associated phenomena have been surprisingly non-useful to coastal zone managers. Toxicologists apparently confine their work to first causes and spend little time on basic causes. Therefore, we don't know exactly *what* has changed in the coastal waters environment to cause these lethal effects.

Clearly, toxic outbreaks have increased exponentially over the past few decades and, therefore, something unusual *is* happening. Some change has taken place. The outbreaks have often reached epidemic proportions—but still very little research has been aimed at discovering root causes, even by rich countries like the United States, England, and Japan.

3.13 CARRYING CAPACITY

Carrying capacity is a way to express the concept that there are limits to resource use. It is often applied to the sustainability of the coastal tourism industry. Borrowed from wildlife biologists and other renewable resource managers, the concepts of *population carrying capacity* and *maximum sustainable yield* have been applied to determining the limits of sustainability. For ICZM, carrying capacity is most often invoked to limit tourism.

Carrying capacity analysis as a method of numerical, computerized calculation for prescribing land use limits and development controls with cold objectivity was born in the 1960s. It did not achieve much success in influencing government policy because of the complexity of the parameters and because politicians and administrators are reluctant to have their judgment preempted by a computer. Nevertheless, a non-prescriptive and more qualitative and normative concept of carrying capacity exists today and *does influence tourism development.* This concept often invokes a quota system to limit visitorship (64, 82).

Carrying capacity analysis—as an approach to prevent overloading of environments by people and their activities—has been around a long time. Regardless of how logical the approach or how good the model, however, it has been virtually ignored by land use planners and environmental management agencies. Ecologically, the concept has found some use in visitor loading in the National Parks and in controlling range herds. But for coastal/marine development control, it is hard to find good examples. (82)

In regard to environmental impact, *carrying capacity* refers to a maximum level of activity beyond which there will occur physical deterioration of the resource or damage to natural habitats (5). There are other terms in use with similar meaning—limits to use, maximum occupancy, sustainable limit, etc.—but all refer to a definable, often quantitative threshold. In the development context, some prefer the term *limits of acceptable change* (LAC), which to some people seems a more flexible concept with a built-in expectation of impact and an admission that development will modify the resource.

Sorenson (323) considers carrying capacity "an alluring concept but putting it into practise is fraught with difficulties." He mentions the problem of proving simple cause/effect correlations of human impact against biodiversity decline as well as the sometimes ease of technical adjustments and change of expectations. Carrying capacity is not fixed but can be reduced by human or natural damage or increased through selected management procedure. With carrying capacity, as with other biological analogies, the social dimension complicates the procedure for estimating limits (117). Some of the key components—such as tourist or user satisfaction—change when the users themselves or their preferences shift. Consequently, skepticism remains about the applicability of carrying capacity as a management technique and disagreement exists over estimation methods (117).

Therefore, the actual carrying capacity limit—in numbers of visitors or any other quota or parameter—is usually a judgment call based upon the level of change that can be accepted, regarding (a) sustainability of resources, (b) satisfaction of resource users, and (c) the level of socio-economic impact (64). As an example of flexibility, the following visitor control measures may be employed to reduce visitor impacts and/or increase carrying capacity at coastal and marine national parks and reserves (64):

Administrative

1. ***Concessions:*** Control numbers/activities by limits on concessionaires (lodging/meal facilities, tour operators, etc.)
2. ***Time separation:*** Includes seasonal closure and time intervals
3. ***User fees:*** Controls numbers by "ability to pay" the fee
4. ***Speed limits:*** Also engine horsepower, size and type of craft
5. ***Activity restriction:*** For example, no spearfishing, jet-skiing, and so forth
6. ***Access restrictions:*** Special places (zones) and time periods
7. ***Zoning:*** For assigning specific uses to specific areas; the most fundamental measure, always to be done
8. ***Quotas:*** Numerical limits on visitors; the most direct method

Physical

1. ***Signage/exhibits:*** For announcement of rules and for providing directions and education/awareness (Figure 3.20)
2. ***Trails/routings:*** Surface or underwater to guide visitors
3. ***Mooring buoys:*** Reduce anchor damage, provide positioning, and control numbers of boats on a site
4. ***Ramps and docks:*** Limit launchings, provide orientation
5. ***Guides:*** Guidebooks, brochures enhance visitor awareness
6. ***Transport:*** Special interpretive craft (subs, glass bottoms)
7. ***Dry observatory:*** View of nature without entering the water
8. ***Artificial habitat:*** Enhances nature and attracts visitors

In the coastal zone, there has been much concern about the carrying capacity of human and natural habitats to support tourism and resort development. In this connection, carrying capacity has been

Figure 3.20. At popular coastal parks and nature reserves, visitor control is essential to avoid exceeding carrying capacity, as is done at Xel Ha (Yucatan, Mexico), with rules announced by prominent signs. (Photo by author.)

defined as the "physical, biological, social, and psychological capacity" of the environment to "support tourist activity without diminishing environmental quality or visitor satisfaction" (221). Control of tourism is a good example of the use of the carrying capacity idea.

According to Miller (247), it is vital to distinguish between two interpretations. In the one sense, carrying capacity refers to the optimum density of tourists for the benefit of their enjoyment—e.g., the density of people on a beach or visiting an historic site or attraction. In the other sense, carrying capacity refers to a certain threshold level of tourist activity beyond which there will occur damage to the environment, including natural habitats, such as coral reefs. The accumulated environmental effects of tourism development include reduction of biodiversity, human health problems, depletion of natural resources, and loss of jobs, income, and hard currency earnings.

A third category should be added—"socio-economic carrying capacity." When the social threshold is exceeded, extreme troubles arise. In the Caribbean, a major cause of social unrest and, therefore, tourist discomfort has been too rapid expansion of mass tourism. The socio-economic capacity should be limited to the maximum visitor loading acceptable to the people living in the surroundings of the tourist destination area and others especially affected. But resort development is usually a foreign investment scheme done over the heads of the local people. (See also "Tourism" below.)

After examination of carrying capacity potential as an analytic tool, Dunkel (117) concluded that "perhaps more important than any other management technique is the attempt to establish upper limits on the volume of tourists." But some tourism specialists believe there can be no mathematical formula for how many tourists are enough or too much and instead favor constant vigilance of the stress on services and the impacts of tourism on the environment (64).

Dunkel's research—plotting the number of tourist arrivals since World War II for the Virgin Islands and Bahamas—yielded semilogarithmic curves that resemble curves of a biological population reaching the carrying capacity of an ecosystem (117). By building more rooms and importing more resources, these islands can enlarge their capacity, but their efforts to attract more tourists through expending more funds on promotion have been only moderately successful. If this artificially expanded carrying capacity collapses, there is virtually no domestic agricultural base left to sustain even the islands' current population.

Sadler (300) believes that carrying capacity provides a frame of reference and is widely used to underline the importance of maintaining a level and mix of development that are environmentally and culturally sustainable. Various measures of environmental and social carrying capacity have been developed. Despite the effort expended on this work, however, operational definitions of thresholds or tolerance levels remain elusive. It is recognized that some carrying capacity criteria are measurable while others are not, in any practical sense.

The Economic Commission for Latin America and the Caribbean (120) reported that megascale carrying capacity levels were tried for Barbados, St. Croix Island (U.S.), and Bermuda. In the case of Bermuda, capacity was established at one-half million visitors annually based on physical carrying capacity levels (e.g., 14,500 maximum hotel rooms). Barbados used limitations on docking space, and the U.S. National Park Service at St. Croix (Virgin Islands) has limited the number of commercial tour boats allowed to go to the Buck Island coral reef daily.

Also, the Seychelles Islands in the Indian Ocean recently declared their tourist capacity reached and announced a prohibition on future increases. Ecuador set a limit for the Galapagos Island National Park of 12,000 visitors/year in 1973; they raised it to 25,000 in 1982 but actually allowed 47,000 in 1990.

The Virgin Islands National Park at St. Croix limited the total number of tour boats allowed to go to Buck Island daily. Many marine parks, particularly coral reef parks, are installing special boat moorings in visited areas and discouraging visitors from free-anchoring. In this way, the visiting boats can be limited in number and located appropriately—when the time comes for control. Looe Key Marine Sanctuary (Florida, USA), with 52 such moorings, is a good example of this approach (Figure 3.21).

After considerable study, Costa Rica set a limit of 25 visitors per night to the turtle nesting area of Nancite Beach in Santa Rosa National Park (17). Similarly, Queensland (Australia) limited vistors to a total of 100 at Michelmas Cay to protect seabirds, as recommended by the Great Barrier Reef Park Authority (139), although regular beach goers, in general, seem happy enough with 100 people on 0.1 ha of beach (Figure 3.22).

While many of the above carrying capacity limits seem to be holding, others have not done so well. For example, Ecuador set a limit of 12,000 visitors per year to the Galapagos Islands National Park but raised it officially to 25,000 in 1982 and actually allowed 47,000 visitors in 1990. But sensitive sites are restricted to a maximum of 12 visitors, while more durable sites have a carrying capacity of 90 (64).

More often, limits to sustainability are recommended but never implemented. For example, the beach at Hanauma Bay near Honolulu now receives over 10,000 visitors per day when the recommended capacity (in 1977) was 1,000 (37). This creates not only visitor chaos (via tour bus) but also degradation of the fish-watching experience on the closely adjacent reefs. (See "United States, Hawaii: Tourism Threat to Haunama Bay" case history in Part 4).

In a study of Samui Island, an important resort in Thailand, Leksakundilok (216) considered resource protection, pollution, socio-economics, water supply, and cost efficiency. He found that each controlling parameter had a different level of capacity for a different management input and therefore derived a family of carrying capacity curves (Figure 3.23).

Like nature reserves and national parks on land, marine protected areas should be developed with specific objectives for managing human uses including carrying capacity (see also "Protected Natural Areas" in Part 2). Whether large with multiple uses or small and highly protected, the desired levels of usage may be achieved through some or a combination of the following, according to Kelleher and Kenchington (203):

- Establishing area boundaries for specific activities (i.e., zoning)
- Enforcing closure during parts of the year critical to life histories of species or for longer periods
- Setting size limits, maximum permitted catches, and harvest limits
- Prohibiting or limiting use of unacceptable equipment
- Licensing or issuing permits to provide specific controls or to limit the number of participants in a form of use
- Limiting access by setting a carrying capacity that may not be exceeded

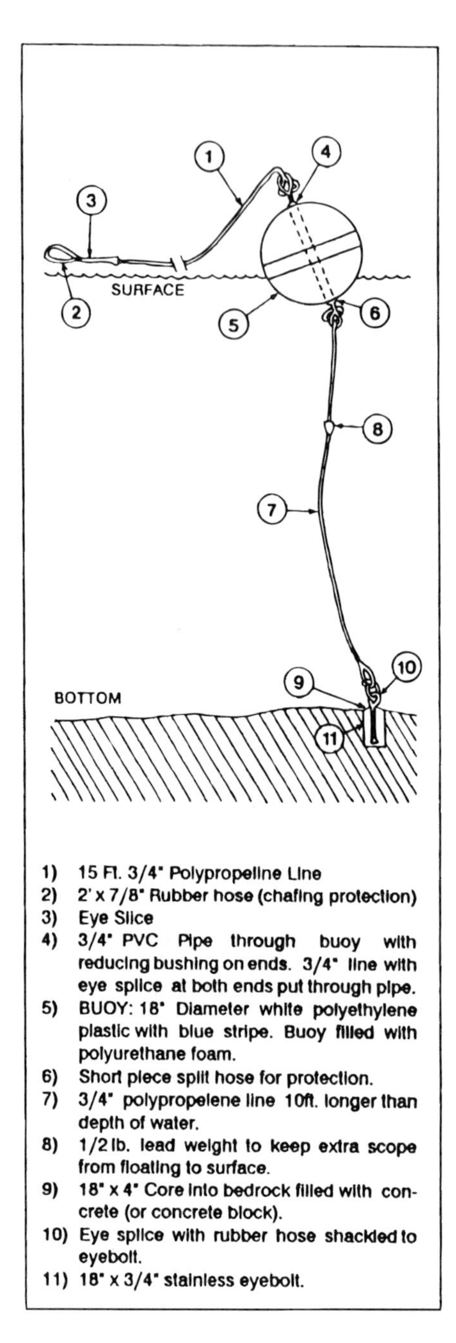

Figure 3.21. Details of the drill-anchored fixed mooring buoy system. (*Source:* Reference 243.)

3.14 CIGUATERA

Ciguatera is a toxic reaction to the eating of tropical fishes. It is caused by a specific nerve poison termed *ciguatoxin.* It is normally taken in by eating large predaceous fishes, particularly barracuda, but also large groupers, amber jack, "king" mackeral, snappers, and so forth. While its occurrence is worldwide in the tropics, it is geographically patchy. For example, on the north side of Cuba (facing the United States) ciguatoxic fish are common, while on the south side (facing Mexico) they are quite rare, according to Dr. F. Castro (62).

Figure 3.22. Beaches are a major tourist attractant and source of income for numerous countries. (Photo by author.)

In the United States, ciguatera claims nearly 10,000 victims per year (3.2), but very few die (<1.0 percent). It accounted for 22 percent of all chemical foodborne diseases in the United States in a nationwide survey (11).

Most cases produce cardiovascular symptoms, such as irregular heartbeat, but also nausea, vomiting, diarrhea, achiness, dizziness, exhaustion, chills, strange skin sensations, and hot-cold reversal. The cardiovascular effects may end in three days, but other effects may continue, even for months or years (3.2). No antidote is known to exist and therefore treatment is only for the apparent symptoms—neural, muscular, vagal, etc. If taken within 48 hours, **mannitol** (an alcohol, $C_6H_{14}O_6$) is said to reduce the effects. There is no practicable method to test individual fish for ciguatera.

The organism responsible for ciguatera is a microscopic "armored dinoflagellate" type of alga that often lives within the beds of the macro-algae (on the sea bottom), where it attaches to leaf surfaces. Herbivorous coral reef fishes that feed on the macro-algae also eat the dinoflagellates and store the toxin within their tissues. Other fish eat the herbivores, and larger fish eat them, passing the toxin up the food chain where it is concentrated by the highest level predators.

Therefore, anything that increases macro-algae growth could expand the base for the symbiotic ciguatera-producing dinoflagellate. The most reasonable management method for reducing macro-algae is reducing nutrient pollutants (nitrates and phosphates) that stimulate algal growth. These nutrients come from sewage, factories, and land runoff (11).

The pattern of ciguatera depends on the intensity of the disturbance, in both time and space. The first fish to become poisonous are the herbivores and detritus feeders, followed a few months later by the carnivores. The various links of the food web are toxic. The unsafe areas increase in space when moving predators are involved. Frequently, only the large predacious species, especially those that feed on fishes that accumulate toxins, are poisonous. Single prey fish (herbivorous) may have not accumulated enough toxin to cause ciguatera if eaten by humans (11).

Maragos (231) claims that ciguatera incidence may also be linked to marine construction. Dredging, filling, and other physical changes to tropical habitats have been implicated as causes for the increased incidence and outbreaks of ciguatera fish poisoning. In Oceania the poisoning is caused by the dinoflagellate, *Gambierdiscus toxicus.* Considerable research has been accomplished by the Japanese and French

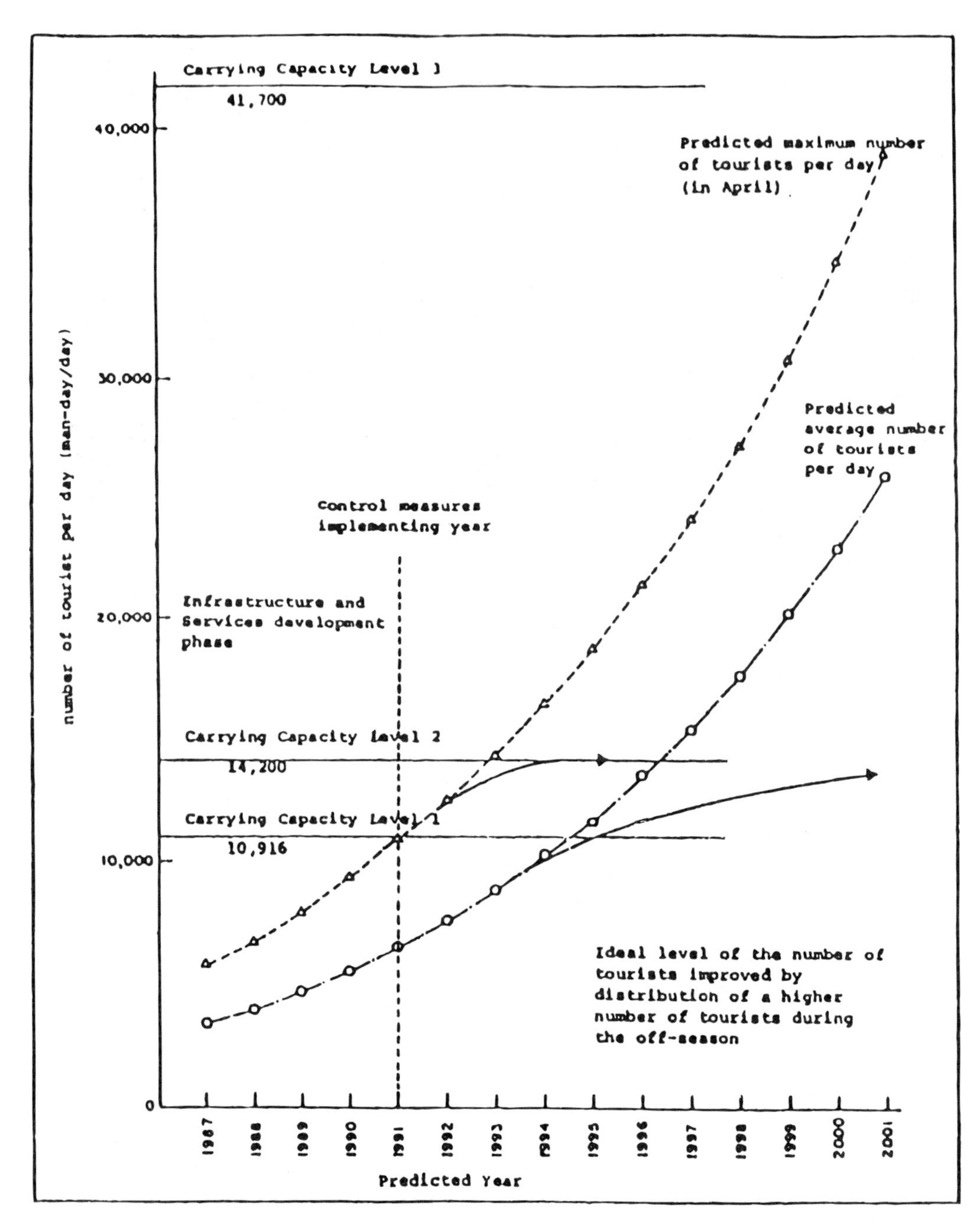

Figure 3.23. Predicted number of tourists per day and the computed carrying capacity of the Ko Samui resort area (Thailand). (*Source:* Reference 216.)

on the causes, symptoms, and distribution of ciguatera, especially in French Polynesia, where it is quite prevalent.

There is also circumstantial evidence for a relationship between ciguatera and construction in South Pacific coral reef areas; specifically, the lack of ciguatera on some atolls prior to construction, followed by heavy outbreaks during and after construction, which occurred at Palmyra, Johnston, and Bikini Atolls (231). Eventually it may be possible to link directly specific types of coastal construction or other types of disturbance to specific outbreaks of the fish poisoning, which in turn could result in modification of project designs, siting, and construction methods or other activity controls to limit ciguatera outbreaks.

For now, there is sufficient circumstantial cause-and-effect information available to ICZM planners to require counter-measures. The management responses should be (a) to reduce the coastal waters spillover effects of landside development to the minimum, particularly nutrient pollution from sewage, industries, and general land runoff, and (b) to prevent physical disturbance to coastal ecosystems, particularly coral reef systems. In future, when science has refined the linkage of development to biotoxin-producing organisms, more specific management actions may be indicated. (See also "Biotoxins" above).

3.15 CITES

The international species protection treaty, Convention on International Trade in Endangered Species (CITES), was concluded on March 3, 1973, and came into force in July 1975; by June 1990 there were 109 parties (nations having ratified or acceded to the convention). It is specifically concerned with those species whose long-term survival is at risk from international trade; species threatened primarily by habitat loss, local exploitation, or any other pressure are eligible only if international trade is also a significant factor.

The convention prohibits commercial trade in seriously threatened species; i.e., trade in species vulnerable to exploitation but not yet at risk of extinction is sanctioned in a regulated manner. Trade in endangered species may be authorized in exceptional circumstances, provided the import is not primarily for commercial purposes. All international shipments must be covered by an export (or re-export) permit and an import permit from the country of destination. For trade in threatened species, an export permit is required but not an import permit. There is no restriction on the specimens' use, but an export permit may be issued only if the exporting state is satisfied that the export will not be detrimental to the species.

Only a small proportion of the taxa covered by CITES are marine, and many of these are higher vertebrates (Figure 3.24). The total listed is 68, of which 42 are in the endangered category (some categories include more than one species).

Special rules apply to specimens that come from ranching or commercial captive breeding operations, and these are important for the many marine species that can be farmed. Theoretically, a well-managed operation could displace trade in wild products and dampen the effects of the latter. However, such trade may also stimulate demand and, if the origin of products is not distinct, make the control of illegal trade more difficult. For a farmed product to displace a wild product in trade by market forces, it must be available in large quantities relative to the wild supply and must be cheaper or of better quality; all of these factors are difficult to achieve in the early stages of the operation.

Captive-bred endangered specimens may be traded commercially if they are born in captivity from parents that mated (or otherwise transferred gametes) in captivity and the parent stock was obtained without detriment to the survival of the species in the wild, is maintained without augmentation from wild populations except occasionally for genetic improvement, and is managed so that it reliably produces second generation captive-bred offspring.

Ranching, in the context of CITES, is distinct from captive breeding in that the stock is obtained from the wild and the specimens are reared in captivity until they are of marketable size. (In aquaculture, ranching usually refers to the rearing of animals in hatcheries and their subsequent release into the wild for harvesting at a later date). Whereas captive breeding runs the risk of being viewed as an alternative to habitat conservation, successful ranching depends on the maintenance of a healthy wild breeding population and can thus encourage habitat protection and conservation of the species in the wild.

The CITES program is struggling in many countries as a result of inadequate national legislation, low penalties, widespread smuggling, fraudulent documents, permits issued wrongly, inadequate training and resources for police, customs, and management authorities, and low priority given to controlling trade in wildlife (Figure 3.24).

In spite of the difficulties, CITES is considered one of the most successful global conservation agreements. Most countries are willing to accept its basic principles, and the permit system through which it operates ensures that it can be more effectively enforced than many other treaties. Producer nations tend to support it because they see controls at the place of import as well as export as an

Figure 3.24. The gray whale of the eastern north Pacific is a success story in international protection of endangered species because it recovered so well under protection that the herd now numbers nearly 15,000 along the North American coast. (Photo by author.)

essential weapon in the fight to control the trade. In general, consumers support its aims of maintaining populations of species at a level at which they can still be exploited for trade; they benefit from being able to obtain wildlife resources of other countries, and the price they pay is cooperation in controlling the trade.

Source of this section: S. M. Wells and J. G. Barzdo (378). Wells is with University College of Belize, Belize City, Belize, and Barzdo is with the CITES Secretariat, Lausanne, Switzerland.

3.16 THE COMMONS

The richest coastal resources are in the shallow littoral waters of the territorial sea, in lagoons and estuaries, and in the intertidal transition zone of beaches, mangroves, and tideflats. In most countries these places are not owned by individuals or private entities, but are "commons" where the public has more or less free access and use. Therefore, the resources of coastal areas are the responsibility of governmental agencies. Depending upon traditions and circumstances, jurisdiction may be shared in any combination of entities: national government, state or provincial governments, tribal chiefs (in some South Pacific countries), and so forth.

According to Dixon (109), given the richness and diversity of coastal resources, it is not surprising that there are many conflicting claimants for common resources. Within the framework of socio-economic analysis, resource use conflicts arise from a number of reasons; these include the nature of common property resources and poorly defined property rights and also the existence of economic externalities and the presence of non-marketed goods and services (see "Economic Valuation" in Part 2).

Common property resources (CPR) or open-access resources belong to a community or a collective. Examples include marine fisheries, groundwater, and certain public coastal areas. No one individual owns the resource or controls its management. The main danger with CPR is over-exploitation (as in fisheries or groundwater) or congestion (for recreational facilities).

Closely related to the CPR question is that of *property rights.* If property rights for a resource are vested in the community, one has a CPR. In many cases, property rights belong to individuals or groups but may be poorly defined or indeterminate. Agricultural lands may be leased out season by season; in other areas, farmers have customary use rights but do not have formal titles. Communities may manage a CPR such as a mangrove in a sustainable way for generations only to find out that they have no legal title and can lose the use of the resource. Political factors may play a role in assigning formal titles to valuable resources. It is usually the poorest members of society who lose when property rights are changed. (109)

A major purpose of coastal management is the caretaking of common property resources of the "wetside", such as coastal waters, coral reefs, or mangrove forests. Management of common property is an important function of government and one that often gets too low a priority; as stated by J. Gritzner (150), "neglect of common property issues is often a principal cause of failure in development projects." Of course, it is recognized that management programs must of necessity include control on use of private lands and private activities even while planning and management strategies are aimed primarily at protection of "the commons" because much of what occurs on land strongly affects the sea. (See "Land Use" in Section 1.6.)

3.17 CONFLICT RESOLUTION

Because the coastal management process operates at the interface between land and water, there is often intense conflict between private (or quasi-private) property-based operations in shorelands, on the one hand, and public (common) property-based activities in the tidelands and coastal waters, on the other. The coastal management process may play an important mediating role between conflicting coastal "wet side" and "dry side" interests. This essential role should not be played out sector against sector. Rather, the mediating/coordinating entity must look at all sectors with legitimate interests to find the most broadly compatible solutions.

As an example, integration may be needed among fisheries, tourism, oil and gas development, and public works where these sectors are all attempting to use the coastal zone simultaneously. Both fisheries and tourism depend to a large extent on a high level of environmental quality, particularly coastal water quality. Both sectors may receive "spillover" impacts such as pollution, loss of wildlife habitat, and aesthetic degradation from uncontrolled oil and gas development. In another example, fisheries may require port services similar to those that tourism depends on—an infrastructure system that supplies water, sanitation, transportation, and telecommunications. Therefore, planning for both should be integrated with that for transportation and public works sectors. (326)

Almost every economic sector has a strong stake in the coastal area. It would be virtually impossible to allocate the coast to a single one of these economic sectors for development or even to give one or two sectors a priority for coastal development. Conflicts that arise over coastal environmental and resource issues can be severe, including loss of life (e.g., the Indonesia fishing "wars" of the 1970s) and destruction of property (e.g., a riot of 50,000 people who burned of a tantalum factory in Phuket, Thailand, in 1986 (see "Thailand: Phuket's Tantalum Riot" case history in Part 4). In fact, it is the intense conflicts over use of the coast that so often arise among the various sectors that makes the ICZM process so necessary.

To take an example, the 30 or so developing nations with extensive mangrove forests in the humid tropics should have a strong incentive for the integrated approach to conflict resolution. Most developing nations with extensive mangrove forests are confronted with conflicting claims and uses that create stresses which threaten mangrove sustainability. Conversion of mangrove forest to aquaculture ponds or croplands often presents a particularly difficult conflict with local people dependent on the mangrove for sustenance (326).

A major role of coastal management is to identify conflicts over coastal land and coastal renewable resources and to find ways to allocate and manage uses for the optimum long-term benefit of the nation by using a multiple-use format. Methods for resolving conflicts include fact finding and executive decision, study commissions, bargaining sessions, informal negotiation, formal mediation, administrative or public hearings, and adjudication. Any method can be included in an ICZM management framework.

One approach to conflict resolution is negotiation or *facilitated dialogue.* Several experiments in negotiated rule making have been organized in the United States to avoid the litigation that oftens accompanies the passage of legal environmental standards. Representatives of key interest groups concerned with specific regulations engaged in facilitated discussion over the precise details of rules. In recent cases, the negotiating group convened a one- or two-day meeting each month. The total negotiation lasted five to nine months, though longer or shorter time frames are equally possible (326).

In negotiations, industry, environmental groups, and state and federal agencies are all represented. Subcommittees of the full negotiating group may be formed as needed to delve into specific issues and report back to the full group. A "resource pool" is made available to help defray costs of attendance and to underwrite the costs of retaining technical experts to help clarify the issues at hand. Each negotiator has an equal voice, and each has a veto, since the objective is to secure a consensus among all parties (326).

Another approach is *mediation.* If communication has broken down and ministries or resource users are frozen into antagonistic positions, a neutral outsider acceptable to all sides can help reopen discussions by serving as a go-between. With appropriate technical knowledge, mediators can help disputants invent ingenious solutions to problems that make joint gains possible for all parties.

Mediators typically take a more active role than facilitators. While they help to set agendas, run meetings, and record minutes, they may also meet individually with key actors in a dispute to better understand the interests underlying the positions of the various actors (326).

Nine steps that might compose mediated negotiation or a policy dialogue are listed below, with questions that should be addressed in designing a mediation process (326):

1. ***Entry of the Nonpartisan Intervenor:*** How is "help" triggered? Who should ask for help? Should the convenor be an expert in mediation techniques, an expert in the issues at hand, or both? Should the nonpartisan party be appointed or come forward as a volunteer intervenor? Who pays for the mediator?
2. ***Choosing Representatives:*** Which parties should participate? Only government agencies with conflicting policy goals? Users of coastal resources? Non-governmental organizations? Corporations? Donor institutions? By what criteria are stakeholders recognized? Which spokespersons should represent the interests?
3. ***Setting Up the Process:*** Is there a specific order in which the issues should be taken up? Are issues linked in such a way that they should be considered together? Schedule? Deadline? What are the ground rules? Who convenes the meetings and who chairs them? How are uncooperative parties dealt with? Can negotiators speak for their constituents? Are they willing to sign a negotiated agreement?
4. ***Joint Fact-finding:*** What dimensions of the natural systems and technology are in dispute? What data and analysis might help clarify these issues? Do the parties each have the capability to understand technical material? How can unequal capability be addressed?
5. ***Options:*** Are there enough issues on the table to make trades possible? Are there interdependent issues under discussion so bargaining can lead to a positive sum outcome? Are parties willing to give something up to get something else in exchange? Should the mediator invent specific compromise proposals? Should parties develop entire competing proposals, or work on single text of an agreement and negotiate each portion of it?
6. ***Selling the Agreement among the Constituency:*** Did each negotiator get an agreement acceptable to the people he or she represents? Will it have to be revised to be acceptable "back home"?
7. ***Ratification:*** Are some last-minute revisions needed to meet the requirements discovered in the previous step? How can a written, but still informal, agreement be linked to more formal

mechanisms? What legislation, contracts, covenants, or interagency agreements need to be signed?

8. ***Monitoring and Evaluation:*** How will parties be held to their written promises? If there were contingent clauses in the agreement, did forecasted events materialize or not? Should the negotiators automatically reconvene after a fixed period of time?
9. ***Remediation:*** Should an updated agreement be negotiated later if conditions change?

If the above fails to produce an acceptable outcome, *arbitration*—the intervention of a respected nonpartisan party—may be useful. *Binding* arbitration represents a higher degree of intervention than facilitation or mediation, since an arbitrator generally has the authority to impose a solution. (326)

China has explored the use of an "arbitration commission" to resolve coastal management disputes, as part of the implementation of the Law of Coastal Zone Management of the People's Republic of China. According to a leading coastal planner based in Nanjing, the commission is to arbitrate the disputes arising between the boundaries of the provinces and counties, and between the developing industries. (326)

3.18 CORAL REEF RESOURCES

Coral reefs occur along most shallow, tropical coastlines where the marine waters are clean, clear, and warm. They are one of the world's most productive ecosystems. Reef systems are highly subsidized by nutrients and plankton brought to them by water currents and by excreta of reef fishes (when they return from off-reef forays). Consequently, the basis for the high productivity of the coral reef ecosystem is a combination of the primary production of the reef and support from its surrounding environment.

"Fringing" reefs—those whose platform is contiguous with the shore—are the most common and widespread of the reef structural types. The reef zone may terminate with a ridge rising a short distance offshore (100 or perhaps 500 m) and running parallel to the shoreline. Although they occasionally extend into the intertidal zone (Figure 3.25), fringing reefs are usually found below the low tide level. Their littoral distribution renders them more susceptible to degradation from coastal activities than other reef types. One of the largest fringing-reef formations lies along Saudi Arabia's 1,100-mile Red Sea Coast; the reef is successful because of an absence of freshwater (and sediment) runoff. (83, 271, 316)

Patch reefs are isolated and discontinuous patches of coral reef, lying shoreward of offshore reef structures. *Barrier reefs* are linear, offshore reef structures that run parallel to coastlines and arise from submerged shelf platforms; the water area between the shore and reef is often termed a *lagoon.* The world's largest barrier reef system, the Great Barrier Reef, occurs off the Queensland coast of Australia, and the second largest lies along the coast of Mexico, Belize, and Honduras. *Atolls* are circular or semicircular reefs that arise from subsiding sea floor platforms as coral reef building keeps ahead of subsidence.

With few exceptions, coral reef development occurs only where seawater temperatures remain above 20°C. As a result, most coral reefs are found in the tropics, with the exception of higher latitudes where there are warm ocean currents. The areas of greatest coral reef development are the Western Pacific and Indian Oceans and, to a lesser extent, the Caribbean Sea, including the Bahama Islands area.

The corals themselves are relatively slow-growing colonies of animals, with growth rates ranging from 1/10 cm to 10 cm in height per year (Figure 3.26). The large and diverse animal populations associated with the reef are supported by the high productivity of the reef and its ability to assimilate and hold nutrient and recycle it rapidly. Even though they live in nutrient-poor waters, their capability for efficient recycling of scarce nutrients (by the whole reef community) helps make coral reefs some of the richest, most complex communities on earth. This nutritional advantage combined with their ability to build physical habitat—the limestone structures built by the corals with the aid of their zooxanthellae—makes the coral community a truly remarkable habitat.

Reef-building corals are a symbiotic association of coral animals and microscopic algae called *zooxanthellae.* In nutrient-poor waters, the only concentrations of nutrients available are those already incorporated in phytoplankton or zooplankton (tiny drifting animals). The coral animal is highly adapted for capturing plankton from the water, thereby capturing nutrients. Corals through photosynthesis (154)

Figure 3.25. Extensive intertidal coral reefs fringe the shores of Palau. (Photo courtesy of World Wildlife Fund.)

metabolize their food and give their nutrient wastes to the zooxanthellae, which use those nutrients to produce more food.

In this remarkable symbiosis, the zooxanthellae are confined to the endoderm of the coral's coelenteric cavity, each algal cell being enclosed in an amoebocyte, inserted between or at the base of the lining epithelial cells. They are found in all hermatypic (reef-building) corals and occur in many other coelenterates as well (Figure 3.27).

The zooxanthellae are the immobile or vegetative stages of dinoflagellates. A zooxanthella has a tough cell membrane through which it is evidently able to take up some of the normal animal waste products, such as CO_2. Guanine and adenine can also be taken up, and PO_4 and NH_3 are utilized from the surrounding waters. (250)

A benefit to the coral's metabolism may be the secretion by the symbionts of small amounts of hormone or vitamin traces, without significant contribution to nutrition. The zooxanthellae may have an important role as removers of the products of excretion, promoting the higher metabolism and faster growth that enables hermatypics to form reefs where their temperate relatives without zooxanthellae cannot. Also, the zooxanthellae have been held to exert a direct effect on calcification and skeleton formation. Calcium passes through the endoderm into the ectoderm; it becomes absorbed on a mucopolysaccharide membrane directly outside the ectoderm, part of the organic matrix that serves for the initial stages of mineralization. In the sequence diagrammed in Figure 3.28, Ca combines with HCO_3 to form first $Ca(HCO_3)_2$ and then $CaCO_3$ (250). From 160 to 800 tons of calcium carbonate per acre are deposited each year on coral reefs (213).

Figure 3.26. Extensive study of coral reefs has shown that they grow vertically at greatly varying rates (0.1 to 10 cm/yr). (Photo courtesy of U.S. National Oceanic and Atmospheric Administration.)

Catches from coral reefs often make up a significant portion of the local fisheries. In western Sabah, for example, reef fishes comprise nearly one quarter of the total catch of the Kudat and Kota Kinabulu areas (391). Artisanal coral reef fisheries have been reported to account for up to 90 percent of the fish production in Indonesia and up to 55 percent of production in the Philippines (equivalent to more than 54 percent of the protein intake of all Filipinos). Reports of fish yields from coral reefs appear to vary widely, from less than 1 metric ton per square kilometer per year to almost 20 metric tons per square kilometer per year.

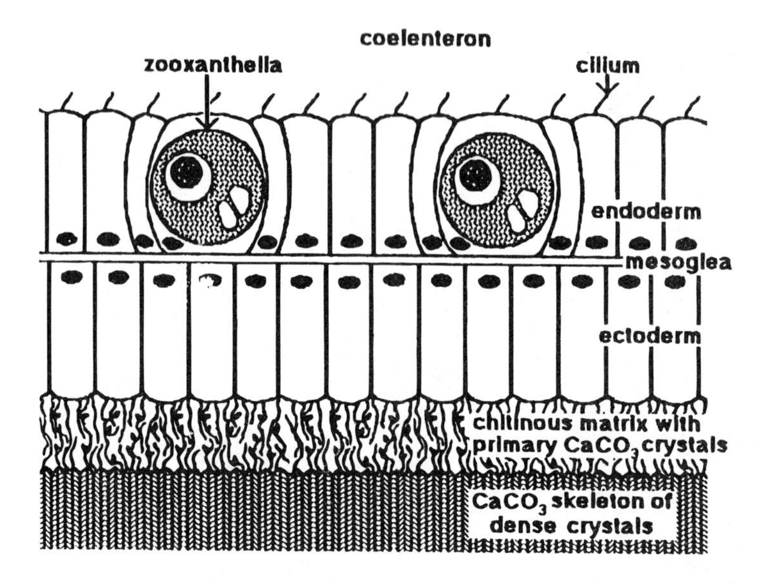

Figure 3.27. Location of the zooxanthellae in the endoderm of hermatypic (reef building) corals. (*Source:* Reference 230.)

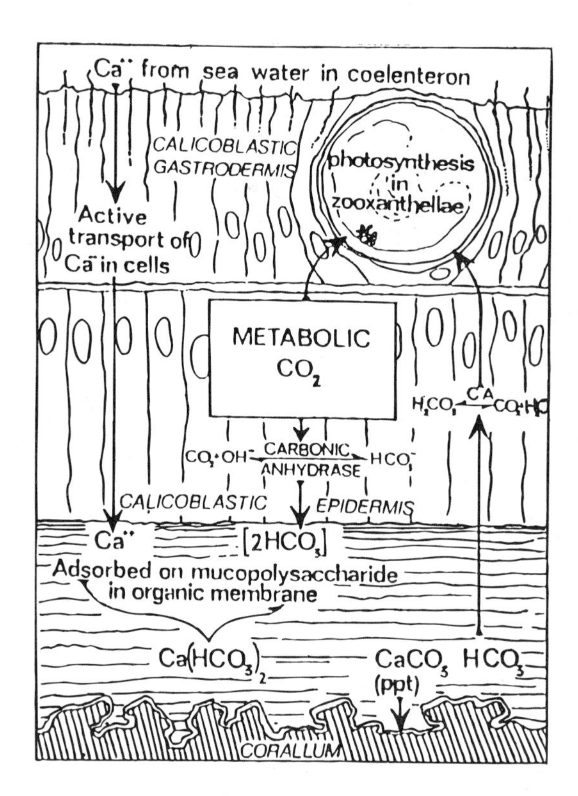

Figure 3.28. The chemistry of calcium carbonate deposition, including the role of carbon dioxide. (*Source:* Reference 250.)

Present evidence suggests that a sustainable harvest of all edible fish, crustaceans, and mollusks averaging 15 metric tons per square kilometer per year could be derived from most shallow coralline areas, those less than 30 meters deep (information from the Caribbean). The theoretical potential harvest on a global basis, therefore, amounts to 9 million metric tons per year, 12 percent of current total world fish production (391).

In addition to the fisheries conducted on the coral reefs, large yields are obtained from offshore marine fisheries supported by the reef system; for example, the diet of tuna caught off New Caledonia consisted of 58–73 percent food items of coral reef origin.

Coral reefs also support booming tourist industries in many countries. Catering to snorkelers, divers, underwater photographers, sightseers, and fishermen, reef tourism produces billions of dollars of foreign exchange earnings annually. One reef system alone—Pennekamp State Coral Reef Park (Florida, USA)—attracts 1.5 million visitors per year. More than half of the foreign exchange earnings of the Cayman Islands are from coral reef-based tourism. There are now 45 protected coral reef areas (including national parks) in Caribbean countries, but in the absence of broad-based coastal conservation programs, many have become seriously degraded (166).

Coral reefs serve as natural protective barriers, deterring beach erosion, retarding storm waves, allowing mangroves to prosper, and providing safe landing sites for boats. While Sri Lanka's comprehensive national ICZM program (see "Sri Lanka" case study in Part 4) is aimed primarily at reducing risks from natural hazards, beach erosion, and coastal flooding, a strong component is coral reef protection, which necessitates the cessation of coral reef mining (7).

Both human and natural causes of coral reef damage are at work to degrade the coral reefs of many countries. An important purpose of coastal management is to prevent further damage and to restore coral reefs. The discussion of the human sources of impact given below is excerpted from material prepared by M. Goodwin (78):

1. **Chemical Pollution**
   - **Agriculture:** A wide variety of pesticides and fertilizers are used in agricultural development projects in the Caribbean, often in excessive quantity. These compounds are carried by runoff to coastal waters and have been associated with coral reef destruction (e.g., in Grenada).
   - **Fishing:** Use of chemicals (e.g., bleach, cyanide) for fishing poses a threat to reef vitality in the Pacific.
   - **Industry:** Some corals are known to be sensitive to heavy metals that are produced by mining, refining, dredging, and manufacturing activities (Brown, 1987). The extent of this sensitivity among species and the prevalence of heavy metal contamination is not known.
   - **Petroleum:** Exposure to hydrocarbons has been associated with abnormal reproduction, growth rate, feeding and defensive responses, and cell structure in corals. These effects are worsened by dispersants used to clean up oil spills.

2. **Mechanical Damage**
   - **Dredging:** Cutterheads, buckets, pipes, and so forth associated with sand and gravel extraction cause obvious damage when they come into contact with living benthos, though sedimentation is probably the most serious impact of dredging.
   - **Fishing:** Adverse impacts associated with fishing include disintegration of reef structure in order to weight traps or to remove hiding places, damage caused by beating coral surfaces to herd fish into nets, destruction by explosives, and anchor damage (all prevalent in the Pacific).
   - **Shipping:** Accidental groundings can cause destruction of corals and other benthos in localized areas; cruise ship anchors can destroy masses of coral.
   - **Tourism:** Snorkel and SCUBA divers do extensive damage by striking the reef with hands, feet, or equpment or by collecting coral for souvenirs. Resort facilities onshore can be exceedingly destructive.
   - **Collectors:** Corals are often harvested for sale as souvenirs to tourists and for export; the market for decorative coral species is often quite lucrative.

3. **Nutrient Loading**
   - **Aquaculture:** Discharge from aquaculture facilities may bring fertilizers, waste feed, and other materials. Organics and fertilizers cause eutrophication and oxygen depletion. Cleaning chemicals may be toxic to reef species.

- **Sewage:** Sewage contains nutrients that cause eutrophication and favors algae, which can overgrow reef corals. Nutrients are particularly damaging to coral communities because they have evolved to live where nutrients are scarcest (35.1) and therefore are easily overcome by algae when sewage reaches them, suffering from, predation, and destruction.
- **Shipping:** Sewage and other wastes from yachts and commercial shipping may pollute crowded harbors, but they typically receive more waste from land sources of sewage.

4. **Sediment Loading**
   - **Construction:** Coastal construction associated with road and airport construction and shoreline development often results in heavy sediment loading.
   - **Erosion:** Inadequate land management in many coastal areas contributes to soil runoff from farms and settlements; this runoff sediments the reefs.
   - **Dredging:** The most pervasive impacts of dredging on coral reefs are caused by suspension of silt, sedimentation, turbidity, oxygen reduction, and release of bacteria and toxic matter (Figure 3.29).

In addition, there are damages from natural causes, such as (a) outbreaks of reef-destroying animals such as crown-of-thorns starfish; (b) diseases like whiteband (which kills elkhorn coral) and blackband (which kills large structural corals); (c) hurricanes that smash the coral and "sandblast" away the living tissue; (d) coral disablement and death from "bleaching" episodes; and (e) die-off or depletion of essential symbionts, such as parrot fish and sea urchins that clean the reef of algae. In the following list Goodwin (78) discusses the natural causes:

Figure 3.29. Dredging operations cause turbidity, which can, if excessive, damage coral reef ecosystems. (Photo by J. Chess.)

1. **Seastorms:** Fragmentation and dislodging of shallow corals is a frequent consequence of hurricanes and other major storms that produce severe wave action. These may be thought of as normal events which contribute to successional and reef-building processes. On the other hand they may be thought of as natural disasters which, as for Jamaica, have devastated much of the reef structure of north and northwest exposures, with serious economic consequences and long-term natural recovery prospects.
2. **Pathogens of corals:** Marine disease epidemics have recently spread rapidly through the Caribbean, often with serious consequences. The extent to which any epidemic is linked to human-related sources, such as pollution, is not known, but some linkages are suspected. Degenerative coral diseases of concern are "whiteband" and "blackband" disease, which have been the subject of scientific scrutiny only in the past five to eight years.
3. **Pathogens affecting symbionts:** The long-spined sea urchin (*Diadema*) has a most important role in the coral reef community. Urchins graze algae from coral surfaces, preventing destructive overgrowth of the coral community. In the early 1980s, urchins were eliminated by disease from much of the Caribbean. This is particularly serious where algae-grazing fish are rare (e.g., parrot fish, tangs) because corals and algae compete for space. Eutrophication from sewage and reduced grazing allow algae to smother corals. However, it has been noted that an over-abundance of urchins could result in their grazing directly on corals (143).
4. **Bleaching:** In the 1980s, corals in parts of the Caribbean were devastated by the "coral bleaching" syndrome in which corals expel their xoozanthellae. The coral structure then turns bright white; months later they may recover partially. The extensive 1987 outbreak of bleaching was associated with an unusually warm sea temperature and a more stagnant ocean. El nino events in the Pacific are associated with extensive bleaching and death. While individual corals may live through an incident, they are believed to be greatly weakened and more subject to other sources of mortality.
5. **Ciguatera:** A toxic reaction in people who eat fish from the Carribbean is called *ciguatera,* tropical fish poisoning. It is included here because it has such deleterious effects on fisheries and fish eaters in many Caribbean countries and may be connected to human-related causes (e.g., eutrophication leading to increased abundance of bluegreen algae).

According to Dizon (112), the most destructive enemy of coral reefs is man. For example, humans take out corals by the boatloads for fill and lime kilns, for house foundations and embankment of streets, canals and fishponds. The destructiveness is multiplied severalfold by dynamite fishing and bottom trawl fishing. Collecting and exporting corals also wreaks havoc on coral reefs, unless adequate and effective control measures are enforced. (112)

In the Philippines, over 70 percent of coral reefs have been degraded and/or depleted; the original inventory was 27,000 km (234). The findings of an intensive coral survey for 1982 revealed that, in 633 Philippine sites investigated over the entire archipelago, only 5.5 percent showed excellent cover; 24 percent, good; 38.3 percent, fair; and 32.1 percent, poor (112). In Indonesia less than 7 percent of the coral reefs could be classified as in "excellent condition" (75+ percent live coral cover).

Sediment loads from deforestation add a large quantity of nutrients, silts, and high turbidity to the Caribbean basin, reducing its suitability for coral reef growth. Jackson (188) noted that there has been 40–90 percent dieoff of corals in less than 10 m (about 30 ft) of water depth at five different areas of the southern Caribbean that have been particularly well studied—Jamaica, U.S. Virgin Islands, Barbados, Panama, and the Netherlands Antilles.

Hallock and Schlager (154) note that coral reefs in good water can grow 10 m (vertically) in 1,000 years but also can self-destruct at the same rate from excess nutrient exposure. Tomascik (346) notes that nutrification hinders calcium deposition, thus interfering with basic coral reef formation. Wells (376) notes that hurricanes, diseases, and sea level changes show that reefs are well adapted to recover from natural stress—"complex" reefs recover in 20 to 50 years.

Coral reef degradation has serious consequences for tourism, fishing, beach stability, and, particularly, coastal/marine parks. For example, most of the 21 countries and 49 parks (or reserves) in the Caribbean with coral resources have problems (166). Coral reef degradation can ruin a national park and cut severely into tourism. Some reefs are virtually beyond repair (those closest to settlements), but many that are degraded could be rehabilitated if techniques were available.

Unfortunately, there are numerous destructive forces at work and important coral resources are being degraded at a rapid rate, some of these forces can be easily controlled through coastal management programs, but others present serious socio-economic and political problems for many countries. For example, Sri Lanka is faced with finding alternative jobs for thousands of coral miners put out of work to protect national coral reef resources. (See "Alternative Livelihoods" above.)

According to Wells, reef management by forming a system of protected areas is advisable. Such protection can help halt further degradation, facilitate the recovery of devastated areas, protect breeding stocks, improve recruitment on neighboring areas, and maintain the sustainable utilization of reef resources (376).

Most of the above types of problems can be effectively addressed through simple management programs. A comprehensive approach would be through an ICZM-type which emphasizes resource management coordinated with development management. (See "Coral Reef Management" in Part 2).

The following impacts are specifically related to tourism development along coral coasts (308):

1. **Anchor Damage**
   - Breaks or damages corals.
   - Some designs, notably plough anchors, are particularly destructive.
   - For small boats a sand bag can be an effective and relatively undamaging temporary anchor.
   - At intensively used reefs, compulsory anchoring areas or compulsory moorings (permanently installed) may be useful.
2. **Diver Damage**
   - Almost all diving results in minor unintentional damage to corals and other reef biota; at frequently dived sites this damage can become significant and can lead to local loss of fragile species.
   - On intensively used reefs, periodic closure to allow recuperation of dive areas may be needed.
3. **Small Boat Damage**
   - Small boats and inexperienced boat handlers grounding on reefs can cause considerable physical damage to shallow areas, particularly at low tide.
   - On intensively used reefs, a system of designated boat channels and moorings to keep boats away from shallow, fragile areas may be necessary.
4. **Reef Walking**
   - Walking on reefs at low tide is a popular method of reef viewing which inevitably causes some physical damage.
   - In areas with a highly developed cover of fragile corals, severe damage to corals can occur.
   - Reef walking should be controlled, and a system of periodic closure for recuperation may be necessary.
5. **Construction of Tourist Facilities**
   - Has immediate mechanical impact.
   - May alter water flow around the reef and thus change a major ecological factor.
   - May shade reef locally, reducing photosynthesis.
   - May become a point source of pollution and littering.
   - Should be the subject of prior environmental assessment.

3.19 CYCLONES, HURRICANES AND TYPHOONS

The most destructive force that coastal residents of many countries face is the combination of a tropical cyclone—wind, waves, surge, and rain. A severe tropical storm is officially called a *cyclone* (or, in the Western Hemisphere, a *hurricane*) when the maximum sustained windspeeds reach 120 kilometers per hour (75 miles per hour or 65 knots). Hurricane winds may reach sustained speeds of more than 240 kilometers per hour (150 miles per hour or 130 knots). Cyclones, unlike less severe tropical storms, generally are well organized and have a circular wind pattern with winds revolving around a center or *eye* (not necessarily the geometric center). The eye is an area of low atmospheric pressure and light winds. The characteristic clockwise (southern hemisphere) or counterclockwise (northern hemisphere) circulation of its clouds and winds about the eye (a cloudless core) is set in motion by the earth's rotation.

Atmospheric pressure and windspeed increase rapidly with distance outward from the eye to a zone of maximum windspeed, which may be anywhere from 7 to 110 kilometers (4 to 70 statute miles) from the center. From the zone of maximum wind to the periphery of the cyclone, the pressure continues to increase; however, the windspeed decreases. The atmospheric pressure within the eye is the best single index for estimating the surge potential of a cyclone. This pressure is referred to as the *central pressure index* (CPI). Generally, for cyclones of fixed size, the lower the CPI, the higher the windspeeds. Cyclones may also be characterized by other important elements, such as the radius of maximum winds (R), which is an index of the size of the storm, and the speed of forward motion of the storm system (V_F) (362).

Many climate factors can encourage or discourage the development of a hurricane. It is these influences that explain a particular storm's behavior. A cyclone is an area of low atmospheric pressure and rotating winds and clouds that draws its energy from a warm ocean surface where the temperature is more than 80°F (26°C). A cyclone's strength depends largely on the degree to which air is allowed to flow unhindered into and out of the central column, or eye, which is defined by towering walls of thunderclouds spiraling around the core. A cyclone draws most of its energy from the heat of the ocean's surface water.

Warm vapor-bearing air from the surrounding ocean, under higher pressure, displaces the lower pressure air at the storm's core in a powerful rush, moving from the bottom of the column to its top. The greater the pressure difference from the periphery to the center of the storm, the more powerful the flow of air and the higher the winds. For the process to be sustained, the rushing air must be vented from the top of the central column, and surrounding climatic conditions can help or hinder this venting.

The wind of a cyclone can be very damaging, particularly when it exceeds 160 km/hr. But the massive surge of seawater the cyclone pushes ahead of its eye is capable of massive damage when it hits shore. The surge may be a mass of water more than 7 meters high, traveling at a speeds up to 45 km/hr. Approximate surge heights for various wind speeds are as follows (152):

Wind speed (km/hr)	Surge height (meters)	Wind speed (mph)	Surge height (ft)
60	1.2	37.3	3.8
80	2.3	49.7	7.6
100	3.3	62.1	10.7
120	4.3	74.6	14.1
140	5.3	87.0	17.2
160	6.7	99.4	22.0
180	8.0	111.9	26.3
200	9.1	124.3	29.7
220	9.5	136.7	31.2

Note: Wind speeds in excess of 120 km/hr or 74 mph are considered to signal an official cyclone (hurricane, typhoon).

Many countries suffer great losses of life and property from cyclones (e.g., Mexico, Bangladesh, some Pacific island countries, particularly the Philippines, the United States, and many Caribbean Island countries).

An example is Mexico's worst hurricane, "Gilberto," which crossed the Yucatan Penninsula in September 1988 with winds of 115–150 miles/hr (185–240 km/hr) (Figure 3.30). Evacuation programs reduced the number of human casualties, but other losses were high: 1.5 million chickens, 6,000 flamingos, and 15,000 hatchling turtles killed; 100,000 beehives and 140,000 ha of corn destroyed; 200 fishing craft wrecked; 15,000 breaks in electric lines; millions of dollars lost from tourist cancellations; 60 percent of corals detached; 60–90 percent of shore vegetation damaged (wind and salt vapor); tourist beaches lost up to 2 meters depth of sand (Cancun); 90 percent of mangroves damaged (or destroyed); 13,000 homes and numerous hotels and other structures damaged; and the whole north coast altered as nine cuts were made through the barrier islands, destroying roads, stranding villages, and flooding the sensitive estuaries with ocean water (Figure 3.31).

Another example is Bangladesh, which is catastrophically vulnerable to cyclones because (a) the Bay of Bengal enhances and aims cyclones directly at Bangladesh and (b) large areas of the lowland, deltaic coast are floodable. The country was hit with 16 major cyclones in 32 years (1960–92) and suffered as many as 300,000 deaths in a single event. The monthly frequency of cyclonic storms in coastal West Bengal is shown below for the period 1891–1970 (152):

Month	Cyclonic storms	Severe cyclonic storms
January	5	1
February	1	1
March	4	2
April	19	8
May	39	26
June	35	4
July	38	7
August	26	1
September	32	10
October	62	26
November	68	33
December	34	14
Total in 80 years	363	133
Annual average	4.5	1.7

With all the cautions taken by the government, the April 1991 cyclone killed 139,000 people and left as many as 10 million homeless. In addition, there was an enormous loss of homes, livestock, vegetation, and animals. (86)

The 7- to 8-meter-high surge hit Bangladesh just north of Chittagong at midnight, with the high tide, and began to subside at dawn. By then, 90–95 percent of the ripe crops (e.g., Boro rice) were destroyed from salt water submersion where the center of the cyclone had passed. Stored paddy and seeds were lost. Saline intrusion spoiled surface water and penetrated much of the ground water. Birds, reptiles, rodents, worms, and so forth were decimated. The death rate for poultry was 90 percent; for goat and sheep, 80 percent; cattle, 60 percent; and water buffalo, 50 percent. Most species of trees were stripped of their leaves. The exceptions were betelnut, coconut, and palms. (152)

Many coastal management lessons were learned from Bangladesh's 1991 cyclone. First, earthen embankments may lead to a false sense of security if they are not located and designed correctly. The typical coastal embankment was "not even able to save itself from the cyclone" (152).

Second, coastal shelter belts, particularly a broad fringe of mangrove (see "Greenbelts" below), appear to work very well for (a) protection of the embankments and other works necessary for property protection and (b) reduction of loss of life from the cyclonic sea surge ("tidal wave"). Also, it seems evident that the sand dunes found on the coastal barrier islands serve a major function in protecting life and property by providing a surge barrier and wave-dampening function (Figure 3.32).

Third, specially built cyclone shelters, able to accommodate all people at risk and their livestock and highly valued transportable possessions, are effective for the people willing to abandon their homes. (As squatters, many landless people fear they will lose their homes if they evacuate.) The elevated evacuation roads are also effective; livestock on such roads have a better survival chance.

Fourth, because well-built structures resist damage, coastal people should be required or encouraged to build strong houses; the coastal management entity should provide specifications and model designs.

Fifth, cyclone tracking is essential. (This was done very well in Bangladesh.) It should include the identification of cyclone formation, tracking of cyclone direction and speed, estimation of cyclone strength, and prediction of cyclone landfall and inland path, storm surge, duration, rainfall, and windspeeds.

Six, a warning system with messengers to go to distant, rural places is essential, including a warning system for ships and fishing boats. This should be followed by timely evacuation of people in high risk zones and of their valued property and by the timely cyclone-proofing of non-transportable property.

The long-term answer to the cyclone threat is integrated coastal management, which can control coastal demography and the locations and types of structures.

3.20 DECENTRALIZED MANAGEMENT

In this approach, the ICZM program is decentralized, with local or regional governments being the administrative entity. Customized planning and management programs would be created specifically

Figure 3.30. *Hurrican Gilberto* (1988) was Mexico's most damaging cyclone, destroying homes, vegetation, livestock, and ships (coast of Yucatan shown above), but human life was spared. (Photo by author.)

Figure 3.31. Yucatan's protective barrier island system was punctured in nine places by *Hurrican Gilberto* (1988), destroying roads, stranding villages, and flooding brackish estuaries with ocean water. (Photo by author.)

Figure 3.32. West Sonadia Island, Bangladesh, where residents told the author their lives were spared from the 1991 (April 21) Bay of Bengal cyclone because the big sand dune diminished the waves and the coconut trees provided safety above the surging waters; the young survivors are shown above at the base of the dune in front of their village homes. (Photo by author.)

for the conservation, economic, and social needs of particular coastal municipalities or regions as provided in "Situation Management" (See in Part 2).

The decentralized approach to implementation enables a nation to obtain experience with ICZM area by area, allowing the more advanced localities to proceed while others get organized. It provides time to develop and recruit expertise and presents later opportunity to make needed mid-course corrections (75).

While it may be appropriate to emphasize a decentralized ICZM program, it should be understood that all *must have* national government participation because national governments usually retain most of the authority for management of water bodies and oceans (the wetside of the coastal zone).

It would be quite feasible to organize an ICZM program to provide both nationwide protocols and standards as well as local program requirements, the former dominated by national interests and the latter by local interests. In fact, the U.S. ICZM program, operated by individual states, allows both statewide and area-specific programs. But all state and local programs must be in conformance with national standards for coastal conservation.

An interesting example is Alaska, one of the 50 United States of America. Alaska operates under the aegis of the U.S. national ICZM program (see "United States, National Program" and "United States, Alaska" case studies in Part 4). Alaska's municipal governments develop coastal management programs under state guidelines and standards. The act sets up a Coastal Policy Council to oversee the development of such programs and to resolve conflicts during their implementation. Approval of local programs, however, requires action by the state legislature in addition to the Coastal Policy Council.

The council has issued, and the state legislature has adopted, guidelines and standards for the following categories of coastal uses and activities (339):

- Coastal development
- Geophysical hazard areas
- Recreation

- Energy facilities
- Transportation and utilities
- Fish and seafood processing
- Timber harvest and processing
- Mining and mineral processing
- Subsistence

In addition, there are also official state guidelines and standards for resource management, including inventorying and managing such habitats and ecosystems as (339):

- Estuaries and lagoons
- Wetlands
- Tide flats
- Rocky islands
- Barrier islands
- Rivers
- Offshore areas

The State of Yucatan, Mexico, has also developed an ICZM-type plan for management of its coastline—"Plan de Manejo Integral de la Zona Costera," which is discussed in the "Mexico, Yucatan" case history in Part 4.

In selecting the boundaries for the local ICZM unit many factors have to be considered, including existing political subdivisions and traditional ways of regional thinking. But to the extent possible, the unit should reflect the natural boundaries of the landscape. Baker and Kaeonian (13) recommend for Thailand that the landforms and ecosystems provide practical units for the identification and analysis of land and water resources. Within each unit, information would be collected on (a) the functions that the component landforms and ecosystems serve, (b) the goods (wood products, fish larvae for aquaculture) and services (nutrient exchange, control of erosion) they provide, and (c) factors that influence their development and which form the elements for management controls and the investments required to exploit coastal resources on a sustainable basis.

Specifically, the national ICZM program can be organized to implement individual regional programs, if it has the flexibility to recognize regional distinction. The national program should be organized to facilitate regional economic development planning activities.

The regional (i.e., sub-national) emphasis arises from the issues orientation of ICZM programs; the issues are often distinctive to particular regions and therefore the resolution of the issues may also be seen as a regional matter. For example, if beach erosion (or oil extraction, or port development, or mangrove cutting) affects only one region of a country, then the search for solutions is seen as meeting a more localized need, not a national one. The regional emphasis is quite compatible with a national ICZM program approach.

Examples of ICZM decentralization are Sri Lanka, India, and United States. Sri Lanka is engaged in a strong campaign to devolve coastal management to its provinces using Situation Management as a main approach (see "Sri Lanka" case history in Part 4). India has delegated responsibility for its current (1994) ICZM planning initiative to the individual states—West Bengal, Tamil Nadu, Gudjarat, and so forth. The United States ICZM program was designed and has been operating for 20 years as a decentralized program (see "United States. A National . . . Program," in Part 4). In all cases, of course, central government participation is required because of the extent of national authority over coastal matters. The central government role is strengthened by its role as funding agent of provincial programs.

3.21 DIVERSITY INDEX

For scientific purposes, mathematical methods are available by which to attempt a numerical measure of diversity of species. The idea is to be able to evaluate complex data sets on the occurrence of various taxa in specific locations and to be able to make comparisons over time or between locations. Examples of these are presented in this subsection.

Shannon-Weaver Species Diversity Index — This index is computed using

$$Hc^1 = -\sum_{i=1}^{S} P_i \text{ Ln } P_i$$

where

Hc^1 is the diversity index,
S is the number of species,
P_i is the proportion (percent cover) of the ith species in a sample, and
Ln is natural logarithm.

This index has the attribute of being influenced by both the number of species present and how evenly or unevenly the individuals are distributed among the constituent species. This information function is sample size independent (except for samples of <100 when species number increases with sampling intensity). Thus, most samples of different sizes can be directly compared. This index is computed for contiguous transects of five and ten.

Margalef's Species Index — An alternative measure of diversity that incorporates S (the number of species) and N (the total number of individuals in all the species).

$$D = (S - 1) \log_{10} N$$

Brillouin's Diversity Index ***H*** — Brillouin's H gives the actual diversity of a fully censused collection of organisms and does not require unrealistic assumptions about the sampled population. This diversity index provides better discrimination when small replicated samples are used, as compared to a single large sample. Brillouin's index H is defined as

$$H = (1/N) \log (N!/n_1!, n_2!, \ldots, n_s!),$$

where N is the total number of individuals in the sample and n_1, n_2, . . . n_s are the numbers of individuals of the constituent species. It should also be pointed out that diversities at a generic level are almost as effective as those at a species level and that, in a complex system such as a coral reef, they save money and time.

Two methods that are particularly suitable for coral reefs were reported at a United Nations workshop, "Comparing Coral Reef Survey Methods" (355). They can be used to compare species composition of two coral communities, as shown below:

(1) Jaccard's coefficient (1908)
$CC = (b + c)^a - a$
where $b + c$ = number of species in communities one and two, respectively.
a = number of species in both communities.

(2) Sorenson's index (1948)
$$CC = \frac{2a}{(b + c)}$$
where $b + c$ = number of species in communities one and two, respectively.
a = number of species in both communities.

Tomascik (345) suggests *Simpson's Measure of Concentration* ($D'n$) as an unbiased estimate of dominance in a fully censused community, as shown below:

$$D'n = \frac{\Sigma\, ni\, (ni - 1)}{N(N - 1)}$$

where ni is the number of individuals on the ith species and N is the total number of individuals of all species in the sample.

$D'n$ can also be computed from coral coverage data as $D'c$:

$$D'c = \frac{\Sigma\, ci(ci - 1)}{C(C - 1)}$$

where ci is the coverage of the ith species and C is the total coral coverage. The coral community will exhibit high dominance if two individuals, drawn at random and without replacement from an S-shaped community containing N individuals, belong to the same species.

3.22 DREDGING TECHNIQUES

The most common excavation techniques include use of a crane with a clamshell, dragline (or bucket) dredging, and pipeline dredging from a floating barge or ship for suction dredging, cutterhead dredging, or hopper dredging (Figure 3.33). Explosives and drilling and shooting are often used to facilitate the removal of particularly hard materials. (231).

Dragline or bucket dredging is most often accomplished from the shore or from fill causeways or dikes built out on shallow flats from the shoreline. The crane swings an open bucket out over the water and drops it to the bottom. As the crane begins to winch in the bucket (which is shaped like a scoop and has teeth to break up hard rock), material accumulates in the bucket as it is dragged along or up slope. The crane then hoists the full bucket and swings it to the stockpile site, where it is emptied.

Clamshell dredging can also be accomplished from the shoreline but more often is accomplished from a floating barge. A hinged clamshell or jaw-like device is opened and then dropped by the crane vertically into the water. Impact on the bottom causes the jaws to dig in the substrate. The jaws are then winched shut, hoisted out of the water, and then swung over to the stockpile site. Barge-mounted clamshells can be used in offshore waters especially for sand mining or collection. Land-based draglines are simple to operate, often requiring only a skilled crane operator, but are limited by the depth and distance to which the bucket can be cast by the crane (usually 10–15 m). However, temporary dikes or causeways can be constructed to increase the range of the crane and allow dredging to be accomplished in broad, shallow areas. But clamshell operations are usually limited to about 460 m.

Pipeline dredging is done with a large suction pump that delivers the dredge spoil to a distant point. The dredging plant is mounted on a floating barge or ship (Figure 3.34). The head end of the pipe is often mounted on a hinged frame so that it can be raised and lowered easily from the surface. Bottom material is sucked in with seawater to create a slurry mixture that is transported through the pipe back into the water at a distance away or is discharged onto the shorelands in a confined basin (Figure 3.35). (231)

Typical discharge rates for pipelines flowing at 12 ft/sec are listed below (59):

	Discharge rate*	
Diameter (inches)	**ft^3/sec**	**gal/min**
8	4.2	1,880
10	6.5	2,910
12	9.4	4,220
14	12.8	5,750
16	16.5	7,400
18	21.2	9,510
20	26.2	11,740
24	37.7	16,890
27	47.6	21,300
28	51.3	23,000
30	58.9	26,400
36	84.9	38,000

* To obtain discharge rates for other velocities, multiply the discharge rate in this table by the velocity (ft/sec) and divide by 12.

Hopper dredges discharge the material into "hopper bins" aboard a ship. After the bins are filled, the ship lifts up the suction pipe, travels to an aquatic disposal site, and releases the material through

Figure 3.33. A hydraulic, or "pipeline," dredge engaged in channel deepening is releasing dredge spoil into the waterway, causing environmental damage through water turbidity. (Photo by author.)

bottom opening hatches on the bins. Hopper dredges are specialized ships normally used for redredging; i.e. to maintain navigability in deep draft commercial harbors and channels.

Cutterhead dredges consist of a suction dredges to which rotating cutterheads equipped with sharp "teeth" are attached to the head end of the pipes. The rotating cutterhead is able to cut through hard rock, removing both soft and hard materials. Cutterhead dredging is a sophisticated operation. The action of a rotating cutterhead on a suction dredge may generate considerable turbidity and sedimentation during the grinding and loosening of hard rock. Variables that affect the quantities of these parameters include size of teeth, rotational velocity, degree of suction, type of cutterhead, and characteristics of the substrate.

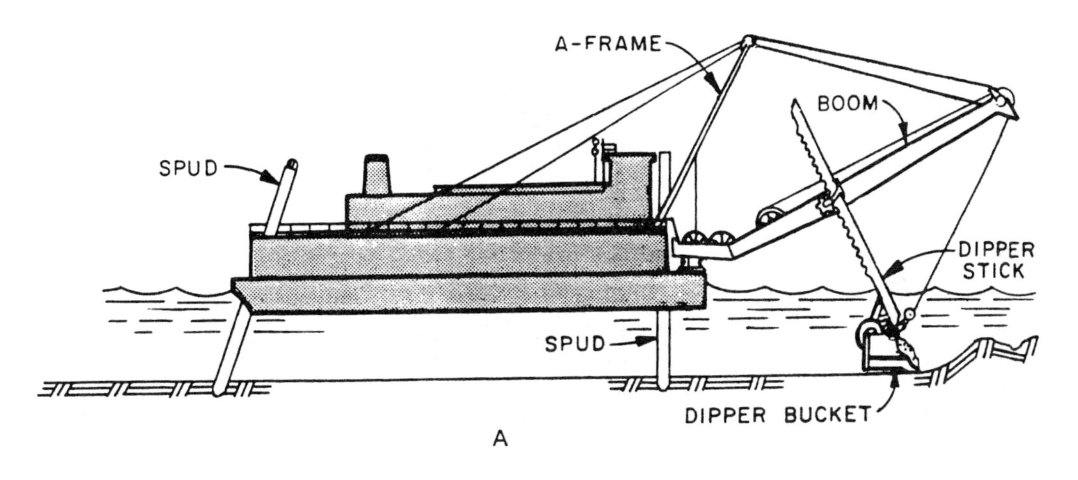

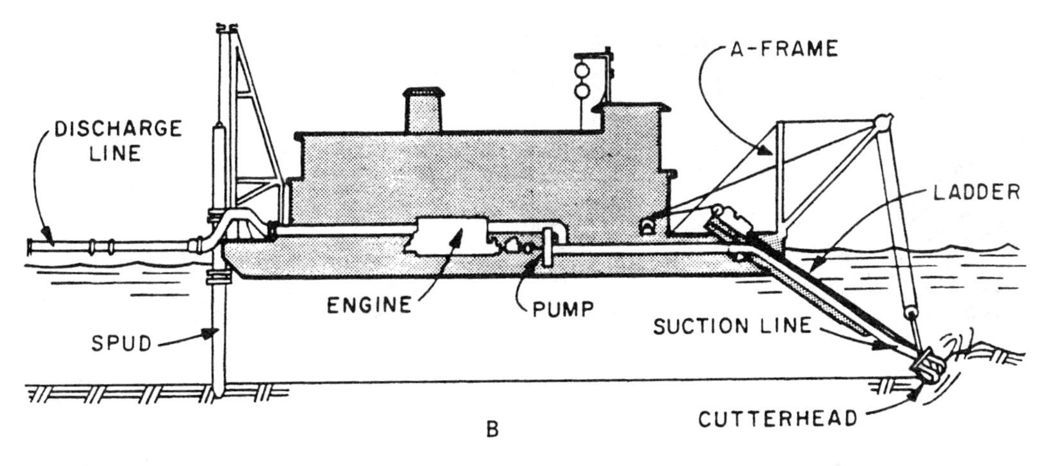

Figure 3.34. Two popular types of dredges: (A) bucket (mechanical) dredge and (B) hydraulic (suction) type. (*Source:* Reference 362.)

For clamshell and dragline dredging, turbidity and sedimentation impacts are more localized and are generated as individual pulses when the bucket or clamshell is dropped and hoisted. In contrast, pipeline dredging can generate turbidity and sedimentation continuously at both the suction and discharge ends of the pipe if the latter involves disposal or spillage back into the water. These impacts are usually localized at the suction end and can be reduced further if a jet probe is attached to the head, allowing it to be buried beneath the sediment surface before suction is applied.

The action of a rotating cutterhead generates additional turbidity and sedimentation during the grinding and loosening of hard rock. Ideally, a slurry mixture of 75–80% water is desired to ensure its rapid transport through the pipeline and discharge. Too much material in the slurry (lower water content) increases the risk of clogged lines, pipeline breaks, and pump malfunctions. Too little material in the slurry (higher water content) reduces dredging efficiency and quantities. (231)

Suspended sediments can have a range of impacts on coral reefs depending upon species, dredging technique, and degree of sedimentation. Some Pacific reef corals (species of *Acropora, Porites, Psammocora, Montipora, Astreopora*) and large-polyped forms (*Favia, Favites, Leptastrea, Lobophyllia, Fungia, Turbinaria, Plerogyra,* and *Physogyra*) are more adapted to withstand suspended and accumulating sediments, compared to other corals (other species of *Acropora, Pocillopora, Heliopora* and *Pavona*). Species common in sandy environments, including back reefs and lagoon floors, are naturally resistant to sediments compared to species adapted to wave exposure on ocean reef slopes, where sediments tend to be less prevalent (231).

Figure 3.35. An hydraulic dredge creating real estate by depositing dredged material on coastal wetlands. (Photo by M. Fahay.)

In one example, slurry from cutterhead dredging at Kwajalein spilled out over a large reef tract, burying coral communities and inhibiting recovery. At Okat reef at Kosrae (see case study), the rate of slurry discharged into a retention basin exceeded the basin's capacity, causing slurry to overflow the walls and spill out over 10 hectares of seagrass and coral habitat and bury it under 0.25 to 0.5 m of fine slurry muds. In this last case the impact could have been prevented by reducing the rate of slurry discharges. (231)

Sedimentation at the dredging end of operations can also cause some impacts to adjacent communities. At Kosrae, sand from cutterhead dredging was transported by strong currents to bury adjacent seagrass beds and, where slurry discharge into a retention basin exceeded the basin's capacity (Okat Reef), slurry overflowed the dikes, spilling out over 10 hectares of seagrass and coral habitat to bury it under 0.25 to 0.5 m of fine slurry muds.

Bacteria counts were measured in relation to dredging activities in a lagoon in Guadeloupe, French West Indies. Total counts of 10 to 100 living bacteria per cm^3 of water were found in clear coral zones whereas 50 times more were found in the dredged coral area. Suspended sandy particulate matter resulting from dredging seems to be the direct cause for the abnormal increase in bacteria counts (309).

Also refer to "Dredging Management" in Part 2.

3.23 DYNAMITE FISHING

The term *dynamite fishing* refers to all uses of explosives to kill or stun fish for the purpose of harvesting. The practice has caused considerable damage, particularly to reef and fishery habitat and poses a significant educational and environmental problem in Oceania. Aside from the devastating effect on coral habitat, the use of explosives for fishing is extremely wasteful. (231)

Although illegal in most countries, blast fishing continues to be openly practiced and tolerated, if not encouraged in some fishing areas. Whole villages benefit from it in remote parts like in the Philippines, for instance. Relatives and neighbors may get fish enough for more than a day's meal. When the catch is bountiful, blast fishermen give away more than 10 kilos to their co-villagers and sometimes to visitors from neighboring villages. This is why, whenever law enforcers sleep in the village, the people will remark, "We will be hungry" (184).

The activity presents a relatively faster, cheaper way of earning money. The fishermen also believe that there is no other effective way of catching some types of fish like anchovies, except through blast fishing. They further justify the practice by saying that it is better than stealing or committing other crimes. Besides, blast and nonblast fishermen alike believe that the harmful effects of blast fishing is limited and confined to a very small area only, where the dynamite was thrown (184).

Explosive devices are constructed by fishermen from available materials consisting of gunpowder taken from bombs, cannon shells and bullets (basic ingredients: TNT and cyclonite, RDX or hexagen) fertilizer and diesel oil, TNT smuggled from legitimate users (such as mining firms), or mixtures of such substances as potassium chlorate, sulfur, and gum resin purchased from local drug stores.

In the Philippines, the variety of blasting devices used locally to kill fish range from handmade bombs to dynamite. However, the most common device is a bottle filled with layers of sodium nitrate alternating with layers of pebbles. The cord-type fuses are usually commercially obtained. Sodium nitrate is sold legally to induce ripening in mangoes and so is difficult to control. (242)

The significant increase in the use of explosives for fishing during the period immediately following World War II was due to the availability of ammunition which illegally fell into the hands of fishermen. When this source was exhausted, fishermen found other sources. Oftentimes the explosives are stolen or otherwise obtained from construction companies and/or military sources at the present time.

The exposive materials are packed in bottles varying in size from coke bottles to gallon glass-jars which are fitted with either home-made pistons and fuses or ready-to-use ones purchased from illicit traders. Occasionally, fragmentation hand grenades are also used.

Dynamite users in the Philippines ("blast fishermen") usually work in pairs or in small groups, using small hand-paddled outrigger boats (bancas). In the case of a pair, one acts as the "dynamiter" and the other as the "boatman." The dynamiter locates fish schools and drops bombs on them; the boatman paddles the banca in the search for fish, quickly escaping from the explosion point. Both men free dive to retrieve the fish from the bottom, where they sink usually. The pair may be accompanied by some divers using their own bancas. Recently, fish dynamiters have used motorized bancas or boats capable of quick get-aways from police (the practice is illegal).

Smaller bombs (coke or beer bottle size) are used against small schools of fish at shallow depths, gallon-sized bombs against large schools or large fish in deeper water. Small bombs are thrown at intended victims, but large bombs fitted with longer fuses are simply dropped, often weighted down with sinkers to ensure their explosion at or near the bottom. A certain proportion of fish killed remains uncollected because of factors such as currents, which may carry fish to inaccessible, deeper water, and limitations of free diving. Consequently, this fishing method is wasteful.

Accidents caused by the premature explosion of bombs in the air have occurred, killing or maiming the dynamiters. While casualty rates are not known exactly, it is estimated that, in three Philippine localities known to the authors, one out of five or six dynamiters has either lost one or two arms or died instantly. The presence of one-armed men in Philippine coastal localities appears to be a reliable indicator of dynamite fishing.

The shock wave of an underwater explosion tends to travel in all directions, diminishing as an inverse function of the cube of the distance traveled. Therefore, to double the blast force, the charge must be increased 8 times (a factor of 2^3) and to triple the blast over-pressure, the charge must be increased 27 times (a factor of 3^3). Air bladders of ray-finned bony fishes are readily ruptured by underwater explosions, but those without bladders survive explosions fairly well. The size of the lethal zone of an underwater explosion depends upon the size of the charge in relation to several other factors. A typical depth charge, for example, may be expected to be lethal to most marine organisms within a radius of 77 meters, within an area of 1.9 hectares and a water volume of 1.9×10^6 cubic meters. For fish with air bladders, these figures would have to be multiplied by 1.16 and 64, respectively.

Fish killed by dynamite blasts are easily recognized by the effects of the sudden, tremendous over-pressure on their bodies, including ruptured air bladders, ruptured blood vessels causing hemorrhages, broken bones, and loss or mutilation of body parts. Since ruptured air bladders and broken bones show up clearly in X-ray films, these characteristic features can be used by law enforcement agencies for identifying fish killed by dynamite blasts.

A bomb the size of a beer or coke bottle exploding at or near the bottom will shatter to pieces all stony corals in an area with a diameter of 3 meters, while a gallon-sized bomb will destroy an area about 10 meters in diameter. A crater-like structure is created at the blast site, over which coral rubble is scattered. Such a structure is roughly circular, as seen from the top, the rim marked by surviving corals. The physical damage becomes progressively smaller as the point of explosion increases in vertical distance from the bottom since, as already stated, the force of an underwater explosion decreases inversely as the cube of the distance. Heavily dynamited reefs lack relief and are practically reduced to rubble. Dynamited reefs may take a very long time to recover even if left alone. About 38 years are thought to be required for a reef to recover to 50% hard coral cover.

Although blast fishermen in the Philippines experience the constant fear of being caught, this is not from being imprisoned but rather from the fact that they would have to pay out a considerable amount of money to bribe the law enforcers to release them. Even if they are released, they are already tagged by the authorities as blast fishermen. From time to time, some of the authorities would return and visit their homes to ask for more bribes. Bribing the law enforcers is a most detested action by blast fishermen. Some of them say that, were it not for the corrupt authorities, their lives would have been a lot lighter (184).

One way in which fishermen create a "better" atmosphere for blast fishing is by establishing "friendly relations" with the law enforcers. This is mainly done through drinking sessions and free dinners with the authorities whenever they are in the village. Another way of avoiding imprisonment is by maintaining links with high military and civilian officials, otherwise known as "godparent" (*padrino*) system. Many blast fishermen are often released through the intercessions of their "godparent" (184).

Source of this section: A. C. Alcala, Silliman University, Dumaguete City, Philippines, and E. D. Gomez, University of the Philippines, Quezon City, Philippines (see Reference 4).

3.24 ECODEVELOPMENT

Ecodevelopment is a process that could be defined as a site-specific package of measures, developed through people's participation, with the objective of promoting sustainable use of land and other resources, as well as on-farm and off-farm income-generating activities that are not deleterious to environmental values. The objective, ultimately, is better conservation, but this in turn should mean better lives for local people and a more satisfying occupation for managers. As a necessary part of resource management, rural development programs must be undertaken to reduce the dependency of local communities on diminishing forest resources in many countries. The programs must promote more efficient use of natural resources and/or provide alternatives where appropriate. This is the essence of ecodevelopment which is closely related to social forestry and other community-based programs.

Many of the current rural development programs are capable of supporting this theme, either in their present form or with some modification. Advantage must be taken of ongoing programs wherever they can be made to serve but many additional programs will need to be launched. And while ecodevelopment was formulated for land resources, the coastal zone has both wet and dry forests.

Ecodevelopment is a joint effort between all agencies; government and non-government, and the affected people. The manager during planning, has to decide, in consultation with other agencies and the people, how much of a particular activity or development is required how much is being already done or is being planned by some other agency and then decide how much augmenting for which resources will have to be found. The management plan document should indicate general or specific roles to be played by various agencies, including the ICZM-type agency.

There is a multitude of rural development measures that can be directed towards ecodevelopment. The special skill of the ecodevelopment officer must be the ability to recognize what measures will be most appropriate in each situation, and to know when and how to employ specialist advice. To this extent he or she is a facilitator rather than a specialist in any particular branch of rural development.

The ecodevelopment planning process, unlike most others, has only a beginning and no end. It begins with the definition of the approach and continues even when implementation has begun, in the shape of annual and site plans. In practice, the process consists of a number of actions, concurrent and sequential, which can be grouped into the following stages:

- Reconnaissance
- Trust building
- Participatory planning
- Participatory implementation
- Monitoring

These steps can easily be meshed with an ICZM-type program for coastal lowlands, forests, and plains, as well as wetlands.

Reconnaissance: Obviously, this aims at obtaining the first impression about the area, people, resources, demand, shortages, current policies and programs, etc. This essentially is a broad survey of the situation, both on the ground as well as in official records and, even history. This will involve field visits, studying maps, reports, gazetteers, working plans, relevant literature and holding informal discussions. This stage should imperceptibly merge with the trust building stage as many activities will be common to both the stages.

Trust Building: This involves such actions/gestures on the part of the planner (who may be a manager as well) which promote mutual confidence between coastal managers and the people. The actual actions involved in trust building are just common sense. It may involve informal discussions, on the wayside or in the villages, small helps as gestures of good will and anything else that can promote rapport.

Participatory Planning: This stage is the real core of the entire planning exercise. Here the collection of information and data becomes more formal and structured and ideas start taking a concrete shape. It is now widely agreed that ecodevelopment planning should be done using rapid rural appraisal techniques—a set of semi-structured exercises designed to facilitate a productive interaction with rural communities. Various exercises such as mapping, transect walk, venn diagram, and timeline are aimed at eliciting people's perceptions about their lives, resources, development agencies, and so forth which can be used for any rural development planning work.

The participatory planning process involves setting up coordinating mechanisms at village and broader levels, preparing comprehensive plans for ecodevelopment, including soil and water conservation, agricultural development, irrigation, cottage industries, apiculture, fisheries, alternative energy programs, etc. It also includes preparing a composite plan document for not less than five years.

Participatory Implementation: Participation of the people does not end with completion of a plan. People are continuously involved in decision making related to implementation. Annual plans and site specific details will have to be prepared in which joint decisions will have to be made. The targets indicated in the plan will have to be examined and readjusted in the light of the experience gained and the resources available. Site-specific micro plans may have to be made annually.

Source of this section: H. S. Pabla, S. Pandey, and R. Badola, Wildlife Institute of India, Debra Dun, India (274, abridged by author).

3.25 ECONOMIC BENEFITS OF PROTECTED AREAS

Since planning should accomplish certain defined goals and objectives, the protected areas planner hopes that these will be clearly laid out by policy makers. However, because such goals are often ambiguously stated, the planner's first job may be to interpret and refine the mandate for the planning process and to request necessary clarifications. In so doing, the planner may find that the program needs substantial and detailed justification, particularly when it is based on administrative rather than legislative action. The justification may be based on political, social, or economic grounds. But economic justification is now fashionable, and the planner should be prepared to provide the necessary analysis of costs and benefits for the program.

Quantitative economic indicators of protected area values are usually numbers of visitors or jobs created, weight or value of fishes landed, money or days spent in hunting and recreational fishing and so on. Examples of monetary and non-monetary benefits that may be measured in one way or another are (308):

- Gate or license fee totals, to indicate the economic value of tourism to the protected area. These are also indicators of the willingness of the public to pay for recreation privileges at the site.
- Total tonnage at dockside or retail value of fish landings to calculate the contribution of a protected area to fishery revenues (i.e., the economic value of the breeding ground of a fishery resource).
- Total income from recreational and commercial equipment, lodgings, and food and transportation to estimate the contribution of a protected area to supporting industries.
- Total hotel catering, product processing and packaging, equipment production (factory) and distribution (outlet), guide, and other jobs in industries linked to the protected area.
- The probable total cost of property damage (to roads, buildings, livestock, and crops) through storm waves and winds multiplied by the probability of storm damage (i.e., after the felling of mangrove, disturbance of dune vegetation, or blasting of coral reefs) to obtain an estimate of the annual benefit of natural storm damage control.
- The number of visiting students or student groups, their range of ages, and the number of teaching institutions represented. These give estimates of the value of the protected area for education.
- The number of researchers, research projects, theses, and publications, to indicate the value of the protected area to research.
- "Head counts" for bus, boat or other groups of visitors to a protected area. The figures can be expressed as a total or a percentage of the state or national population for an estimate of the social value of the site.

3.26 ECONOMIC VALUATION

It is difficult to assess ecosystems adequately with economic techniques normally used in project planning because of two separate phenomena, according to Burbridge (32). First, many of the products and services they provide are unpriced and are therefore not easily measurable in monetary terms. Second, the benefits from many of the products and services often accrue at some distance from the system itself and are thus usually treated as external to the location or site of the physical system.

These two phenomena are illustrated by Figure 3.36, where a mangrove system is used as an example. The figure demonstrates that the products and services produced by the mangrove forest can be divided into four categories based upon whether they are utilized within or outside the physical system (on-site versus off-site), and whether they are priced and exchanged using conventional market mechanisms (32). Using this method of analysis, it is relatively simple to demonstrate that some goods, such as mangrove timber, will be extracted on site and can be easily valued by observing their local market price.

Other goods (such as fish that depend on mangroves for some part of their life cycle) may be obtained at some distance from the site and sold through a market, so their value is also relatively easy to demonstrate. The most difficult category to value are the products or services that the mangrove generate but that occur off-site and for which people do not normally pay. For example, the reduction of storm damage resulting from mangroves acting as a buffer to waves and wind (32).

An alternative but complementary way of examining the benefits (and potential losses from development) relating to ecosystems is to consider all the different uses of the ecosystem in question. These can be separated into three main types. "Direct uses" are where, for example, products such as fish are harvested directly from the ecosystem. "Indirect uses" are where benefit is gained indirectly from an ecosystem usually through support and protection of other economic activities elsewhere. The third category is "non-uses," which refers to values accruing to an ecosystem without any actual current use of that ecosystem. (329)

		Location of Goods and Services	
		On-site	Off-site
Valuation of Goods and Services	Marketed	1 Usually included in an economic analysis (e.g., poles, charcoal, woodchips, mangrove crabs)	2 May be included (e.g., fish or shellfish caught in adjacent waters)
	Nonmarketed	3 Seldom included (e.g., medicinal uses of mangrove, domestic fuelwood, food in times of famine, nursery area for juvenile fish, feeding ground for estuarine fish and shrimp, viewing and studying wildlife)	4 Usually ignored (e.g., nutrient flows to estuaries, buffer to storm damage)

Figure 3.36. Four components of economic valuation of coastal resources; traditional analysis would treat only the one component, "marketed/on-site." (*Source:* Reference 36.)

Non-use values are increasingly becoming acknowledged and accepted as real economic benefits. They include "option value," the benefit gained from retaining the option of using a resource at some time in the future, and "existence value," the benefit accruing to individuals simply as a result of knowing that an ecosystem exists and will continue to exist in the future. Part of the satisfaction for existence value comes from the fact that an individual knows that future generations will be able to enjoy the ecosystem's continued existence. This is often referred to as "request value." (329)

In addition to the above values, which may or may not be valued monetarily, ecosystems have an additional value which will always be impossible to put a monetary value on. This is the "intrinsic value" of an ecosystem. It is based on the premise that other organisms have an equal right to existence regardless of any use of them by man. (329)

To assess the merits of alternative forms of development, it is necessary to consider the economic value of products and services produced by each and how they would be affected by development schemes. The absence of quantifiable market values for many "environmental" goods and services does not present an insurmountable problem because qualitative assessments of their significance can be incorporated into a carefully constructed analysis (34). (See also the subsections, "Environmental . . . ," "Economic . . . ," and "Social Impact Assessment" in Part 2.) A distinction is made between economic analysis which requires consideration of economic factors external to the physical system and financial analysis which normally ignores such factors (32).

Financial Analysis is concerned narrowly with profits and losses of private individual projects. Typically, it relies on market prices to guide investment decisions.

Economic Analysis in contrast, is concerned with the total effect on society and evaluates development alternatives based on changes in social welfare, or the sum of individuals' welfare.

Project Analysis usually focuses on easily measured direct costs and benefits and often ignores environmental "externalities," many of which are damage costs such as respiratory illness from air pollution or loss of fisheries because mangroves were cut.

Many of the issues requiring evaluation will be about multiple-use conflict. In economic terms, the multiple-use concept requires that all actual and potential uses for resource utilization schemes be determined and compared, so as to ensure that the resources are used in a way that maximizes the net benefit to society obtainable from them. It is the concept of the optimum allocation of resources which lies at the very heart of economics.

When comparing different uses of resources, all benefits and costs relating to their uses need to be considered. When a resource is used for one particular use as opposed to another, it has what is known as an opportunity cost. This can be defined as the value of that resource in its next best alternative use. This value should be used in the economic analysis as the cost of using the resource. For example, in determining the allocations of fresh water for either irrigation or fishery maintenance, if an excessive amount of freshwater is taken for irrigation, which leads to a collapse in the fishery, then the use of the water for irrigation will have an opportunity cost equivalent to the income which could have been

Figure 3.37. Fisheries create wide employment onshore in the processing and marketing sector and in boat building and equipment and supplies; here in El Salvador, a dugout canoe is under construction. (Photo by author.)

generated by the fishery (e.g., the Egyptian sardine fishery in the Mediterranean and the hilsa fishery in Pakistan).

Environmental and economic *externalities* exist when the actions of one resource user have an impact, positive or negative, on the welfare of another who is not part of the decision-making process. Changes in freshwater stream flow, perhaps as a result of irrigation development, may damage a coastal mangrove. Mangrove conversion to an industrial site will affect nearshore fisheries and supporting services (Figure 3.37). Coral mining may lead to increased coastal erosion and storm surge damage. In each of these cases, a decision made by one resource user imposes additional costs on others. Since these costs are not taken into account, the level of resource use will be greater than would be the case if all benefits and costs were considered. (109)

Because numerous economic sectors are influencing the coastal area, one must examine the economic "externalities" of each. When any one sector attempts to gain the highest economic yield from its activities, it often attempts to avoid responsibility for its external effects. For example, a factory operator may wish to avoid financial responsibility for the degradation of fisheries or inhibition of tourism caused by his factory wastes polluting a bay. The ICZM process examines the effects of "externalities" of any one sector on other sectors, most importantly the effects of "dryside" (shoreland) private activities, upon the common resources of the "wetside" of the coastal area. Ideally, these externalities should be incorporated (internalized) into the economic analysis.

An instructive case in point is that for an issue of logging vs. coral reef uses analyzed by Hodgson and Dixon (see "Philippines, Palawan" case in Part 4). Here the tourism and fishery values of the reef were shown to exceed that of clear-cut logging of adjacent slopes that threatened the reef.

If left purely to market (or political) forces, a suboptimal result will come about when dealing with coastal resources. This is particularly true if one is concerned with social welfare and long-term sustainable management of coastal resources. However, while some factors can be quantified (e.g., expected changes in income levels), others are not quantifiable (impact upon traditional life-styles). Both types of effects should be researched and brought together with other relevant information as part of the package of factors to be considered.

Unfortunately, the market signals (and market imperfections) favor rapid development and overexploitation of coastal resources. These patterns can be the result of a variety of reasons. Greed-driven resource exploitation by a handful of wealthy and powerful people can be just as destructive as poverty-driven overexploitation by large numbers of coastal residents. In both cases, the results are similar—short-term benefits are extracted at the cost of much greater potential long-term returns. (109)

Economic valuation for coastal management requires a mix of standard economic and financial appraisal methods coupled with special approaches developed for natural resource and amenity valua-

$$NPV = \sum_{t=1}^{n} \frac{B_d}{(1+r)^t} + \frac{B_e}{(1+r)^t} - \frac{C_d}{(1+r)^t} - \frac{C_e}{(1+r)^t} \quad \text{or} \quad \sum_{t=1}^{n} \frac{B_d+B_e-C_d-C_e}{(1+r)^t}$$

Where NPV = net present value

B_d = direct project benefits

B_e = external and/or environmental benefits

C_d = direct project costs

C_e = external and/or environmental costs, including environmental protection costs

r = discount rate

t = year in which costs or benefits occurred

n = number of years in economic time horizon or project lifetime

Σ = summation sign

Figure 3.38. Method for calculating net present value. (*Source:* Reference 41.)

tions. Among the more commonly used project appraisal methods include analyses such as the net present value, internal rate of return, and benefit-cost ratios. Below are various techniques recommended by Dixon (110), which can be used to value the various project costs and benefits which go into the economic appraisals:

Generally Applicable

1. Those that use the market value of directly related goods and services:

- Changes-in-productivity approaches
- Loss-of-earnings approaches
- Opportunity-cost approach

2. Those that use the value of direct expenditures:

- Cost-effectiveness analysis
- Preventive expenditures

Potentially Applicable

1. Those that use surrogate-market values:

- Property-value approach
- Other land-value approaches
- Wage-differential approach
- Travel-cost approach
- Marketed goods as environmental surrogates

2. Those that use the magnitude of potential expenditures:

- Replacement costs
- Relocation costs
- Shadow-project approach

Additional Methods

1. Contingent valuation methods, which use survey-based methods to value environmental impacts:

- Bidding games
- Take-it-or-leave-it experiments
- Trade-off games
- Costless choice
- Delphi technique

2. Macroeconomic models, for assessing regional effects of environmental impacts:

- Input-output models
- Linear programming models

External costs that are often excluded from *financial* analyses—such as the costs imposed by pollution—should be considered by donor agencies and project planners in an *economic* analysis (34). International donor agencies normally have some broad social objective in stimulating development. Therefore, a more broadly based assessment of projects and the ecosystem framework within which they operate is required, one which expresses a development project's impact upon society as a whole.

In the economic analysis of projects, the timing of costs and benefits is particularly important. Since money is valued higher at the present than in the future, a discount rate must be applied so that all future streams of costs and benefits relating to a project can be assessed on an equal basis. The discount rate effectively converts all the costs and benefits accruing over the lifetime of a project to present day values. The net result of each project can then be compared by calculating the *net present value* (NPV), using the formula shown in Figure 3.38 (41).

Present values may also be used for the *internal rate of return* (IRR) and the *benefit-cost ratio* (BCR). These are related to NPV as follows:

NPV = Present value of benefits − Present value of costs

IRR = Discount rate that results in the present value of benefits becoming equal to the present value of costs

BCR = Present value of benefits/Present value of costs

These measures are related in the following way:

NPV	BCR		IRR
If > 0 then	>1	and	$>r$
If < 0 then	<1	and	$<r$
If $= 0$ then	1	and	$= r$

Calculated present values of $100 in future years at various discount rates are shown below (41):

	Discount rate				
Time (yr)	**2%**	**5%**	**8%**	**10%**	**15%**
0	$100.00	$100.00	$100.00	$100.00	$100.00
10	82.03	61.39	46.32	38.55	24.71
20	67.30	37.69	21.45	14.86	7.56
25	60.95	29.53	14.60	9.23	7.05
40	45.29	14.20	4.60	2.21	0.57
60	30.48	5.35	0.99	0.33	0.04
100	13.80	0.76	0.05	0.01	—

Unfortunately, there are not simple "ABC" techniques that can be employed in conservation economics by non-professionals. It will usually be necessary to engage a professional resource economist to assist with economic analyses. If this is not possible, it may be best to use simple arithmetic arguments.

3.27 ECOSYSTEMS

Ecology, like biology or geology, is a field of study, not a popular movement. It is the study of the way in which all the forms of life relate to their environment. The origin of the word *ecology* reaches back to the 1800s, when a word was needed in the scientific community to define the study of how animals interact with their environment. Ecology is the dynamic aspect of nature's system. In marine biology it relates each species to such factors as water conditions, nearby objects, the food chain, and potential cooperating and enemy species.

The basic tenet of ecology is that life does not exist apart from its environment and that each species is, in fact, a product of its own particular ecological system, or *ecosystem,* owing its existence to factors of climate, geography, and geology. The reverse may also be true because many marine species modify their own environments.

Coastal ecosystems are so different from their terrestrial counterparts as to require different and special forms of conservation. For example, such ecosystems as coral reefs, beaches, coastal lagoons, submerged seagrass meadows, and intertidal mangrove forests have no counterparts in terrestrial resources. There are other important differences; e.g., since marine organisms are in closer chemical contact with their surrounding medium than land organisms, they are jeopardized more by pollution (308).

The ecosystems of the sea can achieve great complexity, as in coral reefs, and very high bioproductivity, as in "upwelling" areas where ocean waters rise to the surface. These systems differ from terrestrial systems in many ways. Three-dimensional phenomena are, of course, more marked and important in the ocean, where organisms are less tied to the solid bottom than are land organisms to the earth. But, more important, because of the fluid nature of the seas, whole biological communities exist as floating plankton-based entities distributed horizontally and vertically through broad ocean spaces and are displaced over them as they drift with ocean currents.

Currents are also great mixers, transporting organic nutrients produced at one site to distant locations and carrying planktonic eggs and larvae of organisms to colonize distant habitats. In addition, many marine species actively migrate long distances, like tunas, turtles, whales, and eels. Other sources of variability include the waves resulting from tropical storms. Since marine organisms are in closer chemical contact with their surrounding medium than land organisms, they are jeopardized more by pollution (308).

Although aquatic environments differ latitudinally in temperature regime, illumination, and seasonality, differences also exist between habitats or life-zones at particular latitudes. There is some variation with depth. Thus, coastal habitats in temperate zones are characterized by wide seasonal changes in temperature and light, as well as diurnal changes in temperature in some habitats, such as the intertidal. Tropical environments, on the other hand, may experience only slight seasonal changes in day length and illumination intensity. (305)

Organisms in polar regions must have strategies to cope with prolonged harsh conditions and be adapted to exploit the short periods of favorableness both for growth and reproduction. In the tropics, where seasonal stability and other conditions are favorable growth and reproduction are not so constrained, and other challenges, such as interspecific competition, become the more important constraints on the organisms. (305)

Temperature is an important variable in ecological systems and methods of predicting its effect are useful. The most common method of quantifying rate changes in physiological processes with temperature is by the temperature coefficient (Q_{10}), which is the factor by which a process (or reaction) rate increases over a rise of 10°C. A Q_{10} of 2 signifies a doubling of reaction velocity over the stated 10° temperature interval. Thus,

$$Q_{10} = k_1 + {}_{10}/K_1 \tag{2}$$

or, expressed for any temperature differential,

$$Q_{10} = (k_2/k_1)\exp[10/t_2 - t_1] \tag{3}$$

This form is the most widely used, even though it has long been established that Q_{10} is not a true constant but varies considerably at different temperatures, generally declining with increasing temperature. However, for most biological reactions Q_{10} lies between 2 and 3 and is of descriptive (and comparative) value despite the availability of a range of more sophisticated temperature formulas.

A most important element of the ecosystem is the *food chain.* In its simplest form, there would be four links in this food chain. First are the *primary producers* made up of all plant life. Plants are found either growing in the bottom (mangroves, marco algae, seagrass) or as myriad floating microscopic cells known as *phytoplankton. Plankton* means all of the small life in the water. *Phytoplankton* is the plant portion and *zooplankton* is the animal part.

The second link is the animals that feed directly on plant life. They are known as the *consumers* and they include shrimp, clams, worms, insects, some fishes (e.g., mullet), and zooplankton. The third link includes the animals that eat the consumers. These are the *foragers,* such as croaker, killifish, mackerel, sole, crabs, and pinfish. The final link in the food chain comprise animals that feed on the foragers. This group, known as the *predators,* includes sharks, marlin, chinook salmon, king mackerel, bluefish, tuna, and bottlenose dolphin.

Actually, the food chain is not as simple as these four discrete categories would imply—many species short-circuit the food chain. For example, predators like cod sometimes feed on consumers as well as foragers. Or foragers like seatrout may become predators at times. Because of these short-circuits, some ecologists call the system a *food web* (Figure 3.39) instead of a food chain. Others call it a *food pyramid.*

Of particular importance in the coastal food chain is wetlands detritus, small floating particles of plant matter from decomposing mangrove or cordgrass leaves or other plant tissues. Detritus is consumed by a wide variety of shrimp and other small estuarine life forms which in turn serve as forage for birds and fish.

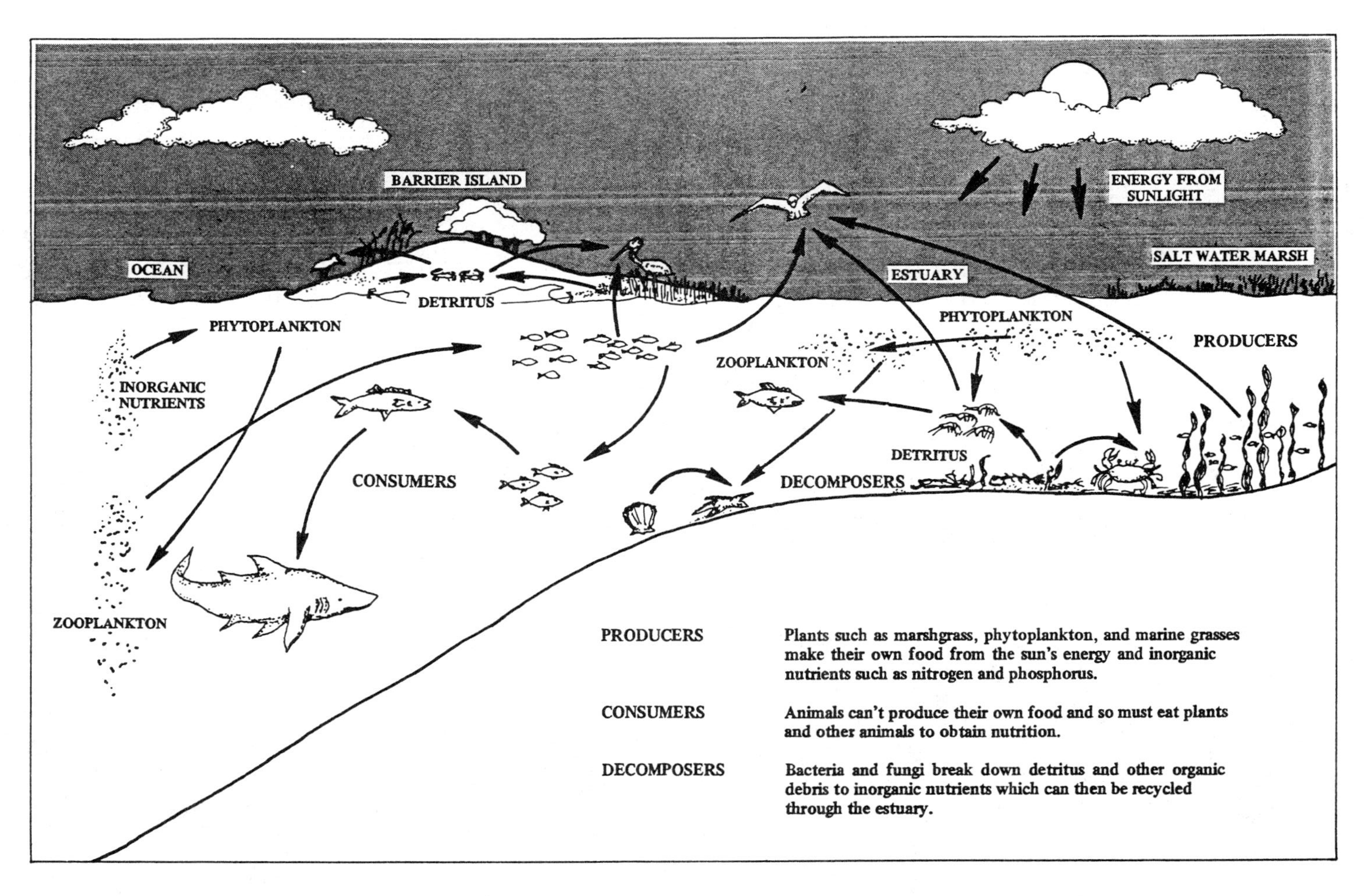

Figure 3.39. The "food chain" or "food web" of the coastal ecosystem needs to be maintained throughout its various parts and processes. (*Source:* Reference 107.)

All aquatic plants (primary producers) are nourished by nutrient minerals dissolved in the water, particularly compounds of nitrogen and phosphorus, which are supplied from within the ecosystem through a continuous internal recycling process. However, nutrients continuously trickle out of the system and are replaced by minerals from land runoff and other sources.

Sunlight is the basic force driving the ecosystem. It is the fundamental source of energy for the aquatic plants. It must be able to penetrate coastal waters so as to foster the growth of both the rooted plants, such as seagrasses, and the suspended algae (or phytoplankton). Increased turbidity, from the addition of suspended matter to the water, reduces light penetration and depresses plant growth. Coastal waters are normally more turbid than open ocean waters, more laden with silt and more rich in suspended life.

Of the various gases that are found dissolved in coastal waters, oxygen is of the most obvious importance to the fauna. Coastal waters need a high oxygen concentration to provide for octimum ecosystem function and highest carrying capacity.

With a limited amount of primary production typifying a body of water, there will also be a limited number of consumers, foragers, and predators. A rule of thumb is that there will be a 10 to 1 reduction through each link of the food chain. The effect becomes apparent when we consider that to produce one 20-pound tuna, it would take 200 pounds of mackerel, that ate 2,000 pounds sardines, that ate 20,000 pounds of tiny invertebrates, that ate 200,000 pounds of algae—100 tons of algae was produced to create a 20-pound tuna. This is a very simplified example, but it does demonstrate why the largest fish are scarce.

Primary productivity measures are useful in diagnosing the condition of an ecosystem because they indicate the potential capacity to support life. By comparing the actual abundance of life with the potential abundance, one can determine whether the system is malfunctioning and needs attention. Because there are great difficulties in measuring the abundance of the variety of life forms in an ecosystem, ecologists often use "primary productivity" as a gauge of the total productivity of the system

Primary productivity measures the growth rates of plants, such as algae, seaweed, and marsh grasses. These values provide an index of the biotic *potential* or carrying capacity of an ecosystem. By an extension of this approach, one can learn the sources of energy that fuel the ecosystem and the rates of energy flow through it, as well as diagnose the present condition of flows and determine whether some correctable blockage is interfering with energy flow and thus lowering carrying capacity. A loss of productivity is to be presumed adverse—vital productivity areas require protection.

A coastal management program must be compatible with ecological theory and fashioned to the specific qualities of the ecosystem to be managed. The productive capacity of a coastal water body is governed by the interplay of the chemical, geological, physical, and biological factors that together govern its productivity. Each type of pollution, removal of habitat, interference with primary productivity, or other disturbance reduces the vigor of the ecosystem. (59)

Each of the numerous components of the coastal water ecosystem must be safeguarded—fringing vegetated water areas, estuarine bottoms, shellfish beds, and the breeding, nursery, feeding, and resting habitats of species. In addition, basic dynamic processes must be maintained—water circulation, nutrient input, and sunlight penetration of water. Only after these vital basic processes and components have been identified, and their vulnerabilities to disturbances are known, can a comprehensive management program be developed. (59)

Lagoons and estuaries are the richest of all coastal waters not only because they produce an abundance of fish and shellfish but also because they serve special needs of the migratory nearshore and oceanic species that require shallow protected habitat for breeding or as sanctuary for their young. Second only to the estuary is the nearshore zone, the band of shallow waters adjacent to the ocean shore, often bounded on the seaward side by a coral reef. (57)

The entire dynamic balance of the ecosystem revolves around and is strongly dependent on water circulation. Vertical and horizontal water circulation transports nutrients, propels plankton, supports and spreads "seed" stages (planktonic larvae of fish and shellfish), flushes away the wastes of animal and plant life, cleanses the system of pollutants, controls salinity, shifts sediments, mixes water, and performs other useful work. The specific pattern of water movement found in the estuarine portion of any coastal system is a result of the combined influences of runoff volume, tidal action, wind and, to a lesser extent, external oceanic forces.

Salinity or salt content of the water is a critical factor for lagoon and estuarine species. Generally, there is a gradient in salinity that starts with a high concentration in the ocean, decreases inward through the estuary, and drops to near zero at some distance up the lagoons and tributaries. Some coastal species tolerate a wide range of salinity, whereas others require a narrow range to live and reproduce successfully. Some species require different salinities at different phases of their life cycles, conforming to regular seasonal rhythms in the amount of land runoff.

The floors of coastal basins are important. They provide the basic form and structure of the basins and govern the flow of water through them, as well as harbor the richest habitat areas of coastal waters—clam beds, coral reefs, submerged grass beds, and so forth. Estuarine floors are usually biologically richer and more vulnerable to adverse impacts than are nearshore ocean floors.

Many commercially or recreationally valuable species depend on the basin floor for habitat and they forage about within the bottom sediments for their food. The community of life of the basin floor is also a major element in ecosystem stability. The bottom species are highly diverse—including worms, lobsters, clams, oysters, shrimps, and fish.

Ecologically healthy estuaries have clean and firm bottoms and undisturbed habitats with a high resource carrying capacity. The system's capacity is reduced when functioning grass beds, shellfish beds, coral reefs, and other vital areas of the basin floor are seriously altered or degraded and when sediments accumulate on the bottom of the basin, causing shoaling and lowered water quality.

3.28 ECOTOURISM

A recent alternative to typical packaged tourism is the *ecotourism* paradigm, whereby upscale, empathetic tourists pay in hard currency to experience nature (e.g., hiking in mountains, birdwatching in wetlands, snorkeling over coral reefs, etc.). The local people often profit by providing tourism support services and suffer minimum disruption to their traditional life. And most important, the host country gets a good lesson in why nature should be protected—it attracts ecotourists, who bring money (Figure 3.40). So it is in everybody's best interest to protect environmental quality and natural diversity, or so goes the paradigm.

Its advocates hope that through ecotourism the more rapacious resource activities—such as clear-cutting of mangroves, mining of coral reefs, and hunting of endangered species—can be discouraged in favor of the lighter footprint of selective, personalized tourism—much of it directed to national parks and nature reserves. But ecotourists usually want to go the "unspoiled places" to see "nature in the raw," putting stress on areas that typical tourists may not visit. In today's travel market, ecotourism is often included in the category *adventure tourism.*

Ecotourism is a leading force for coastal conservation in countries like Costa Rica and Dominica. But it needs controls based on carrying capacity to ensure that natural resources are not overused and that the parks are not overwhelmed by tourists. The point is that ecotourism is no miracle fix for environmental problems—development control based on carrying capacity is still necessary. (See "Carrying Capacity" above and "Tourism" below.)

3.29 EDUCATION

In the coastal management context of this book, *education* is meant to help people accomplish things themselves; *awareness* is meant to alert the people to programs or actions promoted by agencies or special interest groups and which are supposed to have social benefit (see "Awareness" in Part 2). Education would often be used to introduce new eco-friendly technologies or basic understanding of ecology.

A central educational facility is often effective in putting across the basic ideas of conservation (Figure 3.41). But active contact may be more effective in accomplishing specific education tasks. To take an example, in the Dominican Republic, members of the Centre for Marine Biological Research (CIBIMA) of the National University give frequent talks to community groups, charcoal-makers, fishermen and government officials in charge of mangrove management (149). These talks, usually given

Figure 3.40. Ecotourism can yield high profits with a minimum of environmental and social cost to the community if managed properly. (Photo by author.)

in areas where specific problems are identified, have been effective in promoting a more careful and less wasteful utilization.

There are different stages in the process that humans use in considering a new idea, according to Hudson (177):

- Revelation: exposure to an innovative idea
- Interest: seeking of more information
- Evaluation: new idea set beside held beliefs
- Trial: testing of new idea
- Adoption: adoption of idea if it proves successful

Figure 3.41. Public aquariums provide a good opportunity for general education and for creation of public awareness. (Photo by author.)

In close-knit communities, the knowledge that a protected area can increase fish stocks and catches in nearby fished areas can lead to positive attitudes toward the protected area and compliance with its rules (308). For example, this approach (along with some materialistic motivation) was used in a program to encourage fishermen of Discovery Bay, Jamaica, to shift to larger mesh size wire for fishpots to allow escape of small fish and thus to improve sustainability of the fishery. When they turn in one fishpot made with small size mesh (1.25 in. or less), fishermen receive enough large mesh size material to make *two* new pots of equivalent size. (149).

Mass education needs to have popular appeal, which arises largely from the communication method and media used. Three examples are given below. The first (Figure 3.42) is a view of a tropical marine ecosystem used to explain the "food chain" by use of the following text, wherein the numbers match those of the figure (149):

> Sunlight provides direct energy for the ecosystem (*1*). But water currents also bring animals, plants and dissolved food (called "nutrients") to the reef (*2*). The reef plants use the sunlight and the nutrients to grow.
>
> Plants are everywhere on the reef. Some are in the water, such as plankton (*3*). But most are in the sand, on the rocks, and in the coral polyps and skeletons. Algae of varying colors—red, brown, and green—grow over the surface of dead coral, providing food and shelter for small animals and adding color to the living reef (*4*).
>
> There are many reef animals which eat the plants, and these are, in turn, eaten by other animals. Thus energy is passed up a "food chain" from the plants to sharks or grouper.
>
> Animals that eat plants are mostly small, like the crustaceans and worms. But there are also animals like sponges, which filter plankton from the water, and snails and sea urchins, which graze plants on the rock surface (*5*). Many different kinds of animals feed on the tiny plants and animals that live in the sand. Most of them simply eat the sand and digest what food is in it (*6*). Some fish eat plants in or on the coral, including the parrotfish (*7*).
>
> The next link in the chain of energy is the animals that eat the plant-eaters. They are mostly fish, including those that eat large zooplankton, those that hunt small animals close to the bottom, and those that eat the coral skeleton to get the worms and mollusks that live inside it (*8*).

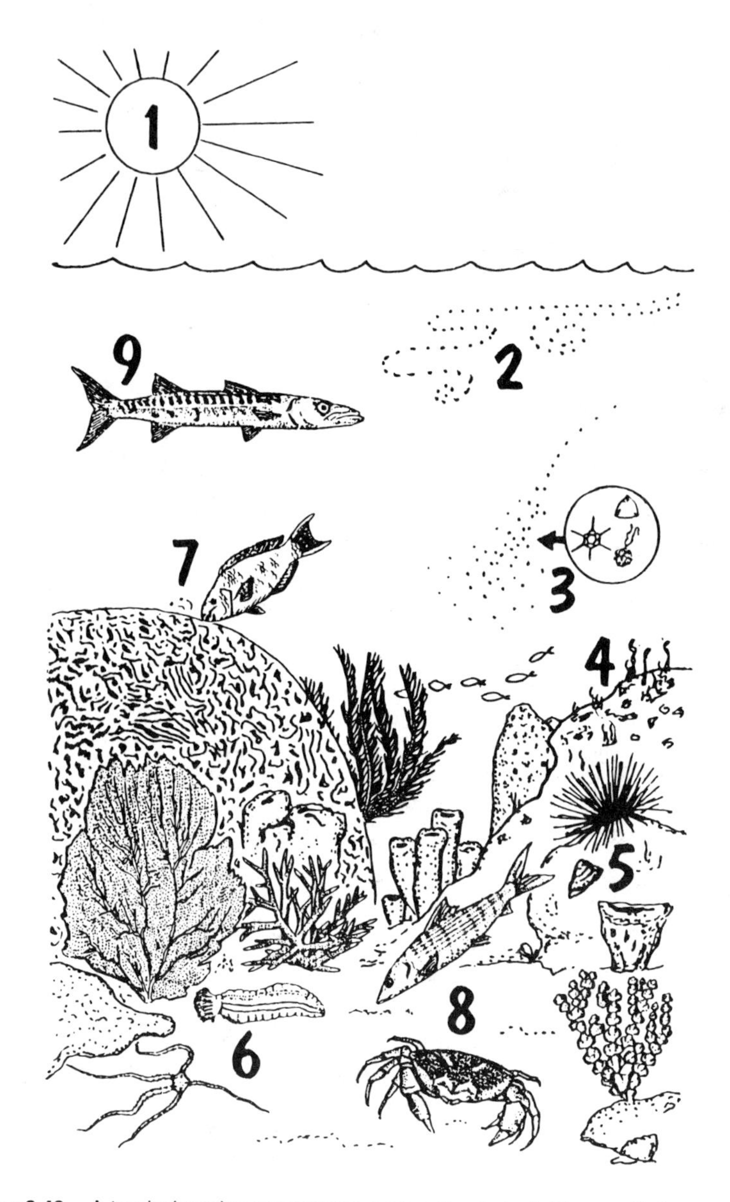

Figure 3.42. A tropical marine ecosystem; explanation in text. (*Source:* Reference 149.)

The chain usually stops when the food energy reaches a "top" predator—unless someone such as yourself is fishing on the reef, ready to catch and eat the fish (*9*). You can see how human beings have become the last link in many food chains.

The second and third examples, in Figures 3.43 and 3.44, explain graphically to a rural audience the dangers of tropical fish poisoning, *ciguatera.*

3.30 ELECTRIC POWER GENERATION

The lessons learned from the past regarding large steam-powered electric plants are as follows: (a) there are good and bad places to put steam plants; (b) there are good and bad ways to design steam plants; (c) there are good and bad ways to operate steam plants; and (d) the sources of damage are well known. The same lessons hold whether the steam plant is fired by oil, coal, gas, or nuclear power.

Both fossil-fuel and nuclear-fired electric generating plants operate on the same general principle. Steam is produced from water by burning the fuel substance and is used to power a turbine, which turns an electric generator. The spent steam is condensed, and the water returned to be revaporized into steam to start the cycle anew. The steam condenser is cooled with water that is either drawn continually from a natural water body or recirculated through a closed-cycle cooling system. The delta T across the condensers—difference in temperature between ambient and exit waters—is around 11°C (19–20°F).

A large nuclear generating unit of around 1000 MW^e (megawatts electricity) with "open cycle" (or "once-through") cooling may be fitted with a row of six 140,000 gpm (gallons per minute) pumps, each with a separate intake. In front of the pumps are various screens and other devices to protect the

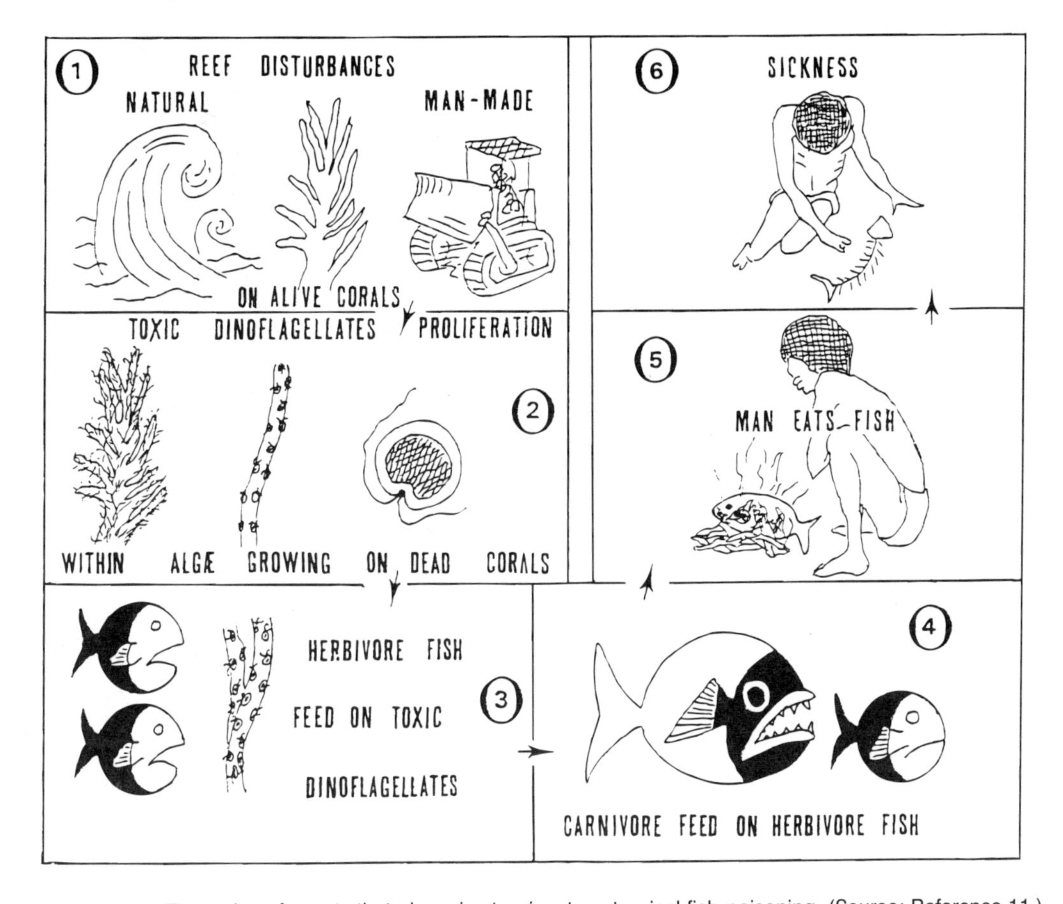

Figure 3.43. The series of events that gives rise to *ciguatera,* tropical fish poisoning. (*Source:* Reference 11.)

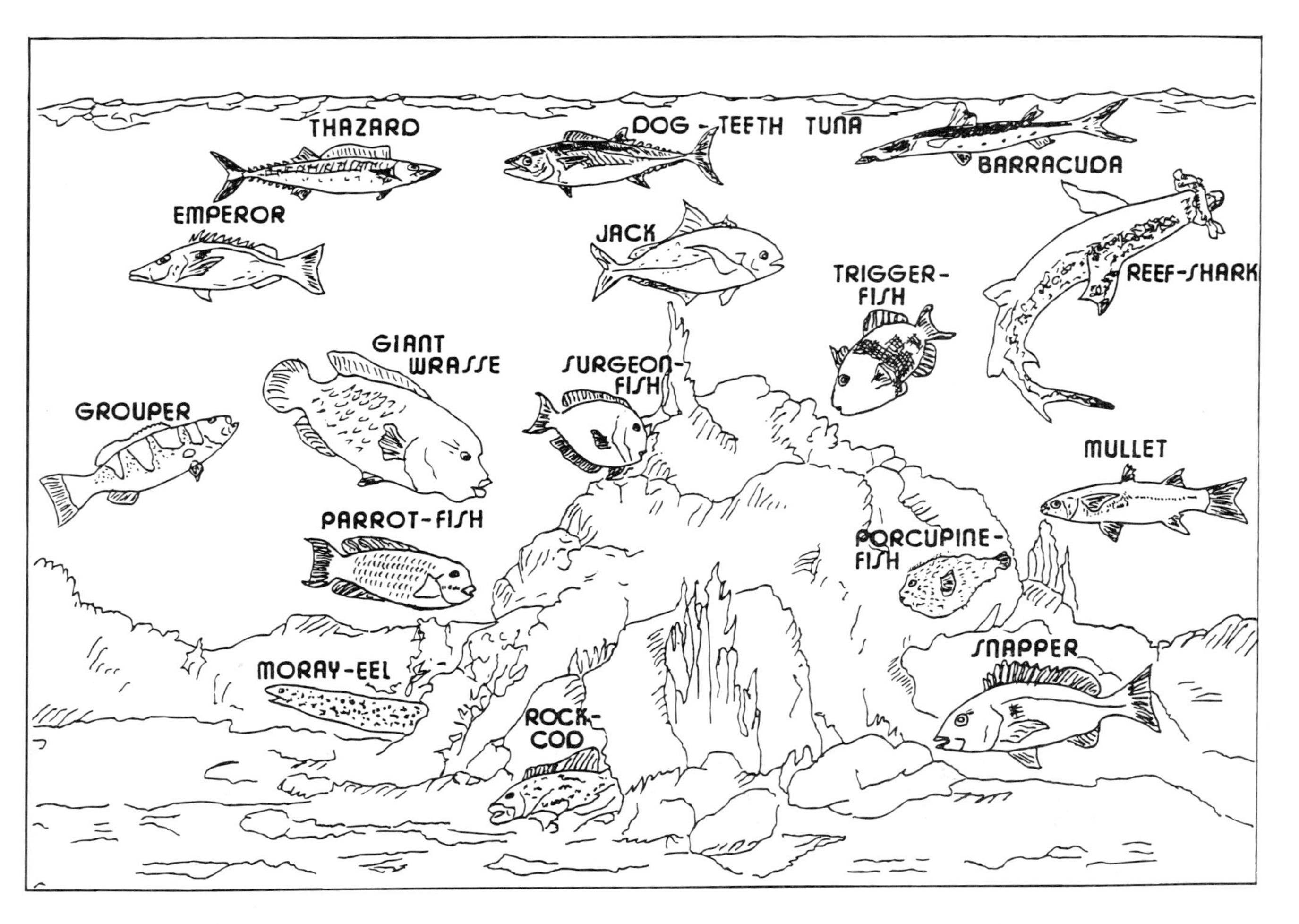

Figure 3.44. The variety of fishes that can transmit *ciguatera*, tropical fish poisoning. (*Source:* Reference 11.)

cooling system from damage caused by floating debris. A typical screen is made of sections of 3/8-inch mesh mounted on drums, so that it may be rotated for cleaning when quantities of debris or dead fish accumulate and reduce the flow of water through it (Figure 3.45).

The amount of water required for open-cycle cooling of steam condensers varies. With a typical modern nuclear plant of 1000-MW^e capacity, water goes through the plant in less than 1 minute, and its temperature is raised by 10 to 34°F (5.6 to 19°C) before being discharged directly back into public waters. Diesel fuel plants are more efficient than nuclear plants and discharge about one-third less waste heat. Whereas a fossil fuel plant might require about 600,000 gpm of cooling water per megawatt of electricity produced, a nuclear plant might require 900,000 gpm of water from an adjacent tidal river, estuary, or coastal waters.

Inside a large (1000-MW^e) nuclear plant, the cooling water passes through a manifold of 35,000 1-inch metal tubes, each about 50 feet (15.2 meters) long. The tubes are surrounded by the steam that is to be condensed. As the cooling water traverses the tubes, it cools the steam, gains heat, and then flows via a junction box out a conduit back to the source water body. Once back in the source body, the heated water may rise, sink, or remain suspended, depending on the relative densities of effluent and receiving waters. (59)

The heated effluent from open-cycle power plants adversely affects the natural patterns of life and the behavior of all aquatic species and thus is called *thermal pollution.* The entire aquatic ecosystem may be degraded by thermal pollution. How pervasive this pollution may be and how damaging depends on the size and flushing characteristics of the public water basin that is threatened with pollution.

Very often a greater threat is the high death rate of organisms suspended in the water and drawn into power plants with the cooling water (Figure 3.46). Aquatic forms drawn in with the cooling water are exposed to heat, turbulence, abrasion, and shock. The potential for environmental damage from massive "entrainment" and death of these organisms—fish, plankton, and the larval stages of shellfish—is so great that it is a major factor governing the design and location of power plants in the coastal zone. In the United States the number of fish impaled on these screens has exceeded five million in a few weeks' time at estuarine-sited power plants with the old open-cycle cooling (59).

The areas of greatest environmental threat are the enclosed waters—estuaries, bays, lagoons, and tidal rivers—which are of critical environmental concern because of their high productivity and the abundance and diversity of life that they support. These areas, which require the highest degree of protection, are also the most attractive as power plant sites for many practical and strategic reasons and therefore are the most threatened.

The principal criteria for locating power plants in the past have been (a) cooling water availability, (b) fuel availability, (c) land suitability, (d) engineering feasibility, (e) cost of land, (f) transmission of power to market areas, and (g) for nuclear plants, nuclear safety. Aquatic resources and ecological considerations have often been ignored, resulting in much serious damage to natural systems. Electric power production can be harmonized with environmental protection in coastal areas if certain traditional practices are modified to respect resource and ecological needs. (59)

The major solutions are to (a) locate power plants along the open coast, where there is deep water nearby for strategic placement of intake and outlet structures if open-cycle cooling is to be used, and (b) reduce the volume of cooling water used by plants on estuaries by requiring closed-cycle systems, which recirculate cooling waters, rather than open-cycle systems, which continuously withdraw from and discharge into the environment large volumes of water.

When open-cycle plants are permitted on the ocean coast, intakes, outlets, and other submerged structures must be designed with care to prevent entrainment, entrapment, and impingement of aquatic life. To minimize adverse aquatic impact, the first requirement for selection of an ocean coast site should be a broad survey to locate all important habitat areas. Such areas should be classified as off limits for power plant intakes and effluent discharges. The second requirement is that the cooling system be designed for minimal disturbance of coastal ecosystems. Specifically, the following should be required: (a) limit the discharge of chemicals, (b) provide for minimal disruption of natural water flow, and (c) locate cooling water intake and outlet for minimal effect on biota.

The toxicants actually or potentially associated with the operation and maintenance of nuclear power plants and cooling tower structures include the following: acids, acrolein, arsenic compounds, ammonia and amine compounds, boron, carbonates, chlorine and bromine, chlorinated and/or phenylated phenols, chromates, cyanurates and cyanides, hydrazine compounds, hydroxides, metals and their salts, nitrates

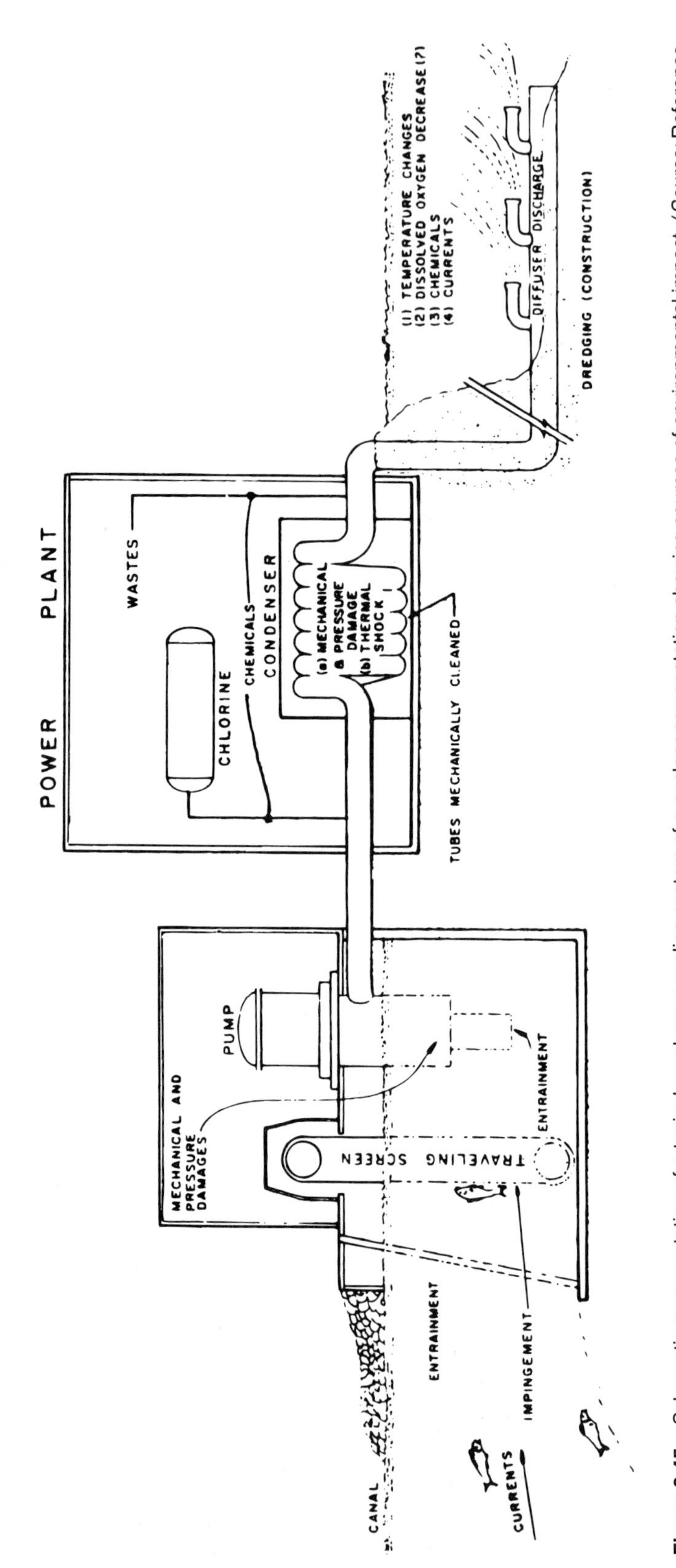

Figure 3.45. Schematic representation of a typical condenser cooling system of a nuclear power station showing sources of environmental impact. (*Source:* Reference 363.)

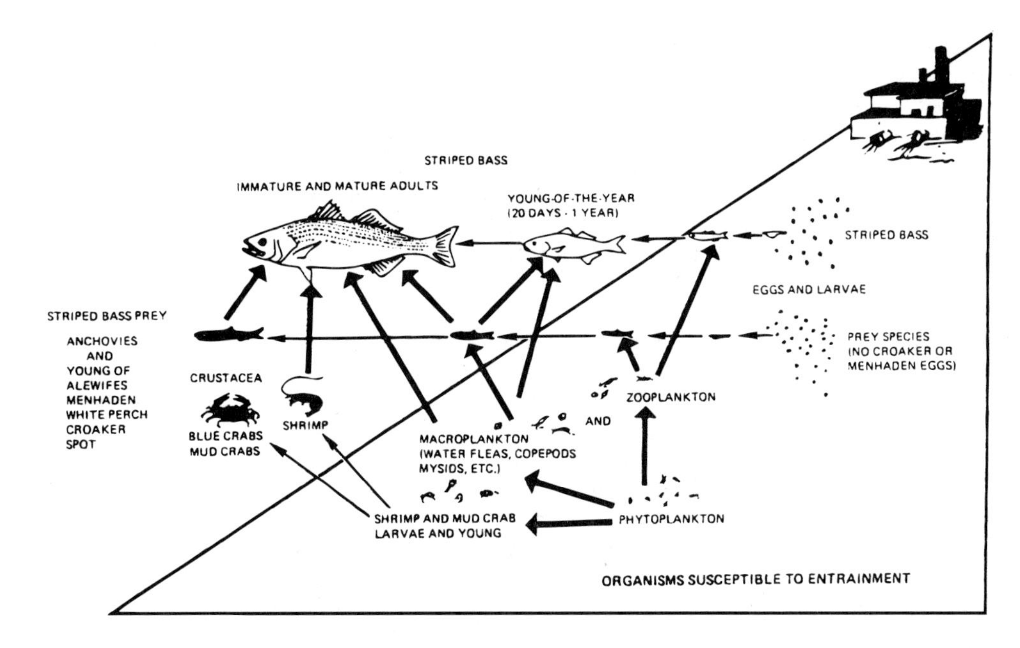

Figure 3.46. Potential power plant impacts on an example estuarine fish, the striped bass, and its food chain. (*Source:* Reference 59.)

and nitrites, potassium compounds, phosphates, silicates, sulfates, sulfides, and fluorides (59). The impact of chlorine should not be underestimated, especially when it is found in company with amines and other compounds with which it can form more toxic and/or longer lasting chlorine-based compounds.

3.31 ENDANGERED SPECIES

Endangered species are those in danger of extinction throughout all or a significant portion of their range; *threatened species* are those likely to become endangered within the foreseeable future. Because much of species loss is due to habitat loss, it is critical to provide a means whereby the ecosystems upon which endangered and threatened species depend may be conserved. One of the unfortunate consequences of growth and development has been the extermination of certain species and subspecies of fauna and flora. Such losses in species with educational, historical, recreational, and scientific value continue to occur, and a key to more effective conservation of native fauna which are endangered or threatened is to encourage conservation programs for such wildlife (59). The care of both species and their habitats is included in biodiversity iniatives (see "Biodiversity" above).

Society must face ethical issues relating to species extinction. Human beings have become a major evolutionary force, lacking the knowledge to control the biosphere, but having the power to change it radically. We are committed to our descendants and to other creatures to act prudently. We cannot predict what species may become useful to us—we may learn that many apparently dispensable species can provide important products, like pharmaceuticals, or are vital parts of life support systems on which we depend (see "Genetic Diversity" below). For reasons of both ethics and economic self-interest, therefore, we should not cause the extinction of a species (308).

A distinction of the sea is its limited endemism: marine species and subspecies are only rarely confined to certain small areas. There is great mixing of the ocean and its species and few sharply defined biogeographic provinces with unique species compositions. Since very few species are confined to narrowly bounded habitats, the chance that any species would be extinguished by human activities is very low. Saving species from extinction is thus not as strong a motivation for marine protected areas as is conserving renewable resources. (308)

Worldwide, the endangered and threatened species lists include mostly big animals, including many of the coastal/marine mammals, such as, dugong, river dolphin, some whales and other cetaceans, seals and sea lions, along with crocodilians and several species of sea turtles (Figure 3.47). Along the coastal beaches and wetlands of the littoral, there are many water birds that are listed or have been, such as terns, egrets, eagles, and pelicans.

To cite some examples, in the Mediterranean, the most endangered species (species on the verge of extinction) are the monk seal (*Monachus monachus*), the audouin gull (*Larus audouinii*), and the leatherback (*Dermochelys coriacea*), loggerhead (*Caretta caretta*), and green (*Chelonia mydas*) turtles. Also threatened is their food supply and the general quality of their habitats. The endangered monk seal is described below along with the threats to their survival—underlining the importance of protecting habitats to save these species.

The monk seal, once found along all shores of the Mediterranean and Black seas, the coasts of northwest Africa, and around Madeira and the Canary Islands now heads many lists of endangered species. Its present population is estimated at about 400, concentrated along the Turkish and Greek coasts and the Aegean islands. Major threats to the monk seal include disturbance at breeding sites, habitat destruction, marine pollution, competition with fishermen for food, and hunting. Recent coastal development has destroyed many areas used for giving birth, forcing the females to more remote spots. Monk seals also present an easy shooting target as they sleep in caves.

To conserve any marine species, careful management procedures must be directed to two goals: conserving species stocks so that the breeding potentials of populations are not destroyed by overharvesting, and conserving the support systems (e.g., nutrient inputs) and critical habitats (e.g., feeding, breeding, shelter, nursery, and migration areas) of these species stocks (308). It is also important ot avoid disturbing or frightening most wild species (Figure 3.48).

The United States has had several successes with endangered species restoration programs. Examples of coastal species that have been declassified as "endangered" because their abundance returned to safe levels include the gray whale, the American alligator (Figure 3.49), and the brown pelican.

Control of international trade in endangered species—as living animals or as animal products—has been achieved by an international treaty with 109 signatory countries. This is the Convention on

Figure 3.47. All seven species of sea turtles worldwide are endangered by human activity, mainly at the beaches where they nest. (Photo by Richard Stone.)

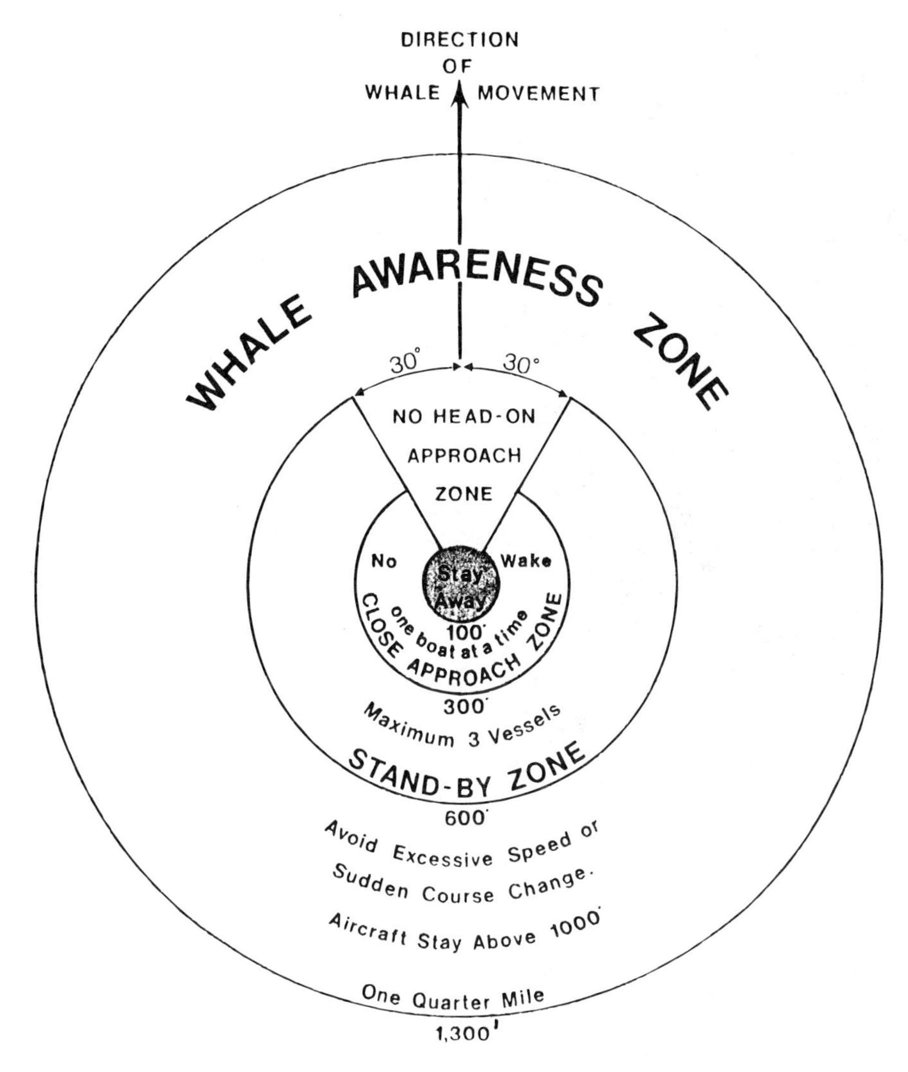

Figure 3.48. The U.S. government has found it necessary to regulate the activity of boats that visit whales in their natural habitat, as shown above. (*Source:* U.S. National Oceanic and Atmospheric Administration.)

International Trade in Endangered Species of Wild Fauna and Flora (CITES), which came into effect in 1975 (see "CITES" above).

3.32 ENVIRONMENTAL AUDIT

A major purpose of ICZM-type programs is the examination of proposed major development projects to determine the impacts they have on coastal resource systems and to recommend design or location changes which can eliminate or reduce any negative impacts. This regulatory function will usually require some form of environmental fact-finding procedure. If the potential negative impacts of a commercial activity are eliminated or reduced to a minor level or if they are appropriately "mitigated" (i.e., offset by compensatory activity such as environmental benefits on or off the development site), the project will be permitted.

The private sector has responded to these concerns and recently has shown concern about how different manufacturing and distribution activities might affect the environment. As a result, various

Figure 3.49. The American alligator, once protected as an endangered species, has fully recovered its former abundance and is no longer listed as endangered. (Photo by Luther Goldman.)

businesses are sponsoring an *environmental audit* of their operations. This usually involves hiring an environmental consultant to examine the full gamit of activities of the business.

The subjects of the audit may include any or all of the following, and other items where appropriate:

- Policy, responsibilities, and organization
- Planning, monitoring, and reporting procedures
- Management and staff awareness and training
- External relations with regulatory authorities and citizens
- Compliance with regulations
- Emergency planning and response
- Pollution sources and minimization
- Pollution treatment and discharge
- Resource savings
- Housekeeping
- Land management

The first level of examination may be a general review of the environmental status of the business, including such subjects as strengths, weaknesses, opportunities, and threats to explore various environmental policies and aspects:

1. What are the environment-related strengths of the enterprise or business unit?
   - Environment-friendly products
   - Processes that save resources and do not cause environmental hazards
   - Corporate image to be a "Green and clean" products
   - Staff and management committed to environmental protection
   - Research and development capacities for "clean" products
2. What are the environment-related weaknesses?
   - Products that cannot be recycled

- Non-recyclable packing materials, bottles, etc.
- Polluting processes
- Hazardous wastes
- "Polluter" image
- Staff and management not committed to environmental protection

3. What are the environment-related opportunities?
   - Entering into new markets
   - Being among the first to offer an "environmentally friendly" version of a traditional product
   - Securing long-term survival by shaping a "Green" business image
   - Raising the performance of collaborators by setting a new goal for environmental protection
   - Saving resources (e.g., energy) and costs
4. What are the environment-related threats?
   - Environmental regulations require additional investments and could render products unprofitable because:
   - Mid-term survival of enterprise is threatened
     a. increased state intervention in and control of business activities;
     b. citizen activist groups take action against enterprises;
     c. competitors gain market shares with "Green products";
     d. identification of staff with the company decreases, retainment and recruitment of personnel becomes more difficult;

Such a review may best be done by assembling an "audit team" from the company to assist the consultant with the review. A side benefit of this approach is that the environmental consciouness of the employees will be raised and an interested inner cadre will be formed.

The audit itself should go through a standard, pre-arranged, procedure such as the one shown in Figure 3.50 that was created by the Canadian Naranda Corporation and adopted by the International Chamber of Commerce's environmental auditing working party. This procedure divides the audit into three parts: pre-audit, site activities, and post-audit.

The type of benefits to accrue to the company from environmental audit might include any of the following:

- Savings due to reduced consumption of energy and other resources
- Savings due to recycling, selling of by-products and wastes, resulting in decreased waste disposal costs
- Reduced environmental charges, pollution penalties, compensations following legal damage suits
- Increased marginal contribution of "Green products," which sell at higher prices
- Increased market share due to product innovation and less performative competitors
- Completely new products opening up new markets
- Increased demand for a traditional product which contributes to pollution abatement
- Improved public image
- Renovation of product portfolio
- Productivity improvement
- Higher staff commitment and better labor relations
- Creativity and openess to new challenges
- Better relations with public authorities, community, and Green activist groups
- Ensured access to foreign markets
- Easier compliance with environmental standards

The environmental audit should lead to a continuing environmental management program for the company. This could make for a significant change in some businesses because, conventionally, managers make the decisions on the basis of internal costs and benefits to their organizations, whereas environmental management takes in account the external positive and negative environmental effects of actions that do not necessarily have a direct economic relationship with the enterprise or action proposed.

Good environmental management can be of economic advantage to an organization, but its justification must be based on the recognition of these wider responsibilities. Creating this environmental

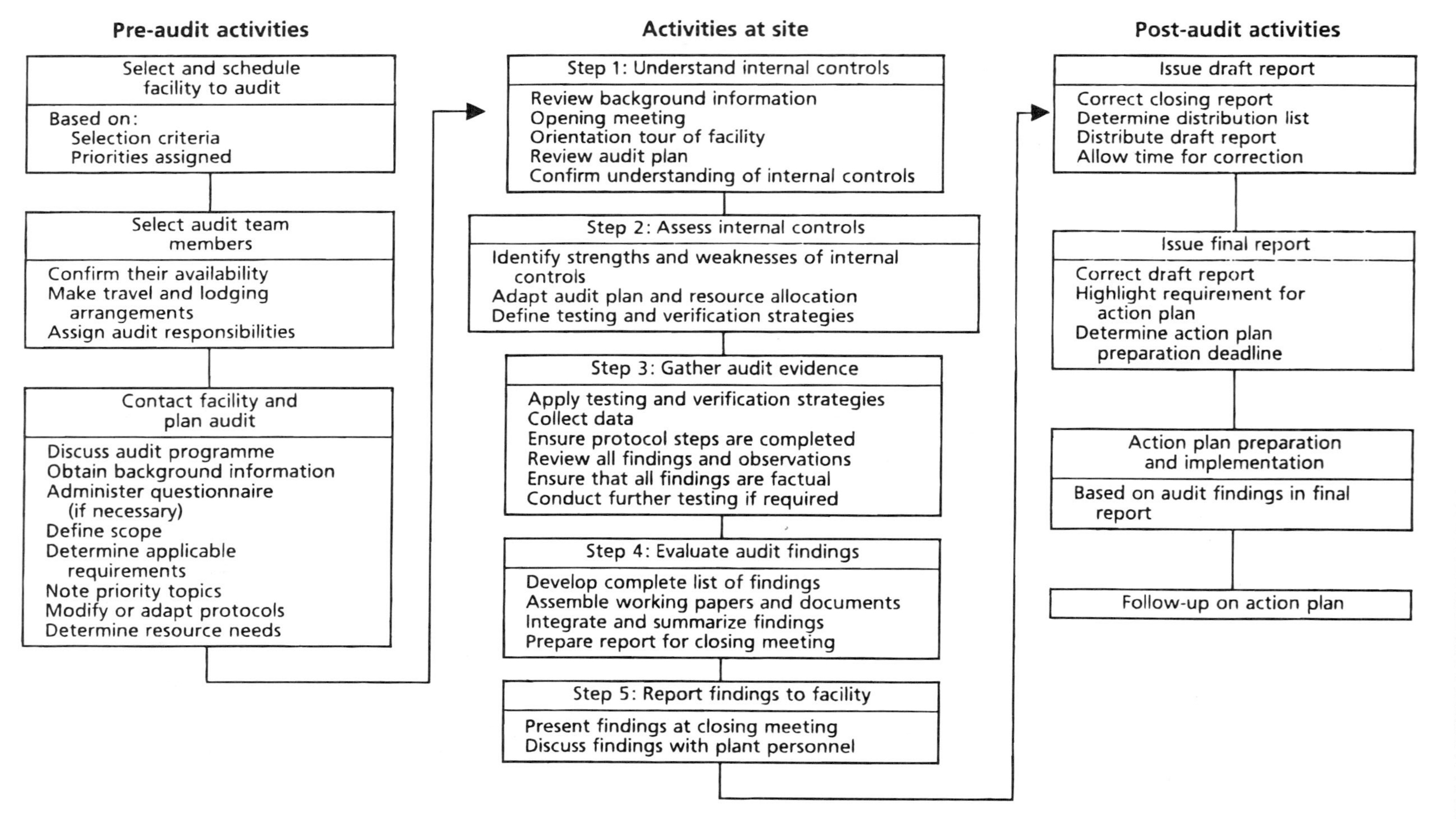

Figure 3.50. Basic steps in an "environmental audit." (*Source:* Reference 261.)

awareness is the greatest challenge in developing environmental management in a company. Successful environmental management, just as the development of positive policies regarding equal opportunities, employment or the creation of better working conditions, require the extensive participation of workers as well as management.

Environmental auditing needs the strong endorsement and active support of top management. The procedure should be clearly communicated together with the appropriate incentives. The auditing team has to gain the confidence of the audited units and make clear that the objective is to identify ways to progress, and not to punish managers. Some companies audit their facilities regularly. One company, for example, audits each refinery every six months, medium-risk facilities every three years, and low-risk facilities every five to six years.

The full audit team may vary from three to eight people. They may be full-time auditors, subject specialists, representatives from the business unit being audited, representatives from other company plants or qualified external consultants. It is also advisable to include worker's representatives who will require adequate training and information on the auditing process.

Source of this section: K. North, OECD, International Labor Office, Geneva (261).

3.33 EUTROPHICATION

When the fertility balance of a coastal system is upset by excessive input of nutrients, the system suffers from a condition known as *eutrophication,* whereby the growth of aquatic plants accelerates to excessive amounts. Algae and seaweed multiply beyond the limit of the ecosystem to respond, causing serious imbalance or even "crash" of the system. The result may be evident in the putrid condition of the water body—algal scum over the water, dead and dying fish and shellfish, turbid water, absence of dissolved oxygen, dense growth of seaweed.

On the other hand, eutrophication may act in far more subtle ways. Without much visible evidence, it can act to retard coral growth, repulse fishes from the area, and reduce biodiversity. Ecologists contrast the "eutrophic" state (nutrient rich) with "oligotrophic" (nutrient poor) and "mesotrophic" (mid-range) states.

In coastal waters, eutrophication is caused by an excess of chemical nutrients—mainly phosphorus (P) or nitrogen (N) or both—which are dissolved in the water and therefore invisible. In ordinary amounts, P and N provide essential ingredients for plant growth and are a vital part of a balanced ecosystem. Only when they are in excess do they cause problems. In coastal waters, these excessive amounts originate on land in most cases, not in the water. When there is trouble with P or N, the source of excessive nutrient is most often human activity.

In water, the common *dissolved* form of P is phosphate and of N is nitrate (often P is present more in particulate organic form). Both P and N are common in agricultural fertilizers and are often washed off farmland by rain to pollute coastal waters. Livestock wastes can also contribute P and N (dissolution during decomposition). Household products such as dishwashing compound may contain high levels of phosphate. But the main source in most areas is human waste from sewers, septic tanks, or dispersed excreta. Only when these sources supply *excessive* amounts is there a problem; that is, only when the byproducts of human progress overwhelm nature.

The net effect of sewage is to stimulate the population growth of both phytoplankton (plant plankton) and zooplankton (animal plankton). It does so by providing dissolved fertilizing compounds of N and P, which the phytoplankton take up when light and other conditions are favorable. This growth-stimulating process is frequently excessive, resulting in an overabundance of phytoplankton, known as a *plankton bloom.*

When cloudy weather develops or the supply of nutrients becomes depleted, the phytoplankton cannot be sustained and they may suddenly die (along with the zooplankton they have been supporting). The decomposing plankton settle to the bottom, where they deplete the available dissolved oxygen of shallow areas, often resulting in reduction of oxygen concentration below that required by fish and other organisms. Fish kills frequently result. (243)

Sometimes P drives eutrophication and sometimes N does. It depends on which dissolved compound is the scarcest. If the water has ample phosphate but limited nitrate, then nitrate is the driver, the triggering agent. If, on the other hand, phosphate is limited, its addition will trigger eutrophication. The normal ratio of elements for growth of living tissue is P:N:C = 1:16:100. In fresh waters, phosphate is most often the triggering agent. In confined coastal water basins (lagoons, estuaries), nitrogen is most often limiting (59, 130). In high salinity lagoons with minimal terrestrial influence, phosphate appears to be limiting (130). Also, coral reef systems under maritime influence are shown to be phosphate limited (20). Notice also that the relative amounts of P and N may vary with season in temperate regions, with N peaking in winter or autumn and P in summer, as shown in Figure 3.51.

At middle level of supply, P or N can stimulate growth of phytoplankton/zooplankton sufficient to seriously reduce water clarity. The plankton creates turbidity, which screens out sunlight, reducing the amount of light that reaches coral reefs on the seafloor (corals must have light to live; see "Coral Reefs" above). Also increased nutrients can encourage bottom-living algae and certain invertebrates to smother coral reef communities (346). Eutrophication is also responsible for troublesome overgrowth of seagrasses (see "Sea Grasses" below).

Among numerous coastal and marine systems that suffer from eutrophication are the following examples (351): the southern North Sea (high winter nutrient levels); northern Adriatic Sea (foul, slimy material on Italian and Yugoslavian beaches); Inland Sea of Japan (phytoplankton blooms); New York Bight (fish and invertebrate kills at mouth of New York Harbor).

From a management point of view, it is unfortunate that politicians, administrators, and managers—who are absorbed with land-oriented human health matters—usually do not pay attention to eutrophication unless it reaches crisis levels, when it may be too late for any easy solution. The most difficult message to get across is that normal sewage treatment plants do not take out these troublesome fertilizers or

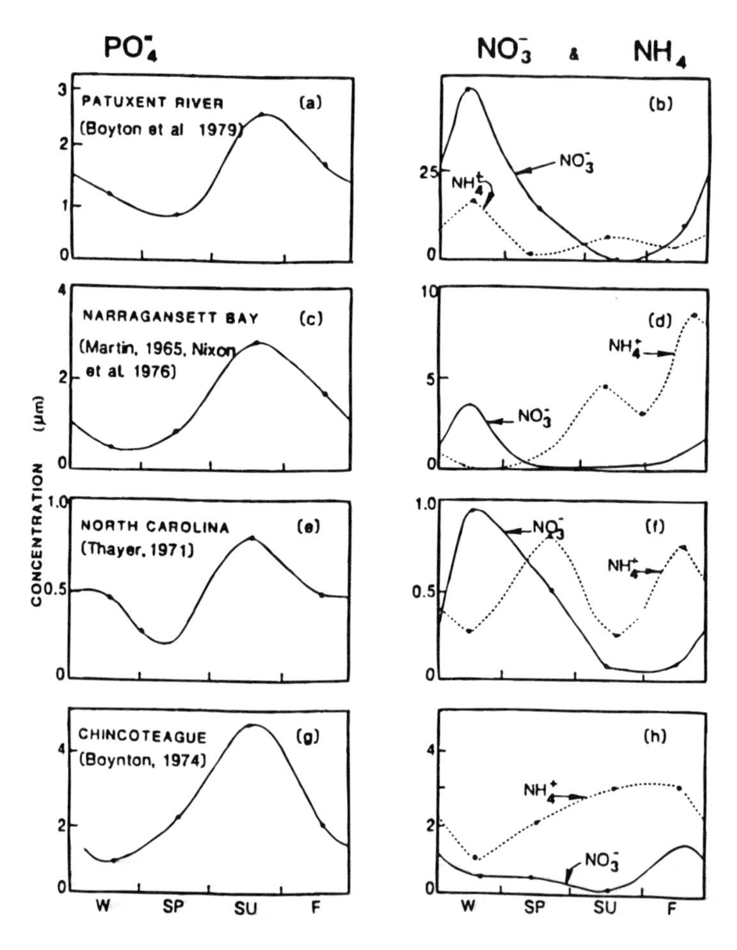

Figure 3.51. Examples of seasonal variations in the amounts of phosphate (PO_4) and nitrate (NO_3) in U.S. coastal waters in Winter (W), Spring (Sp), Summer (Su), and Fall (F). (*Source:* Reference 185.)

else, they take out only a small part. Even when they do function properly—which is the exception for most sewage plants in tropical countries—the removal ratio is slight.

More bad news is the ironic situation that, in fact, "secondary" sewage plants manufacture phosphate and nitrate pollutants. They convert organic matter to the inorganic nutrients phosphate and nitrate. In this way, secondary sewage plants have become *sewage refineries.* They take in organic matter (to reduce its BOD) and refine it to N and P, which are then discharged to coastal waters. Effluents from treatment plants may contain 5 to 10 mg of P per l (371).

Because N and P pollutants can be so damaging when they occur in high concentrations in the plant's effluvium, the method and location of effluent discharge are critical factors. That is why it has become common to build *ocean outfalls* to carry the troublesome effluvium far away from shallow coastal waters. (See "Sewage Management" in Part 2.)

Where P is limiting, a concentration between 0.01 and 0.1 mg/l is "sufficient to promote accelerated eutrophication" (371). Because phosphorus-based household and commercial detergents that are disposed with washwater are also a major source of phosphate in some places, they are now being controlled in some countries to protect the quality of shallow coastal waters. (See "Nutrients Management" in Part 2).

3.34 EXCLUSIVE ECONOMIC ZONE

The Exclusive Economic Zone or "EEZ" is a 200-nautical-mile (320 km) wide marginal zone defined by international treaty whereby the adjacent country has the exclusive privledge for exploitation of marine resources. The EEZ is very important to small countries because it vastly increases their area of economic influence—for example, from 70 to nearly 500,000 square miles in the case of the Marshall Islands in the South Pacific. Countries that cannot themselves exploit the vast EEZ can "lease" it to other nations for fishing or other uses. While such matters are outside the scope of coastal zone managment as it is usually construed, the EEZ could be included in the ICZM planning area.

3.35 EXOTICS

The introduction of exotic (alien) species to an ecosystem is sometimes deliberate and sometimes accidental. In the coastal zone, many exotic plants have survived introduction, including trees like coconut, Australian pine, and Brazilian pepper. Some shellfish have also made the transition, including clams and oysters. The mobile species are most difficult to introduce into coastal/marine waters.

Exotic species may spread to the detriment of endemic species. When not in competition with native species, introduced species are not necessarily harmful. However, there is the danger of importing dangerous pathogens with the target species.

Historically, the object of most introductions was release into natural waters on the chance that results might be beneficial. With this criterion, some of the introductions of exotic marine species were deemed successful, and others were not. Most recorded successes involved moving temperate species of molluscs, crustaceans, and fishes, from one continent to another (Europe, United States, Japan, New Zealand). However, the criterion of success for most of these was simply whether the object species "took hold" in the new environment. The ecological consequences were simply unknown.

In recent decades, considerable initiative by entrepreneurs and governments has resulted in a new world wave of introductions of marine species, this time involving many tropical species. This wave has been met with considerable resistance when the objective was release into natural waters, because of uncertainty as to the full ecological ramifications. However, species intended for containment in "tropical" fish tanks or in aquaculture facilities arouse less concern, even though it must be presumed that the contained species will sooner or later find their way into natural waters. In this way, hundreds of species of fishes, molluscs, and crustaceans have found their way into exotic environments. Some have survived and multiplied, with consequences that are still unknown. It is far more difficult to observe both the spread of exotic aquatic animals and the displacement of other species than to observe terrestrial plants or even exotic aquatic plants.

The concern about disease is that might be uncontrollable in the new environment, where resistance to the disease is absent. This probably has already affected wild communities in unknown ways through escape of disease-carrying individuals or discharge of pathogens to the natural environment. Transfer of species is often a risky business for the operators. For example, in trying to solve shortages of seed stock, Taiwanese aquaculturists import gravid (egg bearing) shrimp from the Philippines and other countries. However, in 1987 and 1988, mass mortality and dwarfism caused a crisis in the industry and many operators had to shut down. These larvae from imported stock are easily infected with viruses common to the Taiwanese hatcheries.

There are other adaptive problems with introductions and transfers that planners should recognize. For example, crayfish imported to the Dominican Republic from Louisiana for aquaculture burrowed through dikes, causing them to collapse and allowing the crayfish to escape into the wild with unknown effects. In an example of a more thoughtful and studied approach, conch from a Caribbean hatchery intended for a research transplant to the Florida Keys were found through pre-planting tests to be genetically different from the local stock and had inferior growth characteristics, requiring that the transplanting be postponed for further study and perhaps terminated.

The planner wonders whether diseases exacerbated in culture facilities could pass into the natural environment and have a major impact on wild populations. For example, the oyster industry of Europe was crippled for ten years by disease and predator problems, possibly associated with transfers of exotic stock.

In light of the uncertainty of success with introduced species and the possibilities of genetic problems and ecological disruption, the planner might also wonder why commercial aquaculturists want to use exotic, rather than local, species. The answer is that local markets are not usually lucrative and, therefore, the international marketplace is the target of the most ambitious entrepreneurs. This market is organized to favor a certain few species that attract premium prices. In addition, certain species have the reputation of being fast growing or having other valued attributes. As an example, the black tiger shrimp (*Penaeus monodon*) fits the profile of ease of culture and premium price (and is also the one that caused the trouble in Taiwan).

The planner's concern is to find a method to evaluate proposals for aquaculture expansions and new starts so as to provide an enlightened framework for decision making. For species introduction, one method that can be recommended is the "ICES code," which has been developed by international cooperation over the past 15 years (ICES is the International Council for the Exploration of the Sea). This code is actually a comprehensive system for assessing the probable consequences of species introductions and transfers and can be utilized successfully for evaluating most proposals involving the use of non-local stocks in aquaculture.

For information, contact the following: ICES Secretariat, Palaegade 2–4, DK-1261 Copenhagen K, Denmark.

3.36 EXPLOSIVES

Explosives are commonly used in coastal construction to remove rock. They may also be used in sea quarrying, seismic survey, and "dynamite fishing." The effects vary according to the charge and the local geophysical conditions.

Underwater explosions introduce noise, localized turbidity, and some chemicals (nutrients or heavy metals) into the water column. It is the shock wave and concussion effects that are the most significant, according to Maragos. Impacts are proportionately greater the closer to the detonation sites, particularly for open water explosions. Within the immediate vicinity of the explosions, substratum material is pulverized and a crater is formed. (231)

At intermediate distances the substratum can be shattered or dislodged. At greater distances, substrate or corals projecting above the bottom are fractured or broken, and at the greatest distances (up to 100 m away from large explosions involving 200 kg of explosives or more), coral heads and rocks may be sheared or dislodged from the substrate. There is some evidence to suggest that operations involving staggered detonation times or explosives with slower detonation speeds cause less impact. (231)

In one test in Hawaii, a series of explosive charges was detonated to excavate completely a shallow draft harbor basin and channel without mechanical dredging. The very large explosions caused massive fish kills and damage to coral reefs. The use of this technique would be unacceptable except in an emergency situation. Nevertheless, explosives are commonly used in conjunction with dredging to shatter, loosen, or fracture hard consolidated rock and facilitate its mechanical removal. Explosives are either laid on the bottom, placed in holes or crevices, or are loaded and tamped into pre-drilled holes (drilling and shooting). The quantity of explosive energy needed is less if placed in holes because more of the explosive energy is transferred to the rock strata to break it. (231)

Explosive energy transferred to the water column can cause considerable ecological damage from shock wave and concussion effects. "Drilling and shooting" is now a standardized technique for quarrying armor stone from hard reef flats in Oceania. The drill holes are carefully sized and placed and the charges set so that fracturing maximizes the production of a desired size of quarry stone. (231)

Explosives used by fishermen to kill or stun fish for ease of collection are discussed in "Dynamite Fishing" above. Most fish killed or stunned by explosives do not float to the surface. Organisms above or to the side of explosions are more likely to be injured or killed compared to organisms deeper than the explosion; hence, deeper water explosions tend to generate higher mortality and casualties. Explosives packed into holes or crevices, or covered with sandbags, etc., will cause less damage from shock and concussion. Fish and mammals with air cavities (bladders, lungs, etc.) are more prone to injury. (231)

It may seem strange that the use of explosives in drilling and shooting operations on coral reefs has so far caused rather few recorded impacts on reef environments. But smaller charges are used and the reef is buffered from shock and concussion because the charge is tamped into pre-drilled holes for detonation.

3.37 FISHERIES

Fisheries play a key role in many, many countries. Nevertheless, decision makers often underestimate the economic role of fisheries. With close to 80 million tons of fish per year harvested worldwide, the sea provides more protein worldwide (16.1 lb) than do beef and mutton combined (12.2 lb). The annual harvest also exceeds world beef production by a substantial margin. (30)

Fisheries supply 23 percent of all animal protein consumed worldwide. In many countries, fish are the principal source of animal protein. Small-scale fisheries are virtually the largest supplier of animal protein (+75%) to several hundred millions of people in Senegal, Tanzania, Ghana, The Gambia, Nigeria, Cape Verde, Sierra Leone, Togo, Indonesia, Burma, Bangladesh, India, Sri Lanka, Guyana, Trinidad, and Yemen (30). Yet economic planners take slight notice, perhaps because seafood production is a mysterious pursuit carried out far from cities and incomprehensible to many persons. Also individual fisheries are variable and largely unpredictable and unprogrammable.

Fisheries also provide livelihood for millions of fishermen and their families and for others in the fishing industry, including boat builders, trap and net makers, packers, distributors, and retailers—all of which enhances social, cultural, economic, and political stability in the coastal areas. A strong domestic fishery promotes self-sufficiency and reduces the outgo of foreign exchange. Also, profitable coastal fisheries reduce rural population migrations to already overcrowded cities.

For all these reasons helping to maintain fisheries is an important function of coastal zone management. It is essential that water quality and critical habitats be protected and that convenient boat landing and fish processing facilities are available to fishermen. In a time of rapid population increase, it is necessary to conserve the productivity of coastal habitats for fisheries, the same as we protect rangelands for livestock, farmlands for crops, and forest lands for wood.

More than 99 percent of the world catch of marine species is now taken within 320 km of land (within the EEZ), and more than half of the total biological production of the ocean takes place in that zone. The presence of the continental shelf directly leads to high production because it concentrates activity into a thin layer of the sea and provides a substrate (solid surface) for fixed plants and benthic animals. In addition, the topography of the shelf stimulates the upwelling of deeper waters carrying chemical nutrients to the surface (308).

Unfortunately, seafood's contribution to national diets and income is diminishing, partly because fisheries are not usually managed for sustainability. Past and present overfishing has led more than 25 of the world's most valuable fisheries into serious depletion. As an example, in the Gulf of Thailand in the early 1960s, only a few hundred trawlers could be counted and the catch rate was about 300 kg/

hour. Two decades later, as the number of far more efficient trawlers had increased dramatically, the catch rate had declined to 50 kg/hour. (279)

In addition to depleting fish, crustacean, and mollusc stocks, overfishing has greatly depleted some species of whales, sea cows, and sea turtles. Many species are under pressure because of incidental exploitation; that is, they are captured along with "target" species, killed, and discarded (Figure 3.52). An example of this wasteful practice is the incidental capture, killing, and discard of sea turtles in fishing nets, which threatens several species. As the commercially valuable fisheries for fish, crustaceans, and molluscs become more fully exploited, the effects of habitat destruction and pollution will become more evident, particularly on those species depending on coastal wetlands and shallows or on inland wetlands and floodplains for nutrients or for spawning grounds and nurturing areas.

The outlook is not bright. As human populations swell and as aspirations for economic growth remain strong, pressures to overexploit coastal resources intensify. Too many people are placing unrealistic demands on these resources, and the results are readily evident—deteriorating water quality, coastal erosion, loss of mangroves, decreasing biological diversity, destruction of coral reefs, water polluted by untreated sewage and, occasionally, disastrous oil spills. All of these problems cause a decline in the productivity of coastal resources.

Of the greatest economic importance is maintaining bioproductivity for fisheries—an obvious example of an ecological process directly supporting people. Naturally productive ecosystems, such as estuaries, provide free of cost what expensive aquaculture can barely match—continued fish production. Continued fish production means continued livelihood for fishermen and for others in the fishing industry, including boat builders, trap and net makers, processors, distributors, and retailers (Figure 3.53). Finally, continued livelihood means continued social, cultural, economic, and political stability (308).

Fishermen are the real day-to-day managers of the fishery resource. Any program that seeks to improve and regulate the fishery sector can only do so effectively by working through the fishermen/managers. In an overcrowded, open-access artisanal fishery, fishermen as individuals feel a sense of helplessness to initiate changes. Assisting them to organize and to implement fishery management activity can change the situation. This approach is called *community-based management.*

Figure 3.52. Fish traps are indiscriminate—their catches often include excessive amounts of undersized and unwanted fish. (Photo by Bill Keogh.)

Figure 3.53. Artisanal fisheries provide onshore employment for numerous family and community members (Thailand). (Photo by author.)

Success in addressing one key management issue spurs a commitment to maintain what has been gained and a willingness to solve other problems, such as halting destructive methods—explosives, muro-ami or fine mesh nets—and creates a less disturbed, less heavily fished environment, which encourages many fish species to return to nearshore waters.

In this book, the focus is on habitat management—protection of mangroves, coral reefs, etc, to maintain the resource potential of fisheries at its highest level. Harvest control to prevent over-exploitation of fish stocks is not a subject of this book. Coastal management, as it is addressed here, is about *development* planning and management, not *resource* management per se. Harvest management (controls on catch, seasons, areas, gear types, etc.) is not typically a subject for development management. But, in some smaller countries with consolidation of agency responsibilities, the two subjects might be combined. The typical ICZM role is to protect habitats and environmental quality along with managing development. The coastal management program should usually be viewed as working in cooperation with the fisheries authority.

M. Gawel (137) takes a different view than this book. He believes that fisheries harvest management should be closely integrated into ICZM programs for some countries. He argues that if the ICZM program is evaluating impacts of habitat change but not over-fishing, and if each of these actions is to be managed separately, this would not be practical for Micronesia. Gawel emphasizes that, in fact, the fisheries agencies of the Pacific States are usually responsible for environmental assessment of development and similar activities.

In any event, there will naturally be a much closer connection between coastal management and the small-scale, artisanal fisheries than with the more industrialized offshore component. This is because the artisanal fisheries are conducted in near-coastal and lagoonal/estuarine waters where coastal management is most relevant. Also they tend to live along the coast where habitat protection is taking place and are the subject of modern participatory management.

The fisheries of developing countries are predominantly small-scale and artisanal in form. They are characteristically labor-intensive operations conducted by craftsmen with low income and production-quality levels, poor mechanical sophistication, limited fishing range, insignificant political influence, constrained market outlets, and restricted social mobility and employment opportunities. This type of fishery (Figure 3.54) contains millions of full-time professional fishermen, grouped in villages along the coasts of developing countries.

There are large variations in the manner in which fishery activities are performed, depending on the gear used in fishing. While fish resources are owned as common property and may give the impression of egalitarian access, the means of production, in general, are privately owned and their distribution is rather askew. It has been estimated that in Asia only one third of all fishermen own a boat, two thirds own some type of gear, and one third own neither a boat nor gear. Marketing of fish is tied to money lending. Most marketing activities are financed by noninstitutional sources of credit, which provide the cash necessary for handling the catch as well as for family needs, such as food, clothing, and medical assistance.

Fishing families have incomes below those of many other groups engaged in rural sector activities and most economic groups engaged in urban sector activities. This is not true for all, but certainly for the large majority of those families involved in small-scale fishing. Even in Malaysia, which has one of the highest incomes per capita among the developing countries in its region, about 75 percent of the total number of fishermen's households live below the "poverty line" (115).

Most of the technology in harvesting is labor-intensive. Although there are fisheries that have motorized their boats, very few have mechanized the process utilizing the gear. The technology used in handling and processing is very primitive, which results in significant harvesting losses. Very few small-scale fisheries use ice or have the required type of shore storage facilities (i.e., cold storage, ice plants). Although there are large numbers of small-scale fisheries operating at a subsistence level, there are also significant numbers of small-scale fisheries which are commercially oriented.

"Sustainable use" has been at the foundation of modern fisheries management for many decades. It has been implemented with explicit goals such as Maximum Sustainable Yield (MSY), Maximum Economic Yield (MEY), and Optimum Sustainable Yield (OSY) as concerns for sustainability spread to cover progressively the physical resource, its value, and all other social benefits from fisheries. This concept implies an assumption about reversibility. Resources will unavoidably be affected by exploitation in terms of abundance, spawning biomass, species composition etc. The need to conserve options for

Figure 3.54. Fishing is the economic mainstay of the residents of Inhaca Island, Mozambique. (Photo by author.)

future generations implies that modifications are reversible within some useful time-frame, perhaps 10 years.

According to Pauly (279), over-fishing may be viewed as having five facets, of which three are recognized forms of over-fishing. The first well-known form is *growth over-fishing,* which is what happens when fish are caught before they have time to grow. After World War II, methods by which growth over-fishing could be diagnosed in practice and remedied by fishery management (for example, through the imposition of appropriate mesh sizes for fishing gears) were created. Research work related to over-fishing consists of estimating the ages and the growth and mortality rates of fish and assessing the (mesh) selection characteristics of fishing gears. (279)

A second recognized form of over-fishing is *recruitment over-fishing,* which refers to fishery-induced reductions of the number of young fish entering the fishing grounds. Recruitment over-fishing can be brought about by (a) reduction of the spawning stock (which may become so small such as to produce a limited number of eggs and hence of recruits) and (b) coastal environmental degradation, which usually affects the size or suitability of nursery areas. Preventing recruitment over-fishing is not a matter of letting each female spawn at least once (since less than one in a thousand anchovy or shrimp larvae grow up) but rather a question of not fishing too much.

Surplus production models are used to manage tropical fisheries. These models do not distinguish between growth and recruitment over-fishing but rather lump the two processes into a single category of general biological over-fishing. Here, *economic over-fishing* is shown to be a fishery at a level of effort higher than that which maximizes the economic rent, i.e., the differences between gross returns and fishing costs. Note that this optimum level is less than maximum sustainable yield (MSY) and that, therefore, maximum economic yield (MEY) is less than MSY (279). Many models have been created to express what *too much* is (Figure 3.55) in the generic sense of biologic overfishing.

Pauly (279) introduced the concept of *eco-system over-fishing* to characterize the process that takes place where fishing is to intense that it alters the balance of species on the fishing grounds, with some species increasing, but failing to replace the depleted ones. This process implies that a part of the system's ecological production now goes into side branches of the food webs (i.e., into nonresource species). The suggested remedies usually involve a mix of management measures (e.g., mesh size regulations, closed areas or seasons, limits on gear sizes or on craft designs, etc.).

By all scientific standards, there are too many small-scale fishermen in Southeast Asia. And because their numbers are rapidly growing, Pauly calls their form of over-fishing *Malthusian over-fishing,* after the Reverend I. R. Malthus (1766–1834), the famous prophet of doom. It is what occurs when poor fishermen, faced with declining catches and lacking any other alternative, initiate wholesale resource destruction in their effort to maintain their output (279).

This may involve (in order of severity and rough time sequence) (a) use of gears and of mesh sizes not sanctioned by government; (b) use of gears not sanctioned within the fisherfolk communities and/or catching of fish "reserved" for a certain segment of the community; (c) use of gears that destroy the resource base; and (d) use of "gears" such as dynamite or sodium cyanide that destroy habitat and endanger the fisherfolks themselves (see "Dynamite Fishing" above).

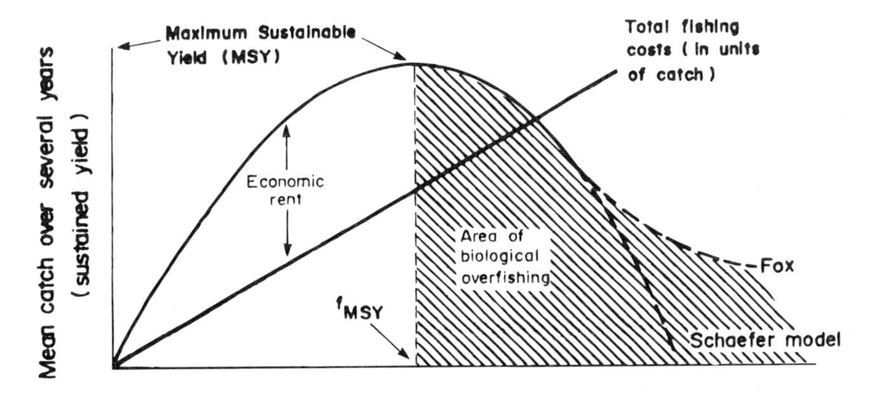

Figure 3.55. The standard fisheries yield model. (*Source:* Reference 279.)

There being quite simply too many fishermen chasing too few fish this situation can corrected only by altering the fishermen/fish ratio. The solution is clearly alternative employment opportunities, in villages and towns and cities, for young and not so young, for literates and illiterates, and stabilization of populations.

The introduction of trawling in ASEAN waters in the late 1960s is another case of resources use conflict. As noted by Chua (51), initial economic benefits led to a proliferation of variously sized trawlers and uncontrollable operation in inshore waters. But, in Malaysia and Indonesia, many lives were lost due to social antagonism between inshore fishermen and trawler operators. Many inshore fisheries were depleted rapidly. Well-known fishing grounds were rendered "over-fished." The Straits of Malacca, Sulu Seas, Gulf of Thailand, Java Sea, and many others have their demersal fish stocks depleted (51).

The situation has become so serious that the Indonesian government found it socially, politically, and economically beneficial in the long run to completely ban trawling in Indonesian waters in 1983. However, with sound planning and effective controls, the above loss and depletion could have been avoided (51).

3.38 GENETIC DIVERSITY

Wild genetic resources are lost through the extinction of a species or through the extinction of individual populations of that species (genetic impoverishment). The first process is final and irreversible. The second is a matter of degree and is to some extent reversible. In the sea, where endemism is low, the problem is less one of species extinction than of genetic impoverishment. No significant or detectable increase in extinction rates of fish species has been observed in the ocean, but populations have been extinguished by over-fishing, pollution, and habitat destruction. Organisms occupying disappearing habitats will probably never again reach their present levels of genetic diversity (308).

It is necessary to recognize the difference between biological diversity, reflected by the number of species, and genetic diversity, reflected by the variation within a species. The land has more species than the sea and hence greater biological diversity. Marine organisms, however, tend to exhibit more genetic variability; thus, they have greater genetic diversity. Both types of diversity are important and of benefit to people.

Genetic material determines how much species can adapt to changes in their environment. Individuals having the most genetic variation (and hence greater tolerance of environmental changes) have been shown to have better survival rates or higher relative growth rates. Variety of form or appearance within a species are expressions of different genotypes (i.e., combinations of genes). Alternative forms of a particular gene (called *alleles*) cause variation among individual organisms, for example, in the background color of shell. The number and relative abundance of alternative forms of a gene in a population is a measure of genetic variation (called *heterozygosity*). The total amount of genetic variation in all populations of a species is a measure of its genetic diversity (308).

Human activities diminish genetic variation and encourage the extinction of species in a number of ways (308):

- Pollution and other environmental changes that stress a population, causing differential mortality, extinction, or both
- Fishing pressure that favors one genotype over another
- Artificial selection and domestication, which can result in conscious or unconscious inbreeding and genetic impoverishment
- Introduction of exotic species and diseases

Broadly speaking, there are three ways to preserve marine genetic resources against such human-caused losses. One is establishing gene banks, "storehouses" that maintain genes for future use (more widely used for terrestrial genetic resources). A second means is preventing the overexploitation of species by managing the harvest, or by supplementing the harvest of wild stocks with cultivated products, or by prohibiting the harvest and trade of depleted and endangered species. A third means is creating protected areas for habitats, since a major threat to the survival of some populations of species is the destruction of critical elements of their habitat. Such coastal and marine protected areas function as in situ gene banks, preserving genetic material within an ecosystem rather than a special storehouse and are promoted by biodiversity conservation programs (see "Biodiversity" above) (308).

Coastal and marine protected areas can help maintain in situ gene banks in a number of ways. They protect rare, threatened, and endangered species and populations or species known or likely to be of value as genetic resources (e.g., the wild relatives of farmed species or other wild species useful to people).

Local extinctions and depletions of stocks have resulted in part from habitat destruction and in part from the high market demand for such species; for example, whales, turtles, dugongs, and certain molluscs and corals. (See "Protected Natural Areas" in Part 2.)

3.39 GEOGRAPHIC INFORMATION SYSTEMS (GIS)

The advent of computers has changed the way society deals with information. The rapid advances in computer technology have reduced the cost of microcomputer systems and made them readily available, according to K. S. Pheng and W. P. Kam, authors of this subsection. The recent development of a class of computer-assisted mapping tools called *geographic information systems* (GIS) has provided the means for dealing with spatial information which are inherent in most resource data. Previously requiring large computers, GIS are now available on microcomputers and work stations. This increasingly puts them within the budget of more institutions which deal with resource data.

In natural resource studies, much of the data is spatial in nature; for example, the distribution of fisheries resources in coastal waters. Decisions in resource use often depend on the spatial distribution of the resource in relation to other factors such as transportation. It is often helpful to examine the spatial relationships among resources, human settlements, sources of employment, and available transportation network than merely producing statistical tables or graphs on them.

GIS, computer-assisted mapping tools, are used in collecting, storing, retrieving, processing, and displaying spatial data. Organized to process spatial data from the real world, GIS provide the tools for a wide range of applications in virtually every discipline that deals with spatial information.

All geographic data can be reduced to three basic entities: (a) points (e.g., the locations of settlements or of water quality monitoring stations); (b) lines (e.g., rivers and road networks); and (c) areas (e.g., districts of the coastal zone as defined for a costal resource study, the boundaries of which are invented, or areas bounded by natural boundaries like river basins and bedrock types).

Most resource data with spatial information are readily inputted as points, lines, and areas with attributes tagged onto these entities. The analyses for GIS are transformations of geographical data and attributes in the form of a map or referenced to a map. For example, for an overlay of two map layers showing the extent of mangrove forests of two different years within the same study area, GIS should be able to show exactly where mangrove areas have increased (e.g., where the shoreline has accreted) or have shrunk (e.g., along an eroded coast or inland where agriculture has encroached).

GIS provide a convenient tool for resource assessment, planning, and management because they carry out analytical functions, are integrative and spatial, and can be updated. GIS are integrative because they can take data of different formats from different sources. These data are converted into a consistent internal format and scale within the GIS. The various map layers for a particular area are geometrically registered with one another and with a base map.

GIS also facilitate the preparation of map "overlays." For example, a map layer showing the delineated coastal zone could be overlaid on a soils map, covering a bigger general area, to mask out areas outside the zone to display only the soils within it.

Calculation of area is another feature of most GIS. A two-map cross tabulation can calculate the areal extents of various soil types under various forms of cultivation (e.g., rubber, oil palm, etc.), an operation that would otherwise require painstaking manual overlay and planimetric procedures.

GIS can also be used to determine areas that are suitable for a specific use. For example, a number of criteria (soil and water properties) are used by the aquaculture expert to determine suitable sites. These are mapped in seperate thematic layers. Overlays of these layers are carried out, depending on the criteria given by the specialist. The "sieving" techniques used to produce suitable aquaculture sites may range form indexing overlays, whereby individual map layers and class values are ranked either as absolute conditions or of relative weight, to more complex overlays involving Boolean logic or mathematical combinations of the input layers.

GIS can take into consideration all these factors and help the resource planner make trade-off decisions in resource allocation. The process can be repeated several times, each time changing the various assumptions and criteria. Different scenarios can then be generated which will assist the resource planner and policy and decision makers to pick on the most suitable one for a particular situation. The database that is already in the GIS can be used to determine the environmental impacts of a particular decision.

The role of the GIS as a tool in resource studies is continous. GIS are open ended and can easily receive new data or change old data. Therefore, the GIS data bank can easily be updated.

The following are six potential applications of GIS in integrated coastal zone management (ICZM):

1. **Cartography:** a study by the United States Geological Survey to produce topographic maps of the nation for use of federal and state agencies, commercial companies, and individual citizens at 1:24,000 scale series.
2. **Land management:** profiles developed for each drainage basis based on GIS inventories of the extent of forest lands, cultivated land, urban areas, stream shore, lake shore, silt soil, sand soil areas of 3–6 percent, slope and other parameters for water resource assessment.
3. **Freshwater habitat management:** a case study on impact assessment of contaminants. Creates databases for habitat potential, attribute file of habitat condition and stream dimensions, watershed boundaries, point file of contaminant discharge. Describes downstream impact in terms of proportion of fish production loss. Analyzes habitat affected by contaminants and converts habitat areas into fish production.
4. **Marine habitat management:** creates database for various attributes, point data, bathymetry, sediment type. Establishes criteria for suitable habitat model by describing relationship between spatial variables. Overlays maps to produce the desired output.
5. **Potential for aquaculture development:** data sets used are salinity requirements, soil characteristics, rainfall pattern, land use (mangrove vs. non-mangrove area) for determining potential area for shrimp farming. Data sets used are environmental parameters, infrastructure, land use, soil types, hydrographic factors, coastal geomorphology, and meteorological characteristics for determining potential area for aquaculture development. Also used are water quality, existing land use patterns, distance from water source, geomorphological features, and distance from existing aquaculture farms for determining potential area for shrimp and fish hatcheries.
6. **Coastal resources study:** Identifies socio-economic variables that might influence developments in a coastal environment. Data sets used are population, employment statistics, income levels, educational background, infrastructure, and public amenities.

The selection of a GIS system should basically be needs-driven. Most of the needs for ICZM have much in common with those for other resource management. A few requirements may be peculiar to the nature of the coastal zone, as described above. Three main criteria that might be considered in evaluating the suitability of a particular GIS for ICZM are:

- The extent to which the intrinsic data model and data structure used by the system are amenable to the wide-ranging spatial scales and resolutions required of coastal zone studies
- The efficiency and effectiveness with which spatial and nonspatial data are captured, edited, stored, retrieved, and displayed
- The kinds and capabilities of spatial analysis tools for modeling the coastal zone properties and processes

Some basic GIS functions for ICZM support, comprising generic GIS functions and database compilation needs, are as follows (from G. Shultink in Reference 86):

- Digitizing and editing of point, line and area data
- Support of UTM, LAT/LON, and other major geographic referencing systems reflected in current and future map data sources

- Relational data base interface with all spatial data sets permitting selective grouping, cross-tabulation, scaling of, and extrapolation of attribute characteristics
- Data capture, editing, display an annotation in raster, and quadtree or polygon format
- Graphic map display, scaling, updating, annotation, and geographic transformation to major geo-referencing systems
- Raster image analysis of satellite digital data and supervised classification using selective control points and three-dimensional geometric correction (rubber sheeting)
- Weighted map overlay, boolean logic, and statistical analysis capabilities
- Three-dimensional terrain analysis with optional point and contour elevation inputs, including slope characteristics calculations such as gradient, length, aspect, and spatial filtering
- ASCII file input and output capabilities
- Area and tabular summary calculations with graphic displays (e.g., histograms and piecharts)
- Chloropleth and color mapping with geographic references, legend, and scaling operations
- Lightweight transportable Global Position System units to verify topographic maps position during field surveys and satellite ground truth verification

The volume and storage capacity of GIS may also be an important consideration for a major ICZM project, but this is largely determined by hardware configuration. It is also essential, a fact which can be time consuming in itself, that the factor analysis and weighting decision rules or algorithms are very carefully defined and developed (86).

It may be good to look at a few practical and philosophical problems in designing and setting up GIS in a project or agency doing coastal planning, such as the five listed below:

1. It is necessary to identify the real needs of the actual user group before designing a system. Such an information system has to change and expand over time. Nor can GIS meet all the needs of users/planners/managers, so it is necessary to set priorities for use.
2. GIS are not meant to replace other data sources but to supplement them. GIS assist in decision making and are not an archive of data; they need a specific purpose. A key feature is up-to-date data that are easy to manipulate and use for the intended goal of coastal area management or objectives therein.
3. GIS not only serve but also shape the planner. This means that unless the GIS user is focused on a useful output, he/she may end up making the tool (GIS) the final output. This is often the pitfall of sophisticated computer systems when operated by people nor fully in command of them. This implies that adequate training of users is a prerequisite to successful GIS application.
4. GIS are not a panacea for coastal area management and planning. They are tools and new technologies to be matched by thoughtfulness and caution about the scope of use, scale of design, and ultimate outputs.
5. GIS's usefulness depends on actual use (i.e., they must be tried and experimented with and continually used so that outputs can be critically appraised and applied).

Note that time dependence, flows, and uncertainty are not effectively represented in maps and that subjectivity is present in producing many types of maps (e.g., soils and vegetation). Resource management requires an understanding of how ecosystems at one scale are related to ecosystems at large and small scales which include the human dimension. This problem, however, is not easily solved through GIS application. (282)

Similarly, suitability maps can be prepared for specific uses of the resource, for instance, the suitability of mangroves, by virtue of its biological and other characteristics, for timber extraction on a sustainable basis: and for strict conservation and other compatible uses like nature recreation and wildlife snatuaries. Each resource sector (fisheries, forestry, agriculture, and resource-based tourism and recreation) may be identified separately as a suitable area for its own use. All of these assessments need to be studied together and incorporated so that coastal resources can be put to optimal use with the minimum of conflicts in terms of impacts on the resource and on people. (282)

Van Classen (88) warns that the organization of a GIS for environmental assessment (EA) is recommended only when all of the logical relationships, attributes, and map bases have been previously developed and are working, and data entry is a routine matter, Computerized GIS are warranted where long-term monitoring of the effects of a project on the environment of a region is contemplated, where data will add to the ongoing maintenance and use of an existing regional database, or where it will be used as an ongoing resource management or regional planning tool. (88)

Many of the real-world complexities of the coastal zone cannot be adequately captured and represented, spatially and temporally, in GIS. Future advances may improve its capabilities. However, current GIS, while employing simplifications of the coastal zone models for spatial display and analysis, are still a useful tool for ICZM planning (283).

The basic components of a typical microcomputer-based GIS system (Figure 3.56) include a computer or central processing unit (CPU) linked to a disk drive storage unit (for storing programs and data) and to one or two video display units or monitors (a monochrome and a color monitor in the case of a two-monitor system). A digitizer is used to convert data from maps into digital form and sends them to the computer. Other optional peripheral devices can be added to increase data storage capacity like high-capacity hard disks or tape-drive units and special hardware which scan aerial photos directly or displays maps quickly. (282)

The basic hardware configuration, shown in Figure 3.57, meets most needs and the cost is not prohibitive. Users increasingly find it useful to carry out spatial and logistical analyses on mapped data which had previously been extremely difficult to do by hand. The flexibility of GIS in a wide range of applications lies in the generic ways by which they deal with spatial or geographical data. These data describe objects from the real world in terms of:

- Their position in relation to a known coordinate system
- Their attributes, which refer to the characteristics of the object (vegetation type, population size, etc.)
- Their topopogical relations (i.e., the position of one object in relation to another)

For further information on the subject see "Database" above and "Mapping" in Part 2.

Source of this section: This section was mostly adapted from GIS articles appearing in the newsletter "Tropical Coastal Area Management," Vol. 4, No. 2, 1989 (ICLARM, Manila) by Kam Suan Pheng and Wong Poh Kam (see Reference 282).

3.40 GLOBAL WARMING

Global warming due to the "greenhouse effect" may significantly change the future of the seacoast, according to many experts. But this is not a new discovery. According to F. U. Mahtab (229), there have been warnings of danger since 1827, when the French mathematician Baron Joseph Fourier described how carbon dioxide warmed the atmosphere and warned that humans could affect the world climate. At the end of the last century, Svante Arrhenius, the great Swedish scientist, coined the phrase "the greenhouse effect." At about the same time, Jules Verne wrote of the consequences of rising sea levels (see "Sea Level Rise" below).

The greenhouse effect describes the outcome of loading Earth's atmosphere with various gases—mainly carbon dioxide—produced by combustion in factories, homes, automobiles, and electric power plants, as well as by natural means. Carbon dioxide remains in the atmosphere for a long time. Gases in Earth's atmospheric blanket act to hold in heat that would otherwise escape to space—like the glass roof of a greenhouse. They let the sun's rays through to the Earth but trap some heat from being radiated back into space. When we add more gas to the atmosphere—as we do when we burn coal, oil, forests, or anything else—we increase the potential temperature of the planet. Other important greenhouse gases are methane, CFCs, nitrous oxide, sulfates, and water vapor.

If there were no CO_2 or other gasses in the atmosphere, the Earth's temperature would, on average, be a bleak 18°C below zero. Natural levels of the gas keep it at a comfortable 15°C and make most life possible. We emit about 5.4 billion tons of it every year from burning oil, gas, and coal, and this increases by about 100 million tons a year. (229)

Each molecule of the CFC gases used in aerosol sprays, fast food cartons, and fridges traps 10,000 times more of the Earth's heat than does a molecule of carbon dioxide! Methane (brewed in garbage dumps, paddy fields, and cow's stomachs), which is steadily increasing, now makes up about 20 percent of greenhouse gasses and is said to last longer in the astmosphere than CO_2. Together, these various

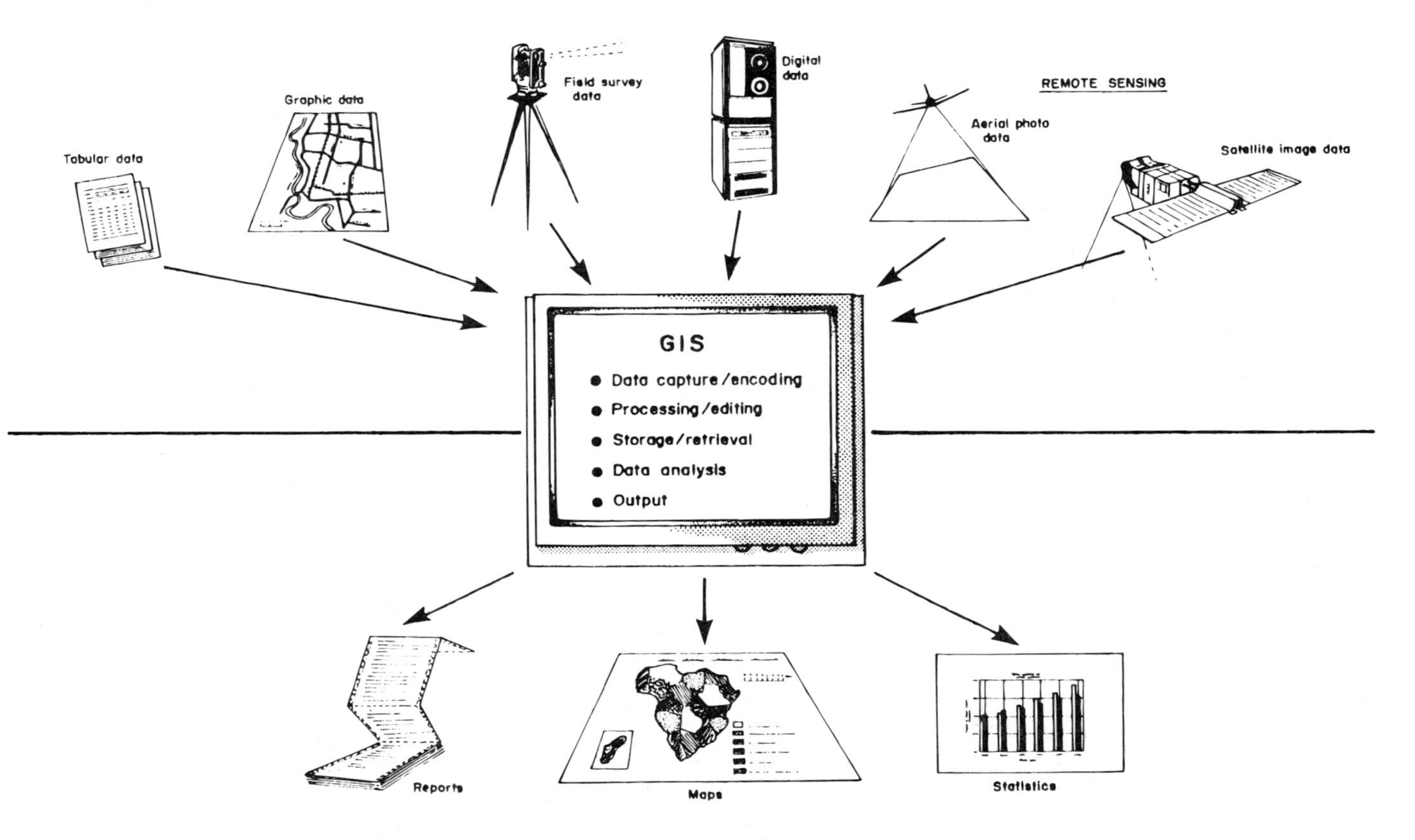

Figure 3.56. The basic components of a typical microcomputer-based GIS system. (*Source:* Reference 282.)

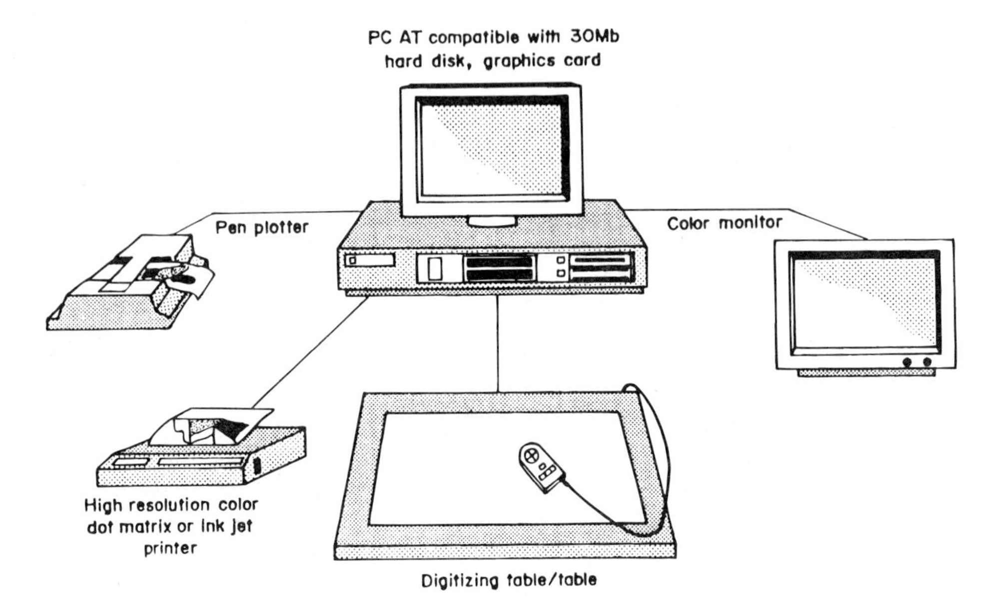

Figure 3.57. The basic hardware configuration for an inexpensive microcomputer-based GIS system. (*Source:* Reference 282.)

gases double the effect of carbon dioxide and have already warmed the Earth by about half a degree centigrade over the last 100 years. Another 1.5 to 4.5°C will be added by the 2030s. At the high end of estimates, Earth's air temperature would increase +4°C (7°F) by the year 2050 and could be the highest rate of global warming ever experienced in Earth history. This extra heat would be enough to thoroughly unbalance our climate. (229)

Cutting down trees, as in clear cutting the tropical rainforests, makes things worse. Living trees "mop up" carbon dioxide; when they are cut down and burned they release it—contributing another 1.5 billion tons a year (229). Vegetation removes the gas responsible for at least half the greenhouse effect—carbon dioxide—and converts it to plant tissue (a good reason to protect rain forests). Also, the planktonic flora in a healthy ocean absorbs enormous amounts of the troublesome gas and helps prevent accelerated global warming.

It appears from recent findings that the greenhouse effect to date has been to elevate nighttime lows rather than daytime highs. Also, there is evidence that, in the northern hemisphere, the warming may be happening mainly in the winter and spring and somewhat in the fall. The atmospheric phenomenon thought to account for the warming at night—increased cloud cover—is keeping daytime temperatures lower (but the extra cloudiness might be caused by the warming itself). Some experts think that increased cloudiness results less from warming, and more from particles of air pollution in the atmosphere—mostly sulfates from the burning of fossil fuels. These have a cooling effect both by reflecting sunlight back into space before it can reach the ground and by acting as surfaces upon which water vapor condenses to form clouds.

Global warming may bring more hurricanes (cyclones) each year to coasts like the Atlantic coast of the United States. It will also increase beach erosion and flooding of coastal communities by elevating sea level (seawater expansion and melting of the south polar icecap). Sea level has risen by 6 inches on the USA Atlantic coast in the last half century (see "Sea Level Rise" in Part 2).

Sea life will be affected by the greenhouse effect in the following ways, according to Bigford (24):

1. Rainfall patterns will shift and freshwater flows alter, leading to shifts in environmental conditions, particularly of lagoons, estuaries, and bays.
2. New hydrological regimes could force an ecological transition with more impact than human development, pollution, and resource exploitation.

Thus, mega-changes are predicted for coastal ecosystems with extremely complex rebalancing of physical, chemical, and bological parameters. Some changes will benefit fisheries and sea life in general, others will jeopardize them, and most will induce some amount of stress, temporary or permanent (24).

Ray (294) believes that, while the most dramatic effects may be seen in polar seas, even tropical coral reefs will experience major impacts. In fact, the world-record hot year of 1987 produced the first serious outbreak of "coral bleaching," a stress reaction in which various corals turn pale and go into a prolonged state of shock when the water temperature exceeds 86°F (31°C).

The spectre of an ocean too warm for coral reefs is disturbing because so many tropical countries depend heavily on their reefs for food, jobs, and hard currency earnings from tourism. Equally serious are the prospects that sea level may increase too fast for the reef-building coral organisms to adjust.

Rising sea temperatures (as high as 2.5°C for some parts of the sea) will convert some temperate water bodies (like the northern Gulf of Mexico, USA) into tropical seas and shift most species northward. This will include fishes and other aquatic species like seals, alligators, and pelicans.

To understand the future of global climate change, scientists have been looking at past events, like the El Nino of 1982–83 during which a large area of the Pacific ocean surface was heated by 7°F, an extreme change in ecological terms. (El Ninos are rapid ocean warmings along the Pacific Equator caused by global climate shifts.) The change translated to a 2–3° increase of water temperature off the coasts of California and Oregon (USA)—enough to have had a major impact on fish. According to Fluharty (129), "The salmons were more dispersed along the coast; their growth was slower and, judging from the lower rate of return to the streams for spawning, their survival was poorer." Fluharty also noted the unusual occurrence of squids, leatherback turtles, and brown pelicans off the north Pacific coast.

3.41 GREEN BELT

Some Asian countries are engaged in strip planting of their shorelines to create "greenbelts" of trees for several purposes. First, the trees control erosion and stabilize the littoral by holding sediments and building up land. Second, a greenbelt of trees will effectively reduce the force of devastating storm surges and waves that accompany cyclones. Third, trees provide an amenity and a source of food and materials for coastal communities. Fourth, trees are beneficial to biodiversity and can create habitat corridors for wildlife.

A greenbelt differs greatly from conventional forestry and should not be considered a source of government revenue, as most national forest reserves are in Asia. Greenbelt projects should be based on concepts of social forestry, ecodevelopment, and participatory planning. Local communities should help decide the details, assist with the nurseries and plantings, and directly receive most of the direct and secondary benefits.

There may well be a gradation of tree species from the water's edge inland, starting with mangroves in the intertidal and going eventually to hydric species, including perhaps betelnut, date palm, coconut, etc, which survive cyclones rather well, and perhaps bananas and fruit trees on higher soils, such as those of a coastal embankment (Figure 3.58).

Bangladesh is a good example. A major campaign of Bangladesh since the 1960s is construction of coastal embankments, or sea dikes. The main purposes are to block salinity intrusion and moderate storm surges. The problem has been to design them so they will not collapse when either overtopped by extreme storm surges and flood backwashes. If they do they will lead to a false sense of security; therefore, they have to be correctly designed and protected.

Vegetation has proved very valuable in securing the Bangladesh embankments. For this and other reasons, Bangladesh has already planted 100,000 ha of coastal fringing mangrove forest and is now planning a greenbelt to infill the existing afforested stretches of the coast to create a greenbelt for the whole occupied coastline from west to southeast; that is, from the edge of the Sunderbans in the west to the southern tip of the country, at Teknaf. The total length of shore line to be covered is 1,874 km. More than 100 nurseries are planned. (71)

All people living in cyclone "risk-zones" would benefit from coastal greenbelts in terms of security and many in terms of improved access to food, materials, and shelter and some in terms of income. But every effort must be made to ensure that the outputs (coconuts, fruits, nuts, thatch, cane, poles, fuel) as well as any land adjustments benefit the common people and do not accrue especially to politically powerful and already rich landholders or government officials.

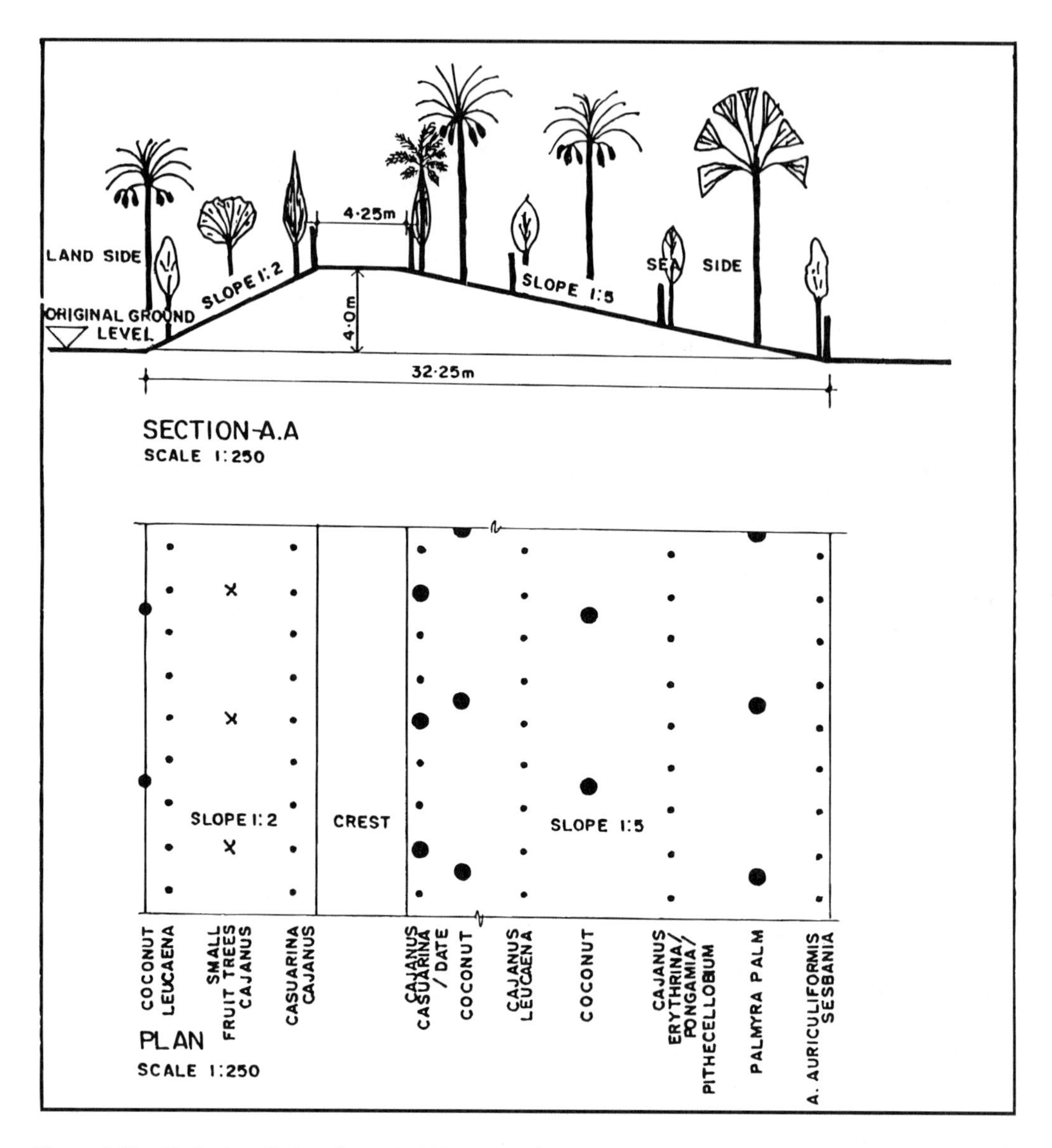

Figure 3.58. Typical profile for a "greenbelt" incorporating the coastal embankments of Bangladesh—purpose is to stabilize the embankments and provide a baffle against storm surge from periodic cyclones. (*Source:* Reference 135.)

3.42 IMPACT TYPES

The potential impacts of various human activities to marine environments may be evaluated on three levels—local, regional, and global, according to Olsen (270). Specific projects may contribute to environmental degradation on one or more of these levels.

The sources of impacts are varied and are categorized and defined in different ways by different sources. However, one can make the following distinctions (65):

1. Naturally caused impacts—Those derived from natural systems such as global temperature changes, hurricanes (typhoons), floods, diseases which usually result in the gradual or sudden death of organisms and in which man has had no discernible input.
2. Non-natural impacts—These are always somehow the result of human activities which have direct or indirect effects on ecosystems.

3. Direct impacts—Coral collecting, trampling, spear fishing, over-fishing, speed boating, trail walking, and tree cutting are examples that have visible immediate effects.
4. Indirect (or external) impacts—Inland terrestrial activities such as deforestation, improper soil coverage, the indiscriminate use of agri-chemicals all contribute to environmental degradation directly or indirectly on other coastal and marine systems.
5. Physical impacts—these can be natural or non-natural in origin, as well as direct or indirect and result in the physical alteration of a particular ecosystem, such as anchor damage, boat grounding on coral reef, or eutrophication by sewage in a lagoon.
6. Ecological impacts—they may be a result of any of the above categories but pertain to a broader scale of changes with more long-term effects and the possibility of permanent alterations to an ecosystem.
7. Socio-economic impacts—these are usually a result of a direct or indirect interaction with foreign visitors on local communities. These can result in changes in the habits of other resource users, competition for scarce resources, and changes in social habits.

Because of the nature of water, dispersal of chemical and physical impacts, such as oil spills, can be extensive, following water circulation patterns or becoming concentrated as they move up the marine food chain. In some cases, ecological community structures can be affected by actions taken on marine coastal ecosystems separated by hundreds of miles (270).

Ten impact types are listed by DuBois (115):

1. ***Loss of mangrove forests*** due to rate of exploitation above that which would sustain the resource.
2. ***Coral reef destruction*** leading to reduced yield of fish species associated with reefs (Figure 3.59); reduced tourism attraction; and loss of rare ecosystems due to:
   - Dynamiting to harvest fish from the reef
   - Harvesting of coral for construction material or, in the case of exotic coral species, for jewelry or souvenirs
   - Siltation and smothering of coral from erosion associated with upstream deforestation
   - Death of coral from deleterious effluents associated with mining or oil spills
3. ***Congestion and intensive use*** of coastal resources such as water, species, and fisheries, due to high population density and continuing growth, expansion of tourism industry, and growth of commercial and industrial activity.
4. ***Depletion of wildlife species*** such as turtles, crocodiles, deer, waterfowl, and manatees.
5. ***Pollution of coastal waters,*** which adversely affects fisheries yield and tourism revenues due to:
   - Industrial wastes
   - Sewage
   - Agricultural pesticide run off
   - Oil spills
   - Bilge discharges
   - Toxic contamination of fish and shellfish
6. ***Beach and coastline erosion*** caused by removal of coastal mangrove forests; construction of coastal installations which alter current and wave action patterns; or mining of beaches for sand.
7. ***Swampland filling*** that reduces fish spawning and "nursery" habitat and reduces fisheries yields.
8. ***Upland deforestation*** leading to erosion and sedimentation.
9. ***Over-fishing*** leading to reduced fisheries yield and in some cases species extinction.
10. ***Salt water intrusion*** leading to loss of coastal agricultural lands and potable water supplies.

Additional impact types are listed in other sections; e.g. see "Environmental Assessment" and "Dredging Management" in Part 2. (See Figure 3.60 for impacts on specific habitats.)

The most serious adverse impacts of human activities on the coastal zone are water pollution and habitat destruction according to Olsen *et al.* (270) as described below.

Pollutants can enter marine ecosystems from direct discharge, runoff from land, or atmospheric deposition. Direct discharges include pipeline discharges ("point-sources") of municipal and industrial wastes into marine environments, discharge into rivers which then flow into marine environments, and discharge from vessels. Runoff (often called "non-point" source pollution) from agricultural and forest lands and from urban areas and areas of coastal building may be the greatest threat to marine ecosystems, as well as the most difficult one to stop. Nutrients, such as phosphate and nitrogen—applied in large quantities in agriculture as fertilizers—can cause eutrophication in coastal ecosystems by stimulating photoplankton, algae, and submerged vegetation which rapidly grow and then die giving rise to anaerobic conditions.

Toxic substances form another category. They include heavy metals, pesticides and hydrocarbons which are harmful to marine species and to humans. The effects of these substances may be lethal or chronic, including maladies as tumors, skeletal deformations and shell diseases. "Bioaccumulation" often results in threats to human health, when toxic substances become concentrated in tissues of marketable fish and shellfish. Measures must be taken to diminish the sources—through recycling, proper disposal, and reduced use of toxic materials (270).

Increased loads of suspended sediment from land erosion and dredge spoils also cause pollution. They cloud waters, reducing the penetration of sunlight, thereby reducing normal levels of photosynthesis. Excessive sedimentation may smother benthic ecosystems, including coral reefs; for example, 70 percent of the coral reefs of the Philippines are said to have been degraded by sediments in the nearshore waters. Suspended sediments may be contaminated with toxic substances (270).

Habitat destruction has been a monumental problem in coastal wetlands, which are filled, dredged, channelized, flooded, and paved over. Because they are based upon a fragile physical structure, coral reef ecosystems are also highly vulnerable to habitat destruction. Direct habitat destruction has been less severe in other marine ecosystems, but there is reason for concern. Dredging and dredge disposal activities have destroyed habitat in many estuaries and more recently in off-shore environments. Proposals for seabed disposal of waste and seabed mining for mineral poses threats to deep-water benthic ecosystems. Large spills of hazardous substances, such as oil, can cause instantaneous habitat destruction with a long recovery period. Habitat destruction can also occur indirectly, as a result of the cumulative effects

Figure 3.59. To prevent damage by tourists to the coral reef, underwater signs are employed by the Virgin Islands National Park. (Photo by author.)

Development Activity	Type of Ecosystem							
	Marshes	Deltas	Estuaries	Mangrove swamps	Seagrass beds	Coral reefs and lagoons	Beaches	Islands
Agriculture and farming		•	•	•	o	o		•
Feedlots, ranching and rangelands	o	o		•	o	•		•
Forestry		o		•				o
Aquaculture and mariculture	o	o	•	•	•	o		
Nearshore-catch fisheries		o	•	o	•	•		o
Dredging and filling	•	•	•	•	•	•	o	o
Airfields	o	o	o	o	•	•	•	•
Harbours	•	•	•	•	•	•	o	•
Roadways and causeways	o	•	•	•	•	•	o	o
Shipping		•	•	o	o	o	o	o
Electric power generation	o		•	o	•	•	o	o
Heavy industry (onshore)	o	•	•	•	•	•		•
Upland mining	o	•	•	o	•	•		o
Coastal mining	o	•	•	o	•	•	•	•
Offshore oil and gas development		o	o	•	•	•		•
Military facilities, training and testing	o				•	•		•
Land clearing and site preparation	o	o	o	o	o	•	•	•
Sanitary sewage discharges	o		•	o	o	•		•
Solid waste disposal	•	o	o	•			o	•
Water development and control	•	•	•	•				•
Shoreline management and use		o				•	•	•
Coastal resource uses	•	•	o	•	•	o	•	•

o Significant adverse effects likely
• Adverse effects possible

Figure 3.60. Types of impacts on coastal natural habitats caused by various human activities. (*Source:* Reference 232.)

of severe water pollution and water diversion (270). Some types of fishing are also destructive of habitat such as: dynamite, muri-ami, kayaka, and otter trawling near coral reefs and other sensitive areas.

3.43 INDICATOR SPECIES

The best general indicator of ecosystem quality may be the biota itself. It is often assumed that any significant reduction in the abundance or diversity of species is in itself *prima facie* an adverse impact. Also, any shift in biotic communities from desirable to less desirable species could be considered adverse. Yet these biotic effects are often surprisingly difficult to pin down at the community level with hard and fast numerical indices. Therefore, individual species are often chosen to represent the ecological community as a whole.

An indicator species is one chosen to represent conditions in an ecosystem usually because it is either especially sensitive to change in the environment or because, as an indicator of ecosystem condition, it is particularly easy to work with. However, choosing one, or even several, species to represent an ecosystem has its pitfalls.

An indicator species might be used in management to monitor the effects of pollution or other degradation of an environment by comparing the present standing crop of the species to the standing crop during a previous undisturbed condition, or to some specified standard. The pitfalls of this method of monitoring lie mainly in sampling errors and in the difficulty of ascribing an observed change to a specific type of disturbance, for example toxicity versus nutrient reduction.

Trained scientists who have the opportunity to study and measure a variety of species are able to make a judgment of biotic conditions but rarely can provide single bellwether species that can be readily inventoried to provide a ready index of trends in carrying capacity. Certain invertebrate species of the bottom would be preferred indicators because they stay in place and they are sensitive to disturbances. Also the analysis of certain rooted aquatic plants represents a promising approach, and it would seem useful to explore the possibility of using such plant indicators in coastal zone ecosystem management programs. For example, the following associations between specific rooted plants and environmental conditions appear to provide useful indicators (145):

Plant	Condition
Widgeon grass (*Ruppia maritima*)	Presence indicates acceptable turbidity conditions
Sea lettuce (*Ulva*)	Dense concentrations may indicate eutrophication due to sewage or other causes.
Water milfoil (*Myriophyllum spicatum*)	In fresher waters, presence is adverse (may indicate eutrophication).
Eelgrass (*Zostera maritima*)	In saltier waters, presence may indicate good conditions.

To the extent possible, potential indicator species should be identified whenever *baseline* surveys (to establish the present condition of an ecosystem) are to be done (see "Baseline Database" above). Under favorable circumstances, change-or-no change of one or more indicator species may be the best measure of development project impact.

There are certain organisms that may be useful as indicators of coral reef stress (145). Fish, being relatively more complex organisms, many aspects of their biology and behavior may be used to gauge the degree of "suitability" of their habitats. The presence or absence of certain species is also highly indicative in certain cases, because, being mobile, fish would be expected to select habitats with more favorable conditions. Examples of possible candidates for indicator organisms are certain species of chaetodonts (butterfly fishes), which are obligate coral predators (Figure 3.61). But the abundance measures must be interpreted in context of the type of reef community under evaluation.

Multiple species or community indicators are probably used more often than single species. The extent of cover by coral and other benthic species of the available reef substrate and the range of species present are generally accepted as significant indicators of reef condition. In general, high cover and high species diversity indicate a healthy reef. A third factor, mean size of coral colonies is often used as an indicator of the stability of a coral community since large average size indicates that several years

Figure 3.61. An accepted indicator of a healthy reef environment is the presence of a variety and abundance of butterfly fishes. (Photo by Bill Keogh.)

have passed since the area was colonized by corals and the presence of a lot of small colonies suggests recent colonization. (145)

In one example, three groups of fishes were used as indicators (snappers, grunts, and hogfishes) to determine the effect of a ban on spearfishing at a coral reef marine reserve in the United States (Looe Key National Marine Sanctuary in Florida waters). In the method used, a diver randomly selects a point on the forereef and censuses all species observed in a 5-min period in a 7.5-m radius (85). A total of 130 baseline samples were collected between 1979 and June 1981, when spearfishing was banned. A total of 160 samples were collected two years after the ban during an intensive fish survey of the sanctuary in the summer and fall of 1983.

The results are shown below for some important species. Five potential target species not observed before the spearfishing ban were observed after the ban. All 15 species that were spearfishing targets increased in abundance, while 14 of 15 increased in frequency of occurrence. Examples of six species follow (85):

	Abundance			
	Before		After	
Species	**Mean**	**SE**	**Mean**	**SE**
Ocyurus chrysurus	3.84	0.88	6.92	0.82
Lutjanus apodus	0.7	0.19	0.81	0.22
Haemulon macrostomus	0.13	0.06	0.35	0.11
Haemulon sciurus	0.67	0.24	1.86	0.46
Haemulon plumieri	0.21	0.05	0.85	0.21
Haemulon carbonarium	0.05	0.03	2.2	1.5

3.44 INDUSTRIAL POLLUTION

Many industries are attracted to the coastal zones where they (a) benefit from access to low-cost marine and inland transportation systems, (b) can use seawater for process or cooling purposes or for waste disposal, (c) utilize marine transportation, and (d) draw directly on the marine environment for raw material. Industries are also attracted by labor availability in coastal population centers (Figure 3.62).

The wastewaters from industries sited in coastal areas may pollute coastal ecosystems. The impacts range from relatively minor disturbances (such as temporary, localized turbidity increase) to major disruptions (major water pollution caused by discharge of toxic chemicals). Thermal pollution must also be considered, along with the uptake of large quantities of cooling water. Both pose a major threat to free-floating planktonic forms of life, including larval and juvenile species of important fish and crustacean species.

Synthetic organics are exclusively the products and by-products of industrial synthesis. Some of the most persistant organic synthetic compounds belong to the chlorinated hydrocarbon group. This

Development projects \ Environmental parameters	Surface water hydrology	Surface water quality	Ground water hydrology	Air quality	Land quality (pollution)	Fisheries	Vegetation	Forests (Resource depletion)	Mineral resources	Aesthetic	Socio-economic	Public health
Food processing	○	⊕	⊕	•	○	•	•	•		•	•	○
Sugar refining	○	⊕	○	•	○	•		⊕		•	•	○
Pulp and paper	○	⊕	•	•	○	⊕		⊕		•	•	○
Fertilizer	•	⊕	•	⊕	○		⊕		•	•	•	•
Cement	○	•	•	•	⊕	•			⊕	•	•	•
Tannery	○	⊕	○	⊕	•	⊕				⊕	•	○
Pharmaceutical	○	⊕	○	•	•	⊕				○	•	⊕
Steel and iron manufacture	•	⊕	•	⊕		•			⊕	•	•	•
Electroplating	•	⊕	○	○	○	•	⊕			○	•	○
Petrochemical	○	•	•	•	⊕	•	○		⊕	•	•	•

⊕ Significant impact.
• Moderate to significant impact.
○ Negligible impact.

Figure 3.62. Potential environmental impacts of various categories of industry. (*Source:* Reference 41.)

group includes biocides such as DDT-types, hexachlorobenzene, hexachlorocyclohexane, dieldrin and toxaphene as well as polychlorinated biphenyls (PCB). They move horizontally with air and ocean currents and vertically through the air/water interface by precipitation (wet and dry) and volatilization (228).

The sources of metal released to the environment are mainly the combustion of coal and oil, pyromatallurgical processes, industrial, agricultural or consumer use, and transport. All these activities contribute to releases to the atmosphere, soil or aquatic ecosystems. Inputs to the marine ecosystem are derived from atmospheric emissions, river transport, and the discharge of sewage and dredged soil (228).

Rivers, sewage, sludge, and "dredge spoils" carry metals, mostly in particulate form, which has less opportunity to escape to the open sea than dissolved components. The transport and trapping mechanisms are (a) movement with coastal water bodies; (b) sedimentation; (c) leaching from, and sorption on particles; (d)) flocculation. From the discharge towards the open ocean both the particulate-to-dissolved metal concentration ratio and the sediment metal concentration decrease as a result of sedimentation. In sea water, a metal can be dissolved or particulate, free or complexed, charged or uncharged, inorganic or organic. The chloride affinity of lead, cadmium, and mercury increases from lead to cadmium to mercury, and affects speciation. When organic lead and tin reach sea water, total decomposition follows entry within a few weeks (228).

A marine species that takes up a metal from water or sediment does so either directly or indirectly by ingestion with food. Diffusion, facilitated transport, or absorption on gill and surface mucus are the mechanisms of uptake from water. Marine algae can accumulate arsenic, lead, mercury and selenium with high (× 1,000) bioconcentration factors. Bioaccumulation of arsenic, cadmium, lead, mercury, selenium and tin from water is the important uptake process for mussels and oysters. The possibility of adverse health effects depends on the toxicity of the metal and on the degree of exposure. Toxicity is an inherent quality but frequently differs with the various compounds of the same metal. Dose or exposure for seafood consumers depends on the concentration of metals in seafood, the chemical form of the metal and the quantity of seafood consumed (228).

Methylmercury is able to damage the sensory part of the nervous system of humans who eat contaminated fish or shellfish. The first symptom is paraesthesia (abnormal sensations), which is followed by uncoordinated movement, constricted visual field, slurred speech, and hearing difficulties in more severe cases. Limiting the number of seafood meals to less than one per day often keeps methylmercury exposure within tolerable limits (228).

In locating a new industry it will be necessary to ensure that its treatment needs are incorporated into the community's long-term plan for environmental protection. Cleaning up industrial effluents may require advanced treatment systems. For example, since the constituents of industrial effluent are usually quite different from those of domestic sewage, separate private systems may have to be constructed by industry, and planned for accordingly. When discharge into the municipal collection network is allowed, private pretreatment units will probably be necessary.

Industrial facilities with difficult discharge problems, such as food-processing, petrochemical, wood pulp, and steel processing plants, should not be located on confined estuarine waters. Moreover, tidal streams, dead-end harbors, small lagoons, and similar small or poorly flushed water bodies should be avoided because of their extremely limited capacity to accept and assimilate even small amounts of contaminants. (59)

The *chemical industry* is vast, and, in general, it can be broken down into three divisions. The first produces the basic chemicals, including inorganic chemicals, acids, alkalies, and salts. A second division manufactures intermediate chemicals, such as plastics, synthetic fibers, fats, and oils. The third group manufactures finished chemicals such as drugs, cosmetics, soaps, and pesticides. Obviously, each industry produces its own set of wastewater toxicants, which may be unique to that industry or shared with other ones. The toxicity of the individual toxicant or of a group of toxicants varies widely and depends on a multitude of factors. Each industry has to be approached independently in regard to the potential or proved effects of its wastewater on the particular biotic community receiving its effluent. (59)

The *pulp and paper industry* has the potential of contributing a variety of chemicals to coastal pollution loads. The kinds and amounts of toxicants vary with the particular process. In addition to solid wastes and carbohydrates, which can reduce oxygen concentrations to toxic levels, heavy metals used as fungicides, acids and bases for treating pulp, chlorine for bleaching, and other chemicals are found in the wastes of this industry. What is disturbing is that, of the 1.9 trillion gallons of water discharged annually by the paper and allied products industry, only 34 percent undergoes any treatment. (59)

The *petroleum industry* produces the obvious oil and oil-coated solids. In addition, the electrical method of crude desalting produces wastewater containing sulfides, suspended solids, phenols, and ammonia—and all this commonly at an elevated temperature. Crude oil fractionation produces sulfides, chlorides, and phenols. The thermal cracking process, hydrotreating, lube oil finishing, and other operations produce wastewater containing phenols, oils, sulfur compounds, ammonia, and stable oil emulsions. There are also leaks, spills, and tanker groundings. (59)

The *steel industry* is plagued with an especially large wastewater problem; waste streams flow at rates of 10,000 to 25,000 gallons per minute. Waterborne wastes of this industry include suspended solids, oils, hot water, acids, plating solutions, dissolved organics, soluble metals, emulsions, and coke plant chemicals. (59)

Brewery and distillation industries produce quantities of both liquid and solid waste. The brewing process (beer-making) involves production of a fermented beverage of low alcohol content from various types of grain (41). Barley is commonly used in the brewing process because it contains about 60 percent starch. Before fermentation, the starch present in barley is converted into sugar by a process called "malting"—moistening barley grains with water to initiate germination. During germination an enzyme (diastase) is formed, which converts starch to maltose. Germination is stopped, the batch is washed and the ground barley is extracted (with hot water) to give a liquid called *wort,* which is boiled with hops and passed to a fermenter where yeast is added to convert sugar in the wort into alcohol and carbon dioxide. Filtration or carbonation, for example, can also be carried out to make beer of the desired type. Liquid waste is basically a product of the barley steeping, while solids consist of the sprouts and rootlets screened out during the malt preparation. Wastes from fermentation and subsequent operations include spent grain, hops and yeast, liquors from spent grain, yeast washing and pressing of yeast wastes, and wash waters (e.g., from the washing of equipment, casks, barrels). (41)

The liquid wastes have a high suspended solid (SS) content, a high biological oxygen demand (BOD), and a high chemical oxygen demand (COD). Typical effluent compositions are (41):

BOD_5 (mg/l)	1200–3000
SS (mg/l)	100–800
COD (mg/l)	3500–4000
pH	4.3–11.0

Hard drinks (high alcohol content) are obtained by distillation (41). The type of liquor produced from distillation will depend on the alcoholic liquid used as a base (e.g., the base for brandy is wine, whisky is distilled from a fermented wort derived from corn, rum is distilled from a product obtained by a fermentation of molasses. (41)

In *cane sugar manufacture* cane stalks are crushed between rollers to extract juice which is screened and treated with lime to prevent sucrose inversion. The extract is heated, insoluble compounds removed by filtration, and the liquor concentrated until a mixture of sugar crystals (raw sugar) and syrup (molasses) is formed. Sugar is separated from molasses by centrifugation. (41). Cane crushing water contains a substantial amount of soil, impurities, and sugar. Disposal of the solid waste may lead to eutrophication problems because of huge amounts of BOD. The return of mill mud and mill ash to the canefields is highly recommended. Effluents typically contain (41):

BOD_5 (mg/l)	500–3000
SS (mg/l)	100–2000
pH	6–9

Metal finishing is used for protecting metals against corrosion, improving their properties, or enhancing their appearance. Pretreatments are usually necessary to remove natural oxide coatings, corrosion products, and protective oils and greases. The oxide films and corrosion products are removed by either mechanical methods (brushing, descaling, polishing, shot blasting) or chemical processes (acid pickling, sodium hydroxide descaling). Oil and grease is removed by organic solvents. The methods used for metal finishing may be grouped as (41):

- Electroplating
- Stripping
- Chemical conversion (anodizing, phosphating)
- Chromating metal coating (e.g., galvanizing, rust proofing)
- Machining
- Final polishing
- Case hardening

Pretreatment by means of acid pickling prior to galvanizing will produce high concentrations of iron salts, which produce insoluble hydroxides. Liquors from pickling of copper contain salts of copper and zinc. Electroplating acid wastes generally contain dissolved copper, nickel, zinc, chromium, cyanide, and sometimes cadmium or lead. Anodizing gives rise to a range of wastes, including both alkaline and acidic solutions, nitric acid and nitrates, phosphoric acid, chromium-bearing acidic solutions, bright anodizing solutions (nitric/phosphoric/acetic acids), and nickel-bearing sealing solutions. Typical electroplating effluents contain (41):

pH	3–8.5
Cu (mg/l)	0.1–0.3
Zn (mg/l)	0.1–8
Cr (mg/l)	0.1–5
Ni (mg/l)	0.01–60
Cd (mg/l)	0.01–0.1

Leather manufacture, both the preparation of the hide and the subsequent tanning, contribute to pollution, particularly organic matter. The first stage is to remove hair, protein, grease, and oil. Hides are prepared for tanning in a number of stages to remove dirt, blood, manure, and nonfibrous proteins as well as to preserve the hides for long periods. This stage is followed by liming and unhairing in which the soaked hides are treated with a mixture of lime slurry and a solution of sodium sulphate and then delimed with weak acids (and/or acid salts) and bated with a proteolytic enzyme (pancreatin, trypsin) which peptizes the protein fibers and removes unwanted protein (elastin). (41)

Tanning produces heavily contaminated wastewater. It may be alkaline or acidic, is highly colored, and has high oil and grease levels and high salt content. Such wastes will generally have high BOD, COD, SS, and odor. Also of concern would be chromium, which is toxic to humans and to aquatic invertebrates and fish. Typical effluent composition is shown below (41):

BOD_5 (mg/l)	500–1500
COD (mg/l)	3000–6000
SS (mg/l)	400–1000
O/G (mg/l)	120–500
Tot Cr (mg/l)	12–100
pH	6–11

Food processing involves the categories of industry grouped as follows: (a) abattoirs and meat canning/packing; (b) manufacture of dairy products; (c) canning and preserving of fruits and vegetables; and (d) canning and preserving of marine food (e.g., fish, crustacea, shellfish, clams). They produce the following pollutants in the waste system (41):

	Abattoir	Dairy	Fruits/veg.	Fish
$BOD_5$5 (mg/l)	1200–3800	1600–2500	1000–3000	3000–3500
SS (mg/l)	450–2400	900–1400	100–400	1200–1600
COD (mg/l)	5600–6900	3000–4200	2000–4000	4000–5000
O/G (mg/l)	20–100	200–400	—	40–60
pH	7–7.2	4.5–10	4–12	6–8

Cement is a complex calcium aluminosilicate material produced by the combination of limestone with material containing alumina and silica (usually clay). After crushing and grinding, the limestone and clay are blended to give the correct composition for the kiln in which they are heated to form "clinker". Then a wet slurry is introduced, followed successively by heating, calcining, drying and pulverizing. The hot gases produced by combustion carry entrained dust with them. Dust emissions are also produced in crushing, grinding, blending, clinker cooling, finish grinding, and in moving finely divided material to silos and in packing. Wastewaters are produced during blending and from the wet scrubbers that remove dust from the kiln exit gases. These waters contain significant quantities of suspended (inorganic) solids (41).

Pollution control and location are the major management tools. As a rule, industries with high waste output, such as power plants with large estuarine water intakes, chemical plants with irremediable toxic discharges, and oil-transfer terminals, should not be located on estuarine water bodies unless there is no practicable alternative—that is, the private and public costs of protection of ecological resources from pollution would be exorbitant (Figure 3.63).

In locating a new industry it will be necessary to ensure that its treatment needs are incorporated into the community's long-term plan for environmental protection. For example, since the constituents of industrial effluent are usually quite different from those of domestic sewage, separate private systems may have to be constructed by industry, and planned for accordingly. When discharge into the sewage collection network is allowed, private pretreatment units will probably be necessary to reduce the industrial waste flow before discharge, in order to protect the municipal facilities and the receiving waters.

Industrial facilities with difficult discharge problems, such as food-processing, petrochemical, wood pulp, and steel processing plants, should not be located on confined estuarine waters. Moreover, tidal streams, dead-end harbors, small lagoons, and similar small or poorly flushed water bodies should be completely avoided because of their extremely limited capacity to accept and assimilate even small amounts of contaminants.

In many countries that have undergone intensive development, only a few coastal locations ideally suited for industrial use (in relation to waste discharge) still remain. Coastal management should identify, inventory, and reserve these prime locations as important industrial real estate resources. Many such sites in growing metropolitan areas have been and continue to be taken over by housing and commercial

Figure 3.63. Factories with major chemical pollution potential should not be built in the coastal zone. (Photo by author.)

establishments, which are not really dependent on waterfronts. To ensure that prime sites with the lowest pollution potential are available when needed for industrial use, special land-use controls are appropriate to restrict the development of these sites to waterfront-dependent industry.

3.45 INFORMATION NEEDS

Recognizing that coastal resources conservation is based upon relevant, reliable, information for decision making, it follows that choosing the right information is critical to the success of coastal planning and management. Information acquisition is especially crucial to the Strategy Planning phase. It is during this phase that important decisions will be made about the future of the coastal management program, or even whether it will have a future. Clearly, these decisions should be made in the most information-rich circumstances so that the consequences of taking or not taking specific actions are predictable.

The following are important categories of information for coastal management programs (65):

1. Physical environment: terrain data (including history), erosional processes, storm surge, winds, tides, air-sea interaction, sediment transport, geological setting, subsidence, sediment supply, and condition of above.
2. Biological environment: Primary and secondary production, distribution and extent of living marine resources, major habitats and ecosystems, ecological relationships that determine productivity, presence of rare, threatened or endangered species, indicator species; and condition of above.
3. Sociological information: Resource dependency; historic use patterns including methods; factors determining historic use patterns; identify current use patterns; identify whether current use patterns are sustainable; demography; socio-cultural information, including social equity; land and sea tenure.
4. Economics: Resources and resource use patterns should be evaluated, including economic output. Qualitative values—resources and resource use patterns that cannot be economically valued are to be considered. Specific economic pursuits can include fisheries, tourism, ports, energy, settlements, transportation, aquaculture, mining, oil/gas, waste treatment and disposal, traditional practices.
5. Issues: It is necessary to identify the problems and issues facing ICZM and the stakeholders in use and management of coastal resources and potential conflicts among them.
6. Institutional mechanisms (national, state, and local): ministries and departments with responsibilities, and their organization and hierarchy; NGOs; legislation on zoning, pollution, resource, utilization; interagency councils; advisory panels; standing agreements with private parties; permitting and other administrative processes to carry out legislation.

Information needs can form a long list; therefore, priorities have to be established for most programs (also see "Database Development" in Part 2.)

One of the urgent requirements for coastal management programs is assessment of the present status of coastal resources. This would include the status of many of the following: eutrophication, occurrence of red tides, ciguatera, coliforms at beaches, nitrification and salinization of groundwater, fish abundance, habitat degradation levels, identification of local and upstream impacting activities (agriculture and others) wetland obliteration, etc.

Information collection, analysis, and presentation can be useful for planning and management or a waste of effort and money. For example, the standard apology—"more information is needed" may be only procrastination. Experience shows that lack of information is seldom, if ever, an adequate excuse for not beginning ICZM planning. More information is always helpful in making decisions. But decisions always have to be made under various conditions of uncertainty. Because obtaining information is *not* costless, there is a limit to data acquisition with available resources.

Environmental information can be prepared in several ways. Each category of data can be analyzed using (a) one characteristic only (i.e., single-factor analysis by progressively adding factors or characteristics according to potential or increasing limitations; added-factor analysis by adding characteristics according to increasing limitation, risk, potential); or (b), sorted-factor analysis by using a stepped, decision-tree approach similar to those used in taxonomic or identification keys (to sort or classify groups of significant characteristics) along the lines suggested; and/or weighted-factor analysis by combining and weighting characteristics to reflect relative importance (88).

Single-factor, added-factor analyses can be readily accomplished using manual overlay techniques. Van R. Classen (88) states, however, the overlay mapping system can quickly become cumbersome as the number of attributes, factors, and characteristics increases. Where large sets of data exist, the number of maps can become unmanageable and computer assistance to handle the data may be required. Such computer-handled data systems are usually termed *geographic information systems* or GIS. (See "Geographic Information Systems" above.)

It must be noted, however, that a GIS does not reduce the need to gather information or to develop the maps for input into the digital computer base. The use of the computer will not reduce the amount of mapping work required in the first instance. It may also delay the completion of the assessment (see "Geographic Information Systems" above), unless the computer hardware and software systems are well developed and running without "bugs."

3.46 INLETS

Any inlet through the shoreline to a port, harbor, bay, lagoon, or estuary can cause problems, particularly if the inlet is mobile or breaks through a sandy beachfront. Inlets are notoriously mobile, changing shape and moving *en toto* along the beach to left or right. Trying to stabilize an inlet through a sandy barrier island system to accomodate navigation, for instance, is a real engineering challenge.

Inlets affect the stability of adjacent beaches by interrupting littoral drift and trapping sand. As sand moves into the inlet, the inlet narrows, causing acceleration of current velocity and thereby increasing the capacity of the current to suspend and to carry sand. On the flood tide, then the water flows in through the inlet, sand is carried a short distance inside and deposited. Such sand creates shoals in the landward end of the inlet, known as the *inner bay* or *flood-tide delta.*

On the ebb tide, the sand is carried a short distance out to sea and deposited on an outer bar or *ebb-tide delta.* When this bar becomes high enough, waves break over it, causing sand to move toward the beach. The net result is the accumulation of sand on the bar and the "starvation" of downdrift beaches. However, this process reaches a state of equilibrium when the inlet bars are fully formed, after which natural bypassing will occur. When inlet channels are artificially deepened by dredging and the deposited sand is regularly dredged and hauled away, their capacity to store sand may be increased, thereby causing narrower beaches on the downdrift side of the jetty by reducing the supply of sand to these beaches.

Jetties are structures similar to groins but used to modify or control sand movement at inlets only. Jetties not only stabilize the location of the channel by controlling sand movement but also shield vessels from waves. They are constructed of steel, concrete, or rock, depending on foundation conditions, wave climate, and cost. Although jetties are similar to groins in that they dam the sand stream, they are usually considerably longer and larger, often extending seaward to a depth equivalent to the channel depth desired for navigation purposes.

To eliminate undesirable downdrift erosion, some projects are provided with a bypass system for dredging the sand impounded by the updrift jetty and pumping it through a pipeline to the downstream eroding beach. These bypass systems ensure a flow of sand to nourish the downdrift beach and prevent shoaling of the entrance channel (Figure 3.64 and Figure 3.65). This may be accomplished in conjunction with an offshore breakwater. (59)

A more recent development provides a low section or weir in the updrift jetty, over which sand moves into a predredged deposition basin. By dredging the basin periodically, deposition in the channel is reduced or eliminated. The dredged material is normally bypassed to the downdrift shore (Figure 3.66).

Experience shows that both inlet deepening and inlet stabilization projects affect the sand supply moving along the beachfront and that either can lead to a major imbalance of the beach system. Therefore, inlet projects must be considered in the context of a comprehensive long-term plan for stabilization and protection of the beach system, which involves both land control and construction elements.

Figure 3.64. Entrance to a small boat harbor is protected from filling with sand by building a parallel breakwater. (Photo courtesy of the U.S. Army Corps of Engineers.)

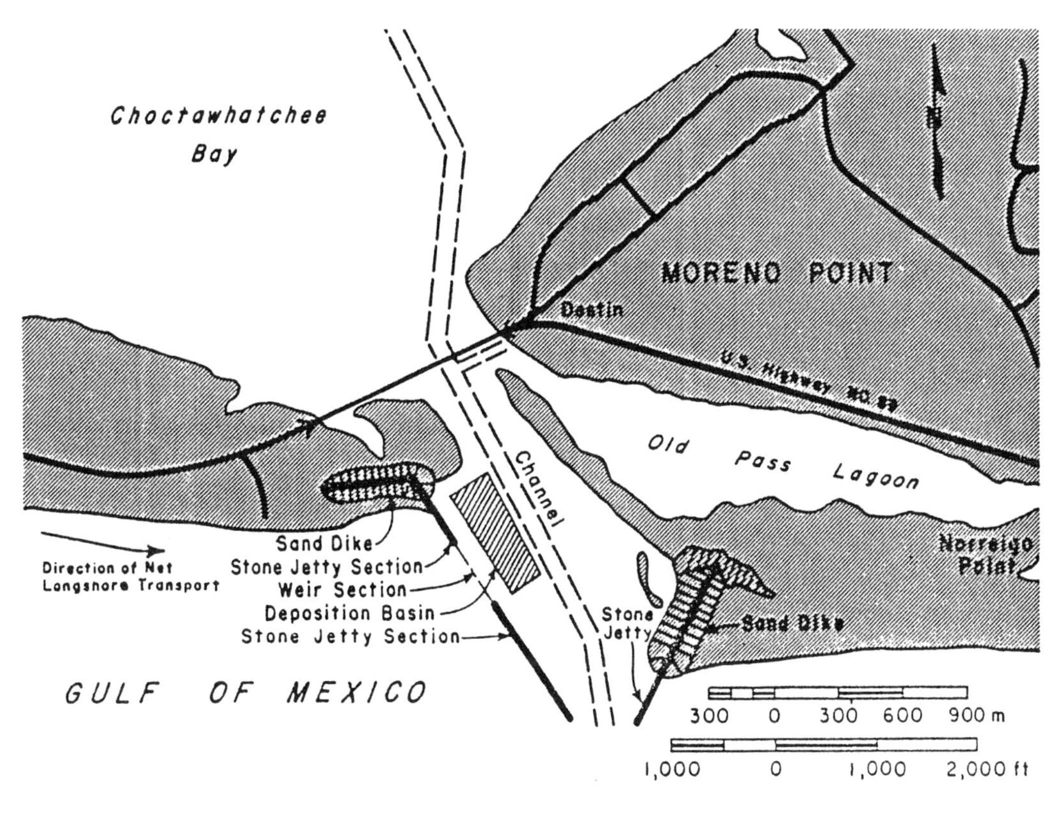

Figure 3.65. A system for passage and collection of littoral sand at a Florida inlet. (*Source:* Reference 362.)

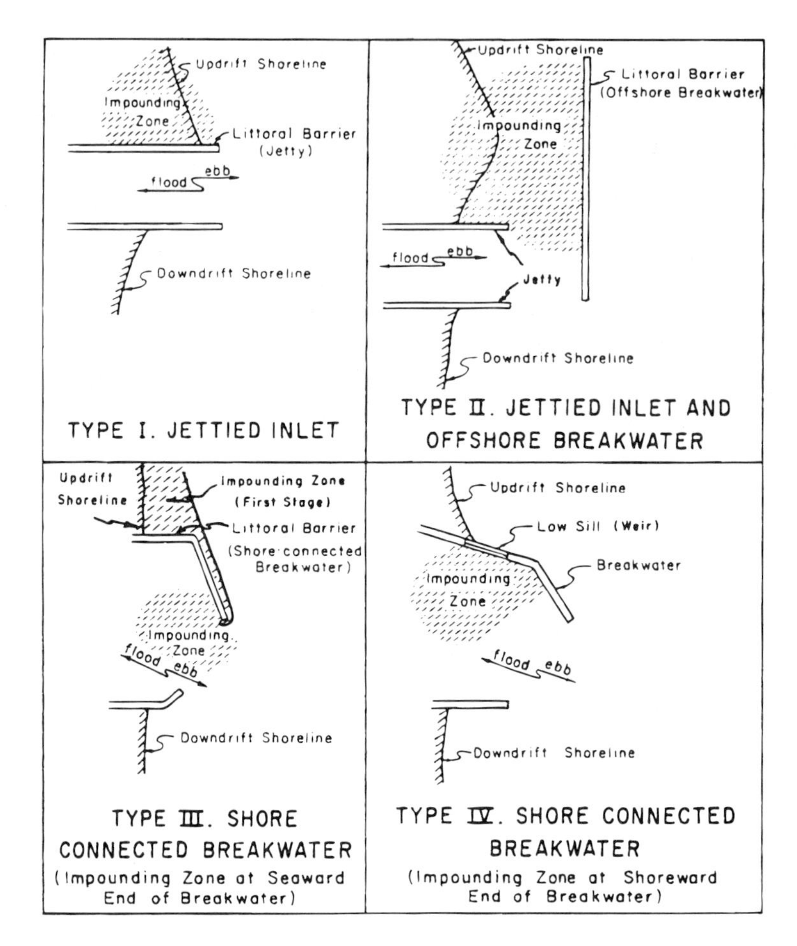

Figure 3.66. Various methods of reducing and managing sedimentation of harbor entrances. (*Source:* Reference 362.)

3.47 INTERNATIONAL ASSISTANCE AGENCIES

Coastal development with major environmental implications is often supported by the big "international donors," the bi-lateral and multi-lateral banks and foreign assistance agencies, like World Bank, Asian Development Bank, International Development Bank (Latin America), United Nations Development Program (UNDP), USAID (USA), CIDA (Canada), DANIDA (Denmark), GTZ (Germany), JICA/OECF (Japan), etc.

Members of the "donor community" routinely require environmental assessment (EA) of major development projects, including socio-economic impacts. Thus, the current requirement is often for a combined "environmental and socio-economic impact assessment" (ESEIA). Most donors have developed guidelines for EA or ESEIA which are incorporated into their project review cycles (see Figure 3.67). It is relevant that ICZM feasibility projects for certain countries have been funded by some donors (e.g., USAID and UNDP).

Most of these agencies now have strong environmental components, often including (a) a central environmental staff, (b) requirements for environmental assessment (EA) of development projects, and (c) funding for special "environmental projects."

For an example, the Asian Development Bank (ADB) has an Office of Environment which plays an important role in the day-to-day operations of the bank. By contributing to policy-making decisions and serving as a resource to bank staff, this office has facilitated the integration of environmental planning at various levels of bank operations as follows (2):

- Providing technical advice: The office of the Environment provides technical advice and assistance to Bank staff at all stages of program and project planning.
- Training bank staff: The Office provides annual training programs at the Bank, introducing staff to the principles and practices of environment and natural resources planning and management. Environmental staff also attend non-Bank training to improve skills in environment-related operations.
- Generating guidelines: The Office has prepared guidelines for non-environmental staff to assist them in assessing the potential social and environmental impact of projects in various sectors (Figure 3.67).
- Reviewing and monitoring: Projects are monitored throughout the project cycle for their compliance with recommended environmental mitigation measures. Environment staff also conduct environmental appraisal during post-project evaluation.
- Performing impact assessment: The office conducts preliminary environmental examination on all projects. A formal Environmental Impact Assessment (EIA) is prepared early in the planning stage for every project where substantial impact is expected.
- Building a database: The Office has developed the Environmental Monitoring Information System (EMIS), a computerized database which contains information about the environmental components of projects, including monitoring and follow-up requirements.

International organizations and donor agencies have a major role to play in mobilizing finances for ICZM managed coastal development and in formulating optimal revenue-generating systems. Conservation-oriented investment opportunities in coastal seas, including environmental rehabilitation, should be identified and encouraged as a useful complement to development-based funding, particularly for innovative resource developments.

Lack of financial support at national and regional level hampers the management of ocean and coastal resources in developing countries. Protection of resources and adoption of more "responsible" development schemes for long-term benefits may have immediate costs which developing countries

Figure 3.67. International financial aid organizations now require environmental scrutiny on all major projects, even the most distant coastal environments (Gentoo Penguins in the Antarctic). (Photo by Philippa Scott, courtesy of World Wildlife Fund.)

might not be able to afford. Therefore, the promotion of ICZM in developing countries might best be in the form of a program which utilizes some proportion of financial support from international funding agencies. But, still the ICZM program must be founded on the national economic planning of coastal countries.

Funding sources need identifying and proper funding procedures should be promoted. In particular, national sources should be identified (fees, fines, taxes, rents) and short-term financing cycles by donors should be replaced by medium to long-term cycles. Various international banks and donors should allocate a significant share of the funding they provide for pre-investment studies and better coordinate their interventions among themselves.

3.48 KELP BEDS

Along higher latitude coasts of certain countries, there are extensive beds of kelp (e.g., Southern California and S.W. South Africa). Kelp is a brown alga that roots on the bottom of the sea and extends, via a long stalk, to the surface, where its fronds spread over the surface of the sea. Kelp often grows in thick stands, creating a submerged forest that provides an important multiple-species habitat. These kelp stands are vulnerable to over-harvesting (for alginic acid), species imbalance (too many kelp-eating sea urchins), and pollution.

Kelp grows best in relatively cool waters where depths are less than 100 feet. The kelp bed breaks the force of the sea and provides a strip of quieter water between it and the shore. It provides food and a favorable habitat for many fish, as well as sheltered nursery areas for their young. Kelp beds are also a favored haunt of sea otters, which feed there on fish, crustaceans, and sea urchins. Because kelp is such an especially important component of the coastal ecosystem where it is found, the kelp bed should be classified as a critical habitat area and given full protection.

3.49 LAGOONS, ESTUARIES, AND EMBAYMENTS

The subject here is shallow, semi-enclosed, and sheltered littoral water bodies, including lagoons, estuaries, small bays or embayments, sounds, fjords, esteros, etc. These basins have especially high natural values. But many have been extensively altered for economic purposes such as conversion to harbors or even large commercial ports, with adverse effects on coastal natural resources (Figure 3.68).

Littoral basins are special habitats for biological resources but also the locus of substantial economic activity (e.g., in the Caribbean, in West Africa, and in the Mediterranean). The pressure on them will increase as the population and economy of a country expands. Fisheries, shipping, commerce, industry, tourism, housing, and institutions all crowd along their shores. If improperly planned and single sector oriented, such development along the shores of littoral basins can cause severe degradation of the biodiversity and loss of natural resources. This creates a variety of short and long-term economic losses and opportunity costs as the result of resource collapse.

The most common forms of littoral basin are estuaries, lagoons, and embayments. There is some ambiguity in the terms "lagoon" and "estuary". *Estuaries* are semi-enclosed basins permanently connected to the sea whose waters are diluted by fresh water drainage, often from rivers. Estuaries exist in some form throughout the humid tropics but not often in arid/semi-arid regions (where major rivers are few and discharge is sporadic) or on small islands. *Lagoons* are basins with limited fresh water supply and high salinity, often cut off from the sea for part of the year by seasonal formation of sandbars, when they often become hypersaline. Lagoons exist widely, occupying nearly 15 percent of the world's coastline. It should be noted that the "lagoons" formed by coral reefs are not usually included in the above definition, because they are quite different water bodies (e.g., not enclosed by drylands). *Embayments* are more open littoral basins, with less restriction to the inflow and outflow of salt water and a rather high salinity (nearer to oceanic). Note that these littoral basins are not the "enclosed seas" often referred to in United Nations parlance, which are very large, deep water bodies such as the Black Sea and the Mediterranean (which are not separately treated here).

Littoral basins play a major role in the life cycles of economically important finfish and shellfish species by providing feeding, breeding, and nursery habitat (391). In many coastal areas, littoral basins

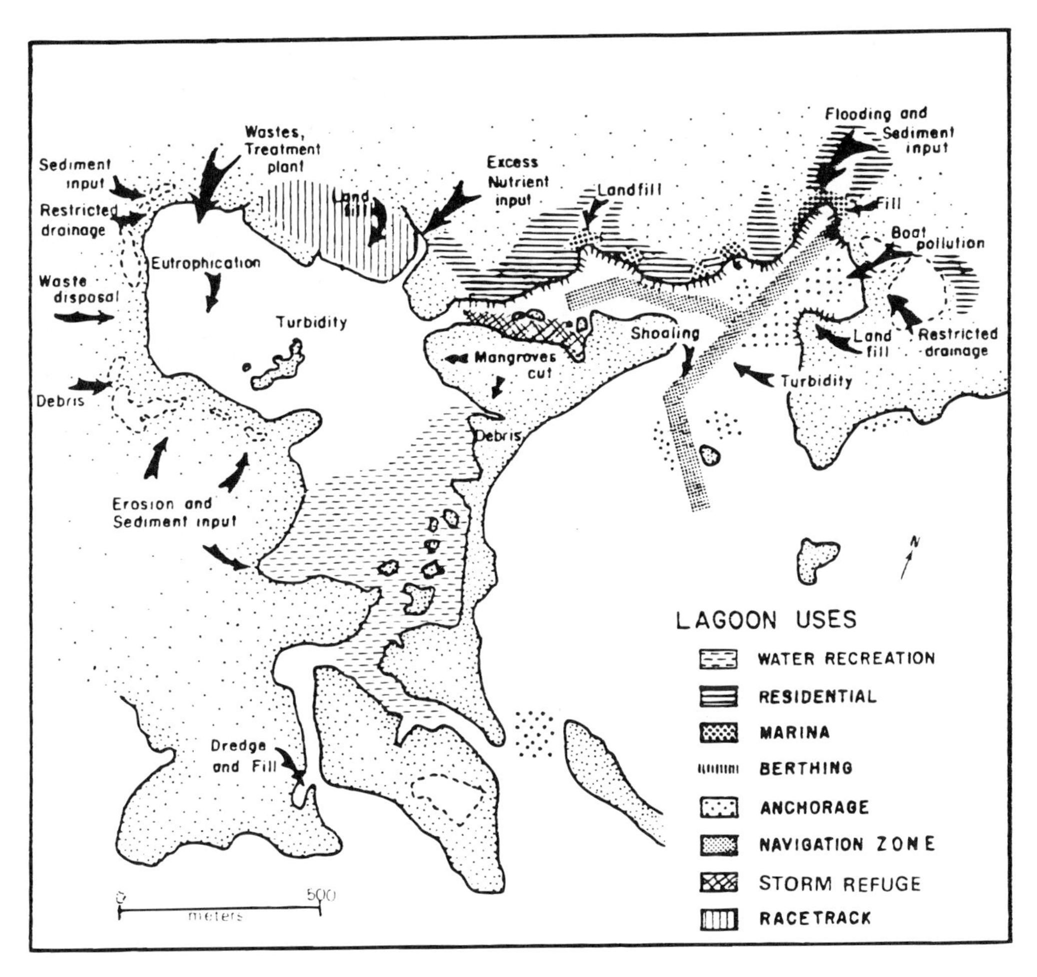

Figure 3.68. The environmental impacts and stresses on a typical mangrove lagoon in the Caribbean region (Jersey Bay, St. Thomas, USVI). (*Source:* Reference 349.)

are essential in the life cycles of most commercial species; for example, over 90 percent of all fish caught in the Gulf of Mexico are reported to be estuarine dependent to some degree (Figure 3.69) (59).

Both estuaries and lagoons maintain exceptionally high levels of biological productivity and play important ecological roles, including (a) "exporting" nutrients and organic materials to outside waters through tidal circulation; (b) providing habitat for a number of commercially or recreationally valuable fish species; and (c) serving the needs of migratory nearshore and oceanic species which require shallow, protected nurture areas for breeding and/or sanctuary for their young. Embayments of various kinds are also particularly rich in sea resources.

A key factor for estuaries is the variable salinity. A salinity gradient usually decreases from near open sea salinities at the mouth to near zero in the fresh waters at the head of the estuary and in its estuarine tributaries. The mid-range of salinities (brackish) in a true estuary can provide a rich feeding zone for fishes, particularly the youngest which often benefit from the special nurturing properties of estuaries (Figure 3.70). These nurture areas are often in the vicinity of the mangrove or grassy wetlands and tideflats around their peripheries.

Among the more important factors explaining the high production levels of estuaries are (a) organic nutrients, and oxygen provided by estuarine flora and marine inputs; (b), the benefits of solar radiation which are enhanced because of the shallow depths; (c) favorable salinity gradient; and (d) high vertical mixing rates which assist gas exchange and other functions. Also significant is the amount of wetlands, seagrass beds, and other features which provide critical habitat for numerous estuarine species.

Figure 3.69. Tidal rivers are an important link in the coastal resource system (Baños del Este, Uruguay). (Photo by author.)

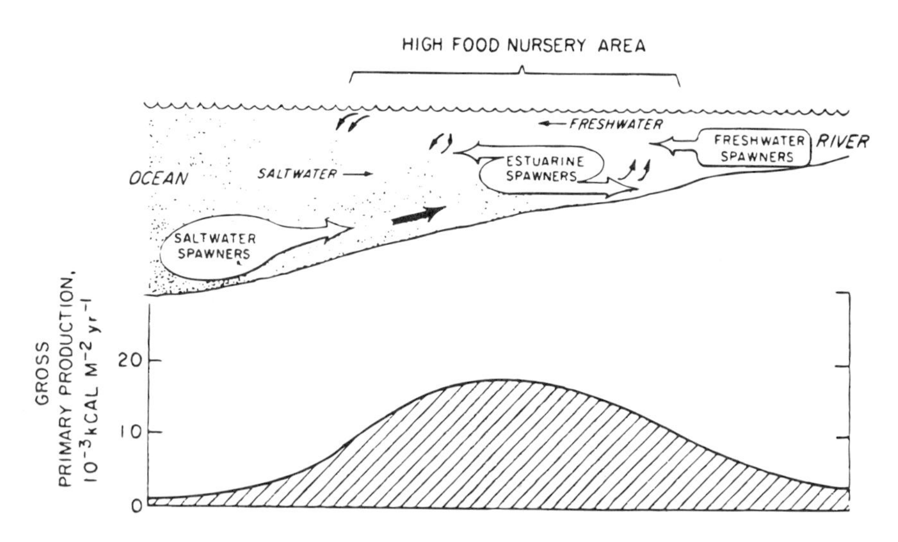

Figure 3.70. Lagoons and estuaries play an important role in the survival strategies of many marine species, especially in nurturing the young stages—estuarine productivity is exceptionally high. The area of greatest richness occurs between the ocean and the freshwater source, where intermediate salinity is found and where larval and juvenile fish and shellfish find optimum survival conditions (richness is measured as gross primary production). (*Source:* Reference 114.)

Regarding horizontal water circulation, the combined influences of freshwater flow, tidal action, wind, and oceanic forces result in the specific pattern of water movement found in any estuarine system. Tide is often the dominant force in water movement. Its amplitude varies within the system, decreasing inward from the ocean through the littoral basins and into the tributaries. Water movement also varies with the shape, size, and even the bottom material of individual parts of the system. Circulation forces tend to be lesser and flushing rates poorer when tidal amplitudes are low and the basin is deep or long. Any disruption of flow that seriously reduces flushing allows a buildup of pollution and also may permit the salinity to increase to levels that will be adverse to the biota.

These coastal water bodies have provided sustenance to sizeable human settlements dating back to prehistoric periods. But in today's world, lagoons and estuaries must accomodate a multitude of purposes including shipping, industrial and domestic waste disposal, mariculture, recreation, and residential development. Large areas of lagoons and estuaries have been "reclaimed" to create port facilities or commercial properties. Some large lagoons have been totally drained and/or filled to create real estate or agricultural land, most notably in land-scarce regions (such as in Japan and in the Netherlands).

Uncontrolled development activity can destroy critical coastal natural habitat and create sources of water pollution. A major source of degradation of shallow littoral basins is their continued use as pollutant discharge areas. Aside from outright fish kills and other dramatic effects, pollution causes pervasive and continuous degradation, evidenced by the gradual disappearance of fish or shellfish, or a general decline in the natural carrying capacity of the system. Lagoons are particularly vulnerable because they have restricted circulation.

Effects of pollution are evidenced by the gradual disappearance of fish or shellfish, or a general decline in the natural carrying capacity of the system. The most likely sources of pollution are agricultural and industrial chemicals and organic wastes. One major source of lagoon/estuary degradation is their continued usage as pollutant discharge areas. Aside from outright fish kills and other dramatic effects, pollution causes pervasive and continuous degradation. Such contaminants in high concentrations create a hostile environment that drives away fish, prevents shellfish from reproducing, and/or undermines the food chain.

In lagoons (shallow basins with little freshwater inflow and restricted openings) wind is often the main force effectively driving horizontal circulation, in both the process of mixing and that of inducing currents. Lagoons are often poorly flushed and quite vulnerable to a buildup of contaminants and to other ecological disturbances.

Generally, the typical stratified estuary is somewhat less vulnerable to pollution than its opposite, the lagoon or other estuary of mixed or unstratified type, because flushing of pollutants is more rapid. However, the bottom flow can recycle pollutants that sink, sending them inward to the upper estuary before they rise again. Also, the saltier water moving inland at the bottom may return sediment toward the land, dropping it at the head of the estuary and thus causing shoaling and restriction of flow.

The increasing usage of many of these water bodies for transport of oil, chemicals, and other toxic materials—whether by ships, barges, pipelines, or railroads—presents a continuous threat to these fragile ecosystems. This pollution is particularly damaging to internal lagoons because of their sluggish circulation which enables the concentration of pollutants to reach high levels.

Because of the importance of water circulation and flushing, activities that alter lagoon configuration can create disturbances with far reaching effects. Major adverse effects stem from construction of causeways and bridges and from dredging undertaken to create navigation channels, turning basins, harbors and marinas. Other problems arise from laying pipeline or excavating material for fill or construction.

"Dredge-and-fill" activities adversely affect the coastal ecosystem in a variety of ways. Among the potential effects of dredging are the following: (a) to create short- and long-term changes in water currents, circulation, mixing, flushing, and salinity; (b) to add to water turbidity, siltation, and pollution; and (c) to lower the dissolved oxygen.

A number of projects and activities in the estuarine basin have the potential to significantly alter the natural pattern of circulation and flushing and would be presumed to be adverse, such as (a) restricting flow through inlets and passes by constricting them with bridges, causeways, or bulkheads; (b) impeding water flow with "spoil banks" of disposed dredged material; (c) diverting water flow by channel dredging; and (d) altering water flow patterns by the encroachment of bridges, piers, trestles, and fills into coastal waters.

Because interruption of water circulation appears to be the most serious of the various physical effects from alteration of the water basin, any change in the configuration of the coastal water basin from structure or excavation that reduces water circulation is presumed to be detrimental and should be avoided. Therefore, excavation of lagoon/estuarine bottom material through dredging to create and maintain canals, navigation channels, turning basins, harbors, and marinas must be controlled.

An increasing threat to the well being of estuaries is the impoundment and/or diversion of rivers at upstream locations. When portions of the coastal watershed system are altered or short circuited, the natural flow pattern is disrupted and estuaries may be subject to surges of fresh water. This not only disturbs the ecosystem, but also increases flood hazards.

In management programs, the most confined embayments (particularly lagoons) need the maximum of protective controls: protection of wetlands, tideflats, and beaches; additional "buffer strips" above wetlands; control of sewage and storm drainage effluents; safeguards against runoff of soils, fertilizers, and biocides from the coastal upland; restrictions on industrial siting; and so forth (Figure 3.71). Such development on the shores of estuaries and lagoons, if not properly planned and safeguarded, creates a variety of short and long-term economic losses and opportunity costs resulting from resource collapse. This complex of problems can best be addressed through an ICZM-type approach.

The catchment of an estuary should be considered as part of the estuarine ecosystem and land use in the catchment coordinated with the overall aims of coastal management (180). Because many of the forces which can detrimentally alter littoral basins have their origins outside the estuarine system itself (e.g., upland agriculture, river diversion), estuarine conservation and utilization can benefit greatly from

Figure 3.71. A productive coastal mangrove lagoon that is protected to maintain biological diversity (Tarpon Bay, "Ding Darling National Wildlife Refuge," Sanibel, Florida, USA). (Photo by author.)

integrated coastal planning and management (ICZM) programs that are linked to upstream management (see also "Watershed and Upland Effects" below).

In summary, the estuarine ecosystem has a natural ability for self-maintenance and self-renewal from human or natural disturbance if its basic configuration and characteristics are maintained. On the other hand, these ecosystems are seriously jeopardized by factors that permanently alter the prevailing salinity, current, and nutrient cycling patterns. Conservation of the ecosystem and its resources can be achieved most simply by preventing any significant change from occurring in these factors. If this is not practicable under the circumstances, then every possible effort should be made to minimize the losses in biodiversity and valuable natural resources, according to the counter-measures proposed throughout this guidebook.

3.50 LITTORAL DRIFT

A common problem with sandy beaches is erosion and net loss of the sand resource caused by imbalance in accretion and depletion. A major force in controlling the balance is *littoral drift,* the term used for the drift, or lateral movement, of sand along a beachfront. Sand transport on beaches is primarily caused by currents generated by breaking waves striking the shore at an angle and then transporting the sand away at a reciprocal angle. It carries sand that has been stirred into suspension by the turbulence of the breaking waves. The strongest movement of sand due to this process is when the waves strike at about a 30-degree angle to the beach.

It is estimated that over 90 percent of nearshore sand transport takes place between the shoreline and the outer line of breaking waves, the zone of littoral drift. This portion of the nearshore area is known as the *surf zone* and can vary in width from less than 10 meters to many hundreds of meters.

On a coast facing to the east, storm waves from the northeast would produce a high rate of littoral transport toward the south. Conversely, mild wave action out of the southeast would result in a much smaller rate of littoral transport to the north. However, if the southeast waves existed for a much longer time than did the northeast waves, the effect of the southeast waves might well be more important in moving sand than that of the northeast waves. (59)

Because the amount of sand that is transported along a beach by littoral drift depends essentially on two parameters (wave height and breaker angle—the angle between wave crest and shoreline), it is possible to estimate the amount of sand transported at any point by measuring these two parameters during representative periods of the year. Sand transport values are generally calculated for a period of one year and given as a volume of sand moved past a shore point in units of cubic meters of sand per year (m^3/yr). The average annual net rate of littoral transport varies considerably from place to place and depends on the local shore conditions and alignment, as well as the energy and direction of wave action in the area, but can exceed 500,000 m^3/year for some beaches.

Littoral drift is an eternal natural process, but it usually attracts our interest only when development occurs too close to the beachfront and when bulkheads or stoneworks are installed to stop the waves and protect the hotels, houses, and other structures that should never have been built so close to the sea edge. Once this mistake has been made, it is compounded by attempts to control the erosion that is actually *caused* by these structures—specifically sand that is reflected away from the beach by them. The usual counter-measure is to try to catch the sand moving along in the littoral drift in an attempt to rebuild the beach that has been lost. This is usually done by building long perpendicular piers, called *groins,* or breakwaters. While they may hold some extra sand for the perpetrator, the downdrift properties are deprived of sand; it is often a case of "robbing Peter to pay Paul." (See also "Beach Erosion" above).

3.51 MANGROVE FOREST RESOURCES

The term *mangrove* refers to any of several dozen species of trees that are capable of living in saltwater or salty soil regimes. About 24 million hectares of mangrove forest wetlands occur in coastal areas of subtropical and tropical countries of the world. Because mangroves are sensitive to frost and freezing temperatures, the latitudinal limit of this type of wetland is determined by temperature. Mangroves are found in the intertidal zones of sheltered coastlines where wave forces tend to be reduced (Figure 3.72).

Figure 3.72. The mangrove forest is a natural habitat of very high ecological value and should always be treated with special care (Ivory Coast). (Photo by author.)

Ecologically, mangrove communities play essential roles in littoral areas. A prominent role is the production of leaf litter, which is exported to lagoons and the nearshore coastal environment (often as partially decomposed detrital matter). Through a process of microbial breakdown and enrichment, the detrital particles become a nutritious food resource for a variety of marine animals. The organic matter exported from the mangrove habitat is utilized in one form or another by the inhabitants of estuaries/lagoons, near-coast waters, seagrass meadows, and coral reefs which may occur in the area. Many of the commercial shrimps and many fish species are supported by this food source. (See also "Mangrove Forest Management" in Part 2.)

Mangrove forests occur in a variety of configurations (Figure 3.73). Some species send arching prop roots down into the water (Figure 3.74), while others send numerous vertical "pneumatophores" or air roots up from the mud. In size, mangroves range from bushy stands of dwarf mangroves, found in Gudjarat (India) or the Florida Keys (USA), to 30-meter or taller stands, found in the Rio San Juan (Venezuela) or the Sunderbans (Bangladesh, India). Mangroves propagate by producing waterborne "propagules", which are not seeds but rather embryonic plants (28).

Mangrove ecosystems also provide a valuable physical habitat for a variety of important coastal species. Waterbirds and shorebirds are well known and highly valued inhabitants of mangrove wetlands, as are alligators and even tigers, deer, and monkeys. Many of the migratory species of birds nest in the mangrove forest. Equally important (but less evident) inhabitants are crabs, shrimp, and the important juvenile stages of commercial and sport fishes, along with numerous forage species of fish, invertebrates and insects (84). Also, mangroves provide the nutrient for rich feeding grounds for many marine species from various trophic levels. Hence, many productive fishing grounds are found adjacent to mangrove areas (280).

Using peninsular Malaysia as an example, Jothy (195) found that an estimated 31 percent of the fisheries (about 200,000 tons) have an association with the mangrove ecosystem. In the west coast of the peninsula alone, 41 percent of the landings (about 178,000 tons) appear to be associated with the mangroves. In the Philippines, 71.9 percent of the total "municipal" catch from 1982 to 1986 was shown to be in association with mangroves). The mangrove forest is also a special nurturing area for

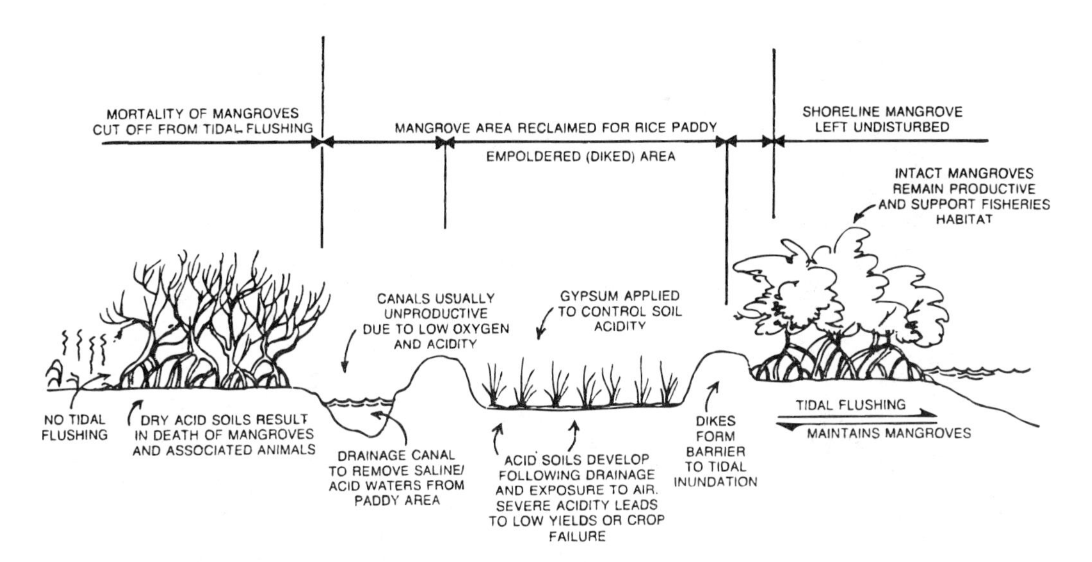

Figure 3.73. In a multiple-use setting, the mangrove forest needs to be carefully managed to avoid harming its ecological values. (*Source:* Reference 316.)

Figure 3.74. Typical South Pacific mangrove stand: (1) *Brugulera gymnorhiza;* (2) *Ehizophora samoense;* (3) *Calphyllum innophyllum.* (*Source:* Reference 250.)

the youngest stages of many important species of finfish and crustaceans. Medicinally, mangroves are good sources for organic compounds made into anti-convulsant, anti-tumor, and anti-inflammatory drugs (189).

Another factor not to be overlooked is the value of a wide mangrove forest fringe in combatting natural hazards. Shoreline mangroves are recognized as a buffer against storm-tide surges that would otherwise have a more damaging effect on low-lying land areas—particularly, they protect life and property from the huge "storm surges" that accompany cyclones (hurricanes). Also, mangroves are noted for their ability to stabilize coastal shorelines that would otherwise be subject to erosion and loss. Two of the major worldwide mangrove types are shown in Figures 3.75 and 3.76.

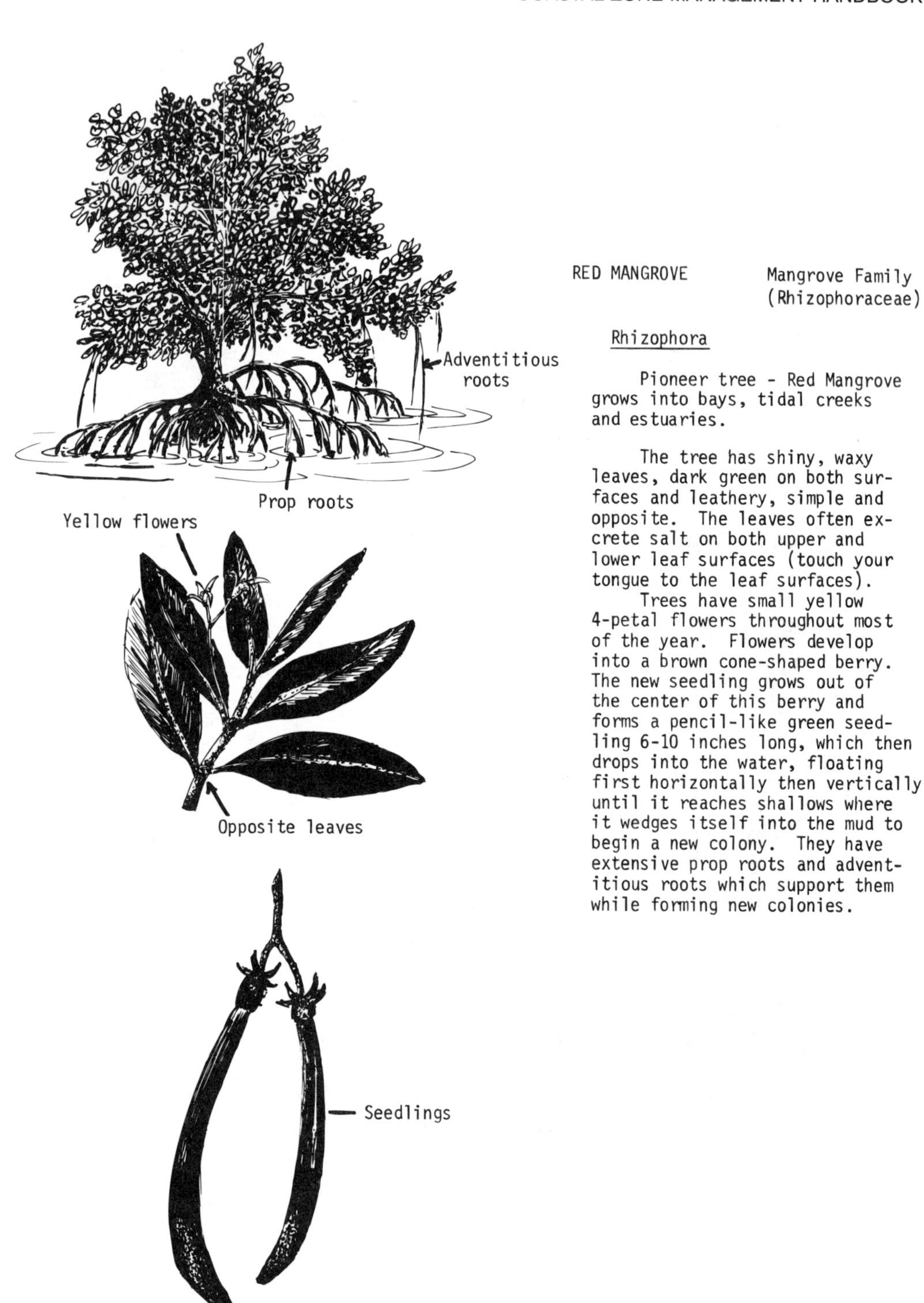

RED MANGROVE — Mangrove Family (Rhizophoraceae)

Rhizophora

Pioneer tree - Red Mangrove grows into bays, tidal creeks and estuaries.

The tree has shiny, waxy leaves, dark green on both surfaces and leathery, simple and opposite. The leaves often excrete salt on both upper and lower leaf surfaces (touch your tongue to the leaf surfaces).

Trees have small yellow 4-petal flowers throughout most of the year. Flowers develop into a brown cone-shaped berry. The new seedling grows out of the center of this berry and forms a pencil-like green seedling 6-10 inches long, which then drops into the water, floating first horizontally then vertically until it reaches shallows where it wedges itself into the mud to begin a new colony. They have extensive prop roots and adventitious roots which support them while forming new colonies.

Figure 3.75. The red mangrove. (*Source:* Courtesy of William Hammond, Reference 59.)

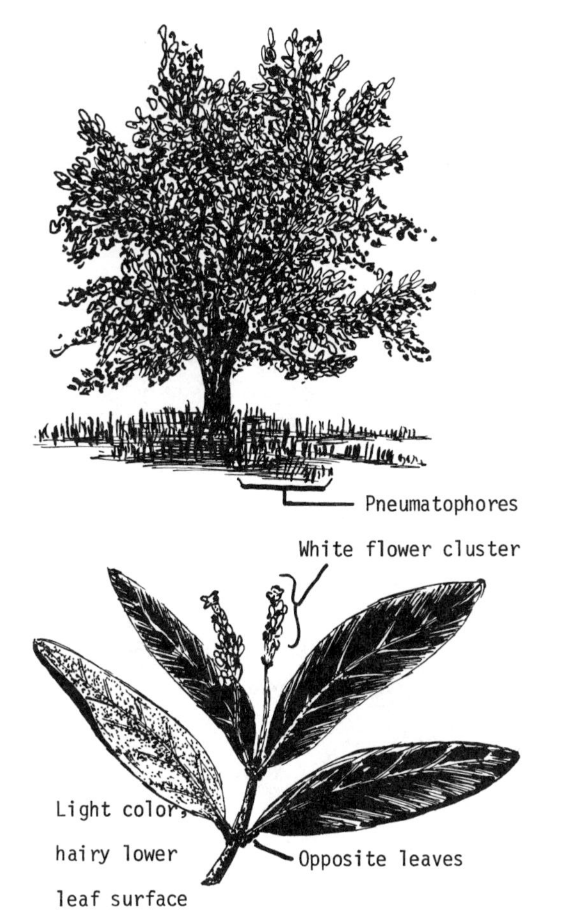

BLACK MANGROVE Verbena Family (Verbenacea)

Avicennia

Black Mangrove is closely associated with Red Mangrove stands on the shoreward borders.

Black Mangrove has opposite oblong to elliptical green leaves that are a lighter color and hairy underneath. It has small white flower terminal clusters with four-lobed flowers.

The seed pod resembles a large lima bean about an inch in diameter.

Black Mangrove is very efficient at excreting salt through its upper leaf surface. In India and Asia this sometimes is used as a source for salt.

The Black Mangroves generally have a surrounding carpet of pencil or finger-like projections called pneumatophores sticking up from the soil surrounding the tree. Their function is thought to be related to support and respiration. In any event, they form ideal habitat for juvenile fish to hide, feed and mature safe from large predatory fish.

Figure 3.76. The black mangrove. (*Source:* Courtesy of William Hammond, Reference 59.)

Mangrove forests are usually common property—owned by all the people—because they are usually intertidal and belong to the sea, not the land. They are often heavily used for subsistence. In many parts of the world, human populations rely heavily upon the variety of products that can be obtained from mangrove forests such as timber, pulpwood and chips, fuelwood and charcoal, honey production, and sundry domestic products. Below is an extensive list of mangrove forest products, provided by Saenger *et al.* (304):

Fuel

Firewood (cooking, heating)
Charcoal
Alcohol

Construction

Timber, scaffolds
Heavy construction (e.g., bridges)
Railroad ties
Mining pit props
Boat building
Dock pilings
Beams and poles for buildings
Flooring, paneling
Thatch or matting
Fence posts, water pipes, chipboards, glues

Fishing

Poles for fish traps
Fishing floats
Wood for smoking fish
Fish poison
Tannins for net and line preservation
Fish attracting shelters

Textiles, leathers

Synthetic fibers (e.g., rayon)
Dye for cloth
Tannins for leather preservation

Food, drugs, beverages

Sugar
Alcohol
Cooking oil
Vinegar
Tea substitute
Fermented drinks
Dessert topping
Condiments from bark
Sweetmeats from propagules
Vegetables from propagules, fruit or leaves
Cigar substitute

Household

Furniture
Glue
Hairdressing oil
Tool handles
Rice mortar

Toys
Matchsticks
Incense

Agriculture

Fodder and "green manure"

Faunal items

Fish
Crustaceans
Shellfish
Honey
Wax
Birds
Mammals
Reptiles and reptile skins
Other fauna (amphibians, insects)

Other products

Packing boxes
Wood for smoking sheet rubber
Wood for burning bricks
Medicines from bark, leaves and fruit
Paper of various kinds

The value of the mangrove resource in terms of its marketed products can be expressed in economic terms. The "free" services provided by the mangroves are difficult to measure and consequently are often ignored. These "free" services would cost considerable energy, technology and money if provided from other than natural sources. Since these values are seldom taken into account in the governmental decision process, the total value of the mangrove resource is most often quite significantly understated (156).

Mangrove forests are too often considered wastelands of little or no value unless they are "developed." All too often, this term means conversion of the mangrove ecosystem to some other form of use that is assumed to be of greater value. In these situations, the basic habitat and its functions are lost, and that loss is frequently greater than the value of the substituted activity, on a long term basis. In general, these kinds of problems are generated by ignorance of the resource value of the functioning mangrove system and the absence of integrated planning that takes these functions and values into account.

The conversion of mangroves for aquaculture and other land uses is detrimental to those marine species whose survival depends upon mangroves. In Ecuador, the large-scale reclamation of mangrove areas for shrimp ponds has contributed to the decrease in shrimp fry (baby shrimp) availability for stocking because most shrimp farms still depend on natural wild fry collection. In Asia, many aquaculture operations rely on the collection of naturally occurring seed stock of penaeid shrimp and finfish like milkfish, groupers, snappers and sea bass that are mangrove dependent. Although the commercial hatchery production of penaeid shrimp and sea bass has been attained, many hatcheries still depend on wild sources of broodstock (280).

Thus, the destruction of mangroves—as in the Ecuador experience—affects the availability of fry and broodstock and, consequently, aquaculture production. In terms of capture fisheries, the same can also be said. Low recruitment will consequently affect production. With fishing grounds throughout the tropics already overexploited, mangrove destruction can only further reduce stock recruitment and production.

There is clear evidence that the April 1991 cyclone—which ravaged the S.E. coast of Bangladesh and killed nearly 140,000 people—would have been far more devastating had there not been mangrove along part of the shoreline (some natural, some planted). In a post-cyclone field trip in 1991, this author was able to confirm that the littoral strip mangroves planted by the Bangladesh government in the 1980s did save thousands of lives and millions of dollars worth of property during this event. To increase protection, Bangladesh is now planning to complete a "greenbelt" of mangrove and other vegetation along 1,760 km of coastline (more than 100,000 ha have already been planted). The primary purpose is attenuating storm surge from giant cyclones—other purposes are to stabilize sediments and "build" land as well as gain the products and conservation benefits of mangroves (see "Greenbelt" above). (303)

While mangrove forests provide life support and income for millions of people, few countries have created effective mangrove conservation programs. Where appropriate laws have been passed, enforcement is often inadequate. The lack of an overall national conservation policy or an ICZM-type program for the coast often weakens the potential for enforcement of laws and regulations.

Then, too, mangroves are socio-economically valuable if just left in place, because they preempt development sites that are at too low an elevation, are hazardous real estate sites, or eliminate valuable ecological resources. For example, one of the strategies of the Gulf of Kutch Marine Park in India is to replant as much destroyed mangrove as possible to "reclaim" the tideflats as a viable part of the park and particularly to prevent the spread of the huge solar salt industry (310), which has already taken over 20,000 ha of mangrove tideflats (see "India" case history in Part 4).

The widespread depletion of the remaining 24 million hectares of coastal mangrove forest of the tropical world from all these causes is a troubling example of rapidly shrinking littoral habitat. Several approaches are available for conservation of mangroves—general regulations prohibiting their removal, establishing single-use (protection) preserves, and managing them as multiple-use areas (protection and use). Multiple-use areas would attempt to optimize all the compatible uses, socially, economically, and environmentally. (See also "Mangrove Forest Management" in Part 2).

In an example of emerging multiple use, Bangladesh allows a yield of timber from the buffer zone of 650 sq km around the Reserve Forest of the Sundarbans, much of which is a nature reserve (42.2). Production is 4,500 m^3/yr and of fuelwood 18,500 t/yr (i.e., 37,000 m^3/yr), the felling cycle being 20 years. However, the forest area of the Sundarbans, with an expanse of 4,264 sq km, has a tiger reserve of 2,585 sq km including 1,830 sq km of the core area. Spotted deer and a multitude of other species also take refuge there. The Sundarbans are endowed with 30 out of 53 species of the "true" mangroves of the world. In another two decades, the demand for the fuelwood is likely to go up by 20 percent which warrants a rapid afforestation program in neighboring West Bengal (India), including enough mangrove area to produce 13.5 mt fuelwood for the state by 2005 A.D. Fortunately, there are patches of barren lands which are available for an afforestation program covering an area of 50,000 ha. (189).

Integrated coastal zone management (ICZM)—which involves simultaneous attention to all sectors and considers the maximum sustained yield of each resource, including fisheries—is an approach that is especially important in the management of mangrove forests. For one reason, an overall national conservation policy and a ICZM-type program for the coast strengthens the potential for enforcement of laws and regulations. For another, it enables maximum sustainable utilization of resources through multiple use approaches.

Therefore, in the face of growing pressures on coastal resources it seems essential to conserve mangrove wetlands which are one of the most important of the world's littoral habitats. It is clear that the thirty or so developing nations in the humid tropics with extensive mangrove forests should have a strong incentive for integrated coastal management (ICZM). Most developing nations with extensive mangrove forests are confronted with similar stresses which threaten the sustainability of this renewable resource. Conversion of mangrove forest to other uses should be considered only for emergencies. (326)

3.52 MARINAS

The environmental impacts of marinas and small harbors depend on site location, design, construction methods, and "housekeeping." Proper site planning can help avoid or minimize many of the impacts that might otherwise result from marina development (Figure 3.77). Designing the marina to take maximum advantage of the natural attributes of a site can contribute significantly to reducing or eliminating potential environmental problems from marina construction. During marina operation and maintenance, implementation of an operations and maintenance plan can contribute significantly to the

Figure 3.77. Small boat harbors, or marinas, are essential facilities of the coastal zone. Careful site planning can help avoid unecessary environmental impacts caused by their construction and operation. (Photo by M. Fahay.)

environmentally sound performance of the marina facility. Alternative measures taken during marina design and construction are directed toward avoiding adverse impacts on water quality and aquatic biota, grassbeds, wetland habitats, and protected species. Following are some guidelines for marina facilities (165):

Habitats: In siting and constructing marinas, great care should be taken to protect critical habitats, such as mangrove forests, particularly valuable tideflats, waterbird nesting sites, oyster bars, etc.

Dredging: Where dredging is necessary, provisions should be made to dispose of dredged material in upland areas away from the marina. Suction head dredges should be used for this work around marina structures to prevent damage that dragline and clamshell dredges may cause. Silt cutains shouild be placed around the work site.

Runoff: Maintaining water quality within the marina basin requires effective management. Paint spraying, sand blasting, engine repairs, boat washing and similar maintenance activities should be performed on shore and where pollutants can readily go into marina waters. Use of non-phosphate detergents can greatly reduce the amount of nutrients entering marina waters.

Boat Wastes: Properly maintained and convenient waste disposal services, including garbage disposal and onboard wastewater collection, are important (facilities for pumping out should be provided). Oily wastes in the bilge water can be reduced if boats are fitted with oil filtering devices on bilge discharge or by pumping bilge water into a shoreside facility.

Recreational boats are a source of water pollution. In one hour 10 modest size pleasure boats can discharge the following: BOD, 14 g; coliform bacteria, 380 million individuals; non-volatile oil, 200 g; volatile oil, 113 g; phenol, 2.4 g; and lead, 1.2 g (80.3).

Other sources of pollutants are: boat maintenance activities, including fiberglass repair, washing, sanding and painting—antifouling paints should not contain significant amounts of PCBs or TBT. Sanding and painting boats add toxic materials to the aquatic environment from antifouling paints. Any sewage that leaches into marina water from malfunctioning (poorly designed, sited or maintained) septic systems contribute to the overall coliform bacteria levels of marina waters.

3.53 MARSHLANDS

Intertidal areas of the coastal edge are sometimes in marshes (grasses and rushes) (Figure 3.78). Marshes, where they exist, serve many of the same ecological purposes as mangrove forests. They assimilate nutrients and convert them to plant tissue which is broken into fine particles and swept into the coastal waters when the leaves die. This plant detritus provides the food base for many important species of fish and shellfish. In addition the marsh provides a special habitat for many species, including the young stages, a nurturing place. (See also "Tideflats" below.)

Furthermore, the tidal marshes contribute organic material (detritus) to the base of the food chain. Substantial portions of such estuaries consist of tidal marshes, coastal wetlands, and streams that feed into the marshes and estuaries. The tidal marshes proper are dependent in great part for their high biological productivity on nutrients from inland and upland sources delivered by the freshwater inflow systems. (45)

Tidal marshes also provide excellent cover (habitat) for the young of estuary-dependent fishery resources and for all life history stages of many wildlife species. In addition, the marsh vegetation helps to weaken the onslaught of storm-generated waves and acts as a reservoir for coastal stormwaters, so that shorelands and nearby settlements are protected. (45)

Even more striking, however, is the ability of estuary-tidal marshes to accomplish waste treatment through the "tertiary" stage of nutrient removal and assimilation. This valuable contribution, unaided by man, is free and has always been taken for granted, or its value has not been recognized (45). The

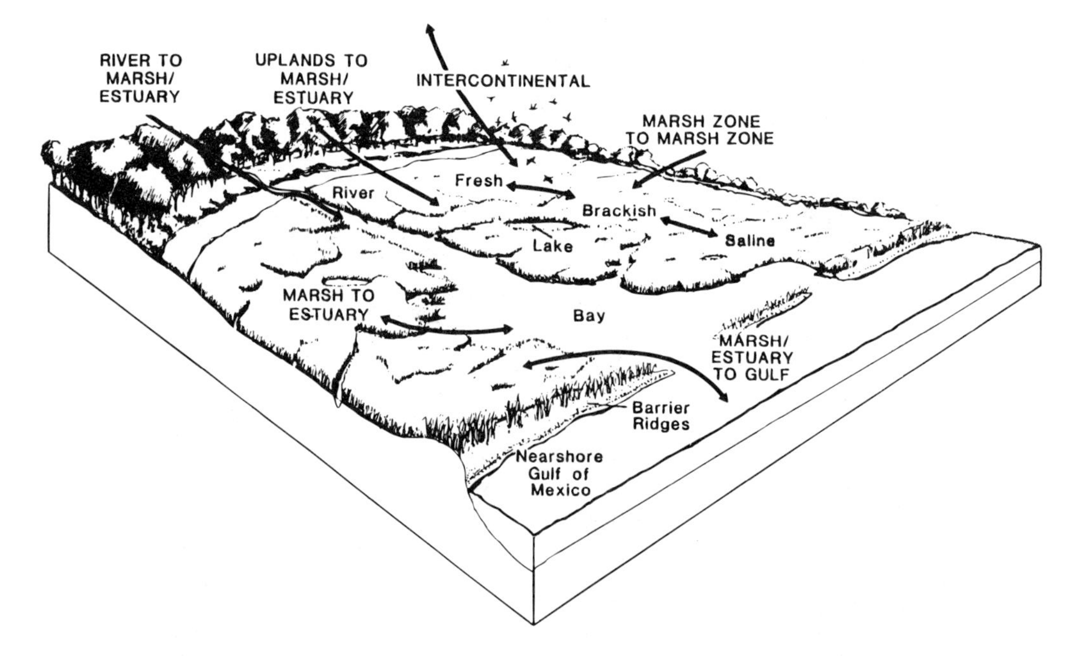

Figure 3.78. Wetlands, like this deltaic marsh, are closely linked to sea and upland ecosystems. (*Source:* Reference 147.)

value of tidal wetlands as pollution filters has been documented: data from the Tinicum marshes near Philadelphia indicate 50 to 70 percent reductions in nitrate and phosphate levels several hours after the waters from sewage and effluent passed over a 500-acre tidal marsh. (259)

In terms of wildlife productivity, the salt marshes play a major role by contributing food, shelter, and nesting sites for thousands of waterfowl. In the United States, there are wading birds, such as the herons and egrets, marsh hawks, ospreys, rails, marsh sparrows, and some 50 other species in the marsh avifauna (Figure 3.79). Among the reptiles, the diamondback terrapin is unique to the tidal marshes. Muskrats are abundant and "harvested" for their fur; a variety of other mammals are frequent visitors and indirectly dependent on these wetlands. (259)

Tidal marshes are among the most productive ecosystems in the world. For example, in the southern marshes of the United States, up to 10 tons of marsh grass may be produced annually. In the Northeast this figure ranges from 3 to 7 tons. Recently it has been documented that tidal wetlands are important pollution filters (see also "Tideflats" below). Data from the Tinicum marshes near Philadelphia indicate 50 to 70 percent reductions in nitrate and phosphate levels several hours after the waters from sewage and effluent passed over a 500-acre tidal marsh. (259)

Geologically, too, the marshes play a significant role. They act as sediment accretors or as depositories for sediments, thereby reducing the frequency of dredging for navigation. This in turn reduces the potential for smothering shellfish and other bottom organisms in the estuary, an indirect result of dredging. During severe storms the marshes exhibit resiliency and thereby act as buffers to protect the developed continuous shoreline. (259)

Salt marshes have suffered greatly from development activities. Filling, dredging, ditching, impounding, and draining, as well as polluting, have greatly reduced the total acreage. In Connecticut over 50 percent of the marshes have been destroyed, and somewhat similar trends have occurred in other highly populated coastal regions. Some human activities have actually obliterated the marshes;

Figure 3.79. Wetlands provide essential habitats for numerous species of birds and other wildlife (Sacramento Wildlife Refuge, California, USA). (Photo courtesy of U.S. Fish and Wildlife Service.)

others have modified their biotic composition and productivity. Ditching has had a profound effect in reducing invertebrate productivity and modifying the vegetational pattern. Impounding destroys the salt-marsh vegetation and results in fewer saline water bodies favoring waterfowl. Causeways constructed for highways and railroads restrict tidal flow, thereby modifying the vegetation and encouraging the less desirable *P. communis.* Tidal gates designed for flood protection also restrict tidal flushing. (259) The modern solution is thoughtful development planning and use of mitigation techniques (Figure 3.80).

3.54 MINING

Several types of mining bring coastal environmental problems. Here we discuss mostly the two types of mining that take place in the water for which the products are mostly construction materials, sand, and limestone from coral.

Sand for construction is a valuable commodity, but it must be remembered that taking sand from any part of the beach or the nearshore submerged zone can lead to erosion and recession of the beachfront. The key to the natural protection provided by the beachfront is the sand which is held in storage and yields to storm waves, thereby dissipating the force of their attack. Consequently, beach conservation should start with the presumption that any removal of sand is adverse and should be prohibited, unless it can be shown to be naturally replaceable; that is, if it is a sustainable use of a renewable resource (65).

Sand mining in lagoon environments in Fiji involved clamshell dredging of holes, which caused slumping of seagrasses into the depressions. Suction head dredging at Keauhou (Hawaii) involved use of a jet probe to minimize sedimentation however, the excavation craters eventually led to the undermining and collapse of fragile finger corals living on the sand within 100m of the craters. In contrast, augering and clamshell operations at Barbers Point, killed only a few adjacent corals, and the remaining halves of corals cut by the auger survived. (231)

Increased turbidity is often caused by excavation in coastal waters. The excess suspended matter tends to settle out on shallow bottoms, where it may be readily resuspended by wind action or boat

Figure 3.80. One form of “environmental mitigation” is maintaining wildlife habitat on wetland project sites; for example, the wetland and beach are preserved at the “Port Liberté” project site across the Hudson River from New York City. (Photo by author.)

traffic (59). But the direct damage potential is expected to be highest where a critical habitat lies adjacent to a donor site (source of material) or disposal site (place of deposit)—whether a beach or a submerged area.

Special considerations are needed to ensure that the environmental impact of dredge mining for sand is minimized (see also "Sand Mining" below). Proper management of the dredging operation is aimed primarily at controlling the effects from the reintroduction into the water column of polluted bottom sediments. Major problems are increased water turbidity and the release of large quantities of trapped nutrients, organic materials, and toxic pollutants in the spoil. Short-term effects that may be expected are clogging of the gills of aquatic species (with silt); reduced light penetration; eutrophication; depletion of dissolved oxygen content; and uptake by organisms of heavy metals, pesticides, or other toxic substances stirred up by the dredging, which may accumulate in their tissues to extremely high concentrations. (59).

Coral mining for building materials takes place in many countries of Southeast Asia. Blocks of coral are used for fences, houses, churches, road surfaces, etc. Burning of coral produces quicklime, a type of mortar. Because coral reef systems provide such great benefits in terms of fish production, tourism, and shore protection, they should be protected from most removal of coral material. (See "Coral Reef Management" in Part 2 and "Indonesia, Bali," "Tanzania," and "Sri Lanka" case studies in Part 4.)

3.55 MULTIPLE USE OF RESOURCES

Multiple use is starkly simple in concept but often difficult in fulfillment. The multiple use premise is that a single resource area can be shared by multiple users with an optimum socio-economic outcome guaranteed through use of an equitable balancing process. But there are many political and bureaucratic obstacles, particularly where the coastal commons are involved (See "Commons" in Part 3).

Multiple use contrasts with *exclusive use,* in which only a single use is permitted for an area, say aquaculture, military activity, or nature protection. Because public benefit should be the main criterion for use of the commons, the mix of uses should be that with the maximum socio-economic benefit. The mechanics can often be accomplished through zoning (see "Zoning" in Part 2).

According to Burbridge (31) the potential benefits that complex resource systems—such as estuaries and mangrove forests—offer to social and economic development depend on maintenance of the functional integrity of these natural systems. Also, factors such as the potential utility of the goods and services, manpower, and market forces play a significant role in the way natural systems are used. The multiplicity of benefits these natural systems offer is often overlooked because most development plans for coastal areas focus upon maximizing single purpose uses. The management systems created may be too narrow and generally fail to harness much of the resource potential offered by natural systems (31). As a result, many opportunities for the improvement of economic and social welfare of coastal people are not realized (Figure 3.81).

In economic terms, the multiple-use concept requires that all actual and potential uses for resource utilization schemes be determined so as to ensure that the sum of the "opportunity costs" is minimal. Opportunity costs represent the value of those lost options (or opportunities) that would otherwise be derived from using other resources, as opposed to committing one resource for an exclusive use. For example, in determining the allocations of fresh water for either irrigation or fishery maintenance, if the use of fresh water is exclusively for irrigation, then this imposes an opportunity cost for fisheries which equals the annual income which could have been obtained but is now lost due to its collapse (e.g. the Egyptian sardine fishery in the Mediterranean and the hilsa fishery in Pakistan).

A major advantage in a balanced multiple-use program is that investment risk is lowered. The multiplicity of uses provides a hedge against failure of any one use of a resource and enables flexibility in face of unexpected change in the market or natural variation in the productivity of the resource. In summary, the maximum flow of natural goods and services from a coastal resource system can be expected in a multiple-use ICZM approach. (31)

There are cases of single-use development of coastal resources that economically, even in the short-term, are quite inefficient according to Dahuri (101). A dramatic example of economic inefficiency in single-use is the massive withdrawal of coastal lowlands and mangroves from common property reserves

Figure 3.81. In a planned multiple-use approach, fishing and tourism, along with many other livelihoods, can be pursued simultaneously; however, others such as mining of beach sand cannot. (Photo by author.)

and their conversion to brackish water shrimp ponds on the North Coast of West Java, Indonesia (see case study in Part 4). Then because production costs often exceeded revenue, many shrimp culture productions in Indonesia collapsed leaving destroyed managrove forests in their wake. Also, many people lured from other occupations—salt production (Figure 3.82), rice cultivation, mangrove related products, etc.—lost their jobs. The loss of mangroves coincided with a significant decline in catches by coastal fishermen who were dependent on species nurtured by the mangrove forest.

One large obstacle to equitable sharing of resources in a multiple use approach problem is that there may be both winners and losers and the outcome is not too often a democratic one in some countries. For instance, it is less than likely that the potential losers will become actual losers *if* they are among the politically powerful, as it is also less than likely that the potential winners will become actual winners if they are of the common people. Nevertheless, some countries will be able to administer balanced multiple use approaches following democratic, egalitarian, principles.

Another very basic problem is that responsibility for complex natural resource systems is given to agencies which represent only one economic sector. For example, because mangroves are trees, mangrove forests are often made the responsibility of a Ministry of Forestry. While alert foresters may become aware of the value of mangroves in controlling coastal erosion and the importance of mangroves to the conservation of fisheries, their primary responsibility is to develop mangrove forests for wood production; therefore they may not be asked to conduct broadly based environmental and economic assessments of mangrove systems or to develop management plans for sustainable uses other than forestry (31). Sustainable extraction of timber and secondary forest products from mangrove is difficult to achieve and provides relatively modest revenues to forestry departments. The limited revenues earned provide little incentive to manage mangrove forests in a manner which will sustain secondary uses such as fisheries or honey production (33). But mangrove forest management clearly needs the multiple-use approach.

Management practices should optimize the conservation of mangrove resources in a multiple use way so as to provide for traditional and contemporary human needs, while ensuring adequate provision

Figure 3.82. Uses such as salt manufacture can be compatible with other uses, such as agriculture, in a multiple-use coastal zone management program. (Photo by author.)

of reserves suitable for protection of the diversity of plant and animal life within them, according to Hutchins and Saenger (180). Being a renewable resource, mangrove ecosystems should be managed on a sustainable-use basis—providing sustainable economic returns while maintaining the ecosystem as close to its natural or original state as possible.

A single use, such as preservation area (nature reserve or national park), may be desirable for certain parts of extensive mangroves to serve as refugia for fauna and flora and as a resource for restoring areas in which management policies have failed or accidents have occurred. Preservation can buffer the area generally and can be an advantageous part of an overall multiple-use management plan.

In summary, integrated, multi-sectoral, resource planning and management are a needed element in the coastal zone. To consider multiple-use potential, decision makers need to know what economic and social uses can successfully coexist in an area and which cannot (Figure 3.83). In most situations the multiple-use approach should be advocated. In some situations, however, exclusive use could be recommended.

To handle the increasing need for multiple-use management, new institutional mechanisms are needed, and some are being tested now. The United States, Austalia, India, and several other countries are exerimenting with different forms of multiple-use balancing (see "Situation Management" and "Multiple-Use Management Authorities" below). It is nearly always a mistake to postpone a decision at one of the early decision-making stages until all the information necessary for a later decision-making stage is obtained.

One suggested sequence or hierarchy of decision making in establishing and managing a coastal multiple-use management area is (203):

Stage 1. Legal establishment of boundaries

Stage 2. Plan for allocation by zoning

Stage 3. Enactment of zoning regulations

Stage 4. Specific site planning

Figure 3.83. Farming is compatible with nature protection in the Baños del Este Biosphere Reserve in Uruguay. (Photo by author.)

Stage 5. Specific site regulation

Stage 6. Day-to-day management

Stage 7. Review and revision of management

At each of these stages of decision making, the following factors should be taken into account. (However, the level of detail in which these factors are presented and considered should increase from Stage 1 to Stage 7) (203):

- Geographic habitat classification
- Physical and biological resources
- Climate
- Access
- History
- Current usage
- Management issues and policies
- Management resources

It is recommended that the above be incorporated into an ICZM plan.

3.56 MULTIPLE-USE MANAGEMENT AUTHORITIES

Historically, conservation of critical marine resources has been accomplished using either: (a) *regulatory means;* that is, using regulatory power (police power) of government to set up rules, each usually affecting a particular user group (sector), and often unaccompanied by specific boundaries; or

(b) *proprietary means;* that is, exercising government's trusteeship (over common property) or ownership rights (to an area having specific boundaries), as a nature reserve, national park, or etc.

Because of significant limitations in this "either/or" approach, there has arisen a new form of resource administration—the multiple-use marine *management authority.* Such authorities have a modern format but use many of the traditional methodologies. However, you would not necessarily find them titled by the generic term "management authority"; instead, terms such as "park authority" or "sanctuary administration" may be used. Examples of existing and emerging authorities include: the Great Barrier Reef Marine Park Authority of Australia and the Florida Keys National Marine Sanctuary of the USA. Both have purview over quite large areas of the sea (see the case studies for each in Part 4).

Replacement of single-use management by multiple-use approaches is also occurring in many resource limited developing countries. A trendsetter is UNESCO's "biosphere reserve" concept which would reserve a modest part of a larger conservation area for strict nature protection ("the Core Zone") but prescribe eco-friendly rural uses for the rest (the "Buffer Zones"). This concept of controlling uses in a buffer zone, rather than prohibiting them, has been named "ecodevelopment" in some countries. But giving the concept a name is much easier than giving it life.

Regarding regulations, if not targeted at specific sites they are often difficult to implement; therefore, it may be a benefit to identify for special attention those particular areas that need conservation management. However, reserving a large area exclusively for nature protection, as traditionally employed for national parks and nature reserves, may be less than appropriate in today's use-oriented world. Also, broad areas managed for multiple use (including nature protection) are superior to the opposite—isolated, highly protected pockets, in an area that is otherwise unmanaged or is subject to regulation on a piecemeal basis—according to Kelleher and Kenchington (203). One solution to these problems is to opt for defining a sizeable multiple use management area to accomplish both nature protection and use-oriented resource management. This could best be administered by a modern multiple-use management authority such as that established for the Florida Keys, USA, (Figure 3.84).

Coastal resource use, particularly, should be based upon sustainable multiple use principles (see "Multiple Use" below). Because of the linkages between marine environments and between marine and terrestrial environments it is important to make provision for the control of activities which occur outside a management area which may adversely affect features, natural resources or activities within the area (203). A collaborative and interactive approach between the collateral jurisdictions is essential. One solution is to establish a coordinative mechanism in a "Zone of Influence" surrounding the area.

Politically, it is most difficult to achieve the optimal balance of uses in a surrounding buffer area, even when the management agency owns the land and has full authority. The park or reserve manager may not be fully supported by politicians or superiors in trying to stop adverse activities—such as, tree cutting, game poaching, cattle grazing, or shrimp culturing—when jobs or food security are at stake.

Basically, multiple-use management requires entirely different approaches than traditional park or nature reserve management. In recognition of this, countries like India are re-training some of their managers in subjects like ecodevelopment, rapid rural appraisal, and social forestry (274). But, the governments for which the managers work, must also change the institutional structure to accomodate new, multiple-use, management approaches. The structure recommended here is the multiple-use management authority, as used to manage the Great Barrier Reef. But to implement this would likely require special legislation.

Such legislation can be justified on the grounds of a worldwide failure of traditional piecemeal protection of marine and littoral areas, which has lead to diminished biodiversity and depletion of seafood supplies. The Management Authority and its powers should be authorized by the highest body responsible for such legislative matters in most countries. Legislation should identify and establish institutional mechanisms and specific responsibilities for management and administration of marine areas. The legislation should contain enough detail for (203):

- Proper implementation and compliance
- Delineation of boundaries
- Providing adequate clarity of authority and precedence
- Providing infrastructure support and resources to ensure that the necessary tasks can be carried out

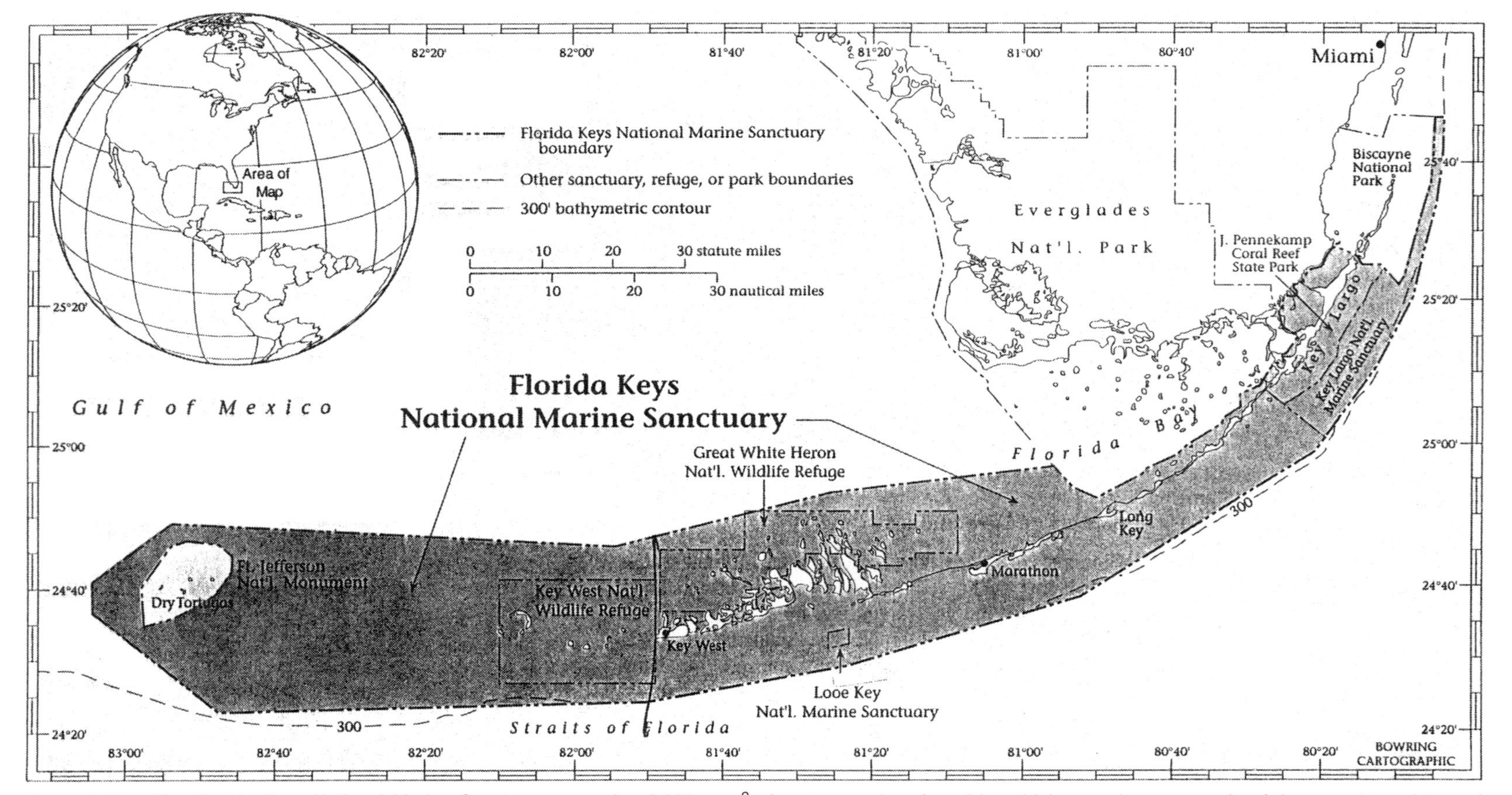

Figure 3.84. The Florida Keys National Marine Sanctuary—covering 2,800 n.m.^{2} of water, coral reef, and intertidal area—is an example of the recent trend toward multiple-use management programs. (*Source:* Courtesy of Center for Marine Conservation and U.S. National Oceanic and Atmospheric Administration.)

The legislation should be guided by these objectives (139):

- Provide for conservation management over large areas
- Provide for a number of levels of access and of fishing and collecting in different zones within a large area
- Provide for continuing sustainable harvest of food and materials in the majority of a country's marine areas

Coordination of planning and management, by all intragovernment, intergovernment, and international agencies with statutory responsibilities within areas to be managed, must be provided within the legislation. Ideally, the legislation should provide the Management Authority with jurisdiction over all marine and littoral resources of flora, fauna, island terrain and overlying water and air. It would be helpful if each Authority were to operate within a broader ICZM program. Full participation of stakeholders should be established in the legislation—resource management programs depend critically on the support of local people (203).

Legislation and management arrangements should grow from existing institutions if possible, following these guidelines (203):

- Creation of new agencies should be minimized.
- Existing agencies and legislation should be involved by interagency agreements where practicable.
- Existing sustainable uses should be interfered with as little as practicable.
- Existing staff and technical resources should be used wherever practicable.
- Unnecessary conflict with existing legislation and administration should be avoided.
- Where conflict with other legislation and administration is inevitable, precedence should be defined unambiguously.

A master plan should be prepared (see "Management Planning" in Part 2). It is particularly important to base the plan on the concept of zoning (see "Zoning" in Part 2). Opportunities should be provided for the public to participate with the Management Authority in preparing the master plan, including: the preparation of the statement of purpose and objectives; the preparation of alternative plan concepts; and the preparation of the final plan.

Management strategy should be consistent with the legal, institutional and social practices and values of the nations and peoples enacting and governed by it. Consideration should be given, where local rights and practices are firmly established, to arrangements for specific benefit to local inhabitants in terms of employment or of compensation for lost rights. Or, in the spirit of ecodevelopment, to helping to improve their welfare in many other ways (infrastructure, health, education, water).

3.57 NATURAL HAZARDS

As many experienced planners and managers already know, the main threat to the shoreline is usually from development on land next to it. Shoreline protection requires coordinated management of the shore edge and the land behind it, as well as a way to limit buildings, control excavation, and limit beach protection and inlet structures (79, 58).

Cyclonic storm surge and wave runup which elevate the sea surface are often more damaging than wind. The short-lived but intensive winds of hurricanes and cyclones exert enormous forces on both natural and built systems. They drive before them masses of rising water—known as *storm surge*—which in very intense storms can elevate normal water levels to 20 feet or more above normal sea level in particular locations. A moderately intense storm may raise water levels 6 to 12 feet above normal (such as Hurricane Alicia which struck Galveston, Texas, USA, in 1983 with maximum winds about 110 mph). The notorious cyclone that struck Bangladesh in April 1991 created a surge higher than 20 feet above mean seal level. Storm winds also build waves on top of the storm surge, which can increase flood elevations by as much as 55 percent over the storm surge level. The huge hurricane "Gilbert" that hit Yucatan in 1988 destroyed houses and perforated the beachfront (Figure 3.85).

A wave's energy is dissipated as the wave strikes the coastline as beaches, dunes, vegetation, and structures built in the wave zone absorb this energy. At the boundary of land and sea, beaches and dunes yielding masses of sand to waves, act as efficient dissipators of wave energy. Where beach sand is deposited in bars offshore, water depth is decreased and storm waves break farther from shore, thus protecting the land area. Sand deposited further landward (e.g., in dunes) increases the effective width and volume of the beach, further reducing storm impact. This beach and dune system needs to be protected.

Coral reefs also protect the shore by reducing wave impact. An example of damage is provided by Bali, Indonesia, where coral mining (removal of coral material from reef ecosystems) caused tourist beaches to erode and millions of dollars to be spent protecting and rebuilding the shoreline (see "Indonesia: Bali" case study in Part 4). So important are reefs that many countries have special reef conservation programs. For example, Sri Lanka organized a nationwide coastal management program with a main goal to protect reefs so as to save their southwest shoreline from serious erosion (see "Sri Lanka" case history in Part 4).

Similarly, mangrove forests serve to dissipate wave energy and to protect the land areas behind them from the erosive forces of storms. Mangrove removal serves to jeopardize life and property because the major shield to dissipate wave energy and to stabilize the shoreline from the erosive forces of storms is gone. The value that these natural resources have for hazard prevention reinforces the need to identify them as critical areas and provide for them strong measures of protection. For example, mangroves are so important to Bangladesh in cyclone protection that the country is planning to plant an extensive greenbelt along its coastline (see "Bangladesh" case study in Part 4).

But, natural hazards damage at the coast can also come from inland. An example of uncontrolled development leading to high property damage and disablement of much of the aquaculture industry, is the major flooding along the Ecuador coast caused by massive runoff from rainstorms from the "El Niño" event of 1983.

Figure 3.85. Large storms can have major ecological impacts; for example, by cutting through the boundary islands, *Hurrican Gilberto* caused major flooding of Ria Lagartos (Yucatan, Mexico) with ocean water. This completely changed the ecology of the 70-km lagoon and interrupted a major salt production facility. (Photo by author.)

In many of the more densely populated nations, the risks of natural disasters to inhabitants of the coastal lowlands are being increased by population increases and development projects. Coastal people become more susceptible to natural hazards such as floods, typhoons, or tsunamis when land reclamation projects encourage settlement in dangerously low-lying areas, or when land-clearing and construction removes protective vegetation, reefs, or sand dunes (326). A particularly disastrous example is Bangladesh, where more than 140,000 people were lost in a cyclone that struck the coast in April 1991 (Figure 3.86).

Because protection against hazards begins with preservation of coastal landforms that provide natural resistance to wave attack, flooding, and erosion from hurricanes and storms, projects that remove or degrade protective landforms—removing beach sand, weakening coral reefs, bulldozing dunes, or destroying mangrove swamps—should be prohibited. For example, if dunes are removed, the risk to coastal development behind the former dunes is greatly increased. Similarly, if mangroves are removed, the land areas behind them are endangered by the erosive forces of storms (84).

Because certain natural disaster protections are so closely related to resource conservation, they should be included in an ICZM program. Specifically, water related disasters—which are virtually inevitable in any coastal nation—result from cyclonic storms, tsunamis, shore erosion, coastal river flooding, landslides and soil liquefaction. The Philippines, Bangladesh, Jamaica and dozens of other countries are affected by severe storms and flooding (163). Therefore, hazards loss reduction should begin with preservation of coastal landforms that provide natural resistance to wave attack, flooding, and erosion from hurricanes and storms.

Natural hazards cannot be addressed solely in single sector plans for, say, public health and safety because natural disasters cut across all sectors. Wind damage from a hurricane, inundation by a tsunami or earthquake (Figure 3.87), or rapid coastal erosion can affect tourism, the fishing industry, port operations, public works, and transportation. Housing and industry are also vulnerable. Natural elements which should be protected include dunes, beaches, and wetlands that protect coastal inhabitants and property against moderate storms and absorb some of the more violent energy of major storms.

The ocean beachfront may be the most hazardous place to build. Keeping development far enough inland to avoid the high risk zone is important in natural hazard prevention efforts. Therefore, a setback

Figure 3.86. Cyclone shelters in Bangladesh can harbor 1,000 to 2,000 people each during storm events. Most of the time they are used for schools, as is this one near Chittagong. (Photo by author.)

Figure 3.87. Integrated coastal zone management programs should consider the effects of all major natural hazards, including earthquakes. (Photo by author.)

line should be delineated at a safe point inland from the beach and all construction kept behind this line. This planning solution is the concept of "retreat" (see "Retreat" and "Setbacks" in Part 2) which requires these three steps:

1. Predict how far back the beach will erode in the future (say 50 years from now).
2. Identify this line on appropriate maps.
3. Prohibit any building seaward of this line.

Continued severe beach recession is certain and predictable along much of the coast. It is unwise to allow development of property that will certainly be lost to the sea, especially when the security of buildings so often creates demands for groins, bulkheads, and other protective works, which may further imperil the whole beach system.

3.58 NATURE SYNCHRONOUS DESIGN

The most cost-effective and environmentally correct approach to coastal development and coastal engineering is one that respects the strength of natural forces operating at the coast. Using this approach, project designers utilize or adapt to natural forces, pursuant to the "nature-synchronous" or "design-with-nature" approach to development, first advocated by the planner Ian McHarg (241). The concept applies to beachfront development and construction, wetland usage, natural hazards protection, etc.

Countries with limited investment capital and a critical need for food and economic betterment cannot afford to waste their money on over-engineered, overly expensive or ineffective structures. Yet this has happened too often with projects engineered to defy natural forces and ecological processes. Natural forces are exceptionally strong in coastal areas—wind, waves, erosion, storm surges, and tides.

For example, the beachfront is a place of extraordinary release of natural energy and a place where mistakes can be very costly. Special risks are attached to development on the ocean beachfront, where buildings are directly in the path of storm-driven waves. Beaches and dunes shift with changes in the balance between the erosive forces of storm winds and waves, on the one hand, and the restorative powers of tides and currents, on the other.

Consequently, along much of its length, the coast is a risky place to maintain habitation. The costs in property losses and human lives have been high. Furthermore, enormous sums of public money are spent on typical "hard engineering" efforts to stabilize and safeguard beaches. These efforts are rarely rewarded with success. There is another approach—nature-synchronous engineering or "soft engineering"—which offers a more cost-effective alternative to "hard engineering" and that works well in many, but not all, situations.

For beach projects, the soft engineering or nature-synchronous approach recognizes that the natural beachfront exists in a state of dynamic tension, continually shifting in response to waves, winds, and tide and continually adjusting back to an equilibrium state. Long-term stability is gained by holding the slope or profile of the beach intact by balancing the sand reserves held in various storage elements—dune, berm, offshore bar, and so forth. Each component of the beach profile is capable of receiving, storing and giving sand, depending on which of several constantly changing forces is dominant at the moment. The storage capacity of each of the components must be maintained at the highest possible level. This can be facilitated in some cases by proper design in locating wave-absorbing structures (rip-rap) rather than wave-reflecting structures (concrete bulkheads).

Many applications of nature-synchronous engineering approaches for beach stabilization can be cited; for example, harbor inlet stabilization, dune management, and beach replenishment. There are other applications such as vegetated stormwater drainage ways and buffer strips (instead of concrete ditches), mangrove buffer strips along channels (instead of bulkheads), and use of flood plains to store flood waters (rather than concrete storm water canals). Most soft engineering approaches have less damaging impact on coastal renewable resources than hard approaches.

Soft engineering for beaches recognizes that beach stability can be facilitated in some cases by proper design in locating wave-absorbing structures (rip-rap) rather than wave-reflecting structures (concrete bulkheads). With beach projects, designing with nature recognizes the existence of tremendous natural forces and works with instead of against the forces that accompany storm waves, littoral currents, and tsunamis, and is realistic about the capablility of natural defenses to protect the shore.

Kana (99) recommends mimicking nature by the use of "sand breakwaters," a technique based on natural sand bar migration. As sand bar migration occurs, the beach will undergo accretion in the area protected and erosion to either end of the bar (Figure 3.88). The extent of erosion and accretion are related to the size of the bar. Once attached, the bar introduces a new sand supply to the beach. This natural nourishment is then spread in both directions away from the center of the bar. This can be used to extend the design life of beachfill projects (see also "Bali, Indonesia" case history in Part 4). The time for this cycle of bar migration and attachment to occur depends on the quantity of sand involved. Small volumes (under 50,000 cubic yards-cy) might be distributed in a matter of months, whereas large volumes (say, 500,000 cy) could require perhaps three years. Sand breakwaters can be pumped in at lower unit costs ($1.50/cy compared to $2.65 cy: 1990 US$ for regular beachfill because little shaping is specified—as the sand breakwaters are pumped, wave action will rework the features into a natural slope. (99)

Inlet stabilization represents a good example of "soft engineering." At Captain Sam's inlet (South Carolina, USA) nature-synchronous inlet relocation was used instead of the typical double jetty construction. Based on historical analysis and continued threat of erosion to a downdrift community, plans were devised to relocate the inlet artificially.

The relocation freed more than one million cubic yards of sand to migrate (under wave influence) to eroding downdrift beaches. The project was completed in February 1983, and performance to date is up to expectations. Total cost of the project, including engineering and monitoring, was US$250,000. (99)

In summary, in both planning and in management (project review/environmental assessment) phases, preference should be given to nature-synchronous options.

3.59 NOISE AND DISTURBANCE

Loud noise may be defined as a form of environmental pollution. In the absence of appropriate safeguards, construction activity can produce high noise levels and cause air quality deterioration (e.g.,

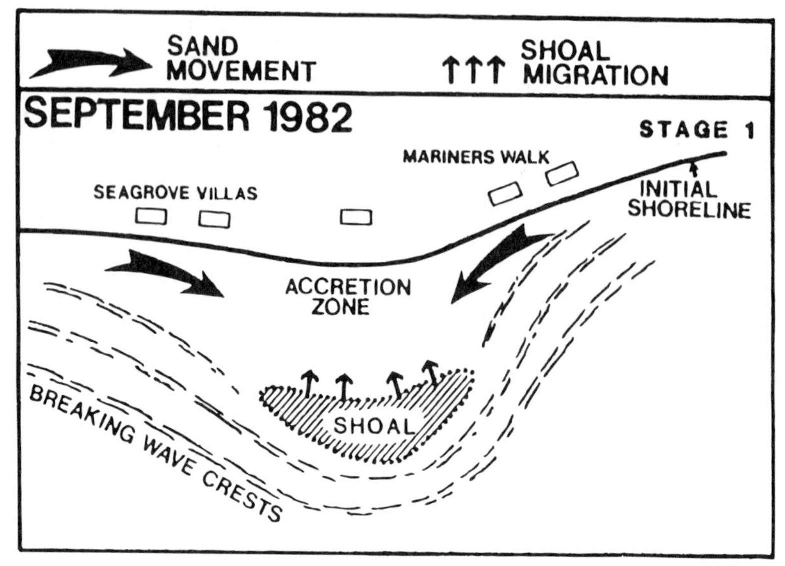

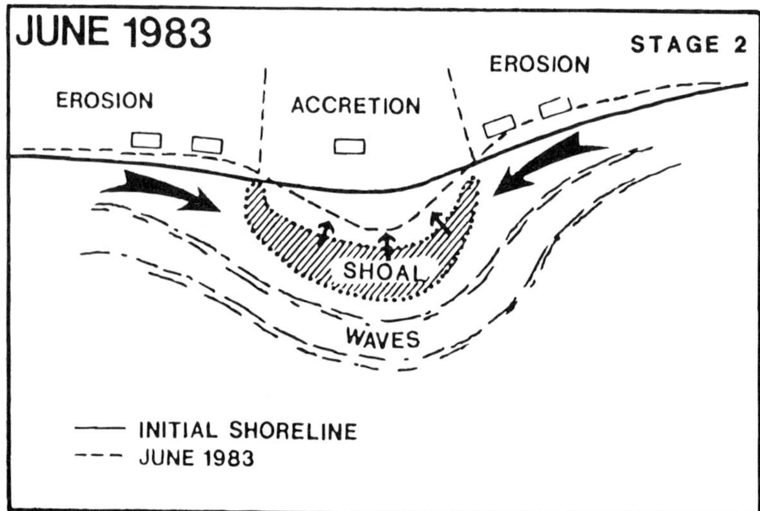

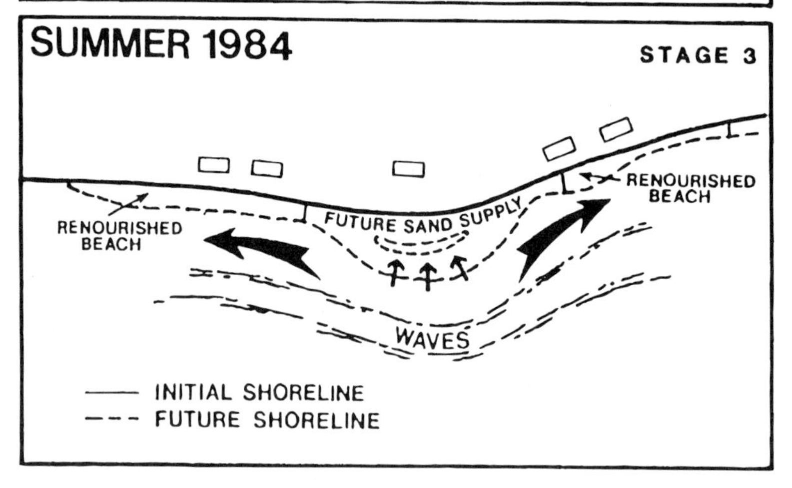

Figure 3.88. Nature-synchronous “soft” engineering is often more effective and less expensive than “hard” engineering solutions. The sand breakwater shown above protects the shoreline by mimicking nature through “shoal bypassing” and “natural nourishment.” (*Source:* Reference 99.)

dust, fumes, odors) during construction and maintenance operations, which would not only disturb the community peace but also disrupt the tranquility and ambiance that attract tourists. Also noise and disturbance negatively affect birds and other species.

During coastal construction, the work yards and work sites are often in the shorelands adjacent to heavily populated commercial or residential areas. Therefore, it is necessary to consider that the construction work could cause considerable disturbance. Coastal construction often requires excavating and moving materials (e.g., sand, rock, concrete) and involves the use of heavy machinery operating close to inhabited areas over a prolonged period and also moving of large equipment and materials through the streets. This can create more general disturbance, smoke, noise, and traffic, and cause air quality deterioration (e.g., dust, fumes, odors). Also there will be stock yards, casting yards, truck parks, etc., that are a major source of disturbance.

Noise can be defined as unwanted or unpleasant sound. Human response to noise is complex, but the following factors appear to determine the objectionability of noise: intensity; pitch or frequency; duration; time of day; background noise; number of exposures; adjustment to exposure; individual sensitivity; and psychological factors. *Pitch* describes the frequency of the sound in cycles per second (or Hertz). The intensity ("loudness" or sound pressure level) is measured with a sound level meter and is expressed in decibels.

The effect of noise on wildlife includes reduced hearing acuity, masking of auditory signals, behavioral changes, and physiological stress responses. In general, noise can reduce wildlife hearing sensitivity; mask social signals; induce panicking, crowding, and aversive behavior; disrupt breeding and nesting habitats and possibly migration patterns; and change blood pressure/chemistry, hormones, and reproductivity. Some animals have been able to adapt to noise sources and to differentiate dangerous ones from others. Boating can be detrimental to wildlife populations if boaters intrude into otherwise secluded habitats. Noise levels from boat motors (outboards) have been reported to reach 80 decibels at 50 feet distance (165).

Such impacts should be mediated, particularly loud, explosive, and irregular noises. Standard, appropriate, safeguards should be imposed to avoid such negative effects. Certainly, noise should not exceed 60 decibels for quiet areas, 70 decibels for residential areas, or 75–80 decibels for commercial areas, at a distance of 30 meters from the source. Disturbances can be controlled by equipment scheduling, routing of traffic, locating yards, and dust, noise, and exhaust abatement. Basic information on noise sources is given in (Figure 3.89).

It is generally accepted that people have a lower tolerance to noise when it interferes with sleep or rest and, therefore, many noise control ordinances specify more restrictive noise level standards during the night. Actual ambient nighttime noise levels in a city generally run 5 to 15 decibels lower than daytime levels.

3.60 NURTURE AREAS

There are special coastal habitats that serve the special needs of coastal species that require shallow, protected areas for breeding and/or sanctuary for their young—"nurture" or "nursery" areas (Figure 3.90). These special nuture areas support large fisheries; for example, over 90 percent of all fish caught in the Gulf of Mexico are reported to be estuarine dependent to some degree (59, 391).

Of the greatest economic importance is maintaining the productivity of fisheries. Continued fish production means continued livelihood for fishermen and for others in the fishing industry, including boat builders, trap and net makers, packers, distributors, and retailers—all of which enhances social, cultural, economic, and political stability. Of particular importance, a strong domestic fishery promotes self-sufficiency and reduces the outgo of foreign exchange.

A healthy, functioning, marine life nurture area depends upon many components. For example, a snook (*Robalo*) nurture area needs to have a shallow intertidal area with mangrove edge and an admixture of fresh water and, outside, a productive feeding and cryptic habitat of seagrass beds. After this initial period (1/2 year or so) the snook move into deeper waters and utilize a variety of habitats in enclosed water of estuaries and around channels, where good water quality becomes important (75). Therefore, to improve the snook's survival it would be necessary to reverse the degradation of habitat and water

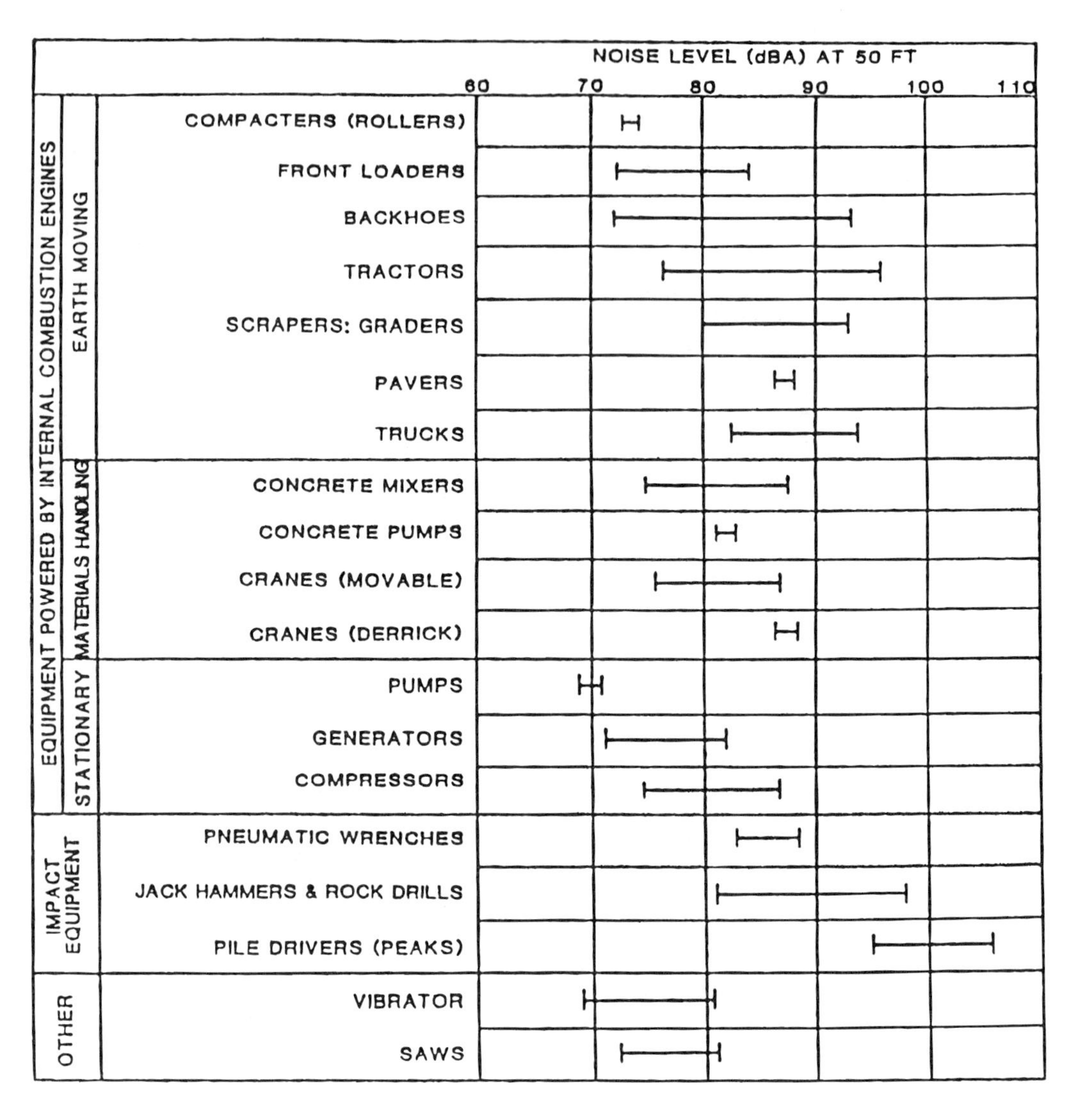

Figure 3.89. Noise levels for various types of construction equipment—any noise greater than 80 decibels is disturbing. (*Source:* Reference 365.)

quality in a multi-habitat aquatic system (see "Restoration and Rehabilitation" in Part 2) including high wetlands, low wetlands, flats, streams, channels, and open shallow waters.

Identification and management of nurture areas are discussed in "Critical Natural Areas Identification" and "Protected Natural Areas" in Part 2.

3.61 OXYGEN

Of the various gases that are found dissolved in coastal waters, oxygen has the most obvious importance to sea life. Dissolved oxygen (DO) is needed to keep organisms alive and to provide for optimum ecosystem function and highest carrying capacity. Coastal waters typically need a minimum of 4.0 mg/l and do better with 5.0 mg/L. Much of the available supply is produced by plants, which remove carbon dioxide from the water (in photosynthesis) and replace it with oxygen.

Oxygen is readily soluble in water. Its solubility rate varies inversely with water temperature and directly with atmospheric pressure. At normal atmospheric pressures, solubility rates of oxygen

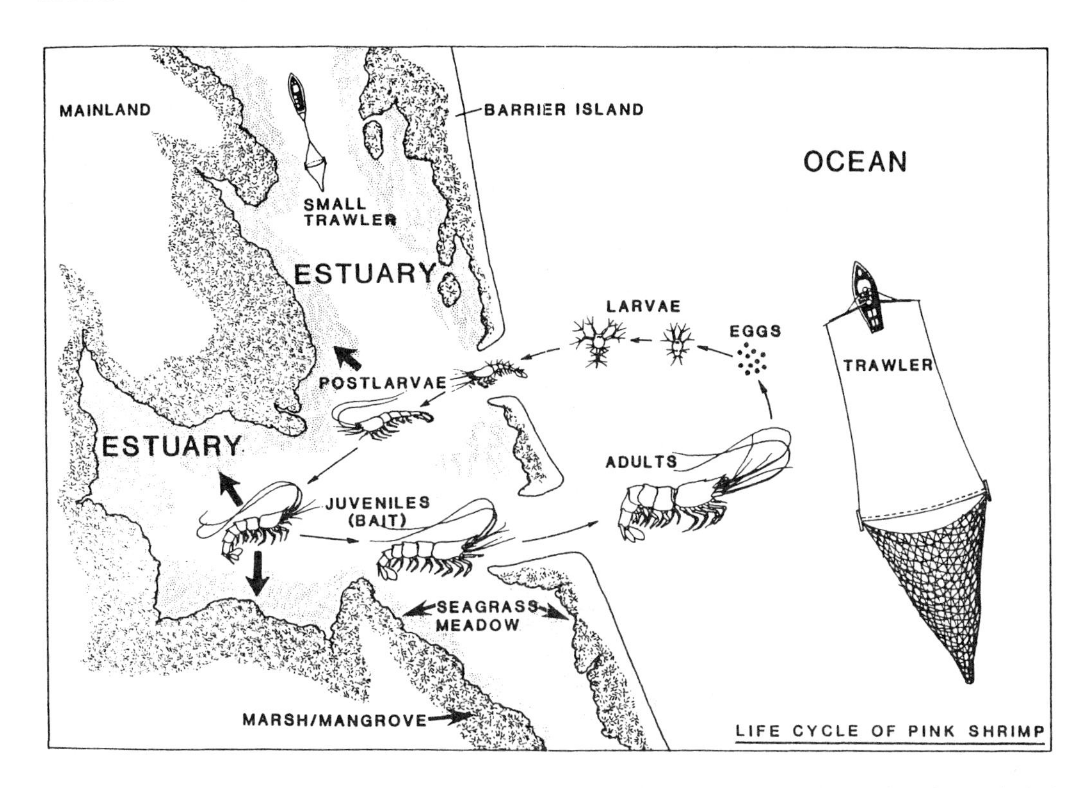

Figure 3.90. Shrimp populations are critically dependent upon lagoon and estuarine habitats for survival of key life stages. (*Source:* Courtesy of Florida Dept. of Natural Resources.)

in water range from 14.5 mg/l at 0°C to 7.8 mg/l at 30°C. Dissolved oxygen concentrations may be expressed in milligrams per liter (mg/l), parts per million of water, or as a percentage of saturation (356).

The main source of oxygen is from the atmosphere. Oxygen is dissolved into the upper layers of the water body through the air-water interface and is dispersed by wind and wave action, vertical mixing, and other forms of agitation. Another important source of oxygen is from the well-known photosynthetic process of plants, which may be expressed by the equation:

$$6\ CO_2 + 6\ H_2O \rightarrow C_6H_{12}O_6 + 6\ O_2$$

Carbon dioxide and water react chemically in the presence of light (solar energy) to produce sugar and oxygen. This oxygen becomes dissolved in the water body and enters the organic cycle of the lake. Often, when lakes are in a stagnant condition, photosynthesis is the main source of oxygen. Photosynthesis is also an important source of oxygen in tropical lakes which have long periods of intense sunlight and infrequent turnovers (356).

The minimum amount of DO required for healthy ecosystem function is maintained by natural processes in undisturbed coastal waters (except some confined estuaries during warm seasons). But oxygen may fall to unhealthy levels when sewage and other wastes (e.g., from food processing) with high biological oxygen demand (BOD) pollute coastal waters and induce high bacterial action. The bacteria involved are common residents of coastal waters and are species that multiply rapidly to reach enormous abundance, thereby depleting the water of oxygen faster than it can be replaced by either plants or the atmosphere. (See also "Oxygen: BOD/COD Measurement" in Part 2.)

DO depletion caused by pollution places stress on aquatic animals, reduces their ability to meet the demands of their environment, and in extreme cases may cause death by oxygen starvation. The amount of DO in the water depends upon salinity and temperature; i.e., cold water of low salinity retains more DO than does warm water of high salinity (Figure 3.91).

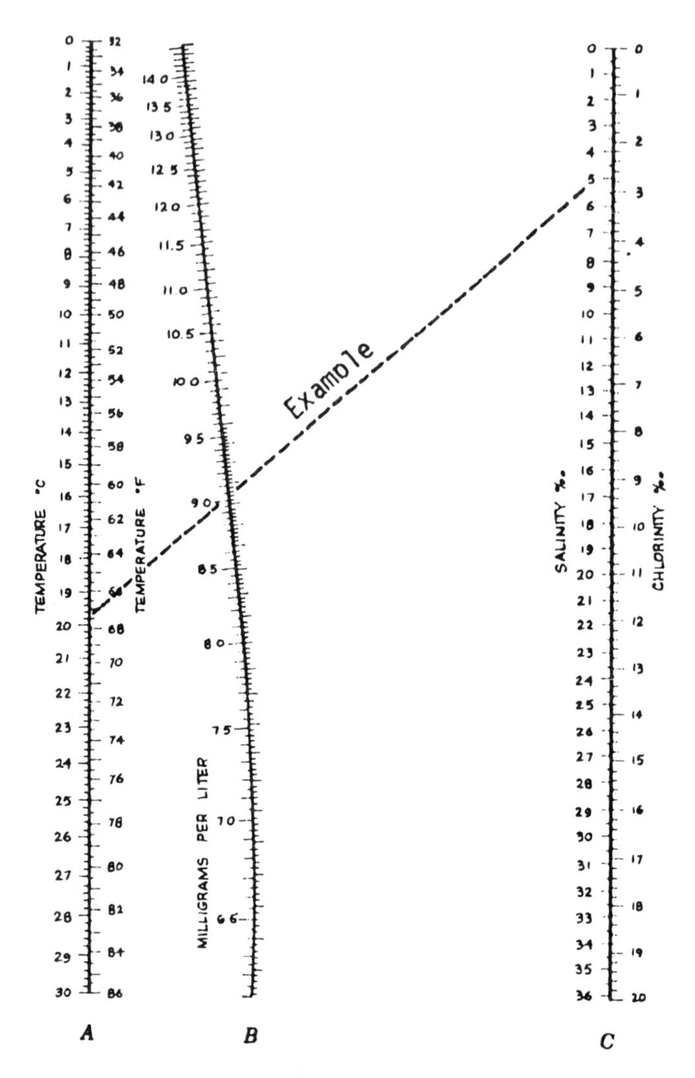

Figure 3.91. Nomograph relates (B) solubility of oxygen in water to (A) temperature and (C) salinity in the example, the water could hold 9 mg/l of oxygen. (*Source:* Reference 19.)

Fish occupying a habitat well saturated with oxygen during daylight hours may be subjected to stress conditions in the early morning hours when the demand for oxygen exhausts photosynthetic reserves. Generally, waters with DO concentrations less than 5.0 mg/l will not support good fisheries. A change in temperature, barometric pressure, and/or contamination could change the oxygen percent saturation. The best conditions vary for each species of fish (59).

Because of variation according to time, weather, and temperature, DO tests should be run during the same period (week and time of day) if yearly comparisons are to be made. Also, in shallow waters, there is usually adequate mixing of water from the surface to the river bottom. However, in impounded deep areas, there may be little mixing of the water. This could cause significant differences in DO measurements from the surface to the river bottom.

The whole ecosystem is degraded by inadequate oxygen. It is therefore essential to include conservation of DO in the coastal management program and to arrange controls for the maintenance of an optimum oxygen environment. The goal should be to maintain oxygen at as near the saturation level as possible and never less than 5 mg/l except when a prevailing natural situation keeps the concentration lower.

3.62 PARTICIPATION

Coastal resource planning and management programs require the highest level of public participation possible or appropriate. The general public of the community, along with the private sector interests and any other stakeholders, should be consulted about major coastal developments—this can be called the "sunshine rule," bringing decision making out "into the sunshine."

People who live along the coast and have traditionally used coastal resources are greatly affected by new conservation rules and procedures. Therefore they must be involved in the formation of new coastal policies and rules on resource use, if they are to support them. (See also "Public Participation in Planning" in Part 2.)

According to Coetzee (93), a major lesson learned in South Africa is that extensive public involvement should have the first priority in organizing coastal management. He believes that the recent emphasis on *integrated* coastal zone management (ICZM) is particularly appropriate as it may also help focus the need for involvment of the public into the process (i.e., at policy formation, programming of action, and implementation stages).

Unfortunately, planners and managers have too often resorted to public participation and involvement as a last means and only because a particular management decision is difficult to enforce or encounters some form of opposition. Participation is not a way to sell *premade decisions* either. It should be a two-way consultation with ideas growing in both directions.

If we were to attempt a description of participation, we would see it as a true dialogue between all parties concerned with a particular resource in order to ensure that there is a sharing of agendas. Participation is not intended to change the views of the fishermen, the government officials, the planners, or the consumers. Nor is it a means to get a particular group or sector "aligned" to the needs of another group. What it does is to ensure an appropriate shift from single sectoral concerns and self-centered concerns to a collective agenda which all parties will be better prepared to address. Participation serves to unite people in the sharing of needs and ideas and in the working of solutions.

Using aquaculture development as an example, participation on the part of coastal communities will require conscious efforts to involve them in the process of aquaculture development. It certainly will not come about without efforts to decentralize control and decision making over the coastal zone itself and the technologies that are appropriate there (326). Nor will participatory development come about without efforts of interested researchers, extension workers, rural bankers and non-governmental community developers to make certain that communities are directly involved and supported over the long term. (See also "Aquaculture Techniques" above and "Public Participation in Planning" in Part 2.)

In answer to the question, "Why should local communities support sustainable and ecologically sound uses?" Alan White (381) answers as follows: "Traditionally, these communities understand the limits to natural marine ecosystems and productivity, but in our modern, over-populated world, this tradition is being lost." White develops the argument further in the following statement about user participation (381):

> Surely coastal people who have not previously participated in management planning need encouragement and guidance to learn participation capability. In other words, they may have to learn how to participate. The idea of speaking up may be quite intimidating to many persons. Also, in order to be effective, they may need to organize themselves into an effective "receiving mechanism". This may be for single villages or for larger geo-political units. The reason for this being favored in many countries is that individually the rural communities may be unable to optimize socio-economic capabilities.

Education and pilot projects can show and remind people what is possible in terms of sustainable use and the value of protected areas. But education is not participation. Participation comes from wanting to support common values to gain some real or perceived benefit for the individual and community. Without it, marine resources can never be conserved and sustained because "enforcement" of laws in such a commons is not practicable. When people decide to "participate," they (as resource users) will make the real difference in protected area management. The solution thus lies in helping people to decide to participate in a constructive manner. Once resource users decide to do so and receive the associated benefits, the process will perpetuate itself.

White (382) notes that small islands provide advantages to coastal resources management because resources are more available to them and less so to "outsiders." This provides an incentive for the islanders to manage their own area to their own advantage. But to acheive this there must be linkages among all participants—community leaders, local politicians, enforcers, government agencies, NGOs, and the private sector (382).

To a particular community, their special problems have importance, and government's inability to see their situation can be frustrating. Communities can resolve many of these special problems themselves by freely exchanging ideas and information and working towards commonly acceptable solutions. Government can assist by being a receptive part of this problem-solving process, working to understand the problems of individual groups, and supplying technical or institutional assistance to encourage local efforts.

While some people will easily give their opinion, others are either shy or cynical. For example, Johannnes reports: "I asked four fishermen independently why they had not raised their concerns about the causeway [a roadbed through the water] with the government. Their replies were almost identical. Each said that the government did things to suit the government, not the people of the atoll [coral island], and that they felt they would not be listened to." The lesson: People need encouragement to participate.

This author, having spent many months in Bangladesh, finds this politically reorganized country's view on participation most relevant. There the central issue of participation deals with redistribution and control of resources and power in favor of those who are supposed to be the end beneficiaries of the economic development activity. People's advocates argue that an ideal project should be designed in cooperation with the beneficiaries and that they should take control of available resources and decide on the best socio-economic status for themselves.

Participation of the beneficiaries at project design and implementation stage has the underlying assumption that these communities have the wisdom and knowledge to distinguish between economic alternatives in terms of their welfare, and that such choices are invariably better than the choices made for them.

Often, governments and donor agencies seem surprisingly reluctant to invest resources in the activities associated with participation at decision-making or implementation stages. One reason appears to be that governmental agencies feel that discussion and consultation with intended beneficiaries about their problems and possible solutions might raise accelerated expectations on what will be done for them. The agencies, therefore, seem to prefer delivering benefits at a time and place of their own choosing in order to keep control over development activities. In the process, however, there is no assurance that they would be delivering the most needed benefits or even the correct ones to the intended beneficiaries.

According to Wells (377), community involvement and citizen participation are proving to be most important in management of coral reefs. In many Pacific nations, this dates back centuries in the form of traditional marine tenure systems, with villagers and family groups controlling fishing on the reefs adjacent to their land. The success of these traditional systems is visible in cases where they have broken down or been lost and replaced by open-access systems which regularly lead to over-exploitation and conflict between user groups, even if western-style fishery and environmental regulations have been introduced. Community involvement has been the key to the success of the community-run Philippine reserves, higher fishery yields providing positive feedback to the community and helping to generate a sense of pride in their efforts.

3.63 PATHOGENS

Disease-carrying bacteria and viruses (or pathogens) associated with human and animal wastes pose threats to humans by contaminating seafood, drinking water, and swimming areas. Eating seafood and even swimming can result in hepatitus, gastrointestinal disorders, and infections. Also, there are pollution-bred pathogens that infect various species of sealife.

There are several sources of bacterial contamination in the coastal area. Leaking septic tanks can pollute both ground and estuarine waters, as can septic tanks that are spaced too closely, placed on porous soils, or located in high water tables (254). Sewage treatment plants and package plants can fail, allowing wastes to enter surface and ground waters. Discharges of human waste from boats can contaminate estuarine waters, particularly in marinas where there is a concentration of boats in a small area. Animal feedlot and stormwater runoff can also cause contamination.

As an example, nearly 20 percent of the shellfishing waters in North Carolina (USA) are closed and the number of acres of closed waters increases each year. Total coliform bacteria and other pollutants are transported into shellfish waters indirectly by runoff during periods of significant rainfall and directly through groundwater. In a North Carolina study (254) direct transport accounted for less than 3 MPN/100 ml for both total and fecal coliform, and indirect transport or surface runoff of contaminated waters resulted in total coliform ranging from 22,100 to 240,000 MPN/100 ml and fecal coliform from 330 to 13,000 MPN/100 ml in shellfish waters immediately adjacent to the runoff. Therefore, it was concluded in these studies that conventional septic tanks appeared to adequately filter out bacteria from the septic tank leachate *under normal conditions.*

The incidence of disease in fishes is relatively high in sewage-polluted waters. Fish afflicted with fin rot have fins partially or totally missing. In the United States, the disease is known from sewage-polluted areas of Narragansett Bay (Rhode Island), New York Harbor, and southern California (particularly in the area affected by discharge from the Los Angeles White Point sewer plant). Infected species include the sole, flounder, bluefish, and weakfish. Both bacteriological and toxic metal associations are implicated. (243)

Pathogens of coastal waters that affect human health usually originate from human and animal excreta conveyed into the waters either accidentally or purposefully. The following pathogen types can be found in coastal waters: bacteria, viruses, protozoa, and parasitic helminths (including roundworms and flukes) (14.1). Waterborne diseases can be conveyed to humans directly, by bodily contact with coastal waters, or by ingesting seawater or indirectly by the eating of contaminated shellfish. Potential diseases are cholera, typhoid, infectious hepatitis, jaundice, and diarrhoea/dysentery. Such pathogens (particularly those from the intestines of warm-blooded animals) frequently persist in coastal waters for a long time and travel rather great distances.

The danger level for waterborne diseases is most often monitored by measuring the abundance of a certain sewage-associated, or coliform bacteria (*E. coli*) in the water. When this abundance is too high, use of the water is restricted). For example, for shellfish growing the water should not contain more than 70 *E. coli* in 100 ml of the water being tested (see also "Water Quality Standards" in Part 2).

Exposure to disease is maximized by eating contaminated shellfish. For example, massive outbreaks of cholera and infectious hepatitis can often be traced to shellfish harvested from waters polluted with sewage. The reason is that shellfish filter large quantities of water during feeding and remove the pathogens from this water. The shellfish are not infected themselves, they just store the pathogens in their bodies to be passed along to any human that eats them. (243)

Exposure is also maximized by swimming at crowded or sewage infested beaches. Gastro-intestinal diseases can be contracted as well as various eye, ear, and respiratory disorders. Many of the World's most prestigious resort beaches are polluted with disease causing pathogens and people are contracting disorders without even knowing the source is the contaminated coastal waters along the beach. Resort owners and local officials are loathe to discover beach contamination and often remain silent when they do so as not to discourage tourists.

Counter-measures include (a) pipe sewage to ocean outfalls; (b) improve functioning of coastal septic tanks and cess pools; (c) provide more consistent and better disinfection of effluent (chlorine application, etc.). (See also "Ocean Outfall Placement," "Sewage Management," and "Septic Tanks" in Part 2).

3.64 PERFORMANCE STANDARDS

Performance standards are environmental requirements for project design and construction. These standards are usually applied through a permit system. Once these standards are agreed, construction of a project is approved with the understanding that specific properties and characteristics of the natural system must still function in a prescribed manner after development. Accordingly, the developer is allowed to proceed with the project if he can show proof that the natural system will still "perform" according to prescribed "standards." In this way, the developer is free to devise any approach to the project which will guarantee full function of the natural ecosystem after construction. A performance standard might be the quantity of fish a pond shall produce, the quality of water that passes from a parcel of land, the amount of water current flowing through an estuary, or the uptake of pollutants by a wetland.

Saenger (301) recommends performance standards to define the maximum permissible impact of specific uses on particular resources. Certain uses (which may or may not have adverse impact on the resources or other users depending on their design, location, the natural values and hazards at the site, adjacent uses, and other factors) require special permits which will be granted if the applicant can demonstrate that the use impacts will not exceed permissible (acceptable) levels. "Threshold levels" or "acceptable limits of change" are often used to set the permissible levels which govern performance standards.

3.65 PETROLEUM INDUSTRIES

All phases of oil and gas production—extraction, transport and refining—can cause serious environmental impacts to coastal waters. In the past, important disturbance of coastal ecosystems has been caused by oil and gas operations in the coastal zone. Now, much activity takes place on the outer continental shelf. Moreover, hazards to marine and coastal environments can be expected to increase as technological improvements permit oil extraction to take place at greater and greater depths.

The major potential environmental disturbances in the extraction and processing of oil and gas are (a) pollution by oil spills from blowouts, pipeline ruptures, and transport accidents; (b) preemption or destruction of vital habitat areas (e.g., shellfish beds, wetlands) and energy flows; (c) general disruption of the coastal environment (e.g., from channel dredging); and (d) primary and secondary environmental and socioeconomic impacts from a wide variety of associated onshsore facilities (59).

Development of oil and gas resources is a complex industrial process that requires extensive advance planning and coordination of all phases from exploration to processing and shipment. Each of dozens of components linking development and production systems has the potential for adverse environmental effects on coastal water resources, and it is necessary to understand these probable effects in sufficient detail so that methods can be devised to minimize environmental damage.

The transport, storage, and handling of diesel, oil, gas, aviation fuel, and other petroleum products have serious potential environmental impacts. These must be assessed and planned for (41).

Tomascik (345) lists the following as major impacting activities:

- Tanker spills and tanker-related accidents (Figure 3.92)
- Oil platform blow-outs
- Tanker ballast discharge
- Release of production water from oil platforms
- Land clearing associated with petrochemical plant and storage tank construction
- Construction of transportation facilities on land and sea
- Platform construction and operation
- Placement of anchor buoys
- Loading and off-loading of materials
- Oil spillage during drilling and operations
- Release of drilling muds, drilling fluids, and drilling cuttings
- Waste discharge such as sewage
- Oil spills during transportation

The deleterious effects on the marine environment and the living resources as a result of a growing frequency of oil spills such as the *Exxon Valdez* have caused public awareness and widespread political attention (Figure 3.93). It is interesting to note that of the total budget of petroleum hydrocarbons introduced into the oceans, 34.9 percent arises from marine transportation, 26.2 percent from river

Figure 3.92. The *Amoco Cadiz* polluted much of the Brittany coast and the English Channel with Iranian-Saudi crude oil when it grounded at Portsall, France, on March 16, 1978. (Photo by author.)

runoff, 9.8 percent each from natural seeps and atmosphere rain, 4.9 percent each from urban runoff, industrial and municipal waste, while oil refineries and production account for only 3.3 percent and 1.3 percent, respectively (16). The total contribution from offshore oil production in terms of blowout of oil-wells or leakage from producing sites is said to be relatively insignificant (Figure 3.94).

It should be noted from the above that 66 percent of oil entering the sea is from marine transportation or local runoff, two sources of particular relevance to coastal management programs. It can be presumed that this relative ranking (for Eastern Peninsular Malaysia) would apply to many countries with the same habitat types and could be used for guidance until specific data are collected for the country. But slick trajectories for spills at sea have to be calculated for each country and each section of coast separately.

Petroleum companies should be required to employ the latest spill prevention technology (e.g., automatic shut-off devices) and have spill-response equipment available. Even so, when spills do occur, the agency concerned will need to be well prepared to deal with them.

To ensure that this preparation is thorough, each community should have an Oil Spill Contingency Plan that is frequently tested and updated (Figure 3.95). The plan should contain a complete list of oil-response equipment (owned by private companies and government), trained personnel, and a 24-hour pollution emergency alerting and response network with designated authorities and their contact telephone numbers. Regional response entities are useful for small countries, e.g., many Pacific island governments participate in the South Pacific Regional Environmental Program (SPREP) Regional Oil Spill Contingency Plan", which provides an assistance network in the event of a major accident and provides oil-spill response training offered by SPREP (41).

All petroleum products biodegrade naturally. Thus, if a spill is being carried out to sea and is not threatening important resources, it is best to allow the oil to be broken down and dispersed naturally. When oil does reach coastal areas, its effects depend on the coastal ecosystem present. For example, oil floating on the water surface will not have a major impact on subtidal corals. Mangroves, however, are particularly susceptible because of their intertidal location and air breathing roots. Most shorelines can be ranked according to a sensitivity and clean-up index (e.g., exposed rocky headlands and hard-packed mud flats are less sensitive and easier to clean; medium-grained sand beaches and cobble beaches

Figure 3.93. An "oil boom" is placed across the mouth of Britanny's Aber Wrach estuary to protect it from oil spilled in the *Amoco Cadiz* grounding. (Photo by author.)

are moderately sensitive; and estuaries, mangroves, and coral reefs are quite sensitive and practically impossible to clean when oiled) (41).

The persistence of oil in the marine environment depends on the amount and type of oil spilled and the wind, wave, current, and weather conditions. Aviation fuel and gasoline (petrol) are "lighter" and vaporize quickly into the atmosphere, creating less of a pollution and clean-up problem. However, because they are more toxic and highly flammable, special care must be taken if a large amount is spilled in shallow areas or is trapped under shore structures (e.g., piers, wharf areas). Diesel, distillate, and refined oils are more of a problem. Heavy crude oils and ship (bunker) fuel are especially persistent.

Figure 3.94. Oil extraction often occurs in coastal waters where special efforts must be made to ensure that spills and leaks are minor and the damage can be confined to small areas. (Photo by author.)

Figure 3.95. Oil pollution is particularly harmful to coastal birdlife. (Photo courtesy of U.S. National Oceanic and Atmospheric Administration.)

When oil is spilled, the response plan decision-making process should be rapidly put into action to prevent the worst damage, facilitate clean-up, and to minimize environmental degradation and financial losses (Figure 3.96). Two decision areas are particularly important. The first is whether to use chemical dispersants or to try and physically contain and remove the oil from the water or shoreline. Some modern dispersants can be used where fish, coral, seagrass, and other benthic organisms are present without doing too much damage to these. However, a trade-off is required between allowing oil to come ashore where its impact depends on shoreline type, and dispersing oil offshore where its impact depends on bottom communities, depth and water movement (41).

Figure 3.96. French Army troops stuggle to clean up the beaches of Brittany, France, during the *Amoco Cadiz* disaster of 1978. (Photo by author.)

A second important decision in oil spill response involves disposal of collected oil and oily debris. After oil is physically contained, usually by floating booms or other barriers, it is collected for re-refining or disposal. This collection process and the shoreline clean-up of spilled oil are very messy operations. In the end, a decision must be made as to how and where to dispose of the oil and oily debris. The relatively clear oil should be stored in barrels for reshipment to refineries. The oily debris must be stored and shipped, buried, or burned. Burial is not always a good solution, as oil can seep into groundwater or streams, and it may be difficult to find available land on small Pacific islands with customary land ownership. Burning can cause local air pollution problems, and there will usually be leftover material that cannot be burned. Thus, some oily debris may also need to be shipped for proper disposal (41).

The laying of buried pipelines on the seabed can cause sediment disturbance during trench excavation. Some inshore sediments are particularly fine (very small particle size) and are easily resuspended and the disruptive effect of such fine sediment on corals, seagrass, fish, crabs, and shellfish in a sheltered body of water can be very marked. The location and nature of all buried pipelines should be carefully planned and mapped. Chronic leaks of petroleum products have polluted coastal waters for long periods (41).

The bilges of most ships and tanks of petroleum tankers need to be cleaned periodically. Emptied oil compartments are also filled to adjust trim and ballast. Subsequent discharge of oil-contaminated ballast is therefore a source of oil pollution. These were once a major source of marine oil pollution as tanker masters took the easy option and cleaned out tanks while underway and out of sight of land—whether in national or international waters. Strict regulations under the International Maritime Organization have curbed much of this abuse of the seas. It still can be done, however, where surveillance of shipping is weak, as in the South Pacific. (351)

Tanks used for storage of petroleum products on shore need to be cleaned occasionally. The sludges removed from the bottoms of these tanks are always potential problems for the environment. Careful arrangements for their disposal must be decided and included in the waste management procedures which should be part of the outcome of the environmental assessment.

With sufficient advance knowledge, both industry and public authorities can assess the environmental acceptability of each of hundreds of options for linking the components of production and delivery systems and thereby choose the approach that will best protect fish and wildlife resources from adverse impacts, including the probable secondary effects from onshore facilities.

Tomascik (345) lists the following environmental protection guidelines for critical natural areas, using coral reefs as the ecosystem in jeopardy:

1. Limit offshore and/or coastal oil drilling operations to 5 km or more from all coral reef areas.
2. Ensure that oil terminals and other oil installations are at least 10 km away from coral reef habitats and must be constructed down-current.
3. Ensure that in the case of an oil spill in a coral reef area, the use of mechanical removal of oil is the *first* recommended technique if possible.
   - Use booms, skimmers and sorbants.
   - Mechanical removal will most likely fail under the following physical conditions: currents greater than 2 km/h; high winds, greater than 12 km/h; low wave action.
   - In cases where mechanical removal fails, non-toxic oil dispersants should be applied at the leading edge of the spill as soon as possible.
4. Ensure that a contingency plan and a strategic deployment of clean-up hardware is in place in all oil installations with a potential to have an impact on coral reefs through spills.
5. Ensure that all oil-related development activities must go through the EIA process.
6. Ensure that all oil installations have appropriate waste water treatment plants.
7. Ensure that waste water discharge from all oil installations is processed through an appropriate treatment and discharged down-current and away from coral reef areas.
8. Ensure that the discharge of production water is limited to deep water, away and down-current of coral reef areas.
9. Ensure that the disposal of drilling muds and drilling fluids is limited to deep water, away and down-current from coral reef areas.
10. Develop and implement strict guidelines for the handling of all products in all oil installations.

11. Do not site oil storage tanks in close vicinity to coral reefs.
12. Implement an environmental monitoring program, for all coastal and platform oil installations, in the immediate operational area and in adjacent coastal and marine ecosystems.
13. Avoid the use of toxic oil dispersants to clean up oil spills in close vicinity of coral reef areas;
    - Chemical oil dispersants should be used only if (a) mechanical equipment is not available; (b) mechanical equipment cannot be used because of physical conditions in the area; and (c) if a more sensitive or economically important resource is being threatened.
    - Use dispersants that have effective range of 0 to 40 percent salinity, and 20 to 37°C temperature.
    - The dispersant/oil ratio should be in the region of one part of dispersant to 20–30 parts of oil, and the rate of undiluted dispersant application must not exceed 110 liters/ha.
    - The dispersants must not contain compounds which could expose the user to an unacceptable toxicological hazard during normal spraying or handling operations.
    - Preference should be given to dispersants containing less than 3 percent by weight of aromatics.
    - The use of benzene, chlorinated hydrocarbons, phenols, caustic alkali, and free mineral acid must be prohibited in all oil dispersants.
    - Suppliers must state the application rates recommended for the products offered.
    - Use dispersants only if the oil type is amendable to dispersant use. Heavier oils are more difficult to disperse than lighter oils.
    - Oil weathered for more than two days is *not* amendable to dispersant use.
    - The area must have an active water exchange rate and preferably be one of high energy input.
    - The area must not contain eggs or larvae of ecologically and/or economically important species (e.g., shrimp, coral).
    - Oil dispersants are generally more damaging to coral reefs than is undispersed oil.
    - Use of dispersants to prevent oil reaching a reef is recommended only if the upstream site is less sensitive than coral reef itself and sufficiently distant to ensure that the dispersed oil does not reach the coral reef.
    - Ensure that dispersed oil does not reach coral reefs.
    - Use of dispersants on oil already over coral reefs is recommended only to stop the oil from impacting other critical habitats, and only if the water depth over the reef is greater than 10 meters.
    - Dispersed oil is generally more damaging to seagrass beds than undispersed oil.
    - Ecologically it may be more desirable to allow oil to beach where it can be cleaned up mechanically rather than to disperse it at sea where it enters the water column (Figure 3.97).
    - Avoid the use of dispersants on beached oil, since it may cause the oil to sink into the substrate creating the potential for a long-term impacts.
        - Use of dispersants is recommended to prevent oil reaching important seabird nesting and roosting sites.

Tomascik (345) lists the following dispersants by level of toxicity to marine life (arranged within each group in increasing order of toxicity):

LOW	MEDIUM	HIGH
Finasol OSR-7	Corexit 9550	ADP-7
Cold Clean 500	Jansolv	OFC D609
Corexit 7664		Corexit 9527
		Conco K
		V-25

3.66 POLITICAL MOTIVATION

Coastal environmental programs are usually initiated in response to a perceived use conflict, a severe decline in a resource, or a devastating experience with natural hazards. Launching a coastal program demands strong motivation. Such motivation can arise from events that dramatize the importance and vulnerability of coastal resources. The potential long term socioeconomic benefits of coastal

Figure 3.97. These razor clams are part of a delayed (two weeks) kill of 16 million shellfish near Lannion, Brittany, caused by a combination of oil from the *Amoco Cadiz,* the chemical dispersants, and sinking agents used to combat the spill. (Photo by Farnum Allston.)

management must be evident for environmental quality and natural area protection to enjoy continued support. Fisheries productivity, increased tourism revenues, sustained mangrove forestry, and security from natural hazard devastation appear to be the four most common and persuasive arguments for ICZM (326).

Olsen (270) believes that putting coastal management on a political agenda which is already overcrowded with economic problems can be a most difficult undertaking. Much of this problem is because coastal resources are not often recognized as a political subject. They do not appear in national economic statistics as a separate sector. Periods of crisis involving coastal areas, such as coastal storm damage, flooding, oil spills, fish kills, or crashes in the population of a coastal species may get political attention but will lead to reforms only if groups within government and the private sector are prepared to take advantage of them. Public education efforts on environmental protection are weak in most countries, and coastal resources appear as only a subset of basic national environmental problems such as water pollution, over-fishing, soil erosion, poor sanitation, and lack of potable water.

Redirecting a policy maker's attention is fraught with difficulties because each is already overburdened and also trying to keep their agenda relatively unchanged, according to Tobin (343). Regardless of the merits of an issue—such as wise and sustainable use of coastal resources—few policy makers are naturally inclined to seek new items for their agenda. For every new issue they place on their agenda, some other issue must be neglected or removed. Moreover, even when policymakers are readily amenable to a new item on their agenda, they must still cope with significant matters they cannot ignore. At the national level, these include defense, foreign relations, international trade, budgets, revenues, tax codes and, normally, programs focusing on health, education, welfare and agriculture. In short, issues related to ICZM compete with scores of other, more familiar ones that claim both categorical precedence and policy makers' attention.

From a political point of view, the distribution of costs and benefits is especially important. On the one hand, environmental regulation frequently concentrates costs among relatively few readily identifiable interests (thus sparking resistance) while distributing benefits among a large class of people who may not even know they are beneficiaries. On the other hand, environmental regulation normally imposes costs in the present in anticipation of future benefits. This is a difficult situation for politicians and regulatory agencies dependent on public support. (343)

But special coastal problems such as deteriorating water quality, decline in species abundance, or a crash in coastal fisheries need attention and may even require documentation in order to prove they exist. The impacts may not be noticed outside of the localized areas where they occur and therefore will be given little attention.

It is clear that the thirty or so developing nations in the humid tropics with extensive mangrove forests should have a strong incentive for integrated coastal resources management. Most developing nations with extensive mangrove forests are confronted with similar stresses which threaten the sustainability of this renewable resource. Conversion of mangrove forest to aquaculture ponds or croplands often presents a particularly difficult conflict of uses (326).

Another area of motivation is coastal natural hazards. These are usually addressed in sectoral plans for public health and safety. But natural disasters cut across all coastal dependent economic sectors. Wind damage from a hurricane, inundation by a tsunami, or rapid coastal erosion can affect tourism, the fishing industry, port operations, public works, and transportation. Other sectors such as housing and industry are also vulnerable. The potentially devastating consequences of development in coastal hazard prone areas, necessitate the use of integrated coastal planning (326).

Public involvement is important. Creating the political demand for institutional reforms requires translating local problems and needs into a national agenda item. The national response must emphasize the need for national governments to channel resources, information, technical assistance and authority to levels of government which are much closer to the resources and conflicts. But the best legal framework will go awry, and even the machines and modern technology will prove ineffective, without the active involvement of the citizenry (270).

National agencies which have at least a portion of the necessary authority for research, issue assessment, planning, decision-making, implementation and monitoring, are normally unwilling or find it difficult to conduct joint action with other units in their own or other ministries. Critical interactions and relationships should be identified and established in the beginning of the project, even if it slows the pace of activities. Agencies rarely have sufficient funding, staff, or expertise to develop policies sensitive to local environmental conditions. Financial assistance distributed among the most important implementing agencies, rather than held in one lead agency may facilitate joint action and collaboration.

Local governments may have limited jurisdiction and little capability to control or guide the pace of urbanization, to influence large publicly funded or private development projects, to conduct assessments and research, effectively mediate disputes among coastal resource users, or to control large-scale over-harvesting or degradation.

The following five points would be useful in discussing the merits of CZM with a national government (212):

1. Emphasizing the *development* aspects of coastal management as a positive approach.
2. The "balanced" nature of the coastal management approach should be emphasized. The emphasis in the program also should be on *management,* that is, the prudent use and development of resources, and not on *preservation per se.* Of course, protection of sensitive coastal areas is a key objective of CZM.
3. Determine early who is potentially opposed to CZM and develop arguments particularly aimed at that sector:
    - Energy and industrial interests can be convinced that CZM will facilitate rational development and increase the predictability of the coastal situation.
    - Local authorities should realize that they can increase their impact on national government coastal policies through a partnership involving local CZM programs.
    - Rival government agencies should be incorporated into the CZM planning process early on and made partners in the program.
4. Clearly, one needs to argue the logic and reasonableness of rational coastal planning and management and obtain a maximum number of allies for this point of view.
5. Governments involved with planning for 200-mile exclusive economic zones should understand the value of a foundation of rational coastal planning and management.

Many rural communities dependent on coastal resources (and any other means useful to scratch out a living) are living at the margin and do not have the cushion to risk management actions that might erode even for short periods their means to a livelihood. Politicians may be willing to put a program in place but not enforce it if it means forcing people beyond the margin.

Political and financial support is dependent on the level of awareness of decision-makers. In educating politicians and economic planners, it is important to use language and concepts with which they are familiar. Carrying capacity studies that utilize cost-benefit analysis can produce figures such as estimates of revenue, sustainable yields, and other quantifiable data that can be used to convince decision makers of the economic and social benefits of protecting ecosystems (64).

3.67 POLLUTION

Being at "sea level" elevation, the coast is the eventual receiving basin for all land runoff water and the wastes it carries. Coastally discharging rivers gather and carry large amounts of pollutants to the coast with their rampaging flood waters (see "Upland Effects" below). Consequently, a large share of the world's pollution ends up in coastal lagoons, estuaries, wetlands, submerged grass beds, and coral reefs.

It follows that pollution of the coastal sea is largely a problem to be solved by management of the land. The reason: land surface sources of pollution are believed to be responsible for more than three-quarters of marine pollution, via rivers, direct discharges, and the atmosphere. The rest comes from such sectors as shipping, heavy industry, waste disposal, and offshore oil production.

Contaminants are transported from land to sea by rivers, direct run-off, or point discharges. They are partly retained in the water and sediments of estuaries, bays, beaches or open coastal waters, and partly transported through these boundary areas to the open ocean. The disposal of domestic sewage and industrial effluents from population and industrial centers located along the coastlines extends coastal pollution beyond the estuaries. Moreover, coastal waters are not only more polluted than the open ocean, they also offer significantly more opportunities for exposure through contamination of the world's most important fisheries and recreation areas (228). Some typical coastal water pollutants and their effects are listed in Table 3.1.

A group of chemicals known as *priority pollutants* have been identified by the United States Environmental Protection Agency (USEPA) as presenting health or environmental risks at certain concentrations in water. The priority pollutant list includes both organic—or carbon-containing compounds such as chloroform and toluene—and inorganic chemicals such as heavy metals. Heavy metals include a variety of metallic elements such as copper, zinc, nickel, chromium, and lead. It has been estimated that the list addresses up to only 1 percent of the organic content of a wastewater sample. Many other organic compounds may be present which have, as yet, unknown potential effects on human health and the environment. (366).

The carrying capacity of the coastal ecosystem is limited by water quality. The effect of any pollutant depends on where it goes, how concentrated it is at the point of discharge, how rapidly it is assimilated or flushed out of the environment, and whether it can be dissolved in the water column or is chemically fixed to sediments. All of these conditions depend on water movement and circulation patterns which, in turn, are governed by the relationship of tide and river flow to estuarine shape and size. In many bays, embayments, lagoons, and tidal rivers, circulation is sluggish and pollutants may build up to a level that can cause damage, even with efficient treatment of effluents.

Excess nutrients received by the aquatic environment from land sources sometimes may enhance fisheries through increased nutrient supply, but the benefits of eutrophication seem to be outweighed, in most cases, by negative impacts such as excessive algae, oxygen deficiency, coral reef destruction, etc. (See "Nutrients Management" and "Sewage Management" in Part 2.)

The persistence of pesticides and other toxic chemicals and heavy metals is of great concern to maintenance of biodiversity and on the suitability of seafood for the human diet. The need to protect human health provides a strong incentive to avoid discharges of such chemicals into the marine environment. (See "Toxic Substances" below).

Pollution is often at its worst in the harbors of large coastal cities and industrial ports where sewage and toxic industrial wastes combine to damage coastal environments and resources as well as to jeopardize human health (see "Ports and Harbors" below).

Table 3.1. Typical water pollutants and their effects.

Pollutant	Source	Effect on coastal waters
Petroleum hydrocarbons	Fuel exhausts Motor oil and grease Power plant emissions Industrial discharges Spills and dumping Leaking underground storage containers Urban runoff	Spills can kill aquatic life, damage beaches, and permanently destroy wetlands. Runoff can be toxic to marine organisms—causing death, disease, and reproductive problems.
Chlorine	Water treatment plants Swimming pool backwash	Kills aquatic life.
Nutrients	Agricultural, forestry, and urban runoff Industrial and boat discharges Sewage treatment and package plants Septic tanks Animal feedlots	Enrichment of rivers and sounds (eutrophication) resulting in algae blooms, which can alter the food chain by decay, depleting oxygen and causing fish kills. Eutrophication is also suspected of causing some fish disease problems.
Fresh water	Water running of impervious surfaces Land clearing Draining wetlands Channelization of streams	Changes salinity patterns in estuarine habitats, causing slowed growth or death of juvenile organisms or poor reproduction.
Bacteria and viruses	Septic tanks that are spaced too densely, placed on porous soils, located in high water tables, or that leak Sewage treatment or package plants Boat discharges Animal feedlots Urban runoff	Contaminates shellfish waters, so consumption of shellfish may cause disease. Contaminate groundwater, so using for drinking or bathing may cause disease. Contaminates surface waters, so swimming may cause disease or wound infections
Sediment	Land clearing Dredging Erosion	Clogs marine waters. Covers marine habitats, smothering some organisms. Causes turbidity in water, shading out producer organisms and altering the food chain.
Temperature	Factories Electric generating plants Urban runoff	Alters reproduction of fish. Reduces dissolved oxygen, which may then cause fish kills. Contaminates fresh water supplies used for drinking, irrigation, and the like.
Heavy metals	Fuel and exhaust of motorboats and automobiles Industrial emissions and effluent Sewage treatment plant effluent Landfill wastes/leachate Urban runoff Naturally in soil Hazardous waste fills and disposal	Accumulate in fish tissues and are passed on to humans Contaminate drinking water, causing brain damage, birth defects, miscarriages, and infant deaths.
Synthetic organic chemicals	Forestry, urban and agricultural runoff Industrial and municipal effluent Spills or dumping	Cause cancer, birth defects, and chronic illness when consumed in contaminated water supplies or seafood.

Source: Reference 254.

Some pollutants that may affect coral reefs include (206):

1. *Herbicides:* May interfere with basic food chain processes by destroying or damaging zooxanthellae in coral, free living phytoplankton, or algal or sea-grass plant communities; and can have serious effects even at very low concentrations.
2. *Pesticides:* May selectively destroy or damage elements of zooplankton or reef communities. Planktonic larvae are particularly vulnerable; may through accumulation in animal tissues have effects on physiological process.
3. *Antifouling paints and agents:* May selectively destroy or damage elements of zooplankton or reef communities; not likely to be a major factor except near major harbors, shipping lanes and industrial plants cooled by sea water.
4. *Sediments and turbidity:* Smother substrate; smother and exceed the clearing capacity of some filter feeding animals; reduce light penetration, may alter plant and animal vertical distribution on reefs; even slight changes in level may influence this distribution; may absorb and transport other pollutants.
5. *Sewage/detergents:* May interfere with physiological processes.
6. *Sewage/nutrients and fertilizers:* May stimulate phytoplankton and other plant productivity beyond the capacity of reef animal grazing and thus modify and overload the reef system.
7. *Petroleum hydrocarbons:* Have been demonstrated to have a wide range of damaging effects at different concentrations.
8. *Heated water from power station and industrial plant cooling:* Will locally change ecological conditions; water temperature is a key factor in distribution and physiological performance of most reef organisms (Figure 3.98).
9. *Hypersaline waste water from desalination plants:* Will locally change ecological conditions; salinity is a key factor in distribution and physiological performance of many reef organisms.
10. *Heavy metals;* Metals (mercury/cadmium, etc.) may be accumulated by and have severe physiological effects upon filter feeding animals, reef fish, and by accumulation up the food chain in higher predators.
11. *Radioactive wastes:* May have long-term and largely unpredictable effects upon the genetic nature of the biological community.

The most dramatic threat to the estuarine ecosystem is a catastrophic spill of oil, or release of other hazardous materials. The large volumes of petroleum and chemical products transported through the estuarine zone by ships, barges, pipelines, and railroads present a continuing potential for accidental bulk spills of oil or chemicals. Oil pollution may originate during exploration, production and transportation phases of the oil industry. With reference to ASEAN countries, Chua and Charles (53) emphasize that, "in recent years, oil pollution of the marine environment has been an issue of considerable national and international concern" (see also "Petroleum Industries" above).

A United Nations report (351) concluded that the "major causes of immediate concern" in the global marine environment upon entering the decade of the 1990s were "coastal development and the attendant destruction of habitats, eutrophication, microbial contamination of seafood and beaches, fouling of the seas by plastic litter, progressive buildup of chlorinated hydrocarbons, especially in the tropics and subtropics, and accumulation of tar on beaches."

Not all pollution comes from pipes or other specific points. Sources of "diffuse" (or "non-point") pollution that affect coastal areas are septic tanks, dumps, landfills, concentrations of boats, and, particularly, storm-water runoff from adjacent watersheds. These sources may cause serious eutrophication or toxicity where pollutants concentrate in confined lagoons or estuaries. Clearly, dumps, sanitary landfills, septic tanks and similar sources, should be placed away from watercourses and, to the extent possible, out of floodplains, to prevent leaching of pollutants into coastal waters. Also, standards may be needed to prevent pollution from boat and marine wastes (59).

Among the most important causes of nonpoint source pollution are land-alteration activities, principally those associated with site preparation for development and for cropland, as well as shoreline and water-basin alterations. Specific constraints should be imposed on project location, design, and drainage engineering throughout the coastal watershed including stormwater systems. Large-scale, stormwater sewer systems that collect runoff and pipe it directly into coastal waters not only introduce high loads

Figure 3.98. Heated wastewater from the cooling system of a nuclear power plant rises into the ocean off the California coast. (Photo by author.)

of pollutants (if not treated) but cause accelerated discharge to the coast. The flow of runoff in storm sewers may be stopped or reversed by storm surges. Consequently, with runoff obstructed, low-lying areas may flood with damage to shops, homes and other structures. (59).

For protection of coastal waters, the best storm-water system is one that most nearly simulates the natural system; that is, one that has features to detain storm runoff and to provide the maximum of soil filtration for natural purification of pollutants. Ideally, management should preserve and enhance the use of existing natural drainageways—creeks, sloughs, swales and so forth—to the maximum. (59).

Over the past several decades, environmental groups and regulatory agencies in the developed world have been effective in creating the perception that marine systems cannot accommodate societal discards without resource loss. On the other hand, social and natural scientists and engineers have shown that comparisons of disposal options (air, sea, and land) for a specific waste in a particular region do not exclude the ocean and, in fact, in some cases favor it.

3.68 PORTS AND HARBORS

Becoming involved in port planning is very important for conservation interests. Numerous sources of habitat alteration and environmental pollution need consideration (Figure 3.99).

According to Velsink (368), a differentiation should be made according to the time horizon of planning, on the one hand:

- *Long-term:* A period of about 20 years—port infrastructure and strategic Master Plan
- *Medium-term:* A period of 3 to 5 years—for instance, the planning, design, and execution of new terminals or the first-phase implementation of a new port plan
- *Short-term:* A period of about 1 year—procurement of equipment, small infrastructure improvements and adaptations

Figure 3.99. Each harbor facility needs a strategy to prevent pollution of coastal waters from fuel, various chemicals, and human wastes. (Photo by author.)

A differentiation should also be made according to scope, on the other hand:

- National or regional port planning
- Planning new individual ports
- Extensions or adaptations to existing ports

In the port planning process, the following stages or steps can be distinguished in more or less chronological order (there will be overlaps and feedbacks) (368):

- General definition of the objectives
- Delimitation of the coastal zone where the port has to be situated if it is to satisfy its requirements: centers of activity/connections with the hinterland
- Collecting existing data
- Rough forecast of the cargoflows: origin/destination, definition of the hinterland
- Study of the likely shipping patterns; number/type/sizes of the ships for the various trades, i.e., translation of the cargoflows into a shipping forecast
- Provisional determination of the required areas of land and water and the required waterdepth (primary program of requirements including nautical aspects
- A broad investigation into the local environmental conditions: geological and geotechnical, demographic, and sociological, etc.
- Generation of different outline plans for different locations, thus developing basically a complete overview of feasible solutions
- Screening of alternative locations and plans on basis of criteria like:
  - The available areas of land and water (plus extension possibilities)
  - Oceanological considerations (waves and currents)
  - Nautical considerations

  - Coastal engineering considerations
  - Geotechnical aspects
  - Environmental aspects
  - Connections with the hinterland
  - Sociological considerations (urbanization, potential workforce)
  - Industrial engineering aspects (situation of the area and elevation, cooling water, foundations, safety)
  - Cost aspects (construction and maintenance)
  - Accessibility (i.e., the frequency with which ports in alternative locations would have to be closed because of waves, currents and/or winds)

- Pre-selection of the most promising alternatives
- The drawing-up of provisional masterplans for these alternatives
- Evaluation procedure plus final selection
- Location-orientated site investigation
- Optimization: hydraulic model investigations, navigation simulator investigations, land use, costs
- Detailed masterplan and costs estimate: cost/benefit analysis

It will be clear that many divergent disciplines are usually involved in this type of planning operation. Some of these are (368):

- Oceanography and coastal engineering
- Hydraulics
- Hydro-nautics and nautical technology
- River engineering (sometimes)
- Traffic engineering and road engineering (roads and railroads)
- Transport engineering
- Maritime engineering
- Maritime engineering
- Structural engineering
- Dredging technology
- Geology, geotechnology and seismology
- Industrial engineering
- Safety engineering
- Macro-economics
- Business economics
- Transport economics
- Econometrics
- Organization and management
- Physical planning
- Sociology
- Ecology and biology
- Environmental impact assessment

A maritime protocol, or set of operational guidelines, for the more important impact generators should be created (Figure 3.100), including no bilge pumping in the area (even when filters to take out oil are installed); no garbage or other solid waste "over the side" disposal; and no disposal of waste oil or other liquid wastes.

A protocol to require onshore disposal of all plastics waste would be a positive benefit from the project. For many reasons littering and polluting of the sea should be stopped.

Most chemicals entering aquatic ecosystems rapidly bind to particles suspended in the water and accumulate in sediments on the bottom of the water body. This strong bond to particles initially decreases concentrations of chemicals in the water above the sediment. However, these sediment pools later serve as a continuing source of toxic chemicals to the water, and also to plants and animals inhabiting the area. Thus, chemicals present in all components of an estuary—its sediments, water, and biota—play

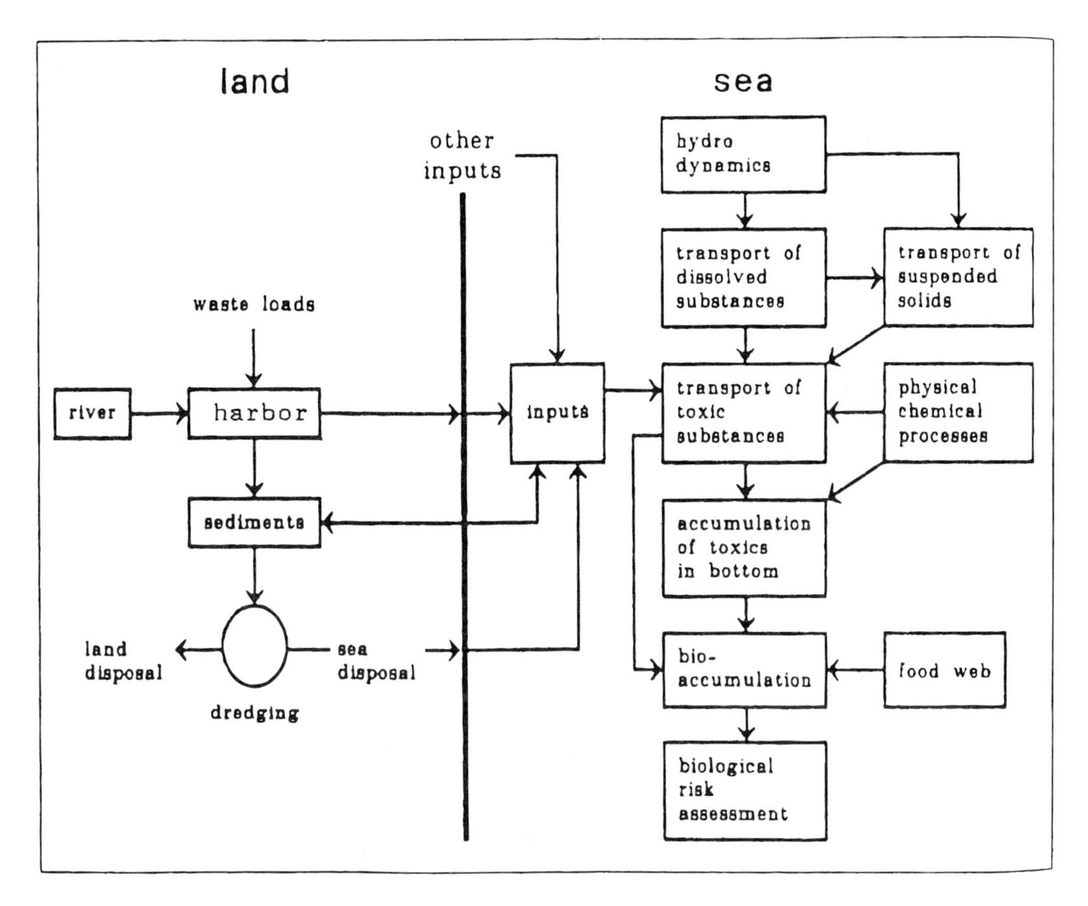

Figure 3.100. Impact pathway for disposal of dredge spoil in a harbor channel deepening project. (*Source:* Reference 233.)

an important role in determining the human health effects and ecosystem damage generated by chemical contamination.

Squibb (330) found that mercury, dieldrin, DDT, PCBs, phenanthrene, and pyrene concentrations exceeded acceptable concentrations in water, biota, and sediment samples from New York Harbor. They pose a hazard both to organisms inhabiting the harbor and to humans who consume fish and shellfish from the area. Concentrations of six metals (cadmium, copper, lead, nickel, silver, and zinc) in both water and sediments were above levels considered to be safe for marine life. In addition, sediment concentrations of PAHs were above safe levels.

Many ports rapidly become silted in and unusable, "erosion control structures" that accelerate erosion rather than slow it and river mouth entrainment schemes that unleash a host of unpredicted impacts.

3.69 PRINCIPLES AND PREMISES

This entry articulates the principles and premises underlying the comprehensive approach—integrated coastal zone management (ICZM)—which is the theme of this book. The goal of ICZM should be to *manage the coastal area as a unit* (shorelands plus coastal waters) and to *integrate* the conservation process with all appropriate economic sectors (shipping, mining, housing, transportation, aquaculture, etc.). The main targets of ICZM programs are conservation of resources, security of coastal communities, and control of coastal economic development.

What appears below is a listing only. However, the substance of the principles and premises is incorporated into the various parts of this book. A detailed elaboration of the principles and premises can be found in the United Nations/FAO guidebook, "Integrated Management of Coastal Zones" (see Reference 66):

1. The coastal area is a unique resource system that requires special management and planning approaches.
2. Water is the major integrating force in coastal resource systems.
3. It is essential that land and water uses be planned and managed in combination.
4. The edge of the sea is the focal point of coastal management programs.
5. A major emphasis of coastal resources management is to conserve common property resources.
6. Prevention of damage from natural hazards and conservation of natural resources should be combined in ICZM programs.
7. Coastal management boundaries should be issue based and adaptive.
8. All levels of government within a country must be involved in coastal management and planning.
9. The nature-synchronous approach to development is especially appropriate for the coast.
10. Special forms of economic and social benefit evaluation and public participation are required for coastal management programs.
11. Conservation for sustainable use is a major goal of coastal resources management.
12. Multiple-use management is appropriate for most coastal resource systems.
13. Multiple-sector involvement is essential to sustainable use of coastal resources.
14. Traditional resource management approaches should be encouraged where their results are positive.
15. The environmental impact assessment approach is essential to effective coastal management.

3.70 RAMSAR CONVENTION

Since its inception (in 1971) the Ramsar Convention has provided the principal intergovernmental forum for the promotion of international cooperation for wetland conservation. The principal obligations of signatory nations include:

- To designate wetlands for the List of Wetlands of International Importance (Article 2.1), to formulate and implement planning so as to promote conservation of listed sites (Article 3.1), to compensate for any loss of wetland resources if a listed wetland is deleted or restricted (Article 4.2), to use criteria for identifying wetlands of international importance.
- To formulate and implement planning so as to promote the wise use of wetlands (Article 3.1), to make environmental impact assessments before transformations of wetlands, and to make national wetland inventories.
- To establish nature reserves on wetlands and provide adequately for their wardening (Article 4.1), and through management to increase waterfowl populations on appropriate wetlands.
- To train personnel competent in wetland research, management and wardening (Article 4.5).
- To promote conservation of wetlands by combining far-sighted national policies with coordinated international action, to consult with other contracting parties about implementing obligations arising from the Convention, especially about shared wetlands and water systems (Article 5).
- To promote wetland conservation concerns with development aid agencies (Figure 3.101).
- To encourage research and exchange of data (Article 4.3).

3.71 RAPID RURAL APPRAISAL

Rapid rural appraisal (RRA) is a process for gathering and analyzing information from and about rural communities in a brief time period (weeks). In terms of structure, it is somewhere between formal survey and totally non-structured interviewing. Unlike other methods of social investigation, RRA sets out to create a dialogue with project clients (beneficiaries). This aspect of the method can reveal information on values, opinions, objectives, and local knowledge as well as "hard" data on social,

Figure 3.101. While the main purpose of "Ramsar" wetlands is nature protection, other uses have to be accommodated; for example, this pipeline can be installed without significant ecological harm if the wetland is properly restored after construction. (Photo by author.)

economic, agricultural, and ecological parameters. The quality of data produced depends largely on the team's skill and judgment.

For the purpose of "ecodevelopment," the involvement of the people in the subject area is essential. For this reason a participatory approach, rather than a researcher-subject relationship, is preferable even at the stage of information gathering. Therefore, we call the method herein discussed "Participatory Rural Appraisal" (PRA) as one form of RRA. Some of the specific advantages of the PRA method, then, are:

- It is cost effective in terms of money, time, materials, and manpower.
- It is interdisciplinary and can include decision makers as well as researchers.
- Non-sampling errors are reduced.
- The interactive methods allow close discussion with locals so that researchers and interviewees can see things from a shared perspective.
- There is such flexibility that hypotheses can be re-evaluated even during the fieldwork.

The PRA method is:

- Rapid and comparatively inexpensive
- Eclectic—using a variety of survey and interview techniques as the need arises

- Holistic—building up a multidisciplinary picture of the situation
- Interactive—generating dialogue between researchers and subjects

The general tools or strategies of PRA include:

- Interview and question-design techniques for individual, household, and key informant interviews
- Group interview techniques, including focus-group interviewing
- Interactive data gathering
- Cross-checking
- Use of pre-existing and secondary data sources
- Methods for obtaining quantitative data in a short time

The specific applications include the following:

1. Participatory Mapping:
    - Social Mapping: Village layout, infrastructure, population, households, chronic health cases, family planning, size of family, etc.
    - Primary Resources Mapping and Modeling: Land, water, tree resources, land use, land soil types, cropping patterns, land and water management, productivity watersheds, degraded land, treatment points, etc.
2. Transects
    - Observational walks to study natural resources, photography, indigenous technology, soil and vegetation, wildlife, farm practices, problems and opportunities which are cross tallied with the resource mapping and modeling.
    - Historical Transects: Pictorial-graphic representations of the area of different points in time to give evolutionary trends in land use, vegetation, erosion, population, etc.
3. Time Line:
    - Time and events, historical evolution of a village, agricultural practices, health care practices, etc.
4. Seasonality Diagramming:
    - For obtaining seasonal patterns of rainfall, employment, income and expenditure, debt, credit, food and nutrition, diseases, fodder, milk production, marketing, etc.
5. Ranking:
    - Matrix Ranking
    - Preference Ranking
    - Scoring Ranking
    - For ranking items such as trees, crops, varieties/types of breeds of livestock, wildlife, fodder, supplementary income-generating activities.
    - Wealth Ranking (status or economic order of members of community), such as, better off, medium, poor, poorest, etc.
6. Diagrams:
    - Venn diagrams ("Chapati diagrams")—Used as a means of identifying relationship between a village and its environment in order of relative importance.

There are numerous guide books on the subject of PRA (or RRA) for the practitioner. Here, we just introduce the topic in order to show that the methods of PRA could apply to coastal zone management. They are relevant to fisheries, mangrove forest use, coral mining, etc., as well as to the typical rural situation involving agriculture, grazing and forest cutting.

Source of this section: H. S. Pabla, S. Pandey, and R. Badola, Wildlife Institute of India, Debra Dun, India (274).

3.72 RED TIDE

The phenomenon known as *red tide* carries the threat of sickness (or even death) from paralytic shellfish poisoning (PSP) to those who are unfortunate enough to eat bivalve shellfish (mussels, oysters, clams, scallops) infested with certain microscopic organisms that float in the sea (uni-cellular dinoflagellate planktonic algae of the genus *Pyrodinium*). The risk from PSP—a nerve poison—is particularly high when the sea becomes visibly red or brownish red from the spread of these dinoflagellates throughout the water—it is almost as if dye had been put into the sea.

When eaten by bivalves, the dinoflagellates break up and the poisons are released and "stored" in the digestive gland of the shellfish which then are eaten by humans. Small filter-feeding fish like sardines and anchovies may also take in the dinoflagellates but they can be eaten safely by humans if the innards are first removed. But some red tides are toxic to fish and massive kills occur, often with windrows of dead fish on the beaches. The whole foodchain may be involved with high mortalities of seabirds, and even marine mammals.

Pyrodinium bahamense are often the causative organisms of paralytic shellfish poisoning and represent major red tide danger in many tropical countries. These tiny unicellular algae multiply rapidly in suitable conditions in bays and along coastlines of tropical countries. They produce a number of similar poisons (neurotoxins) which are collectively known as paralytic shellfish poisons (PSP) or toxins (226).

Eaten by bivalve shellfish—mussels, oysters, scallops, and cockles—the dinoflagellates break up and the released poisons are stored in the digestive gland of the shellfish, which may then be eaten by humans. Small fish, such as sardines and anchovies, also eat the dinoflagellates, but the fish can be eaten safely by humans if the "guts"—the intestine, gills, and other viscera—are removed carefully.

To give an example, PSP deaths occurred in Papua New Guinea in 1972, in Borneo for the first time in 1976 (causing deaths and many illnesses), in central Philippines in 1983 (claiming over 20 lives with 200 reported illnesses. Also PSP cases have occurred almost every year since 1984 in Sabah and in 1987 and succeeding years PSP made shellfish toxic on the Pacific coast of Guatemala (226).

What exactly triggers these massive "blooms" of microscopic algae is not known although works-of-man are often suspect, meaning that there is a management issue here as well as a medical one. For these organisms to bloom, a large supply of nutrient is needed, mostly phosphorus (P), although nitrogen (N) could enter in. Both P and N can be supplied by common polutants—sewage effluents, urban or farm runoff, or industrial discharges. Nutrients can also be released by navigation dredging and other construction disturbance. But until researchers are able to make more exact linkages, the management response should be to discourage nutrient pollution, uncontrolled navigation dredging, intertidal landfills, and other activities which could supply a red-tide bloom with the essential nutrients.

3.73 REGIONAL DEVELOPMENT PLANNING

It is useful to compare integrated coastal management to other types of regionally organized development and conservation programs, such as river basin planning or watershed management. Such activities have been conducted for many years in many countries, and this experience is useful in evaluating ICZM type programs. For example, river basin planners may organize their programs at any one of the four levels below:

1. *Water supply planning:* The program focuses on development of water supply to urban centers only and does not consider other uses or economic sectors.
2. *Multipurpose water planning:* The program considers all uses of water from the source at hand, but not other sectors.
3. *Integrated water planning:* The program focuses on water uses but coordinates with other economic sectors.
4. *Comprehensive water planning:* The program incorporates water use planning into an evenhanded general economic development program where all sectors are given equal priority.

The ICZM process, in the form proposed in this book, is analogous to the third level above, where the subject is conservation of coastal natural resources but where planning is fully integrated with other economic sectors as appropriate (e.g., mining, fisheries, housing, industry, tourism, agriculture, etc.). We do not believe that Level 1 or 2 analogs provide the multi-sectoral interaction necessary to plan and implement a balanced program of sustainable multiple use of coastal resources. For most countries it seems appropriate and possible, at present, to elevate coastal natural resources to the priority position they would have in a "Level 3" type approach in the above analogy (Figure 3.102).

ICZM-type programs can benefit from the regional planning approach in at least two different ways: (a) the national ICZM program can be organized to implement individual regional programs, if it has the flexibility to recognize regional distinction, and (b) the national ICZM program can be organized to facilitate input into presently ongoing regional economic development planning activities. If either the first option (specific regional ICZM program) or the second (facilitated input into regional development planning) is chosen, it is appropriate for coastal interests to understand what regional planning is and how it operates. [Note: One of the best reference documents on regional planning is "Integrated Regional Development Planning" produced by the Organization of American States (OAS) in 1984 (263), in cooperation with the U.S. Agency for International Development (USAID) and the National Park Service of the United States Department of the Interior.]

For professional planners, understanding the benefits and mechanisms of ICZM-type *programs* should be easy, as we have mentioned previously. While on one hand, these programs are distinctive because the coast is such a different landform with such different resources and unique problems, on the other hand, the ICZM *process* of program development is a familiar one. Regional economic development planners and resource planners, particularly should feel comfortable with ICZM because it deals with designated areas, the variety of resources and economic sectors therein, and other compatible subjects.

In general, the nationwide ICZM program should pick up the essence of regional planning (i.e., over a broad area economic opportunities should be identified, evaluated, and prioritized). Development strategies should be evolved that integrate the interests of various economic sectors, that optimize the benefits of resource uses, and that assure that development initiatives are sustainable and socially equitable.

Once a regional planning document has been created, it will serve as an effective guide to the ICZM manager in project review and assessment. Hanson (159) makes the point that "what is needed, in general, is a sharper focus on the definition of sustainable development and how alternative approaches to projects will help to attain the objective."

Figure 3.102. Freshwater swamps are linked to the coastal zone through shared wildlife and through water quality conditioning of runoff water before it enters the coastal waters. Regional planning can assist to protect freshwater in such wetlands for benefit of the coast. (Photo by Charles Wharton.)

The regional development approach provides a means to at least partially overcome the limitations of uncoordinated project planning and implementation (263). Yet development will still be heavily dependent upon individual project analysis, and within this, some form of extended cost-benefit analysis is likely to dominate. Inputs from environmental assessment may be incorporated into the cost-benefit analysis if appropriate. It is important to utilize the variety of ways in which environmental inputs may be incorporated into the planning cycle.

Environmental impact and social impact are specially important considerations in planning regional development. A human ecology perspective takes into account traditional uses, rights, and special needs of tribal minorities. Migration and population expansion also must be considered (263). Regional development is premised on the concept of increased social equity.

The Organization of American States (OAS) regional planning unit has evolved a simplified, regional approach, which includes the following three elements (263):

Diagnosis—A rapid analysis to determine the principal problems, potentials, and constraints of a region. The development diagnosis can include evaluation of natural resources and socio-economic conditions; delineation and analysis of subregions; identification of critical institutions, sectors, and geographic areas; generation of new information; and assembling of ideas for investment projects.

Strategy—Selection of pressing issues and opportunities for addressing them with the resources available. These opportunities suggest actions that are politically feasible within a time frame short enough to maintain momentum. (Less critical issues can be left for another round.) Alternative strategies can be presented so the government has a choice.

Projects—Preparation of interrelated investment projects to implement the selected strategy. The projects provide a balance among infrastructure, production activities, and services.

In the planning model OAS uses, the resource management specialist has three main important tasks in the development process: identifying the natural goods and services available from the regional ecosystems, identifying potential conflicts in the use of these goods and services, and helping to resolve those conflicts given the socio-economic policies in force in the region. If the potential conflicts are identified early in the planning process, before much money is spent or positions are hardened, they tend to be easier to resolve (263).

Integrating conservation into regional development plans calls for continuous liaison between the various planning and management authorities and local communities. Thus, the conservation authority must seek good working relationships with the other agencies concerned. Clear formulation of local conservation needs and objectives can provide a useful input into the development of a regional plan and its implementation. Such strategies and plans should set goals and standards for development and conservation of natural resources that will optimize sustainable production from a multiple-use system, without foreclosing options for future use. (77)

Conservation must be justified in both biological and socio-economic terms because government officials and the public generally undervalue environmentally sound development. Conservation must bring real socioeconomic benefits, such as the following (77):

- ***Stabilization of hydrological functions.*** Natural vegetation cover on water catchments in the tropics plays a valuable role, acting like a "sponge" to regulate and stabalize water runoff. Deep penetration by tree roots or other vegetation makes the soil more permeable to rainwater so that runoff is slower and more uniform than on cleared land. As a consequence, streams in forested regions continue to flow in dry weather and floods are minimized in rainy weather. In some cases, these hydrological functions can be of enormous value, worth many millions of dollars a year per catchment, particularly where reservoirs and coastal zones are protected. Watershed protection has been used to justify many valuable reserves that otherwise might not have been established. Coastal ecosystems can benefit from such protection.
- ***Protection of soils.*** Exposed tropical soils degrade quickly due to leaching of nutrients, burning of humus, laterization of minerals, and accelerated erosion of topsoil. Good soil protection by natural vegetation cover and litter (especially significant in grassland ecosystems) can preserve the productive capacity of the reserve itself; prevent dangerous landslides; prevent costly and damaging

siltation of fields, irrigation canals, and hydroelectric dams; safeguard downstream coastlines and riverbanks, and prevent the destruction of coral reefs and coastal fisheries by siltation. Prevention of siltation of reservoirs is a key benefit.

- ***Stabilization of climate.*** There is growing evidence that undisturbed forest actually helps to maintain the rainfall in its immediate vicinity by recycling water vapor at a steady rate back into the atmosphere and by the canopy's effect in promoting atmospheric turbulence. This may be particularly important in the production of dry-season showers that are often more critically needed for settled agriculture than are the heavier monsoon rains. Forest cover also helps to keep down local ambient temperatures, with benefits for surrounding areas in agriculture (lowered transpiration levels and water stress) and for human comfort.
- ***Natural balance of environment.*** The existence of a protected area may help maintain a more natural balance within the ecosystem over a much wider area. Protected areas afford sanctuary to breeding populations of birds that control insect and mammal pests in agricultural areas. Bats, birds, and bees that roost and breed in reserves may range far outside their boundaries to pollinate fruit trees in the surrounding areas.
- ***Regional pride and heritage value.*** The development of sources of regional and national pride provides benefits that are not easy to evaluate, but are nevertheless truly valuable. Individual and organizational donations of money for the preservation of their local heritage is a clear indication that they appreciate the value of the protected areas in their region.

The above benefits are of different scales of magnitude, accrue over different timespans, and fall to different groups in the local community, but they are additive and the total value to the region as a whole can be considerable. All benefits may not be derived from all reserves.

The best set of land uses for a particular area will depend on comparison of the sum of the benefits obtained through conservation versus the values of benefits attainable if the area were designated for alternative use. It is easy to find socio-economic justification for conservation of non-inhabitable marginal lands, but much harder to justify reserves in areas of high agricultural potential, which are often biologically the richest and the most valuable for conservation (77). Most often the best economic solution is a good multiple-use mix.

3.74 REMOTE SENSING

Data generated remotely may be recorded in either digital or photographic form. Remotely sensed data that are not camera-based are usually in digital form and typically are stored in computer-compatible tapes. These are processed through the use of image-processing systems (IPS), which are particularly used in conjunction with satellite imagery.

The IPS consist of hardware (computer) and software that transform the digital data into images. The IPS preprocess the data by making the necessary radiometric (e.g., effect of haze) and geometric (e.g., effect of earth's rotation) corrections. The preprocessed data are displayed in a computer and further processed using enhancement and spectral classification techniques to match the spectral signals with ground information (e.g., vegetation, landforms), according to Pheng *et al.* (283). The main purpose of image processing and analysis is to extract relevant information, which is then represented in "thematic" maps and interpreted for various purposes such as environmental assessment, cartography, meteorology, military, and management.

Remote sensing technology has been widely used for land resource evaluation, especially with the launching of satellites such as LANDSAT (USA) and SPOT (France). The high spatial resolution of multiband radiometers on LANDSAT and SPOT has also proven useful not only for land-based studies but also for research on shallow-water bathymetry of coastal areas. They are a useful source of data for coastal zone studies, which include both the land and coastal waters. Coupled with the use of digital mapping and GIS technology, they offer considerable potential as tools for coastal zone planning and management (283).

LANDSAT's "thematic mapper" has a ground resolution of 30 m and uses 7 spectral bands (violet-blue to infrared) with coastal applications. In contrast, SPOT has 3 multispectral bands (for color images) with ground resolution of 20 m, and in the panchromatic mode (for black and white images), the resolution is 10 m. In essence, SPOT has a better spatial resolution, but poorer spectral resolution than LANDSAT. Multispectral data from LANDSAT and SPOT have been used in land use assessment, urban planning and coastal studies, particularly in the intertidal zone; e.g., bottom substrates and algal species differentiation (283).

Information on suspended sediment in the water column, topography, bathymetry, sea state, water color, chlorophyll-a, sea surface temperature, fisheries, oil slicks and submerged or emergent vegetation including mangroves, has been provided by available remotely sensed data. Remote sensing has also been used in inventory and assessment of various coastal resources as well as production of maps (283).

Coastal applications, band designations, and spectral ranges of LANDSAT are shown below (283):

Band	Spectral range (μm)	Color	Application
1	0.45–0.52	Violet-Blue	For water body penetration, useful for coastal water mapping, also for differentiation of soil from vegetation
2	0.52–0.60	Green	Measures visible green reflectance peak of vegetation for vigor assessment
3	0.63–0.69	Red	A chlorophyll absorption and for vegetation discrimination
4	0.76–0.90	Near infrared	Useful for delineation of water bodies and determining biomass content
5	1.55–1.75	Middle infrared	Indicative of vegetation and soil moisture content; differentiation of snow from clouds
6	10.40–12.50	Thermal infrared	Used in vegetation stress analysis, soil moisture discrimination, and thermal mapping
7	2.08–2.35	Middle infrared	Discriminates rock types (geological applications); for hydrothermal mapping

For marine vegetation studies, imagery can be enhanced to highlight dense seagrass either on exposed intertidal banks or in clear, shallow (less than 2 m) water—in *large scale studies* remote sensing is more efficient than traditional (in-water) methods. Remote sensing has been utilized to map atolls, coral reefs and islands. Remote sensing used for reef fisheries planning in the Maldives was able to correlate water color and other marine environmental factors with the amount of phytoplankton in a reef lagoon to estimate marine primary productivity (283).

More comprehensive information on water quality parameter patterns in bays or lagoons can provide a better understanding of the ecology, biology, and dynamics in these ecosystems which are important inputs to management. In areas such as turbidity, temperature and chlorophyll and transparency, remote sensing leads to better understanding of the hydrodynamic condition of the system, particularly on the productivity of marine waters and indirectly on fisheries biomass. While the same water quality parameters can be investigated through conventional survey techniques, these are time-consuming and expensive. Remotely sensed data provide a synaptic view and thus, may be applicable to modeling, mapping and monitoring of water quality.

With respect to coastal management, remote sensing may provide the most feasible means for mapping, particularly for insular (archipelagic) countries. For example, in countries with many scattered islands (e.g., Indonesia and the Philippines), land inventory is usually difficult and expensive using conventional survey methods, even with the use of aerial photography. The synoptic coverage of satellite remote sensing makes it less costly in the long run, especially for mapping wide areas of land and vegetation cover (283).

Despite the undoubted potential of remote sensing, there are still limitations and practical problems encountered in their use. For a dynamic system like the sea, which varies on time scales of seconds, minutes, hours, and days, the frequency of data acquisition by LANDSAT or SPOT becomes more limiting than it would be for land applications. This is particularly so in the case of coastal waters which are tidally dominated, whereby changes occur over short time spans of minutes and hours. Earth resource satellites, with a revisit time interval of 16 to 26 days, cannot provide the temporal resolution to capture the dynamism of coastal waters. Furthermore, the use of such data of low temporal resolution gives rise to difficulties in interpretation of imagery taken at different tidal times, as encountered in the mapping of reefs in shallow waters, where the spectral response pattern of the same reef can differ vastly depending on the water depth, which fluctuates with the tide (283).

Perhaps the most serious constraint faced in the use of satellite data from passive remote sensing systems is cloud cover. It has been estimated that at a latitude of 50° N in Europe, there is only a 5 percent chance or less of obtaining two consecutive satellite images of the same area containing less than 30% cloud. The situation can be expected to be far worse in the humid tropics, especially in the coastal zone, where it is not uncommon to find a band of cloud hugging the coastline while the landward interior might be cloud-free (283).

Another problem facing the use of remotely sensed data for the study of waters is the limited penetration of electromagnetic radiation. Thus, most of the coastal and marine phenomena sensed remotely are surface phenomena. While light penetration *per se* may not be as serious a problem in shallow coastal waters, a different problem is faced in the interpretation of remotely sensed data for coastal waters. This is the confounding factor of suspended and dissolved matter with water depth in the reflectance of electromagnetic radiation by the water body (283).

Potentially, airborne remote sensing allows for greater flexibility in the deployment of sensor type, time and frequency of data acquisition, and selective areal coverage. However, such facilities are not easily available in many parts of the world, and are also very costly. In many countries in this region, CZM studies still resort to the more conventional use of aerial photographs. But the poor light penetration and the absence of spectral resolution in conventional panchromatic aerial photographs limit their use in coastal water studies (283).

Geographic data in the form of aerial photographs, other airborne imagery, and maps are among the most important information for both project-specific or long-range environmental planning. Such information collected at the same place but at different times gives a historic record of environmental change, or stability. Geographic data provide the basis to calculate the area of mappable resources and impacts; to help site settlements, roads, ports and agricultural areas; to formulate land use and coastal zone plans; and to facilitate impact analysis using overlay techniques. Photographs and maps are easily read, interpreted, and transcend language or cultural barriers to communication and analysis. Aerial photographs can also be used to establish field surveys and sampling strategies to reduce cost, improve efficiency, and ensure adequate sampling of all relevant habitats and environments (41).

Most modern large-scale maps now rely in part on aerial photography (Figure 3.103). Photographic interpretation and ground-truthing (spot checks in the field to verify photointerpretations) are also important components of mapping. Aerial cameras with 9″ × 9″ negative format provide among the best quality photographs. Film is readily available in black and white, visible color, near infrared color, print negatives, and positive transparency types, as compared in the table below (41).

Type	Advantages	Disadvantages
Narrow band (black and white)	High resolution and good water depth details (up to 50 ft)	Discrimination between different hues of same shade is not possible
Near infrared (false color infrared)	Excellent discrimination among vegetation, crops, and other photosynthetic organisms	Poor depth penetration in lakes, ocean waters (maximum of 3 to 6 ft)
Thermal infrared	Detects temperature anomalies (especially in water bodies), such as cold water or warm water plumes and discharges	Poor resolution; normally requires use of a scanner (emitting device) to bounce the thermal IR waves off the land surface
Radar wavelength	Highest resolution due to short wavelength	Rapid attenuation; low subsurface penetration often requires use of scanner
Visible color	Great discrimination at depth in the ocean; good discrimination between most landcover categories	Some color attenuation at high altitude; cloud and haze interference

Vertical aerial photographs are usually more useful than oblique aerial photographs. Cameras are mounted in the floor of a plane or jet aircraft. The aircraft follows predetermined flight lines, velocities,

Figure 3.103. Aerial photo of Appalachicola, Florida (USA), used in environmental and land use surveys. (Photo courtesy of Appalachicola Commission.)

and altitudes, usually along straight courses and constant altitudes, and the camera takes a series of overlapping frames of the land or seascape.

Although the camera faces straight down for vertical photographs, only the central part of each photograph is truly vertical and free of distortion. At increasing distances from the center (or vertical) distortion results from increasing side angle and distance between the camera and the subject (ground surface). This distortion is greater for wide-angle photographs. Most distortion is predictable and can be optically corrected to produce orthogonal aerial photographs. Other calibrations involve locating specific surveyed landmarks on the photo to establish distance scales and map coordinates (e.g., longitude and latitude). Standard geodesic survey techniques are used to make and calculate the exact locations of the ground control points. Although corrected aerial photographs are needed for map making, uncorrected photos can still provide many important types of environmental information and serve as the basis for rough maps (41).

Low-altitude imagery offers the greatest resolution and detail but has limited field of view (or swath). Hence, more photographs and flying time are needed per unit area. Common scales range from 1 inch = 20 ft to 1 inch = 1,000 ft (or 1:240 to 1:12,000, respectively). Common purposes for low-altitude photography include surveying proposed road alignments, settlements, ports, and preparing master plans, site plans, or facilities plans. Low-altitude photography is also important for flooding or inundation analysis, for the monitoring of beach or shoreline changes, the stability of stream courses, and changes in agriculture or forest cover or conditions. Low- and medium-altitude aerial photography is also important for accomplishing coastal resource inventories.

There are some clear advantages of using remote sensing and Geographic Information System (GIS) computer technology in a complementary manner. Remotely sensed data, and the products of image processing, are already in digital form with more or less standard formats. Although raw remotely sensed data are not geographically correct, preprocessing using well-established algorithms in image processing software can bring the data to acceptable levels of geographical accuracy, which can register with other conventional map data. There are several ways to integrate remotely sensed data with GIS computerized data handling (see "Geographic Information Systems" above) (283):

- Aerial photographs and photographic output of satellite images (subjected to some preprocessing and image enhancement) are manually interpreted, and hand-drawn thematic maps, e.g., vegetation cover, are digitized into GIS.
- Digital remotely sensed data are analyzed or classified digitally, the outputs are produced in hard copy as conventional maps, which are then digitized into GIS.
- Digital remotely sensed data are classified or analyzed using automated digital methods, and the output is transferred digitally into GIS.
- Raw, digital remotely sensed data are entered directly into GIS, where all processing is done.

According to van R. Classen and Pirazzoli (89), the initial capital cost of satellite remote sensing is high and includes satellite design and launching, the establishment of ground control and receiving stations and, or users, the development of computer systems and software for analysis of data. Once established, the cost per processed image is relatively low.

The trend in the development of earth resource satellites has been to decrease the size of the minimum area of resolution (the pixel) and to increase the number of sensors designed for specific, often narrow, wavelengths. These developments have resulted in increasingly large amounts of data which have to be analyzed. To extract the maximum amount of information from the data, the sophistication of a satellite has to be matched by sophistication in the user's computer equipment and software. On the other hand, however, a hard-copy photographic or graphic (black and white or color) plotter image of an area can be used to great effect by a good interpreter (89).

The cost of "hardware" items can be alleviated to some degree by sharing facilities either intra-nationally, with many national or provincial agencies using a single resource inventory/remote sensing analysis center, or internationally on a regional basis, e.g., the Economic and Social Commission for Asia and the Pacific (ESCAP) Regional Remote Sensing Program; South Pacific Commission; ASEAN nations; etc. New developments in low-cost, stand-alone work stations for image analysis can however provide a more localized capability for US$23,000–35,000 (89).

By contrast, the capital cost of aerial remote sensing is relatively low although, depending upon the type of sensor used, there may be significant computer and software costs for analysis. Unless they can be partially absorbed in another function, such as routine surveillance, the operational costs of obtaining imagery from aircraft can be high. This is particularly true if aircraft and crew have to remain on standby for long periods waiting for suitable weather conditions. Airborne cameras and sensors are, however, necessary for work at scales less than 1:25,000, which are generally beyond the capacity of present satellite systems. It should also be noted that air photograph interpretation costs per 1,000 square kilometers for large areas are higher than those incurred in interpreting satellite imagery (89).

Suffice to say that before undertaking operational use of remote sensing data a program investigating and developing specifications is necessary to ensure that the precision and cost of data collected and analyzed is justified by the usefulness of the information obtained (89).

3.75 RESEARCH NEEDS

Coastal planning gives rise to a whole range of questions requiring scientific expertise. Development activity anywhere in the coastal zone—watershed, floodplain, shoreline, or water basin—is a potential source of stress to the coastal ecosystem. There are many critical natural components of this ecosystem: fringing wetlands, estuarine bottoms, shellfish beds, coral reefs and the breeding, nursery, feeding and nesting habitats of coastal species (Figure 3.104). Each component should be evaluated and designated for protection or for limited use. In addition, the basic ecosystem processes—water circulation, nutrient input, and sunlight penetration of water—must be considered, and thresholds of permitted stress established.

The variety of uses envisioned for the coastal areas and the effective utilization of the resources present numerous challenges, including the tests of modern ICZM techniques. Much of the future development will depend on having the necessary capabilities and tools. It will be necessary to:

1. Understand ocean and coastal processes, their relationship and impact upon the resource endowment
2. Evaluate, develop, and conserve the resource endowment
3. Evaluate safety and environmental protection requirements in the marine areas as well as those related to the protection of life and property in the coastal areas
4. Evaluate information needs for monitoring the effects of present and planned activities on the marine and coastal environment
5. Assess new or improved technology to support coastal area EEZ surveying, mapping, and research programs
6. Develop approaches for multiple uses and identify, avoid, and resolve potential and current conflicts between the various uses of the sea
7. Establish management structures for planning and coordinating the development activities in the coastal zone and its EEZ component
8. Develop a regulatory framework and coordination machinery needed to implement the national EEZ plan

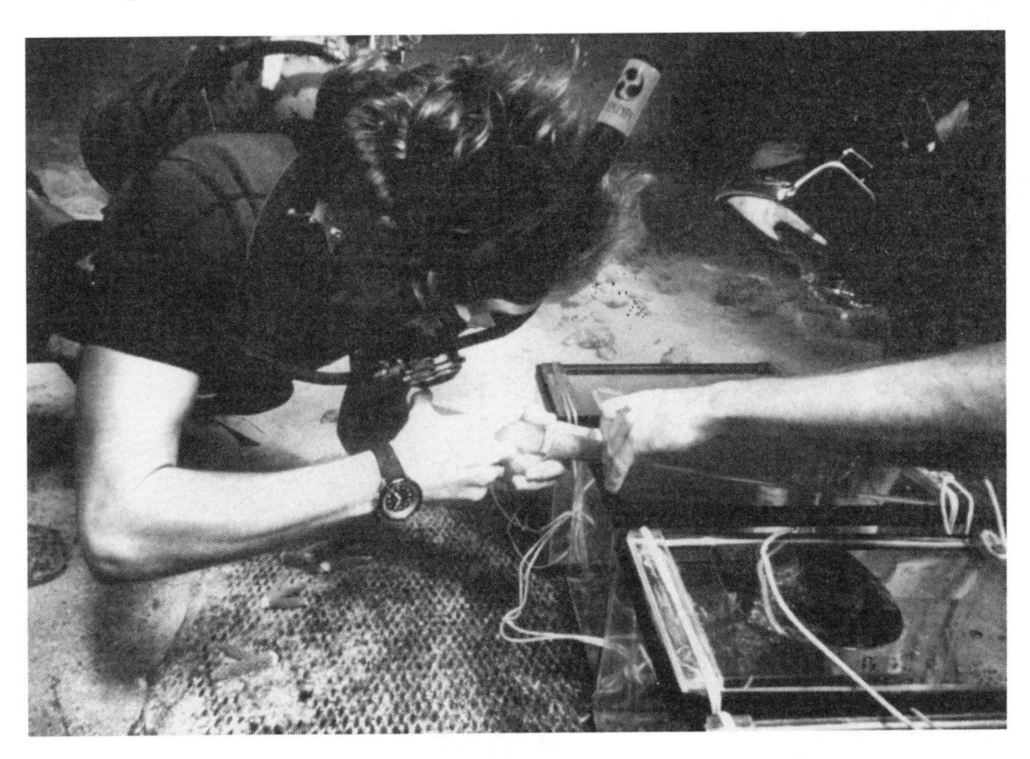

Figure 3.104. Because coral reefs are so valuable, they have been the subject of extensive environmental research. (Photo courtesy of U.S. National Oceanic and Atmospheric Administration.)

The amount of damage that may result from any disturbance depends on the characteristics and vulnerabilities of the specific ecosystem; this is where the expertise of natural scientists is so vitally needed. The services that natural scientists can perform for planners fall into five broad categories:

- ***Description:*** Inventory and evaluation of resources, including critical areas; description of components, processes, and functions of the natural systems involved (Figure 3.105)
- ***Diagnosis:*** Analysis of past environmental damage and the present condition of natural systems; identification of problems, including causes and consequences of ecological disturbance
- ***Prediction:*** Identification of probably ecological effects of specific land and water uses; delineation of the capacity of natural systems to support development
- ***Prescription:*** Recommendation of requirements to maintain natural carrying capacity and to preserve resources
- ***Implementation:*** Advise in formulating criteria and standards for management implementation; design of environmental monitoring programs; and interpretation of data and reevaluation of standards

Planners generally decide which of these functions scientists are required to perform in a planning process. Scientists can participate actively in planning, as originators of their own recommendations, or they can simply be passive reviewers of documents generated by non-scientists.

Regardless of how scientists participate, their task is not a simple one. The ecological consequences of coastal development are so complex and so imperfectly understood that it is often extremely difficult to formulate specific development standards for protection of natural environments and conservation of resources.

National research capacity is required to (a) understand ocean and coastal processes, their relationship and impact upon the resource endowment; (b) evaluate, develop and conserve the resource endowment; (c) evaluate safety and environmental protection requirements in the marine areas as well as those

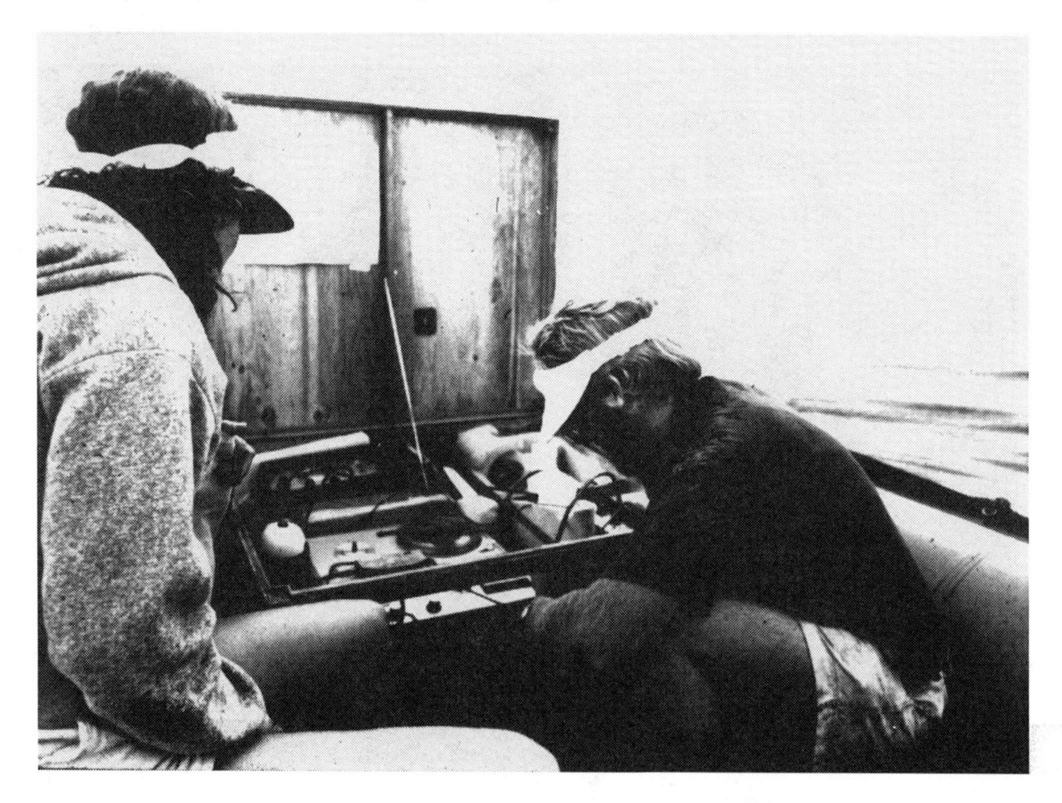

Figure 3.105. Using an underwater recording device, these technicians make tapes of the voices of gray whales off Lahaina, Oahu, Hawaii. (Photo by author.)

related to the protection of life and property in the coastal areas; (d) evaluate information needs for monitoring the effects of present and planned activities on the marine and coastal environment; (e) assess new or improved technology to support coastal area and EEZ surveying, mapping, and research programs; (f) develop approaches for multiple uses, and identify and resolve potential and current conflicts between the various users of coastal seas; (g) establish management structures for planning and coordinating the development activities in the coastal zone, including the EEZ; and (h) develop a regulatory framework and coordination machinery needed to implement a national ICZM plan.

But in addition, social research is needed to understand the situation of the people to be affected by coastal development. This is often done by a technique known as Rapid Rural Appraisal (RRA). RRA research techniques can be used to uncover the actual situation and conditions confronting the inhabitants of the coastal zone (see "Rapid Rural Appraisal" subsection above). Preliminary data and information on resource use productivity, stability, sustainability and equitability can be collected. And also, first-hand knowledge on basic demographics, socioeconomics, technology, production and cropping systems are gathered. Problems confronting and opportunities opened to the inhabitants can be looked into and elaborated upon with select groups of key informants and knowledgeable persons. (47)

Scientific research has provided a general sense of the values supported by natural systems and of the effects that development can have on these values. But widely applicable "cookbook" standards for development have yet to emerge. In the absence of these standards, the analysis of area-specific scientific data on natural systems may be necessary for each planning program so that the range of potential environmental impacts can be known. Guiding principles such as "protect wetlands" or "preserve water quality" simply do not do the job. These principles must be translated into specific development controls.

Moreover, although scientists can often state the optimum conditions for ecosystem function, they are not equipped to determine what constitutes socially acceptable or unacceptable levels of resource protection. In effect, scientists can establish criteria upon which public decision making can proceed, but they are not qualified, by virtue of their expertise, to make those decisions.

The special role of science in coastal planning can be an uncomfortable one for the scientist. The role does not involve basic research, which many scientists relish. Rather, it involves information transfer, including data interpretation and consultative services, a role which many scientists disdain. The planners' main need is not for new research, but to get more of what scientists already know applied to the problems of planning and management. Unfortunately, this kind of information transfer has many advocates but few professional practitioners.

3.76 RESTORATION AND REHABILITATION

It is axiomatic that the most developed coastlines have the most degraded renewable resource systems. Pollution, habitat conversion, and interference with water circulation, among other effects, have seriously reduced the economic and environmental benefits of most natural systems. While it is an important function of ICZM to constrain further ecological losses, it is also important to plan for the repair of natural systems damaged in the past.

The following definitions can be used to clarify two separate restorative aproaches:

Restoration involves putting back what was there as exactly and completely as possible;

Rehabilitation involves repair or substitution of function; no attempt is made to restore the ecosystem to its original configuration but rather to *replace* the original by a different but valuable one.

Exact *restoration* is an engineering activity with little design input. On the other hand, *rehabilitation* begs the questions: What design would yield the highest socioeconomic benefit, considering regional needs for natural goods and services? If waterfowl habitat is critical, then a relatively shallow open water area would be most appropriate. If fish nurturing, shorebird habitat, shoreline stabilization, or run-off water purification are the priority needs, then different designs are indicated.

Given the money, environmental engineering can restore almost any type of damaged coastal ecosystem. For example, a degraded wetland can be regraded, reshaped, rewatered or replanted—that is, the topography of the soil can be reshaped, or the supply of water restored, or the proper balance of vegetation cultured (75). A beach can be rebuilt, a mangrove forest replanted, an estuarine channel reopened, a seagrass bed replanted, a polluted lagoon cleaned up. Most often the priority need is for hydrological restoration.

It is evident that much of the world's coral reef habitat is in poor condition and in need of repair. Are there management actions which might accelerate recovery? The answer seems to be that if a coral reef has been damaged by pollution, hurricanes, mining, or boat anchoring, it may be difficult, but not impossible to rehabilitate it given sufficient political will and money, some degree of restoration can be accomplished. A notable example is the reduction of sewage discharge into Kaneohe Bay, Hawaii, as a result of which the reefs are recovering from extreme algal overgrowth. (See "United States, Hawaii: Success with Kaneohe Bay Ocean Outfall" case history in Part 4).

While local "point sources" of pollution can be detected and corrected most easily (with money), the bigger problems are those of more diffuse origin (e.g., runoff pollution from deforestation and agriculture and overfishing). Passive rehabilitation is considered to be a slow process, taking 20 to perhaps 50 years or more for relatively mature regrowth of corals.

A United Nations report (351) cites many cases of ecosystem recovery after water quality cleanup. Examples include the removal or reduction of phosphate, methyl mercury, pulp mill waste, DDT and PCB, and petroleum.

Critical habitat needs such as these can be identified and mitigation goals specifically stated in terms of fish and wildlife or ecologic targets (75). The shift from the reactive to the strategic approach would bring a shift from "supply-side" to "demand-side" thinking about ecosystems.

A regional restoration strategy can be organized to respond to "cumulative impacts" and to provide "offsets" for any environmental damage in degraded ecosystems. Effectiveness in aquatic habitat restoration requires understanding regional ecosystems, their present condition (how far degraded), and what values are most important and should be given priority for rehabilitation (plant productivity? bird habitat? nursery area?). Given this, one can formulate goals and advanced criteria for permit review and mitigation and even reverse the trend of negative cumulative impacts and bring about a positive cumulative impact sequence.

In summary, the ICZM program should include an ecosystem rehabilitation planning element. This will require an investigation of the condition of coastal resources, an economic evaluation of losses and benefits of rehabilitation, and recommended priorities for rehabilitation projects.

3.77 RISK ASSESSMENT

All data pertaining to environmental effects possess some degree of uncertainty and many involve random, stochastic processes, according to Carpenter and Maragos (41). For example, an oil spill from a tank farm could damage nearby habitat and aquatic birds but, often, no information is given about the likelihood of the spill or the magnitude of the impact on the bird population. For significant impacts, a quantitative probabilistic risk assessment may be useful.

Some definitions of risk terminology are (14.1):

- A *hazard* is a danger, peril, or a source of harm.
- *Risk* is the chance, possibility, or probability of adverse consequence, loss, or injury.
- *Uncertainty* is lack of knowledge about an outcome or result.
- *Assessment* is appraisal or evaluation in order to judge.
- *Analysis* is detailed examination or thorough study in order to understand.

Risk assessment addresses three questions: What can happen; what can go wrong? How likely is it that the event will actually occur? If it does happen, what is the range of consequences?

Environmental impact assessment (EIA) and accompanying engineering and economic analyses will generate most of the scenarios for the first question and also will predict the nature of the consequences in Question 3. What remains is to answer questions about the likelihood of hazardous events and to express the severity of impacts statistically.

Risks are measured in terms of both frequency and severity. Nuclear power plants, for example, may present the risk of a catastrophic but rare failure. Pollution discharges may present the risk of frequent events but with minor damage. Highly sophisticated techniques have been developed for quantifying risks from cancer from exposure to toxic chemicals. It is possible, however, to estimate risks in a practical way by using technical judgment and experience and thus to give decision makers more information about impacts.

An example is a geographic information system (GIS) used to help pick the site for a hazardous waste landfill. The systematic examination of the movement of electroplating wastes from the point of generation, through transport, storage, treatment, and disposal, revealed that the biggest risk was in contamination of surface water or groundwater. The risk could be reduced by siting the landfill well away from any surface water, irrigated fields, fishponds or wetlands.

It is not necessary to have complete probability distribution data in order to perform a useful risk assessment. It is often helpful to the decision maker to have some idea of the frequency and relative scale of adverse consequences.

3.78 ROADWAYS, CAUSEWAYS, AND BRIDGES

Roadways in the coastal zone have complex side effects. A major effect is the replacement of mangrove and other tidal wetland areas with the roadbed. A second major effect is the blockage and disruption of normal circulation patterns, both tidal and land drainage. Roadways built parallel to the coastline and through coastal zone water areas have often acted as dams to fresh water draining down toward the sea or as blockages to tidal flows within estuaries. All roadways in coastal areas should be located so that they conform to existing topography, require a minimum of alteration of soils and vegetation, and are provided with ample drainage channels or culverts (Figure 3.106).

A major environmental problem involves the effects of roads that are partly devised to "open up new territory" for development. In such cases, unless a complete projection of effects is available,

Figure 3.106. Because roadways and bridges can have major environmental impacts their routes and designs should be carefully planned. (Photo courtesy of U.S. Army Corps of Engineers).

significant environmental deterioration will ensue. In locating a roadway network, it is necessary to consider the land and its resources as a total natural system. Land-use planning then aims at identifying and setting priorities for various uses and amenities and concurrently assessing environmental and cultural effects.

Factors requiring consideration include community, institutional, and residential values; the value of land for recreation; surface water and groundwater values; and wildlife and fisheries resource values. With these values considered, roadway location can then be best determined to serve simultaneously the purposes of preservation of natural systems and need for urban expansion.

The construction of a number of highways through tidal wetland areas has effectively separated the upper reach of a wetland from the lower reach, completely altering the circulation patterns of the wetland system. Besides obstructing tidal flushing, causeways also act as water barriers during severe storms and hurricanes. The blockage can prevent the normal circulation of waters and exchange of nutrients and organisms and lower the biodiversity of the coastal system. The effects of such blockages can be avoided by elevating the roadway instead of using solid-fill causeways.

The basic circulation pattern of the wetlands or tidal marshes should be preserved by appropriate use of culverts under filled causeways where the roadway routing crosses upland tributaries (Figure 3.107). Also, when abutments or fill areas must impinge on water areas, it is necessary to reduce the encroachment to the minimum. As a rule of thumb, the cross-sectional area of a waterway should in no case be reduced to less than that which can adequately pass the 100-year maximum flood waters, defined as not raising the floodwaters more than 1 foot above the natural level.

Also, solid-fill construction usually requires extensive dredging for fill and the excavation of construction access canals, which result in additional loss of wetlands and tidelands and soil discharge into the estuary.

When building a causeway over wetlands, construction activity should take place on the causeway structure and off the wetlands to the maximum extent possible. That is, heavy equipment needed to place the roadway pilings—cranes, dredges—should be operated from the roadbed. This topside construction approach is recommended to avoid the need for barge canals.

Bridges should be designed so as not to impair tidal flow in respect to volume, velocity, or direction. Abutments should be built back from the water edge, and clear spans used rather than piers. Most simply, the cross-sectional area of a watercourse should not be reduced by abutments, support piers, pilings, and so forth.

According to Maragos (231), road causeways can have substantial impacts on coral reef ecosystems if they block water circulation or degrade water quality (Figure 3.108). At Palmyra Atoll, road causeways completely blocked exchange of lagoon waters to the open ocean causing the catastrophic collapse of lagoon ecosystems (see Figure 6, of "South Pacific" case study in Part 4).

Ecological impacts often include reduction of water exchange, aggravated water pollution from sewage discharges, and blockage of migratory pathways used by reef fishes. Causeways are now being planned on Kwajalein Atoll to include measures such as adequate sideslope protection, bridge openings, and culverts to reduce impacts on water quality. Such projects are extremely expensive and require considerable maintenance and repair, however (231).

Maragos also describes a large roadway constructed across a reef flat at Moen, Truk (see Figure 5 of "South Pacific" case study in Part 4) blocked off circulation to Pou Bay, the most important estuary and subsistence fishery habitat on the island. Raw sewage discharges from villages at the shoreline of the bay aggravated water pollution, potentially contaminating shellfish and generating public health hazards. Improvements to the causeway in 1977 included enlarging the culverts, significantly improving water exchange and bay water quality, but it took considerable effort to convince reluctant engineers and government officials of the value of installing the expensive culverts. The large causeway to connect the port and airfield to the town of Kolonia, Pohnpei has blocked circulation to much of Kolonia Bay, has decreased water quality, has increased sedimentation, and also choked off one mangrove area. A similar causeway at Kosrae (see Figure 9 of "South Pacific" case study in Part 4) to connect the main island to Lelu, a populated offshore island, reduced circulation and water quality in Lelu Harbor, substantially reducing fish and shellfish catches in what used to be Kosrae's most productive lagoon and reef habitat (231).

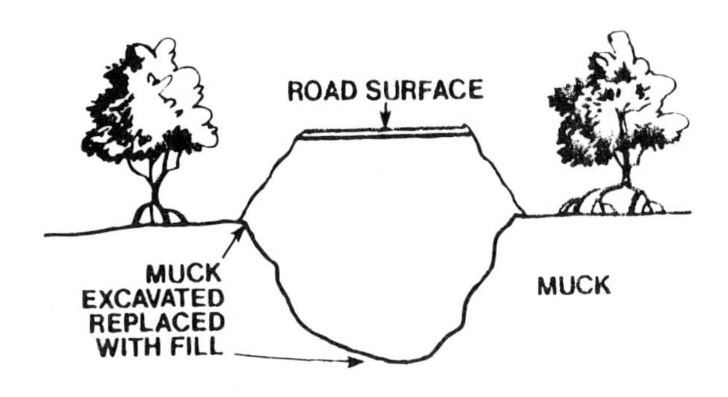

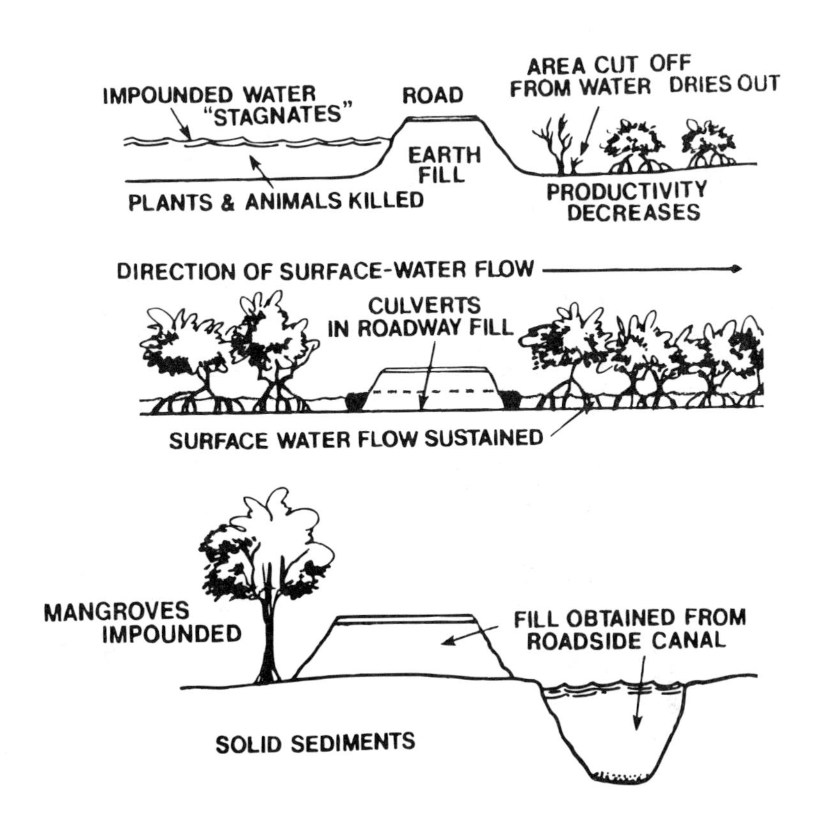

Figure 3.107. Four different road construction techniques are shown along with the environmental impacts typical of each. (*Source:* Reference 316.)

3.79 SALINITY

The salinity of coastal waters reflects a complex mixture of dissolved salts, the most abundant being the common salt, sodium chloride. Salinity throughout the coastal ecosystem fluctuates with the amount of dilution by precipitation and by land drainage or river inflow. Typically, there is a gradient in salt content that starts with a value of about 35 ppt (parts of salt per thousand parts of water) offshore, drops to about 30 ppt in nearshore waters and in the seaward ends of estuaries, and then falls to less than 0.5 ppt at some distance up the tributary rivers. In estuaries, where more saline and heavier water is entering beneath fresher runoff water at the surface, a powerful two-layered circulation pattern is established (Figure 3.109).

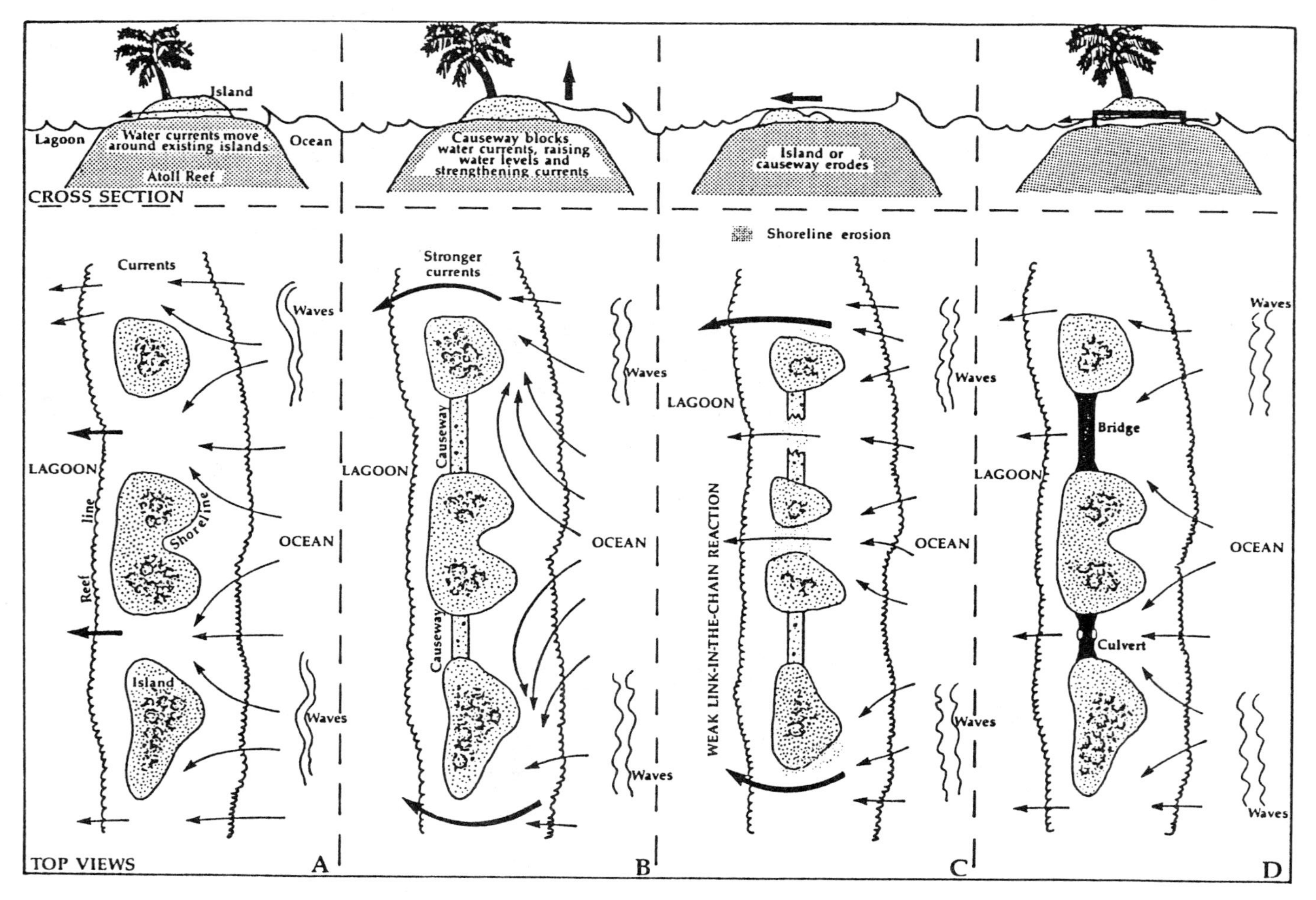

Figure 3.108. How road causeways can block water circulation and degrade coral reef ecosystems: (A) in natural state, waves push water into lagoon; (B) causeways block currents and raise water levels over land; (C) eventually, water cuts through island or causeway; (D) either culverts under or bridges over will allow water to flow through naturally. (*Source:* Reference 213.)

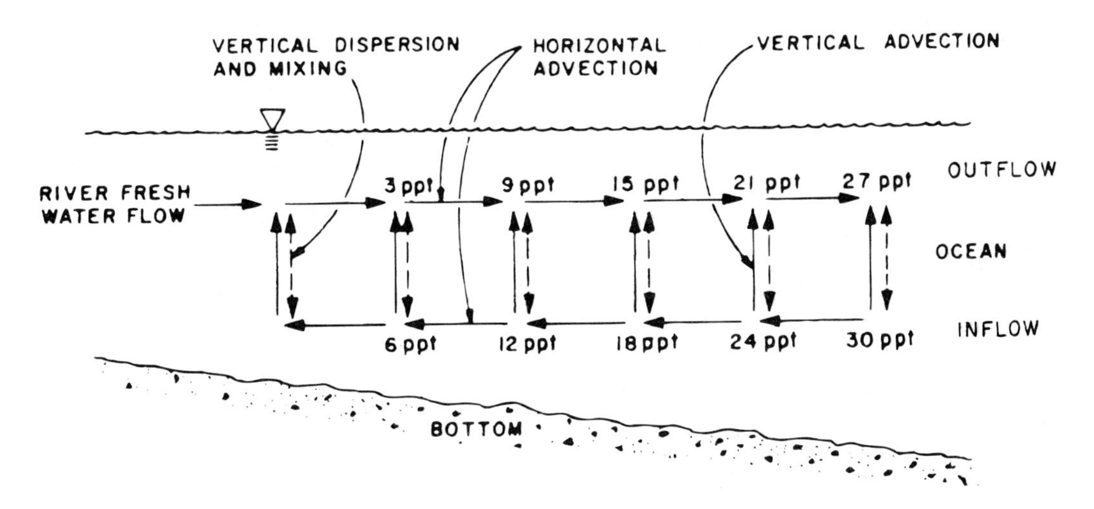

Figure 3.109. Salinity variation and current flows in typical stratified estuaries. (*Source:* Reference 57.)

Some coastal species can tolerate a wide range of salinity, whereas others require a narrow range to live and reproduce successfully. Some species require different salinities at different phases of their life cycles, such as are provided by regular seasonal rhythms in salinity in estuaries from spring runoffs and summer drought. Most of those with a narrow tolerance can nevertheless withstand short-term exposure to a much wider salinity range, depending on the rate of change. As with other environmental factors, coastal species have evolved over the years in harmony with their salinity environments and have become adapted to the natural regime.

Development activities that alter the salinity balance of coastal waters, particularly estuaries, can reduce productivity and biodiversity. For example, any type of solid-fill causeway can disrupt the natural tidal flow in an estuary sufficient to upset the delicate salinity balance essential for the survival of many estuarine organisms. In another example, diversion of river water flow away from an estuary can raise the salinity to unfavorable levels.

Salinity may be measured chemically (ppt) or electrically (millimho/cm) (the conversion ratio of chemical to electric measure is 6.76 ppt = 1,000 micromos). One can convert from one measure to the other if temperature of the water is known. Rough conversions can be made from Figure 3.110. See Table 3.1 for the relation of temperature and conductivity.

3.80 SALTWATER INTRUSION

When groundwater resources are overexploited, more water is used than the aquifer can replenish through infiltration. Without sufficient recharge of the groundwater, wells can go dry or "pull" saltwater from the estuaries into the aquifer. Salt-contaminated water cannot be used for drinking or crop irrigation, posing a serious problem for local communities and farmers.

For this reason, it may be necessary to designate "capacity use areas"—aquifers where withdrawals must be regulated to prevent the groundwater from being overused. Such areas may encompass a large portion of the coastal area.

The big users of groundwater in coastal areas are municipal water suppliers and industry. The time has passed in most areas where such uses can go unmanaged in heavily populated areas. For example, in Long Island (New York, USA) overpumping for municipal supplies and industrial operations caused the freshwater head to drop as far as 35 feet (10.7 meters) below sea level. The resulting intrusion of seawater forced many communities to abandon their wells. Along California's populated coast there has been seawater intrusion in at least 12 localities; 7 others are known to be threatened, and 15 others are regarded as potential intrusion sites. (59)

The solution to seawater intrusion is comprehensive water management. To prevent saltwater contamination and land subsidence requires a comprehensive management based on a thorough under-

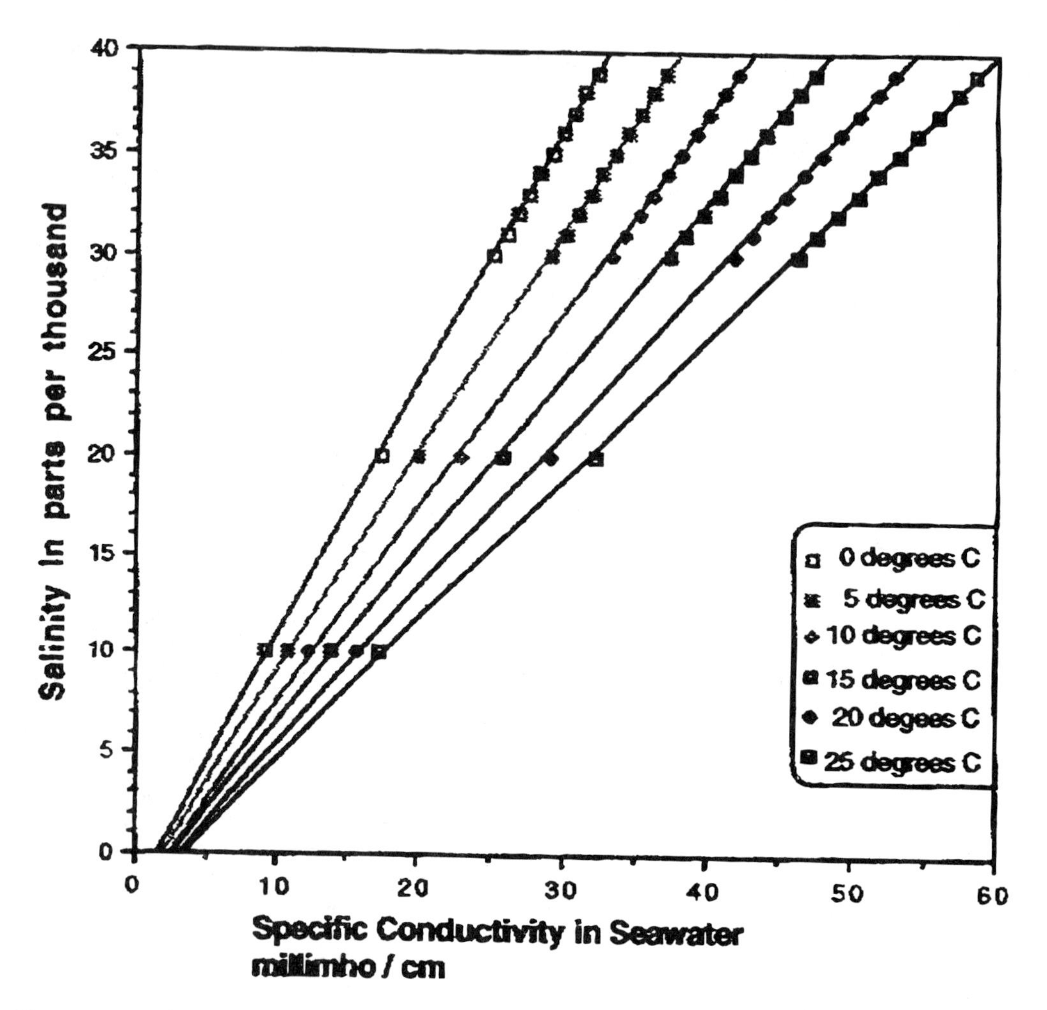

Figure 3.110. Standard conversion system from conductivity to salinity. (*Source:* Reference 59.)

standing of the groundwater system—its supply limitation, its replenishment rate, the present and anticipated future demands, and any special pertinent characteristics. A total management program provides for groundwater, surface water, and reused water supplies to be inventoried and utilized in a coordinated plan.

The management program might identify the number and location of wells, together with their pumping rates and annual limitations on total extractions. Upper and lower groundwater levels would also be defined. Water quality objectives would be set, and sources and causes of pollution carefully controlled. Recharge using storm flows, imported water, or reclaimed water could be involved. In one California case where extensive withdrawal caused saltwater intrusion in the fresh water supply, a water control program was implemented that reduced the excessive amounts withdrawn and allocated the water resources according to historical uses.

The recharge potential of groundwater aquifers should also be protected in the management program. Aquifers are naturally recharged by rain percolating downward from the land surface or laterally from a lake or stream. In most cases, the aquifer will be recharged when the permeable formation is near the land surface. The area of recharge, where rainwater or seepage actually enters the aquifer, may be a long distance from the wells, thereby requiring regional management.

Surface aquifers vary in level with seasonal changes in rainfall, sometimes rising above the land surface and flooding it, but usually remaining below at a depth of several inches to many feet. There may also be a number of deeper confined aquifers, which are separated from each other and the surface aquifer by more or less horizontal layers of impermeable rock or clay (Figure 3.111).

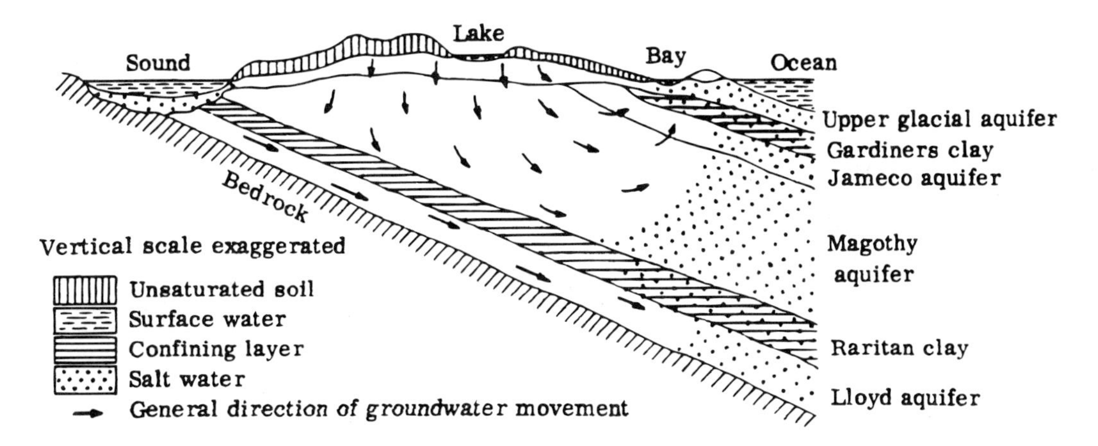

Figure 3.111. Typical subsurface geo-hydrology of the coastal zone, shown by a geological profile of Long Island (New York, USA)—the type of aquifers shown are vulnerable to salt water intrusion when pumped. (*Source:* Reference 240.)

Typically the confined aquifers are sloped seaward with fresh water flowing out of them where they intercept the seafloor. The head pressures on the aquifer normally prevent salt water from intruding into the fresh water but overpumping may cause intrusion if the water is drawn from either surface or deeper aquifers. If a confined aquifer is overpumped, the head pressure will fall and the result is the same: salt water will intrude.

3.81 SAND MINING

Mining of beach sand has been a common practice in many countries. As an example, the mining of sand and/or aggregate has been identified as a serious problem in the islands of the East Caribbean (40). Throughout the Caribbean islands there is legislation controlling sand mining; however, in many cases it is the Ministry of Public Works who has to implement this legislation, and all too often it is that ministry that has been the advocate of sand mining. In addition, there is always the problem of monitoring and implementation; in some cases this is carried out by the local village councils (see "West Indies" case study in Part 4).

Many communities and states recognize the serious consequences of removal of sand from the beach *per se* but do not understand the effects of dredging sand immediately seaward of the beach (Figure 3.112). When sand is mined from the nearshore zone of the beach system, a submerged depression is created. Nature's response is to replace the lost material via wave and current transport. This takes sand from the beach, eroding its structure and depleting its stores. The result may be a perpetual cycle of dredging sand offshore and placing it on the beach, while the new supplies are carried back into the sea to fill in the depressions caused by the mining. (59)

For these reasons, large-scale excavation of sand from any part of the active beach system, whether above or below water, should be prohibited. This means from the backshore of the beach seaward as far out as strong wave energy penetrates to the bottom, which could be as far as a mile or more offshore (often to depths of 40 or more feet).

Obviously beach sand mining should be stopped—unless shown to be sustainable. But until an alternative can be offered at a reasonable cost, this may not be politically feasible. Nevertheless, viable alternatives for beach sand should be actively sought. (See "Mining" above and "Beach Management" in Part 2.)

The following guidelines for sand removal in the coastal zone were prepared by the Coast Conservation Department (CCD), who manage the Sri Lanka ICZM program (43). CCD recognizes that removal of sand from estuaries, beaches, spits, dunes, and sand bars will contribute to coastal instability. However, as these have been traditional sources providing sand essential for building and other activities, the following guidelines are applied by CCD in the issue of permits pending the location of alternate sources:

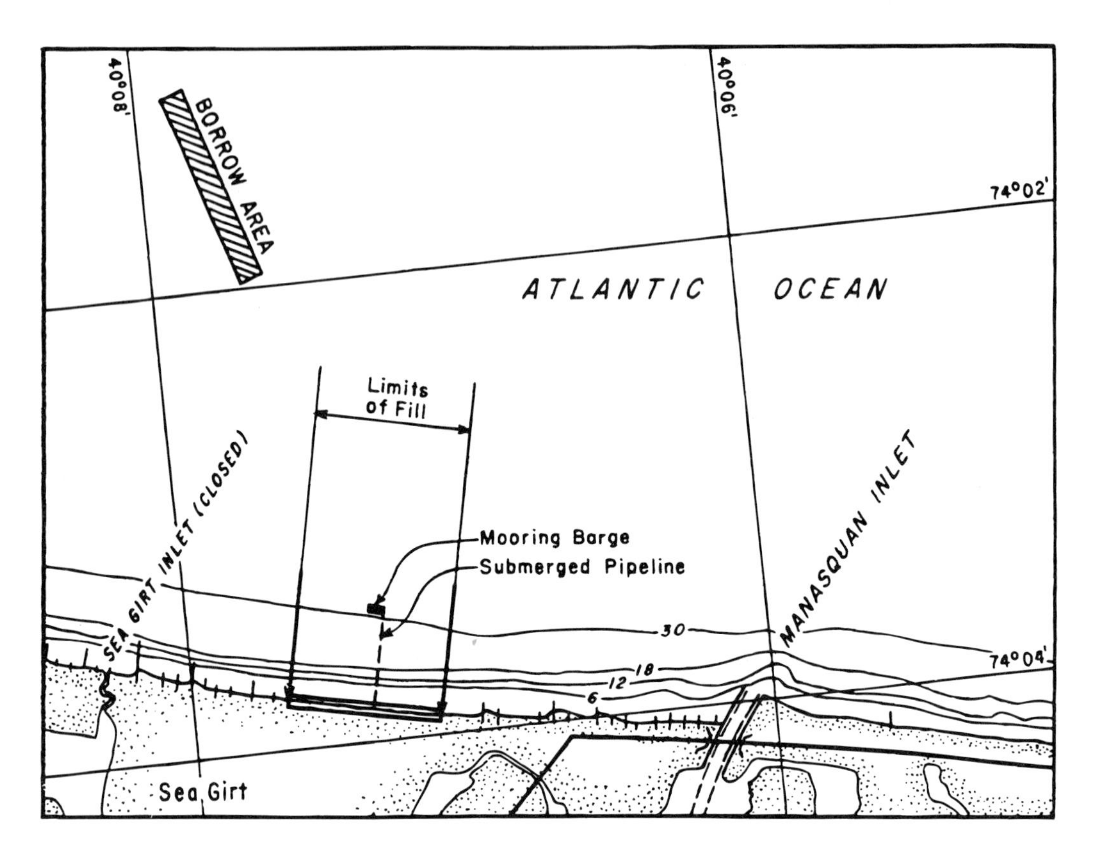

Figure 3.112. Arrangement of dredge and piping system for coastal sand mining (New Jersey USA), an operation that needs to closely controlled. (*Source:* Reference 362.)

1. ***Riverine estuaries.*** Removal of sand may be permitted where the downstream beaches do not indicate signs of deficient sand supply. A minimum distance that must be maintained between the mining site and the river mouth will be specified in each permit.
2. ***Beaches, barrier beaches, and spits.*** Removal of sand from accreting beaches may be permitted if such removal will not cause adverse environmental impacts in adjacent sites.
3. ***Multiple dune systems and stable reservoir areas.*** Removal of sand from multiple dune systems or stable reservoir areas known to be the termini of sand circulation cells may be permitted. When sand mining is completed, the developer must re-plant the dune with appropriate vegetation to prevent further sand loss due to wind erosion. Backfilling will be carried out by the developer as specified by CCD.
4. ***Sand bars.*** Sand removal may be permitted if (a) downdrift areas are not already eroding; (b) such removal will not contribute to erosion of the shore or in downdrift areas. Sand bar breaching may be permitted to allow free flow of backwaters during flood periods. If the adjacent areas or the littoral cell concerned indicates a deficient sand budget, sand so removed shall not be removed out of the littoral cell.
5. ***Offshore areas.*** Offshore sand mining will not be permitted within the 5 fathoms contour or 1,000 meters seaward from the low water mark, whichever is furthest. Where there are fringing coral reefs or sandstone reefs, sand mining in the area between the reef and the shore will not be permitted. Offshore sand mining may be permitted in other locations if the following conditions are met:

- The developer carries out, at his cost, investigations deemed necessary by the CCD and submits to CCD all required documentation for evaluation of the permit application.
- The mining site is located as far as possible from any living coral reefs. The CCD will specify the minimum distance between the mining site and the nearest coral reef.

- The mining is restricted to a depth, determined by CCD for a particular site.

Methods of mining, stockpiling of mined sand, method of transfer from the mining site to storage site, and the storage site to retain/disposal sites are specified by CCD.

3.82 SEAGRASS MEADOWS

Submerged seagrasses are often abundant in the shallow waters of temperate and tropical coastal zones of the world. Seagrass beds, or meadows, are highly productive and valuable resources, which enrich the sea and provide shelter and food for some of the most important and valued species of fish and shellfish. Their broad distribution is attributable to the fact that seagrasses can tolerate wide salinity ranges (from nearly fresh water to hypersaline seawater). They prosper in the quiet, protected waters of healthy estuaries and lagoons, often in beds, or meadows, that are easily delineated for classification as critical habitat areas. They are normally found in areas where light can readily penetrate (clear, shallow water) and enable photosynthesis to occur. Seagrass beds may extend shoreward to the point where wave action prevents their establishment.

High productivity of seagrass habitats is associated with both seagrass abundance and the production of "epiphytes" attached to the leaf surfaces. For example, primary productivity (amount of plant production) for two common seagrass species has been shown to be higher per acre than for average corn and rice cultivation in the United States. Abundance is related to depth and other variables—salinity, temperature, turbidity, currents, and so forth.

Seagrass meadows may meet either mangrove or reef communities at their boundaries, and seagrass frequently provides the link between mangrove and coral reef ecosystems. The migration of animals at various life stages from one ecosystem to another for feeding and shelter, coupled with currents that transport both organic and inorganic material from runoff and tidal flushing, links the offshore coral reefs to nearshore seagrass meadows and the seagrass meadows to mangrove estuaries.

Seagrasses provide a substantial amount of nourishment, nutrients, and habitat, vital places of refuge not found in the open ocean. They attract a diverse and prolific biota and serve as essential nursery areas to some important marine species. Seagrass meadows are also known to trap and bind sediments which otherwise would move freely with the water as particulate pollutants.

Although seagrasses are a relatively hardy group of plants, they are vulnerable to damage from excessive siltation and turbidity, shading, and water pollution, as well as fishing practices that use bottom trawls which scrape and plow the meadows. Turbidity and shading reduce ambient light levels in the water, resulting in a lower rate of photosynthesis or, in extreme cases, completely inhibiting it. Certain pollutants in seawater not only have toxic effects on the growth and development of seagrasses, but also on many of their animal associates.

Seagrasses are sensitive to hot-water discharges and, for example, are usually eliminated from areas closest to cooling water discharge points of coastal power plants. Other major threats are dredging and filling operations in seagrass meadows; for reasons that are not clearly established, seagrasses only slowly, if at all, naturally revegetate areas that have been dredged or seriously disturbed. Typically, when a seagrass meadow is eliminated, its fauna also disappears from the area. The disappearance of seagrass communities may go unnoticed because, unlike mangroves and coral reefs, seagrass communities are not visually obvious to most observers.

3.83 SEA LEVEL RISE

The World Ocean is rising under the influence of global warming (see "Global Warming" above). There is not good agreement in the scientific community as to the rate of rise that can be expected in the next 100 years or whether such rise will influence all coasts the same. And yet this information is vitally needed for coastal zone planning and resources management because some scenarios show a potential for worldwide catastrophic damage from flooding, erosion, storm surge, and associated effects.

To complicate matters, *tidal amplitude* is also increasing in some parts of world (e.g., tidal amplitude is, increasing at 20 cm/100 years along the coast of Netherlands—the same rate at which average sea level has been rising there) (233). This means that high water events will be even more extreme.

Physicists and climatologists believe that a doubling of the emission of greenhouse gases could lead to a +1.5°C to +4.5°C rise in the average atmospheric temperature and that, because of the phenomenon of thermal inertia, an accelerated sea level rise in the short and medium term is "inevitable" (181). These estimates have grave consequences for the evolution of the coastal environment, and it is essential to make allowances for them in any coastal development and protection plan. The most significant consequences for accentuating the littoral risk are of the following types (181):

- Erosion and modification of the coastal morphology with retreat of the line of the coast
- Flooding of humid coastal zone
- Undermining of the effectiveness of coastal defense structures, with higher property damage risks
- Rise in the level of groundwater tables and saline intrusion, especially in estuaries
- Fluvial sedimentary deposits at the level of present river mouths
- Damage to sub-surface littoral networks for drainage, run-off, and services

Sea level rise is commonly believed to be one aspect of the general global warming caused by release of carbon dioxide into the atmosphere. There are three major causes of sea level rise: melting of glaciers, heat expansion of the ocean water mass, and meltdown and disintegration of the Antarctic ice cap. Scientists seem comfortable with predicting the first two but not with predicting the third.

Many different predictions for sea level rise exist, varying between 50 and 250 cm/100 years. United Nations panels have estimated 50 to 95 cm/100 years—but, of course, the rate will not be linear because global warming is expected to accelerate. According to Revelle (65.4), in the next hundred years, glacier melt will elevate the sea by 40 cm and heat expansion by 30 cm. To this rise of 70 cm ($2\frac{1}{4}$ ft), one should add Antartic meltdown—perhaps 100 cm—bringing the total rise to 170 cm ($5\frac{1}{2}$ ft). The Antarctic meltdown has a long delay factor and therefore the 100-cm rate would not be reached for many years. But a dramatic increase in the melt of the polar ice (atmopheric temperature increase of +3°C.) could raise the sea level as much as 1.5 m by 2030 (229)

Many coasts of the world, including those of the United States, have already been undergoing long-term increases in relative sea level. For example, the average for Atlantic sites (along open ocean) of Eastport, Newport, New York, Sandy Hook, Charlestown, and Miami Beach was 2.77 mm/yr or 27.7 cm (10.92 in) in 100 years (data for 1903–1970). The trend for this century is shown in Figure 3.113.

This rise has been associated with serious recession along many miles of United States beachfront. The particular sections of coast that have suffered much higher than average sea level increases have often experienced the greatest recession of the beachfront. In some coastal areas, such as the Texas

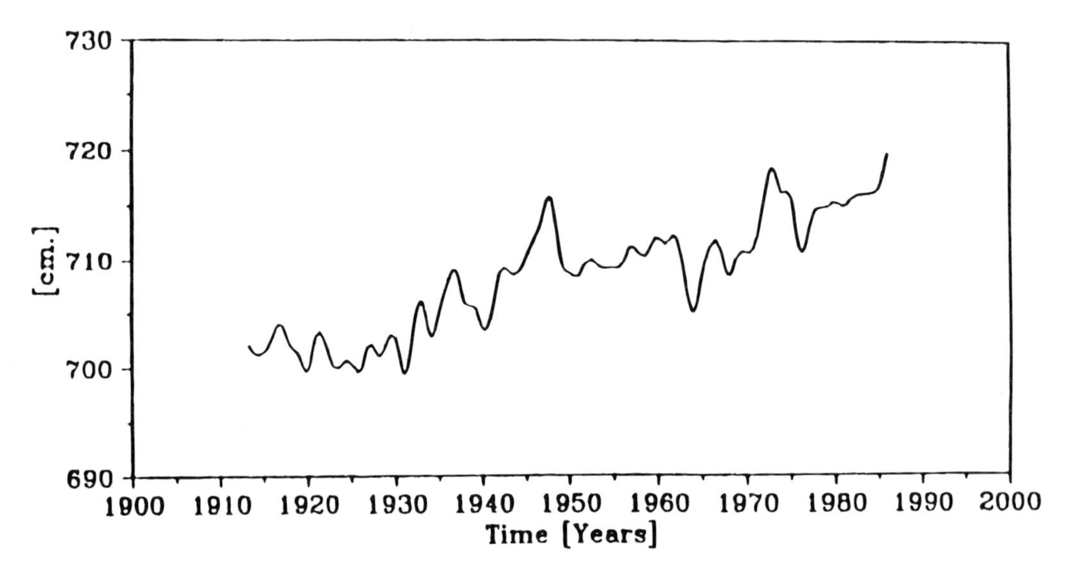

Figure 3.113. Annual mean sea level (Key West, Florida, USA) showing a rather steady rise, 1930–90 (Revised Local Reference from Permanent Service for Mean Sea Level). (*Source:* Reference 238.)

Figure 3.114. Completed 100-m by 2-km beachfill at Corpus Christi (Texas, USA). (Photo courtesy of U.S. Army Corps of Engineers.)

(USA) coast, recent rates of increase have been more than 1.4 feet (43 centimeters) per century, requiring dikes for lowland areas and beachfills for some shorelines (Figure 3.114).

The worrisome problem is that, if the Antarctic ice cap disintegrates at a higher than expected rate, sea level will increase dramatically and catastrophically. The potential rise with total global-warming-induced disintegration is additional 500 cm (195 in)—enough to submerge permanently massive areas of coastal lowlands in dozens of countries (e.g., more than 40% of Bangladesh) and to threaten much more land during storms. While such an increase may sound unreasonable, 125,000 years ago sea level was 7 meters (23 ft) higher than now (294).

Several factors will determine the severity of impacts from sea level rise on natural resources and their habitat, including the following (24):

1. Physical obstruction to inland habitat shifts as sea level rises—marshes, mudflats, and coastal shallows must be able to shift horizontally without interruption from natural or human barriers; species adapting to new environmental conditions must have the same freedom. Any physical constraint will increase wetland loss, affect zonation under the new tidal regime, and probably decrease resource populations and seafood harvests.
2. Resilience of species to withstand new environmental conditions (light, salinity, water flow, sediment abrasion, etc.) during periods of erosion-induced transition—through behavioral changes, genetic selection, or ecological adaptation, species must respond to habitat changes at a rate at least as swift as the environmental changes.
3. Rate of environmental change—slow shifts in climate and inundation should allow species to adapt or relocate. However, if the changes are expressed through severe storms, habitat migration and species adaptations may not keep pace with rising waters, high wave energies, and associated physiological stress.

When all factors are considered, local impacts to fish habitats will vary greatly, but significant wetland loss, with many losses in valued finfish and shellfish stock, is certain. These changes will alter fish stocks and productivity.

Conservative estimates of sea level rise indicate that rising waters threaten to affect about 60% of the 4.3 million ha (11 million ac) of the saline and coastal freshwater wetland in the United States. Impacts to living resources and their habitat depend largely on the rate of inundation, shoreline geology (slope, material, orientation to waves, etc.), the dominant vegetation species, and society's decision on whether to protect the shore, each of which affects habitat stability.

Effects on coral reefs will vary according to the rates of rise and of coral growth. In the Caribbean, reefs normally grow vertically at the rate of one quarter to one half inch per year. If sea level rise is faster, then the Caribbean coral reefs might be overwhelmed, according to Ray (291), and reefs may be drowned. Conversely, some high level reefs—e.g., the Great Barrier Reef of Australia—might benefit from a modest sea level rise because the reefs are already high and their tops are exposed to wave action. A sea level rise over the presently depauperate reef flat will encourage corals to grow on the flats, making them ecologically productive as well as pleasing to tourists—a rare benefit from sea level rise.

But all sea level change can not be attributed to increased ocean height—potential land subsidence should also be taken into account. Subsidence most usually takes place in zones of sedimentary accumulation (deltas, swamps, lagoons) where subsidence rates of 0.5 to 2 cm are typically recorded. (181)

Subsidence also occurs near geological fault zones and areas of uplifting and where water or oil is being extracted at high rates near the shoreline. The eastern part of Lake Maracaibo (Venezuela) has subsided at a rate of 80 cm per 10 years over 50 years (4 m total) and the Niger Delta at 25 cm per 10 years. (181)

Thus, it would seem to be necessary to take into account, as of the present, this future "sea level rise" factor in any study of the protection, development, and planning of coastal zones, without waiting for significant refinements in the predictive models and scenarios, which could make it possible to reduce the ranges of values currently available for average sea level rise (181).

Many countries are now very seriously planning defenses against sea level rise (e.g, Bangladesh, Netherlands, Maldives). Where retreat is not practicable, physical defenses—seawalls, dikes, etc.—are being evaluated. In Netherlands, the design parameters for sea dikes include (233):

- Wave runup
- Margin for
  - Seiches
  - Gust bumps
- Sea level rise
- Settlement of subsoil

These parameters are incorporated into the general dike design developed for the Netherlands shown in Figure 3.115.

According to Bigford (24), coastal zone planning for rising waters offers affected parties two basic choices: entrench with dikes or retreat to higher ground. Broadening their management perspectives, states must also consider basic changes in definitions and initiatives. State coastal zone boundaries and floodplain zones should shift with higher waters. Other coastal management strategies should include siting standards and development limitations, with the dual intent of protecting existing structures and preserving remaining space for water-dependent users.

The influence can be expected along most beaches during times of rising sea level because the effect of increased sea height is to force the beach inland as the shore profile adjusts to equilibrium (59). Setback standards can limit construction in threatened areas and perhaps preclude rebuilding if a structure is damaged. Such zones should be redefined as the shoreline moves. A recession line to govern the setback distance is important wherever a beach is receding significantly from erosive forces.

Ecological buffer strips could yield similar benefits. Building codes should restrict bulkheading and encourage movable (or expendable) structures. One local planning option is to develop marine enterprise or economy reserve zones. If located on high ground and properly engineered, fishing and other shore-dependent industries might be able to operate in these special waterfront parcels despite rising waters.

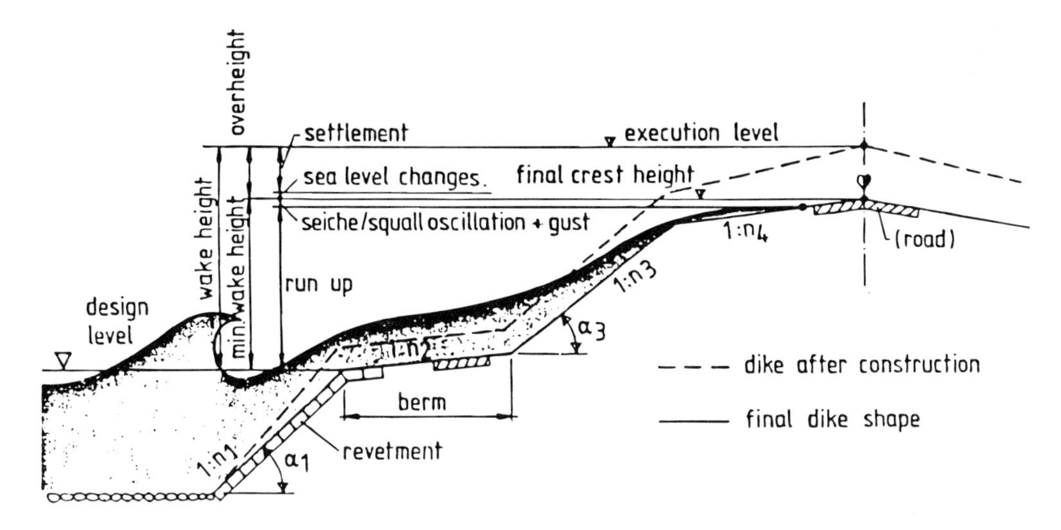

Figure 3.115. Sea defense arrangement combining revetment and sand dike (as utilized in Netherlands). (*Source:* Reference 233.)

To protect shoreside infrastructure, these industries must eventually choose whether to retreat or entrench. Retreat represents a conscious decision to abandon waterfront land and seek alternate space. Along most port and harbor shorelines, space is at a premium, but one or more sectors of the least water-dependent industries (supply, processing, etc.) might find suitable space. New space will probably bring economic inefficiencies, perhaps by fragmenting activities or moving away from an established infrastructure. Still, retreat may be the most sensible option, coastal geology, infrastructure, and real estate values permitting. (24)

For the purposes of coastal zone management and planning, it seems advisable to choose an arbitrary rate of sea level rise—150 cm/100 years is suggested—and modify it if necessary, first, as local conditions merit and, second, as future studies indicate. From a non-linear (accelerating) rise of this amount, one could use either of the two planning horizons below:

40 years (to 2035) — 50 cm (19 $\frac{1}{2}$ in.) rise
60 years (to 2055) — 80 cm (31 $\frac{1}{2}$ in.) rise

3.84 SEDIMENTS AND SOILS

The "soils" of concern in the coastal zone are mostly sediments—materials deposited by water action. These include those of both marine and terrestrial origin. Knowledge of sedimentary dynamics is important to the process of coastal zone management because they influence water quality, beach stability, estuarine life forms, plankton blooms, and even the fate of coral reefs (see "Suspended Particulate Matter" below).

Sediments of human origin, whether from liquid waste disposal (sewage and industrial organic waste) or excessive soil erosion (land clearing and working), are usually environmentally harmful. The lighter particles—finer mineral silts (<0.1 mm) and organic particles—are particularly troublesome (Figure 3.116) because they go easily into suspension and also because they may carry toxins (adsorbed), stimulate plankton blooms, or make the water excessively turbid (see "Toxic Substances below and "Nutrients Management" in Part 2).

The sediments produced and transported naturally in the coastal sea are usually a lesser problem than those of terrestrial origin. For example, coral reef systems produce a considerable amount of calcareous sediment but are found in water with only slight amounts of suspended matter, perhaps only 2 or 3 mg/l and thus very clear water, because the sediments they produce do not easily suspend and tend to be incorporated into the reef structure. Wetland elements such as mangrove forests salt marshes produce fine organic material ("organic detritus"), much of which is retained by the root mats.

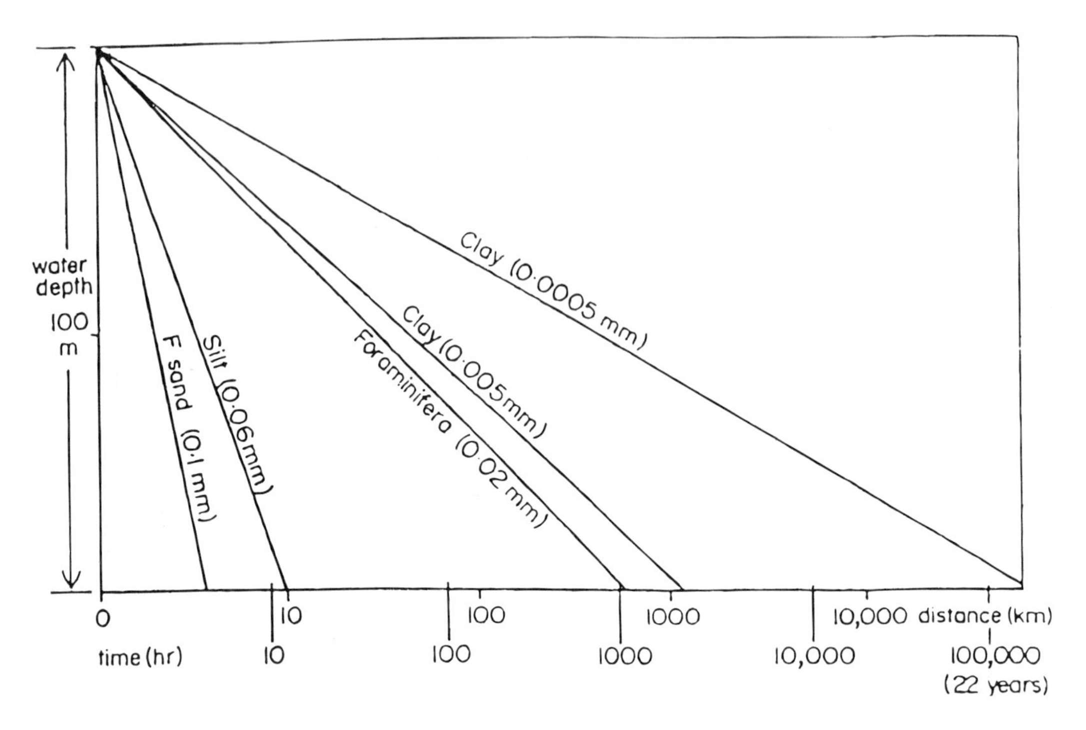

Figure 3.116. To calculate the drift parameters of toxic sediments, one can use particle size and approximate settling rates such as those above. (*Source:* Reference 239.)

Lagoons and estuaries often do accumulate fine sediments that can easily resuspend. Consequently, suspended matter of 50 to 80 mg/l would not be uncommon in such water bodies. But, the species that occupy bays, estuaries, and coastal lagoons have evolved to prosper in these conditions, so long as the sediments do not become excessive or laden with toxic matter.

Regarding coral reefs, reduced light intensity associated with increased turbidity is probably less damaging than smothering effects of settled particles according to Salvat (309). Waters over dredged areas can contain massive amounts of bacteria; there have been reports of 50 times more bacterial biomass than in non-polluted seawater. These conditions were associated with the disappearance of 20 fish species out of 29 in a dredged lagoon in Guadeloupe, and almost total disappearance in neighboring areas which were moderately or extremely disturbed (146). A. Smith (pers. comm.) suggests that a sedimentation rate of 10 mg/cm^2/day is "normal" for coral reefs.

3.85 SETTLEMENTS

The special environmental problems of settlements in waterfront and flood plain areas demand special precautions in planning and construction. It would be useful to have special land use allocations and regulations and systems of review for waterfronts and flood-prone areas, including subdivision reviews, building codes, zoning actions, and other land and water-use control programs, to take into account the unique coastal conditions and problems of these areas.

The environmental disturbances typical of shoreline development include problems not only with dredging but also bulkheading and attempts at landfill or erosion prevention along occupied shorelines. Additional problems of waterfront occupancy arise from natural storm and flood hazards, the lack of suitable soils for foundations or drainage, the shortage of freshwater supplies, and the accumulation of wastes (Figure 3.117).

The proximity of settlements to the shore also means that polluted runoff may go directly to the coastal water basin with little time for natural purification through vegetation and soil. Land clearing and site preparation cause direct discharge into coastal waters of polluting sediment and nutrients that

Figure 3.117. Bad housekeeping practices lead to bad health problems—public education is needed. (Photo courtesy of World Health Agency of the United Nations.)

can drastically reduce water quality even from a distance inland (Figure 3.118). Serious and often irreversible pollution also may result from sediments dispersed in channel-dredging operations. Buffer areas are mandatory, and special provisions must be made to handle land grading and drainage during site preparation.

Waterfront development by dredging wetlands, tidelands, and estuarine bottoms and using the "spoil" to fill and elevate the land is one of the most disturbing types of coastal development, particularly when canals are dredged and the dredge spoil is piled on adjacent wetlands or low lands to gain elevation and to create lots for canalside homes. The canals collect pollutants so rapidly that there is often inadequate time for natural purificating action. Canalside settlement affects the ecosystem adversely in three major ways: (a) lower water quality, (b) habitat loss, and (c) reduced ecosystem productivity.

The planning stage of a settlement provides an excellent opportunity to ensure that access to the public shoreline is facilitated and that special efforts are made to preserve the drainage systems of floodplains because these areas have high water tables and direct hydraulic connections to coastal waters. The lower floodplain can serve as a buffer area to protect vital water areas that lie along the shoreline, such as wetlands and tidal flats. The floodplain filter or buffer strip protects the wetlands and tidelands from runoff surge and heavier loads of sediment and other pollutants than they can assimilate. All existing residential development should be brought into conformance with a community plan or ICZM special area plan.

Regardless of the type of development, a buffer strip of unaltered land should be required between the water's edge and a stipulated setback line. The purpose of a buffer zone is to separate land clearing and construction from the shoreline and to intercept water runoff and to provide for purification of the water by soil infiltration and vegetative "scrubbing" before it enters any water areas.

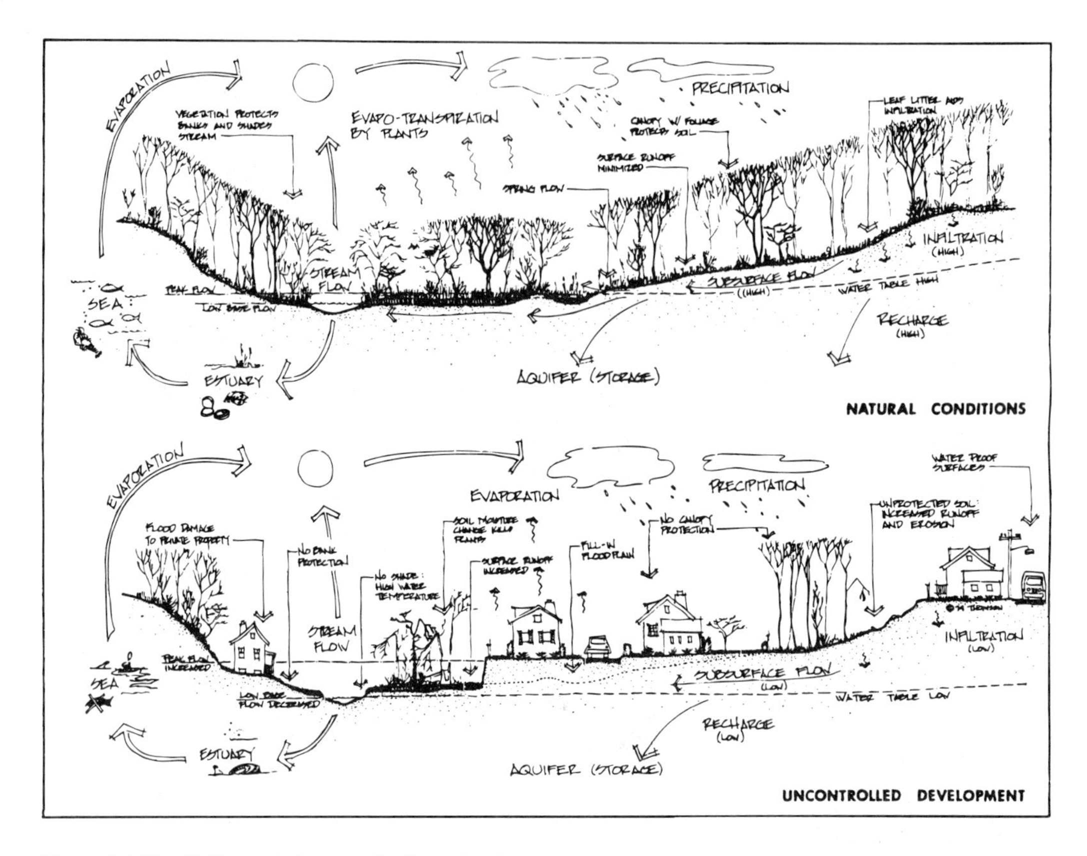

Figure 3.118. Settlements in coastal valleys alter hydrological conditions; negative effects should be countered by appropriate water/wastewater management. (*Source:* Drawing courtesy of S. Hammill, Reference 157.)

Roadways can be a problem and the nearer they are to water, the greater is their potential for adverse impacts. Location is the most critical aspect of roadway planning. Roadways should be located so that they conform to existing topography. They should not be constructed in vital habitat areas, wetlands, and other water areas. But if they must, they should be elevated on open pile supports installed over existing grade.

Plans should avoid the location of bridges and causeways across wetlands, tidelands, and water basins. Roadways should not impinge on floodprone areas except under strict environmental constraints. Roadway design should not require landfill, nor should it impede the natural water flow. Neither excavation nor filling of wetlands is an acceptable activity. Nor should canal dredging to drain lowlands be encouraged. (See "Roadways, Causeways, and Bridges" above.)

Water pollution is often concentrated in coastal waters, particularly near big cities, estuaries or littoral industrial complexes. Industrial, agricultural and urban effluents disrupt the natural trophic equilibrium of all coastal marine ecosystems through the process of anthropogenic eutrophication. The transition to dystrophic situations usually results in a loss in biological diversity. Considerable amounts of sediments, due to harbor or artificial beach construction or local erosion, increase water turbidity, which reduces the photosynthetic ability of seagrasses. Sediments also trap toxic pollutants which then form a deposit on the sea bottom, conducive to the contamination of demersal species (142).

Sediment and other settleable substances resulting from dredging and associated activities for shoreline developments form objectionable deposits that have an adverse effect on valuable marine bottoms such as grass beds and communities of marine organisms. Deleterious and/or toxic substances discharged from storm sewers or overtopping seawalls and discharges from other sources often contain excessive loads of dissolved nutrients, pesticides, fungicides, herbicides, heavy metals, and mineral and organic residues. Nutrients from fertilizers are a recognized threat to bays and estuaries. The nutrients stimulate the growth of algae to the extent that the original plant-animal balance is affected. Many of the pesticides, fungicides, herbicides, and heavy metals are accumulative in marine plants or animals and therefore, even a low levels, pose a hazard to plant and animal life and to human health.

Oxygen levels are depressed to low values in confined water areas where process of eutrophication is initiated by an excess flow of nutrients—particularly nitrates and phosphates—to surface water bodies (142). This excess causes an over-development of a few species of algae and phytoplankton, leading to a disruption of ecosystem balance. These algae "blooms" are generally noxious and consume enough available O_2 to limit the growth of other species. Decaying biomass also contributes to oxygen depletion and perpetuates the cycle by releasing additional nutrients to overlying waters. Although all coastal areas are affected, this situation is particularly acute in enclosed environments, such as lagoons, which receive large quantities of nutrients. While the run-off of nitrates due to modern agricultural practices is a significant source of eutrophication, urban wastewater and industrial pollution play an undeniable role in the phenomenon.

Considering the impending expansion of cities, pollutants directly discharged into shallow coastal seas are likely to reach crucial concentrations if either no appropriate sewage and wastewater treatment systems are installed (142) or the sewage is not disposed offshore. As an example, the predicted growth of the urban population from 75 million in 1985 to 241 million in 2025 in southern and eastern Mediterranean countries (Morocco to Turkey) will lead to a considerable increase in wastewater, solid waste, and air pollution. Even in the best of situations, the proportion of wastewater discharged into the sea that is treated remains low; in France, it only attains 47 and 49 percent in 1990, respectively, for Provence-Cote d'Azur and Languedoc-Roussillon. The limited water circulation in the Mediterranean Sea renders it particularly sensitive to the pollutant influxes (142).

Because of the extreme scarcity of good shorefront land and its importance to coastal communities, there is good reason to allocate coastal land uses. In general, waterfront land in highest demand is that which extends to and/or across the intertidal zone and carries such values as ship berthing, access to fishing, access to the beach for tourists, and unobstructed view of the sea from home or apartment. With such demand for sites at the water's edge, the development planning focus tends to narrow to a linear approach; that is, it focuses on the mean high water line and the "reclamation," or alteration, of properties that lie along that line. Unfortunately, much of the edge is in wetlands.

Wetlands, tidelands, and seagrasses provide a critical link in the basic food chain energy system, assimilating, converting, storing, and transferring nutrients. Wetlands themselves typically provide about half the basic nutrients for estuarine life systems. When the productivity components are eliminated or short circuited, estuarine ecosystems are seriously degraded.

Wetlands are also key habitat areas. Dredge-and-fill projects to create shoreline real estate from wetlands, by their very nature, obliterate or disable wetlands as well as tidal flats, sea grass beds, and estuarine bottom habitat. The damage is high because much of this habitat serves as critical breeding and nurture areas for fish, shellfish and water birds.

The urban setting presents distinctive problems for waterfront development as well as special opportunities for restoration and enhancement. Currently, the limited waterfront property that exists in most urban areas is often being aggressively sought for residential and commercial development. Wilderness is not found here. Most of the urban shoreline edge has been "hardened" and dredged or otherwise altered, changing its ecological character to the detriment of fish and wildlife. The result is an overall reduction in biological diversity and carrying capacity for desirable species within the adjacent lagoon, estuary, bay, or tidal river. (80)

For all these reasons, the use of shoreline areas for settlements should be carefully planned to prevent unnecessary environmental damage. The density of urban population creates demand for factories, power plants, warehouses, homes, docks, streets, automobiles, water supplies, resource depletion and waste disposal. It results in competition with the needs of Nature, threatens ecosystem integrity and resource productivity, and impinges on the sensibilities of mankind by reducing biological diversity. Although impacts are an unavoidable consequence of human survival, they often can be greatly reduced by appropriate planning and management.

3.86 SEWAGE TREATMENT

Sewage commonly contains nitrogen as nitrate and ammonia, phosphorus as phosphate, oil and grease, detergents, phenols, cyanide, arsenic, and trace metals—cadmium, chromium, copper, iron, lead, manganese, mercury, nickel, silver, and zinc—in concentrations that vary with the source of the waste. Runoff into sewers from farmlands or landscaped areas contains pesticide and herbicide and fertilizer residue (243).

Coastal waters can also be loaded with viral, bacterial, and fungal agents, many of them pathogenic to humans. The source of most of these pathogens is feces (human and non-human), and the carrier that transports it to rivers, estuaries, and then coastal water is domestic sewage. Even a clinically healthy individual excretes in 1 gram of feces more than one million infectious particles consisting of more than 100 different viruses and bacteria. In raw sewage, the concentration can be as high as 500,000 particles per liter. (228)

The key design factor for treatment plants is more likely to be reduction of "basic oxygen demand" (BOD)—which can be handled with aeration lagoons—than phosphate and nitrate which cannot. BOD reflects the demand of decomposing organic biosolids for dissolved oxygen which they remove from the water during oxidation. Excessive BOD causes dissolved oxygen to be severely depleted creating water conditions that cannot support animal life.

Three levels of sewage treatment—primary, secondary, and tertiary—are recognized. However, much sewage is discharged too near the coast and receives little treatment beyond comminution and maceration, which reduces the size of solid particles but does little to remove bacteria and viruses. Chlorination of effluent is commonly employed to kill pathogenic organisms. The extracted solid material (sludge) is burned, spread on land or buried. (243)

Primary treatment removes much floating and solid material by skimming, screening, and desedimentation. Primary treatment removes most of the solid material through a settling and flotation process. The remaining effluent is then disinfected (chlorination, etc.) to kill pathogens. About 40–70% of the suspended solids are removed by primary treatment.

After primary treatment, the following are examples of the changes in composition of the waste which can be expected (59):

	Raw wastewater	Primary treatment
BOD in mg/l	250	175
SS in mg/l	220	60
P in mg/l	8	7

Secondary treatment eliminates nearly all remaining solid materials by sedimentation and biological treatment, but leaves much dissolved phosphate, nitrate, organic materials, and trace metals (some metals are removed quite effectively, whereas others exhibit small or no reduction). (243)

Secondary treatment further neutralizes the primary effluent before chlorination by utilizing biological processes to remove much of the BOD (biochemical oxygen demand) in the effluent. The resulting effluent is then again clarified by settling and is chlorinated before release to the environment. Secondary treatment removes at least 85 percent of the suspended solids and BOD, and almost all of the pathogens.

Secondary treatment has two main methods: (a) percolating filter and (b) activated sludge system. Removal of oxygen demand is undertaken by populations of microorganisms. These systems are designed for the microorganisms to make maximum use of the available "food," resulting in a high degree of treatment. These "extended aeration" systems are used widely for isolated sources of waste such as hotels and small developments. Little excess biomass is created and little sludge is disposed, thus saving in operating costs (59).

See Figure 3.119 for a schematic portrayal of a typical secondary sewage plant layout. Following is a typical pathway sequence for a modern secondary treatment plant (353):

Bar screening
Grit removal
Aerated lagoons
    Complete mix lagoon
    Semi-mix lagoon
Polishing pond
Chlorination
Post aeration

Sludge management
- Drying (lagoons)
- Stockpiling
- Transport offsite

Tertiary treatment is an advanced process, designed to remove nutrients remaining in the effluent—phosphates and nitrates—as well as some other dissolved materials. Tertiary treatment is still in the process of development to make it more effective and less expensive.

Sewage treatment systems can be complex and expensive and have "intensive" maintenance and operational manpower requirements. In addition, they are characterized by a low removal of pathogens and, also, unless "extended aeration" can be used, "sludge" has to be removed constantly from the system and a separate means of disposal identified. Modern "package" plants that serve a particular development have overcome some of these difficulties (59).

Waste stabilization ponds, which are an alternative, especially for smaller communities, reduce sludge problems. Waste stabilization keeps the wastewater in ponds where natural biological processes reduce the oxygen demand, suspended solids, and pathogenic bacteria. Maintenance costs are low, and only basic skills are required of those workers responsible for maintenance/operation. With a retention time of 2 or 3 weeks in an open pond exposed to sunlight, significant removal of BOD and pathogens can occur (59).

Depending on the amount of wastewater, availability of space for ponds, and end use for the effluent, a combination of pond types can be utilized. Such ponds can be used for aquaculture if carefully designed and appropriate monitoring of water quality undertaken. Anaerobic ponds require desludging every 2–3 years, while facultative ponds may need desludging only every 20 years. A disposal route for the sludge has to be found, but the quantities produced are often less than those produced in central or "package" wastewater systems using complex technology. The effluent from waste stabilization ponds can be used for irrigation because it is nutrient rich and low in pathogens (59).

The degree of treatment is important ecologically. Untreated sewage discharged in water bodies can cause bacterial and viral contamination, high turbidity, depletion of dissolved oxygen, elevation of plant nutrients, and accumulation of toxicants when present.

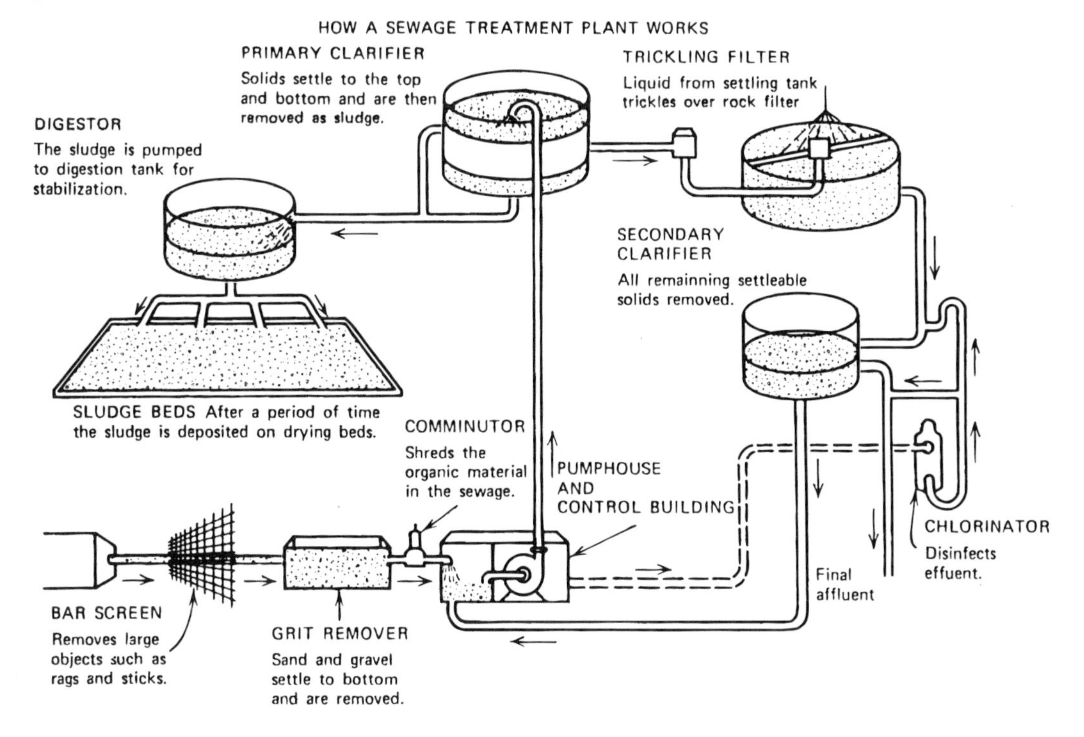

Figure 3.119. Components and processes of a typical secondary sewage treatment plant. (*Source:* Reference 252.)

Nearly all of the basic wastewater treatment processes, including clarification, biological filtration, activated sludge, and chemical precipitation, convert pollutants into the concentrated form called sludge. As wastewater treatment systems become more complete, both in municipalities and in industry, the column of sludge produced and the concentration of undesirable impurities in it will dramatically increase (235). All this sludge must be treated so that it can be disposed of easily and economically without further pollution of water, air, or land.

The basic aim of sludge treatment is to reduce volume and destroy or stabilize sludge solids before final disposal. There are methods for processing sludge (235). One method involves either concentration, digestion, and land application or dewatering and disposal. Sludge concentrators and thickeners increase the solids concentration from 1 to 4 percent. The sludge is then subjected to anaerobic digestion, which makes it easier to dewater and converts some of the organic matter to gaseous end products. This digested sludge can then be used as a soil conditioner and fertilizer alone or in combination with secondary sewage treatment effluent.

Another method involves further dewatering by drying beds, filters, or centrifuges. The dried sludge can then be applied as fertilizer-conditioner or disposed of in landfill. Ocean dumping of sludge is now severely limited. Landfill disposal of sludge cakes is a waste of a valuable resource—the concentrated sludge nutrients—and is impracticable in areas where landfill space is scarce.

Still another approach utilizes heat drying and combustion to handle the dual tasks of volume reduction and solids destruction. Heat drying can be used to produce sludge cakes for fertilization or landfill but is usually a preliminary step to incineration. Operational costs for incineration are high compared to those for other sludge disposal methods. Moreover, sludge incineration presents other disadvantages, such as air pollution, ash disposal, and operational problems due to flue gases.

The use of sludge to fertilize and condition recreational areas, agricultural fields, and strip-mined land is highly recommended. An existing drawback is the presence of residual concentrations of toxic metals and pathogens in the sludge. However, soil and crop management practices allow modification of environmental conditions detrimental to plants or animals (168); such control options are absent in ocean disposal and incineration practices.

It must be recognized that sludge disposal is never easy and sometimes quite difficult. Dumping it in coastal waters is always a bad practice because sludge can blight coastal ecosystems. Use as a soil amendment is better than incineration or burying it in a landfill. Biologically digested sludge can be used to enrich soils or reclaim strip mines—it has the advantage (to some crops) of slow release of nitrogen. Also, there may be opportunities to use it as a fuel source.

When given secondary treatment, the discharged effluent's content of plant nutrients is of primary concern because secondary plants act as "sewage refineries". They convert organic matter to mineral fertilizers like phosphate (P), nitrate (N), and so forth, which cause outbreaks of all forms of algae (see "Nutrients Management" and "Sewage Management" in Part 2). In practice, however, treatment plants frequently operate below design capability; virtually all effluents exhibit some or all of the characteristics of untreated sewage at times. (243)

In summary, the discharge of untreated sewage to an adequate depth of water is the most cost-effective disposal method when (a) the configuration of the coast is such that the required dilution can be achieved and (b) currents prevent the contamination of bathing beaches. In the absence of such favorable conditions there is a case for treatment in sewage plants by mechanical (including filtration), biological (to speed up degradation), and chemical (to precipitate and flocculate) methods (228).

3.87 SHELLFISH POLLUTION

A major problem on a worldwide scale is the existence of pathogenic organisms discharged with domestic sewage to coastal waters. The reason why shellfish is the most important vehicle for the transmission of marine-borne microbial pathogens is that (a) shellfish are frequently cultivated in the vicinity of urban centers with inadequate wastewater treatment and disposal systems; (b) shellfish are usually eaten either raw or after gentle cooking; (c) the feeding activity (filter feeding) of bivalve organisms, such as mulluscs, clams, and oysters, involves the circulation of large volumes of sea water through the shell cavity. Owing to this feeding activity, shellfish concentrations of pathogenic micro-organisms are above concentrations in the overlaying water. For example, when 100 g oyster meat contained 6 to 224 viruses, 70 coliforms and 2 fecal coli, the respective numbers in 100 ml water over oyster beds were 4 to 167, 3, and less than 2. Experimental studies indicated that the capacity of molluscs to concentrate micro-organisms is high but finite. (228)

Epidemics of oyster-associated typhoid fever were reported from France as early as 1816, but the importance of shellfish in the transmission of this disease was recognized only at the end of the last century. Between 1894 and 1896 large outbreaks were reported from the United States, the United Kingdom and France. Typhoid bacillus (*S. typhi*) remained the leading cause of shellfish-associated disease until the middle of the present century when non-specific gastro-enteritis became the prevalent seafood-borne malady. Nevertheless, shellfish-associated typhoid fever remains a hazard wherever the disease is endemic and sewage reaches the vicinity of shellfish harvesting grounds. (228)

Infectious hepatitis (jaundice) is known to be associated with the consumption of oysters, clams, mussels and cockles. The largest outbreak occurred in Sweden in 1955–56 when 629 people developed oyster-associated hepatitis. The spread of the disease in the U.S.A. is demonstrated by an outbreak in 1973 when 293 people developed hepatitis A in New Mexico, Texas, Oklahoma, Missouri, and Georgia after eating Louisiana oysters supplied from a single source. (228)

There are four possible ways to prevent seafood-associated illnesses: (a) to prevent shellfish contamination by the regulation of coastal sewage disposal; (b) to prevent harvesting from polluted grounds; (c) to remove pathogens from the shellfish by efficient depuration; (d) to educate the public in the handling and preparation of such products.

The causal relationship between the pollution of shellfish harvesting areas with domestic sewage and the outbreaks of shellfish-associated enteric bacterial infections led to the application of bacterial indicators for the assessment of the sanitary quality of shellfish and shellfish-growing waters. The use of fecal coliforms as indicators provided useful information when the fecal pollution was moderate to excessive. It became less and less reliable when the numbers were below $70\ 10^{-2}\ ml^{-1}$ in water and $130\ 10^{-2}\ g^{-1}$ in shellfish. The cause of discrepancy between coliform standards and viral concentrations is due to the difference between the resistance of fecal coliforms and viruses to chlorination and natural inactivation factors in both sea water and shellfish. (228)

3.88 SITE MANAGEMENT AND HOUSEKEEPING

Coastal conservation also involves private initiatives, both personal and corporate, to keep the environment clean and productive. The following is a verbatim example of "housekeeping rules" devised for the CALICA Corporation's port in Quintana Roo, Mexico. Calica ships bulk earth materials from the port, mostly to the United States for construction purposes (Figure 3.120).

The eastern coast of the Yucatan peninsula is primarily dedicated to tourism which is supported, in the main, by its marine environment—the white sand, clear water, corals and exotic tropical fish. The pressures being put on this environment by the tourist development excesses are well documented and are currently the subject of discussion among knowledgeable environmental activists.

CALICA'S position in the midst of this tourist area is sensitive and its management has evidenced a serious concern for environmental stewardship on the lands over which it has control.

The geology and hydrology of the area are extremely important considerations in maintaining ecological equilibrium in the area. Along this part of the Yucatan platform the groundwater accumulated during the rainy season migrates to the east and southeast, and ultimately discharges into the Caribbean. CALICA has gone to extreme measures to ensure that the water flowing under its property reaches the sea without chemical or bacteriological contamination.

The following are the controls mandated by the company's management in order to establish sound "housekeeping" practices.

- No human habitation is allowed on the CALICA property. Human wastes and kitchen waste water are treated before being released.
- No fuel bunkering is permitted in the harbor, not even for small boats. There simply are no marine fueling facilities available.
- All ships owned by CALICA or chartered by CALICA are equipped with internal fuel tanks, safely constructed within and separate from the hull. An accidental rip in the hull will not release fuel.

- In the unlikely event of an oil spill in the harbor, there is on hand cleanup equipment and the necessary boats and trained personnel to deploy it.
- For the many land vehicles on the project and maintenance shop is provided where used motor oils are held for recycling.
- The fueling tanks for these vehicles are contained within leak-proof containment walls.
- Solid wastes (wood, plastic, metal, etc.) are separated and sold for recycling or removed to the municipal land fill.
- Garbage from the kitchens is removed several times a week and taken to the municipal landfill.
- Used auto and truck batteries are held and returned to the vendor for recycling.
- No discharge of raw sewage is permitted from the ships in the harbor.
- Ballast tanks are purged at sea in conformance with United States Coast Guard provisions to avoid the discharge of chemical or biological pollutants into the harbor.
- Specially designed loading facilities preclude the dumping or accidental discharge of product into the harbor while ships are being loaded.
- Wash water used to rinse stone before shipment is recycled, the fines (limestone sludge) are dewatered and stored.
- When modifications in the configuration of the harbor are required, a silt curtain is used to control turbidity and regular monitoring of the turbidity is practiced.
- Dust is controlled on roadways and transfer points within the plant area.
- Speed limits are posted and enforced on the company property.
- Reforestation of disturbed areas is planned for those areas no longer needed for operational purposes.
- No fishing or hunting is permitted on the company property.
- A greenbelt is maintained about the periphery of the property to screen it from public view.

Source of this section: Peter Wiese, Vulcan Materials Co., Birmingham, Alabama, USA (387)

Figure 3.120. CALICA Corporation operates its earth materials mining and handling operation according to a formal set of "housekeeping" rules. (Photo by author.)

3.89 SOCIAL EQUITY

In tropic regions, the coast, the sea and the people are often culturally inseparable. The coastal populations of these nations look "outward" from the shore for their livelihood and spiritual well-being. The people of coastal Asia and of Oceania can be characterized as coastal cultures. Remote coastal communities, particularly in Indonesia, the Philippines, Sri Lanka, and the outer islands of Kiribati and the Solomons, are directly dependent upon coastal and marine resources for their livelihood. Often these communities operate as subsistence economies and are especially vulnerable to the degradation of the resource base upon which they depend. Some major, socio-culturally important uses of coastal resources include (148):

Food. One measure of the significance of the coastal resource base is the proportion of protein consumed that is from marine products. Available data for Asia show that people in this region obtain more than 50 percent of their animal protein from fish and shellfish. In nations such as Thailand, abundant, inexpensive seafood is closely linked to the perceived quality of life.

Fuel. Where mangroves are present, they are important as a source of fuel wood for cooking. Fuel wood gathering and charcoal making is a major reason for the loss and degradation of mangrove habitats.

Cultural and Religious Sites. In Sri Lanka, the destruction of religious sites brought by hotel resort development created ugly incidents with strong political repercussions. This prompted the government to inventory and protect all cultural and religious sites along the coast of the island. In other nations there is also a high incidence of cultural and religious sites in the coastal zone. Pressure on these sites increases as development pressures of all kinds grow more intense.

Recreation. Western influences and the pressures of rapidly growing coastal populations make shorefront recreation sites of increasing economic and social importance to the people of the tropic regions.

Capacity. This had both micro and macro aspects, involving changes in the individual, the community, and the nation, including the capacity to develop political and social institutions responsible for production and allocation of resources.

Equity. On the one hand, long-term economic development is stimulated by increasing the human resources in a country and by equalizing the ability to consume. On the other, ensuring more equality in access to benefits is a value in itself.

Participation. Certainly, development should mean increasing the capacity of rural people to influence and control their future, a goal that some believe can be achieved by meeting the following objectives (148):

Empowerment. If powerlessness is to be directly addressed, then the poor must have some political leverage in order to correct grossly unfair decisions regarding the allocation of development resources and distribution of the ensuing benefits. This can be achieved through strengthening livelihood security.

Sustainability. Development includes a long-range concern for the future, and the principal objective of development initiatives should be to generate self-sustaining improvements in human capability and well-being.

According to Gow (148), there is a growing body of knowledge that can help the Third World and the international development community work toward the new paradigm briefly sketched here—of capacity, equity, empowerment, and sustainability. Equally important are the political will and the personal commitment to act on this knowledge. Gow believes that people are seeking an experience of being alive, so that our life experiences on the purely physical plane will have resonances within our innermost being and reality, so that we actually feel the rapture of being alive.

The following was derived from studies of aquaculture, but its conclusions are much broader. In the context of working with coastal communities to develop appropriate aquaculture systems, the following aspects of community structure and institutions appear to be the most important, according to I. Smith (312):

- Informal and formal institutions, especially those of a legal nature that govern property or use rights
- Sources of wealth (productive assets) and degree of concentration of ownership
- Male and female labor use patterns and availability
- Extent of collective action and strong leadership
- Previous experience with and reaction to technological change in aquaculture or other community activities
- Present skill levels, both technical and managerial
- Extent of linkages with external institutions, including credit, extension, and markets

In most instances, technology is a two-edged sword, Smith concludes. While it can potentially liberate and add to general community welfare, it frequently does so at the cost of established socio-cultural values, community structures and institutions. In the case of coastal aquaculture, however, two major factors must be kept in mind with respect to this issue:

First, the vast majority of residents in coastal communities are desperately poor. They are poor because of their lack of access to alternative employment opportunities and because existing community and national structures and institutions most often allow local elites to capture the bulk of any benefits that come from more productive technologies introduced to or adopted by such communities.

Second, the common-property nature of the coastal zones's resources, especially mangrove areas, is being rapidly eroded by the conversion of much of these areas to private fishpond use. This use and misuse of the coastal zone is made possible through subsidized financing and institutional arrangements that favor the large-scale private or corporate investor over the small-scale, perhaps communal, investor.

The above two factors imply that for the majority of residents in the coastal zone there is nothing particularly beneficial in existing community power structures and institutional arrangements, according to Smith's analysis (312). Most often, those communities have experienced only the negative aspects of this technology; for example, in the form of large-scale trawlers that have led to the over-exploitation of may coastal fishing grounds. What has been missing in much of their experience to date with technological advance is an element of community control over its development and use.

3.90 SOCIO-ECONOMIC FACTORS

The human (social) sciences are often asked to help solve complex socio-environmental problems that arise from coastal economic development schemes. It has become popular to include social scientists in the teams that study and weigh and balance development/environment issues. The range of questions social scientists deal with is as wide as the human paradox itself. But usually they want to predict whether a development proposal will adversely (or beneficially) affect the surrounding community. This is not an easy question because it is so difficult to find a *satisfactory* baseline of past or present human condition against which to compare the future with the proposed project built and operating.

Burbridge (32) recommends that ecological, economic, *and sociological* impacts be jointly evaluated. The social science input may involve anthropology, sociology, ethnoscience, or other social science disciplines depending upon circumstances. This recognizes that international donor agencies normally have some broad social objective in stimulating development. Therefore, a more broadly based assessment of projects and the ecosystem framework within which they operate is required, one which expresses a development project's environmental impact upon society as a whole and its economic underpinnings. (Note that example method for these assessments are given in Part 2 under separate headings of "Social . . . ," "Economic . . . ," and "Environmental Impact Assessment.")

Social sciences, such as demography and sociology, depend on official records, censuses, and surveys for information. A census, a comprehensive collection of data about a country's population, obtains factual information from people about their ages, family relationships, residence, work, income, and land to assist policymakers in social and economic planning. This information could be used, for example, to help determine how many people would be displaced by a hydroelectric project, or whether the work force needed for a new factory is locally available. These data can also be manipulated, using statistical and modeling techniques to predict answers to the more complicated questions.

Such a combination of techniques is often used in anthropology and sociology, the studies of human culture and society. These disciplines provide means of understanding a society's values—social integration, religion, relations of respect or familiarity, taboos—essential for minimizing undesirable social disruptions. A large luxury beach hotel, for example could have unacceptable social impacts on an isolated, traditional community. An aquaculture scheme to culture tilapia for local consumption would be doomed to failure if the fish was not suited to local tastes.

Socio-economic assessments can prevent inadvertent damage to the economic base of rural communities. For example, of the greatest socio-economic importance for rural coastal areas is maintaining bioproductivity for fisheries. Here is an obvious example of an ecological process directly affecting people's well being. Naturally productive ecosystems, such as estuaries, provide free of cost what expensive mariculture can barely match—continued fish production. Continued fish production means continued livelihood for fishermen and for others in the fishing industry, including boat builders, trap and net makers, packers, distributors, and retailers (Figure 3.121). Finally, continued livelihood means continued social, cultural, economic, and political stability (308). Therefore, the benefits of any development project which creates negative impacts on fish habitats should be critically weighed against its long term detriments to fisheries.

One difficulty is in romanticizing the status of ethnocultures that exist in a clearly historic condition and that visitors might consider ideal and in balance with nature. W. Clarke (87) articulates the situation in the following paragraphs:

> First, those Pacific peoples—living as they did (or in some cases as they still live) immediately dependent on land, forests, reefs, and seas around them—possessed intimate and exact knowledge of the uses, locations, and habits of an immense variety of plants and animals (terrestrial and marine) and of the potential of their lands for diverse kinds of agricultural production. Based on their knowledge,

Figure 3.121. Fishing is the socio-economic driving force for large coastal communities in many countries, such as this one in Bangladesh. (Photo by author.)

they worked out many effective forms of subsistence production that possessed the quality of sustainability, now so much sought after.

Second, those same people—like all people—were always in dynamic interaction with their environments. Some of their actions were destructive in ways that would today excuse great concern in an EIA. That is, they "impacted" their environments and then had to adjust to the environmental degradation or resource depletion that their own actions had initiated. No time can be found in the history of Pacific peoples before which people were perfect conservationists living in a steady-state of harmony with their environments. Early Pacific Islanders or present-day noncommercial Pacific peoples shared or share the universal human trait of altering environments, often destructively, and then seeking to redress the disadvantages brought by the change. Thus, though the modern world has much to learn from the traditional resource manager, we need not romanticize the past as a conservationist's Garden of Eden from which only the modern commercial-industrial world has strayed. Present-day technologies and population pressure do, of course, increase the magnitude of changes in the environment from development.

To return wholly to past ways of production is not a feasible solution to today's environmental problems. The past was not always a happy place for conservationists in any case. But it can be argued that certain traditional management approaches could be advantageously applied today, perhaps in combination with modern scientific knowledge.

This sort of "progressing with the past" seems feasible in the Pacific where much traditional knowledge of resources remains (though it is disappearing fast), where rural peoples are not generally so pressed against the edge of survival that they cannot risk innovation, and where small islands as well as small-scale tenure systems encourage or make possible experimentation on a less than agroindustrial scale. At the very least, traditional patterns of management should be available for consideration for possible use in the mitigation/ameliorative measures brought forward in EIA.

In social science, as in natural science, simulation models seek to simplify complex realities so that relationships between different factors can be seen. The danger of models is that they will oversimplify human actions to conform with preconceived assumptions or patterns. If these pitfalls can be avoided, models can be extremely useful tools for understanding the social impacts of the changes brought on by development.

People may choose to speak through their leaders or as individuals. Where large numbers of people are affected, necessity usually dictates obtaining information from a representative sample of the people reflecting the economic, social, and cultural diversity of the population.

Often, surveys and similar techniques are inappropriate or inadequate for predicting the impact of a project on a community. Local people may lack the level of education or exposure to external influences to respond adequately or accurately to written or formal questions. Then, different approaches such as observation, reliance on written records, and informal personal interviews must be used.

There appears to be a "critical mass" in terms of involvement and resources which leads to appropriate socio-economic development. Some of the elements of this mass are a technically sound program with clearly perceivable objectives, adequate resources and the freedom for the participants to modify individual program elements based on local needs. When these elements are combined appropriately, program success appears to result.

There is as yet inadequate empirical information about quantity and quality of each of these inputs to achieve success. It is possible to conceive of this critical mass as being a function of the socio-economic condition of the participants, their level of knowledge, skills, existing pattern of technology, extent of incremental improvement sought to be achieved, asset availability and expected resource input. Level of technology and skills available with the members appear to be other vital ingredients.

3.91 SOLID WASTES

Dumping of solid wastes often takes place on the border of settled/developed areas and causes environmental problems in many places, particularly if the dump is at the water's edge. Using the Turks and Caicos as an example (374), the main concerns relate to pollution of surface water and groundwater, dust, odor, breeding of flies and mosquitoes, rodent infestation and unsightly accumulation of rubbish around the perimeter of the dumps due to windblow and animal/bird foraging. Generally, disposal follows a specific sequence. As rubbish is dumped it is burnt to reduce the volume of the waste. (374).

Ecologically, the most worrisome impact is water pollution (Figure 3.122). Water falling on a landfill site percolates through the waste and a number of chemicals dissolve into the water forming a polluting liquid called leachate. That contaminates groundwater and often runs into the sea. Restoration of contaminated groundwater is difficult and expensive, so it is preferable to control contamination. The following strategies to mitigate water pollution are feasible (374):

- Reduce/prevent rainwater entering the landfill site by an impervious "csf" or engineered cover
- Reduce/prevent leachate flow by using an impermeable liner to line the dump site
- Site surface water control
- A combination of the above

Surface water control measures are designed to minimize the amount of surface water flowing onto a site, thus reducing the amount of potential infiltration by provision of diversion berms and drainage ditches using standard civil engineering techniques. Capping of a site is designed to minimize the infiltration of any surface water and precipitation into a site. Impermeable liners provide groundwater protection by inhibiting downward flow of leachate. Old mineral and aggregate extraction pits and natural shallow depressions or small valleys can be used for disposal of solid waste. Increasingly, dumps are managed throughout the world so that industrial waste is incorporated with domestic refuse. Such mixing can be a useful, cheap way of dealing with certain categories of industrial waste. But the nature of any industrial waste needs to be determined prior to any mixing with domestic refuse and subsequent disposal to a landfill. (374).

For existing dumps, two strategies for reducing movement of leachate from a site are grouting and slurry walls. Grouting is the process of injecting a liquid, slurry, or emulsion under pressure into the soil to create a "dam" within the soil. Grouts can be constructed by continuous injecting around the boundaries of the disposal site. The costs are high, and they are only applicable to special cases. (374)

For slurry walls, a trench is dug around an area and it is backfilled with an impermeable material. Slurry walls can be positioned either upgradient from a waste site to prevent flow of groundwater into a site or around a site to prevent movement of polluted groundwater away from a site. Construction is generally simple, not requiring complex engineering equipment. (374)

Figure 3.122. Poorly managed solid waste dumps endanger human health but also pollute coastal waters through runoff and leaching via groundwater. (Photo by author.)

3.92 STORM SURGE

The term *storm surge* is used to measure the elevation over normal water level due to the action of storms. Severe storms may produce surges in excess of 8 meters (26 feet) on the open coast and even higher in bays and estuaries (362). Determination of storm surge water elevations during storms is a complex problem involving interaction between wind and water, differences in atmospheric pressure, and effects caused by other mechanisms unrelated to the storm.

Abnormal rises in water level in nearshore regions will not only flood low-lying terrain, but also provide an elevated base on which high waves can attack the upper part of a beach and penetrate farther inland. Flooding of this type combined with the action of surface waves can cause severe damage to low-lying land and backshore improvements. Wind-induced surge, accompanied by wave action, accounts for most of the damage to settlements, coastal engineering works, coral reefs and beach areas. Dismantling of jetties, groins, and breakwaters; scouring around structures; accretion and erosion of beach materials; cutting of new inlets through barrier beaches; and shoaling of navigational channels can often be attributed to storm surge and surface waves (362).

The extent to which water levels will rise during a storm depends on several factors. The factors are related to the following (362):

- Characteristics and behavior of the storm
- Hydrography of the basin
- Initial state of the system
- Storm intensity
- Astronomical tides
- Direct winds
- Atmospheric pressure differences
- Earth's rotation
- Rainfall
- Surface waves and associated wave setup
- Storm motion effects
- Overwater duration

Winds are responsible for the largest changes in water level. Onshore winds cause the water level to begin to rise at the edge of the Continental Shelf. The amount of rise increases shoreward to a maximum level at the shoreline. Storm surge at the shoreline can occur over long distances along the coast. The breadth and width of the surge will depend on the storm's size, intensity, track, and speed of forward motion, as well as the shape of the coastline and the offshore bathymetry. The highest water level, neglecting the contribution of astronomical tide, reached at a location along the coast during the passage of a storm is called the *peak surge.*

3.93 SUBSIDENCE

Even people who understand the dangers of hurricanes, floods, earthquakes, and landslides are often unfamiliar with land subsidence. Broad regional subsidence can significantly affect a region's vulnerability to other hazards—especially to riverine and coastal flooding.

The costs associated with such subsidence are only poorly known. Damage is almost always attributed to an immediate cause—the flood produced by a hurricane, for example, even though subsidence occurring over a period of years before the flood may have significantly affected the depth of flooding.

The phrase *land subsidence* describes vertical displacement of the land surface caused by removal of subsurface support. The character of the displacement varies broadly, including both catastrophic collapse of the land surface and regional lowerings of the land surface that may take decades to become problems, as in Venice, New Orleans, and Houston.

Development activities that cause land subsidence include (a) subsurface mining of both metals and nonmetals, especially coal; (b) withdrawal of both groundwater and petroleum from unconsolidated sediment; (c) drainage of peat and muck soils; (d) groundwater withdrawal from limestone; (e) solution mining; and (f) surface application of water to undercompacted sediment (hydrocompaction). (173)

Withdrawal of underground fluids (petroleum, water) from unconsolidated sediment has caused major subsidence problems in the United States. This subsidence is caused by sediment compaction. Areas of the port city of Long Beach, California, sank as much as 3.3 m during development of the Wilmington oil field. Control of subsidence by repressuring the field cost US$120 million.

In Houston, Texas, groundwater withdrawal caused portions of the coastal area to subside as much as 3 feet (2.4 meters) below sea level. This subsidence caused outright inundation of low-lying land, drastically increased the flooding danger and made the area especially vulnerable to hurricane disaster. Structural solutions have been attempted. Dikes have been built and pumps installed to help ward off flooding problems. Unfortunately, such structural protection measures treat only the "symptoms" of unmanaged groundwater pumping—the increase in relative sea level and flooding—and do not solve the problem. Subsidence in the Houston area not only caused inundation but also increased susceptibility to storm surges generated by hurricanes.

Subsidence from drainage of organic soils may be the most costly type of subsidence. Drainage of these soils for development inevitably results in subsidence. The processes causing the compaction are both physical and biological. Rates of subsidence depend on climate, type of peat, and depth of water table, but may exceed 8 cm/yr.

In this kind of subsidence the soil is actually being consumed, and future cultivation is threatened. This is an important consideration in the Florida Everglades and in the Sacramento Delta in California. In the delta the problem is compounded because the land has subsided below sea level and must be protected by extensive levees.

The state of Louisiana in the United States contains the delta of the Mississippi, a huge area whose elevation had traditionally been elevated by natural deposit of sediment. But when the river channel was stabilized with solid dikes to stabilize its position and depth for navigation purposes, the sediment was then flushed directly to sea and did not spread over the delta. The delta land began to sink beneath tidal waters at the rate of 155 sq km per year. Wetlands sink first but, as they do so, the waterline moves onto the land. (341)

In subsidence areas where structures are already at risk, cost-effective methods for controlling subsidence need to be developed and refined. This requires not only technical advances but, in some cases, changes in institutions. Legal and institutional obstacles often prevent those affected by subsidence from taking action.

Unregulated development of resources and urban growth lie at the heart of some subsidence problems. If the geographic areas with potential subsidence problems can be identified, land use controls may provide a method for reducing the net cost to society.

Private and public entities (petroleum companies, water suppliers) commonly benefit from an activity that causes subsidence damage, yet the costs of damage are not distributed equally among the benefactors. For example, in the Houston, Texas, area, the whole region benefitted from lower-cost groundwater, but only people in the coastal area and on faults incurred the costs resulting from subsidence. Institutional and legal changes are required to redress these situations including setbacks, pumping limits, etc. (173)

3.94 SUSPENDED PARTICULATE MATTER

Suspended particulate matter (SPM) is a general term for the particulate or non-soluble content of the water. When present in large amounts suspended matter causes coastal waters to lose their clarity. Visually, the water in this state is usually less attractive and less productive (biologically) than clear water. Turbid waters signal the presence of pollution from various sources such as organic waste from sewage, silt from rainwater runoff, excess nutrients (plankton bloom), or disturbed beach or bottom sediments.

SPM includes both inorganic (sand and clay) and organic (particulate material, bacteria, and plankton) substances in the water that are retained on a glass fiber filter (59).

Water that contains an excess of SPM is described as *turbid.* The parameter called *turbidity* measures the *degree* to which water has become turbid and thereby indicates the amount of suspended matter and planktonic life that it contains. The suspended matter may be organic detritus, fine inorganic particles, various colloids, or exotic matter. The planktonic organisms may be any of numerous species of microscopic flora or fauna. (For details see "Suspended Particulate Matter" in Part 3.)

In the tropics, an oligitrophic coral ecosystem in its natural state and essentially free of land influence may have only 2 or 3 mg/l suspended matter and should not be forced past 4.5 mg/l by pollution (see "Water Quality Management: Coral Reefs" and "Nutrients Management" in Part 2). Halleck states that the life functions of corals can be disrupted in water as shallow as 2 meters because of turbidity induced by excessive suspended matter (153).

Sunlight is the basic force driving the ecosystem. It is the fundamental source of energy for plants, which in turn supply the basic food chain that supports all life. Sunlight must be able to penetrate coastal waters so as to foster the growth of both the rooted plants, such as sea grasses, and the suspended algae (or phytoplankton). Increased turbidity from the addition of excess suspended matter to the water reduces light penetration and depresses plant growth. Estuarine waters are normally more turbid (cloudy) than ocean waters, being both more laden with silt and more rich in suspended life (59).

Finely divided SPM may kill fish directly by interfering with respiration, inhibit growth or egg and larval development, interfere with natural movements, reduce the availability of food, and prevent successful reproduction by blanketing spawning sites. Suspended particles also serve as a transport mechanism for pesticides and other toxic substances readily adsorbed onto fine clay particles.

Concentrations of SPM less than 25 mg/l usually have no harmful effect on fisheries. Waters containing concentrations of 25 to 80 mg/l should be capable of supporting good to moderate fisheries (356). Where concentrations exceed 80 mg/l, good freshwater fisheries are unlikely.

In its undisturbed and virginal state the ecosystem responds and adjusts to the existing levels of turbidity caused by the suspended matter. The organisms in turn adapt to these levels, and the system comes into equilibrium. However, when the system is disturbed and degraded by a sudden addition of suspended matter, its carrying capacity is reduced and its general function declines below the optimum condition.

A moderate suspended matter load is typical of healthy and productive coastal water bodies such as estuaries, however. For temperate estuaries, the suspended load varies considerably depending upon many factors (243), but commonly will fall in the range of 20 to 80 mg/l (ppm, parts/million). Estuarine species have adapted to a naturally more turbid environment and here siltation from the fallout of the suspended matter may have more serious effects than the turbidity itself.

A serious environmental threat lies in the side effects of coastal construction operations, both along the beach and offshore in the lagoon and at critical habitats like coral reefs, particularly with the handling of soil and sedimentary materials. Dredging and disposal or filling in the marine environment results in suspension of silt and clay along with any toxic substances therein. This fine material causes turbidity and sedimentation which have negative impacts on water quality and sealife and do serious damage to coral reefs.

A major ecological impact generator is dredge-and-fill, whether for channels, land creation, or "beach nourishment" because of the fugitive fine particles (inorganic and organic) that escape from the work site to create turbidity and cause particle fallout. Also, operation of marine construction equipment can directly (physically) damage the environment (see "Construction Management" in Part 2).

Suspended particulate sediment in estuaries is also contributed by rivers, shore processes, agricultural runoff, industrial effluents (Figure 3.123), sewer and storm water outfalls, marine processes, biological activity and resuspension during periods of excessive turbulence. Maximum concentrations of suspended sediments are anticipated during spring freshet and periods of heavy rainfall with associated erosion. It is believed that SPM provides one of the major pathways for the transport of trace metals from the watermass to bottom sediments in most estuaries. However, dredging with accompanying dumping immediately outside an estuary entrance may increase the suspended load by 100× to 300×. It is believed that present dredging practices of dumping in regions immediately adjoining the mouths of estuaries have contributed significantly to development of baymouth shoals and increased harbour siltation.

It may be necessary to estimate the settling rate of suspended solids so as to determine how long the particles will remain suspended, the progressive change in turbidity, and where and when the particles will settle out (e.g., see "Indonesia: Ujung Pandang" case study in Part 4).

Fugitive silt from the construction operations (borrowing, transport, and filling) can be expected to have the following effects if not controlled:

Figure 3.123. Industrial wastewaters are a primary source of suspended particulate matter (SPM), which reduces productivity of coastal waters. (Photo by M. Fahay.)

- Smother coral reefs and other habitats with a layer of silt
- Debilitate fauna and flora by forcing them to use much energy in trying to combat the silt
- Cloud the water so that sunlight does not reach down to them
- Interfere with their food source (microscopic plankton)

These effects are enough to severely retard ecological processes for the duration of the borrowing and filling operation. Using the coral reef as an example, a decrease in water transparency from turbidity would diminish coral growth, (by reducing sunlight penetration and therefore photosynthesis) and allow the reef to deteriorate under attack from bio-eroders. Fauna would then be depleted (including fishes), the forams (that make most of the beach sand) would diminish, and the reef's beauty and productivity would be greatly reduced. In the extreme case, the reef's function as a natural breakwater could collapse.

Turbidity varies greatly with the seasons and with irregular environmental changes (freshets, winds, plankton blooms). Consequently, there is no one base turbidity value but rather a complex pattern of variation that characterizes an ecosystem. This pattern is difficult to pin down without extensive research. Therefore an appropriate management plan is to eliminate the sources of excess turbidity. (59)

Uncontrolled excavation, transportation, and deposition of sediment creates and disperses large quantities of suspended matter as silt and debris that settles on the bottom of coastal water basins. Silty estuarine sediments act as a pollutant trap, of "sink." Many kinds of pollutants, including heavy metals and pesticides, are adsorbed onto the sediments. Dredging operations resuspend these in the water column, increasing the hazard of exposure to plants and animals. The silt suspension may also increase nutrient release, leading to eutrophic blooms. As silt deposits from dredging accumulate in the estuary, they form a "false bottom," characterized by shifting unstable sediments.

In shallow coastal basins, these semi-liquid silts are quickly resuspended after sufficient intensity and duration of wind. Once resuspended, many factors determine the duration of suspension (59). The frequency, direction, velocity, and duration of the wind, water temperature, CO_2 content, water salinity, rainfall, exposure to the site, depth of the water, currents, vegetation abundance, etc., all interact to determine degree and duration of turbidity.

Management procedures for suspended matter are difficult to summarize because of the extreme ecological variations that exist. However, one example can be stated (98): For beach renourishment, the material used for beachfill should not exceed 5 percent "fines" (silts, muds, clays) by volume (see "Sediments and Soils" above).

The overall solution is to maintain the natural condition of the environment by which the ecosystem has evolved and has naturally prospered. Therefore, it is necessary to prevent the addition of unnatural amounts of silt that would block light penetration, or of nutrients that would stimulate excessive plankton growth and lead to the same condition.

Again, it is a matter of balance and usually the management response should be to maintain the natural level of suspended matter (suspended solids or suspended particulate matter) as far as possible by discouraging major increases or decreases. This is particularly true for oligitrophic ecosystems, those that demand especially clear water like coral reefs.

3.95 SUSTAINABLE USE

The planning and management of economic development of the coastal zone should explicitly address one major guiding principle and development strategy, namely achieving a balanced, maximum, and resource-constrained flow of benefits which is economically and environmentally sustainable. This is the concept of "sustainable development" which has been called "conservation" or "wise use" since the early 1900's.

An important ICZM premise is that coastal renewable resources should be managed to produce benefits on a long term, sustainable, basis. Sustainable use is the alternative to resource depletion that accompanies excessive exploitation for short term profit. A corollary is that development projects should be designed to sustain themselves without continued subsidy.

The Brundtland Commission of the U.N. formalized the concept and defined sustainable development as "development which does not compromise the future." However, little practicable guidance was offered by the commission or since then by any other source on how the concept should be translated into specific plans and projects.

Nevertheless, pursuing development on a sustainable use basis must be recognized as an absolute necessity to sustain progress in health, food security, housing, energy, and other critical needs. Resources should not be harvested, extracted, or utilized in excess of the amount which can be regenerated over the same time period.

Specifically, economic development objectives should be formulated to meet the basic needs and improve the quality-of-life of the coastal population without compromising the long-term productive capacity and utilization efficiency of the resource base. In practical terms, this means that:

- Current and future resource uses should be assessed in a systematic and consistent manner to assure that economic and environmental impacts of land use alternatives are carefully identified and evaluated
- Resource limitations and use constraints are identified to quantify the carrying capacity of agro-ecological zones and other vital coastal ecosystems to define population supporting capacities for subsequent use in land use planning and the formulation of realistic development policies

This process is especially challenging in a country like Bangladesh where major environmental disasters occur frequently and where enormous population pressure is exerted on a limited resource base. This means that meeting the most basic of human needs becomes a paramount concern. Moreover, frequent catastrophic flood events occur—resulting from drainage patterns of the major river systems and by cyclone-induced tidal surges—causing great human losses, physical destruction, and environmental impacts, severely reducing the near-term productive capacity of the resources base.

When governments are unprepared for these events, public policy is forged in the crucible of crisis management rather than that of advanced planning and more fully informed and systematic decision making.

Sustainable development requires maximum efficiency in the use of the resources and in the application of capital investments and recurrent inputs. It also requires a maximum of cooperation between the various economic sectors and agencies of government.

There is a growing effort to ensure sustainable resource use in development projects funded by multilateral donors or international development banks. Both the donors and the banks are requiring more intense scrutiny of the conservation and environmental consequences of development projects as well as social and long term economic effects. The simple "benefit vs. cost" financial approach in monetary terms alone is becoming less appropriate as time goes on. Using an integrative and broad planning process such as ICZM should more than satisfy the needs of international banks and donors for multi-sectoral assessments and programmatic impact analyses.

As explained by Snedaker and Getter (316), in most parts of the world, renewable coastal resources tend to be economically limited. Under the pressure of population and economic growth, the demand for a given resource will commonly exceed the supply over time, be it arable land, freshwater, wood, or fish. Sustainable use management ensures that renewable resources remain available to future generations. The criterion for sustainable use is that the resource not be harvested, extracted or utilized in excess of the amount which can be produced or regenerated over the same time period. In essence, the resource is seen as a capital investment with an annual yield; it is therefore the yield that is utilized and not the capital investment which is the resource base. By sustaining the resource base, annual yields are assured in perpetuity.

According to Burbridge, there are major physical, biological and socioeconomic factors involved in determining levels of reasonable resource exploitation (34). This involves conscious economic choices by individual societies on the allocation of scarce capital, manpower and technology. In essence, the resource is seen as a capital investment with an annual yield; it is therefore the yield that is utilized and not the capital investment which is the resource base. By sustaining the resource base, annual yields are assured in perpetuity.

The difficulty of sustainable use management—or conservation of coastal resources—rises year by year because of increases in coastal populations, desire for foreign exchange earnings, the availability of harvesting technology, and so forth. Restrictive allocation of the resources of the coastal commons tends to come only after excessive exploitation causes serious depletion of resources and resulting shortages of food and materials or social conflicts. However, some coastal communities have solved this problem by traditional management mechanisms and consequently are ensured of sustained yields of seafood and other coastal products. ICZM-type programs encourage the continuance of effective traditional management systems operated at the local level.

According to Caddy (38), the objective of sustainable development in fisheries, forestry, and agriculture can be defined as the achievement of systems of production which are compatible with the natural productivity of the local ecosystem, and in balance with the biosphere as a whole. It is evident that, in order for balance to be achieved, the human population and its demands must not exceed that level at which an equilibrium can be maintained between the use of part of the land and water for production while setting aside other areas for controlled use, which preserves the species and genetic heritage of those organisms. That this balance is far from being achieved in many parts of the world is a matter for great concern, and requires immediate measures to alleviate the human misery that such ecological imbalance inevitably leads to. (39)

In the long term, however, it is essential that human populations be brought into balance with productive capacity. Long-term strategies should be developed and followed to achieve this end, and to restore the productive capacities of these ecosystems. The difficulties of achieving these ends are recognized as enormous, particularly at a time when international sources of credit are reduced, and the world is facing the real possibility of climatic change and its effect on all planetary life, due to uncontrolled developments in all sectors of the world economy (39).

Such an ambitious objective can only be approached by setting aside certain areas which should remain qualitatively unchanged in their fauna and flora by development, while still providing economic, nutritional and recreational benefits to human beings, according to Caddy (39). A strategy which combines restoration of productive natural ecosystems with, where possible, an improvement of the productive capacity of the main producing areas, requires of course integrated planning across many sectors of human activity. And the best way to do this is by creating marine management authorities for particular sites.

3.96 TERMS OF REFERENCE

The outline of work to be performed by a consultant or other contract employee is often called the *terms of reference* (TOR) or perhaps the *scope of work.* It is important to both the client and the consultant that the TOR is properly prepared—complete, accurate, and unambiguous.

It is essential that the major environmental concerns be identified and a search for other likely consequences of development be specifically requested in the TOR. For competitively bid contracts, the consultant will seldom add tasks for fear the resulting higher costs will keep the firm from being awarded the job. During negotiations, however, additional studies may be suggested. Hence, the client must have an understanding of just what is needed in order to avoid paying for unnecessary services. It is also important to request the form of analysis and presentation appropriate to the use of the EIA (i.e., extended benefit-cost analysis, comparative risk assessment, and cost effectiveness) (41).

The example TOR instructions given below are for environmental impact assessment (EIA) preparation; the technical contents typically include (41):

1. The objective of the EIA: What decisions will be made, by whom, timetable, what kinds of advice are required, what is the stage of the project?
2. Components of the project: Sites, technologies, inputs of energy, and materials anticipated
3. Preliminary scope of EIA: geographic, region, lifetime of project, externalities
4. Major anticipated concerns about environment changes and consequences
5. Major anticipated impacts on human health and welfare; on ecosystems
6. Mitigation measures possible. Reasonable alternative project designs for achieving development objective
7. Estimated monitoring necessary for feedback to operations, for detecting environmental consequences and judging whether mitigation measures have been implemented
8. Type of study required (e.g. benefit-cost analysis, land-use plan, pollution control regulation, simulation model, comparison of sites or technologies, risk assessment)
9. Staff level of effort, skills required, cost estimate, and deadlines for completion of tasks

Nontechnical requirements in the TOR may include any of the following (41):

1. Stipulation of references and data to be provided by the client
2. Frequency and subject of meetings and progress reports
3. Opportunities for review and comment on draft reports
4. Prior approval of changes in contractor personnel
5. Payment schedules

6. Liability, insurance
7. Printing, distribution of reports
8. Coordination requirements

3.97 TIDEFLATS

Some coastal basins—estuaries, lagoons, bays—have extensive areas of tideflat which are important in processing nutrients for the ecosystem and providing feeding areas for fish at high tide or birds at low tide (Figure 3.124). In many estuaries and lagoons, tideflats also produce a high yield of shellfish. Mud flats and sand flats are often important energy storage elements of the estuarine lagoon ecosystem. Where mud flats are not present, vital dissolved chemical nutrients (such as phosphates, nitrates, nitrites, and ammonia) can be swept out with the ebbing tides, reducing the energy supply to the food chain. The mud flat serves to catch the departing nutrients and hold them until the returning tide can sweep them back into the very productive vegetated tideflats or wetlands (marsh or mangrove). There appears to be an optimum balance between the proportion of wetland (swamp or marsh) to mudflat area which is vital to the stability and the continued existence of both systems.

Where mud flats are not present, vital dissolved chemical nutrients (such as phosphates, nitrates, nitrites, and ammonia) can be swept out with the ebbing tides, reducing the energy. There appears to be an optimum balance between the proportion of wetland (swamp or marsh) to mudflat area which is vital to the stability and the continued existence of both systems supply to the food chain. (316)

Intertidal mudflats, sand flats, and coral flats are often productive habitats in their own right supporting a large and diverse fauna, including clams, sea cucumers, crabs, or ect which are valuable food sources.

3.98 TIDES

As used here, the word *tide* refers to the "astronomical tide"; that is, the regular water level changes caused by changes in the relative positions of sun, moon, and earth. Tide is a periodic rising and falling of sea level caused by the gravitational attraction of the moon, sun and other astronomical bodies acting on the rotation earth. The earth and the moon constitute a single system revolving around a common

Figure 3.124. Ibis and egret feeding on a productive tideflat that has been protected to ensure a high level of biological diversity (Rookery Bay, Florida, USA). (Photo courtesy of U.S. National Oceanic and Atmospheric Administration.)

center of gravity. The system is in equilibrium, because centrifugal forces exactly counterbalance the gravitational attraction between the two objects. But for objects not at the center of the two objects, this equilibrium does not exist. Although the amount of unbalance is not large, there do exist forces that, at certain regions of the earth's surface at certain times, tend to move objects toward or away from the moon. These are called "tidal forces" (59). Astronomical and meteorological influences on tides are expressed in different tide levels at the shoreline.

Ocean water, relatively free to move under the influence of tidal forces, forms two bulges, one on the side toward the moon, the other on the far side. The earth turns under these bulges and thus there arrive two tides a day, separated, however, not by 12 hours but by about 12 hours and 25 minutes because, while the earth is turning, the moon is also pursuing its orbit around the earth (Figure 3.125). Tides occur in different forms depending upon many factors—diurnal and semi-diurnal tides being the main forms (see Figure 3.126).

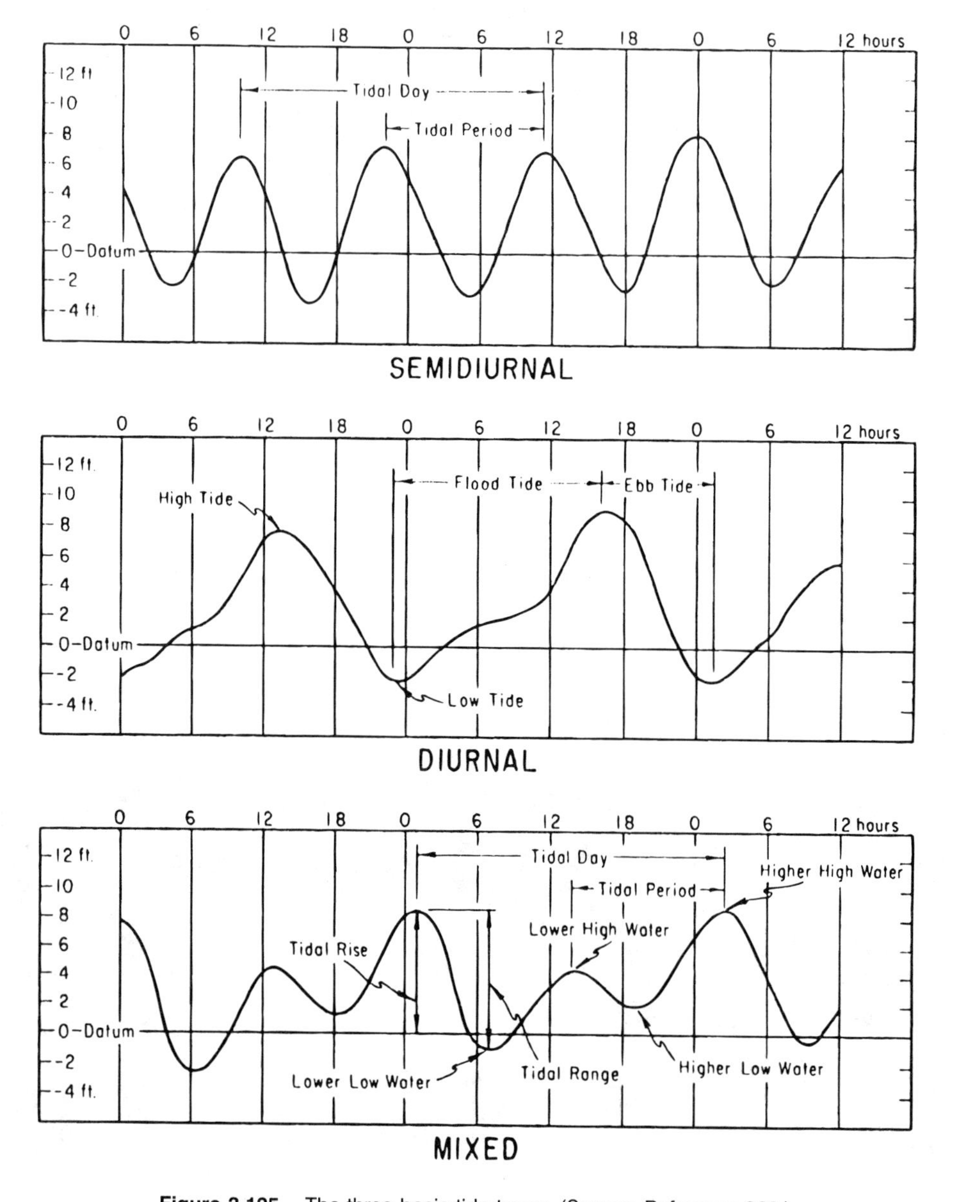

Figure 3.125. The three basic tide types. (*Source:* Reference 362.)

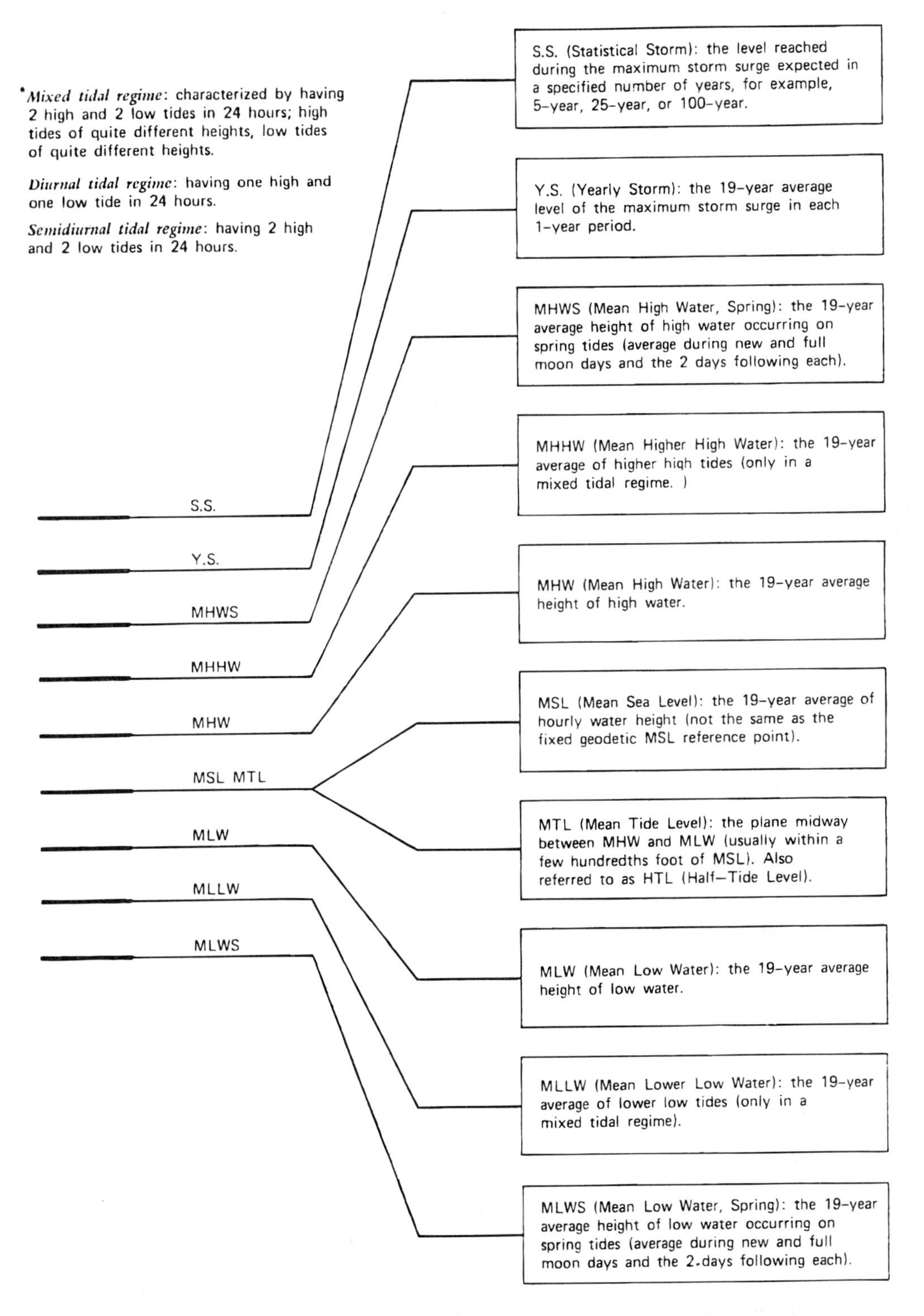

Figure 3.126. Reference heights for different tidal stages. (*Source:* Reference 59.)

Open-ocean tides are small—perhaps 1 or 2 feet in height. Significant tidal range occurs only when local conditions produce it. These conditions involve resonant effects; the surging of water in enclosed areas, some of whose natural periods of vibration may be near the periodicities of the tidal forces (59).

The sun also produces tidal forces on the earth, but it has a little less than half the effect of the moon. When the sun and moon forces cooperate, we have the situation every 2 weeks of *spring tides* with maximum range in height. When the sun's forces are at right angles to those of the moon, there occur *neap tides* of minimum range. Spring tides occur roughly at the times of full moon and new moon; neap tides, when the moon is in first and third quarters. It is typical that where the average range of tides is 9 feet, say at Boston, the difference between the range at spring and at neap tides can be as much as 6 feet. In addition, there are yearly high and low average sea levels driven by yearly tidal cycles (59).

It is of interest to establish and to maintain a network of tide-recording stations located at stable points in secondary ports, coastal lagoons, and estuaries. This second order network would be aimed at allowing for regional extrapolation of data from the basic network, and local estimation of short-period components due to atmospheric and oceanographic factors; if automatic tide recorders are not available for this second order network, the tide can be observed visually by means of graduated staffs. Of course, tidal observations must be referred to some fixed datum to be of full use. For this purpose, a bench mark (BM) is used as the primary reference point. (181)

For illustration, let us consider the simplest type of tide gauge, the visual tide pole. Depending on the location of the BM, the number of instrument stations can be determined (e.g., if the BM is far, more stations are required). The instruments required are an engineers level with tripod, a steel chain and two levelling staffs. In a situation where the BM is very close to the erected tide pole as shown in Figure 3.127, the following stages are followed while leveling the tide pole (181):

1. Set up and level the instrument close to the BM
2. Put a leveling staff on the BM as well as on the erected tide pole
3. Read off leveling staff on BM, as a_m
4. Read off leveling staff on tide pole as b_m

The computational procedure based on Figure 3.127 is as follows:

1. Elevation of top face of tide pole (i.e., from top of BM upward $= a - b$
2. Elevation of BM with respect to ordnance datum $= x$
3. Elevation of BM with respect to chart datum $= x + y$
4. Height of tide pole with respect to ordnance datum $= x + y + (a - b)$
5. Zero of tide pole $= [x + y + (a - b)]$ $-$ length of tide m pole.

3.99 TOURISM

In developing countries, rich in natural attractions, tourism is often the shortest route to broad economic growth and prosperity. Because international tourism is so competiive, countries often target specific market groups using whatever competive edge they have—historic, cultural, adventure, nature. For example, they may focus on young, single women, the silver-grays (senior citizens), or double-income middle age families (Figure 3.128).

A powerful tool for national development, tourism is one of the fastest growing areas of international trade, particularly for smaller coastal countries and island countries with limited development options. It is also a potential area of conflict, particularly in developed countries caught in the typical dilemma of tourism; they want the income from tourism while at the same time they deplore the negative social and environmental impacts.

Typical environmental damage from resort development leaves shorebird habitats destroyed, mangrove wetlands filled for real estate, fisheries depleted, swimming beaches eroding away, coral reef tracts poisoned, harbors polluted. Loss of mangrove forests is particularly damaging because they serve important roles in the coastal/marine ecosystem as well as in the general functioning of tropical forests.

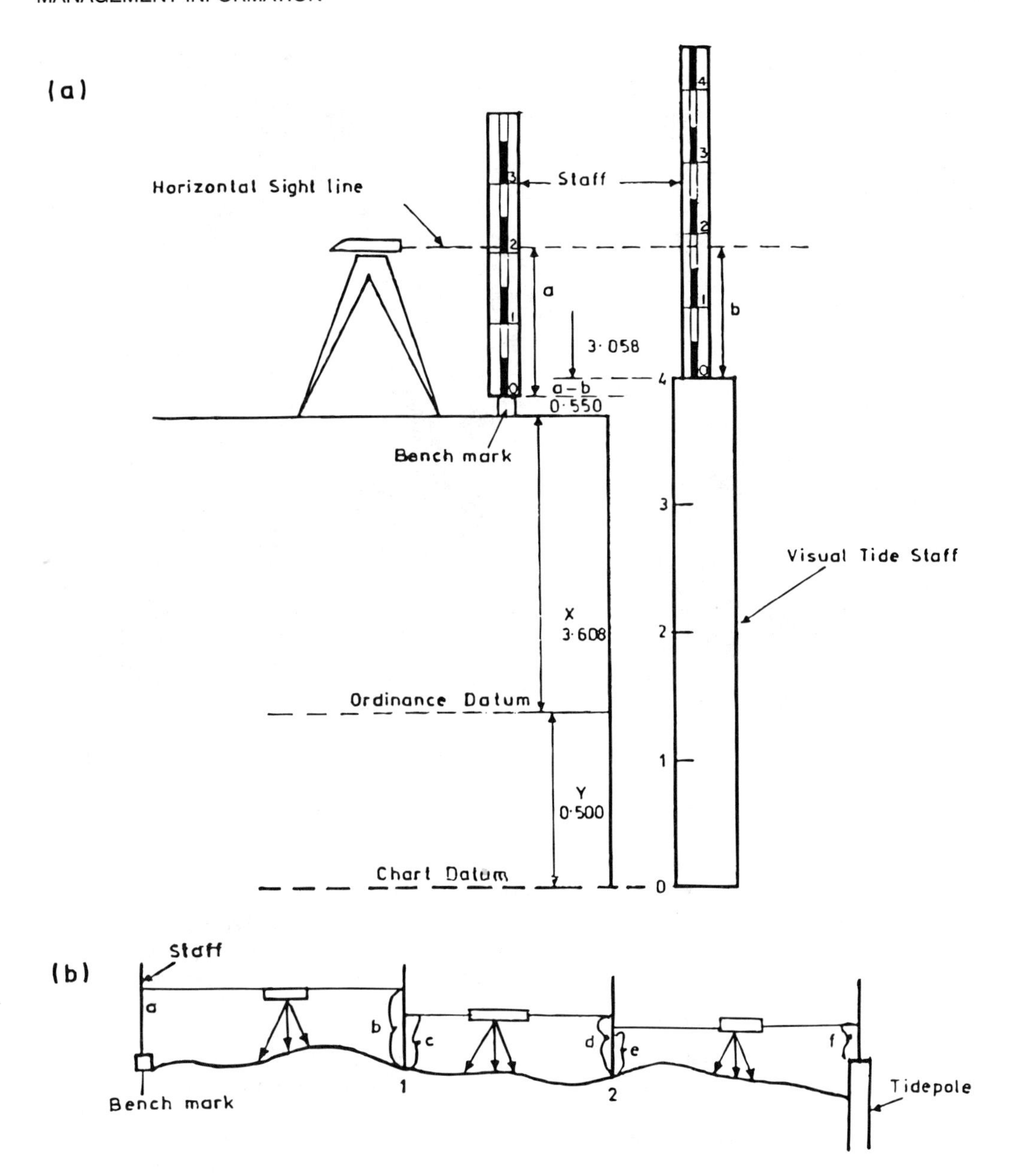

Figure 3.127. Use of a visual tidepole for determining tidal height. (*Source:* Reference 362.)

Saenger (302) identifies sewage as the main culprit but points out that in coastal waters, habitat losses result from dredging or reclamation works associated with the landward facilities or from changed circulation patterns as a result of water-based facilities—dredging, reclamation, and siltation are known to adversely affect most inshore ecosystems.

Sewage also tops Plogg's list of impacts from uncontrolled or poorly planned tourism development in the Caribbean, which include (285):

- Near-shore pollution due to untreated or insufficiently treated sewage which is often washed onto the beaches
- Damage and erosion of beaches as a result of breakwater construction and mining of sand to build hotel and related facilities
- The breakdown of coral reefs directly by stripping and indirectly through pollution

Figure 3.128. A major tourist attraction for sub-tropical and tropical countries is snorkeling or scuba diving on coral reefs. (Photo by Bill Keogh.)

Smith (314) studied the potential damage to coral reefs from cruise ship operations at Grand Cayman. He recorded 3150 m^2 of previously intact reef destroyed by one cruise ship anchoring on one day. Recovery periods of more than 50 years are postulated by Smith.

Coastal areas that have suffered large-scale environmentally damaging tourism developments include: the government sponsored city of Cancun (Quintana Roo, Mexico), the tourist buildup at Buccoo Reef (Trinidad and Tobago), the hotel strip of the Barbados leeward shore, the sprawl of Puerto Galera (Mindinoro, Philippines), and the retailing of the Ilha Comprida (Sao Paulo, Brazil) and the resort city of Montego Bay (Jamaica). The worst of the damage could have been avoided by recognizing sustainable limits and adhering to some simple environmental guidelines and standards, to the extent that they had been available and that the institutional mechanisms to implement them had been in place.

Clearly, in small countries dependent upon coastal tourism, disturbances extend into the social and socio-economic scene. Oftentimes resource plundering and environmental deterioration are root causes of social disruption (Figure 3.129). Dunkel says that mass tourism can "subvert the indigenous culture" and often the economic benefits of tourism have been inflated (117).

There has developed over the years a standard scenario sometimes called the *self-destruct theory* of tourism (the first version is attributed to Stanley C. Plogg, (285)). A recent version by Sobers 1988 (318) follows:

1. A remote and exotic spot offers peaceful rest and relaxation and provides an escape for the rich, who live in isolation from the resident population.
2. Tourism promotion attracts persons of middle income who come as much for the rest and relaxation, as to imitate the rich. More and more hotel accommodation and tourist facilities are built to attract and accommodate more and more tourists. This transforms the original character of the place from an "escape paradise" to a series of conurbations with several consequences:
   - The local residents become tourism employees and earn more than ever before.
   - The rich tourists move on elsewhere.

- The growth in tourist population makes interaction between tourist and resident population inevitable, leading to a variety of social consequences.
- Increased tourist accommodation capacity leads to excess supply over demand and a deterioration in product and price.

3. The country resorts to mass tourism, attracting persons of lower standards of social behavior and economic power. This leads to the social and environmental degradation of the tourist destination.
4. As the place sinks under the weight of social friction and solid waste, tourists exit, leaving behind derelict tourism facilities, littered beaches and countryside, and a resident population that cannot return to its old way of life.

For tourism, the environment is a significant part of the product which a country has to offer. A successful tourism strategy will, therefore, seek to maximize the profits while preserving the natural environment. When the situation deteriorates sufficiently, there is also loss of jobs, income, and hard currency earnings, factors that lead to socio-political instability. Such problems can be alleviated by appropriate planning mechanisms and development controls.

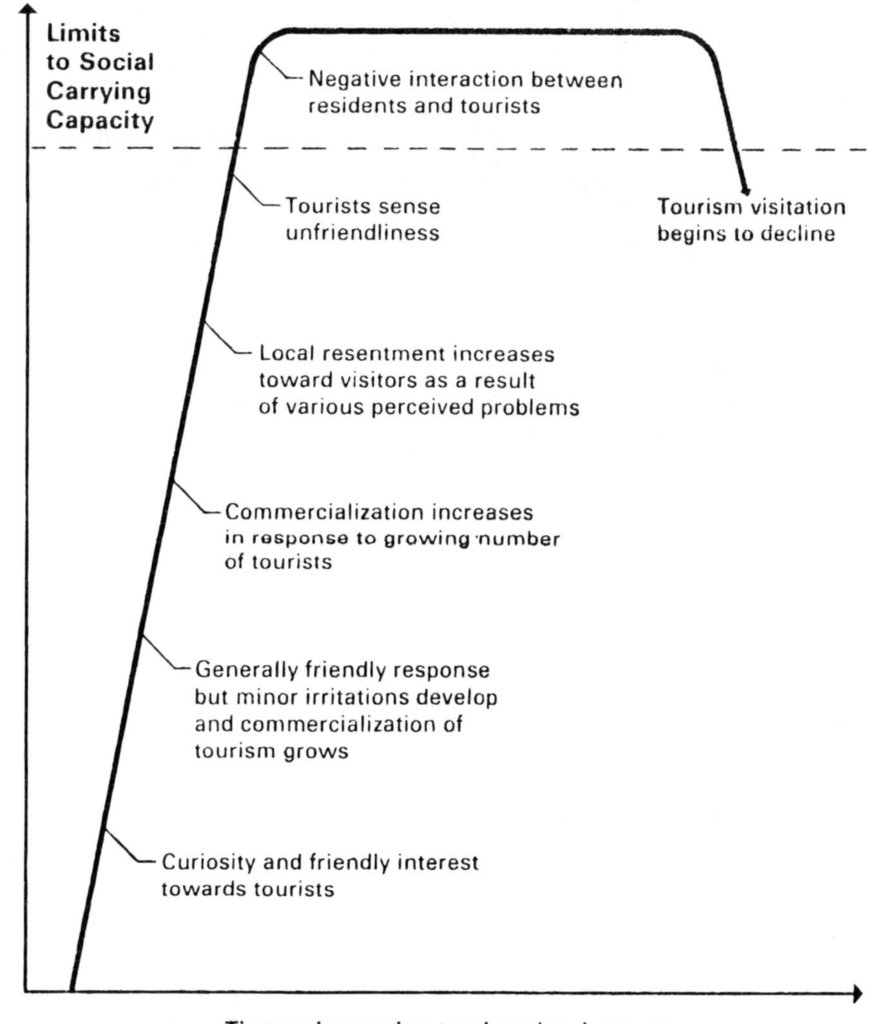

Figure 3.129. Generalized tourist-resident interactions at various stages of carrying capacity. (*Source:* Reference 251.)

Figure 3.130. Ecotourism has become a welcome substitute for mass tourism in many coastal zones, as here in Palau. (Photo by author.)

As Holder (171) points out, the economic importance of tourism as a generator of foreign exchange has grown dramatically as prices for certain export agriculture, oil and bauxite have fallen greatly, reducing the foreign exchange earned by these sectors. According to Holder, the creation of new jobs, in a situation of chronically high unemployment, is the difference between social order and social chaos.

If tourism development is to be controlled, plans have to be formulated, guidelines and standards derived, parks and reserves have to be created, and rules have to be written, implemented, and enforced by governments. These should be based on knowledge of social and environmental carrying capacity and proved methods of visitor management (64). *Carrying capacity* refers to the capacity of an ecosystem to sustain specified resource uses.

Sadler (300) reports consensus from a 1985 conference on introducing the environmental dimension into the tourism industry, from the planning stage through to implementation and management, as follows:

- Tourism planning should be multi-sectoral, part of comprehensive development and land-use planning.
- Policy directives, based on national needs and resources should give direction to the planning process.
- Environmental impact assessment (EIA) should be applied to large-scale tourism projects.

Success comes when tourism planning assumes a holistic approach by explicitly considering a broad range of economic, social, financial, environmental and physical aspects. However, for planning or management to be meaningful, a number of constraints need to be addressed, such as those listed below (27):

1. At the political level more attention needs to be given to the setting of clear objectives as regards the kind and impact of tourism desired (Figure 3.130).
2. At the same time environmental considerations need to be given more emphasis at the political and decision-making levels.

3. Tourism goes across many sectors. Therefore it is rather difficult for any single agency to prepare or evaluate tourism plans and development proposals—in many countries there is no established mechanism for co-ordination and integration of the planning process among the various government ministries, departments and statutory bodies, along with closer collaboration with the private sector.
4. Planners of the region need to create a better understanding that planning can be positive and development oriented rather than restrictive.
5. Planners should also ensure that the social and economic perceptions of the various groups within a society are included through continuous consultations with the population.

Miller (247) notes that planning for tourism has traditionally been oriented to the needs of tourists, but future planning must alter its emphasis and take account of the welfare of the indigenous population. The development plan should seek to discover the most effective means, whether institutional, regulatory, legislative, or economic of integrating environmental and social policy not only with tourism but also with the policies for other economic sectors.

It is not enough just to create awareness among the public and their political leadership—specific actions have to be taken. Even the most compelling arguments for nature protection are only a means to an end. If tourism development is to be controlled, plans have to be formulated, guidelines and standards have to be created, and rules have to be written, implemented, and enforced by governments. These should be based on some knowledge of social and environmental carrying capacity.

It is quite possible for a country to start with simple controls and move later into more complex programs. For example, beach/dune setbacks for hotels or delineation and protection of the most critical ecological areas could be a starting point.

According to Smith (313), present governmental approaches are rather ineffective in directing and managing the development of coastal tourism. New comprehensive approaches are needed, including, for example:

1. Comprehensive planning for the coastal zone requires the preparation of tourism plans at the regional level. These plans need to identify objectives and priorities and should set aside the best tourism resources. Zoning of intended uses is necessary to spatially define areas of concentration, dispersal or absence of touristic, residential, commercial and other functions. In selecting coastal sites, due consideration should be given to visitor access, infrastructure, visitor safety, residential areas, conflicting uses, coastline stability, and visitor information (Figure 3.131).
2. Development control guidelines should establish beachfront setbacks, development density, permitted built form, maximum building heights and landscape conservation. Environmental protection measures should designate areas of protection on land and sea, retain natural rates of fresh water runoff and prevent silt entering the sea. Infrastructure planning is also necessary for the provision of solid waste disposal sites, waste water treatment plants and drainage systems.
3. To guard against unwanted practices, relevant institutions need to have strong powers of regulation. But regulation is not enough. Institutions also need powers of enforcement as otherwise regulations will be ignored.
4. Much planning and management for coastal tourism is likely to originate at national and regional levels. Of equal importance, especially for the operation of tourism ventures and programs, is the local level. Community input, involvement, and cooperation is necessary for successful project programming, implementation, and operation.

A protocol for gauging the tourism potential of a coastal site can be developed using the criteria below. Each evaluated aspect can be more specified and developed by identifying its elements and sub-elements depending on type, characteristic and specific condition of the nature-based tourism area being evaluated. Some of them become critical factors. Generally, those elements are as follows (modified from Reference 357):

Figure 3.131. Tourist information services provide a good opportunity to promote environmental strategies. (Photo by author.)

1. General Attractiveness
   - Beautiful scenery
   - Diversity of prominent resources (asset) as attractive tourism object
   - Uniqueness
   - Originally and balance
   - Diversity and alternative of recreation opportunities
   - Environmental cleanliness (air, water, location)
   - Maneuver space and carrying capacity in hectare (intensive use)
   - Sensitivity
2. Attractiveness of Marine Park (if one)
   - Safety
   - Diversify of fauna
   - Beautiful scenery of the Ocean Park
   - Originality
   - Uniqueness and sensitivity
   - Water clarity
   - Number of location and depth
   - Scenery and comfortability
   - Intensive use area
3. Attractiveness of the Beach
   - Beauty of the beach
   - Type of sand
   - Width and length of the beach (Figure 3.132)
   - Variation of activities
   - Comfortability
4. Infrastructure
   - Road network condition
   - Transportation system, frequency and its intensity

- Reliability of water, power supply
- Supporting facilities
- Special facilities
- Institutions

5. Environmental Situation
    - Land use patterns
    - Population density
    - Ecological impact
    - Natural resources
    - Natural hazards
    - Nuisances (e.g., mosquitoes)
6. Climate condition
    - Rainfall
    - Temperature
    - Sun radiation
    - Wind
    - Humidity
    - Storm frequency
7. Potential Market
    - Domestic tourist (market share: numbers, income, length of stay)
    - Foreign tourist (market share: expected origin, numbers, expenditure, length of stay)
    - Existing tourism facilities
    - Promotional activity
    - Service quality
8. Management
    - Service quality
    - Language capability
    - Maintenance and service facilities
    - Plan and mitigation of environmental impacts
    - Leadership
    - Local participation
    - Employee welfare: health, education, security
9. Impact on Regional Economic Development
    - Business and job opportunities
    - Improvement of income
    - Productivity and efficiency of resources utilization
    - Economic and business information
10. Impact on Cultural inheritance
    - Ethnic
    - Archaeological site
    - Religion and belief
    - Custom, tradition, and solidarity
11. Local Community Effects
    - Community attitude
    - Education
    - Existing media
    - Community-related activities
    - Level of participation (active, passive)
    - Existing local use of tourism site.

To find maximum relevance to the most important aspects, different numerical weights can be assigned and the items ranked as to importance. The weighting factors to be assigned could vary from 1–6, 1–10, or any other number set. These factors will, of course, be different for each situation and each site depending upon the importance of the item to the geo-specific situation.

Figure 3.132. High quality coastal tourist resorts usually have wide, safe, uncrowded beaches that need protective management. (Photo courtesy of Caribbean Tourist Organization.)

3.100 TOXIC SUBSTANCES

A most complex threat to coastal waters is from toxic substances such as heavy metals, petroleum hydrocarbons, and manufactured products like pesticides and PCBs. These pollutants are found in consumer products in daily use and enter both estuarine and ground waters in a variety of ways. They can cause both long-term and immediate damage to human and aquatic life.

There are more than 70,000 synthetic compounds in industrial use in advanced countries—and a thousand more introduced every year—but their effects on human and aquatic life are barely known. Only a small percentage of the chemicals have been tested to determine whether or not they cause health problems. The predominant effects that are known, however, are cancer, birth defects, and both acute and chronic illness. (254)

Traces of toxicants can be highly concentrated through the food web (food chain), and the nondegradable heavy metals and organic toxicants (such as DDT) would be gradually magnified. For example, the "Minamata disease" in Japan (that killed hundreds of people) was caused by methyl mercury concentrated through the food chain. The so called "ouch-ouch disease" is brought on by cadmium concentrated through the food chain of plants. (41)

The main toxicants causing water pollution can be classified by the following four categories (41):

- Nonmetallic inorganic toxicants (e.g., cyanide salts and fluorides)
- Heavy metals and submetallic inorganic toxicants (e.g., mercury, cadmium, lead, chromium, and arsenic)
- Easily degradable organic toxicants (e.g., phenols, aldehydes, organophosphates, and benzene)
- Refractory organic toxicants (e.g., DDT, BHC, dieldrin, aldrin, polychlorinated biphenyl, polycyclic aromatic hydrocarbons, and some aromatic amines)

Heavy metals are given special attention because they are elements that are not degradable and can be quite toxic. As a matter of fact, if not chemically bound in the estuarine environment so as to become unavailable to aquatic life, the heavy metals such as lead, chromium, and mercury can recycle through food chains indefinitely. They are found in pesticide formulations, exhaust of automobiles and motorboats, industrial emissions and discharges, sewage treatment plant effluent, landfill wastes, stormwater runoff, and in soil. Particularly high concentrations of metals are often found in sewage sludge—copper and zinc may occur at 600 and 2,500 ppm when amounts of 0.1 and 10 ppm, respectively, are toxic to marine life. Also, large amounts of metal toxicants are found in the bottom sediments of points and harbors. (219)

Land-disturbing activities such as construction, mining, forestry and agriculture can release heavy metals (that occur naturally in the soil) into coastal waters. Urban stormwater runoff is a particularly significant cause of heavy metal contamination of ground and estuarine waters (254). Copper, found in about 90 percent of urban runoff samples, is associated with liver damage.

Heavy metals can affect human health and natural systems in a number of ways. Lead exposure, for instance, can cause mental retardation and brain damage in children and has been linked with miscarriages, birth defects, and infant deaths. Groundwater supplies may be polluted by leaching from landfills or treatment lagoons, chemical spills, or from movement between ground and surface waters.

Some metals accumulate in the tissue of fish, and so may be consumed by people. As organisms at lower levels of the food chain are eaten by higher organisms, the concentration of metal present in the tissue is magnified. This is particularly a problem with some metals, such as mercury, which remain for a long time and are readily absorbed by fish and shellfish (254). Levels of mercury in fish from unpolluted areas are generally less than 0.1 ppm (based on wet weight), whereas comparable samples from polluted areas have higher values. In one example, levels of heavy metals in commercial crabmeat, were found to be 21 ppm iron, 46 ppm zinc, 466 ppm magnesium, and 15 ppm copper.

In order of generally decreasing lethality to aquatic organisms, mercury, silver, and copper are at the top of the list, followed by cadmium, zinc, lead, chromium, nickel, and cobalt. However, the biological effects of metals in the estuarine ecosystem are quite varied, depending on the particular metal, the organism, and such modifying factors as the presence of other toxicants, the environmental conditions, and the age or condition of the organism (219). Heavy metals such as zinc, mercury, lead and cadmium in their pure state usually are not particularly hazardous to marine life. However, these metals easily combine with other compounds, particularly organic ones, to become quite toxic. Every organism has a certain tolerance level for heavy metals, and metal concentrations as low as 0.5 ppb may be toxic to some organisms. (See "Water Quality Management: Coastal Waters" in Section 2.)

In coastal sediments, heavy metals tend to be associated with the organic fraction of the estuarine bottom sediment and seem to be concentrated to a greater degree in the surface layers than in the deeper sediments. Dredging of such materials in harbors can be troublesome because the excavated sediment may have large amounts of heavy metals; for example, cadmium may occur at 130 ppm in harbor sediments. When these sediments are disturbed by dredging, the toxicants are freed to spread through the water column at often lethal concentrations. (219)

Coral reef communities may be particularly vulnerable to heavy metal toxification because corals rapidly take up metals dissolved in seawater and sequester it in their tissues—Howard and Brown (176) showed 25 different metals in the tissues of five major families of hard corals. It is part of the normal life of these corals to store and use small amounts of metals; it becomes a problem when the metals are excessively available. Interference with important life functions of corals and their zooxanthellae starts at chronic concentrations of 1.0 or 0.1 ppb for some metals (176).

In summary, while mercury continues to be of paramount concern, several other heavy metals, such as lead, cadmium, and zinc, can be troublesome. However, in their pure state "they are not particularly hazardous to marine life" (366). However, they are said to combine with other compounds, particularly organic ones, to become quite toxic, which case, metal concentrations as low as 0.5 μg/l may be toxic. Zinc, copper, nickel, and chromium seem to be less toxic (in lethal effects per dose) than lead, arsenic, cadmium, and selenium to a broad variety of marine life (366). The management reponse to heavy metals pollution is to control the quantity and quality of industrial wastes that are discharged into coastal waters.

Pesticides or biocides (any chemical that kills an organism identified as "pest") are ecologically dangerous, including insecticides, fungicides, piscicides, herbicides, miticides, etc. *Insecticides* are commonly classified into three groups: (a) chlorinated hydrocarbons (organochlorines), like DDT, aldrin, dieldrin, heptachlor, and chlordane; (b) organophosphates, like malathion, parathion, diazinon, and guthion; and (c) carbamates, like Sevin and Zectran. *Fungicides* include such substances as dithiocarbamates (e.g., Ferbam and Ziram), nitrogen-containing compounds (e.g., phenylmercuric acetate), triazines, quinones, heterocyclics, and inorganics like the heavy metals. Herbicides are quite varied, the most common being the phenoxy acids like 2,4-D and 2,4,5-T (219). Frequently used aquatic herbicides include endothal and diquat, often applied in combination with a surfactant (like a detergent).

Some organochlorine pesticides, like Mirex, a chemical commonly used to control the imported fire ant, *Solenopsis saevissima,* in the southern states, are particularly toxic to estuarine organisms. For example, juvenile shrimp and crabs died when exposed to *one* particle of mirex bait; and 1 ppb mirex in seawater killed 100 percent of the shrimp exposed. (219)

Although toxic to crustaceans, Sevin is fairly nontoxic to fish and mammals. In a toxicity test that included 12 insecticides and seven species of estuarine fish, the descending order of toxicity was as follows: endrin, DDT, dieldrin, aldrin, dioxathion, heptachlor, lindane, methoxychlor, Phosdrin, malathion, DDVP, and methyl parathion. Pesticides in coastal waters are particularly harmful to invertebrates. For example, shrimps are killed at concentrations of 1.0 μg/l (ppb) or less of the following: chlordane, DDT, dieldrin, endrin, heptachlor, lindane, and mirex (219).

Pesticides are generally insoluble in water but are readily adsorbed on particulate matter. Uptake of pesticides by estuarine organisms may take place by ingestion of particulate matter (detritus), from water passing over the gills or by diffusion through the skin or exoskeleton.

To marine life, there may be many *sublethal* effects of pesticides, including elevation of metabolic levels, prevention of maturation or molting, reduced ability to osmoregulate, reduced ability to endure temperature and salinity changes, and lethargic behavior. All of these sublethal effects reduce the ability to avoid predation.

Pesticides, paints, household cleaners, and hundreds of other products in daily use contain synthetic organic chemicals—manufactured compounds that can be extremely toxic and may remain in the environment for long periods. These chemicals run off cultivated fields, plant nurseries, lawns, streets, and golf courses. They also are discharged in industrial and municipal effluent and can be spilled accidentally or dumped illegally, contaminating both estuarine and ground waters.

Perhaps the most well known of the industrial toxicants are *polychlorinated biphenyls* (PCBs)—chlorinated compounds that find use in almost every sector and have recently come under close scrutiny (1). They are used in such diverse products as printer's ink and the paint applied to swimming pools. *Phthalate esters* (used as "plasticizers") are recognized as having a potential for serious harm. Pulp mill wastes and acids and chemicals from other industrial operations can also be considered serious threats. (See "Industrial Pollution" above.)

Like the heavy metals, PCBs can occur in high concentrations in sediments (351). Estuarine organisms like fiddler crabs and shrimp readily pick up PCBs from the sediments while filter-feeding oysters accumulate these chemicals from the water (219).

Chlorine, an inorganic chemical, poses a particular problem in the coastal area. It is commonly used as a disinfectant for water treatment plants, swimming pools, sewage effluent. Chlorine is quite toxic to aquatic life and so is a serious threat to coastal water quality (254).

Oil, gasoline, and coal release petroleum hydrocarbons into coastal waters in a variety of ways. Automobile and boat exhaust, motor oil and grease, power plant emissions, industrial discharges, accidental spills, illegal dumping, and leaking underground storage containers all bring petroleum hydrocarbons into ground and surface waters. Spills on land or in the water can damage coastal wetlands and kill or cause long-term damage to marine life and cause a loss of tourist revenue if beaches are oiled. Although oiled sandy beaches can be cleaned (albeit at great expense), oil-contaminated marshes, mudflats, and estuarine bottoms cannot be cleaned.

Protection of coastal water quality involves more than just avoiding lethal concentrations of toxic pollutants. There are lower limits of water quality below which mobile animals either desert an area, or survive in reduced health and abundance, failing to grow or reproduce properly. Coastal/marine migratory fish may be particularly affected by chemical non-suitability of the water and abandon coastal areas with "bad" water. (219)

A variety of potentially repelling substances come from industrial discharges or sewage effluent—aromatic hydrocarbons, synthetic organic substances, and so forth. These repellents may be responsible for the virtual abandonment of many estuarine and nearshore ocean waters by water-suitability-sensitive sport fish species. The result is failure of a fishing area or depletion of the general carrying capacity of the ecosystem. Elimination of the discharge of repellent substances is essential in the restoration of many ecosystems to optimum function. (219)

In the future, industrial and domestic waste loads will increase and so will the potential for accidental spills and dumping.

Safe disposal of toxic wastes is a necessity. Current approaches to waste disposal are too often inadequate. Effluent from wastewater treatment plants or sanitary landfills or the vapors from waste incineration, can contribute toxic substances to the environment. Under most methods of waste disposal, only a small fraction of the toxicants is removed before the effluent or vapors are returned to the waters or atmosphere. Nor can it be expected that improved technologies will soon be employed. Therefore, the accumulation of these substances in coastal waters may continue to cause long-term damage unless source reduction is employed. The idea of source reduction is to stop toxic matter from entering into the air, water, or the wastestream.

In many countries, regulations strictly govern the use and discharge of chemical toxicants. In view of the complicated characteristics of the chemical toxicants and the fact that they are difficult to remove, great attention must be paid to keeping them out of water resources. (For specific standards see "Water Quality Management: Coastal Waters" in Part 2.)

3.101 TRADITIONAL USES

Often a socially desirable and efficient approach to resource management is allocation according to traditional approaches worked out for specific cultures in specific coastal locations over hundreds or thousands of years. The yields obtained from coral reefs and lagoon fisheries in Oceania, *tambaks* in Indonesia, *acadjas* in West Africa, *vallicoltura* in the Venice (Italy) region, "fish crawls" on the Mediterranean and Atlantic coasts, and small-scale traditional fisheries all over the world have been maintained over centuries.

Typically, where traditional fishing rights exist, villages, chiefs, clans or individual fishermen possess the locally acknowledged right to control the exploitation of designated fishing grounds. The existence of such customs has important implications for reef and lagoon management, for they constitute a type of "limited entry" and, as such, a form of fisheries conservation.

Johannes notes traditional coral reef fisheries have formed a vital component of coastal economies in some regions. Where political upheaval has not destroyed local customs, systems of traditional fishing rights often exist in reef and lagoon waters. Such customs are more often found in the Indo-Pacific, where many ancient fishing cultures survive today, than in the Caribbean where aboriginal fishing cultures have been supplanted in relatively recent times.

To strengthen traditional management of coastal resources and to preserve "customary fishing rights," protection has been planned for the Marovo Lagoon in the Western Province of The Solomon Islands. The recommended management authority is the "Area Council." Continuance of the local culture (population 5,500), which is oriented toward the sea and sea fisheries, depends on maintaining fish stocks. This requires not only control over the fish harvest, but protection of habitats against adverse impacts of forest clearing and logging, mining, coastal agriculture, and exploitation by cruising yacht tourists. A protection program can be most effectively advanced in this type of situation by acquiring (if necessary) the appropriate amount of shorelands, transition area, and open water as a protected area, such as a "resource reserve," and delegating its management to the Area Council. (12)

Where fishing rights exist, it is in the best interest of those who control them not to overfish. The penalty of doing so—reduced future catches—accrues directly to the owners. Self-interest thus dictates conservation. Where such resources are public property, in contrast, it is in the best interests of a fisherman to catch all he can. Since he cannot control the activities of other fishermen, the fish he refrains from catching will most likely be caught by someone else. In a fishery open to all, therefore, self-interest dictates overfishing—this is the well known "tragedy of the commons" defined by Garrett Hardin.

A corollary to the above, is that the incentive to protect the resource base—physical habitats and water quality—is stronger where access to the area is limited to traditional users who still maintain control. Thus, in countries where customary fishing rights exist, serious consideration should be given to the maintenance of these rights. However, as Johannes points out (190), "It would be a mistake to romanticize traditional island fishermen, to view them as ideal conservationists living in perfect harmony with nature and one another."

For the most part though, it appears that communities that have gained exclusive rights to their adjacent marine resources have typically evolved effective conservation programs. Recently, the potential of "traditional" conservation approaches (customary fishing rights, marine tenure, etc.) has been increasingly explored. R. E. Johannes explains (190):

> Throughout most of Oceania the right to fish in a particular area was controlled by a clan, chief or family. Generally this control extended from mangrove swamps and shorelines across reef flats and lagoons to the outer reef slope. It would be difficult to overemphasize the importance of some form of limited entry such as this to sound fisheries management. Without some control on fishing rights, fishermen have little incentive not to overfish since they cannot prevent others catching what they leave behind. This is a central tenet of fisheries management science.
>
> Under modern conditions the government must assume the sole responsibility for placing and enforcing fisheries conservation measures. This is a difficult and expensive responsibility under the best of circumstances; it is close to impossible in most tropical artisanal fisheries. Typically there are far more boats, more distribution channels (both subsistence and market) to monitor and regulate than in high latitude fisheries of similar sizes. And there is usually much less money and expertise available with which to do it.

But traditional fishing rights sometimes create difficulties for governments or individuals wishing to stimulate new commercial fisheries. One of the most common of such difficulties in the tropical Pacific relates to the taking by outsiders of tuna baitfish from waters under traditional control. Permission to do so must first be obtained from the "owners." This is not likely to be forthcoming if local ownership customs are not understood and respected and prospective bait fishermen are not prepared to negotiate. Problems have arisen in this regard in a number of different Pacific island countries. In every instance, local authorities and tuna fishermen were not familiar in any detail with traditional fishing rights.

There has been a tendency among some reef fisheries managers to try to invalidate traditional systems of fishing rights because of such problems. But such a traditional system of resource management need not be perfect to be better than the alternatives. Moreover, in some areas such as Papua New Guinea, government attempts to invalidate traditional fishing rights would generate strong adverse public reaction.

Legislation that nullifies traditional fishing rights prevents traditional owners from policing reef resources—something they often do voluntarily if their rights are secure. Invalidating these rights thus increases the fisheries manager's responsibilities and places a heavy additional burden on understaffed and underfunded fisheries departments; the fisheries manager disposes of a service he gets for free and assumes new responsibilities his department is often ill-equipped to handle.

Legislation has been introduced in some countries to protect traditional fishing rights. The goal of these efforts is commendable, but one effect may be to freeze what originally were flexible systems of resource allocation and management; fishing boundaries and rights often changed in the past in order to accommodate population shifts or new political or economic conditions.

It is desirable to study traditional fishing rights before any efforts are made to introduce new forms of fisheries management of new types of fishing. The degree to which fishing rights are elaborated and enforced depends upon their perceived value. As coastal populations swell and the demand for fish increases, the value of traditional fishing rights also increases. It is not surprising under these circumstances, to find that fishermen will "invent" traditional fishing rights if there are advantages to be gained from doing so. Information obtained from fishermen on traditional fishing rights and customs is thus liable to be more reliable if it is gained prior to the development of new fisheries. This is also true in the context of developing marine parks in areas encompassing a traditional reef fishery.

In summary, managing a resource involves regulating the behavior of the people whose activities affect that resource. In order to do so effectively, it is necessary to study not only the resource itself, but also the local methods, traditions and knowledge associated with it and its use. Conservation laws and resource management schemes that are compatible with local customs will gain more public acceptance than those perceived by users as alien. Public support is especially important in developing tropical countries because money and personnel available for enforcement are generally minimal.

3.102 TRAINING

Special training is needed in the skills required for ICZM, particularly for fact-finding and managerial personnel who formulate policy and organize program elements involving skills in law, economics, and development management. The place where the land ends is also the place where the knowledge and experience of most administrators ends. Planners, managers, engineers, and politicians alike are usually not well informed about the sea, the sea coast, and sea life.

Coastal resources management is recognized as an interdisciplinary pursuit, not as a distinct discipline. But those who do have knowledge are usually experts in particular subjects—navigation, fisheries, science, or hydrodynamics—who know the sea and seacoast from a specialist's point of view rather than from a systems perspective. Those few who have prepared themselves to know the sea and seacoast as a natural system, who understand the interactions of sea and coastal land, and who know the requirements of management and governance are exceptionally important as trainers.

Training programs should be developed on various aspects of coastal planning at national and regional levels, including seminars and workshops for policy-makers and planners, short-term courses, development of new academic programs aiming at the development of multidisciplinary competencies (e.g., law of the sea, environmental economy, socio-economy, sociology, demography), preparation of training materials (manuals, simulations, audio visual aids) for various strata of the public, including the rural communities (65).

Where so many different types of activities and technologies are involved. Training would range from short courses and on-the-job training to formal university degree courses. International assistance is available to most developing countries to meet a variety of training needs. As in so many fields, the case study method may offer the best training approach (e.g., see Part 4).

In one example, the Coastal Area Planning and Management Division of Trinidad and Tobago (located in the Institute of Marine Affairs) listed its priorities for staff training in ICZM as (52) (a) assessment of sociological impacts; (b) assessment of economic impacts; (c) cost benefit analysis; (d) simulation analysis; and (e) resource management.

Particularly important for ICZM programs are orientation courses for decision makers and agency staff in the various ministries or departments that are affected by ICZM. Seminars, clinics, field trips, and other dialogues will help these agencies to understand the ICZM program and assure their cooperation. Lack of sufficient data and/or trained personnel should not be seen as a barrier to the initiation of an ICZM program. In practice, there may be normally sufficient manpower with skills in basic ecology, economics, planning, and administration to start the ICZM initiative.

The basic requirement is for a small group of 5 to 10 persons from different disciplines to form the nucleus of an ICZM team. Staff from existing agencies can be secunded and assigned to this team on either a part- or full-time basis without undermining the normal work of these agencies. Additional manpower can be made available by placing emphasis in the ICZM program on in-service training and recruitment of trainee staff who will attend ICZM courses before taking their appointments in government.

On-the-job training for in-country counterparts to expatriate experts should always be included in planning and development aid projects. Making this relationship effective as a training experience requires a sincere recruiting effort by the recipient country to provide a qualified candidate who will gain from his/her experience. The expatriate advisor should help establish the goals in training and set up mutually acceptable measures of progress (296).

The independent foreign study tour is often a good approach. Various disadvantages and constraints should be noted. Such study tours are expensive and lose their value if poorly organized, or too general in focus. They are too often inefficient if arranged for a single "trainee" who requires substantial personal attention from host staff who (depending on protocol) must be of approximately equal rank (296). Language problems can be significant; English will generally be the necessary language.

Selection of a particular country or countries to visit will depend on objectives of the study tour and the willingness or ability of a host country to provide access to its areas and staff. Sometimes host countries are willing to provide some contributions in terms of transportation and accommodation (296).

Temporary international transfers and assignments is another approach to ICZM training. A temporary assignment of one to six months in a particular planning office, research group, university faculty or actively managed coral reef reserve, can be the most effective "on-the-job" training method for mid-level staff (296).

However, such transfers must be very carefully considered and planned in advance so that adequate preparations can be made on both sides. For example, language training may be necessary for the "trainee" and a suitably qualified host area and host staff members must be identified. A realistic list of training objectives must be agreed on with a schedule or sequence for reaching those objectives (296).

Usually it is more effective if the transfer can include a specific short-term project. For example, the trainee could join a planning team for the completion of a new coral reef area management plan, or for the duration of a research survey in the new area. The trainee should be asked to make definite contributions to these projects, not just act as an observer. Transfers to an area just to observe "operations in general", or to "hold discussions with key staff" are not very productive to the trainee and of relatively little value to the host staff either, according to Robinson (296).

Another alternative is to return well-qualified candidates within existing staff for a university degree or diploma in marine science or resources management. If a respected curriculum is available within the country, this may well be adequate and is certainly less costly. Overseas university degree programs involve much more cost (US$25,000/year), a foreign language capability, long residence away from home and perhaps less relevant field experience in non-tropical environments. Successful students may, however, be more useful employees of higher status on return. (296)

Formal graduate level training is now available to those who would choose to make ICZM a career. The list below is not of formal curricula; rather it points to places and persons of opportunity for interdisciplinary training in ICZM. Fully satisfactory generic training courses are very limited. Therefore, the student and professor must cooperate in creating a customized training program and each student candidate may have to negotiate his/her own training program with the institute and professor.

1. Duke University
   Department of Geology
   Durham, N.C. 27008

   Contact: Dr. Orrin Pilkey 919-684-4238

   Comment: Dr. Pilkey specializes in coastal geology, erosion, storm impacts, etc.; graduate studies can be arranged with emphasis on the above along with beach conservation and coastal planning and management; environmental theme. Masters and Doctoral programs.
2. Southern Cross University
   Centre for Coastal Management
   PO Box 157
   Lismore NSW 2480
   Australia

Contact: Dr. Peter Saenger
Telephone: 066-230-631 Fax: 066-212-669

Comment: Offers a three-year Bachelor of Applied Science degree in "Coastal Management" (since 1987); complete curriculum covering all important CZM topics; good facilities for both lab and field work; strong faculty.

3. Nova Southeastern University
Oceanographic Center
Institute of Marine and Coastal Studies
8000 North Dania Drive
Dania, Florida 33004 USA

Contact: Dr. Richard E. Dodge
Telephone 305-920-1989 Fax: 305-947-8559

Comment: Offers a Master's degree in Coastal Zone Management (1 1/2 years) with a science orientation.

4. Oregon State University
College of Oceanic and Atmospheric Sciences
Marine Resources Management Program
Oceanography Admin Bldg 104
Corvallis, OR 97331 USA

Contact: Student Services Office
Telephone: 503-737-5118 Fax: 503-797-2064
E-mail: student advisor @ oce.orst.edu

Comment: Master's degree program designed to train professionals for careers in the wise development and management of marine and coastal resources; core courses in basic oceanography complemented with resource mgmt, econ, policy, communications, geography, land use planning, and special courses in coastal and ocean management (degree program since 1974; 13 associated faculty).

5. University of Delaware
College of Marine Studies
Newark, Delaware 19716 USA

Contact: Dr. Robert Knecht or Victor Klemas
Telephone: 302-451-2336

Comment: Dr. Knecht is very experienced in all aspects of ICZM and can be consulted; Dr. Klemas is a top expert in remote sensing and has been guiding foreign students since 1976.

6. University of Miami
Division of Biology and Living Resources
Rosenstiel School of Marine Science
4600 Rickenbacker Causeway
Miami, Florida 33149 305/361-4624

Contact: Dr. Samuel C. Snedaker
Telephone: 305-361-4624 Fax: 305-361-9306
Easylink: 62845425 Telex: 317454

Comment: Master and Ph.D. programs in major branches of the marine sciences; hard science curriculum but conservation elements can be incorporated.

7. University of Newcastle upon Tyne
Centre for Tropical Coastal Management Studies
Department of Marine Sciences & Coastal Management
Newcastle upon Tyne NE1 7RU
United Kingdom

Contact: Dr. Barbara E. Brown or Dr. Alasdair J. Edwards
Telephone: +44 (0) 191-222-6659
FAX: +44 (0) 191 222 7891

Comment: Special one-year Master's Degree program in Coastal Zone Management (started in 1987); a broadly based course incorporating a multi-disciplinary approach to coastal management that includes environmental assessment, socio-economics, and resource management—strong emphasis on case studies in the Central Caribbean, SE Asia, and Indian Ocean where faculty have worked extensively during the last ten years. Solid base in ecological sciences.

8. University of North Carolina at Chapel Hill
Department of City and Regional Planning
CB# 3140, New East Building
Chapel Hill, North Carolina 27599-3140

Contact: Dr. David Godschalk
Telephone 919-962-5012
David J. Brower
Telephone: 919-962-4775

Comment: Coastal Planning specialization within Land Use and Environmental Planning at both Master's and Ph.D. levels. Faculty expertise includes growth management, hazard mitigation, coastal law and policy, environmental processes, and carrying capacity analysis; core coastal environment course taught in Marine Science Department.

9. University of Rhode Island
Marine Affairs Program
Narragansett Bay Campus
Narragansett, Rhode Island 02882

Contact: Dr. Lawrence Juda
Telephone: 401-792-4041

Comment: URI offers two-year and one-year (mostly for mid-career persons) Master of Marine Affairs degrees with a set curriculum. One area of specialization is Coastal Management. There are opportunities to interact with the Coastal Resources Center, which has an international CZM program (write to Stephen Olsen).

10. University of Washington
School of Marine Affairs, HF-05
Seattle, Washington 98195 USA

Contact: Dr. Marc J. Hershman
Telephone: 206-685-2469

Comment: The school offers a broad two-year Master of Marine Affairs program (themes: Coastal Zone Mgmt., Marine Environmental Protection, Marine Policy, Port and Marine Transportation Mgmt, and Fisheries Mgmt.); Ph.D. students associate with the school and interact with other university departments.

11. University of the West Indies
Faculty of Natural Sciences
Department of Zoology
St. Augustine
Republic of Trinidad and Tobago

Contact: Dr. Peter Bacon
FAX: 809-645-7132
TEL: 809-663-2060, 663-2007, Ext. 2094, 3275

Comment: Dr. Bacon, Prof. of Zoology, coordinates a course in Coastal Management. He is senior author of the workbook *Practical Exercises in Coastal Management for Tropical Islands*. He is regional coordinator of the IWRB specialist group on Wetland Restoration and a member of the UNESCO Task Team on the Impact of climate change on Mangrove Ecosystems.

12. Dalhousie University
Marine Affairs Program
Weldon Law Building
6061 University Avenue
Halifax, Nova Scotia

Canada B3H 4H9

Contact: Coordinator
FAX: 902-494-1001
TEL: 902-494-3555

Comment: One-year interdisciplinary program leading to a masters Degree; established in 1986. Subjects include: ICZM, fisheries, law of sea, shipping, environment, tourism, enforcement, resource development.

Additional sources of training in ICZM can be explored by contacting: Ms. C. Casullo, United Nations University, UNESCO, Room M132 - 1, rue Miollis, 75015, Paris, France. Ms. Casullo operates a database on training in coastal and ocean management.

3.103 TRANSPARENCY OF WATER

Sunlight is the basic force driving the ecosystem. It is the fundamental source of energy for plants, which in turn supply the basic food chain that supports all life. Sunlight must be able to penetrate coastal waters so as to foster the growth of both the rooted plants, such as seagrasses, and the suspended algae (or phytoplankton). The limit for plant growth is the depth at which light intensity is reduced to 1 percent of its surface value. Such a limit is close to 70 m in clear oceanic waters near Pacific atolls. Increased turbidity from the addition of excess suspended matter to the water reduces light penetration and depresses plant growth. Estuarine waters are normally more turbid (cloudy) than ocean waters, being both more laden with silt and more rich in suspended life.

In its undisturbed natural state the ecosystem responds and adjusts to the existing levels of turbidity caused by the suspended matter. The organisms in turn adapt to these levels, and the system comes into equilibrium. However, when the system is disturbed and degraded by a sudden addition of suspended matter, its carrying capacity is reduced and its general function declines below the optimum condition.

Increased turbidity may be caused by excavation in estuarine water basins or by the discharge of eroded oil with the runoff from shorelands. Other increases may be caused by excess nutrients derived from land runoff, sewage, or industrial waste discharges that stimulate algal growth and lead to plouding of the water. The excess suspended matter tends to settle out on shallow bottoms, where it may be readily re-suspended by wind action or boat traffic.

The overall solution is to maintain natural condition of the environment by which the ecosystem has evolved and has naturally prospered. Therefore, it is necessary to prevent the addition of unnatural amounts of silt that would block light penetration or of nutrients that would stimulate excessive plankton growth and lead to the same condition.

There are many other reasons to be concerned about transparency, including the following: (a) coral reefs prosper best in the clearest waters—transparency of 13 m, or 40 feet, and better (sometimes more than 50 m), and they will deteriorate in turbid water; (b) clear waters are aesthetically pleasing, usually healthy, and attractive to visitors; (c) sport divers and underwater photographers are attracted to the clearest waters; and (d) seagrasses grow better in clear water. (See also "Suspended Particulate Matter" and "Sediments and Soils" above and "Turbidity Measurement" in Part 2.)

3.104 TSUNAMIS

Long-period gravity waves generated by such disturbances as earthquakes, landslides, volcano eruptions, and explosions near the sea surface are called *tsunamis.* The Japanese word *tsunamis* has been adopted to replace the expression *tidal wave* to avoid confusion with the astronomical tides. Most tsunamis are caused by earthquakes that extend at least partly under the sea, although not all submarine earthquakes product tsunamis. Severe tsunamis are rare events (362).

A tsunami may be compared to the wave generated by dropping a rock in a pond. Waves (ripples) move outward from the source region in every direction. In general, the tsunami wave amplitudes decrease but the number of individual waves increases with distance from the source region. Tsunami waves may be reflected, refracted, or diffracted by islands, sea mounts, submarine ridges, or shores. The longest waves travel across the deepest part of the sea as shallow-water waves and can obtain speeds of several hundred kilometers per hour. The traveltime required for the first tsunami disturbance to arrive at any location can be determined within a few percent of the actual traveltime by the use of suitable tsunami traveltime charts (362).

Tsunamis cross the sea as very long waves of low amplitude. A wavelength of 200 kilometers (124 miles) and an amplitude of 1 meter (9 feet) is not unreasonable. The wave may be greatly amplified by shoaling, diffraction, convergence, and resonance when it reaches land. Seawater has been carried higher than 11 meters (36 feet) above sea level.

The tsunami appears as a quasi-periodic oscillation, superimposed on the normal tide. The characteristic period of the disturbance, as well as the amplitude, is different at each location. It is generally assumed that the recorded disturbance results from forced oscillations of hydraulic basis systems and that the periods of greatest response are determined by basin geometry (362).

On April 1, 1946, a tsunami struck the Hawaiian Islands with runup in places as high as 17 meters (55 feet) above sea level. Because a tsunami is of short duration, extensive beach changes do not occur, although property damage can be quite high (362).

3.105 TURTLES

Seven of the world's eight species of sea turtle are in trouble, and human activity seems to be their number one problem. Many human activities on the beach and along the seacoast are quite harmful to sea turtles.

At nesting time the female hauls herself high up the beach to find a patch of coarse, dry sand. Here she digs a hole about 2 feet deep and deposits within it about 100 small white eggs. She then covers the nest with sand and returns to the sea. She may repeat this 5 or 6 times per season.

About 50 days after deposit, the young break through their shells and dig their way up and out of the sand. Once on the beach, the 2-inch hatchlings orient to the night sky and scramble toward the horizon of the sea. But bright street lights or house lights can turn them 180 degrees around to scramble in the opposite direction—into human settlements—never to reach the sea.

Conservation needs for sea turtles are well known. *First,* every effort should be made to prevent disturbance to the nesting females. *Second,* do not allow heavy vehicles to be driven over the nests. *Third,* discourage the removal of eggs. *Fourth,* prohibit killing of turtles. *Fifth,* during the nights when hatchlings dig out and scramble to the sea, encourage commercial operators, municipal authorities, and householders to redirect their outside lights or shield them (with rimshades or tree plantings) so they don't shine toward the beach. *Sixth,* control curiosity seekers during egg laying times. In tourist areas, the beaches may be crowded with curious people at night who wish to observe the nesting phenomenon. Their behavior must be controlled.

Seventh, reduce plastic litter going into the sea—hundreds of turtles die each year, their intestinal tracts obstructed by plastics they have ingested. Many more hundreds appear to die invisible deaths at sea because of plastic floatable litter such as plastic bags and sheeting, styrofoam, synthetic rope and line, balloons, and sixpack holders that collect in oceanic "driftlines" or "weedlines." The three main solutions to the plastics problem are: (a) switch to non-plastic packaging; (b) develop new, truly biodegradable, plastics; or (c) prevent littering the ocean with plastics. The first two are up to corporations and government, but the third is up to individuals.

Eighth, preserve ocean beaches where turtles nest, shoreline development has reduced available nesting beaches in many countries by altering or eliminating nest sites (see "Beach Management").

Turtle conservation is a complex matter, involving as it does, controlling development and use of coastal beaches. ICZM is a useful approach because it deals simultaneously with a whole variety of coastal issues, protecting turtles while accomplishing many other goals (Figure 3.133). But regulatory means must also be combined with field conservation activities such as rescuing and protecting turtle eggs where people or natural enemies (e.g., racoons) might harm them (Figure 3.134).

Figure 3.133. El Salvador is planning an integrated coastal management program that includes turtle conservation. (Photo courtesy of Environmental Guidance Group.)

3.106 UNDERWATER FISHING

Each of the several types of underwater fishing has conservation problems. One example is the capture of small fish for personal aquariums. The use of sodium cyanide (NaCN) for the capture of marine aquarium fish became popular 30 years ago. Using cyanide squirted from plastic detergent bottles, divers in the Philippines douse coral head refuges from dawn to dusk to get live aquarium fish. About 50 percent of the exposed fish died on the reef. Other fish were stunned, but recovered if transferred to clean water. About 10 percent of the exposed fish are the colorful species in demand by aquarists. Cyanide kills fish fry and invertebrates. Sea anemones appear to be scarce in areas where the use of NaCN is common (297).

Wafar (373) suggests the following guidelines to limit the collecting of "decorative" fish ("aquarium fish", "tropical fish") in the Lakshadweep Islands of India:

- Marking off the reef areas for this use. This shall not cover more than 10 percent of the whole reef, so as to provide for other users.
- License may be issued to only one operator for each island, considering that the reef area is not greater than a few km^2 in each case. The permit is subject to the condition that the operator has all infra-structural facilities for holding the aquarium fishes without mortality prior to shipment.
- Permit may be issued to one operator to employ not more than 5 field collectors, a figure that is again related to the size of the reef. This is also subject to the field collectors being instructed properly on the collecting techniques least destructive to the reefs.
- Banning of all destructive fishing methods (poison, dynamite) with provision of penalties for infringements.

- Listing of the species that can be caught, and the quotas for the numbers and size of fish in each species. These are to be determined based on species occurrence and abundance surveys and the biology of individual fishes. As a general rule, up to 40 percent of the fish can be allowed to be harvested in one season. This figure is tentative, and is based on the consideration that recolonization of the vacant territories by fish from adjacent zones, and new recruitment can easily replenish the stock.
- Ban on entry of other users in this zone.
- Closure of fishing for any one species when it corresponds with the spawning season of the fish.

In the Philippines, destructive fishing methods such as blast fishing, muro-ami, and kayakas are also destroying the reefs. It is estimated that 10 percent or more of the 600,000 artisanal fishermen use explosives in the Philippines.

About 2,500 fishermen use NaCN to capture aquarium and food fish. NaCN exposure causes internal damage to the fish. Cyanide causes liver and kidney damage and degeneration of the reproductive organs. It can damage gametes, which can affect the survival and growth of progedy. It is likely that cyanide in the blood blocks the action of enzymes involved in respiratory metabolism and causes neurological damage. HCN is quickly absorbed across the gills into the blood, where it is converted to thiocyanate ion (SCN) by an enzyme called rhodanese. Fish dosed with as little as 0.3 mg/l of HCN for five minutes die several weeks to months later. Fish which escape capture after cyanide exposure may also die on the reef (138).

Interviews with fish collectors, exporters, importers, wholesalers, and retailers have shown high mortality. More than 80percent of the fish collected die before the survivors are sold to marine fish hobbyists. Most of the fish collected with cyanide die within 2 months (138).

In the hands of an experienced diver, a spear can be a highly selective and efficient method of collecting reef fishes (Figure 3.135). This selection operates in three ways. First, it can target large individuals of species which may ignore baits at certain times. In some circumstances aggregations of

Figure 3.134. A turtle hatchery on the beach at Terengganu, Thailand, in 1987—each stake represents one laying of eggs (about 100) by a leatherback female turtle. But, by 1992, fewer than 30 nesting females were seen here, and local extinction is now an expected outcome—the result of a conservation initiative that came much too late. (Photo by author.)

Figure 3.135. Spearfishing provides a welcome source of income for artisanal fishermen of Palau. (Photo by author.)

carnivorous reef fishes have been observed which did not take baits but could be collected by spearing. Second, it can target species which are rarely, if ever, taken by line. Third, unlike line fishing, spearing targets fish almost exclusively in shallow waters. These properties of selectivity and the efficiency of many divers make it an excellent collecting method, but one that does not mix readily with the activities of other reef users such as tourists (138).

Artisanal spearfishing guns are often carved from wood, and powered by large rubber strips released with a trigger. The spear is often a metal rod sharpened at one end. The spearfishers use small round goggles made of window glass, wooden frames for each eye and rubber strips. These goggles can cause considerable eye damage when used at greater than a few meters depth because they cannot be equalized through the nose to compensate for rising and falling external pressures. In the Philippines, the divers often use a single, rigid, wooden, shield-like paddle, attached to one foot, to assist them in swimming. Divers traditionally use rocks to assist them in sinking to great depths (30–60 m) rapidly, often resulting in considerable ear damage (242).

By actually diving in the underwater environment in which they fish, a spearfisherman grows to appreciate the beauty and variety of the reef and its inhabitants. Like line fishing, spearfishing is now generally recognized as a legitimate form of recreational fishing and is regulated in many countries. But spearing can clearly have a detrimental effect on the type and quality of experiences enjoyed by other reef users and is, therefore, inappropriate in areas supporting large numbers of recreational divers and other users of the shallow water environment.

Fishes in many areas are already subject to high levels of exploitation. Spearfishing provides another level of exploitation in addition to line fishing and it can target groups not normally caught by line. More importantly, spearfishing is restricted to shallow waters and is often carried out in sheltered reef areas with good visibility. Under these circumstances it will overlap with many other types of reef activity. These include a range of things done in shallow water, including research programs, photography and exploratory diving, especially by newcomers and tourists. Most of these activities involve long periods of passive observation of fishes. Under these circumstances, large fishes become habituated to the presence of divers (138). Direct experience has shown that even a few incidents of spearing in such areas results in characteristic diver-avoidance by fishes.

One solution is to assign separate zones for spearfishing and for visual exploration. In one recorded case (Looe Key Sanctuary, So. Florida, USA), the various species increased by two- to fivefold or more. The fish on Looe Key reef are also tame and approachable now (85).

A growing number of underwater fishers use air compressors with long hoses to facilitate underwater harvesting. The most prominent uses are for spearfishing, lobster gathering, aquarium fish catching and, recently, for sea urchin gathering. The air compressors are of the type commonly used for filling tires at gasoline stations (242). The compressor should involve an air intake extending several meters upwind of the compressor and a series of filters to remove particles from the air.

These precautions are necessary because concentrations of gases such as carbon monoxide and carbon dioxide, which have negligible effects at sea level, can become fatal if inhaled under pressure during a dive. Both gases are produced by the compressor itself, as well as by boat engines. They cannot be filtered out of the air under most conditions, and must be carefully avoided. Oils which enter the compressor can cause lipo-pneumonia as they accumulate in the lungs (242).

Ascents and descents must be carefully regulated to avoid ear and sinus damage and similar injuries (Figure 3.136). Nitrogen narcosis often leads to diving accidents, particularly in waters below 30 m, because euphoric feelings it induces cause judgment to be altered. The use of medical drugs, alcohol, or tobacco smoke in the 12 hours prior to a dive can lead to difficulties during the dive. Underwater times and depths must be strictly limited to prevent decompression sickness, leading to paralysis or death. Frequency of diving is limited to avoid degradation of the bone marrow. Divers must not hold their breath during emergency ascents, which leads to lung bursts, producing emphysema and air embolisms (a frequent cause of death). (242)

The compressor divers of Bolinao (Philippines) often dive in depths of 30 to 60 m for a few hours at a time. They are susceptible to all of the above-mentioned diving hazards. Death or paralysis from decompression sickness is common because the divers routinely exceed the so-called "no decompression" limits used by knowledgeable divers. The air from compressors is laden with oils and dangerous gases. Most spearfishing on the reef slope involves the use of air compressors, such as those used in vulcanizing shops and gasoline stations. The unfiltered air passes a small reserve chamber which provides a final breath of air when motor trouble stops the compressor. The air then passes through a long tube to the diver, who uses the air without a regulator. The divers frequently stay at depths below 30 m for hours at a time and are frequently crippled or killed by decompression sickness and other diver-related maladies. (242)

Banning compressor diving would have a beneficial effect on fisheries ecology in the Bolinao area. There is currently a rapid decline in the number and diversity of fishes reaching adult sizes on the reef slope. This decrease includes a 50 percent drop in species richness and an 80 percent drop in abundance in three years. A ban on compressor diving would serve to curtail this decline and help safeguard the

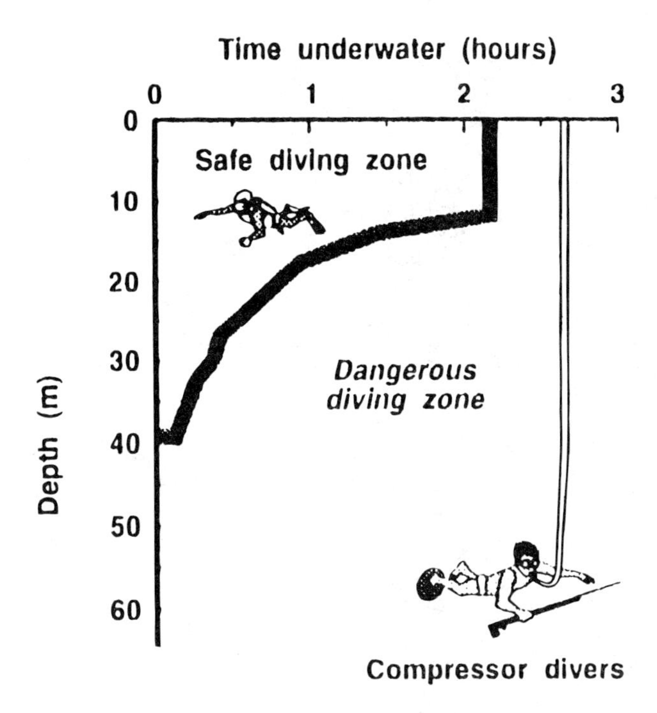

Figure 3.136. "Compressor divers" (without air regulators) often dive too deep and stay too long—the above curve suggests some limits to safe diving. (*Source:* Reference 242.)

breeding populations, which help to supply the coastal reefs along Western Luzon with annual recruits. Thus, a ban on compressor diving would be beneficial for both resource management and humanitarian reasons. (242)

3.107 WATERSHEDS AND UPLAND EFFECTS

A watershed is a geo-hydrologic basin defined by the boundaries of its water catchment area. It is nearly always drained by a river course. Watersheds include all the land, water, and biotic resources within their hydrologic boundaries and behave as ecological units. A watershed's boundaries can easily be demarcated and all its components linked, just like islands.

Water is the factor that integrates the function of a watershed. However, its quality and quantity are themselves regulated by other components of the watershed. A watershed has many subsystems that function within the context of a larger system. All subsystems are connected to each other somehow. When one subsystem changes, all others are affected. In some instances changes are insignificant, but on occasions they can be radical.

Major disturbance of coastal ecosystems from development in the hinterlands includes drainage water pollution, alteration of the runoff water cycle, and loss of coastal wetlands. Sedimentation, nutrient enrichment, and inflow of toxic substances—the principal pollutants—are caused by soil fertilizers, animal wastes, biocides and waste oil carried into tidal waters with surface runoff and stream flow from coastal watersheds from watersheds that drain to the coast. Soil erosion is often the leading problem.

In the United States, the average erosion rate (in English tons/acre) for various land capability classes is cropland, 4.8; rangeland, 3.4; pastureland, 2.6; forestland, 1.2. A relative ranking of sediment yields per year from seven source classes in the U.S.A. shows that agriculture yields four times more sediment to public waters than any other erosion source (59):

Class	Rank
Cropland	200
Construction	50
Disturbed forests	40
Active surface mines	40
Abandoned mines	20
Grassland	10
Undisturbed forest	1

These eroded materials carry excess nutrients and toxic substances into water systems, causing water turbidity and the sedimentation of streams and bays. Many bays and harbors have been seriously degraded by sedimentation, and some require continuous dredging, at a high annual cost to the nation having this problem (Figure 3.137).

Modern farm practices that depend heavily on the use of fertilizers and biocides have greatly increased the potential for disturbance of coastal ecosystems. On the other hand, the accelerating cost of nitrogen fertilizers has decreased nitrogen use. Animal wastes runoff may also be a problem.

Disruption of the natural runoff pattern—its quality, volume, and rate of flow—is caused by diking, drainage, irrigation works, dams, and clearing of natural vegetation. Proximity to coastal waters and to tributary streams is a major factor controlling the severity of the impacts.

Reducing irrigation needs is often beneficial, because diversion of river water can alter the river flow rate or reduce water supply to coastal water basins, thereby upsetting salinity, nutrient and circulation patterns.

Certain pesticides and other biocides have a well-documented potential to harm particular species of animals—fish, birds, and mammals—and to degrade ecosystems. Among the most troublesome of pesticides are the following 12: chlordane-heptachlor; chlordimeform; DBCP; DDT; the "drins" (aldrin, dieldrin, endrin); EDB; lindane; paraquat; ethylparathion; pentachlorophenol (PCP); toxaphene; and 2,4,5-T. As defined by Lugo (225),

> The simplest management practice is to prevent accidental or deliberate introduction of pesticides into surface waters and to restrict the use of chemicals known to be toxic to estuarine organisms. Controlling soil erosion and storm runoff will also help to prevent the introduction of polluting pesticides into the ecosystem.

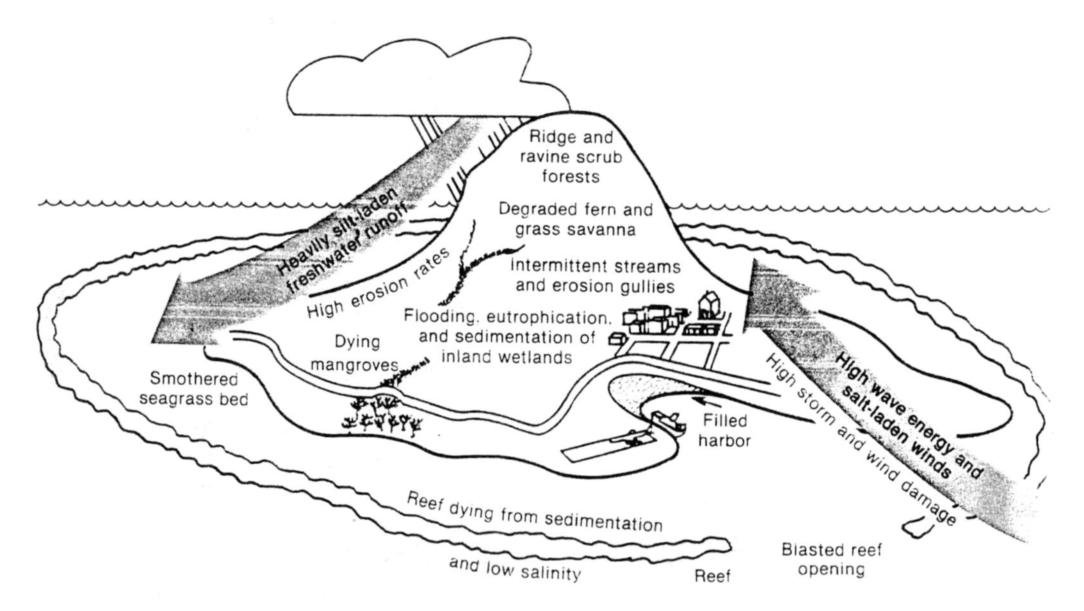

Figure 3.137. Excessive soil erosion is a most severe problem in many coastal zones. (*Source:* Reference 272.)

Conservation practices, including water management measures and soil erosion controls, are needed to alleviate pollution from contaminants carried into water bodies with sediment. Eutrophication (nutrient overfertilization from nitrates in runoff) and soil erosion have introduced vast layers of ooze (suspensible sediments) into the bottoms of many coastal water basins. Channelization and drainage of low-lying shoreland areas, particularly, cause the discharge of large quantities of fine silt into coastal waters (Figure 3.138).

There is a tendency on the part of some growers to over-fertilize farmlands to be sure that a nutrient deficiency does not limit yields. The elimination of excessive application of nutrients is a primary measure recommended for the control of nutrient pollution. Environmentally, nitrogen is usually the key nutrient for algae production in brackish water. Estuarine system are therefore nitrogen sensitive. It follows that nitrogen pollution must be controlled to keep estuarine ecosystems functioning correctly.

In agriculture, soil erosion can be controlled by technical improvement such as improved soil treatment, tillage methods, and timing of field operations or control practices such as terracing, contouring, or water control.

Direct surface runoff volumes can be reduced by measures that (a) increase infiltration rates; (b) increase surface retention in retention storage, allowing more time for water to infiltrate into the soil; or (c) increase the interception of rainfall by growing plants or residues. In many situations erosion can be controlled with agronomical practices that make better use of crop residues or by improved crop systems, seeding methods, soil treatments, tillage methods, and timing of field operations.

Infiltration rates, as well as detention storage, are raised by dense vegetation cover, increased mulch or litter, a high level of soil organic matter, good soil structure, and good subsurface drainage. Improvements in fertility levels and management practices that increase vegetative cover result in lower direct runoff.

Surface storage can be substantially increased by contouring or contour furrowing, graded terracing, and level terracing. Generally, farming parallel to the field contours will reduce erosion. However, when slope length and steepness are great or the area from which runoff concentrates is excessive, such

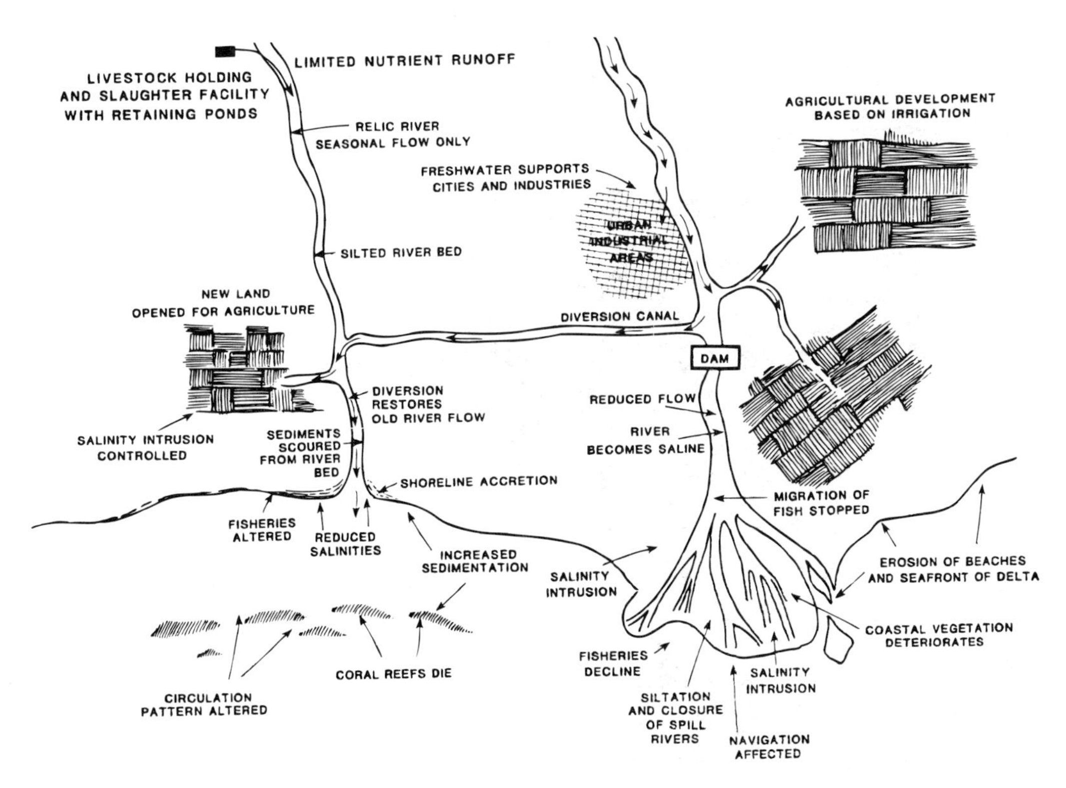

Figure 3.138. The various impacts of agriculture in the hinterlands on coastal zone environments. (*Source:* Reference 316.)

control practices become ineffective and must then be supported by others, such as terrace systems, diversions, contour furrows, contour strip cropping or water control.

Forest managers must understand the complexities of watershed function if they are to play their role as watershed managers properly (Figure 3.139). Unfortunately, there is incomplete understanding of watershed function, particularly in the tropics. It is critical that forestry programs maintain as strong a commitment to watershed research as they do to management. Only through research and a willingness to change management prescriptions when necessary, will we be able to live up to the responsibility bestowed upon us by society (225).

Foresters, through the management of forest lands, can affect the value of a large number of watershed subsystems (e.g., water, wildlife, soil, agriculture, economic, social, etc.) (Figure 3.140). Managing lands with a watershed perspective uses different criteria than managing the same lands for

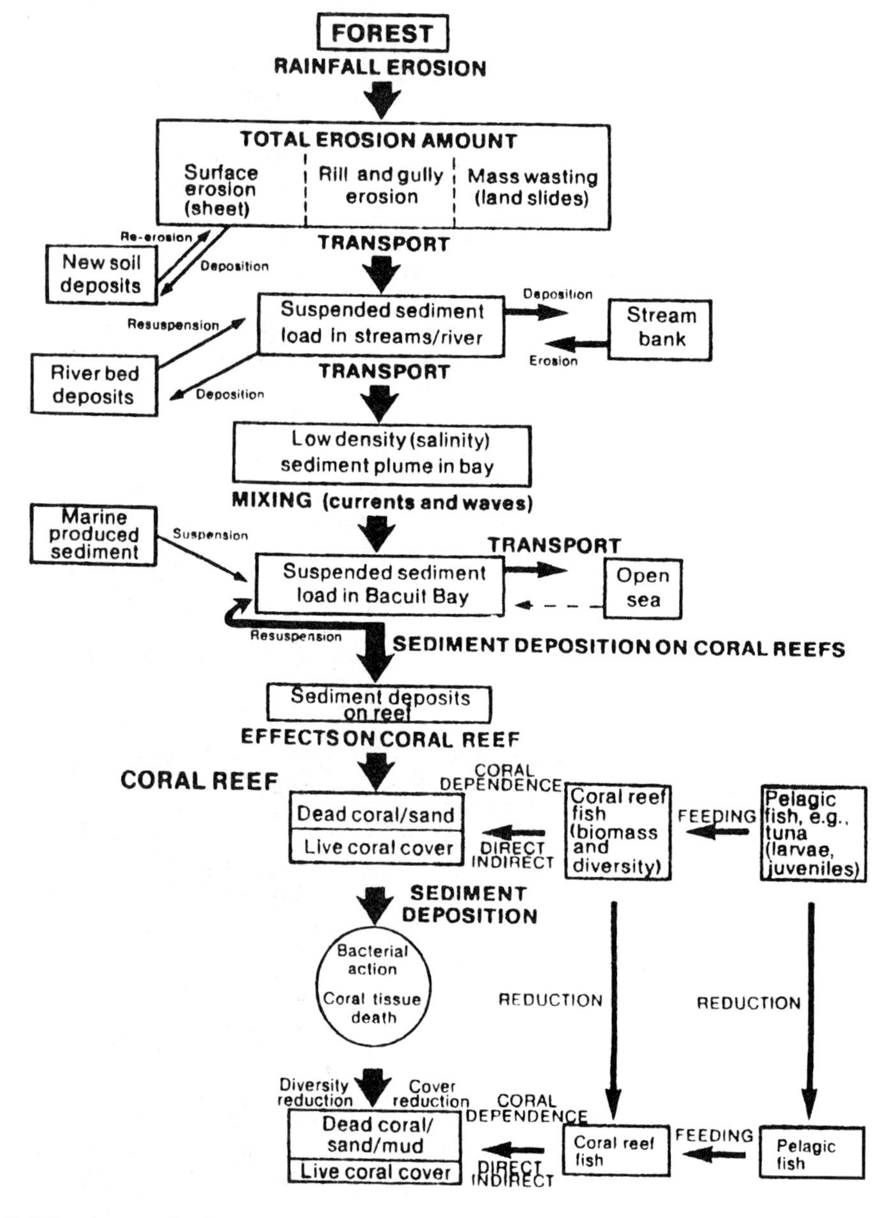

Figure 3.139. A generalized pathway of soil eroded from the forest floor as it is transported to coastal waters. (*Source:* Reference 170.)

Figure 3.140. Coastal managers may find it necessary to persuade other authorities to control uses in coastal watersheds and the hinterlands in general. (Photo by Steve Forsyth.)

a wood yield only. It does not follow, for example, that maximum wood yield will also produce maximum water quality and quantity, or maximum abundance of wildlife. In fact, maximizing wood yield may reduce water quantity, could deteriorate water quality, or reduce the diversity of wildlife.

In forest-harvest activities, both clear-cut areas and logging roads cause increased rates of water runoff and soil erosion. In clear-cut areas, terracing, composting, mulching, and fertilizing help species planted for erosion control to prosper and, by aiding the restoration process, reduce sediment output. Logging trails and roads should be properly located and designed and immediately reseeded to speed the restoration process. (58)

In the recent past, water development specialists tended to allocate water usage based on the most obvious economic and domestic demands, but failed to take into account the key role of fresh water discharge in the maintenance of coastal estuaries and their fisheries (e.g., the serious depletion of fisheries after damming the Nile in Egypt and the Indus in Pakistan). Large dams and their storage reservoirs are major interventions into the land drainage system and are usually adverse to the biodiversity and productivity of coastal ecosystems (Figure 3.141).

The amount of sediment eroded from areas undergoing urban development can be far greater than from any other major land use. Urban construction can increase the natural process of soil erosion in two ways: by removing vegetation whose roots, leaves, and litter stabilize soil; and by removing topsoil, grading slopes, and leaving large expanses of bare land exposed to wind and running water.

Construction sites generally pose a higher potential threat of sediment runoff than the sites of other major land activities. Runoff flow from construction sites often carries enough sediments, toxic materials, nutrients, coliform bacteria, and other undesirable matter to pollute coastal waters.

Urban runoff can be very polluted and damaging to estuaries and shallow coastal waters. The following are typical constituents of urban runoff water amounts in mg/l (366):

Suspended solids	227.0
BOD	17.0
Nitrogen	3.1
Phosphate (hydrolyzable)	1.1

Vegetated buffer strips and such systems as sediment basins can provide sound erosion control for on-going construction operations by detaining runoff, trapping sediment, and preventing increased turbidities in adjacent water bodies. Erosion-control techniques can be divided into three functional types: (a) entrapment of eroding sediments with vegetated buffer strips and sediment-detention ponds; (b) diversion of runoff from likely erosion areas through grading, diversion cuts, and grassed waterways (swales); and (c) prevention of soil movement and erosion, including the use of such methods and reseeding, mulching, and placing of special netting over exposed soils. Your community should implement controls of this sort for all watercourses. (58)

In integrated watershed management, what appears to be common sense is not necessarily correct, and there is a wide gap between available knowledge and practice of watershed management. The following general guidelines are recommended (155):

- Protect a buffer strip at least 50 m on each side of stream banks because these buffer zones are critical areas in the control of water quality.
- Identify and protect areas susceptible to massive erosion (e.g., steep slopes of a watershed).
- Protect cloud forests because they provide a major water input to the watershed.
- Focus attention on roads, ditches, and logging tracts; these are the uses that promote the most erosion in watersheds.
- Minimize non-absorbing surfaces.
- Re-vegetate critical areas first.
- Build management and maintenance into changes of land use.
- Always do research to learn what you are really doing and its consequences.

Figure 3.141. Coastal managers should try to influence the design and operation of dams for flood control, irrigation, and hydropower. (Photo by author.)

Management of watersheds must be well integrated. Wildlife, trees, agricultural crops, water, and urban subsystems within a watershed are all managed by different entities working in isolation from one another, but significantly affecting each other's job. Society, which owns all the subsystems of a watershed, benefits or suffers from the consequences of uncoordinated watershed management (Figure 3.142).

Two simple guidelines designed to make current land clearing and land use practice compatible with the preservation of coastal ecosystems perpetuation of coastal resources are (a) freshwater should enter estuarine waters with a quality, volume, and rate of flow equivalent to the pre-existing natural pattern; (b) for cleared land, sediment loss should be held to a maximum of 15 percent greater than its natural amount; and (c) wetlands are vital habitat areas that require protection and should be exempted from cultivation, leveling, drainage, and impoundment and from roadways and structures.

Management of watersheds must be well integrated. Wildlife, trees, agricultural crops, water, and urban subsystems within a watershed are all managed by different entities working in isolation from one another, but significantly affecting each other's job. Society, which owns all the subsystems of a watershed, benefits or suffers from the consequences of uncoordinated watershed management.

The landward side of the coastal zone which will be likely to have impacts on the coastal zone may already be subject to existing controls. The preferred solution for the coastal agency is not to burden itself with direct control of watershed activities. But the terrestrially oriented authority controlling watersheds should cooperate with the coastal authority to protect the coast from impacts originating upland. Even if it were politically acceptable to intercede directly, the coastal agency would be unlikely to have the staff and skills to deal with the wider range of environments and management demands involved.

At the least, the coastal management agency must interact with terrestrial authorities at the coastal zone boundary. Even if a wide "buffer zone" is placed along the boundary, coastal waters can be influenced by cross-boundary environmental impacts (e.g., polluted runoff) or biological effects (e.g., migrating species). Therefore coastal managers must get involved in inland matters and be aware of activities that could affect the areas of their jurisdiction. This would include the outer "Zone of Influence" (beyond any formal buffer zone) in which coastal managers must exert a positive input into decision making.

Figure 3.142. Wildlife prosper in natural wetlands, often the only habitat where species can survive. (Photo by author.)

3.108 WAVES

Waves are not only a major factor affecting the coast, but also a mysterious force. The size and form of waves are not so easy to predict once they reach the shallow waters of the coastal edge, the coral reefs, beaches, and tidal flats. But in the deeper sea, waves obey simple natural laws so well that one can easily predict their characteristics, as shown in the following:

Wave period (seconds)	Wave length (feet)	wave velocity	
		ft/sec	miles/hr
6	184	30.5	21
8	326	40.6	28
10	512	51.0	35
12	738	61.0	42
14	1,000	71.5	49
16	1,310	92.0	56

The change in form begins when the wave reaches a depth that is less than half the wave's *length.* The final change occurs when the wave reaches the shallows—where depth is less than 1.3 of wave height—when the wave is "tripped" and it break against the shore (Figure 3.143).

As storm waves travel from deep water into shallow water, they generally lose energy even before breaking. They also change height and direction in most cases. The changes may be attributed to refraction, shoaling, bottom friction, percolation, and nonlinear deformation of the wave profile. Waves arriving at the shore are the primary cause of sediment transport in the littoral zone. Higher waves break farther offshore, widening the surf zone and setting more sand in motion. Changes in wave period or height cause sand to move onshore and offshore. The angle between the crest of the breaking wave and the shoreline determines the direction of the longshore component of water motion in the surf zone and, usually, the longshore transport direction. For these reasons, knowledge about the wave climate—the combined distribution of height, period, and direction through the seasons—is required for an adequate understanding of the littoral processes of any specific area (see also "Littoral Drift" above).

Refraction is the bending of wave crests due to the slowing down of that part of the wave crest which is in shallower water. As a result, refraction tends to decrease the angle between the wave crest and the bottom contour. Thus, for most coasts, refraction reduces the breaker angle and spreads the wave energy over a longer crest length.

Shoaling is the change in wave height due to "conservation of energy flux". As a wave moves into shallow water, the wave height first decreases slightly and then increases continuously to the breaker position, assuming friction, refraction, and other effects are negligible, but bottom friction is important in reducing wave height where waves must travel long distances in shallow water. In general, bottom hydrography has the greatest influence on waves traveling long distances in shallow water. Because of

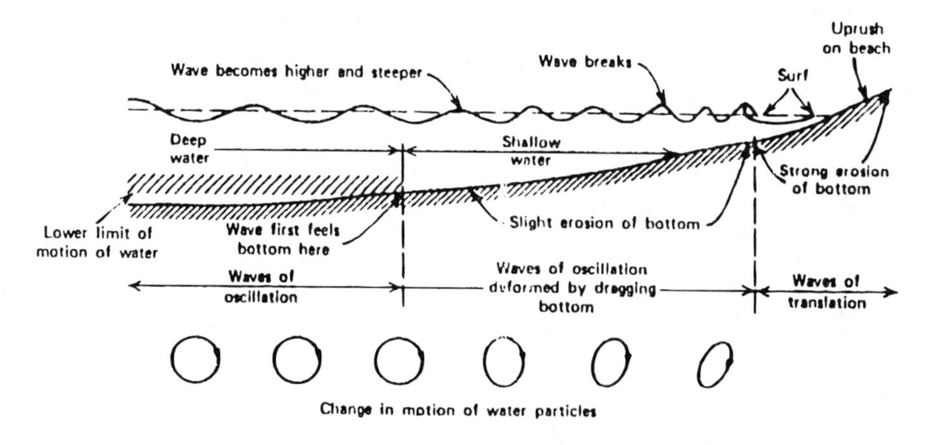

Figure 3.143. The theory of breaking waves. (*Source:* Reference 362.)

the effects of bottom hydrography, nearshore waves generally have different characteristics than they had in deep water offshore (362).

Nonlinear deformation causes wave crests to become narrow and high and wave troughs to become broad and elevated. Severe nonlinear deformation can also affect the apparent wave period by causing the incoming wave crest to split into two or more crests.

The winds build waves on top of the storm surge, which as they strike the coastline can increase flood elevations by as much as 55 percent over the storm surge level. In an open ocean setting, wave heights of 40 to 60 feet and higher have been observed by seamen and operators of offshore oil rigs, but these wave heights are achieved because of the great depths of water beneath them (a 3-foot wave needs at least 4 feet of water depth beneath it to be sustained; a 6-foot wave needs 8 or 9 feet, and so on). As the wave enters shallow water the sea bottom both slows the submerged portion of the wave and reduces the wave height that can be sustained, finally resulting in the breaking of the wave.

While the wave energy is quickly dissipated as the wave strikes the coastline, beaches, dunes, vegetation, and structures built in the wave zone absorb this energy. At the boundary of land and sea, beaches and low-lying dunes may be scoured by waves and the scoured sand washed overland hundreds of feet, or the sand may be deposited seaward, where in decreasing water depths the beach acts to trip the waves.

By their yielding to waves, beaches are efficient dissipators of wave energy: if the sand is deposited seaward, water depth is decreased and storm waves break farther from shore; if the sand is deposited landward as overwash, the beach profile is steepened, potential water depths reduced, further reducing the size of destructive waves.

In this context, dunes contain very important sand supplies. If removed by sand mining or to remove obstructions to ocean views, the risk to coastal development behind the former dunes is greatly increased, for the reasons previously stated. Similarly, mangroves serve to dissipate wave energy and to stabilize the land areas behind them from the erosive forces of storms. Reefs, beaches and dunes, and mangroves are among the most important natural defenses against ravages of wave action on the coast. Reefs also act to trip waves.

The reach between the reef and the shore frequently will be too short and too shallow to permit waves to build to the heights they reached prior to striking the reef. So important are reefs that many countries have special reef conservation programs (e.g., Sri Lanka organized a nationwide management program to protect reefs and save their S.W. shoreline) (see "Sri Lanka" case history in Part 4).

To predict waves one must usually predict wind. Wind waves grow as a result of a flux of momentum and energy from the air above the waves into the wave field. Most theories of wave growth consider that this input of energy and momentum depends on the surface stress, which is highly dependent upon windspeed and other factors that describe the atmospheric boundary layer above the waves. Winds for wave prediction are normally obtained either from direct observations over the fetch, by projection of values over the fetch from observations over land, or by estimates based on weather maps (362).

In predicting winds, much more meteorological data than wind speed and direction are shown for each station on an actual chart. This is accomplished by using coded symbols, letters, and numbers, placed at definite points in relation to the station dot. A sample plotted report, showing the amount of information possible to report on a chart, is shown in Figure 3.144. Not all of the data shown on this plot are included for each prediction, and not all of the data are plotted on each map.

3.109 WETLANDS

This section deals with concepts controlling the management of *vegetated intertidal habitats,* mangrove forests and marshlands (detailed descriptions in specific sections). The purpose is to clarify the definition of wetlands and implications of the concept. Other specific littoral habitats, such as, sea grass beds, coral reefs, and tideflats. Some of the most difficult resource conflicts of the coastal zone arise from conversion of coastal wetlands and lowlands from natural habitats to other uses. The natural values of wetlands are such that preservation of their functions usually precludes development within them (Figure 3.145).

Taking agriculture as an example, the upland economic planner it may think it logical to drain and plant the lowlands and wetlands, until he discovers that the potential of soils in coastal lowlands is

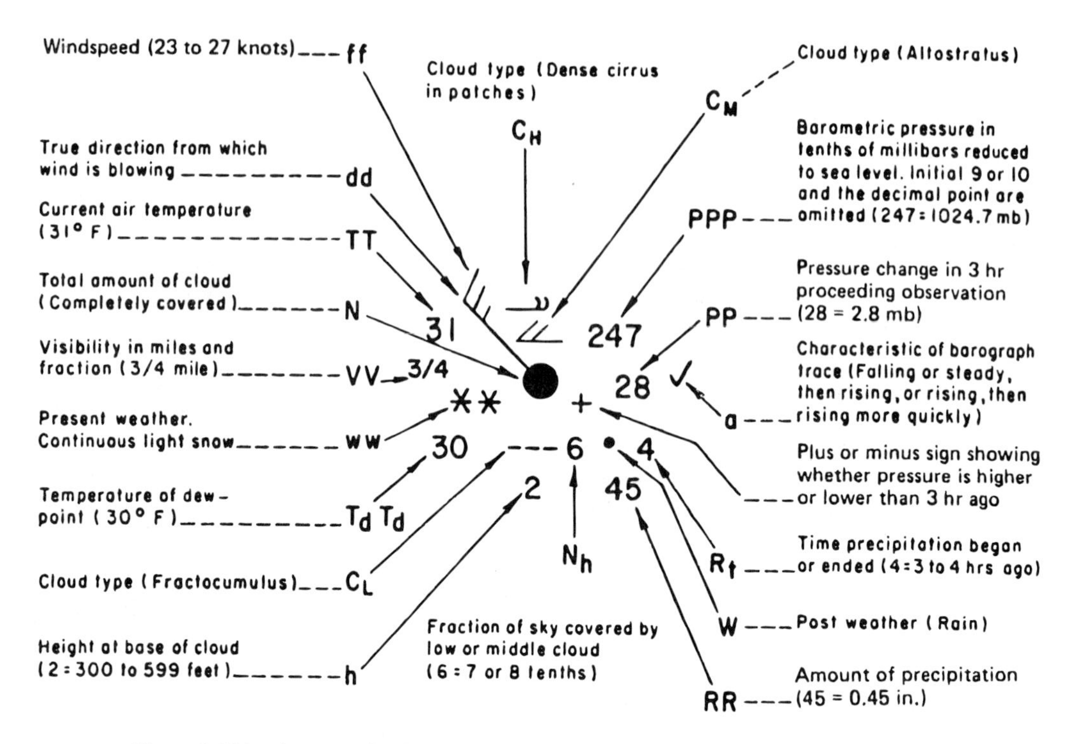

Figure 3.144. An example plotted report for wind conditions. (*Source:* Reference 362.)

usually quite low because peaty conditions so often cause acid-sulfate soils and because there is a high incidence of flooding. One must be very cautious with mangrove swamp or brackish transition swamp/marsh conversions to agriculture. For example, Burbridge and Stanturf (36) see agriculture on the Southeast Asian coast as a risky business, particularly in making careful decisions about using soil sites that require continuous high levels of management to achieve socio-economic benefits. Such floods pose serious risks to farmers (36) and can overwhelm an estuary or lagoon with pulses of sediment, organics, and chemical pollutants.

A variety of difficulties will be faced with wetland conversions to other uses such as industrial, commercial, or residential. For example, large investments are needed to overcome the natural forces which have created these lowland, wetland areas or the natural hazards which will continue to influence their use. And there is also a danger that hazards such as periodic flooding will be exacerbated by poor management of clearing and draining in coastal lowlands leading to flash floods (36). Also, lowland/wetland developments have potentially harmful side effects (e.g., the release of agricultural chemicals into riverine and coastal waters).

Coastal wetlands are found in the transition zone between land and sea. They lie within that part of the land-sea interface that is alternately flooded and exposed by tides and storm action. Some wetlands exist along exposed shorelines while others are found back in lagoons or on up tidally influenced rivers (Figure 3.146). They may connect with non-tidal dry coastal lowlands.

Some persons still consider wetlands as wastelands. This attitude is a holdover from olden times when people needed to believe there were special places on earth where evil was bred—what better candidate than an impenetrable and mysterious mangrove swamp. But in this century, wetlands have been more the victims of technological bravado than superstitious belief. Places that were neither land nor water challenged our inventiveness. We couldn't drive across them or boat through them with any ease. We couldn't easily farm them nor build upon them. The solution was to convert them to either land or water or both by draining, diking, or filling (Figure 3.147). But now the mood has changed. Wetlands are recognized as among the most important biological resources we have. (74)

Events of the last ten years have shown us that there are, and should be, limits on technology. As a nation we are just beginning to practice restraint, to not do some things we are technologically capable

	Estuaries (without mangroves)	Mangroves	Open coasts	Floodplains	Freshwater marshes	Lakes	Peatlands	Swamp forest
Functions								
1. Groundwater recharge	○	○	○	■	■	■	●	●
2. Groundwater discharge	●	●	●	●	■	●	●	■
3. Flood control	●	■	○	■	■	■	●	■
4. Shoreline stabilisation/Erosion control	●	■	●	●	■	○	○	○
5. Sediment/toxicant retention	●	■	●	■	■	■	■	■
6. Nutrient retention	●	■	●	■	■	●	■	■
7. Biomass export	●	■	●	■	●	●	○	●
8. Storm protection/windbreak	●	■	●	○	○	○	○	●
9. Micro-climate stabilisation	○	●	○	●	●	●	○	●
10. Water Transport	●	●	○	●	○	●	○	○
11. Recreation/Tourism	●	●	■	●	●	●	●	●
Products								
1. Forest resources	○	■	○	●	○	○	○	■
2. Wildlife resources	■	●	●	■	■	●	●	●
3. Fisheries	■	■	●	■	■	■	○	●
4. Forage resources	●	●	○	■	■	○	○	○
5. Agricultural resources	○	○	○	■	●	●	●	○
6. Water supply	○	○	○	●	●	■	●	●
Attributes								
1. Biological diversity	■	●	●	■	●	■	●	●
2. Uniqueness to culture/heritage	●	●	●	●	●	●	●	●

Key: ○ = Absent or exceptional; ● = present; ■ = common and important value of that wetland type.

Figure 3.145. Generalized wetlands funcitons and values. (*Source:* Reference 142.)

of doing. Rejecting massive wetland conversion projects as a restraint on technology is becoming a reality. (74)

True coastal wetlands are easily distinguished from other ecosystems. They are shallow enough for a person to wade through and are vegetated with rooted plants like mangroves and marsh grasses. Wetlands are called "wetlands" because they are "land" that is "wet." Lakes or sounds that are 3 meters deep are not usually termed wetlands because they are not wet land (land that is wet)--they are simply aquatic systems. As aquatic systems they have quite different ecological characteristics than true (intertidal) wetlands.

One problem is that aquatic conservation enthusiastists, noticing that wetlands are politically popular to have tried to extend the wetlands definition to include everything from offshore coral reefs to deep sea grass beds to the entirety of large lagoons. The strong image of wetlands then gets lost in ambiguity.

Figure 3.146. All parts of Earth's hydrological system should be considered in a coastal zone resources management program. (Photo by author.)

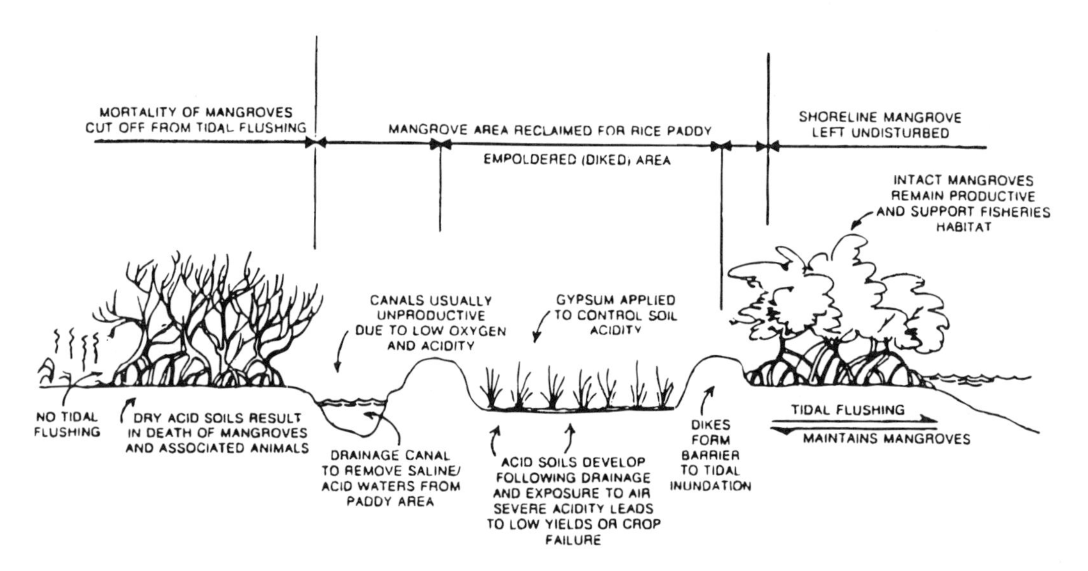

Figure 3.147. The consequences of converting mangrove forest to rice fields by removing vegetation and building dikes and water control structures. (*Source:* Reference 316.)

For example, Larson *et al.* (215) comment that, "depending on which definition is used, wetlands may include . . . rivers, estuaries, oceans, swamps, mires, springs, and flood plains."

The kind of confusion that claims oceans to be wetlands can only dilute wetland conservation efforts and cause socio-political backlash. Supporting conservation of lakes and nearcoast waters for their special values is commendable, but calling them wetlands to gain advantage is not. Use of the traditional, narrower definition is a superior strategy in the long term.

Figure 3.148. The governor of Bali (Indonesia) ordered leases for shrimp aquaculture in south Bali canceled and the ponds replanted with mangrove. (Photo by author.)

In wetlands conservation, it is most necessary to discriminate between "preservation" ("wilderness") wetlands and "multiple-use" ("utility") wetlands. Preservation wetlands are those that should be protected, preserved, and held in their historic condition for all the values that flow from wilderness areas. A goal worth considering is to preserve any wetlands which are in pristine condition and performing their historic ecological functions. No uses would be permitted which would alter a preservation wetland. Wilderness wetlands not already in public ownership (or under guaranteed private protection) should be identified and acquisition and management programs developed and budgeted.

Preservation, or wilderness, wetlands should not be expected to produce shrimp or stop floods or filter sewage—those are utilitarian/economic parameters ascribed to multiple-use, or utility, wetlands. Such wetlands should be conserved (which implies wise use) for economic use, often multiple uses, by keeping them in good functional condition. Major restoration programs are often required (Figure 3.148).

It is natural to think of preservation wetlands more in an area context; that is, to regard their values as associated with structure and expanse, factors that go with scenery, heritage, open space, and such values. Preservation, or wilderness, wetlands include those already designated for perpetual preservation and also candidates for such designation that meet criteria relating to size, location, importance to consequential species, ecological quality, and scarcity on a regional or national basis.

It is equally natural to think of multiple-use wetlands in a functional context; that is, to value them for production of species, processing of water, conversion of materials, or cycling of carbon. Multiple use wetlands should be considered an economic resource to be maintained as appropriate to meet regional wetland priorities for natural goods and services.

3.110 ZONA PUBLICA

In several countries the immediate shorelands are held to be owned by the sovereign or are in a public trust. This creates a most useful linear nature reserve and coastal setback. In Costa Rica, an ICZM system has evolved from the *zona maritima,* which was defined as a 200-meter-wide strip of land (measured back from the high water line) in the 1976 national legislation that authorized it. The 50 meters closest to the high tide line cannot be owned or built upon (*zona publica*), but the remaining 150 meters are under the jurisdiction of the local municipality (*canton*), which may lease it for various uses (*zona restringida*) after a use plan has been approved by the national government. Wetlands (mangrove forests) are given special protection (management by the national Forestry Administration) and areas of national tourism interest (including beaches and facility sites) are set aside by the national Tourism Institute (see "Costa Rica" case history in Part 4). Illegal buildings can be demolished by government, and pre-existing legal ones can be removed if the owner is compensated. (46, 322)

The English legal system offers concepts that lend themselves to absorption into the coastal zone regime such as the English concept of Crown Lands which is commonly found in Commonwealth countries and which, in a way, resembles the Latin American "zona publica." However, these legal conditions may seldom be large enough by themselves to constitute a coherent planning area and implementing legislation will have to be developed further to clarify the applicable area. Although the extent of this area may vary with the situation, it will be most realistic if a degree of flexibility is permitted to reflect the dynamic nature of coastal systems and changing patterns of usage. (134)

The base area is declared to be a 150-foot coastal zone in the British Virgin Islands. Such coastal exclusion zones act as statutory shoreline setbacks. A *Beach Control Ordinance* vests all rights in and over the foreshore and the "floor of the sea" in the Crown and confers on the Minister [of Agriculture, Fisheries and Lands] the right to regulate the use of this area and adjoining land to a distance of "not more than fifty yards beyond the landward limit of the foreshore" and to license its use "for any public purpose, or for, or in connection with, any trade, business or commercial enterprise to any person, upon such conditions and in such form as he may think fit." (104)

PART 4

Case Histories

4.1 AUSTRALIA, CLARENCE ESTUARY: UNGUIDED TOURISM DEVELOPMENT

Background

Maclean Shire is a small rural shire of 14,000 people on the mid-north coast of New South Wales (NSW), Australia. Its 47-km coastline is dominated by the Clarence River estuarine system, with a catchment area of 22,400 km^2. Tourism has long been important to the area, but it has generally been small scale and family oriented. Larger tourist developments have been proposed over the last ten years but none have eventuated. This case history examines the latest proposal for a tourist, recreational, and training complex to be developed by a Japanese company for its employees on an island in the Clarence River.

Land use within the shire is predominantly agricultural (mostly sugar cane) and protected natural areas, along with some large tracts of state forest. Economic activity in the shire is based on fishing, tourism, and agriculture. The prime appeal of Maclean Shire as a tourist destination is its relatively undeveloped character. The predominantly rural aspect of the shire is broken only by the largely unspoiled waterways of the Clarence River estuary, with extensive mangrove swamps (5.2 km^2) salt marshes (2.0 km^2), and seagrass beds (19.1 km^2), and by the surrounding national parks—Bundjalung to the north and Yuraygir to the south.

The Clarence estuary has recently been assessed as of outstanding value in terms of biodiversity, vegetation, wildlife, and fisheries. It contains 29 species of endangered fauna, high diversity and density of wading birds, and breeding pairs of ospreys, sea eagles, and kites. It was found to meet the criteria of a RAMSAR wetland and was recommended for inclusion on that list, even though many of the areas already are protected as "wetland reserves." The estuary also yields 802 tons of fish, 290 tons of crustaceans, and 100 tons of oysters, yearly.

Problem

Tourism associated with the estuary dates back to the turn of the century with the construction by rural landholders of holiday shacks around small fishing ports. This initial phase was followed by the construction of hotels at these now thriving ports in the 1930s and 1940s. Tourism activity in the shire has been small scale, family oriented, and domestic, with the majority of visits to the region originating from NSW country areas (40 percent) and Sydney (33 percent). International tourists account for only 6 percent of visits to the region. At present, the small towns and villages boast nothing larger or more sophisticated than local motels, caravan parks, and pubs, built from the 1950s.

However, since the late 1970s, a series of major developments have been proposed for the Shire (Figure 4.1.1). At least six of those proposals involved overseas interests. The Japanese developer Ipoh Gardens was involved in the $200m "Yamba Waters" proposed for a parcel of Crown land between Yamba and Angourie. The Minister for the Environment rejected the concept in the early 1980s.

In 1983–84 another Japanese company, Kumagai, invested some $9m in the proposed $60m "Wandering Star" development, only to be thwarted once again by the Minister for the Environment for

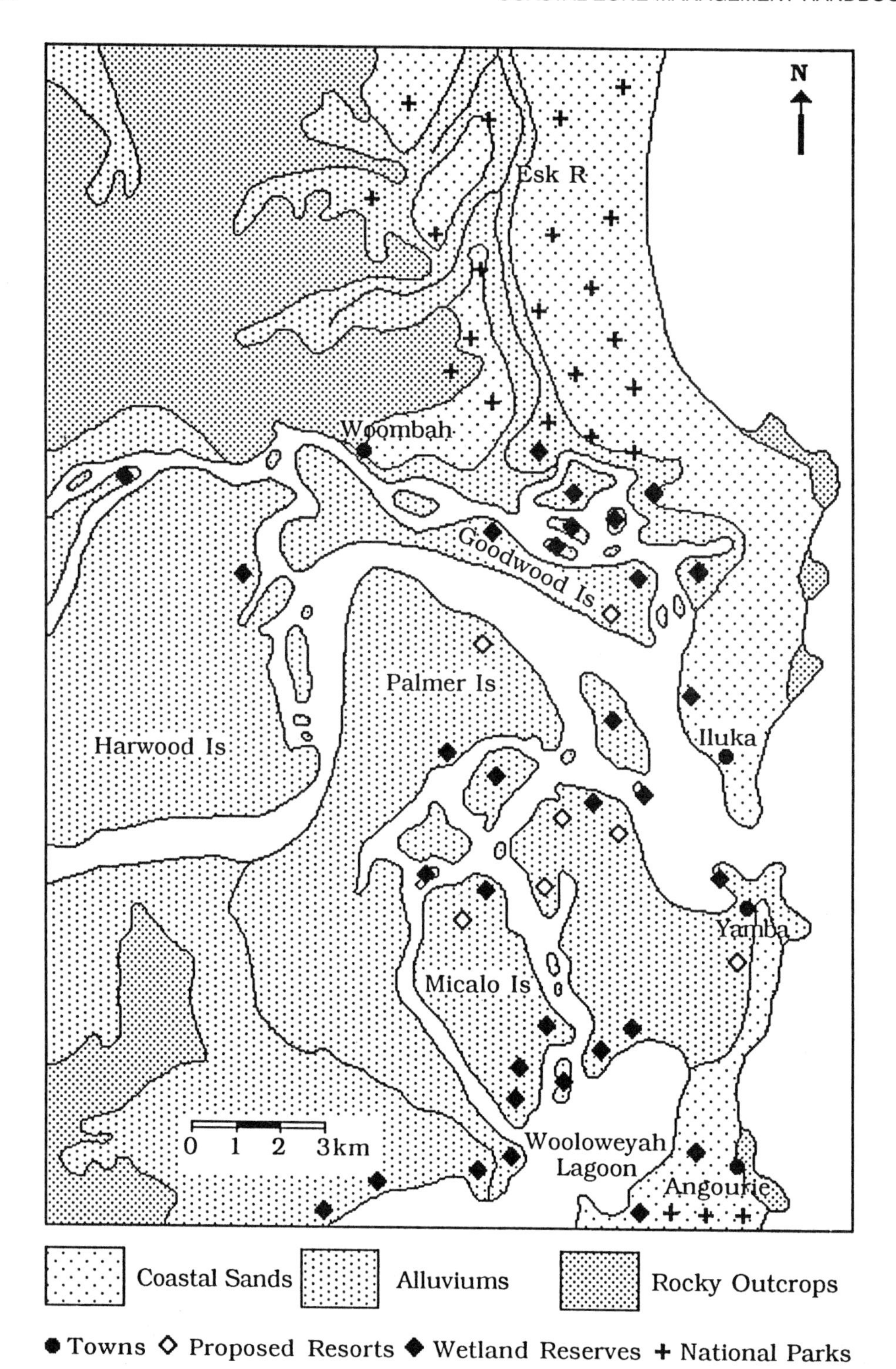

Figure 4.1.1. Lower estuary of the Clarence River (29°25 S, 153°22 E) showing geomorphology, anastomosing waterways, wetland reserves, and national parks, as well as existing towns and proposed resort sites.

contravening the development application. As a consequence, the Maclean Shire Council had some of its planning powers removed by the Department of Planning, and the developer sold its interest. Since then this site has changed hands a number of times, been re-named "Oyster Cove," and has gone bankrupt.

Around the same time, Sabena (Hong Kong backers) commenced a development on Palmers Island North. After driving some 20-odd foundation pylons, work on this project ceased and the site has seen no activity since.

In the early 1990s, an Australian company, Glenpatrick Pty. Ltd., proposed a golf course, club house, and airport for a fly-in/fly-out resort on Goodwood Island with 60 mobile homes. The proposal was rejected as environmentally undesirable (wetlands adjacent) and the land was sold—with no further proposals to date.

For the most recent (and controversial) proposal, land was acquired by the developer, Shin-Ei Co. Ltd., in 1989 and, early in 1990, the council was advised that a request for rezoning would be submitted to allow the development on Micalo Island of a resort to be known as "Dreamtime under the Southern Cross!" The proposal included an 18-hole golf course and associated club house; an accommodation lodge with 52 suites; administration building, plant nursery, maintenance depot, and manager's residence; equestrian center; educational training institute for culture and language; sporting facilities; landscaping and wetland regeneration; 25-berth boating facility; restaurant; general store; boat maintenance facilities; and a heliport. The proposal had an estimated capital cost of $46 million(Aus) and the facility would accommodate 213 visitors at any one time. The development application was lodged with the council in July 1991 and produced immediate expressions of concern from local groups.

Solutions

An environmental impact statement was requested by council, which had originally supported the proposal with an 8–1 vote. However, council elections in 1991 saw the development become the focus of local controversy and, following the elections, the new council voted 5–4 to reject the proposal.

Aspects of the development (loss of wetlands, acid-sulfate soils, etc.) led to state government intervention with the Minister for Planning instituting a Commission of Inquiry, provided for under the "Environmental Planning and Assessment Act." This inquiry commenced in October 1991 while the state government was becoming increasingly concerned that major developments were being held up in the approval process by local governments.

At that stage, the proposal became caught up in a three-way tug-of-war between the state government, the Maclean Shire Council, and the local community:

- *The Council* has the nominal power to approve the development but was divided and ambivalent. Lack of a firm strategy for tourism was seen as a major stumbling block.
- Significant sections of *the community* were voicing concern for environmental reasons (impacts on the waterways, fish population, and wetlands, as well as the larger ecosystem) and social reasons (a Japanese cultural enclave and foreign ownership of Australian real estate). Others believed the development was vital for the local economy.
- *State departments* expressed mainly environmental concerns and suggested a range of conditions that must be imposed on the development. The NSW Tourism Commission did not comment or make a formal submission to the Commission of Inquiry. At the same time, the state government was becoming increasingly concerned generally about local government's obstruction of this and other development, which it considered may have economic benefit for the whole state.

The matter was finally decided by the Commission of Inquiry recommending approval of the project with 85 prerequisite conditions attached and the Minister for Planning over-riding the local council's decision-making authority on the matter to allow the project to proceed.

But the project has yet to proceed. However, the council has responsibility for overseeing the next phase—ensuring consent conditions are adhered to—but is already suggesting that, given the state government's interference to date, it should carry some, if not all, of the responsibility for implementation of the development. The council had indicated to the Commission of Inquiry that its concerns were loss of agricultural land; the potential adverse environmental impact on water quality; fauna and flora within the estuary system; cumulative impact on flood level; economic benefits possibly not being experienced in the local community; and potential social impacts, including demand for community services, increased traffic, and changing life-styles. These concerns reflect those of many in the wider community.

The commission considered that its "conditions of consent" would overcome the potential negative environmental impact of the proposal. After considering the social and economic issues brought before it, the commission found no reason why the proposal should not proceed on these grounds.

Both council and developers are critical of what they regard as unnecessary duplication of state and local effort during the approval process for the Micalo Island proposal, resulting in time being wasted while a "seemingly endless number of approvals were sought and gained."

Lessons Learned

This case study exemplifies a situation typical of a region where tourism is promoted as a source of employment and economic growth in communities that are increasingly concerned about the potential impact of tourism development on the quality of local living environments. In this context, the largely "data-free" debate about how much of the benefits of tourism development flow to the local economy, how much to the state generally, and, in the case of foreign-owned investment, how much to overseas interests, all assume great significance.

Local government generally does not have a clear understanding of planning beyond the prescribed statutory base that is adhered to in New South Wales. Although the "Environmental Planning and Assessment Act" allows for and encourages "strategic planning," most council planning does not adequately cater for large-scale tourism development and unanticipated development. Council planning continues to be dominated by land use zonings and colored maps when what is required is a greater commitment to community consultation and statements of agreed policy and directions.

Most significantly, this case study illustrates the difficulties experienced when there is a lack of a shared vision between local and state government and the community about tourist development in a local area.

Contributed by — P. Saenger and G. Prosser, Centre for Coastal Management, Southern Cross University, Lismore, Australia.

4.2 AUSTRALIA: CORAL REEF SURVEY METHODOLOGY

Introduction

This section focuses on techniques to monitor coral reefs, from planning through to underwater techniques, with brief discussions on data analysis. It is aimed particularly at obtaining statistically reliable baseline data for repeat montoring and reflects lessons learned from extensive field work.

Most remote coral reefs around the world remain unsurveyed and are not being monitored. Implementation of monitoring programs is most pertinent. Establishing permanent sites at a constant depth and running replicate line-transects to record adult communities and narrow belt-transects to record recruitment remains a highly effective tool to monitor a coral community. A permanent record via video monitoring is optional although useful, since further information can be extracted at a later date.

The techniques outlined in this section are a guideline for establishing monitoring sites on coral reefs. In projecting their use to environmental impact assessment, an emphasis should then be placed on depth stratification and pilot studies should be undertaken to determine optimal sampling size, levels of precision, impact detectability, and appropriate levels of replication.

Problem

The chronic effect of humans on the recovery process of coral reefs is our primary concern. Reef flat and deep communities are first to suffer degradation. Coral recruitment is suppressed in polluted areas (17, 18), and community recovery is slowed. Recruitment can be deterred via influences on coral larvae before settlement on the substrate, by overgrowth, or by interference on the juvenile after settlement. By establishing monitoring sites and recording juvenile and adult colonies over time, we may begin to understand the nature of coral reef systems from a global perspective.

Monitoring coral reefs is only a recent scientific endeavor (1970s), and most methods have been adopted from terrestrial botanists. Due to the restrictions placed on sampling effort using SCUBA, time-savers have been identified (4, 9, 19). The techniques most commonly used in the Great Barrier Reef Marine Park are line- and belt-transects, used within permanently marked study sites. These techniques are simple to implement, require minimal equipment, and are widely applicable to coral communities worldwide. (See "Australia, The Great Barrier Reef: Multiple Use Management" below.)

Solutions

Manta towing is probably the most effective technique to gain a first view of the location (5). Although prone to variability and observer bias, the technique is effective for reconnaissance work (8). A person is towed behind a small craft at low speed (1.5 knots) holding onto a manta board containing a recording sheet. Every 2 minutes the boat momentarily stops and the boat driver and the person being towed record approximate position and a qualitative description of the reef, respectively. Large regions can be quickly characterized, and transitional boundaries between different community types are identified. Effective use of the manta tow technique is dependent on good water clarity. Poor water clarity generally requires SCUBA reconnaissance.

When planning a monitoring program, one should divide the area of interest into geographic locations. Charts and high resolution aerial photographs can be used to define topographic features. Windward and leeward locations, or outer and inner islands, often demarcate regions that are subject to different environmental parameters. Once these gross geographic entities are established, the area should be ground-truthed, as follows.

The physical environment is highly influential in determining what type of coral community a reef slope can support—for example, wave exposure, turbidity, and depth greatly influence community composition (6, 10, 15, 18).

Establishing Permanent Sites

Identifying representative locations from the manta-tow data leads to the establishment of permanent sites. There should be at least two sites within each location. Replication of samples within sites is essential (Figure 4.2.1) as what may appear as a change within one location may be merely a local phenomenon, for example, the death of a large coral colony. Fringing reefs vary considerably with

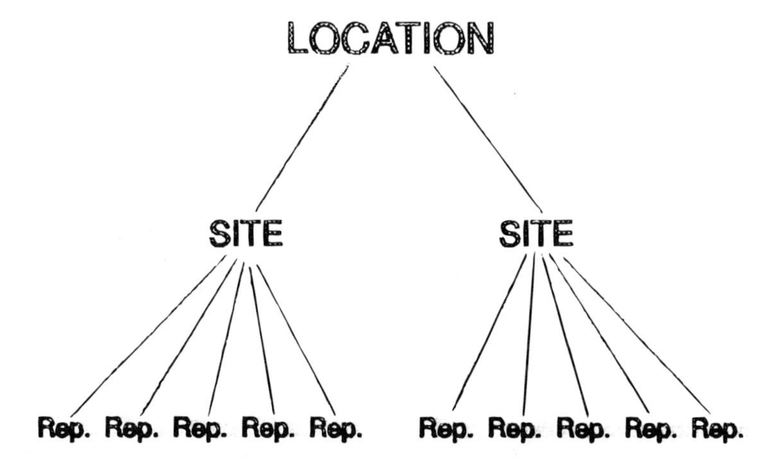

Figure 4.2.1. In coral reef survey, replication of samples within sites is essential, as what may appear as a change within one location may be merely a local event—for example, the death of a large coral colony.

depth, and stratification of each site is ideal but not always logistically possible. Monitoring is a comparison of many geographic locations over time; if a consistent depth is monitored—for example, 2–4 m below low water datum—a program may be still effective without multi-depth sampling.

Steel stakes are often used on the Great Barrier Reef to relocate sites. They last underwater for years except on reef flats where they are continually exposed and rusting is rapid. These stakes should be sufficiently embedded in the coral substrate using a special driver or large hammer so as not to interfere with other reef users. The stakes should be clearly labeled using longitude and latitude as coded numbers. If possible each site should be accurately marked on a reliable map for relocation purposes, with Loran or Global Position System (GPS) fixes or sketch maps with at least three compass bearings from distinct landmarks.

Sample Size

Coral reefs are extremely heterogeneous. In sampling, the goal is to reduce the standard error (variance) of the mean number of organisms. This can often be accomplished better by more sampling of smaller plots and by having a high number of replicates. Estimates of abundance are more precise with a small standard error and statistical tests become more powerful. However, the size of plots should be proportionate to the size of the organisms being measured and be sufficient in size to include the majority of species present. Therefore, a small plot may be appropriate on a reef flat (perhaps 2 m^2) and a large plot on a reef slope (perhaps 100 m^2) where the corals are larger.

A general rule is that the plot size should be approximately 15–20 times the surface area of the organisms being sampled. Although some coral colonies may cover tens of meters under favorable environmental conditions, many colonies do not cover more than 1–2m in horizontal extent. Therefore, replicated 20-m line-transects and 20 m^2 belt-transects are the most common sampling units within the Great Barrier Reef Marine Park. For areas dominated by large massive corals, larger permanent sites and belt-transects have been used (1, 18). Although they are time consuming, these methods provide more detailed information on the coral community.

Line-transect

The line-transect method, adapted to reef surveys by Loya (12), has been widely applied. Cover estimates are made for coral species using a fiberglass tape measure. The tape is laid along the substratum, and edges of each colony (points of transition) under the tape are recorded for a predetermined distance (e.g., 20 m). From replicates of these transects the number of species, their abundances, and relative percent cover can be derived. Line-transects are most successfully used on reefs supporting corals, where the differentiation of individual coral colonies is difficult; for example, communities dominated by *Acropora* species (Figure 4.2.2). The number of replicated lines should be high to cater to the heterogeneity so inherent to coral reefs (13). Five replicates at each site are usually sufficient.

Figure 4.2.2. Line-transects are most successful where the differentiation of coral colonies is difficult, for example, those dominated by *Acropora* species.

Because line-transects are so narrow, there is considerable variability between observers due to "parallax error" (therefore a likelihood of confounding small-scale variation with real temporal change). To reduce this problem when recording with more than one person, all recorders should record information from one line. In this manner an inter-observer error can be incorporated into the calculations. Furthermore, multiple intercepts from the same coral colony should be indicated on the recording paper, as what may appear as abundant recruits may in fact be remnants from a large colony subjected to partial mortality (2), possibly caused by environmental degradation.

Belt-transect

Application of this technique presupposes that individual colonies can be distinguished from one another, which is problematic in the case of branching and encrusting forms such as colonies of *Acropora* and *Montipora* species. Narrow belts (0.5 m by 20 m) have been successfully used to identify recruitment (e.g., colony diameter < 5 cm) (3, 7). Narrow belt-transects are most easily defined under water by using a fiberglass tape to define the length and a T-shaped hand-held PVC pole to define the width of the belt (e.g., 1 m). If larger belt-transects are required, several fiberglass tapes are often used, one to define the perimeter of the transect and others to subdivide the transect (to facilitate recording accuracy).

A permanent belt-transect obtains quantitative data, which can be extracted on coral abundance, colony shape, size, and spatial relationships. One major disadvantage is that taxonomic resolution falls short of what is possible by a trained observer in the field. However, in the 1990s, video monitoring is one of the principal techniques used to monitor permanent sites within the Great. Barrier Reef Marine Park.

Video Techniques

With advances in video technology one must emphasize their importance in underwater use. Video monitoring is economically feasible in terms of time spent underwater and relatively large areas can be monitored. Video monitoring is equivalent to obtaining a permanent record of a belt-transect. The size of the belt-transect will depend on the distance of the video from substratum and the camera lens. Recording height will be restricted by water clarity; however, standardizing the video transect to a 1-m width is appropriate. Once the exact distance above the substrate is identified, to give a 1-m-wide belt, a flexible distance rod, slightly shorter than the required length, can be permanently attached to the camera housing to ensure reasonable consistency. A fiberglass tape measure is laid along the substrate, and filming proceeds along the tape.

To obtain estimates of relative abundance, one divides each species count by the total number of frames counted. One can obtain not only estimates of coral cover, but also estimates of colony size and colony damage. With more sophisticated support facilities, small squares can be simulated on the screen and the video monitor stopped at random intervals; such software is available from the Australian Institute of Marine Science (J. Carleton and T. J. Done). Eventually, the images can be electronically digitized to produce accurate coral community statistics.

A major program being co-ordinated at the Australian Institute of Marine Science, initiated in 1992, involves the video monitoring of permanent sites within the Great Barrier Reef Marine Park. The Marine Park has been divided into 11 sectors for this purpose. Reefs are being targeted within each sector. On each reef, recording takes place on windward and leeward slopes between the depths of 6-m and 9-m (low water datum) as consistently heavy wave action on windward slopes precluded shallower sampling. Transects 100 m long are being recorded. Steel stakes are placed at every 5 m to facilitate recording precision. This is the largest, most comprehensive coral reef monitoring program in the world.

Frequency of occurrence and percent cover estimates (percent of reef covered with live coral) should be calculated for each taxon at each transect. Change in percent cover (line-transects) and number of coral colonies (belt-transects) are usually analyzed via a one-way analysis of variance, with sites nested in locations (e.g., Reference 3). However, there are many ways of assessing temporal change and evaluating whether, or how, this change occurred. This is the essence of statistical analyses, as described in many texts (16, 20).

Physical Environment

For minimal additional effort a database on the physical nature of the reef environment can be established. Characteristics of the reef can be recorded on site—for example, site depth, slope angle, water temperature, salinity, and clarity (Secchi disk). Other information can be collected from maps, charts, and appropriate sources—for example, distance from nearest active volcano, tidal fluctuations, shelf depth, distance from nearest river, catchment size, agriculture type in nearest catchment, distance from city, geology of adjacent land, distance to primary industry (e.g., extraction of minerals), distance to secondary industry (e.g., manufacturing of waste outputs), types, and volumes. Some valuable insights on the distribution of coral species are elucidated when two data matrices, biological and physical, are compared using multi-variate analyses (13).

Lessons Learned

No technique is foolproof. Identification of many coral species is problematic using video monitoring, as for example many species of *Montipora, Acropora,* and *Porites.* Recording is only ideal in clear water, and deriving accurate colony sizes can be difficult due to the high rugosity of the reef substrate. Probably the main disadvantage of video in remote places is that the batteries need to be charged regularly with AC power.

If resource managers are interested in assessing the effect of a development, collecting baseline information and recording information in accordance with the before/after/control/impact (BACI) experimental design is most useful. However, monitoring existent point-source discharges (e.g., sewage outlet from a city or river system) is best achieved by examining at least seven sites moving away from the point source. In this manner, managers have great predictive capabilities using the x-axis as the predictor variable (independent)—for example, distance from the source—and the y-axis as the response variable (dependent), as, for example, coral cover or number of recruits (14).

Steel stakes can mark a site permanently, but the line- and belt-transects should be positioned haphazardly within each site because fixed lines violate the assumption of independence when using analysis of variance techniques in temporal studies. If one wishes to make the lines permanent, thereby increasing re-recording precision, and not violate the assumptions of independence, the difference between sampling times can be analyzed by subtracting abundances of Time 2 from Time 1, rather than analyzing the actual recorded data.

Contributed by — R. van Woesik, Department of Marine Sciences, University of the Ryukyus, Okinawa, Japan.

REFERENCES

1. Cameron, A. M., R. Endean, and L. M. De Vantier. 1991. Predation on massive corals: are devastating population outbreaks of *Acanthaster planci* novel events? *Marine Ecology Progress Series,* Vol. 75. Pp. 251–258.
2. De Vantier, L. M. 1986. Studies in the assessment of coral reef ecosystems. In: Human induced damage to coral reefs, Brown, B. E. (Ed.). *UNESCO Report in Marine Science,* Vol. 40, Ch. 8. 23. Pp.
3. De Vantier, L. M., R. Van Woesik, and A. D. L. Steven. 1992. Monitoring study of the coral communities of Middle Reef, Townsville. *A Report to the Great Barrier Reef Marine Park Authority.* 59 Pp.
4. Dodge, R. E., A. Logan, and A. Antonius. 1982. Quantitative reef assessment studies in Bermuda: a comparison of methods and preliminary results. *Bulletin of Marine Science,* Vol. 32, No. 3. Pp. 745–760.
5. Done, T. J. 1981. Rapid, large area, reef resource surveys using a manta board. *Proceedings of 4th International Coral Reef Symposium,* Manilla. Pp. 299–308.
6. Done, T. J. 1982. Patterns in the distribution of coral communities across the central Great Barrier Reef. *Coral Reefs,* Vol. 1. Pp. 95–107.

7. Done, T. J., L. M. De Vantier, D. A. Fisk, and R. Van Woesik. 1991. Analysis of coral colonies, populations and communities: interpretation of outbreak history and projection of recovery. *A Report to the Great Barrier Reef Marine Park Authority,* 22 Pp.
8. Fernandez, L. 1990. Effect of the distribution and density of benthic target organisms on manta tow estimates of their abundance. *Coral Reefs,* Vol. 9. Pp. 161–165.
9. Goodwin, M. H., M. J. C. Cole, W. E. Stewart, and B. L. Zimmerman. 1976. Species density and associations in Caribbean Reef corals. *Journal of Experimental Marine Biology and Ecology,* Vol. 24. Pp. 19–31.
10. Goreau, T. F. 1959. The ecology of Jamaican coral reefs: I. Species composition and zonation. *Ecology,* Vol. 40.
11. Loya, Y. and L. B. Slobodkin, 1971. The coral reefs of Eilat (Gulf of Eilat, Red Sea). *Symposium Zoological Society of London,* Vol. 28. Pp. 117–139. In: *Regional Variation in Indian Ocean Coral Reefs.* D. R. Stoddart and C. M. Yonge (Eds.). Academic Press, New York.
12. Loya, Y. 1972. Community structure and species diversity of hermatypic corals at Eilat, Red Sea. *Marine Biology,* Vol. 13. Pp. 100–123.
13. Mundy, C. N. 1990. Field and laboratory investigations of the line intercept transect technique for monitoring the effects of the crown-of-thorns starfish, *Acanthaster planci.* Australian Institute of Marine Science, Townsville, 42 Pp.
14. Peters, R. H. 1991. *A Critique for Ecology.* Cambridge University Press, Cambridge, England.
15. Sheppard, C. R. C. 1982. Coral populations on reef slopes and their major controls. *Marine Ecology Progress Series,* Vol. 7. Pp. 83–115.
16. Sokal, R. R. and F. J. Rohlf. 1981. *Biometry.* 2nd Edition. WH Freeman and Co., New York.
17. Tomascik, T. 1991. Settlement patterns of Caribbean scleractinian corals on artificial substrata along a eutrophication gradient, Barbados, West Indies. *Marine Ecology Progress Series,* Vol. 77. Pp. 261–269.
18. Van Woesik, R. 1992. Ecology of coral assemblages on continental islands in the southern section of the Great Barrier Reef, Australia. Ph.D. Thesis, James Cook University of North Queensland, Queensland, Australia. 227 Pp.
19. Weinberg, S. 1981. A comparison of coral reef survey methods. *Bijdragen tot de Dierkunde,* Vol. 51, No. 2. Pp. 199–218.
20. Zar, J. H. 1984. *Biostatistical Analysis.* Prentice-Hall International, USA, 718 Pp.

4.3 AUSTRALIA, THE GREAT BARRIER REEF: MULTIPLE-USE MANAGEMENT

Background

The Great Barrier Reef Marine Park Authority is an independent statutory body with responsibility for the care of the Great Barrier Reef in Australia. The Marine Park covers an area of 240,000 square miles along the northeastern coastline of Australia (Figure 4.3.1). Its 2,900 reefs, vast range of inter-reefal habitats, and almost 1,000 islands make up one of the most diverse ecosystems on earth.

The Great Barrier Reef Marine Park is also one of Australia's richest assets, supporting, among other things, a growing tourism industry, a long-established fishing industry, a major east coast shipping route, a wealth of recreational opportunities, and a livelihood for locals and is of unestimatable value to the Aboriginals and Torres Strait Islanders who have traditionally inhabitated this area of Australia.

Tourism is the most profitable commercial activity, with 24 islands on the reef currently wholly or partly developed for tourism purposes. Reef-based activities associated with tourism include diving, fishing, and boating.

Problem

During the late 1960s and early 1970s, concern had been raised about the changing intensity of human use of the Great Barrier Reef. Of major concern at this time were proposals to establish oil drilling and to mine reefs for limestone.

Historically, Australian Aboriginal people have used the nearshore areas of the reef for at least 20,000 years and continue to do so as an important part of their subsistence, culture, and life-style.

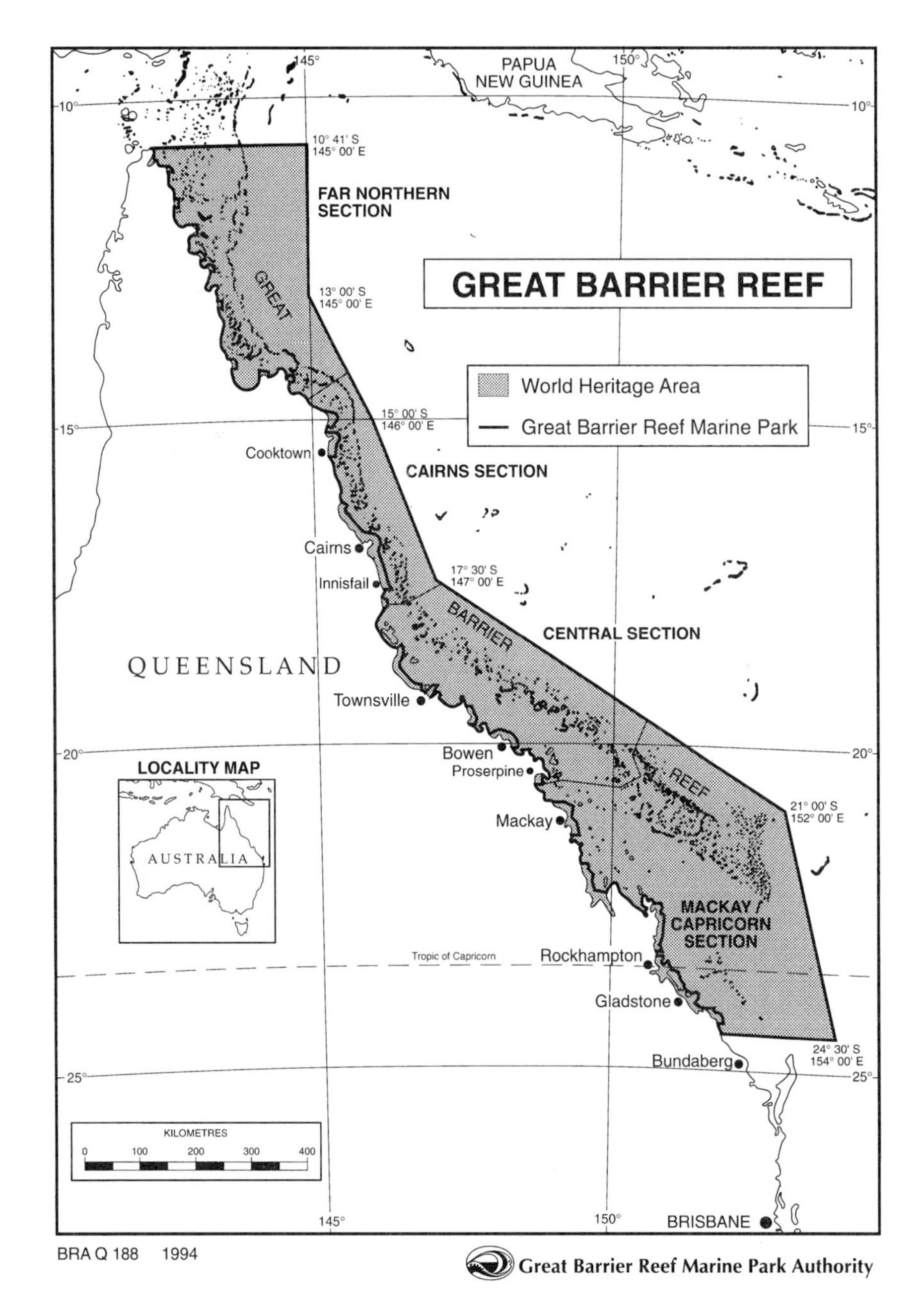

Figure 4.3.1. The Great Barrier Reef Marine Park covers an area of 240,000 square miles along the northeastern coastline of Australia.

Intensive commercial use of the reef commenced in the latter part of last century with the beche-de-mere, trochus, and pearl shell fisheries but has increased dramatically since World War II with tourism, prawn (shrimp) trawling, and reef-line fishing being the dominant activities.

Today, reef-related commercial activities have extended human use to all parts of the reef and have an estimated value in excess of AUS$1 billion per annum. Reef visitation has increased to the extent that some 2.2 million people now visit the reef region each year. Modern-day fishing activities include recreational line and spear fishing, shell and ornamental fish collection, and commercial trawl, line, and net fishing. The commercial fishing industry alone annually generates $128 million a year and directly employs nearly 4,000 people (3).

This growing use pattern caused public concern that the reef—one of Australia's great assets and the largest coral reef in the world—would be rapidly degraded unless conservation measures were imposed.

Solution

The Great Barrier Reef was declared a marine park in 1975. The establishment of a conservation regime that encompasses the entire reef ecosystem and provides for multiple uses is a special feature of the Marine Park. The empowerment of an independent statutory authority to manage the entire area is a unique feature for marine protected areas around the world and has proven a critical factor in the success of the Great Barrier Reef Marine Park.

The Great Barrier Reef Marine Park was established by an Act of Federal Parliament to cover the reefs and coastal waters. A unique feature of the act was that it assigned responsibility for management of the Marine Park to a three-person independent authority comprising a permanent chairman, a representative of the State of Queensland, and a third appointed member who to date has been from a scientific background.

Very importantly, the act also established legislative precedence over all federal and state legislation (with the exception of some aspects of the regulation of shipping and aircraft, which are subject to international agreements).

This is not to say there are no checks and balances on the authority. The act establishes a formal system of zoning, requires extensive public participation in decision making, and provides an independent appeal system for administrative decisions made by the authority.

The Marine Park is managed to provide for multiple uses where this use can be managed to be consistent with the requirements of nature conservation. The act itself banned oil drilling and mining; both activities were considered to be too threatening to the coral reefs. Management of the vast array of remaining activities is a very complex balancing act.

Zoning plans provide the flesh for the skeleton established by the act. They form a framework for management that includes:

- Establishment of "representative areas" of protected habitats as flora and fauna refuges and scientific reference areas
- Protection of sensitive habitats and species from activities that might threaten them (e.g., trawl fishing is precluded from coral reef and seagrass communities, and species that may be particularly vulnerable to exploitation, such as dugong and turtle, receive appropriate protection—Figure 4.3.2)
- Provision for environmental impact assessment for new activities that may have significant environmental impacts, detailed management planning for high use and sensitive sites, and development of conservation strategies for threatened species

Perhaps a measure of the success of the management practices adopted by the Great Barrier Reef Marine Park Authority is the extent to which the authority is now reinforced by the State Government of Queensland. The Queensland government has established marine parks to cover the state's inter-tidal waters and has included these and the state's island parks under a management agreement administered by the authority. Although the operations of the authority itself are funded primarily by federal government appropriation, funding for day-to-day management in the parks (e.g., enforcement, surveillance, extension activities, etc.) is provided equally by the federal and state governments.

Public involvement is the cornerstone of the Marine Park. A formally constituted Consultative Committee was established by the act and advises the authority and the responsible federal and state ministers. The act also requires the authority to seek public input into the development of zoning plans. Specialist advisory committees are also established where appropriate, for example, to advise on monitoring and research programs or to develop more detailed plans for management of intensively used areas.

However, the management of the Marine Park has concluded that it has not yet adequately addressed the needs of Aboriginal people, whose life-style and culture have evolved over many thousands of years of co-existence. This is now a major focus for the authority, and will hopefully provide useful lessons in the future.

Figure 4.3.2. Species that may be particularly vulnerable to exploitation, such as dugong (taken by Aboriginals) receive appropriate protection to ensure sustainability of supply.

Lessons Learned

In a complex area involving local government, state, federal, and international jurisdictions, the establishment of an authority with over-riding powers facilitates a holistic approach to management.

The importance of any authority's status as an independent statutory body supported by comprehensive legislation cannot be overestimated. Many marine protected area management agencies around the world have powers limited to small reserve areas or lack the power to address key impacts because they are the responsibility of a range of other agencies. In contrast, the Great Barrier Reef Marine Park Authority has been empowered to address all activities occurring in the park that might adversely affect the reef.

Any marine resource authority would continually face limits to its management ability. In the case of the Great Barrier Reef, however, these limits are often the result of inadequate management information. Answers to critical questions remain evasive. Such questions include What is the ecologically sustainable level of fishing? How much development is appropriate? What impact are human activities having on nutrient levels, and what will be the effect on the reef?

But one important lesson the authority has learned is that to delay making decisions until perfect knowledge is obtained is to invite ecological catastrophe. The authority makes important management decisions based on the best expert scientific advice *currently available,* taking into account both ecological and social impacts and applying the precautionary principal.

There must always be some limit to the area incorporated within the boundaries of a marine protected area. In this case, many of the problems affecting the park are generated outside its boundaries; nutrient runoff and siltation from coastal areas are of considerable concern. To address the external influences and to strengthen the other existing management arrangements, the authority has recently initiated the development of a 25-year strategic plan.

Wide participation is the key to success with marine resource management. In this case, development of the Strategic Plan has involved all other key agencies, local government authorities, native groups, and other stakeholders. The plan establishes five-year and one-year objectives to be achieved jointly by all groups and agencies. In particular, the ICZM-type plan ensures that the needs of the Marine Park are adequately addressed by integrated catchment management strategies and planning for adjacent coastal areas.

In summary, lessons for marine and coastal management that might be learned from the Great Barrier Reef Marine Park include:

- The advantages of adopting a holistic or whole ecosystem approach
- The advantages of establishing an independent authority with a very strong legislative mandate to focus solely on management of the protected area
- The importance of establishing formal collaborative management arrangements between all relevant levels of government and stakeholders and of establishing processes necessary to control uses of the area
- The realization that decision making should not be put off awaiting "perfect information" but should be supported by the best available scientific information and application of the precautionary principal
- The demonstration that successful management of a marine protected area requires support by the community, who should be involved in the processes of establishing management.

Contributed by — Peter McGinnity, Director of Planning and Management, Great Barrier Reef Marine Park Authority, Townsville, Australia.

4.4 AUSTRALIA, PORT PHILLIP BAY: A FAILED AUTHORITY

Background

Port Phillip Bay is a large marine embayment in southeastern Australia, which hosts the city of Melbourne (population, 3 million) on its northern shores, the city of Geelong (population, 200,000) on the southwestern coast, and many seaside suburbs and smaller coastal towns. The bay has a surface area of about 1,920 square kilometers and a coastline 256 kilometers long; its average depth is about 10 meters (maximum, 24 m).

The coastline is varied, with some cliffs (Figure 4.4.1), extensive beaches, and small areas of mangrove and salt marsh. A Crown Land Reserve—a band 30 meters wide—was declared in the 1870s

Figure 4.4.1. The cliffed coast of Port Phillip Bay at Red Bluff. The coastal reserve is here up to 130 meters wide, backed by a main road and then extensive Melbourne suburbs.

to provide a 100-foot setback along the entire shore. Only a few small sectors of the bay coastline remained in private ownership. Subsequently, about the half the coastal fringe has come under the jurisdiction of government agencies for port operations, navigational facilities, military purposes, and wastewater treatment. The rest remains as a reserve for public uses, including facilities to support waterborne recreation, car parks, picnic areas, and nature reserves.

Problem

Port Phillip Bay and its coastal fringes are intensively used for recreation by the people of Melbourne and Geelong and also by those who live in bordering areas and by visitors from other parts of Australia and overseas (1).

Major recreational activities are both water based (swimming, power boating, water skiing, fishing, and sailing) and land based (camping, picnicking, walking, beach games). Such activities require planning and management to avoid conflicts and maintain the environmental quality and the scenic and scientific interest of Port Phillip Bay and its coastal fringes. Over past decades much political activity, public debate, and media involvement has been spent on management planning but, despite attempts to devise an appropriate management agency, results are few.

Solution

A special Port Phillip Authority was established in 1966. It was allotted the zone extending 200 meters landward and 600 meters seaward from the low tide line (2). It was multiple use oriented with a mandate to assist the Victoria government by coordinating development in the Port Phillip area, preserving its beaches and natural features, preventing deterioration of the coastal environment, and improving facilities "to enable the full enjoyment of the area by the people." It was a good example of Situation Management in the ICZM context.

The authority consisted of an appointed chairman and delegates from four government departments (Department of Crown Lands, Soil Conservation Authority, Ports and Harbors Division, Town and Country Planning Board), with a Consultative Committee including representatives of bayside councils, other departments and agencies, and the general public. Its charter stated that no structure be erected and no works undertaken without consent of the authority, whose recommendations required approval by the Minister for Conservation.

During the ensuing decade the authority addressed such problems as beach erosion, nearshore water pollution, re-vegetation of eroded reserves, and public access to the shore. It devised a permit procedure to approve development projects and other changes in utilization in the coastal fringe and nearshore areas. In its first decade, it received 734 applications for permits, of which 90 percent were approved. The authority also consulted hinterland and offshore management agencies over such problems as nearshore pollution from rivers or drains and discharges from ships at sea.

A typical issue was the proliferation of artificial shoreline structures, such as sea walls, groins, and breakwaters, designed to combat coastal erosion and provide mooring facilities for boats of various kinds (Figure 4.4.2). Depletion of adjacent beaches, a major problem, was partly offset by artificial beach restoration projects. Breakwaters and marina proposals prompted vigorous opposition from opponents of exclusive use by boat people and from environmentalists who were seeking to maintain a natural setting for coastal recreation.

The work of the authority was impeded by limited funding and by existing agencies' unwillingness to share authority. Soon the authority became essentially a permit-issuing agency, was branded an unnecessary bureaucracy, and disbanded in 1980.

As a sequel, coastal management was put in the hands of three agencies: the Ministry of Planning and Environment, the Department of Conservation, and the Melbourne Port Authority in cooperation with other government agencies interacting with 18 municipal councils around Port Phillip Bay, which delegate the work of day-to-day management to no less than 39 committees of management!

Lessons Learned

Interagency rivalry is the burden borne by all integrated coastal zone management (ICZM) programs. In this case, the attempt to achieve ICZM for Port Phillip Bay by establishing the Port Phillip Authority

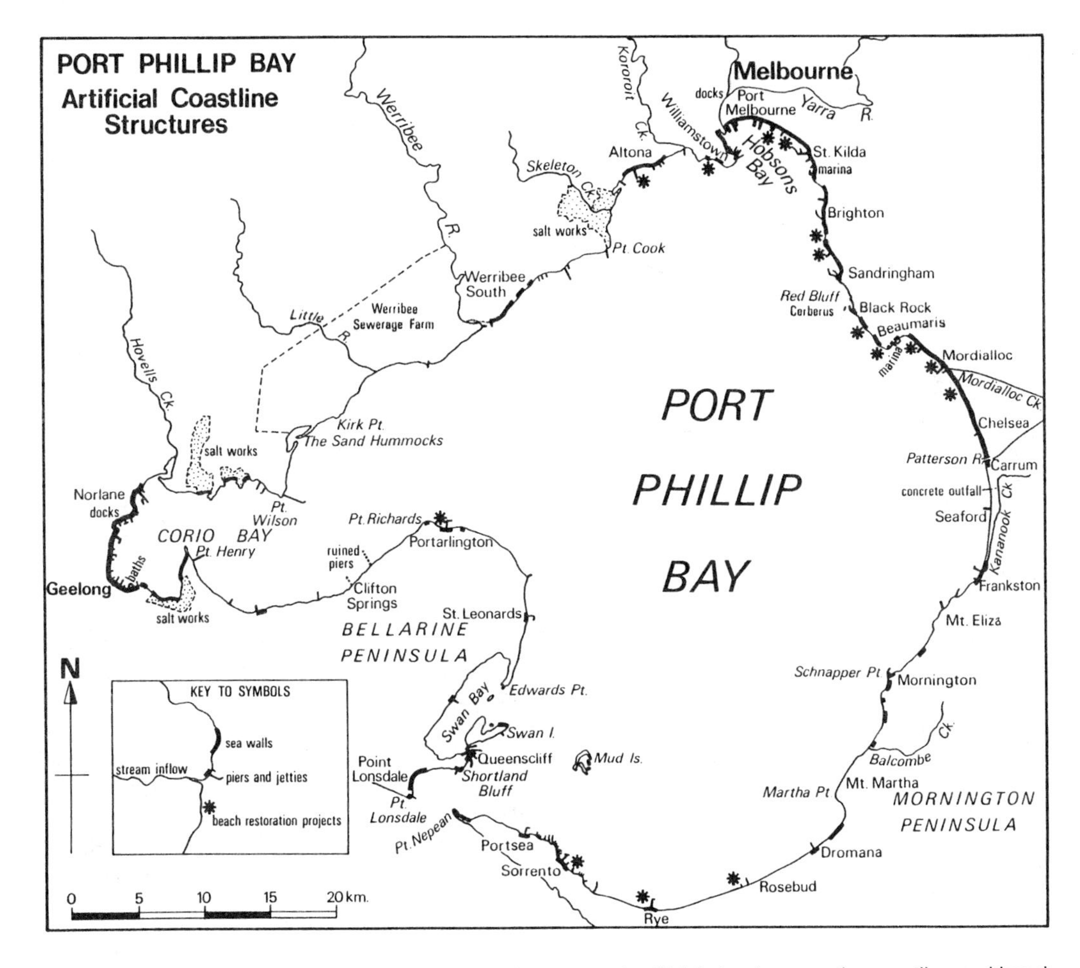

Figure 4.4.2. Port Phillip Bay, Australia, showing the extent of artificial structures on the coastline and beach restoration projects.

failed largely because of this rivalry and other politics. Government departments were reluctant to share authority. Individuals and organizations whose proposals were disallowed opposed the intrusion of this "bureaucracy." Instead of granting the authority sufficient power and sufficient funds to deal fairly and quickly with development proposals, the government was persuaded that it was an unnecessary bureaucracy that could be disbanded as a cost-saving exercise.

Unfortunately, conservation groups and the public often fail to support ICZM of the kind the Port Phillip Authority had achieved. Coastal planning and management in Port Phillip Bay has reverted to the old-style political resolution of competing demands by government agencies with inevitable disputes.

Too many ICZM decisions result more from political expediency, electoral maneuvering, or media pressure than from any scientific or economic consequences. This confused and disintegrated approach to coastal management leads to increasing dissatisfaction and tensions between government departments and agencies, local councils, conservation groups, and the general public, as happened in this case.

More success might come from larger jurisdictional units. For example, it has been suggested that there should be a single integrated ICZM authority for the whole of the state of Victoria, with regional branches along the coast (one would be Port Phillip) and sufficient power and sufficient funding to plan, develop, and conserve the coastline. Unfortunately, this is not yet politically viable because of dissenting agencies and pressure groups intent on implementing their development projects against the public interest.

Note that a similar failure of a regional ICZM program occurred in the Fraser River estuary of British Columbia, Canada (See reference 167 for main text).

Contributed by — Eric C. F. Bird, Environmental Change Unit, 1a Mansfield Road, Oxford University, United Kingdom.

REFERENCES

1. Bird, E.C.F. and P.W. Cullen. 1990. Recreational uses and problems of Port Phillip Bay, Australia. In: *Recreational Uses of Coastal Areas*, P. Fabbri (Ed.). Kluwer Academic Publishers, The Netherlands. Pp. 39–51.
2. Cullen, P. W. 1977. Coastal management in Port Phillip. *Coastal Zone Management Journal*, Vol. 3. Pp. 291–305.
3. Bird, E.C.F. 1990. Artificial beach nourishment on the shores of Port Phillip Bay, Australia. *Journal of Coastal Research*, Vol. S1, No. 6. Pp. 55–68.

4.5 BARBADOS: AN EXAMPLE OF INCREMENTAL COASTAL MANAGEMENT

Background

Within the Eastern Caribbean Islands, tourism is often the major and fastest growing industry and subsequently the pressures on the coastal zone are numerous. Despite this, integrated coastal zone management (ICZM) is still a fairly new concept in these islands. Although some islands have attempted to set up ICZM, this has been done in an incremental manner, starting on a small scale and usually as a response to a particular and serious problem. In this scenario, one particular agency will take the lead role in responding to the problem and will coordinate with other agencies. This leads to the investigation of other problems in the coastal zone and so an ICZM unit evolves. The entire ICZM process may take as long as 30 years; this allows time for training of local personnel and for the public and political levels of society to understand the need for ICZM. Such an approach is also attuned to a particular island's economic situation.

Problem

Barbados is a small Eastern Caribbean Island with an area of 450 km^2 and a population of 250,000. The economy is based on tourism, agriculture, and small-scale manufacturing. The most important sector of the economy is tourism, which is heavily dependent on coastal resources ("sun, sea, and sand")—more than 90 percent of tourism facilities are located along the coast.

Until the 1950s, the coastal zone was sparsely inhabited because of a prevalence of seastorms and flooding by the sea. However, with the advent of tourism in the 1960s, the coastal swamps were cleared and developed with hotels and residential infrastructure. This was particularly evident on the south and west coasts. The east and north coasts, which are exposed to high wave energy and are often rocky and cliffed, have not been so intensely developed.

During the 1970s, coastal erosion, which was related to both natural and human-induced factors, became a serious problem. Most of the coastline on the west and south coasts had become permanently anchored by hotels and other infrastructure; thus, as the erosion continued, the beaches narrowed and steepened and in several instances the hotels themselves were endangered. However, the government had no expertise in the field of coastal erosion, and, although sea defense structures required planning permission, in most cases coastal protection measures were left up to the individual property owner. The challenge in Barbados is not only to protect the coastal infrastructure, but also to conserve beaches, which are so important to tourist industry.

Solutions

In 1983, the government received assistance from the Inter-American Development Bank (IDB). A team of consultants was hired for a 12-month period to conduct a diagnostic and pre-feasibility study of the erosion problems on the south and west coasts. At the same time a government agency, the Coast Conservation Project Unit (CCPU), was established to provide counterpart services. The project also included on-the-job and post-graduate degree training components. Once the project was completed, the government maintained the CCPU, although as a temporary agency without established, pensionable posts. Thus, the CCPU became the nucleus of a future CZM agency.

Between 1984 and 1991, the CCPU continued in the same capacity as a research and advisory agency. It consisted of some 12 persons, and 50 percent of the staff were professionals. The agency was funded entirely by the government. The main functions of the CCPU were environmental monitoring, coastal development control (with the Town Planning Department), advice to the public and other government agencies on sea defense measures, management of beach accesses (with other agencies), and the design and implementation of small-scale erosion measures such as re-vegetation. The environmental monitoring consisted of beach profiles and wave and current measurements. In addition, water quality surveys were conducted in conjunction with other agencies.

In 1991 a three-year feasibility and pre-design project started, funded by the IDB. A major component of this project is institutional strengthening and the development of legislation to assist the Coast Conservation Unit (CCU) to become a permanent agency. In addition, the three-year project designed and tested (through a series of pilot projects) various ways to protect the coast of Barbados. Also, a formal ICZM plan is under development. The next stage planned is a major investment project.

Lessons Learned

From the Barbados experience, it appears that an integrated ICZM agency can evolve gradually and incrementally if it is moving in the right direction (see also "Sri Lanka" case study). In Barbados, coastal management has evolved as a response to a specific problem, namely, beach erosion. This is the case in most of the small islands of the Eastern Caribbean, where islands do not set out to establish an ICZM program; rather, they set out to solve a specific problem. The solution and responses to problems can evolve into an ICZM program. This was also the case in the British Virgin Islands (BVI), where there were two major problems, both of which stemmed from increasing levels of coastal development—loss of mangrove habitats and beach sand mining. Concern about these problems led to the establishment of a Conservation and Fisheries Department, which is now the nucleus for an ICZM program (1).

In the Barbados case, one important lesson learned during the first phase of the IDB project in 1983 was that ICZM cannot so easily be achieved on just one coast of a *small* island. In the small island context, the entire coast must be considered most probably, although more emphasis can be placed on the "problem" situation (see "Situation Management" in Part 2).

Another important lesson to be learned from this case study is the time involved in establishing an ICZM program. It may take as long as 30 years to establish one fully if starting from scratch. This time span is necessary to allow adequate training of local personnel and to provide for public and political levels of understanding of the need for ICZM. For instance, in Barbados, the groundwork was laid in the 1970s, as the erosion problem became apparent and the pressure mounted for the government to take some action. Eventually, the government did take some action and established the CCU in 1983 and also obtained assistance from the IDB. However, it is anticipated that it will not be until the end of the 1990s that Barbados will have a fully integrated program. The same time factor can be seen in the development of ICZM in the BVI (1).

While recognizing that the full establishment of ICZM may take many years, much can be achieved on an informal basis before the legislation and the ICZM plan are in place. Inter-agency committees are a very important part of the process of integration. In addition, it is possible, by working with other agencies, to implement certain measures through existing legislation. Critical agencies in this respect are the planning, ports, fisheries, and public works agencies. While coordination is the key to successful coastal management in these small islands, it is clear that *one government agency must take overall responsibility for ICZM.*

The slow evolutionary process of ICZM is the prime method by which it can be achieved in these small island states. An important factor that must be considered is that in very small islands, where the population is around 100,000 persons or less, every development decision, however large or small, is taken at the political level. Thus, whether the application is for one mooring buoy or a large marina, the decision will be taken by the political directorate. The coastal manager/committee do not take the decision but merely give advice.

Contributed by — Gillian Cambers, COSALC I, Sea Grant College Program, University of Puerto Rico, Mayaguez, Puerto Rico 00681.

REFERENCE

1. Cambers, G. 1993. Coastal zone management in the smaller islands of the Eastern Caribbean: an assessment and future perspectives. *Proceedings of the Eighth Symposium on Coastal and Ocean Management, New Orleans, Louisiana.* American Society of Civil Engineers.

4.6 BERMUDA: TOURISM CARRYING CAPACITY AND CRUISE SHIPS

Background

After the success of 1980, when tourism arrivals broke the 600,000-visitor mark for the first time, the tourism industry in Bermuda experienced a general decline. In 1984, in response to a business community request to breathe new life into the tourism industry, the then-Tourism Minister relaxed a conservative cruise ship policy that allowed only four ships in Bermuda's waters at any one time. That policy changed to allow up to seven cruise ships during a given week and thus allowed the introduction of regularly scheduled weekend cruise calls, which were previously a rarity in Bermuda (3).

The result was that in 1985 there was a 28 percent increase in cruise passengers to 142,903. In 1987, after a minor decrease in 1986, cruise arrivals soared to 153,437 with total arrivals reaching a record 631,000. The pressure on the infrastructure, particularly during the peak months of July and August (Figure 4.6.1), led to calls from the community to cut back on arrivals. As a result, the Department of Tourism implemented a new policy that restricted the capacity of peak season cruise arrivals. Certain factors, however, caused the policy to be short-lived.

Problem

Total arrivals of nearly 610,000 visitors in 1980 consisted of 492,000 staying visitors, a record that is unbroken today. Commercial hopes in 1981, therefore, were quite high. However, a strike by hospital

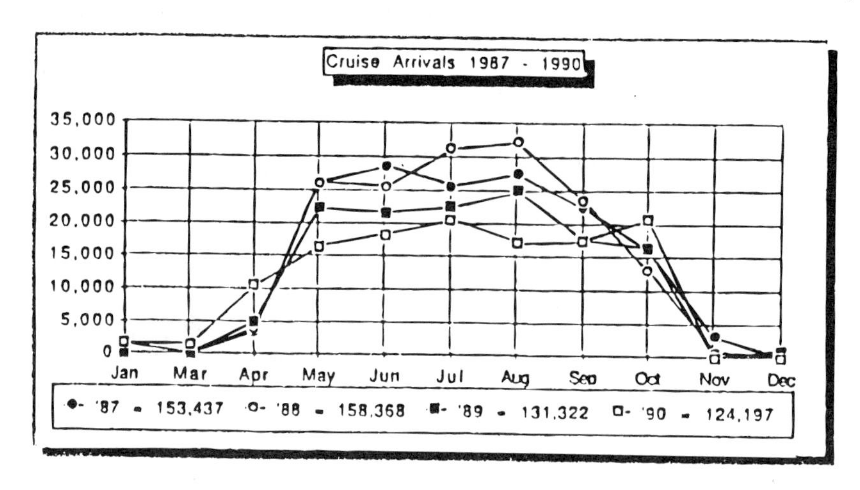

Figure 4.6.1. Cruise ship arrivals at Bermuda ports peak in the summertime. Averages for 1987 to 1990 are shown.

employees in May of that year spread to other sectors of the economy and eventually became an island-wide strike. Hoteliers, fearing that they could no longer provide the level of service that visitors had paid for, closed their doors and sent their guests home. The strike and the negative international press that it generated significantly contributed to the overall decline of 12 percent in tourist arrivals for 1981.

In the two years that followed, only moderate increases were made and it was clear by 1984, when arrivals fell by 7 percent, that Bermuda's problems went beyond the strike of 1981. The United States, which provides 86 percent of Bermuda's visitors, was experiencing a recession. Tourism on a global scale was becoming more competitive, and the cruise vacation was increasing in popularity. Perhaps the main factor contributing to the island's woes, however, was that the Bermuda product, and in particular its physical plant, had become *mediocre* to the discerning traveler.

Also in 1984 there was a change in tourism ministers, and this led to a relaxation of the cruise policy. The policy allowed up to seven cruise ships in the island at any one time, although in reality it seldom reached this number. Ships sailing at that time did not have the capacity of ships today and thus the impact of even seven ships would have been less. The relaxation of the policy resulted in additional ships with larger capacities and weekend calls, which ensured a seven-day-a-week cruise presence. In 1985 there was a 28 percent increase in cruise arrivals and then a decline of 8 percent in 1986.

In 1987 there was an unexpected boost in tourism to Bermuda. Terrorism in Europe caused North Americans to vacation at home or closer to home. With a new advertising campaign in place and a weak dollar, the potential for a successful year was great. By the end of the year, tourist arrivals to Bermuda, both air and cruise, had increased dramatically to 631,314 arrivals, a record even unto today, with cruise arrivals breaking new ground at 153,437. With this increase in arrivals, however, there was a concomitant increase in the level of visitor complaints.

The central concern *now* was that Bermuda was attracting more visitors than it could handle. And there was some truth to this argument. In 1987 there were four months (May, June, July, and August) when the number of arrivals had exceeded 80,000—1.5 times more than the resident population. On any given day of the week, there could be as many as 20,000 visitors on the island's 21-square mile land mass (Figure 4.6.1). Residents wrote letters to the editor of the daily newspaper expressing critical views, and Bermuda's hoteliers were also not happy because the bulk of the increase in visitors was from cruise ships. In a statement for the Tourism Minister, they stated that the sheer numbers of cruise passengers endangered the island's advertised image as a quality destination because the infrastructure could not support such crowds. On the other hand, the retail business community, with very little exception, wanted only more and more.

In 1993 the 1987 capacity was achieved when two cruise lines serving Bermuda replaced their ships with newer and larger ships, something that was permitted in their contracts. In addition, the business community in the east-end town of St. George's, and others, wanted a ship solely for that region. With 1993 being an election year, a dedicated ship for St. George's became an election issue. Before the year's end, a deal was struck for the 1,056-passenger Royal Majesty to sail from Boston beginning in May 1994. Cruise capacity for 1994 now exceeds 183,000, and passengers are likely to top 175,000.

Solutions

The three basic issues of tourism had once again come to the fore: its economic, social, and environmental impacts. If the Ministry of Tourism was to carry out its mandate (3) effectively to promote and control the tourist industry for the benefit of Bermuda (4), it had to take into account these issues. A commissioned study among visitors during the peak period in 1987 found over-crowdedness to be only a minor issue. Another commissioned study, to determine resident attitudes toward tourism, was mostly favorable. Major concerns, however, were the impact on the community of the increased number of visitors and the quality of accommodations. With regard to the environment, ships were required by law to dispose of their sewerage by connecting to the local facility. Laws were already in place to deal with ship emissions and other environmentally unsound practices.

On the economic front, tourism's contribution to Bermuda's economy is strong:

- Forty percent of public revenue is derived from tourism.
- Total business revenue generated exceeds $1,225 million.

- Total income is over $542 million.
- Impact on the balance of payments is nearly $150 million.
- Employment is nearly 13,000, accounting for just under half of the workforce, with another 6,650 jobs being affected (4).

After taking the three impact issues into consideration, the government in 1988 released its new cruise ship policy, which was to take effect from 1990. The main planks of the policy were:

- Generally to limit cruise passengers during the peak season of May to October to 120,000, with 4 regularly scheduled ships and a maximum of 12 occasional callers
- To schedule no weekend cruise calls
- To encourage scheduled and occasional callers to visit during the November–April period.

The policy was expected to have at least four benefits: (a) to reduce the demand on the infrastructure, (b) to spread arrivals throughout the year, (c) to retain the quality of life for Bermudians, and (d) to maintain Bermuda's reputation as a premier resort destination among vacationers.

Lessons Learned

It is most difficult to find general agreement on limiting tourism. Powerful merchant politics usually force open ended tourism—*more is better*—with little thought to the environmental or social consequences. When in 1990—the year of the Persian Gulf crisis—cruise arrivals fell some 7,000 passengers short of the stated goal but air arrivals increased by 4 percent to 435,000, merchants were quick to point out that the carrying capacity policy had failed.

In 1991, when the Gulf crisis became the Gulf War, air arrivals slumped to a 20-year low of 387,000. Cruise arrivals, however, rebounded to 128,000. The lingering effects of the war, combined with the worldwide recession, caused air arrivals in 1992 to fall further to 375,000. In addition, tourism expenditure fell from $490 million in 1990 to $443 million in 1992. The merchant community, reeling from the effects of the decrease in expenditure, put mounting pressure on government to have cruise arrivals increased to the 150,000 level.

Conflicts over tourism never seem to go away. Bermuda could again be faced with the social carrying capacity problems experienced in the summer months of 1987. The Ministry of Tourism is aware of this and is working with the Chamber of Commerce to ensure that no adverse effects of the increased level of tourists are experienced.

Contributed by — Cordell W. Riley, Bermuda Department of Tourism, Hamilton, Bermuda.

REFERENCES

1. Archer, B. 1989. The Bermudian Economy—An Impact Study. Ministry of Finance, Government of Bermuda, Hamilton, p 71.
2. Bermuda Government, Approved Estimates of Revenue and Expenditure for the Year 1990/91, 1990, Hamilton, p 271
3. Riley, C. W. 1990. Bermuda's Cruise Ship Industry-Headed for the Rocks? In: Miller, M. L. and J. Auyong Eds, Proc. of the 1990 Congress on Coastal and Marine Tourism, Vol. II (NCRI-T-91-010), pp. 159–164.
4. Riley, C. W. 1991. Controlling Growth While Maintaining Your Customer Base. Proceedings of the 22nd Annual Travel and Tourism Research Association Conference, University of Utah. p 65–73.

4.7 BONAIRE: CARRYING CAPACITY LIMITS TO RECREATIONAL USE

Background

All of Bonaire's inshore waters have been managed as the Bonaire Marine Park (BMP) since the early 1980s to balance resource protection with recreational use of the island's fringing reefs (5). (1).

With a steady increase in the number of visiting divers to the park (the number of dives made in BMP increased from about 50,000 in 1981 to 180,000 in 1991), there was growing concern with the impact from recreational uses on the reefs and the long-term sustainability of such uses.

Problem

Bonaire is one of the five islands making up the Netherlands Antilles. It has an area of 288 km^2 and a population of just under 11,000, and it is situated 100 km off the Venezuelan coast. Rainfall is 500 mm per year, and the absence of permanent freshwater outflows has allowed the development of a continuous fringing reef around the entire island and its sister island, Klein Bonaire. The quality of the reefs and their easy accessibility has made Bonaire a diver's paradise. Some 40 percent of all visitors to Bonaire are divers, and dive-based tourism generated an estimated gross revenue of US$23.2 million in 1991 (1, 3).

To protect the inshore waters and reefs, the non-governmental Netherlands Antilles National Parks Foundation (STINAPA) in 1979 embarked on a three-year project to create the Bonaire Marine Park, funded by World Wildlife Fund-Netherlands and the Dutch and local governments. Since the project failed to secure a long-term source of funding for management, day-to-day management of BMP gradually became less effective, and by 1986 the park had essentially become a "paper park"—a park that existed only in documents with no management or enforcement.

Solutions

Rising concern over the lack of management caused the local government to commission an evaluation and a revitalization plan for BMP in 1990. The revitalization process took place during 1991 with funding provided by the Dutch government, and by 1992 the park was under active management again and core funding was secured through a system of divers' fees (2).

While the revitalization of BMP was taking place, the World Bank, in an attempt to analyze the success of marine parks in protecting marine biodiversity while at the same time producing economic benefits from development, selected Bonaire as the site for a case study on the ecological sustainability and economic feasibility of marine parks. The study, which addressed (among others) the impacts of recreational use, was quite timely, considering the existing concern over possible degradation of sites as a result of heavy recreational diving.

The major problem for the ecological assessment of BMP was the absence of a time component in data collection on recreational impacts. Monitoring stations had been set up when the park was established, but monitoring was discontinued because of the financial problems outlined above. Fortunately, an extensive reef-mapping project conducted in 1981–82 (4) provided good baseline data for comparison with the 1991 situation.

The World Bank study focused on a comparison of two heavily dived sites and one intermediately dived site, with controls that were selected on the basis of similarity with the test sites in reef structure, coral cover, and exposure, but which received little or no recreational pressure. Four of those sites coincided with the original monitoring stations identified in 1980. The study involved a simulated time component by virtue of its working hypothesis that, if significant differences in coral cover were to be found between a heavily dived site and a control site otherwise exposed to the same impacts, they would have to be attributed to the impact of recreational activities.

The results of the study indicated that the percentage of live coral cover (as measured by tracing and digitizing live coral colonies in 15 photographs per dive site, each photo representing 1.17 m^2 of reef area) was significantly lower in the heavily dived sites compared with the controls, but not in the intermediately dived site. These findings are supported by a comparison of the 1981–82 estimates of coral cover with the 1991 results of the photoanalysis. The comparisons between sites and over time clearly indicate that there is an impact from recreational use.

To address the questions of how extensive the impact might be in relation to the park as a whole and of what is considered "acceptable impact," the study looked at the extent of impact at a heavily dived site. The study arbitrarily concluded that the decrease in coral cover observed at the two heavily dived sample sites was not aesthetically acceptable and in fact exceeded the carrying capacity of those particular sites. The linear extent of "unacceptable" impact was determined by sampling intermediate sites between two heavily dived sites and appeared to be more than 100 m but less than 260 m from the mooring (anchoring in Bonaire is essentially prohibited, so all diving takes place from permanent moorings).

One effect of placing permanent moorings is to concentrate diving at certain spots, an effect that specialists in marine conservation generally agree is beneficial as compared to spreading activity widely over the marine landscape. But any such conclusion has to take into account the specific vulnerabilities of habitats at the subject site.

The comparative dive statistics of the heavily and intermediately dived sites suggested that the recreational impact does not become significant until a site receives consistently 5,000 dives per year or more for a number of years. There seems to be a critical level of visitation above which the impact becomes significant and aesthetically unacceptable.

If the critical level theory can be confirmed by further studies, it would have very important consequences for the carrying capacity of the BMP for recreational diving. Not only did a number of popular dive sites already receive more than 5,000 dives per year in 1991, thus exceeding the presumed carrying capacity for individual dive sites, the theory also represents a handle on determining the carrying capacity for the park as a whole.

The total "diveable" coastline was estimated at 52 km, and if moorings would be spaced 600 m apart the park could have a total of 86 dive sites. At 4,500 dives per year per site (just below the critical level), the maximum theoretical carrying capacity would be 387,000 dives per year. This estimate incorrectly assumes an even distribution of dives over all sites (this is incorrect because of differences in popularity of sites and accessibility). A more practical estimate for maximum carrying capacity was therefore set at about 200,000 dives per year (1, 3).

In view of a recommended total tourist carrying capacity of 100,000 visitors—40,000 of whom would be divers (2)—the 20,000-dive limit would allow 5 dives per visitor. Because the carrying capacity study that identified the 100,000 limit was aimed at socio-economic factors and because our reef use limit is in agreement, one must assume that the ecological and socio-economic aspects are in accord.

The conclusions and recommendations of the study have been incorporated in a study on the integrated socio-economic development of Bonaire (2), which report has been endorsed by the Island Council of Bonaire. If the policies adopted by the Bonaire Island Government are implemented, Bonaire could be an example of the significance to limit growth of the tourism industry, with particular reference to limiting impact of recreational use of the marine environment.

Lessons Learned

The following lessons can be derived from the Bonaire case:

1. There *are* definite tradeoffs between protection of marine resources and using those same resources for economic gain.
2. The impacts of recreational use in the Bonaire Marine Park are limited in extent and concentrated in high use areas. They are decreasing the aesthethic appeal of certain areas but are not yet threatening the biological integrity of the reef system.
3. Effective management of resources that are used to produce economic benefits is extremely important to minimize the tradeoffs (effective management could have contributed to an early diagnosis of recreational impacts in Bonaire and to limiting those impacts).
4. Ongoing monitoring of the resources and collection of statistics on recreational activities are key elements for documenting and mitigating potential impacts from recreational use and must be an integral part of any resource protection and management program.
5. Where tradeoffs between protection of marine resources and use of these resources for economic gain are likely to occur, it is essential to assess the carrying capacity in an early stage of resource development (even if it is only a preliminary assessment or if it only concerns social carrying capacity) and to refine the assessment as more information becomes available.
6. If the use of marine resources for tourism and recreation is to be sustained in the long term, the public and private sectors involving tourism must be made aware early on that resource use may have to be limited at some stage.

Contributed by — Tom van't Hof, Consultant, Marine and coastal Resource Management, The Bottom, Saba, Netherlands Antilles.

REFERENCES

1. Dixon, J. A., L. F. Scura, and T. van't Hof. 1993. Meeting ecological and economic goals: marine parks in the Caribbean. *Ambio,* Vol. 22., No. 2–3, May 1993.
2. Rapport van de Commissie Sociaal-Economische Aanpak van Bonaire. December 1992. 112 pp.
3. Scura, L. F. and T. van't Hof. 1993. The Ecology and Economics of Bonaire Marine Park. The World Bank, Environment Division. Divisional Paper 1993–44.
4. Van Duyl, F. C. 1985. Atlas of the living reefs of Curacao and Bonaire (Netherlands Antilles). Uitgaven Natuurwetenschappelijke Studiekring voor Suriname en de Nederlandse Antillen, Utrecht. No. 117, 1985.
5. Van't Hof, T. 1983. Guide to the Bonaire Marine Park. *Stinapa Documentation Series no. 11.* Stinapa, in cooperation with Orphan Publishing Company, Curacao, 154 pp.

4.8 CANADA: OFFSHORE SEWAGE OUTFALL AT VICTORIA, B.C.

Background

A marine offshore sewage disposal system has been developed over the past 25 years by the city of Victoria, B.C., on the Pacific coast of Canada (Figure 4.8.1). This has involved many public health, engineering, environmental, and financial decisions of a type that concern sewerage authorities in any country (2).

Victoria has opted for a system that screens *but does not treat* sewage before discharge. The environmental impact is limited to within a few hundred meters of the discharge point. The system has been financed by local property taxes, not by major capital funding from outside the region. There is no public health problem from water-borne infectious diseases.

Problem

In Victoria, B.C., as in most cities, human wastes must be disposed of in some way that does not carry diseases, does not impact on other resources, and is not offensive to the community. Sewage is known to carry pathogens and to be toxic, oxygen demanding, and loaded with waste nutrients. It is smelly and looks awful.

Victoria's sewerage system has to remove and dispose of the wastes of about 300,000 people. It is a coastal region of moderate rainfall, alongside a well-flushed strait with a net offshore drift. The physical oceanography is reasonably well documented (7).

The issue in developing the current solution centers on whether, in addition, conventional sewage treatment is necessary. There is some local feeling that discharge of untreated sewage to the sea is in principle environmentally undesirable, in spite of documentation of limited impact since 1971 (5). Some regulatory agencies have stated that the Victoria system is in contravention of regulations, although no formal charges have ever been laid.

This regional issue has been elevated to national and international exposure by the news media of eastern Canada and the United States, including a *New York Times* "exposé" (April 20, 1993).

Solutions

At Victoria, the initial solution 100 years ago, in 1894, was the discharge of untreated sewage just below the low tide line. Gradually, more sewers were built to the low tide level and beyond. In 1971 one of these low tide sewers (Macaulay Point) was extended to a deep offshore outfall and the currently preferred solution was initiated.

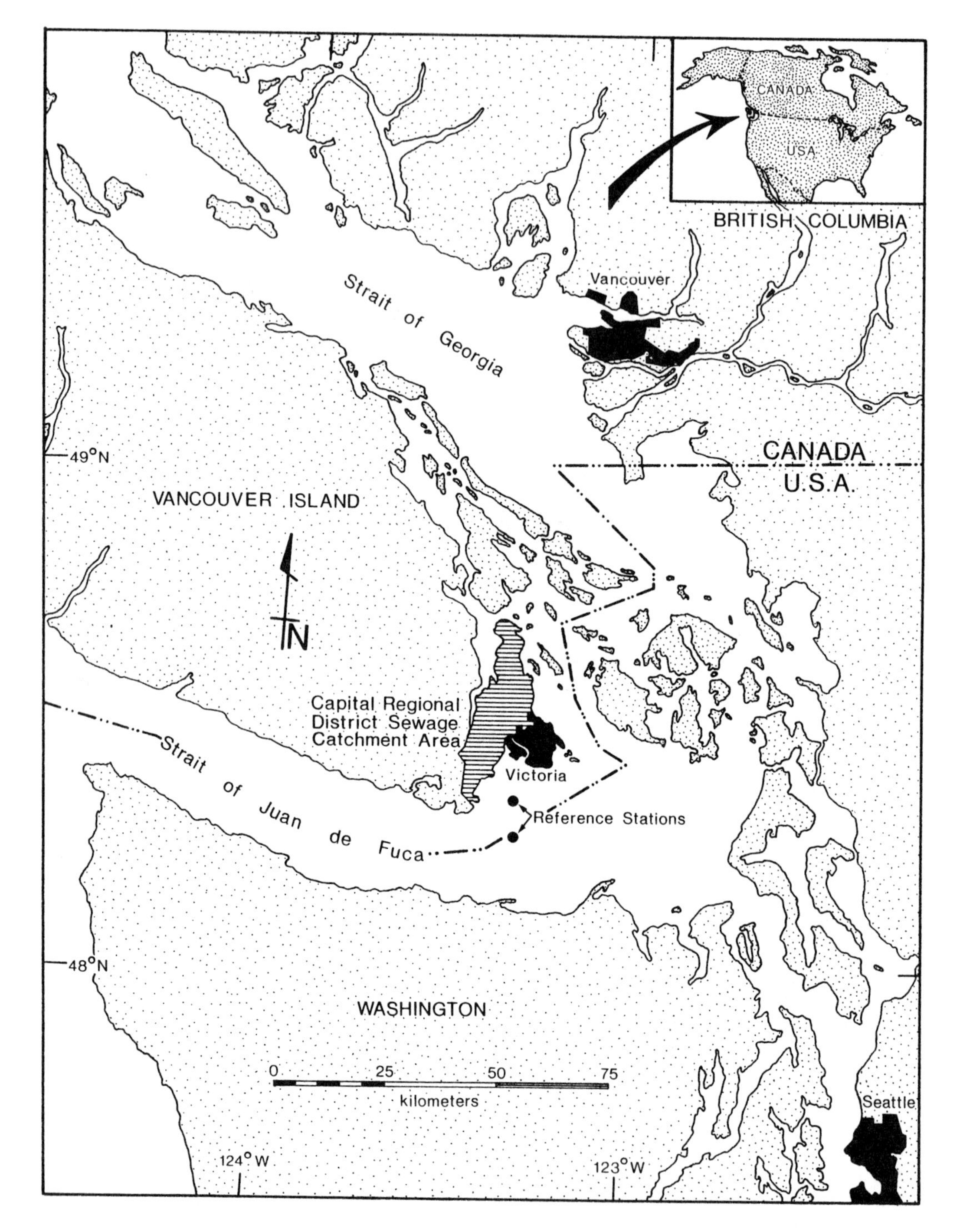

Figure 4.8.1. The location of the City of Victoria on the west coast of North America and the extent of the sewerage authority. Reference stations for the offshore monitoring of sewage dispersal are shown.

The options available to any sewerage authority using a water-based waste-transporting system range from no treatment, through simple mechanical procedures such as screening, to progressively more complex treatment systems customarily labeled primary, secondary, and tertiary treatment. *Primary treatment* consists essentially of ponding the sewage for a few days. *Secondary treatment* involves processing the sewage mechanically or chemically to facilitate biological decomposition (e.g., by air bubbling). *Tertiary treatment* involves extra processing often to achieve recycled, drinking quality water.

All treatments reduce pathogens, suspended solids, BOD (biological oxygen demand), and bad odor. But all treatments also produce sludges, which may be infectious, loaded with toxins, and with a high BOD. These sludges must then also be disposed of in some way that is not polluting.

Screening does not produce a sludge but removes coarse solids (especially the non-decomposables), and only these must be disposed to a landfill. At Victoria, oceanographic flushing is such that chlorination as a final pathogen-killing treatment (which has toxic effects on biota) is unnecessary.

The effluents of treatment systems may still contain high (ecologically damaging) levels of nutrients and some pathogens and solids. At Victoria, sewage treatment is not needed for the protection of public health, and it is not clear how it would benefit other resource uses. A management option of spending millions of dollars on such treatment is troubling to the taxpayers.

Seawater kills pathogens and dilutes, disperses, and recycles nutrients. The dispersed solids deposit below resuspension depths. The 25-year plan started in 1971 has reduced the urban overall system to two offshore outfalls, each with a large catchment area. The design criteria for both was essentially that the sewage plume should rise only to, and disperse at, a trapping depth below a sharp density gradient (the pycnocline or holocline). This same solution was also employed at Larnaca, Greece (see 2.26 "Ocean Outfall Placement" in Part 2). In addition, the dispersing layered field should not drift to shore. System details are provided in Figure 4.8.2 and in the tabulation below:

Item	Macaulay Point	Clover Point
Pipe		
Diameter	914 mm	1067 mm
Materials	Steel	Steel
Offshore	1,717 m	1106 m
Depth	61 m	67 m
Diffuser	Yes	Yes
Overflow pipe	Yes	Yes
Processing	Screens	Screens
1980 flows (m^3/day)	58,000	27,000
Design flows (m^3/day)	100,000	180,000
Permit flows (m^3/day/yr)	54,552 (1975)	63,000 (1980)
Orig. built to low tide	1913	1894
Year extended offshore	1971	1980

Note: In 1991 McMinking and Finnerty Cove outfalls were intercepted and now flow to the Clover Point outfall (see Figure 4.8.2), adding approximately 25,000 m^3/day to the discharge there.

Environmental impact assessment was started, as the 25-year plan was developed, with basic physical oceanography. This allowed calculating by conventional formulae (4) the discharge depth and distance from shore within the known current regime and the predicted future sewage discharge rates.

In 1970, one year prior to commissioning of the Macauley Point outfall, biological and chemical investigations were started and provided baseline data to determine if impact would be as predicted for the design criteria (1). Subsequently, these investigations concentrated on seabed parameters such as sediment chemistry (trace metals, pesticides), coliform bacteria, bioindicator species such as mussels, and biodiversity measured by the variety of species and their abundances (3).

In spite of fish being an important commercial and recreational resource in the area, these were not good bioindicators of impact. Local fish, such as Pacific salmon, are migratory. They remain in the area only briefly and can be subjected to pollution from other areas. The resolution has been that a mixing zone has been permitted around the discharge point. In practice, there is little mollusc harvesting in the Victoria area, especially at 60-m depth.

In 1988, the sewerage authority started a contemporary style re-think of its environmental program by forming a Marine Monitoring Advisory Group (MMAG) of scientists from regional government and university research centers and from regulatory agencies (environmental and fisheries). This group meets 4–6 times a year with the sewerage authority's environmental and public health scientists,

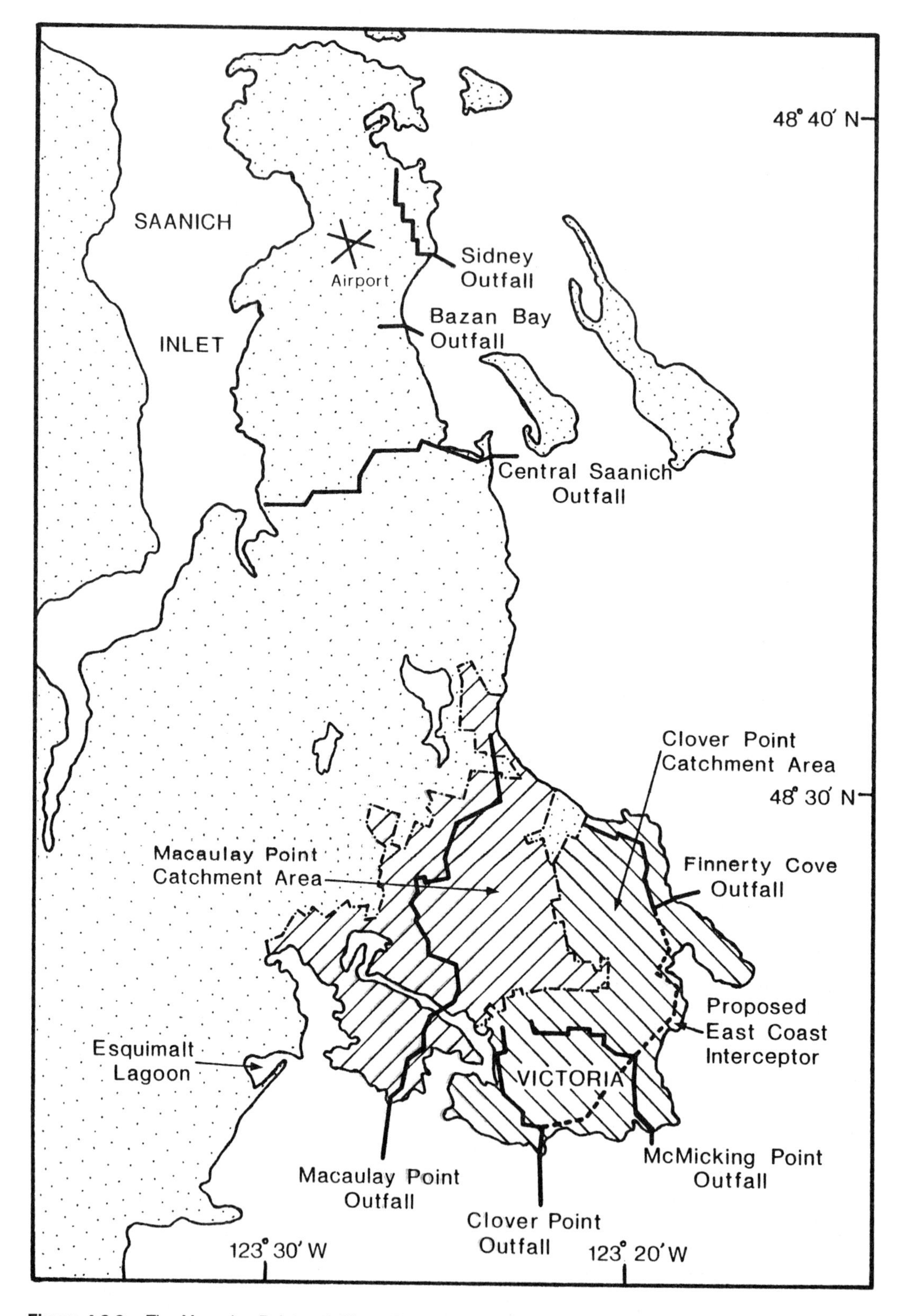

Figure 4.8.2. The Macaulay Point and Clover Point outfalls, discharging screened sewage, have catchment areas encompassing the greater part of the Victoria urban area. There are smaller sewage catchment areas, with secondary treatment systems, in the northern part of the area.

engineers, and physicians. They have completed a review of the "State of the Receiving Environment" following 20 years of offshore sewage discharge (5).

The sewerage authority, on the recommendation of the MMAG, has extended its monitoring program by specific projects. These have been designed to demonstrate the extent and nature of chemical contamination of the seabed and impact on the seabed biological resources (3). Chemical contamination is detectable by many parameters to 400 m and by fewer parameters at lesser levels to 1,600 m maximum distance. Biological effects are even more restricted. There is no detectable effect on the shoreline. The reference stations in Figure 4.8.1 never showed impact from the discharged sewage.

The MMAG is now engaged in designing a new routine monitoring program to meet the more comprehensive contemporary needs for understanding where and how existing impact might occur and what remediation might be appropriate. It is accepted that the frequency of monitoring will vary with the parameters (e.g., it may be appropriate to monitor sediment coliform bacteria annually and benthos every third year). In some cases, such as mapping the biological resources of the area, a review once in 20 years may be sufficient.

In 1992 the sewerage authority made a decision to introduce intensified programs of source control (e.g., controls on toxic effluents of light industries in the area) (3). But in a political reaction, it is also currently required by one of the regulatory authorities to acquire land to house sewage treatment within the next 20 years.

In summary, the Victoria system has been successful in collecting and disposing of human wastes and preventing epidemics of water-borne infectious diseases while producing very limited environmental change. But the system has not avoided conflict with activist perceptions and with some existing government regulations.

Lessons Learned

The most important lesson is that a sewage disposal system for a coastal city may be able to achieve public health protection with little environmental impact for low cost by discharging screened sewage offshore. In so doing, however, it may clash with pre-existing government regulations that have not contemplated the possibility of such systems.

Such regulations should be challenged by appropriate action where they unreasonably preclude the use of cost-effective, environmentally protective systems. It is worth noting that unreasonable regulations in the United States have recently been criticized by the prestigious National Research Council (6).

A lesson confirmed is that it is important to undertake adequate environmental impact assessment prior to the design of such systems. The assessments must be continued indefinitely, as monitoring, in order to check that the design criteria are being maintained. The assessments must be reviewed periodically, ideally by a group of independent experts (such as the MMAG) who can advise on continuation or changes in the monitoring) and on any remedial engineering needed to maintain design criteria.

Other lessons learned include expeditious public release of information about the system and funding sufficient to maintain the environmental assessment program. Both have, at times, been insufficient in the Victoria area. Sewerage authorities need public support that comes from an understanding that their system is environmentally protective and cost effective and that they are being good neighbors to the surrounding communities.

Contributed by — Derek V. Ellis, Biology Department, University of Victoria, Victoria, B.C., Canada.

REFERENCES

1. Balch, N., D. V. Ellis, J. Littlepage, E. Marles, and R. Pym. 1976. Monitoring a deep marine wastewater outfall. *Journ. Water Poll. Control Fed.*, Vol. 48, No. 3. Pp. 429–457.
2. Ellis, D. V. 1989. Sewage—Victoria (Canada). In: *Environments at Risk: Case Histories of Impact Assessment.* Springer-Verlag, New York. Ch. 6. Pp. 126–154.
3. EVS Consultants. 1992 Sediment and related investigations off the Macaulay and Clover Point outfalls. Rept. to Capital Regional District. 193 pp.

4. Grace, R. A. 1978. *Marine Outfall Systems: Planning, Design and Construction.* Prentice-Hall, USA. 600 pp.
5. MMAG. 1993. Scientific consensus on the state of the receiving environment around the Clover and Macaulay Point sewage outfalls: knowledge and concerns. Marine Monitoring Advisory Group Rept. to Capital Regional District. 20 pp.
6. N.R.C. 1993. *Managing Wastewater in Coastal Urban Areas.* National Research Council, USA
7. Thomson, R. E. 1981. Oceanography of the British Columbia Coast. Canadian Spec. Publ. Fish. and Aq. Sci. 56. 291 pp.
8. Thurber Environmental Consultants Ltd. 1992 Capital Regional District, Source Sampling Project. Rept. to Capital Regional District, Victoria. 65 pp plus App.

4.9 CANADA: NATIONAL EXPERIENCE WITH COASTAL ZONE MANAGEMENT

Background

The prospects of a national approach to ICZM for the extensive Canadian coastal zone has been debated, without resolution, for the past 20 years. Although Canada does not have a coordinated nationwide approach to coastal zone management in place, there are successful area-specific applications of the practice in each of the four Canadian coastal regions. This case history examines the reasons why a national approach to ICZM has not been employed and discusses the current conditions, which bode well for its development from the bottom up.

Although one of the largest continental nations, Canada possesses a diversity and abundance of coastal resources and ecosystems along its Atlantic, Arctic, Pacific, and Great Lakes shores which provide significant social and economic benefits to the nation. In terms of the contribution of coastal resources and activities to GNP and the way of life of coastal communities, Canada certainly can be considered a nation with significant coastal regions.

Problem

At 244,000 km, Canada has the longest coastline of any country in the world. However, with vast areas of largely undeveloped and sparsely populated coastal areas and coastal environmental quality conditions that are perceived to be in a relatively healthy state overall, the need for a nationwide response to the planning and management of this vast area has not secured the issue of coastal zone management a prominent position on the national public policy agenda. Moreover, long-standing conflicts over jurisdictional responsibilities in coastal areas have further complicated the task of establishing a cooperative approach.

With increasing trends of contamination of coastal waters, loss of valuable wetlands, and growing resource use conflicts, compounded by the virtual collapse of several major fisheries, the recognition of the need for a more coordinated and integrated response to these issues is being rapidly accepted at the community, bureaucratic, and political levels.

The difficulty is recognized for environmental and resource agencies to work cooperatively among themselves and also with the social and economic agencies and to make direct links to national and provincial economic development plans. Some mechanism has to be created to force cooperative action.

Solutions

The political climate for coordination, harmonization, and partnership between the three levels of government and with other stakeholders in the coastal zone has changed dramatically over the past few years. The 1970s and early 1980s in Canada were characterized by federal-provincial jurisdictional conflict, and conditions were not conducive for cooperative and integrated management of coastal resources and environments. The realities of the 1990s—crisis conditions in many of our coastal areas, vastly diminished administrative resources at all levels of government, a matured federal-provincial relationship, and a growing acceptance of the need for local involvement and community action toward sustainable development—have created a positive climate of cooperation and willingness for joint action.

Coastal zone management in Canada today is being implemented in a number of areas in each of our coastal regions. A broad intergovernmental-industry-community partnership approach is in place in the Fraser River Basin in British Columbia on our Pacific coast, a large ecosystem approach is being pursued for the coastal resources of the Canadian Arctic, both basinwide and "areas of concern" approaches are demonstrating success in the Great Lakes, the St. Lawrence river in Quebec is being managed cooperatively by the federal and provincial governments, and 13 community-based management programs are in place in priority harbors and estuaries in the Atlantic provinces.

Although still not formally referred to as coastal zone management, these area-specific ("Situation Management") approaches are applying the principles and achieving the successes prescribed for coastal management efforts elsewhere. Each approach is attempting to address the critical issues that the region is facing, identify solutions, and work toward their resolution through broadly based and fully cooperative working arrangements.

As each program matures, the scope of the partnerships and issues continues to expand and they are moving from their initial remediation and environmental quality focus to embrace the broader objectives of sustainable environmental and economic development. Probably the most significant development has been the recognition of the inextricable link between the sustainability of our coastal resources and environments and the social and economic well-being of our coastal communities.

Lessons Learned

The Canadian experience has demonstrated important lessons. The first and most challenging has been to have the coastal zone recognized, at the political, bureaucratic, and community levels, as a distinct and valuable national resource in urgent need of comprehensive management. Beyond this, it has been necessary to promote ICZM as an efficient, cost-effective, and non-threatening approach to the problems faced by the multiple jurisdictions that have responsibilities in the coastal zone.

The political motivation for ICZM has been generated by rapidly growing crises in the coastal zone—ones that existing administrative structures could not adequately address—and the demonstration of the success of its application on the ground. All of the technical skills and capabilities have existed, but the willingness to share and apply them in a coordinated manner has taken time to develop.

Administratively, we had to demonstrate, by small-scale and non-threatening case studies, that various agencies and stakeholders could work together and that broad participation was not only possible but essential to success. Only by complementing this bottom-up strategy with top-level federal and provincial policy and program support could national ICZM get a foothold. There is now a growing call to link the existing small-scale coastal management initiatives and to guide their further development with consistent national policy and direction.

This *scaling-up* approach from community-based and regional case studies has demonstrated that we can benefit from sharing and joint action both administratively and economically and that full involvement, from the three levels of government, the private sector, and coastal communities is essential.

Canada now has several fairly mature local and regional coastal management initiatives under way; they are demonstrating the success of this integrated type of approach. Several of the Canadian provinces are currently developing or considering provincial coastal management strategies, and the federal government has organized itself to play a central role in this cooperative partnership approach through an Interdepartmental Committee on Oceans, which is mandated to coordinate and guide marine programs and policies at the federal level.

Although Canada is yet to institute a formal approach to coastal zone management at the federal or provincial levels, the essential working relationships and demonstrated successes of action at the sub-national level have put us at the brink of a nationally recognized program.

Contributed by — Lawrence P. Hildebrand, Environment Canada, Queen Square, Dartmouth, Nova Scotia, Canada.

4.10 CHINA: PROSPECTS FOR INTEGRATED COASTAL ZONE MANAGEMENT

Background

The mainland coast of China is about 18,000 kilometers. The island coastline is around 14,000 kilometers. There are gravelly, sandy, and muddy coastal zones. Some of the largest cities in China, such as Shanghai, Guangzhou, and Tianjin, are in coastal zones. Thus, integrated management of coastal resources (ICZM) is of vital importance to China. China has a long history of coastal resources utilization and has achieved a great deal, especially since the nationwide Coastal Resource Investigation in cities and provinces began in 1977.

Problems

Although more and more people have come to realize the importance of coastal resource management, there still are many problems.

1. The level of coastal resource utilization is very low in China. On average, there is only one medium or large harbor for every 1,000 kilometers of coastline, while in the United States there is one every 120 kilometers. The total volume of goods handled in China's harbors is only one-eighth that of Japan's.
2. In recent years, the focus of coastal utilization shifted from agriculture to aquaculture, such as the raising of shrimp and various kinds of shellfish. Extinction of some fish species is imminent due to over-catch. The area of salt fields is also diminishing.
3. Pollution has worsened along some part of the coastlines. Both the quantity and the quality of the coastal resources has deteriorated.
4. There has been a lack of overall planning. The lack of coordination among different governmental departments and geographic regions has resulted in ever-increasing conflicts.
5. The lack of an authoritative organization in charge of the overall planning resulted in the accumulation of various conflicts.

Solutions

The problems discussed above must be solved effectively to conserve the renewable and nonrenewable resources. The importance of the overall management of coastal resources cannot be overemphasized. The overall management should at a minimum include the aspects listed below.

1. The requirements of different governmental departments are properly balanced.
2. The conflicts among different geographic locations are coordinated.
3. Conservation is kept in pace with utilization so that the ecological system is not in a vicious cycle.
4. Both short-term and long-term benefits are taken into consideration and emphasis is put on the long-term interest.

Based on the present status in China, the following measures must be taken in the overall management of coastal resources.

1. Investigation should be enhanced and optimal planning according to the characteristics of various zones should be worked out. And modifications must be made to the original planning as the natural and socio-economic conditions change with time.
2. Rules must be made or completed.
3. An authoritative organization must be formed to supervise the overall management.

Lessons Learned

Attention has been paid to the above aspects and further effort has been put into the overall planning of coastal resources nationwide. Experimental work will be carried out in some parts of the coastal zones, and the results will be applied to other zones. In the last few years, we obtained the following experience:

1. It is necessary to organize the nationwide Coastal Resources Investigation to get a better understanding of the present status of coastal resources management and to provide basic information for future work.
2. The aim of coastal resource investigation is to obtain information and to use it. Therefore, the investigation results should be made easily available to relative governmental departments when requested. In this way strong support can be gained from these departments in the investigation work. A good example is the Shanghai Jingshan Oil Refinery Plant's expansion project. The information on the coastal resources in Shanghai played an important role in the decision making and construction of the project.
3. When studying coastal resource problems, mathematical and physical models should be employed besides the information from the fundamental investigation work in order to achieve optimal results. We have worked out the Yangtse River Mouth Planning in such a way. Now the problem is that there is not yet an authoritative ICZM organization to ensure the implementation of this plan. This is the key problem to be solved.

Source — Shen Huan Ting, Inst. of Estuarine and Coastal Research, East China Normal University, Shanghai, China (See Reference No. 65).

4.11 COSTA RICA: CONTROLLING THE ZONA PUBLICA

Background

The issue of having a public zone status in the coastal area has been a conflicting topic since the first days of overall coastal planning and legislation. Many countries historically have, declared portions of the coastal area as *zonas publicas*—public zones—in governmental areas based principally on the importance that these areas had for military security or public interest purposes. In so doing they have declared that all coastal frontages were government land, not subject to be registered as private lands. On the other hand, the inhabitants who traditionally have lived and benefited from the resources in the coastal areas have basically considered this dynamic area as an "open area" free for their use, one not subject to regulation.

In the public zone issue there have been different interpretations. Some of the basic ones are:

1. The coastal zone is an area of national heritage.
2. The coastal zone must be accessible to all levels of society.
3. The resources of the coastal zone are not for individual usage but for a more broad national interest.

Problem

The principal problem has been that official boundaries have often not been implemented to guarantee public status. However, an exception is Costa Rica, where there is an official demarcation of the public zone specified by law and marked on the ground. The public zone in Costa Rica extends 50 m back from the mid-tide line.

In the implementation of a public zone, there has not been a uniform application in what should be the areas and locations in which such premises should be applied. Some countries (such as Costa

Rica, Venezuela, and Colombia) have declared that all coastal area frontage (principally, the fringe of land between tides) is public zone.

Other countries may declare only parts or segments of their coastal frontage as public zones. For a majority of the countries, the areas that extend up to the mid-high tide line are public zone. But it is important to emphasize that too often the premise is not kept. Each aspect has required specific legislation and determination of the public zone in coastal areas. These have been subjected to occupation by many activities.

Solutions

In the perspective of ICZM, the implementation of the public zone issue is a conflicting one and the answers are still not clear for many countries with coastal frontages. From the different experiences there is a set of pros and cons in the premise of declaring or instituting public zones. There are many recognizable gains in establishing public zones in the coastal area, principally in the intertidal and supertidal areas. Some of these gains are:

1. Public use is ensured for present and future generations, principally of the beaches and adjacent littoral waters.
2. Freedom of transit along the coast is established as a right.
3. The establishment of a public zone is equivalent in many cases to a setback line for constructions or general development.
4. The recreation and general usage of an important part of the coastal zone is available for all the population, without social discrimination.
5. Coastal areas are not included in private ownerships.
6. Fewer structures are built in public zones compared with privately owned coastal land.
7. The economic value of the land is stabilized and not subjected to the highs and lows of demand.
8. The integrity of the coastal zone is maintained for both scenic and natural values.

On the other hand, there are negative effects in the implementation of public zones if not tightly controlled, including degradation from intensive tourism and recreational use of coastal resources:

1. Large concentrations of people cause deterioration of the beaches, dunes, and vegetation.
2. Extensive public use of intertidal areas and adjacent lands at one particular time causes damage.
3. There is contamination of coastal areas, including plastics and other solid wastes.
4. Conflicting uses are instituted because of the broad public zone status.
5. Conflicts and competition are developed between present inhabitants, who exploit coastal resources, and temporary occupants, who are seeking leisure and recreation.
6. More direct impact and pressure are felt in coastal areas that have been declared reserves and protected areas.

Lessons Learned

In the overall context mentioned above, should public zones be instituted? If the experience is properly planned, definitely yes. The coastal area, principally the fringe of land affected by the tides and land directly adjacent, is a national heritage. The following are some specific lessons learned:

1. Being such a dynamic area, the coast should not be subject to static regulations on development.
2. Being a continuously changing zone, it should be treated as a buffer area between the littoral waters and the continent.
3. Having such a restricted width, each country should guarantee its usage to all levels of society for present and future generations.
4. Coastal planning and carrying capacity studies can determine the ideal level of exploitation of the resources.

5. The negative actions derived from public zone declarations can be perfectly well controlled by ICZM.

Source — Robert Chaverri, Consultant, San Jose, Costa Rica (See Reference 46 in Part 5).

Note — For a detailed review of this subject, refer to the paper by J. Sorensen; Reference 322 in Part 5.

4.12 ECUADOR: INTEGRATED COASTAL RESOURCES MANAGEMENT PROGRAM

Background

Ecuador is a small nation that straddles the equator on the west coast of South America. When the coastal region is defined to include the four provinces that extend from the foot of the Andes to the Pacific Ocean and the Galapagos Islands, this region contains nearly all of the nation's industry and export agriculture. Here the focus is on the mainland coast since Galapagos Province is a special case requiring separate treatment.

Population in the coastal region has increased fourfold since 1950 and attained 4.2 million by 1990, which is half of the country's total population (1). Population growth has been accompanied by boom-bust development that has exploited the region's rich resource base. A series of rivers bring nutrient-rich waters to estuaries that support rich fisheries and abundant mangrove wetlands. Offshore upwelling supports pelagic fisheries.

The most recent boom has been in shrimp mariculture. This has transformed almost all Ecuador's estuaries to make it (1993) the world's largest producer of farmed shrimp. In 1992 the 133,336 hectares of ponds generated some $526 million in exports, making shrimp second only to petroleum as a source of foreign exchange (2).

Ecuador has developed its integrated coastal management program through a USAID-funded project that began in 1986. The project has adopted an incremental, adaptive approach to program design with an emphasis on building local capacity and constituencies for resource management initiatives both at the community level and within central government.

Problem

The mariculture industry is a major problem. It has grown in an anarchic, gold-rush-like atmosphere and in some regions cannot be sustainable. Dense ponds, constricted water flow in estuaries, destruction of mangroves, increased use of agrochemicals, and flows of untreated sewage are all combining to reduce water quality, and this is affecting the productivity of shrimp farms. This pattern of overshooting the capacity of the ecosystem is repeated in agriculture and forestry.

The informal economy and governance systems described by DeSoto (1) prevail in the coastal region. Many shrimp farms are built before permits are negotiated. Along most of the 2,859-km coastline, there is no effective governance structure for resolving the mounting conflicts among user groups or considering the long-term consequences of the development process.

Where government has passed laws to halt or mitigate the overuse and misuse of natural resources, there has been little success in implementation. Destruction of mangrove forests to install shrimp ponds actually increased between 1984 and 1991 despite adoption of more stringent laws protecting this important ecosystem. Similarly, laws and programs designed to control deforestation, overfishing, and point sources of pollution have had negligible impacts. In this context the biggest challenge is to build constituencies for improved resource management and to design governance systems that are adapted to the realities of the socio-political traditions of the country.

Solutions

Progress toward a functioning coastal management program has been made through a collaborative project by the Government of Ecuador and the U.S. Agency for International Development (USAID), administered by the University of Rhode Island. The project began in 1986 and was extended through mid-1994. The project adopted a two-track strategy of building constituencies and governance structures simultaneously within central government (Track 1) and at the community level (Track 2). The Programa de Manejo de Recursos Costeros (PMRC) was formally enacted by executive decree in 1989.

The PMRC features for Track 1 an interministerial commission chaired by the Office of the President and an Executive Directorate that functions as a semi-autonomous agency. Track 2 features five Special Area Management (SAM) Zones and a ranger corps led by each of the seven Navy Port Captains that administer Ecuador's intertidal zone (using principles of Situation Management). The Ranger Corps draws together local representatives of central government agencies with enforcement responsibility.

The SAM zones were selected as microcosms for the combinations of problems and opportunities typical of coastal urban and rural areas. Each SAM zone has a local coordinator and a small staff and works with a Zonal Committee representing local user groups and municipal government. By 1994, three of the seven ranger corps units had been formally assembled and were making progress toward more effective implementation of existing resource management statutes.

The PMRC has been built on the principle of decentralization—i.e., that a local capacity for integrated resource management and constituencies at all levels of society are the foundation of a sustainable program (4). The PMRC has placed a major emphasis on working to respond to needs for both development and conservation and to involve the stakeholders and the interested public in each step of the management process.

The program began with a highly participatory process of issue identification and analysis of the four coastal provinces that featured public workshops and wide coverage in the media, led by the Fundacion Pedro Vincente Maldonado, an NGO based on the coast. This enabled implementation following the 1988 presidential campaign.

Since 1988 the program has focused upon developing detailed ICZM plans for the five SAM zones. These were formally adopted at the community level and then endorsed by the National Commission in 1992. The SAM governance process is proving to be effective in making progress on the major resource management issues. Each plan addresses five generic issues: mariculture development and fisheries land use in the immediate watershed, use of the shorefront, environmental sanitation, and mangrove management.

The planning process has featured "practical exercises in integrated management"—such as latrine building and beach cleanup—designed to produce tangible experience in the implementation of the policies and actions contained in the SAM Plans. The PMRC is also experimenting with arrangements for the co-management of mangrove areas.

In 1993, the Government of Ecuador identified the implementation of the five plans as a national priority and negotiated a $16 million loan with the Inter-American Bank that will fund the program through 1999. The funds will be used for education, training, sanitation systems, research, and bureaucracy and administration.

Lessons Learned

1. The socio-political context within which any coastal management program is initiated must be understood and respected. Experience and management tools being used in other nations must be adapted to fit the realities of a given country and must match the capacity and will to implement new policies. This requires an incremental and adaptive approach to program design.
2. Local capacity in resource management and constituencies should be built simultaneously at both the community level and within central government.
3. The many forms of integration required by coastal management occur when the program is issue driven and geo-specific. In a country that has seen little success in implementing resource management initiatives, a focus on selected special areas is an effective strategy for accelerating the learning process.

Contributed by — Stephen Olsen, Director, Coastal Resources Center, University of Rhode Island, Narragansett, R.I., USA.

REFERENCES

1. De Soto, H. 1989. *The Other Path: The Invisible Revolution in the Third World.* Harper and Row, New York.
2. Egas, E. 1993. The Ecuadorian shrimp industry: its history and future outlook. Paper presented at the *Latin American Seafood Conference,* Fort Lauderdale, Fla.
3. Epler, B. and S. Olsen. 1993. *A Profile of Ecuador's Coastal Resources.* Regional Technical Report Series—TR2047, Coastal Resources Center, University of Rhode Island.
4. Olsen, S. 1993: Will integrated coastal management programs be sustainable: the constituency problem. *Ocean and Coastal Management,* Vol. 21, Nos. 1–3, Pp. 201–116.

4.13 EGYPT, SINAI: THE LAKE BARDAWIL SITUATION

Background

Situated on the Sinai Peninsula, the Bardawil lagoon is a haven for migrating Palaearctic waterfowl and a nuturing wetland that supports a rich fisheries industry. The area has been sparsely inhabited over the last millennia, with the indigenous *Bedouins* practicing their traditional nomadic life-style. Large-scale hydrological interventions and commercial developments now planned, will probably have a profound impact on the natural resources.

With an area of 67,000 ha, Lake Bardawil is the largest lagoon in the Arab Republic of Egypt. It is situated in the north of the Sinai Peninsula between Port Said and El Arish. With an approximate length of 85 km, it occupies more than half the length of the Peninsula's Mediterranean coastline (Figure 4.13.1). The man-made hypersaline lake lies in a region that has a very arid desert climate (less than 100 mm per annum), with the exception of a narrow zone of a few kilometers along the coast which enjoys a more moderate climate. Because of the largely arid conditions, the area has long been under-developed and only sparsely inhabited by Bedouins.

Located at the edge of the Sinai desert, the lagoon is a haven for migratory and resident waterbirds. Included on the list of wetlands of international importance of the Ramsar Convention, the lagoon is classified as one of the most important wetlands in the entire Mediterranean Basin. Next to its adherence to this international wildlife treaty, the Republic of Egypt is party to the Barcelona Convention for the Protection of the Mediterranean Sea against Pollution. Under this treaty parties have agreed to protect coastal areas that are important for the safeguarding of the natural resources and endeavor to maintain protected areas including buffer zones.

During the migratory season large numbers of Palaeartic waterbirds frequent the lagoon; white pelican, gargane, and little stint are of particular conservation importance. The area supports one of the largest breeding populations of little stern, and it acts as a wintering site for significant numbers of Palaeartic birds. With other deltaic Egyptian lakes progressively polluted and sustaining sizeable contraction for urban and agricultural developments, Bardawil will increasingly become a more significant stepping stone for migratory birds.

The lagoon provides significant breeding grounds for Mediterranean fish species. Not surprisingly, the lagoon supports an important fisheries industry. Annual yields reach some 2,500 t of fish. In 1987, some 1,200 t of fish, mainly mullet and sea bream were harvested for the export market, with a value of US$12 million. Some 15 percent of the population of the Northern Sinai Governorate is reported to be involved in the fisheries sector (Figure 4.13.2).

The fish production (and species composition) depend largely on maintaining an optimal hypersalinity level, which ranges from 45 to 55 p.p.t. The water exchange flows and thus the salinity depend on the geo-morphological processes that maintain the low and narrow barrier, which also has a bearing on

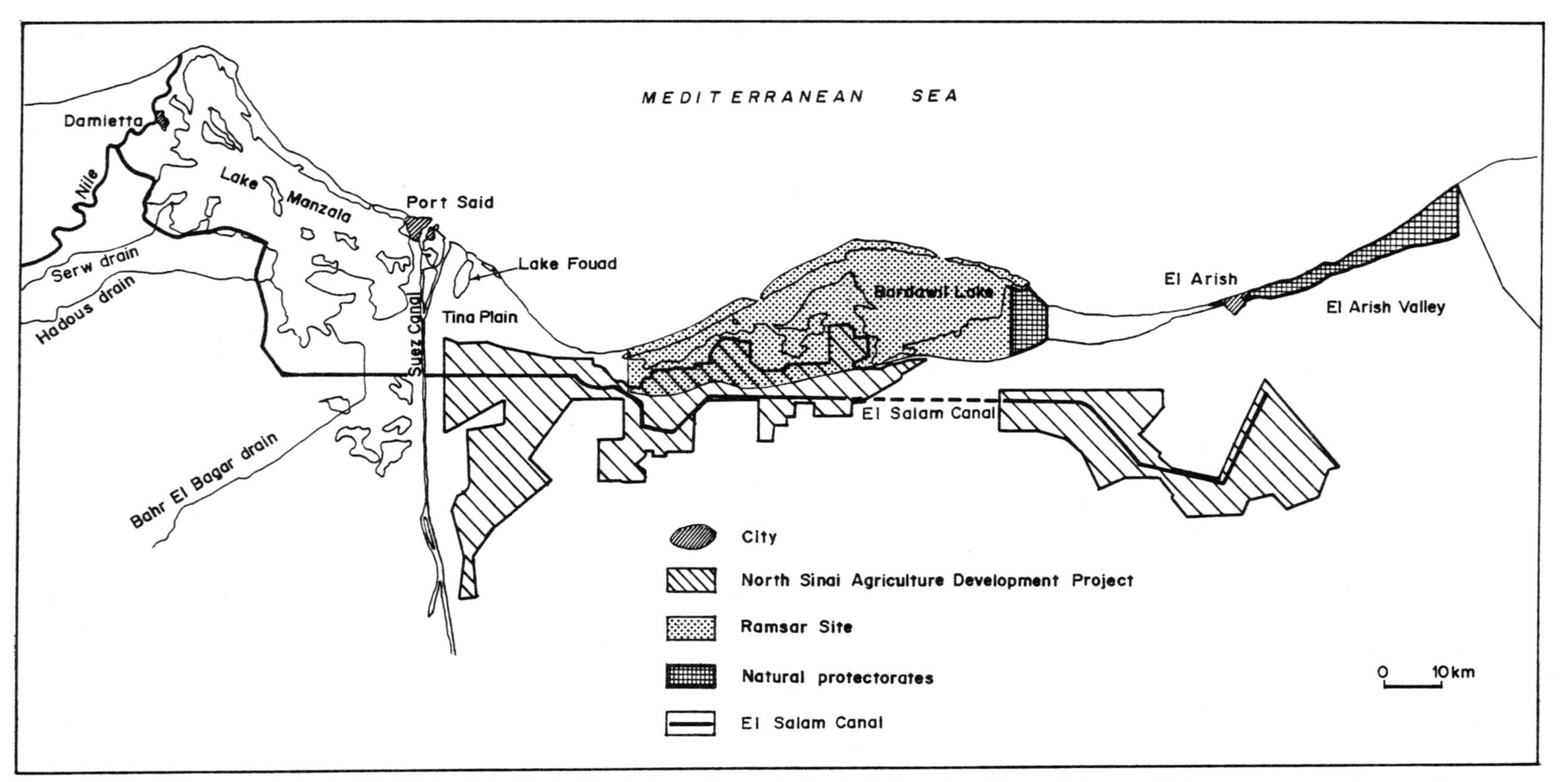

Figure 4.13.1. With an areal extent of 67,000 ha, Lake Bardawil is the largest lagoon in the Arab Republic of Egypt—it is about 85 km long and occupies more than half the length of the Peninsula's Mediterranean coastline.

Figure 4.13.2. Lake Bardawil supports important fisheries—about 15 percent of the population of the Northern Sinai Governorate is involved in the fisheries sector.

opportunities for fish recruitment from the Mediterranean basin. Dredging and coastal engineering works have been applied over the past 60 years to provide favorable environmental conditions for the lagoon's fisheries industry.

In addition to the site's fisheries and biodiversity importance, the area around the lagoon is famous for its archeological finds. Over the last five millennia, Northern Sinai and especially the coastal area around the lagoon, has served as a cultural land bridge, linking the Eurasia continent with the cultivations of the Nile. The coastal land is therefore littered with hundreds of yet unrecorded artifacts and monuments.

Problem

In the absence of spatial plans or regional planning systems, a number of incompatible land use activities have been carried out or proposed; if all are realized, they will seriously affect the sustainable utilization of Lake Bardawil's resource base. These activities will undoubtedly result in a severe conflict between development options and the preservation of Bardawil's function as a major fishery nursery ground and indispensable stopover and winter site for migratory birds. The existing and proposed land use activities include mining, tourism development, fisheries, and agriculture development. The low position of Lake Bardawil makes the site extremely vulnerable to pollution by contaminated surface and groundwater.

Although the entire lake was designated in 1988 under the Ramsar Convention, only the easternmost section is legally gazetted as a natural protectorate. Protection from infringements is far from adequate in absence of legal documents specifying the Protectorate's exact boundaries and lack of strong and well-equipped protected area management. Since 1991, a salt factory has been producing sodium carbonate, and its operation is associated with large evaporation ponds, which partly infringe upon the reserve.

The low elevation of Lake Bardawil makes it extremely vulnerable to pollution. Without any sizable shore vegetation, it is sensitive to outside disturbance. The land and resources, traditionally managed under customary law, will need a broad and comprehensive planning system to guide environmentally benign and sustainable land use options.

Starting in the early 1980s, tourism development boomed, with star-rated hotels and resorts centered around the Governorate's capital town of El Arish, less than 25 km of the lake's eastern section. More recently, tourism developments continued westward along the main transportation artery to El Qantara, which passes just south of the lake's shore. Some 30 kilometers of tourist villages and hotels are to be built along this stretch of coast (2).

The over-riding problem regarding land use activities is the absence of an overall resource management plan or spatial master plan. For example, the Bedouin have practiced their nomadic life-style for centuries. There is still a strong consciousness among the Bedouin tribes in Sinai, which is reflected in the preservation of values and traditions vested in their customary law. Land rights and uses are regulated in long-established traditions and customary law, which is still applied with the consent of the Egyptian government. This century-old system is at the brink of eroding in the wake of the large influx of Nile Valley Egyptians, incomplete environmental legislation, and the lack of an established ICZM mechanism.

A development likely to have a most profound impact on regional land use and the lake's status is the construction of the el-Salam irrigation canal. This intervention seeks to pick up freshwater in the Nile Delta, run underneath the Suez Canal, and feed subsidiary canals in the Sinai desert, thereby providing irrigation to some 170,000 ha of desert land. The proposed irrigated area is mainly situated along the Mediterranean coast, engulfing the Bardawil lagoon. The project, known as the Northern Sinai Agricultural Development Project (NSADP), would require the re-settlement of some 22,000 families from among the rural population of the over-populated areas of the Nile valley and delta.

The various ongoing and planned developments will gravely change the ecological conditions of Lake Bardawil. Implementation of the planned NSADP will strongly influence the fisheries situation and the general environmental status of the lagoon. Whereas uncontrolled tourism developments will indirectly have detrimental impacts on the lake's water quality through disposal of untreated sewage to the sea, close to the lake's inlets. The main impacts are summarized below:

- Construction of water supply systems, which will cause an inflow of freshwater (seepage), thus lowering salinity and increasing the levels of nutrients and pollutants
- Construction of a large irrigated area causing a flow of nutrients and agro-chemicals (and their residues) toward Lake Bardawil
- Construction of settlements, increasing human waste discharge along the lake
- Desert reclamation leading to the loss of important arid habitats for flora and fauna communities
- Increase in (illegal) bird harvesting, especially of migratory species
- Displacement, loss of traditional land rights, and loss of cultural heritage in an area under Bedouin tribal life-style

Solutions

Feasible and cost-effective measures have been identified to reduce significant environmental effects of the NSADP to acceptable levels. Protective measures recommended aim at reducing groundwater seepage toward the lake and thus avoiding agro-chemical contamination. This could be achieved by seeking to match actual supply of irrigation water with actual crop water demand to minimize drainage surplus, especially in the sandy soils (1).

In addition, it is crucial that an Environmental Management Committee is set up to develop a consensus on environmentally benign and economically viable integrated development options for the coastal zone. With Egypt's 59 million people crammed into just 4 percent of the country, a series of government-led population relocation program has been initiated, some of which have already targeted the Sinai. Without a functional planning structure, various governmental and private entities, with their intrinsic aspirations, could start to exploit the coastal resources at any time.

Lessons Learned

Community-based coastal resource management systems have a long history in Sinai and have proven thus far to be effective in managing coastal resources (Bardawil's fish production) and also in fairly allocating access to these resources among the Bedouin users. These management systems are based on in-depth local knowledge and consensus of their utilization among Bedouin tribes.

Like coastal zones in many coastal nations, Northern Sinai has recently seen some major land use developments and will increasingly be the focus of large-scale central government interventions. Such countries should set up all-embracing spatial planning regulation, commission land master plans, as well as institute some type of Board for Planning and Building. The master plan should allow clear designation of land to be set aside for strict preservation of wetland resources and antiquities, shoreline protection for beach recreation, and areas suitable for residential, commercial, industrial, fisheries, and agricultural developments.

Next in importance to identifying appropriate resource uses, the level of development intensity needs to be allocated in coastal developing areas. Incompatible and conflicting space uses can thus be gauged against the master plan and controlled. Any plans not conforming to the master plan should then be vetoed.

Contributed by — Wim J. M. Verheugt, Euroconsult, P.O. Box 441, 6800 AK Arnhem, The Netherlands.

REFERENCES

1. Euroconsult. 1992. Northern Sinai Agricultural Development Project. Environmental Impact Assessment. Euroconsult in association with Pacer and Darwin Engineers.
2. Ramsar Convention Bureau. 1992. A directory of wetlands of international importance. Gland and Cambridge.

4.14 GRENADA: BUILDING SETBACK POLICY

Background

Grand Anse is the most important beach in Grenada for the resident and tourist population. It is therefore very important that the beach itself be conserved. But when the land behind the beach was developed the beach got narrower, and ultimately the buildings were jeopardized by erosion. When owners tried to protect their properties with seawalls, the storm waves were reflected, removing the sand and causing loss of the beach.

Problem

Between 1970 and 1984, the average rate of erosion along the length of Grand Anse averaged 2 feet (0.73 meter) of beach width per year. In addition to the presence of beach structures, there were two major causes for the erosion. The first is pollution of the nearshore water. Pollutants include sewage, pesticides, fertilizers, and high sediment loads. The pollution causes stress and ultimate death of the coral reefs. The reefs are important to the beach because the reefs represent the main source of sand, and they may also act as offshore breakwaters. The reefs from St. George's to Ross Point are dead; this may well have contributed to the erosion at Grand Anse. The reefs off Grand Anse itself are in the main still alive. Another major cause of erosion relates to land and sea level changes. Sea level is rising on a worldwide scale at a rate of 30–150 cm per century.

Another cause often cited is sand mining. Fortunately, large-scale sand mining has been stopped for many years at Grand Anse. Thus, it is not considered a major cause of the present erosion at the site.

Solutions

Due to the erosion problem, a development or building "setback policy" was implemented for all construction at Grand Anse beach. Basically, the setback policy means that all new development at Grand Anse must be placed a minimum of 150 feet (50 meters) inland from the present high water line. The setback will leave room for the average high water mark to retreat inland naturally during the economic life of the property and still leave an adequate buffer zone to absorb the effects of storm or hurricane waves.

Lessons Learned

While it is a relatively simple matter to strengthen and protect buildings in the coastal zone, it is very difficult to conserve an eroding beach. The best way to conserve the beach at Grand Anse is to create a buffer zone between the land and the sea or, to put it another way, to implement a building setback policy.

No country like Grenada can afford expenditures such as those required in Miami Beach, Florida, where US$93 million was spent to replace sand on the beach to protect investments, with no guarantee of how long the sand would last.

Strict monitoring is necessary to ensure that no beach sand mining occurs. Provision of adequate sewage treatment plants should be mandatory in all developments, and the capacity of the sewage system in terms of fecal load must limit development.

Source — Charles H. Francis (See Reference 132 in Part 5).

4.15 INDIA: MANGROVE PLANTING TECHNIQUE FOR THE GULF OF KUTCH

Background

The Gulf of Kutch (northwest India coast) is endowed with a great abundance of biological resources—extensive intertidal mudflats, coral reefs, and mangroves. Mangroves play a vital role in maintenance of bio-diversity and productivity. But the intertidal forests have been destroyed in the coastal wetlands of the Gulf. "Cher" (*Avicennia* spp.) is the dominant mangrove. Techniques of fixation of seeds in muds and plantation of polypot seedlings on mudflats have been adopted. Therefore, it became necessary to reafforest the large mudflats.

Problem

Natural germination of *Avicennia* spp. is adequate in the Kutch, but most of the regeneration is destroyed due to various anthropogenic pressure before its establishment. A large intertidal area in the Gulf and other coastal wetlands could not be regenerated naturally due to various reasons like tidal current and inadequate dispersal of seeds. Artificial regeneration (i.e., sowing of seeds or plantation) is necessary to afforest suitable inter-tidal mudflats in the region.

Solutions

Artificial regeneration of Cher in old mangrove areas and mudflats has been taken up, according to plantation techniques standardized by the State of Gujarat Forest Department for Cher in the intertidal mudflats of Jamnagar and Kutch districts. This started with fixing of seeds, sprouted seeds, and seedlings in mud to facilitate regeneration. Fallen seeds with emerging radicles are ideal for fixing properly in mud. The emerging radicle produces a number of small side roots, which should come in proper contact with mud at the time of its fixation. Ten thousand sprouted healthy seeds should be fixed in a hectare, at spacing of 1 m × 1 m. This number of seeds would be adequate for afforesting good intertidal mudflats.

The afforestation program started in 1983–84, and a total of 3,460 ha were planted by March 1993. Seed sowing of *Avicennia* spp. on mudflats was mainly done at the initial stage; at present, both techniques (seed sowing and plantation of polypot seedlings) have been carried out.

The shape of *Avicennia* spp. leaves makes the front bend with the water current and permits the tide water to flow past easily so that, even with the strongest of tidal flow, the young plants are little affected. Cher seedlings can colonize those mudflats where the new leaves get exposed to sunlight above the water level for as little as a couple of hours a day. At this initial stage, the main danger to the seedlings is from floating algal strings, which get entangled around the seedlings and strangle them.

Three types of extensive mudflats are available in the Gulf of Kutch: high tidal, intertidal, and low tidal. High tidal mudflats receive seawater for a few days a month. The salinity of such mudflats is high, and they are not suitable for regeneration of *Avicennia* spp. But in high rainfall areas high mudflats can support mangrove species because the flow of freshwater during rains reduces the salinity of mud. Such areas can be regenerated by other marsh vegetation, including *Salvadora* spp. and *Suaeda* spp.

Low tidal mudflat remains under water for long periods and is exposed only for a few days in a month. Muds of this zone are very loose and do not provide stability to regenerated plants. Mangrove seedlings often die on such mudflats even after germination.

Intertidal mudflats are ideal for the regeneration of *Avicennia* spp. Those receiving adequate seawater more than 10 days out of 15 days can be selected for afforestation, but brown clayey mud getting regular water from the sea is ideal for regeneration by mangrove species, including *Avicennia* spp. The criteria may vary in other coastal wetlands of the country due to the nature of tides and climatic conditions. This species is a primary colonizer of tidal vegetation. Thick regeneration of *Avicennia marina* and *Avicennia officinalis* has been observed along the creeks and in the mouth of creeks spacing tidal water. Mudflats receiving average 1.0 to 1.5-m tides daily are good for afforestation (Figure 4.15.1).

Seeds of *Avicennia* spp. normally ripen in August and September in the Gulf of Kutch. Abundant seeds are available from August to October on coasts and islands. It would be convenient to collect seeds either from trees directly or from those fallen on the sandy bank. In fact, it is best to collect seeds that are washed up on the sandy beach.

Avicennia spp. are raised in August or September in the Jamnagar district. Seeds or pre-germinated seeds are sown in polythene bags. Sowing of pre-germinated seeds without seed cover provides better results than sowing seeds with their cover. Nurseries are prepared on the intertidal sea bed (Figure 4.15.2). Scattered nursery beds of size 1 m × 10 m are prepared in the mud to reduce transportation cost, and one bed supplies seedings for planting a 0.4-ha area. Average distance between two beds should be 60 m to 70 m (Figure 4.15.3).

Polythene bags, 15 cm × 25 cm, are filled with good quality mud and arranged in rows. Brown clayey mud from the mudflat having the maximum number of holes of mud creatures is good for filling

Figure 4.15.1. Mudflats receiving on average 1.0-m to 1.5-m tides daily are ideal for planting mangrove (*Avicennia*).

Figure 4.15.2. Scattered nursery beds of size 1 m × 10 m are prepared in the mudflats to reduce transportation cost, and one bed supplies seedings for planting a 0.4-ha area.

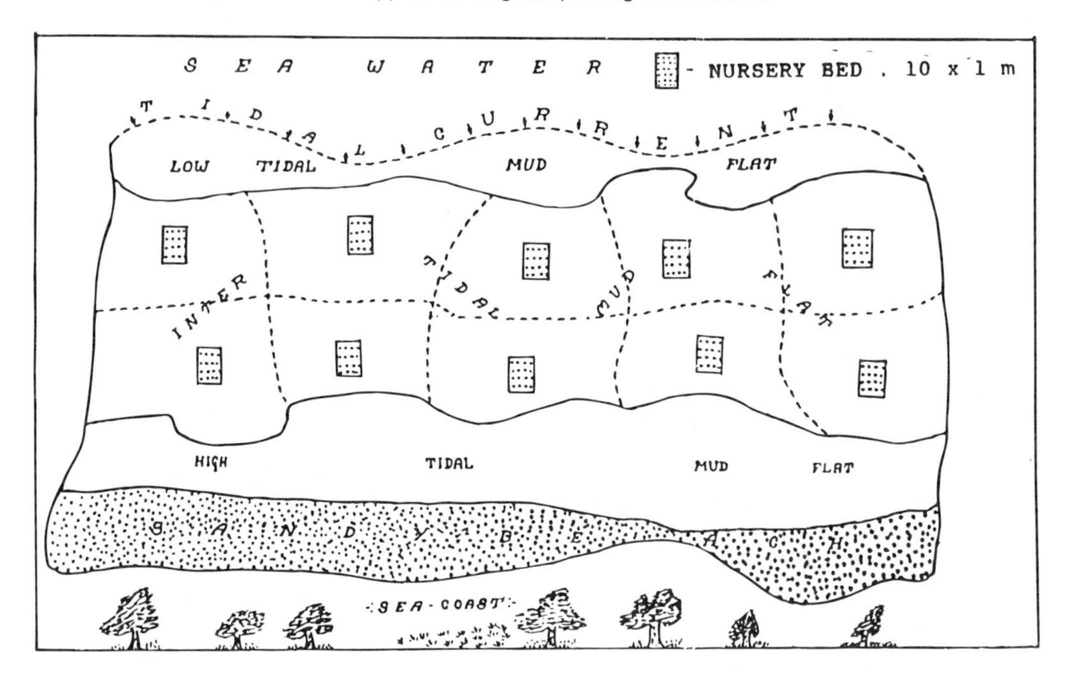

Figure 4.15.3. The average distance between two nursery beds of *Avicennia* should be 60–70 m.

the polythene bags. One third of the length of each bag should be below the mud surface, and the mud dug out should be arranged around the bed as a dike.

Ripened seed germinates within 10 days from the day of seed sowing and attains a 25–30-cm height in six months. Taller seedlings can be swept away by waves or damaged by algae before their roots penetrate and anchor in the sea bed. Five- or six-month-old seedlings should be planted on the site. Plantation work is carried out during low tides. Usually, 2,500 seedlings per hectare at a spacing of 2 m × 2 m are planted in the Gulf of Kutch and take at least two years to become established. Once established, the plants send radial pneumatophores or aerial roots out to all sides and finally anchor themselves in the sea bed.

Lessons Learned

Results of planting intertidal mudflats under these techniques are encouraging in Jamnagar. There is a possibility of high casualties, but this is an appropriate technique to propagate *Avicennia* spp. in the Gulf of Kutch.

Many adverse factors in the intertidal zone cause the death of seedlings. Plants are sometimes washed away by waves or pulled down by algae. Crabs and other sea animals also damage young plants. They should be replaced in the first or second year to fill up the gap. Some marine life, like barnacle deposits on stems and leaves of mangrove trees, retard the growth of plants.

The greatest danger to young seedlings comes from the mass of algae being washed on the shore. The algae floats close to seedlings and their growing tips and gets entangled firmly. During low tide, when the algae dries, it kills the growing tip of the seedling. When the mass of algae is large, it presses the seedling down to the ground. It is, therefore, essential to remove algae regularly for the survival and success of plants in the early stage. The laborers should physically remove the algae without damaging the growing tip and dispose of it on the island or coast. Throwing it back in the sea would bring it back to entangle seedlings.

Manual removal of algae is necessary for the first two years before seedlings or the plants grow taller than the tidal level. Fixing a stick near the plant, toward the coast side, helps in arresting algae.

Source — H. S. Singh, IFS, Conservator of Forests, Marine National Park, Jamnagar, Gudjarat, India.

REFERENCES

1. Arun, P. 1988. Bio-ecology of the Gulf of Kutch, NIO, Goa.
2. Chavan, S. A. 1985. Status of Mangrove Ecosystem in Gulf of Kutch, Marine National Park, Jamnagar.
3. Coastal Environment, May 1992, A Report from Space Application Centre, Ahmedabad.
4. Hamilton, S. and S. C. Snedaker. 1984. Handbook for Mangrove Area Management.
5. Mangroves of India, 1990. Ministry of Environment and Forests, Government of India.
6. Rashid, M. A. 1985. Gujarat's Gulf of Kutch—A Marine Paradise, Forest Department, Gujarat State.
7. Singh, H. S. 1994. Project for Development of Mangroves, Gujarat State, February 1994.
8. Untawale, A. G. 1986. How to Grow Mangroves? NIO, Goa.
9. Untaale, A. G. and W. Sayeed. Distribution of Mangroves along the Gulf of Kutch.

4.16 INDONESIA, BALI: BEACH DAMAGE AND REHABILITATION

Background

When coral reefs that protect beaches are mined for building materials, the beaches may be subject to severe erosion because they are exposed to increased forces of waves and currents. Many countries have had major disruptions caused by mining; for example, the Maldives, Sri Lanka, India, and Indonesia (Bali), all of which have received management attention. Of these three examples, Indonesia's counter measures seem to have been the most effective, although some problems still remain.

Sanur Beach in Bali is an example of this situation (Figure 4.16.1). The beach is about 6 km long. Virtually its whole length is developed for tourist hotels and other resort facilities. The beach and the hotels are naturally protected from erosion by a nearly continuous parallel ridge of coral lying just offshore (Figure 4.16.2). In the 0.5-km (more or less) space between the coral ridge and the beach, there is a very shallow tidal lagoon (Figure 4.16.3), which is rich in sealife and which dampens the force of waves coming over the reeftop (1).

The most active zone of reef building is the upper reef slope lying just seaward of the reef ridge. Here the hermatypic coral (stony reef building coral) is encountered just beyond the algal rim, which is the wavebreak zone of the reef. The abundance and diversity of corals increase rapidly down the reef slope until a depth of 4 or 5 meters is reached when they begin to thin out as a new and different zone of deeper coral species is reached.

The upper reef slope, as the prime coral zone, has the highest percent of coral cover, with up to 80 percent of its surface covered in living coral. This abundance came as a surprise to the author (during an environmental assessment in 1992) because previous information reported that South Bali reefs had deteriorated seriously. Yet this zone had a great diversity of species, including much *Acropora* coral. But here much of the coral is low profile and there are many encrusting species. This basic reef-building zone needs protection, mainly from sewage and agriculture pollution (little coral mining was ever done at the reef ridge, nor is there evidence of anchor damage).

It is noteworthy that the beach is not made of ordinary mineral sand but mostly of tiny organisms (average about 1 mm diameter) called *foraminifera* (*Baculogypsina*), which live in the breaker zone, or "algal rim," of the reef barrier, where they attach to short, tough algae for the few months of their lives. Then their skeletons are swept shoreward by water movement across the lagoon and onto the shore, giving the beach a slightly orange cast. In this way the coral reef a half kilometer seaward serves as a "sand factory," continuously supplying the tourist beaches with a layer of tiny carbonate globes (1). But if pollution or reef destruction depletes the supply of these "forams," the beaches will be diminished (Figure 4.16.4).

Figure 4.16.1. Sanur Beach on Bali is an example of a beach damaged by coral mining and mistaken protection projects. The beach is 6 km long and is mostly developed for tourist hotels and other resort facilities. In the 0.5-km (more or less) space between the coral ridge and the beach, there is a shallow tidal lagoon.

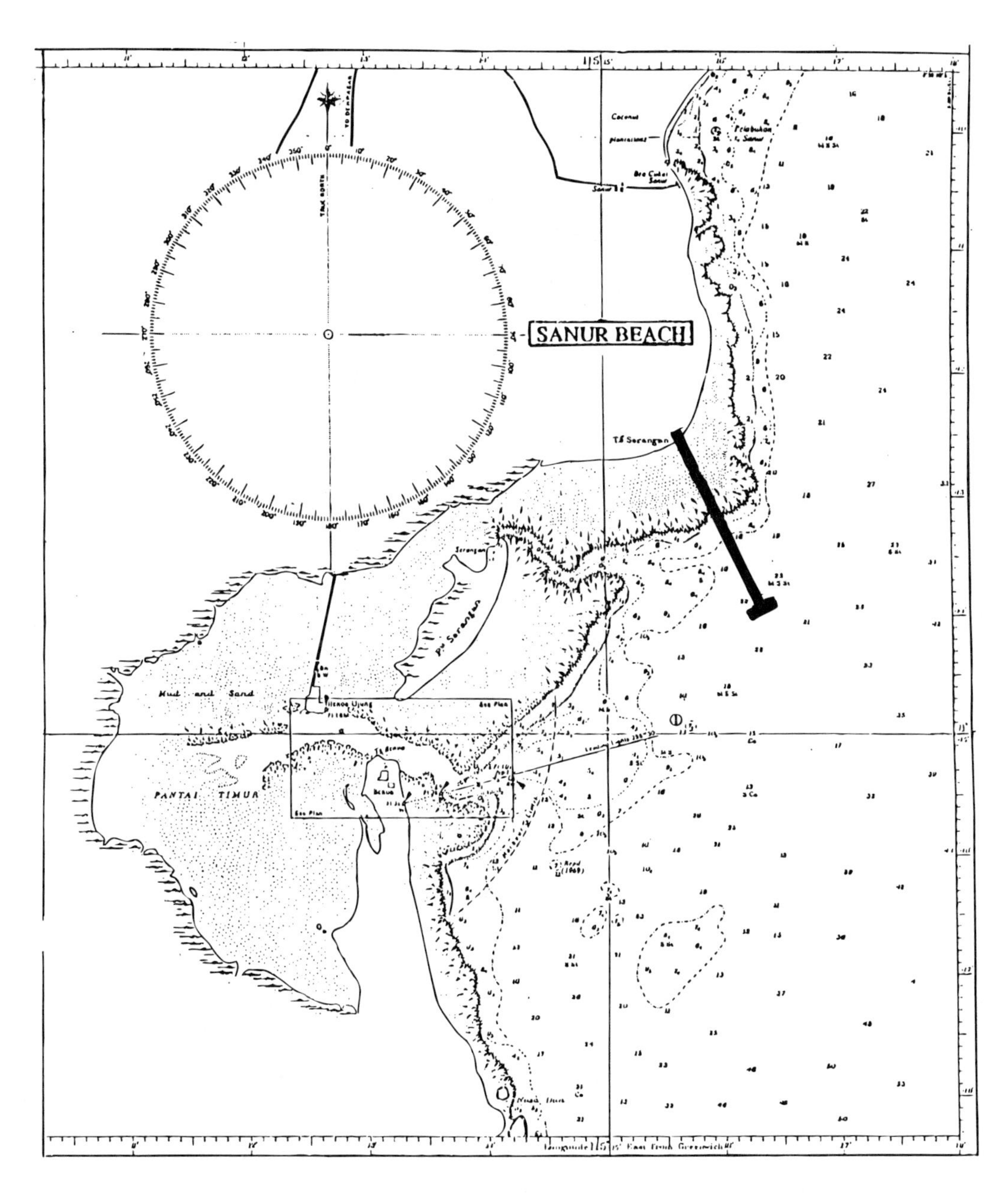

Figure 4.16.2. Sanur Beach is protected from large wave action by a parallel ridge of coral lying just offshore—the *dark bar* is a proposed, ecologically safe location for sewage outfall.

Problem

It is widely accepted that South Bali's beach erosion problem began with "coral mining"; that is, stripping coral from the sea floor for building materials (note that the *reef crest* was never mined, just the lagoon). This age-old practice accelerated dramatically in the mid-1960s as Bali was being transformed into a major international tourism destination. The coral reef ecosystem, which had served as a natural breakwater, tamed the waves, and created calm conditions conducive to wide, deep beaches, was now at risk (Figure 4.16.5). The most damaging change appears to be that the shallow lagoon was altered in configuration and deepened by artisanal mining (Figure 4.16.6). When the coral was removed,

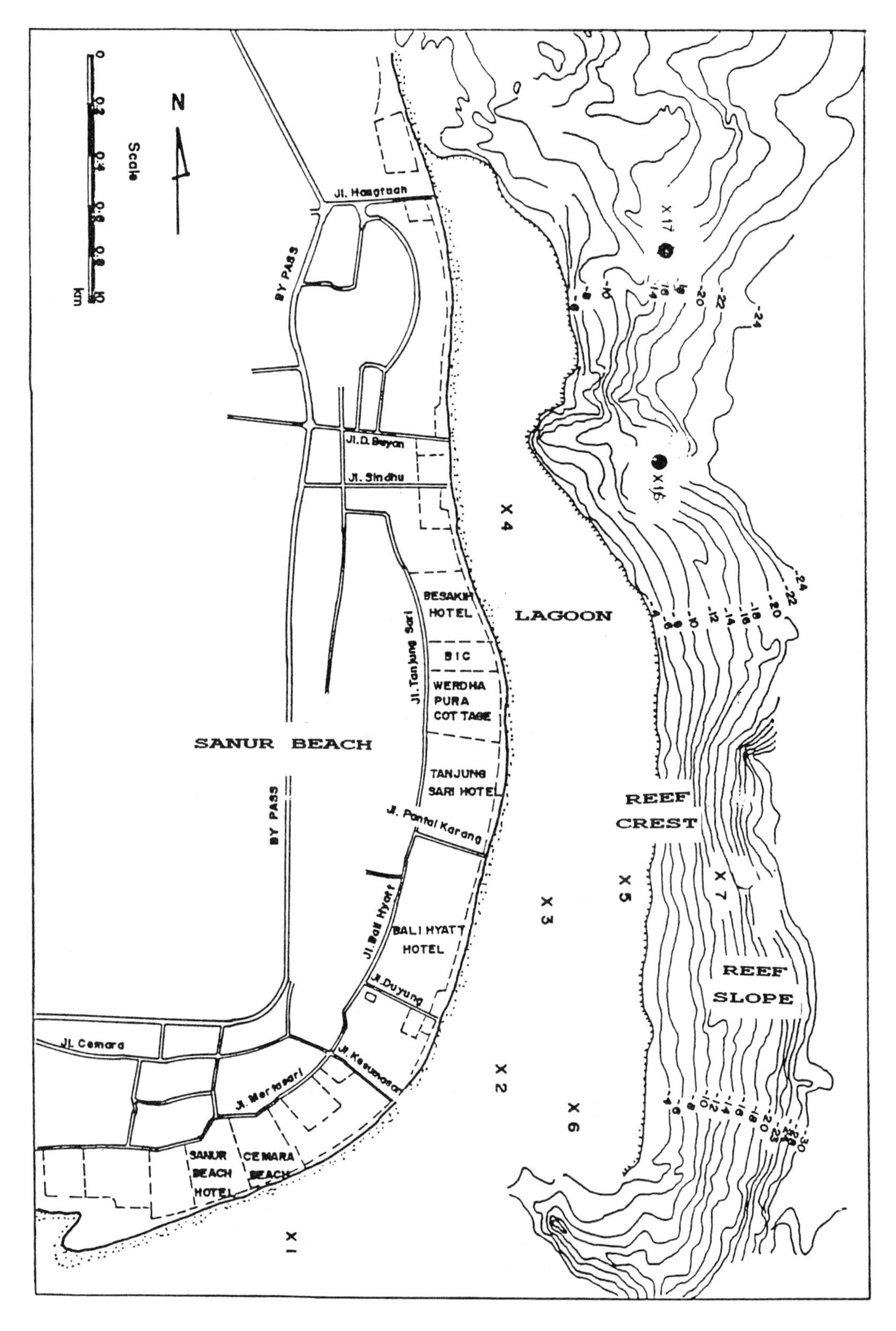

Figure 4.16.3. The Sanur Beach "hotel zone" is separated from the reef and outer slope by a broad shallow lagoon (X followed by a number indicates a water quality sampling station).

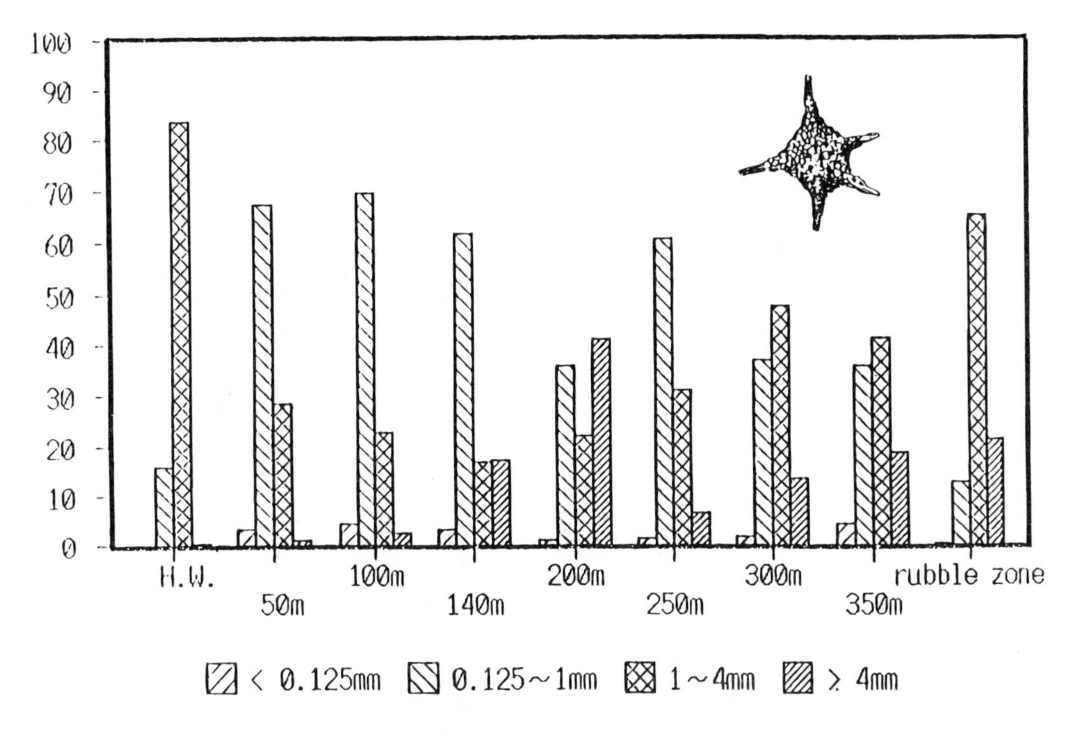

Figure 4.16.4. Sanur Beach is not made of ordinary mineral sand but mostly of tiny foraminifera organisms that average about 1 mm diameter (genus *Baculogypsina*) which live in the breaker zone, or "algal rim," of the reef barrier. Size distributions at different distances outward from the beach (high water to rubble zone of reef) are shown.

stronger waves came through the lagoon, presumably eroding the beachfront (10 to 30 m horizontally) and making certain parts of the beach unsuitable for tourist use.

Coral removal from all of Bali's beaches totaled about 975,000 cubic meters from the mid-1960s to 1980, according to a government engineer (R. Sularto; see Reference 68). This would result in an average lowering of the lagoons of about 0.31 m and is believed to have had the following effects (1):

1. Wave height across the lagoon increased because of the lagoon's increased depth, allowing higher energy waves to strike the beach.
2. The short, bushlike coral growing on the lagoon bottom (*Acropora*) that is a good wave attenuator was stripped away, allowing the waves to pass more easily across the lagoon.
3. An increased beach slope caused some slumpage of beach sand into the lagoon.

There is reason for this author to believe that the problem was exacerbated by one or more quick storm strikes in the late 1970s. Sanur Beach is such a calm and sheltered place under normal circumstances that it is doubtful that the larger particles of sand move around much. Therefore, it is likely that the major movement of sand (including forces of erosion) occurs in response to occasional major events—e.g., high winds and large waves coupled with spring tides (1).

A very infrequent event—a storm that would occur only once in 50 years, equal to a 2 percent chance of occurring in any one year—could create a wave 5 to 6 m high in the deep waters to the South of Bali. A wave this large moving north could devastate South Bali tourist areas except that the shallower water and general shore configuration greatly reduce wave size and force. Nevertheless, storm waves from the south/southwest arrive with sufficient residual force to be of concern (1). We calculated a 2 percent chance (50-yr return period) that a series of waves of 2.25 m (about 7 ft) would strike south Bali in any one year.

Such forces could easily disrupt the lagoon ecosystem and strip enough sand from the beaches to have caused major concern among property owners, particularly if coral mining in the lagoon had

Figure 4.16.5. Balinese coral miner at work in the coral lagoon sawing and removing blocks of coral for building of temples (the only use permitted in Bali).

already reduced wave attenuation potential. The immediate reaction of many tourist hotel owners was to armor their beaches with concrete and stone; while partially successful, this caused new and different problems, like total loss of adjacent and downcurrent beaches at some sites. Had the owners waited, nature would have repaired the damage without human intervention, because beaches are known to yield to storms and then rebuild to their previous state over the next season or few years.

Overall, our project's investigations showed that (a) increasing depth of water in the lagoon is the key factor in lessening wave attenuation; (b) beach erosion was generally most severe adjacent to the lagoon mining sites; (c) a high current speed (ca. 0.4–0.5 m/sec) would be required to lift sand particles of 1 mm or larger; (d) sand is moved mostly by lateral currents, which are related to water *entering*

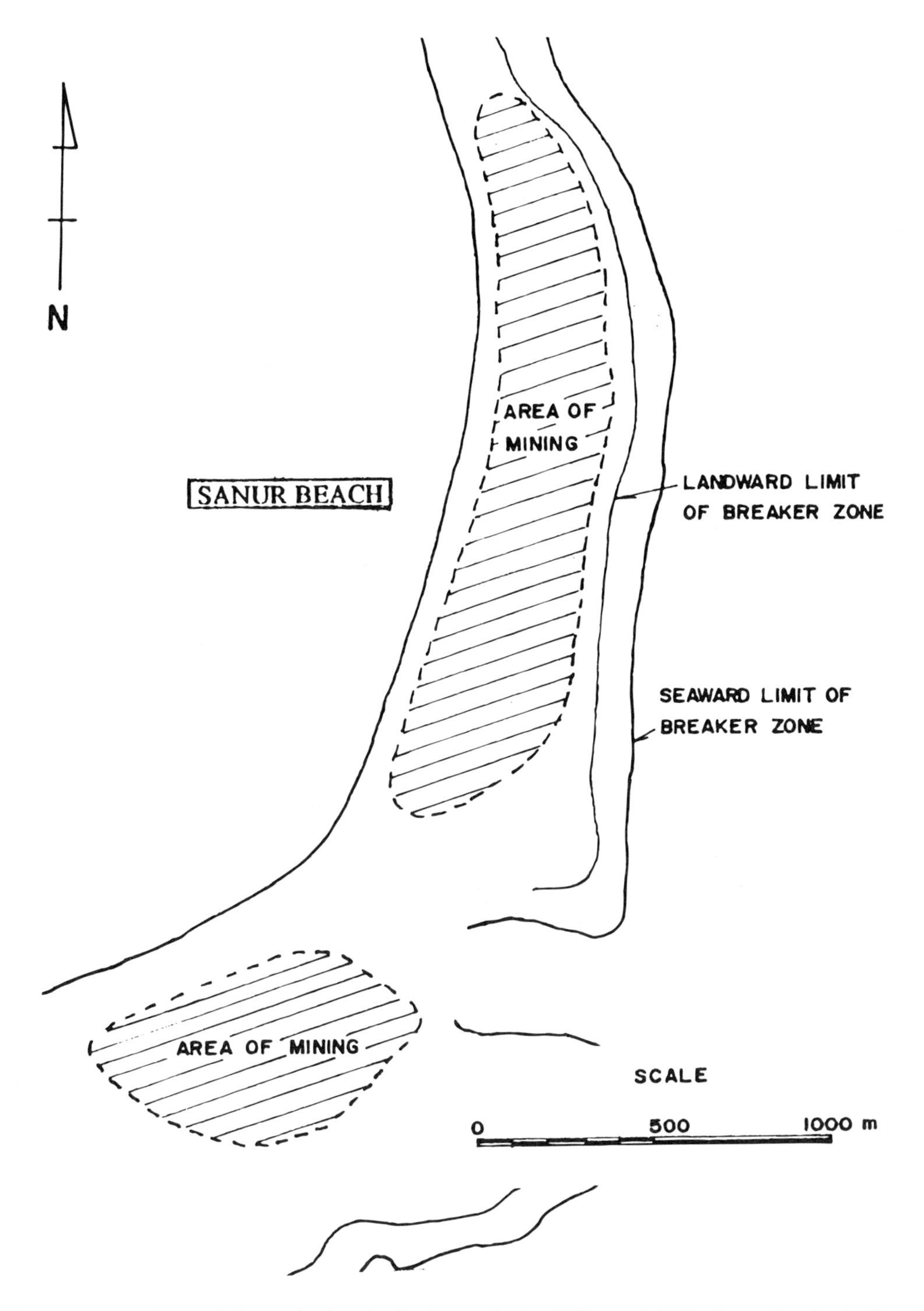

Figure 4.16.6. Heavy mining took place in the lagoon in the 1960s and 1970s, increasing its depth and reducing its wave attenuation ability.

the lagoon over the reef top and *exiting* at two reef gaps; and (e) beach sand is composed mostly of forams from the reef. (1)

Probably the major ecological threat to the coral reef ecosystem of Sanur is actually the poor quality water issuing from Benoa Port just to the west (see Figure 4.16.2). The problems from Benoa discharge are (a) pollution originating in the watershed and carried by streams to discharge into the Benoa Harbor area; (b) pollution originating in Benoa Harbor from ship and shore operations, fish wastes, bilge waters, petroleum discharges and spills, solid wastes, and chemicals originating from other processes such as cold storage operation; and (c) sewage being discharged in Benoa from the Nusa Dua settlement and the international airport.

Now the level of ecological jeopardy has increased because of a major new sewage outfall planned to take all of the sewage from the capital city of Denpasar and the Sanur Beach community and discharge it into Benoa Harbor, after secondary treatment. Benoa Harbor cannot contain the sewage effluent for long because it empties on each tide, disgorging all of its pollutants into the coral reef ecosystem.

It seems that the South Bali coral reef ecosystem could already be near a *pollution threshold.* High concentrations of organic matter are discharged from Benoa Harbor with CODs of 20 ppm reaching some of the reef areas and the amount of dissolved phosphate exeeding ecological limits (1 mg/l) in some places. Increased sewage or increased amounts of fertilizer runoff from paddies could stimulate algal attack of the coral reef. When this happens (as it did at Kaneohe Bay, Hawaii; see case study in Part 4), the algae overgrow the reef, killing the corals. Then the corals are worn down by bioerosion (drilling, grazing, and burrowing animals) and chemical dissolution and the coral reef begins to disappear.

There is a great ecological risk to discharging secondary effluent to the shallow waters along beaches or to lagoons, bays, harbors, or estuaries because secondary sewage plants are really *sewage refineries,* which mineralize organic matter and release large amounts of phosphate and nitrate fertilizers in their effluents.

Solutions

Having been convinced that the mining of the coral lagoons of Bali caused the erosion of the hotel beaches, the governor of Bali issued an order in 1982 prohibiting coral mining. The only exception was an allowance to mine some coral material for the building or restoration of Hindu temples, including bricks sawn from calcium carbonate blocks pried up from the lagoon bottom and kiln-processed quicklime from coral rubble raked from the lagoon bottom (Figure 4.16.7).

Investigations by the author and the design engineers of Indonesia's "Urgent Bali Beach Conservation Project" (68) indicated that the governor's prohibition was being enforced and that, 10 years later, the coral reef ecosystem appears to be recovering from the effects of mining and storms.

An important part of the governor's coral reef conservation program was to create alternative employment by re-training the unemployed miners so they could join the workforce serving international tourists. This appears also to have been a success, with many of the miners becoming marine tour guides, diving guides, beach workers, and hotel workers.

While parts of the Sanur beach have been showing "recovery" over the past several years, full ecosystem recovery would take two decades perhaps, because the *Acropora* needs time to replenish all of the areas where it was devastated and the lagoon needs time to restore its elevation (by 1–1/3 m) gradually through filling with sandy carbonate materials.

Underwater investigations by the author and others, have been most encouraging, in that all marine resources—coral reefs, foram beds, *Acropora* thickets, seagrass meadows—appear to be healthy and recovering from past coral mining and storm problems. The reef itself appeared quite clean and healthy and not yet excessively bothered by algae. Nor did our daytime observations reveal abundant bio-eroders (reef-destroying animals like urchins, which were seen mostly near the reef).

In addition to the governor's prohibition on coral mining, the Indonesia Department of Public Works hopes to replenish the sand "lost" from Sanur Beach by erosion, in the amount of 98,500 cubic meters, along with 29,140 cubic meters of groins, offshore breakwaters, artificial reef, jetties, and revetments. This final engineering design—developed in concert with an environmental impact analysis—was greatly reduced in volume of sand (by a factor of 12) as well as rock and concrete over the original design, benfiting the environment and saving considerable money (1).

Figure 4.16.7. Typical kiln for burning fragments of coral to manufacture quicklime (Nusa Dua, Bali, Indonesia).

This was done by creating a *nature-synchronous* design, which worked with, rather than against, nature. Specifically, the design called for the construction of three major sand ridges in the lagoon parallel to the beach. These were to serve, in the short term, as wavebreaks and, in the mid-term, as feeder systems to add sand gradually to the beach. This reduced the need for large groins and seawalls. In fact, the project design calls for removal of many of the existing inappropriate structures that may have been built in a kind of desperation response to erosion in the late 1970s (1).

The biggest problem remaining is where to find deposits of surplus sand of the right type to match the existing sand on the beaches (this is a strict engineering requirement). Foram shells of 0.5–1.5 mm are usually found in abundance only on beaches behind coral reefs. Stripping other beaches to supply Sanur's needs is a difficult concept to promote (2).

Through the Environmental Management Plan required by Indonesia's statutory impact management system (AMDAL), strict standards for construction and sand filling were prescribed in the required "Environmental Management Plan" to prevent physical damage to the coral from project water craft operation and from release of fine sediments into the water (2).

The beach repair project would move a great amount of sandy materials including very fine silt, which is a polluting substance; therefore, the project had to be designed with extreme care. Marine construction craft are to be kept away from corals. Also construction of concrete and rock groins and seawalls must be done with care, along with operation of storage and casting yards and heavy equipment on the beach and on the local roads.

Specific standards were set for the construction program as are shown in the following (2):

Standard 1: Wash out fines to the extent possible before sand is placed on the beach for the nourishment project.

Even after washing, the sand source used for beach nourishment will contain some amount of fines. But the situation is controllable. The following are recommended to deal with this problem:

Standard 2: Do not use any source that contains more than 1 percent fines (grain size < 60 microns).

Standard 3: Surround work area with silt curtains.

Standard 4: Use methods that cause the minimum release of fines (pipe outfall diffuser, minimum water in slurry, etc.).

This approach requires that the dumping be continually monitored so that undesired effects are not achieved.

Standard 5: The operation shall cease if turbidity, measured as directed, exceeds 10.0 NTU in the water body at distances greater than 200 meters from the point of excavation or discharge, and it will remain in cessation until the problem is corrected.

A serious scenario is that the pollution and subsequent algal growth may interfere with the supply of beach sand from the forams. On the "algal rim" of the reef crest, the forams form a virtual blanket of spheroids over some parts of the surface—right where the full force of incoming waves hammers away at them. They seem tough but are as sensitive to water quality impacts as the corals; for them to continue to prosper and make sand for the beach, they must be protected, particularly from water pollution. Sewage is probably the worst threat to the forams. But an ocean outfall installed only 1.75 km offshore of Sanur Beach would solve the problem (2).

Changing the *location of disposal* from shallow harbor to deep ocean (the "ocean outfall") is favored because there will be deeper water for greater dilution and more favorable currents to sweep the effluent away if located 1.75 km offshore of the south point of Sanur, as shown in Figure 4.16.2. Once dispersed into the sea and diluted, the effluent becomes a benefit, not a liability, because in diluted form it will serve as a nutrient resource for life in the sea.

Sewage engineers proposed a 7-km-long ocean outfall as a "straw man" and then rejected it as too expensive and recommended discharge directly into Benoa Harbor (see Figure 4.16.2). However, careful study of the options shows the following:

1. An ocean outfall at *1.75 km* (not 7 km) offshore would be adequate, ecologically and technologically.
2. With a 1.75-km-distance ocean outfall, primary treatment would be sufficient.
3. A combination of ocean outfall and primary treatment would cost the same amount of money as the present plan or only a little more.
4. The engineers, wanting to build a treatment plant, may have proposed an impractical ocean outfall option, a dummy, to reinforce an argument for its rejection because, if the ocean option were accepted, the plant would not be needed.

Lessons Learned

The Bali beach case is still unfolding, and several more years must pass before we can expect any final outcomes. Yet we have learned some lessons already:

1. Coastal mining destabilizes shallow ecosystems, particularly coral lagoons, and sets the stage for real trouble when extreme natural events occur, like sudden storms.
2. In protection of beaches by coral ecosystems, the lagoon plays a large role. The reef is not just a biological seawall. At Bali, the outside coral ridge usually protects the beach, but in storms the waves pass across and must be attenuated within the lagoon. A lagoon deepened and denuded cannot serve this function well.
3. If a major storm event does erode the beach, beachfront interests shouldn't panic or consider the situation permanent—nature can usually restore the damage, given time. If the beachfront interests cannot wait, then nature-synchronous methods of beach rehabilitation should be used. Most of the structures built at Bali were counter-productive to the tourist industry in the long term.
4. What seems the obvious answer is not always the correct answer, particularly in dealing with the environment. At Sanur there is usually not enough wave energy to move the sand around (waves < 1 m, wind < 10 kt). Therefore, studying the normal situation does not help—the

real forces at work shaping the beach occur during storm events.

5. Coastal development projects, like beach restoration, must be designed holistically; that is, the designs should recognize a very broad spectrum of environmental parameters and understand the ecosystem they are working within. For example, at Bali, the beach protection works of 10 to 15 years ago did not fit the natural forces of the lagoon ecosystem and were often counter-productive.
6. Beaches of organic origin must be treated as such and the source of the material protected. At Sanur, the beach, made of foram shells produced biologically on the reef is the key factor. Pollution (especially eutrophication from sewage) could seriously diminish this resource.
7. A sound approach to controlling erosion of lagunal beaches is to place artificial sandbars in stategic locations parallel to the shore.
8. The idea of secondary treatment of sewage has been so promoted that the often better option of ocean disposal may be overlooked.

Contributed by — Author.

REFERENCES

1. Clark, J. R. 1992. *Urgent Bali Beach Conservation Project: Environmental Study Report,* Vol. 1: *Main Report.* Report to Nippon Koei, Tokyo, Japan. 88 pp.
2. Clark, J. R. 1992. *Urgent Bali Beach Conservation Project: Environmental Management Report,* Vol. 2. Report to Nippon Koei, Tokyo, Japan. 127 pp.

4.17 INDONESIA, JAVA: MULTIPLE USE OF A MANGROVE COASTLINE

Background

The East Java coastline extends south along the Strait of Madura for about 100 km from the delta of the Solo River, includes the delta of the Brantas River, and extends east through the District of Pasuruan. Dominating the coastline is Surabaya, the second largest city in Indonesia, with its neighboring city, Gresik (Figure 4.17.1).

For perhaps centuries, this coastline has been used for aquaculture. The traditional product has been milkfish (*Chanos chanos*), but shrimp and other species of fish have also been raised. At present there is an estimated 39,000 hectares of ponds. The ponds extend up to 10–15 km inland and encompass a range of aquatic environments from tidal brackish to river or rainfed.

The system has been sustainable because of the virtual absence of PAS (acid) soils and a high rate of coastal accretion. North of Gresik, the largest part of the existing tidal pond system has emerged from the sea in the last 100 years since the rechanneling of the lower reaches of the Solo River. In the delta of the Rejoso River, the coastline has expanded by over 3 km since the 1950s. Similar rates occur in the Brantas delta. In parts of the Solo delta, the coastal accretion approaches 80 meters per annum.

A mangrove fringe extends along much of the coastline, and this is converted into new ponds from the rear as the coastline expands. Traditional pond culture involves the establishment of plantations of *Avicennia marina* and *Rhizophora mucronata* along the bunds (dikes). These trees provide protection to the bunds and at the same time serve as a significant source of fuel wood. The pond culture system has created a tidal wetland system many times larger than would have existed naturally.

The Directorate-General (DG) responsible for Protected Forests and Nature Conservation (in the Department of Forestry) wished to investigate options for conservation management of the mangrove vegetation and the populations of waders and other seabirds present in large numbers, including potential sites for reserves. Also, the DG was concerned about the potential for a mangrove "greenbelt" stipulated under recent legislation. The goal of the planning work was to determine a management approach that

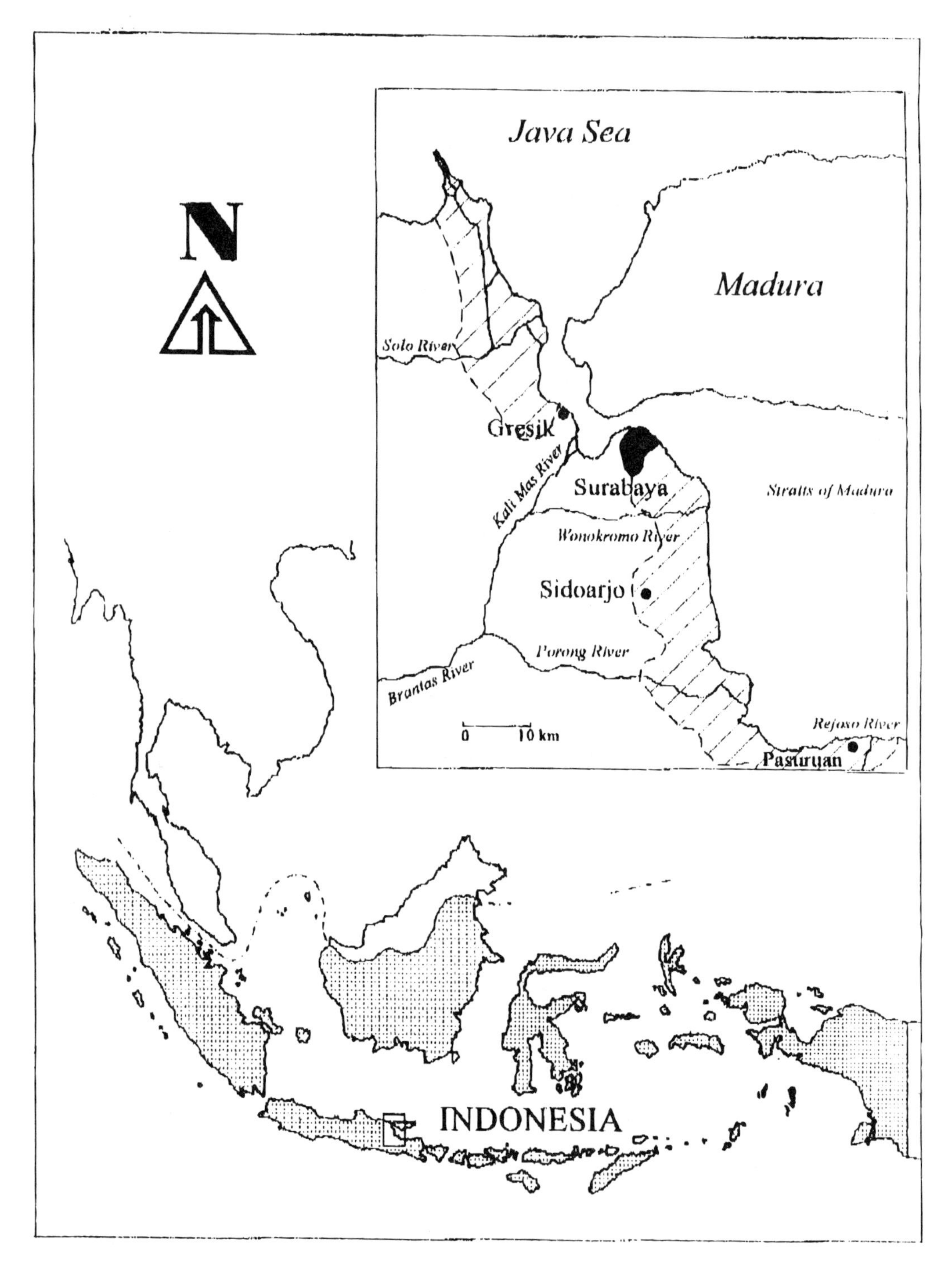

Figure 4.17.1. Project site location for the East Java multiple-use study.

would achieve sustained multiple uses, benefiting nature conservation and social and economic development.

Problem

The major threat to the mangroves of the Solo-Brantas deltas is the growth of Surabaya, resulting in the conversion of coastal land to the north and south of Surabaya for residences and heavy industry. Apart from the loss of productive land and the increased pressure that this places on the remaining aquaculture areas, the conversion creates further sources of water pollution, which is also causing a decline in environmental quality along the entire coastline.

There are no longer new areas of coastal land into which the aquaculture industry can expand to make up any losses to residential and industrial purposes. The only new land is that which is arising through coastal accretion. The pressure to maintain employment in the fisheries sector can be expected to produce pressure for a faster rate of mangrove conversion than wise practice has determined in the past. As a result, a reduction in the present width of the natural mangrove fringe could be expected in many locations along the coastline.

Increased water pollution from expanding urban and industrial activities on the coastline is expected further to increase the risk to successful production of shrimp and to diminish total fisheries production rates. Unless carefully planned and managed, reduction in the quantity and quality of fisheries and fisheries products from the ponds and the offshore stocks can be expected.

Water pollution is a serious issue for the Solo-Brantas coastline, although quantitative data are insufficient to define clearly the scope and scale of the problems. The increasing risk of pollution to shrimp production may be a factor in the slow and patchy development of intensification in the aquaculture industry in the region.

Solutions

Seven major issues are either driving change or being affected by it: (a) physical changes to the coastline (accretion and erosion), (b) administrative responsibility for land management of the mangrove area, (c) water pollution, (d) industrial expansion, (e) urban expansion, (f) the future of aquaculture, and (g) the future of the offshore fisheries.

The physical changes to the coastline, while not entirely natural, are themselves the result of geomorphic processes responding to engineered changes in coastal and estuarine hydrodynamics initiated during the past 100 years for water management programs in the province. These changes set in motion changes to the pattern of supply and transport of sediments in the Straits of Madura which have had profound influence on the shape of the coastline and on the location and rate of accretion and erosion.

The issues of user rights and land tenure are handled by the Department of the Interior, who acknowledge formal or informal village practices derived from earlier customary land administration practices. There is presently no ecological or social argument to alter the existing land administration system as far as the land used for tambak is concerned. Refinement and enforcement of a consistent policy is called for to regulate the progression of short-term land use rights to long-term ownership of newly emerged land (termed *tanah oluran*).

Any attempt to enforce a specified greenbelt through government directive may be impracticable without a profound restructuring of government activities and responsibilities. An alternative approach may involve monitoring of conversion rates by fisheries extension officers with a view to reinforcing best practice among the farming community based upon proven land management by successful operators.

A close examination of the administration of the land is called for as it relates to change in land use and the conversion of tambak land to urban and industrial use. This is required because of the ecological and social impacts of such change but also because of the broader issue of sustainable multiple use and the regional and national values ascribed to aquaculture-based fisheries production.

The provincial planning board has a coordinating function in promoting the development of interdepartmental committees consisting of the line agencies responsible for resource management decisions. Planning, budgeting, and management structures and legislation exist to allow this to happen. If key people can be encouraged to coordinate their efforts in producing consistent policies and using these policies for the basis of resource management decision, the attitude of the resident community and their attachment to the land are sufficient to hope that this model coastal land management system will persist.

Lessons Learned

1. The community must have a central role in identifying and developing appropriate resource use options. Until recently, traditional patterns of resource use in the tidal wetlands and the offshore waters were probably sustainable and met the joint objectives of nature conservation and the conservation of renewable mangrove resources, while at the same time producing fish, prawns, timber, and a range of agricultural products for domestic and export purposes.
2. The apparent sustainability of the system and the sophisticated level of management found here is based on a long-established land tenure system directed toward sustained multiple use and is integrated into the social, cultural, legal, and administrative fabric of the region. That this exists clearly reinforces the need to ensure that options for sustainable multiple use are explored first with the local communities. Unless resource use options are allowed to grow organically from within, it is unlikely that they will be adopted successfully. This is not to deny public sector agencies and their staff a role in this process but rather to emphasize that this role may be supportive, not preemptive.
3. Proposed changes in land use should be based upon an adequate knowledge and understanding of the biophysical, social, and economic environments. For example, analysis of geomorphology, tidal currents, and sediments calls into serious question the location of heavy and polluting industry along the coastline north of Surabaya.
4. There is a need to ensure that management is based upon a good knowledge of the facts rather than upon perceptions. Thus, the accepted truth that the value of mangrove ecosystems is incompatible with aquaculture needs to be re-examined and the question asked in another way. Can coastal production systems be designed to ensure that they provide both ecological and production objectives?
5. Spatial planning in the coastal zone is of limited value when applied on its own. Planning involves the design and regulation of function as well. The essential role of flowing water in linking elements of coastal landscapes together means that sites cannot be isolated and managed alone. The essentially "integrated" nature of coastal ecosystems calls for cross-sectoral integration of agencies.

Contributed by — Jim Davie, Department of Management Studies, University of Queensland, Lawes, Australia.

4.18 INDONESIA, SULAWESI: OCEAN DISPOSAL OF HARBOR SILT

Background

The rehabilitation and expansion of Ujung Pandang Harbor on the Straits of Makassar in Sulawesi, Indonesia, required that 1,000,000 m^3 of bottom material be excavated from the harbor. Special care was necessary to prevent sediment pollution from the marine excavation operation. Standard countermeasures were required (sediment traps, silt curtains), as was appropriate construction equipment (e.g., suction dredge rather than bucket or drag line). But the most difficult part was choosing a silt disposal method.

The project's final design in 1992 was guided by Indonesia's national impact control program, AMDAL, which has three major components: assessment (EIA), management (termed *RKL*), and monitoring (termed *RPL*). In the RKL environmental management program, a variety of efforts are recommended for environmental management, including prevention, reduction, mitigation, enhancement, restoration, education, outreach, and coordination. These all applied to the Ujung Pandang Port Urgent Rehabilitation Project.

Problem

The biggest problem was to devise counter-measures to reduce environmental problems and impacts from disposal of silt. During the feasibility and design studies, there were considered a variety of possible solution, including both land and sea disposal of silt. For land disposal, the material is usually placed in a contained site (diked) on land and allowed to de-water, after which it may be used for soil supplement or left as landfill. For sea disposal, the material is barged or piped to a place where it can be dumped and cause the least environmental problem. Very contaminated material may be placed in special land or underwater containments.

In this case, land disposal and/or possible commercial use of the silt was deemed not feasible. This left only sea disposal as a viable option, and there were two main approaches for this—containment or dispersal. Containment should be favored for badly contaminated silt, but we had no evidence that the silt was contaminated.

Solutions

The objective of open sea dispersal is to spread the material in as efficient a manner as possible so that dilution occurs rapidly as the material is carried with water currents, thereby rendering it as harmless as possible as quickly as possible. To the extent possible the method should simulate the natural process of sediment eroded from the land and carried by rivers to the sea. In this case, it was recognized that some level of impact will occur even though a proper dispersal method minimizes impact. Therefore, the disposal site should not be an area of high biodiversity or of a high level of recreation or commercial fishing activity.

It was clear from published sources of macro-current data that the main current in nearby Makassar Strait is southerly. This told us that the silt disposed at sea will move generally south as long as it remains suspended. We assumed that the micro-currents in and around the disposal site are also generally south moving when tidal oscillations are averaged out (south oscillation was found to be the stronger and would produce a net movement south). The azimuth of southerly flow is about 215°.

An exception is that the current which goes south in the dry season (East Monsoon) may go north for a time in the wet season (West Monsoon) close inshore, perhaps from November to December. Our very limited study with a current drogue (November) confirmed that there *was* a northward flow in the wet season.

To test the idea theoretically, we used a scenario of dispersal of material with a median size (D_{50}) of 15 microns (0.015), a typical silt from the harbor bottom. With a median of 15 microns, 75 percent of the dredged material would range between 5 and 40 microns. It can be calculated that, for *still water* conditions, 75 percent of the material would sink to bottom (30 m) over a period of 4 hours to 25 days at settling rates ranging from 2.1 mm/sec for 40-micron particles to 0.015 mm/sec for 5-micron particles. The smaller size fractions would, of course, remain in suspension much longer, while the larger would sink more rapidly.

Of course, Makassar Strait is not a stillwater body and therefore one would expect a longer duration of suspension of the silt, caused by natural agitation and turbulence. How much longer is not calculable with present data and methods. However, there appears to be no vertical obstacle to the sinking of particles—i.e., we found no thermal stratification in our field work. During the essentially wind-free hours of 6 p.m. to 9 a.m. when wind-caused turbulence is minimal, sinking might proceed closer to the theoretical, stillwater rate.

In the theoretical case, about 25 percent of particles (the heavier fractions) would sink to bottom in $\frac{1}{2}$ day, and a total of 60 percent (40 to 10 microns) would sink in 2 days (see "Sediments and Soils" in Part 3). Doubling the length of suspension to account for suspension factors would allow one day of suspension for particles of 40 microns and four days for those of 10 microns. If the current transported the silt to the south southwest at a net non-tidal current drift rate assumed conservatively to be 0.05 m/sec (about 0.1 knot), then the silt discharged would move south southwest at 4.3 km/day (2.2 n.m.) in suspension.

At the rates estimated above, most of the heavier fractions of silt (>40 microns) would settle out within the inner ellipse (of Figure 4.18.1) during the first day after discharge. The lighter fractions would settle progressively further to the south, carried by current-driven transport (as shown by successive

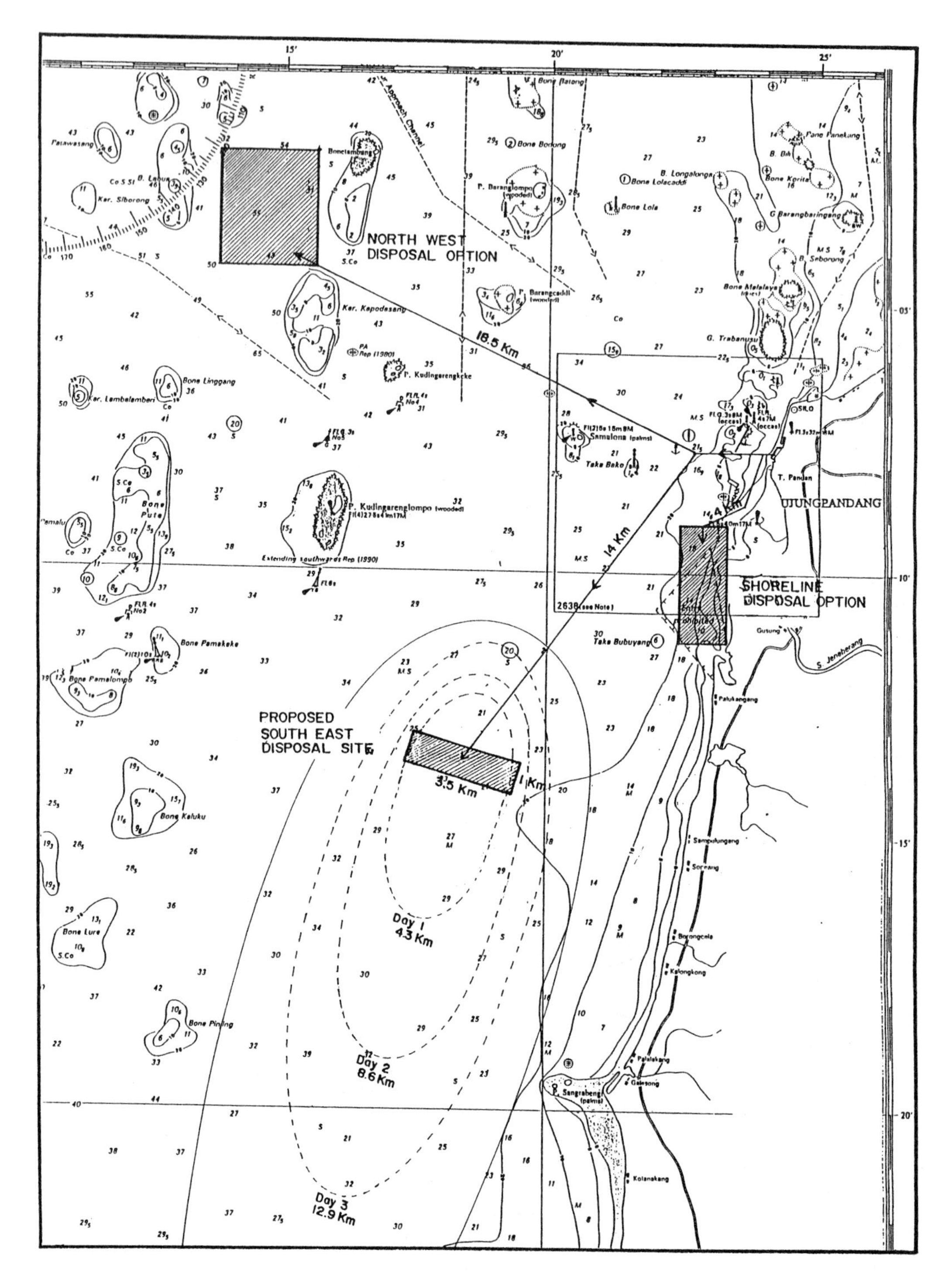

Figure 4.18.1. Disposal options for 1,000,000 m^3 of dredge spoil to be removed from the harbor at Ujung Pandang (Sulawesi, Indonesia). Because the shoreline site and the northwest site were deemed unacceptable ecologically, the southeast site was recommended.

ellipses in Figure 4.18.1), with all materials larger than 10 microns falling out of suspension by the fourth day (eighth tidal cycle). According to this scenario, 60 percent of the silt (600,000 m^3) would have been deposited on the bottom over an area of approximately 90 km^2 to an average thickness of $\frac{2}{8}$ cm ($\frac{1}{4}$ in.).

The <10-micron fraction of silt discharged into the designated rectangle is expected to remain suspended more than 4 days and to be carried south beyond the ellipse of Figure 4.18.1. At this point the material would be expected to be diluted to nearly natural background levels for the southwest Sulawesi coastal sea, as calculated below:

1. Each day a mass of water 5 km wide and 25 m deep would pass through the discharge rectangle (Figure 4.18.1) at a current-driven rate of 0.05 m/sec (4.3 km/day) = 537,500,000 m^3/day.
2. The silt would be discharged into this stream, and the part remaining suspended would be assumed to spread gradually vertically throughout the water column during its southerly drift.
3. With an expected daily discharge of solids of 5,000 m^3/day from project dredging (3 loads @ 5,000 m^3/barge @ 33 percent solids), the dilution achieved would be 1:107,400.
4. This would add 9 mg of suspended matter per liter (reading of <2.5 NTU by turbidity meter) to the background level, minus the amount of fallout, after maximum dispersion has been achieved.

If spread properly over as much as possible of the 4-km^2 area of the discharge rectangle—and assuming that there will be some enhanced mixing by wind in the top five meters of the surface layer and also that the silt will be injected somewhat below the surface upon discharge and will maintain an inertial component—the suspended solids concentration at the boundary of the discharge rectangle (which serves as the permitted "mixing zone") might be around 150 mg/l if each barge load were discharged at around 150–200 m^3/minute over a half hour period. This assumes 10,000 m^3/sec available for dilution (tidal flow $\approx$ 0.5 m/sec, discharge over 3.5 km cruise track, and rapid initial mixing to depths of 5–6 meters).

This 150-mg/l level confined to a small area would result in some quite localized ecological disruption (light blockage, silt fallout on inhabited sea bottom, repulsion of schooling fish, lessening of phytoplankton blooms), but this is considered acceptable under the circumstances because a small area would be involved and the effects would not be permanent. Starting with 150 mg/l at the boundary of the work zone (disposal rectangle), by the end of one day (two tidal cycles) the level of suspended solids added should have dropped to less than 80 mg/l and maybe nearer to 50 mg/l. By the second day the level should be down to under 50 mg/l and maybe 25 mg/l or less, as natural dispersal and fallout processes continue. Consequently, it was proposed to allow a "mixing zone" of 5 km downstream from the disposal site within which the national standard would be temporarily exempt.

According to this scenario, within a few days the silt discharged would become diluted to less than natural background amounts, even for oligotrophic systems, and less than that indicated in the Indonesia official guideline for suspended solids: <20–25 mg/l *desirable* for all uses ($\approx$3–4 NTU) and <80 mg/l *permissible* for fishery and marine park ($\approx$7–10 NTU). Adding the 9 mg/l from silt disposal to the existing background level of $\approx$8 mg/l would give a total of 17 mg/l, which is well below the maximum.

With the approach suggested above for a broad downstream mixing zone, the constraints would be as follows:

1. Turbidity not to exceed 150 mg/l (20 NTU) outside the work zone or disposal rectangle
2. Turbidity not to exceed the national "permissible" standard of 80 mg/l (8–10 NTU) further away than 5 km from the boundary of the work zone

Although there are brisk and important fisheries in the coastal areas surrounding Ujung Pandang, the expected area of major induced turbidity was selected because it is one of the least heavily fished areas (Figure 4.18.2). It is also an area of relatively low biological diversity.

Principal commercial shipping lanes were avoided for the most part in placement of the discharge rectangle, as can be seen in Figure 4.18.2. The occasional ship setting course due south from Taka Bako should not pose a major problem because the discharge rectangle will be buoyed and any project tugs and barges operating at night in the area will be well lighted.

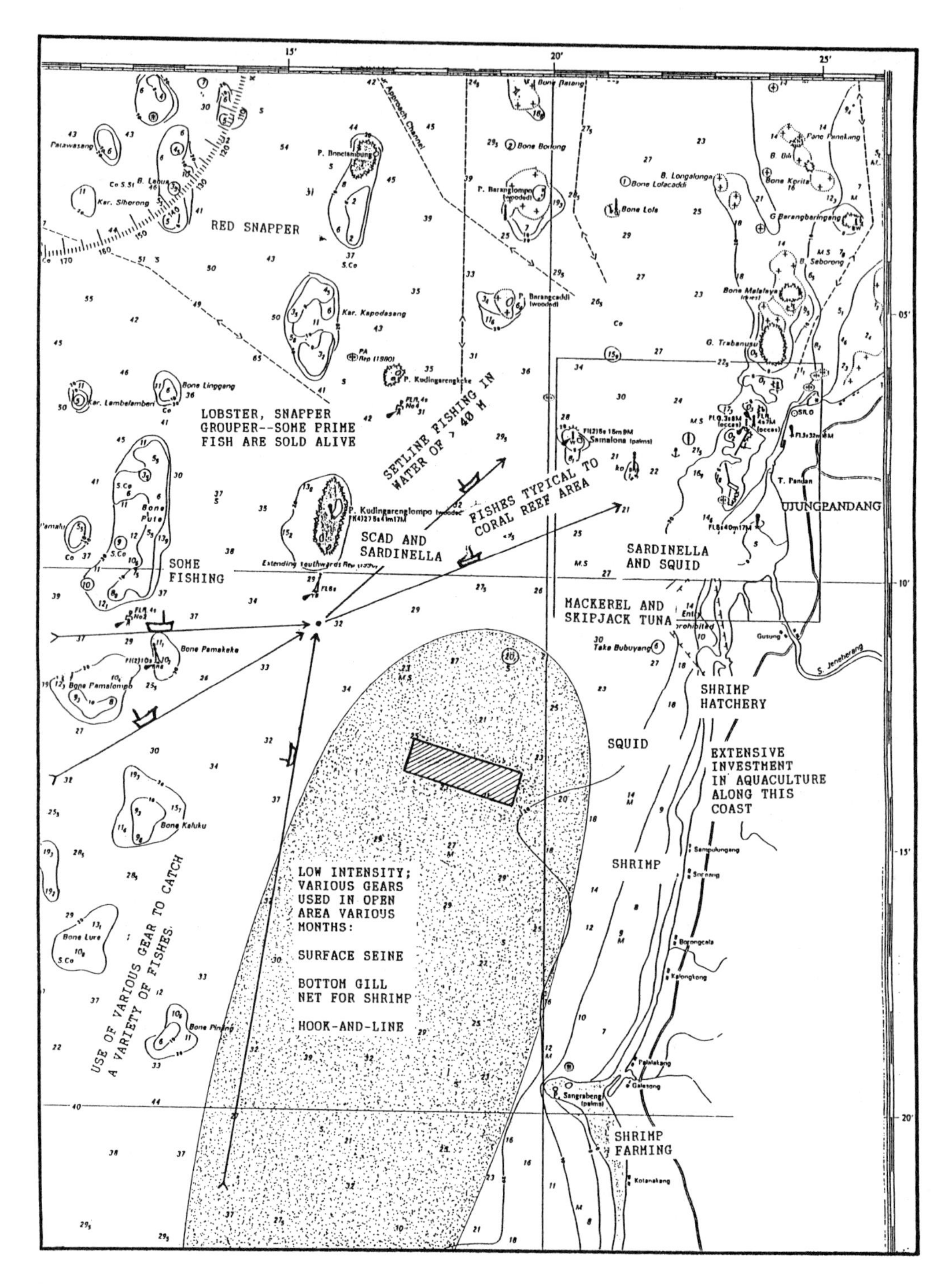

Figure 4.18.2. Examples of fishing locations and other resource user information utilized in choosing the disposal site.

A major location criterion was to avoid the movement of silt discharged by the project onto any coral reefs because the coral reef is the most sensitive of any marine habitat to turbidity and silt fallout. The discharge rectangle, therefore, was located such that the coral reefs of Kudingarenglompo and Samalona and other islands to the north would not be in the path of silt drift in the recommended season of disposal—April to October. To the south no major coral reefs are encountered until the vicinity of Dayangdayangan Island, which we do not expect to expose to more than 20 mg of suspended solids per liter.

Re monitoring, the method recommended required that dumping be continually monitored so that undesired effects would not be acheived. This meant that evironmental management of the Ujung Pandang Port Project should involve real-time monitoring of the suspended solids (SS) load to detect whether the construction work is creating unacceptable impacts, according to proposed standards. Control is to be acheived by maintaining a maximum level of water clarity or, said another way, minimum water turbidity. The plan for oversight will mainly involve real-time monitoring of the suspended solids (SS) load to detect whether the construction work is creating unacceptable impacts. The water quality standards established by the government (KLH)—including limits of suspended solids—are not considered to be controlling so much as guiding. However, the government standard (maximum, 80 mg/l, desirable <25) was used here as the limit to permissible pollution by the project.

The standard would be that water turbidity shall not exceed the national standard of 8 NTU (nephalometric turbidity units) = 80 mg/l suspended solids) as measured by an electronic turbidity meter in any area of the sea influenced by the disposal of dredged material from the Ujung Pandang Port Urgent Rehabilitation Project, except within a permitted non-compliance "work area" and the permitted "mixing zone."

Lessons Learned

A defined "mixing zone" is necessary. While government pollution agencies usually set numerical standards, they often do not set specified working areas, or "mixing zones," outside of which to apply the standard. Clearly, where dilution/dispersion is the working concept, the standard cannot apply at the exact point of discharge (or of dredging) because dilution/dispersion is time linked. In this case we established non-compliance, "work areas" for each situation with the objective that the project will conform to the national KLH water quality guidance standards. The standard is that turbidity at a distance of 250 m from the point of dredging may not exceed 30 NTU. If the limit is exceeded, work would be halted until corrections are made.

Another lesson is that oversight monitoring has to be done by meter so there is real-time, direct readout of the condition being monitored. This is best done by an electronic turbidity meter (simple, inexpensive). Yet this instrument does not read out in mg/l of suspended solids as the standard is expressed; therefore, a conversion scale is needed. In this case, we computed such a conversion by calibrating the turbidity meter against standard solutions prepared from silt taken from the harbor bottom in the particular place of dredging (see 2.42 "Turbidity Measurement" in Part 4).

A major lesson learned is that with knowledge of physical configuration, oceanography, species distribution, and particle size, it should be quite possible to arrange for ecologically safe offshore disposal of non-contaminated silt excavated from harbors. Silt of small particle size (say <25 microns) has little resource value (for landfill, soil augmentation, or other products).

Contibuted by — Author (See Reference 67 in Part 5).

4.19 ITALY, VENICE: A CITY AT RISK FROM SEA-LEVEL RISE

Background

The historic city of Venice is located in the middle of a lagoon, along the Italian coasts of the northern Adriatic Sea. If a sea-level rise occurs in the next century, as predicted by certain climatic models, the flooding situation will rapidly become dramatic; the mobile gates proposed by Italian government experts to protect against storm surges would rapidly prove inadequate to protect the city and its lagoon.

The present dangerous situation results mainly from human-induced processes. Throughout its history, the Serenissima Republic had considered the lagoon as a unitary system to be defended primarily for its strategic and military importance. River courses were diverted from the lagoon, and any construction liable to cause deposition was either strictly regulated or prohibited.

However, since the beginning of the 20th century, priority has been given to the development of an industrial harbor in Marghera and to the setting up of industries (mostly chemical and metallurgical) on newly reclaimed land. Tidal flats have been considered more and more as land resources easy and inexpensive to reclaim and the muddy lagoon basins as ideal places in which to dredge deep channels for navigation into which wastewater could be discharged.

This trend in land use culminated in the 1960s with the construction of an oil terminal in the middle of the lagoon, followed by excessive deepening of the navigation channels (some of them are over 18 m deep, whereas the average depth of the lagoon is only 0.67 m!) and the reclamation of wide new areas of tidal flats for industrial use (Fig. 4.19.1). Dredging and diking were carried out without any concern as to their potential consequences either on Venice or on the hydrodynamics of the lagoon. Among the consequences is a drastic increase in the frequency of floods (3).

Problem

When the city of Venice was constructed, the altimetric situation was better than it is today, though episodic flooding events have been reported during its long history (1). The frequency of flood events has increased dramatically since the 1920s and especially in the 1960s (Figure 4.19.2). This is due to two main causes: (a) an easier penetration inside the lagoon of storm surges arising out at sea or resulting from tidal changes, owing to the increased depth of the passes and navigation channels, and (b) a recent rise in the relative MSL (Figure 4.19.3).

The elevation of street levels in Venice is as low as a half meter above sea level. There has been a dangerous increase in the frequency of flooding during the past few decades. The mean tidal range is about 60 cm and the mean spring tidal range is about 80 cm; exceptional tidal ranges may reach

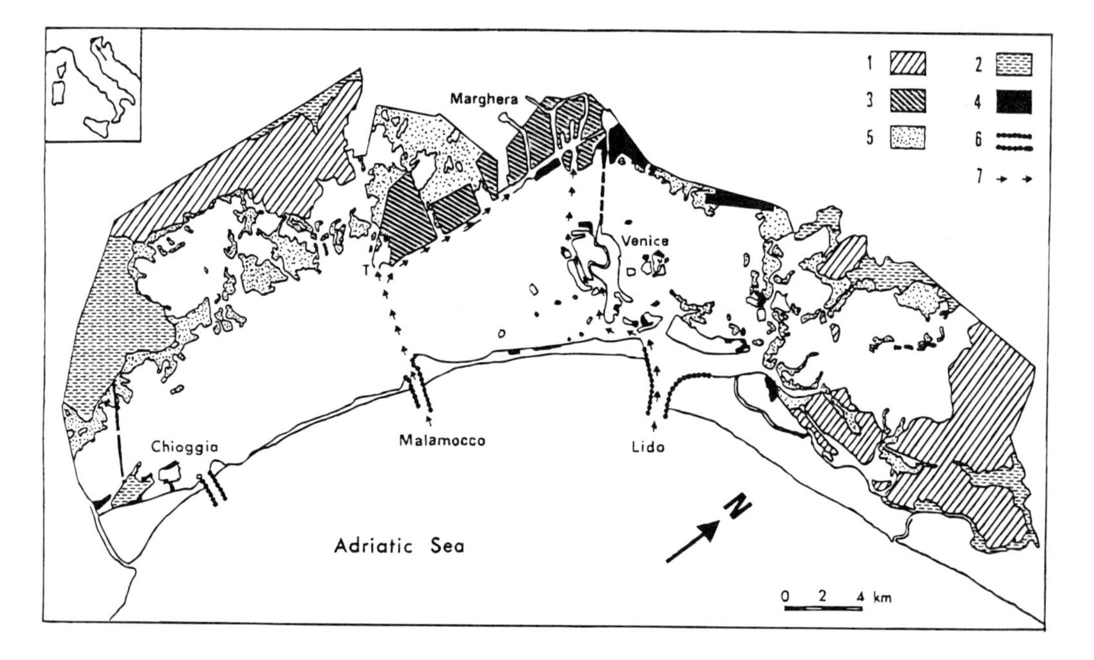

Figure 4.19.1. The lagoon of Venice—1: diked fisheries; 2: reclamation areas diked for agricultural use; 3: tidal flats diked for industrial use; 4: urbanized areas, dumping grounds (since 1900); 5: tidal flats currently open to tide; 6: breakwaters at the inlets; 7: main artificial navigation channels (OT: the oil terminal).

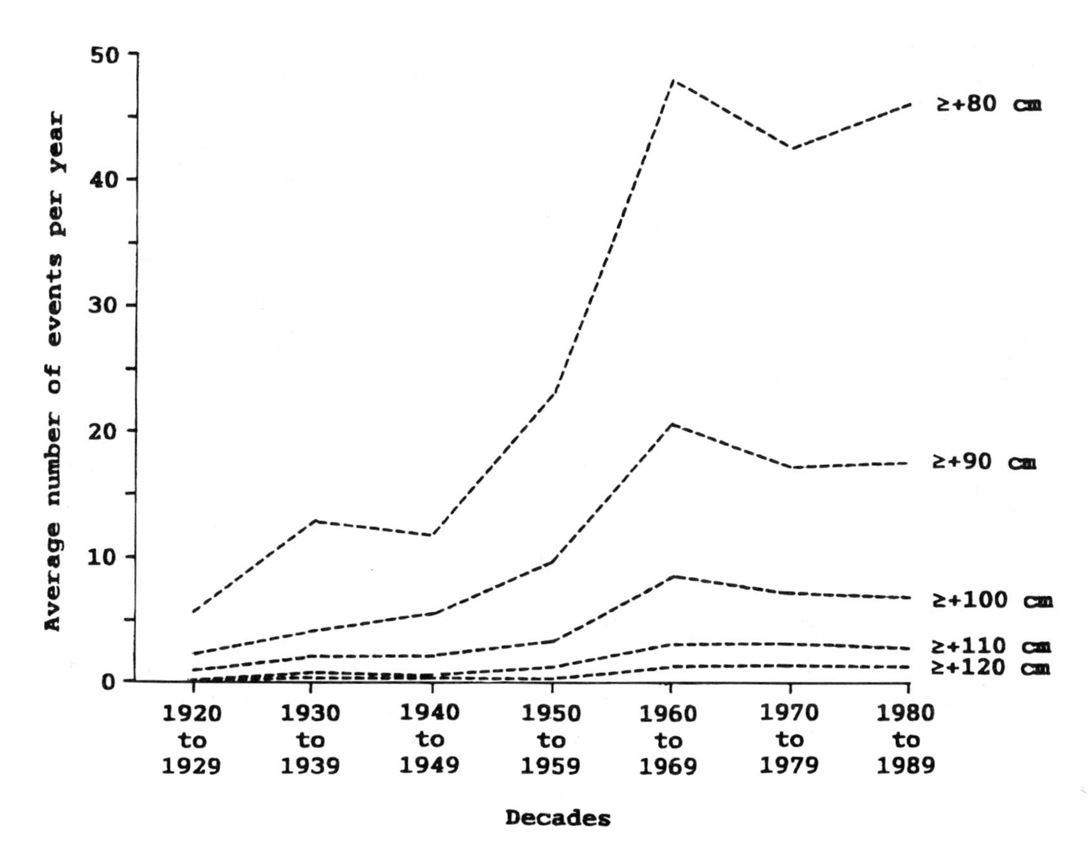

Figure 4.19.2. Average number of flooding events per year in Venice, from 1920 to 1989, with sea levels ≥ +80, +90, +100, +110, and +120 cm above the local datum (adapted from References 4).

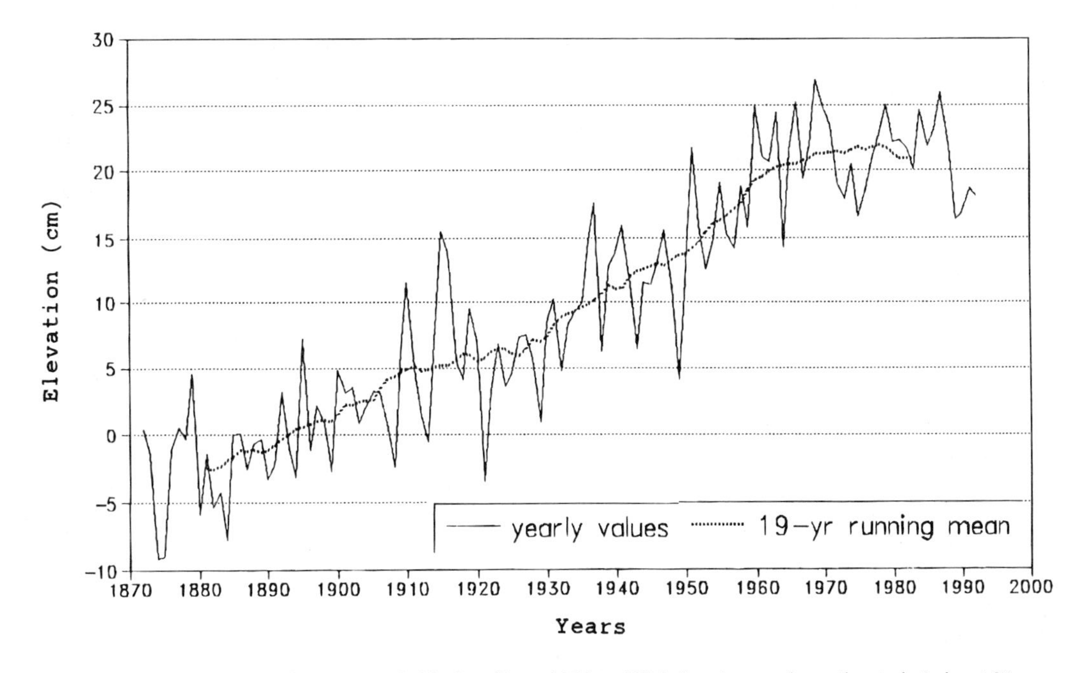

Figure 4.19.3. Relative MSL changes in Venice. From 1872 to 1992, the rise can be estimated at about 25 cm.

Figure 4.19.4. During moderate floods, people can still walk in certain parts of Venice because a boardwalk elevated by 60 cm is provided.

over 160 cm. The lowest point of the city, at +0.67 m above the local datum (i.e., 0.45 m above present MSL), is observed in St. Mark's Square, just outside the Basilica. In 57 percent of the city, the street level is situated at elevations between +1.1 m and +1.3 m above the datum, and 90 percent is below +1.4 m (5).

Since the 1880s, geological subsidence may be estimated to be between 3 and 7 cm. More important is the effect of pumping from underground sources, mostly for industrial use in the port of Marghera. Between 1910 and 1969 levelings measured a subsidence of about 14 cm, with 8 cm during the period 1952–68. Since 1975, when the industrial water supply was ensured by a new canal providing water for Marghera, underground pumping has ceased and human-induced sinking seems to have stopped. The difference between the relative MSL rise recorded in Venice and that reported from the apparently more stable tide-gauge station in Trieste (located in the northernmost part of the Adriatic Sea) is 12 cm since the beginning of this century; i.e., it is of the same order as human-induced subsidence in Venice.

The effects of a climatically induced, near future, sea-level rise would be dramatic in the Venice area. With a new rise in MSL of only 20 cm (which is predicted within a few decades by most climatic models), the return period of a +1.4-m flood would decrease from 4.1 years (at present) to 1.5 years; *sea water would be washed into St. Mark's Square by 55 percent of the high tides* (i.e., each day on average) and stagnate there for 1,283 hours per year (15 percent of the time).

In spite of important fishing and aquaculture activities, during the last decades the lagoon of Venice has received heavy industrial, agricultural and urban wastes. For the northern part of the lagoon, the average loads of nitrogen and phosphorus released into the waters are estimated at about 13,300 and 1,660 tons/year, respectively (2). The situation is already critical at certain periods of the year, with the occurrence of devastating algal blooms; it could rapidly become intolerable if the exchanges with the sea were reduced.

Solutions

At present, almost 30 years after the great flooding of 1966 (1.94 m above datum), in spite of more than 85 further floods higher than +1.1 m in Venice, no real action has been taken to solve the problem and no credible long-term solution is yet in view. Only one palliative has been adopted: from October to April, when storm-surge events are more likely, trestles and planks are placed across some streets of Venice. When the occurrence of flooding is forecast by the special service of the local administration, warning sirens sound in the city a few hours before the event; trestles and planks are then quickly deployed to build a raised path for pedestrians along certain routes (Figure 4.19.4). However, in most places where no emergency measures are taken, activities come to a stop for several hours. Besides, trestles and planks do nothing to prevent ground floors from being flooded and subsequent damage to furnishings and buildings.

Several engineering projects to remedy the present situation have been proposed during the last decades (4). The project preferred by the Italian government seems to be a plan for the construction of a series of mobile gates at the lagoon passes, which will be closed when the tide heights threaten to reach a given level. Each gate, 20 m long, would lie on the floor of the lagoon passes in normal times but would be raised by injecting compressed air to seal off the openings during storm surges. This project, if carried out, would certainly contribute to solving the flooding problem in the short term, but it presents four main weaknesses:

1. The construction and maintenance of underwater structures would be very expensive, and the hinges connecting the gates to these structures would be very vulnerable, particularly with the shock of a boat colliding into the gates.
2. Series of independent gates oscillating with the waves would be incapable of forming a watertight barrier; they would protect the lagoon from most of the effects of a single surge but would be inadequate when more prolonged periods of closure were needed (e.g., in the case of a permanent rise in sea level). Indeed, this would require the transformation of the mobile barrier into a protective system of solid dikes, permanently separating the lagoon from the sea, yet this possibility does not seem to have been taken into account in the project.
3. The present depths in the lagoon passes and in the navigation channels are indeed excessive for a muddy lagoon and incompatible with its hydrodynamic and sedimentary equilibrium. For political reasons this project does not envisage the possibility of decreasing these depths for fear of penalizing maritime traffic. However, since excessive depths facilitate the penetration of storm surges into the lagoon, the gates would have to be closed more frequently than if these depths had been decreased. Thus, maritime traffic will be penalized anyway and the hydrodynamic and sedimentological balance in the lagoon will not be restored.
4. If exchanges between the lagoon and the sea have to be reduced by episodic closure of the passes, pollution levels will increase inside the lagoon, the extent depending on how long the gates remain shut. In the case of the 1966 flood, for example, they would have had to be closed for two consecutive tidal cycles, which would have increased pollution levels by 65 percent, and 6 or 7 tidal cycles (3 to 4 days) would have been required before pollution concentrations returned to normal.

Even in the case of a small near-future sea-level rise, the survival of the historical city will be greatly endangered. To cope with the ongoing environmental deterioration, protection against storm surges is urgently needed. This may consist at first in the construction of mobile gates at the lagoon entrances but, in the case of a continuing sea-level rise, a rapid transformation of these mobile gates into a permanent barrier would be necessary, possibly within a few decades. In all events, a strict control and monitoring system of the water quality in the lagoon should be established before any construction works involving a decrease of the exchanges with the sea are undertaken.

Lessons Learned

Coastal engineering works decided purely on the basis of political choices, without taking into account possible consequences for the environment, can cause serious environmental problems and economic damage, as in the Venice area. Those responsible for all of the mistakes of the past will probably never have to account for their errors; in any event, most of them are no longer alive. Errors made in the lagoon of Venice, which have endangered a unique historical city belonging to the cultural heritage of humankind, should be viewed as a classical negative example, to be avoided here and in other parts of the world.

REFERENCES

1. Camuffo, D. 1993. Analysis of the sea surges at Venice from A.D. 782 to 1990. *Theoretical and Applied Climatology,* Vol. 47. Pp. 1–14.
2. Orio, A. O., and R. Donazzolo. 1987. Specie tossiche ed eutrofizzanti nelle Laguna e nel Golfo di Venezia. Rapporti e Studi, Vol. 11, Istituto Veneto di Scienze, Lettere ed Arti, Venezia. Pp. 149–215.
3. Pirazzoli, P. A. 1983. Flooding (*acqua alta*) in Venice (Italy): a worsening phenomenon. In: *Coastal Problems in the Mediterranean Sea,* E.C.F. Bird and P. Fabbri (Eds.). International Geographical Union, Commission on the Coastal Environment, University of Bologna. Pp. 23–31.
4. Pirazzoli, P. A. 1987. Recent sea-level changes and related engineering problems in the lagoon of Venice. *Progress in Oceanography,* Vol. 18. Pp. 323–346.
5. Pirazzoli, P. A. 1991. Possible defenses against a sea-level rise in the Venice area, Italy. *Journal of Coastal Research,* Vol. 7, No. 1. Pp. 231–248.
6. Rusconi, A., M. Ferla, and M. Filippi. 1993. Tidal observations in the Venice lagoon—the variations in sea level observed in the last 120 years. International Meeting, "Sea Level Changes and Their Consequences for Hydrology and Water Management," Noorwijkerhout, Netherlands. 22 Pp.

Note: — The above is a contribution to the project "Relative sea level changes and extreme flooding events around European coasts" of the European Union "Environment" Program.

Contributed by — Paolo Antonio PIRAZZOLI, Laboratoire de Géographie Physique, Meudon-Bellevue, France.

4.20 MALAYSIA, PULAU BRUIT: THE DISAPPEARING NATIONAL PARK

Background

A major challenge for ICZM is allocation of coastal areas for different uses. Such uses can be designated following broad use zones: protection areas, sustainable use areas, and development areas. Protection areas may encompass strict nature reserves, fish breeding/nursery areas, and buffer zones. Sustainable use areas may permit non-destructive fishing, mariculture or floating cage culture, and managed extraction of timber or non-timber forest products. In development zones, uses may include ports, harbors, hotels, and agricultural or industrial development. But planning for such spatial segregation of activities frequently assumes that the coastal environment is static rather than dynamic.

Problem

In parts of Southeast Asia, there appear to be cyclical coastal erosion phenomena. In some areas the coastline alternately accretes and erodes in a cycle lasting 5 to 30 years. Although this phenomenon is poorly understood and disputed by some, it plays an important role in spatial allocation of coastal resources. This process may be caused by variations in sediment deposition, river flow, wave or wind action, or coastal currents. The erosion can affect apparently stable or accreting mudflat areas and can erode wide belts along long stretches of coast. Although such cycles have been little documented in the literature, evidence can be found in the soil and vegetation profiles in a number of sites in the region.

In 1985, surveys were conducted along 800 km of the coast of Sarawak, Malaysia, by the National Parks and Wildlife Office of the Sarawak Forest Department in conjunction with the Asian Wetland Bureau to identify suitable sites for nature conservation. The major criteria used for site selection were (a) extensive accreting mudflats/mangrove forests and (b) large concentrations of migratory waterbirds and resident fauna.

As a result of extensive ground and aerial surveys, one site in the Rajang Delta was identified as the most important. This site was along the northern and western shores of Pulau Bruit, a 40-km-long island in the west of the delta. This site held more than 70 percent of the migratory waterbirds recorded along the Sarawak coastline and a 1–3-km-wide belt of high quality mangrove forest along the coast. Following the survey, this site was proposed as a new national park.

Between 1985 and 1989, the gazettement process was initiated to declare the site as a park. After gaining the support of the local community, a formal proposal was prepared. This gave recommended boundaries, which included the most extensive areas of accreting mangrove and the most important sites for migratory birds. Boundaries of the reserve were surveyed and agreed with the local community. Mangroves within the park were to be totally protected, and access/disturbance from fisherman would be restricted. The local people were allowed to harvest mangroves and fish from areas outside the park, and any land conversion activities were restricted to the center of the island.

However, in 1987 and 1988 an erosion phase began to affect the main part of the reserve. Over a period of about 12 months, the intertidal mudflats in the proposed park were severely eroded and a 2-meter-high erosion scarp developed along much of the coast. The mangrove zone was eroded landward by a width of at least 500 meters along a considerable distance of the coast. The most valuable parts of the reserve were lost.

Surveys in March 1989 indicated that the majority of the mudflats in the park area were of little importance for migratory waterbirds and that most of the birds were to be found outside the reserve area.

Solutions

Fortunately, the final gazettement of the reserve had not been finalized and it was possible to consider expansion of the reserve to include additional areas of intertidal mudflat and mangroves in the park.

In 1989, there were some signs of fresh deposition of mud and accretion of mangroves in areas that had been eroded earlier, indicating that the cycle may be reversing. However, it is likely to be several years, if ever, before the mudflats regain their former importance.

Lessons Learned

The major lesson to be gained from this example is that parks or protected areas should be sufficiently large to include sections of coast with different erosion/deposition regimes. This will ensure that a variety of suitable habitats are always available in a protected state, whatever the erosion status. It is thus very important to consider the dynamics of the coastal systems before determining boundaries of coastal protected areas. Boundaries should also, if possible, be able to adapt to changing coastlines.

Source — Duncan Parish, Executive Director, Asia Wetlands Bureau, Kuala Lumpur, Malaysia (See Reference 277 in Part 5).

4.21 MALDIVES: AN INFORMAL APPROACH TO COASTAL MANAGEMENT

Background

The Republic of the Maldives may be an example of a country that has accomplished integration in coastal zone management sufficient to meet its needs through informal means. Therefore, it does not, at present, seem to need a formal integrated coastal zone management (ICZM) process to provide a framework for addressing multiple-use issues in the coastal zone. The republic is a small country with concentrated government within which informal mechanisms seem to provide coordination adequate to serve the country's perceived needs.

The Republic of the Maldives comprises an oceanic chain of 26 coral atolls supporting approximately 800 small, sandy islands, as well as many unvegetated cayes. It lies 600 km to the west-southwest of Sri Lanka in the Indian Ocean. The republic stretches along the 73° meridian between 7° 06′N and 0°

42′ S. The area of the country is about 90,000 km^2, but only about 298 km^2 is land. The largest island is only 6 km^2 in area, and most of the islands have an area less than 1 km^2. Only 25 islands have more than 1,000 people and support 53 percent of the population of 214,000. The four cornerstones of the Maldivian economy are fishing, tourism, limited agriculture, and shipping. The remaining 100,000 people are scattered over 177 administrative islands. Most of the uninhabited islands are used for small-scale forestry and agriculture. About 70 islands are exclusively allocated to tourism.

Problem

The "integrated" process of coastal management (ICZM) tends to assume that the existing institutional structure cannot deal with the coastal conservation issues or vested interests that may inhibit management. The response is to recommend a novel framework, which in certain situations may be unnecessary or even counter-productive. If tried where not appropriate, it may be ignored, at the best, or lead to conflict, confusion, and long-term rejection of the practicality of ICZM, at the worst. In view of these possibilities, the ICZM process should involve an early stage analysis of the need for a novel framework in case a workable framework already exists.

In the Maldives an informal traditional and non-innovative approach to coastal management seems to exist already, but it could be strengthened. In this case replacement by a novel ICZM system may at best be counter-productive. The framework includes an Environment Commission containing representatives of various ministries and government departments. The commission identifies policy and is supported by a secretariat in the form of the Environment Section within the Ministry of Planning, Human Resources and the Environment. Government also contains departments dealing with sectoral activities such as physical planning and design, public works, and fisheries research. Most of these departments are represented on the commission and can input to it.

The particular circumstances that support an existing framework that can address multiple-use issues in the coastal zone of the Maldives are a reflection of these basic biological, physical, demographic, and economic features. Central administration and evident multi-sectoral dependency on both coastal activities and the coastal resource base all lead to an integrated approach in dealing with coastal issues.

Despite this, the existing framework may not be operating as effectively as is necessary for two main reasons. First, a local, appropriately educated government cadre does not exist to distinguish between significant and insignificant issues. Making such a distinction has not helped the country to develop. Second, people tend to accept deterioration of their environment as the cost of improving other living standards. For example, poor groundwater quality and other population pressure–related effects on the Capital City (Malé) are accepted as the cost of better education and medical care.

Both failure to identify significant issues and investment of time, energy, and political will in relatively insignificant issues are defeating to the ICZM process. Issues that were first considered to be significant and are now considered to be less so include transient coral bleaching events, localized crown-of-thorns starfish outbreaks, and "reef cracking."

Conversely, the issue of coral mining to provide building materials—which has great social and environmental significance—still has to be resolved. At present, the action is remedial rather than preventive; that is, resolution is concentrated on areas that have already been mined rather than on those that are at risk from mining. An analogy is trying to put out a forest fire after it has died down rather that putting in firebreaks to control it.

Solutions

Efforts to minimize environmental degradation through innovative systems of environmental management are considered to be particularly necessary and urgent in societies undergoing rapid change. This consideration assumes that the adaptation of existing systems is impractical because these systems are inappropriate but cannot be adapted quickly enough. It also assumes that innovative mechanisms can be assimilated quickly or that they can be imposed. These assumptions are simplistic, and the strategies developed from them may not work.

It may well be that in some countries the traditional and informal approach to coastal management can provide solutions in time to meet the threats posed by rapid change and particularly so in a society that can evidently undergo rapid change in other sectors. However, this is not necessarily guaranteed.

The priority must, therefore, be to create an educated cadre of locals who can balance tradition with innovation to meet the ICZM challenges posed by rapid change.

In the meantime there will be a period during which traditional mechanisms will be unable to cope with change and innovative mechanisms will not be accepted. Environmental degradation may occur. Local counterparts supported by expatriate experts might provide limited opportunity for the introduction of innovative management in a traditional context during this period.

Educating the public about development priorities is a more difficult matter and may be achieved only indirectly through further environmental deterioration. There is also a need to educate the public that a deteriorating environment need not be an acceptable cost of improving other living standards.

Lessons Learned

In small countries the deficiencies of the existing coastal management framework may not justify the introduction of a novel ICZM program. One reason is that, in a small country with concentrated government, integration may be achieved de facto, at least to the extent that existing leaders want it, because they are in nearly continuous communication.

However, deficiencies do need to be addressed. There may be a pressing need both to do the research and to provide the educated cadre necessary to interpret the significance of a coastal zone issue before it is considered and dealt with within the existing political framework.

At the simple level it can be suggested that societies undergoing rapid change require innovative management because traditional systems do not have the experience to cope with new issues. However, it can also be suggested that societies that are capable of undergoing rapid change are, by definition, innovative and adaptable because they have assimilated the skills necessary to support change. The problem may, therefore, be the seed of its own solution. However, this assumes a homogeneously changing society and, more critically, that the value systems that drive change are the same as the ones that will accept innovative management. Neither may be the case, at least during the time available during rapid development.

While traditional systems may have their failings, innovative systems may be perceived by a rapidly changing society as inappropriate. The coastal zone manager is, therefore, faced with a dilemma. As with most dilemmas the problem is simply one of perspective. Tradition and innovation are all part of the same historic continuum, with today's innovation being tommorrow's tradition and today's tradition being yesterday's innovation. Identifying management mechanisms is not, therefore, a matter of deciding whether to use traditional or innovative mechanisms. It is first a matter of deciding whether the issues to be addressed through management are soluble and important and then deciding what mixture of tradition and innovation will achieve the desired objective within the most relevant time scale. Objectives chosen through consensus are the strongest.

The coastal zone manager should use a mix of traditional, innovative, impositional, and conceptual management mechanisms to achieve the desired objective but with the caveat that management must ultimately be achieved through consensus. For example, innovative impositional management may best be achieved using traditional value systems within an accepted, rather than a novel, hierarchy. Conversely, pure consensus management might better be achieved through novel systems and not through traditional systems.

Ultimately, it is more than likely that the best strategy for supporting ICZM in any country will be based on answers to the following questions:

1. How rapid is the change that needs to be addressed?
2. How significant is the change?
3. Can the change be managed using traditional mechanisms?
4. Can impositional management be avoided in the available time?
5. Can consensus management be achieved in the available time?

Contributed by — Alec Dawson Shepherd, Hunting Aquatic Resources, Hertfordshire, England.

4.22 MEXICO, CANCÚN: COASTAL TOURISM THREATENS A PRIME LAGOON

Background

Cancún is a world-famous resort on the eastern side of Mexico in the Yucatan Peninsula. Mayans are the indigenous inhabitants of the region surrounding Cancún. Their civilization flourished from the fourth to the tenth centuries. Their architectural monuments and advanced culture still create wonderment. According to one Mayan dictionary, the name Cancún derives from the word for "place full of snakes." According to another source it means "golden serpent." The latter interpretation appropriately describes Cancún as a mixed blessing (Figure 4.22.1).

One could believe that God created Cancún to be an international tourism site. The Cancún geography has been seen as a divine gift to Mexico. The scope and quality of tourist attractions in conjunction with the proximity to the world's largest tourist market (USA) is a tourism opportunity unsurpassed by any other area on the globe. Consider the following list of superlative tourist attractions stated in 1970: perfect weather in the winter months, 19 kilometers of wide white powder beaches fringed with coconut palms and fronting on clear turquoise ocean waters rich with sport fish, a clear 50-km^2 lagoon system (Nichupté Lagoon) that is perfectly protected for water sports, the world's second largest barrier coral reef starting about 30 m offshore, bountiful groundwater, and close proximity to the Mayan archeological sites.

Cancún's most distinguishing natural feature is the Nichupté Lagoon System (NLS). Originally, it was the largest clear lagoon with a white sand bottom in the wider Caribbean and Mexico (and a depth of 2 meters). The NLS is divided by shallows into three basins (Cuenca Sur, Cuenca Central, Cuenca Norte). Two smaller lagoons (Rio Ingles and Bojórquez) connect through narrow channels to the main water body. The whole system is connected by the Cancún and Nizuc Inlets to the open sea (Figure 4.22.2).

An 8-meter-high dune system is found along the southern two thirds of the island's east side. In 1970 the dune was covered by vegetation, which reduced wind drag and prevented the dune from migrating inland. On the seaward part of the island, the dominant vegetation was a tropical dune-beach

Figure 4.22.1. The hotel strip of Cancún was built on a sandy barrier island enclosing an ecologically fragile lagoon.

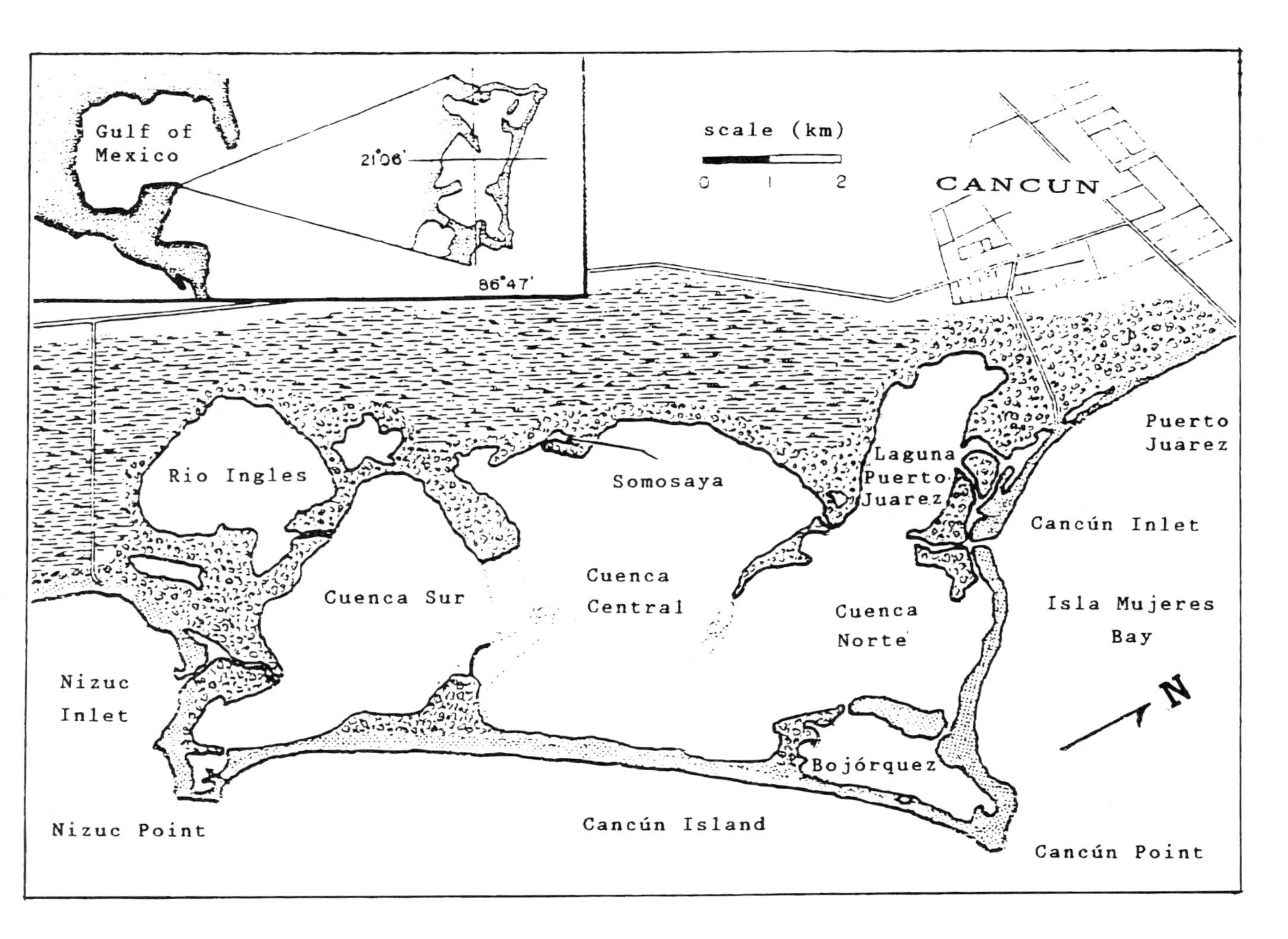

Figure 4.22.2. The Nichupté Lagoon system is enclosed on all sides and has only two small openings for seawater to enter and flush contaminants from the lagoon—the result is a costly example of polution of a tourist amenity.

association of indigenous plants. The central portion of the island was in coconut palm plantations. Mangroves extended along the lagoon shoreline.

In 1969 the Inter-American Development Bank loaned Mexico $US21.5 million to provide half the funds for planning Cancún and developing the infrastructure. Then the planning and development of Cancún occurred through a progression of four master planning efforts and products in 1971, 1982, 1985, and 1988.

Problem

The main problem of Cancún was profiteering at a level that drove resort development beyond any reasonable limit and corrupted any forthright planning attempts. Huge fortunes were amassed while the beginnings of a tourist ghetto were laid down.

The number of hotel and condominium rooms proposed in the 1988 Master Plan exceeded the 1971 Master Plan by a wide margin. The 1971 plan was for about 2,000 units at the end of stage 3, whereas the 1988 plan (3) provided for almost 15,000 units. A comparison of the 1971, 1982, and 1988 master plans (2, 3) shows an increase in the area used by hotels from 25 percent in 1971 to 68 percent in 1988.

Now it appears that Cancún may become another Acapulco (west coast of Mexico). The tourism development of Cancún, like Acapulco, has both degraded the original spectacular attractions of the site and motivated tourists, as well as residents, to go elsewhere.

It is typical that the literature generated by the Mexican National Tourism Agency, FONATUR, strongly suggests that Cancún is a success. But their measure of success is the number of rooms constructed, number of tourists per year, employment numbers, and regional income. Nothing about *quality.*

A Cancún case study by the authors (5) identified ten major problems that were caused by FONATUR's plans and development:

- Degradation of the Nichupté Lagoon system
- Groundwater management
- Borrow pit and strip development along highway 307
- The loss of ocean beaches and the dune system
- Congestion of the highway along Cancún Island
- The chaotic visual impression of resorts on Cancún Island
- The slums and lack of housing in Cancún City
- Beach access and use
- The lack of diverse employment opportunities
- The dependence on the tourism economy

The first four issues—those that pertain directly or indirectly to the environmental quality of the Nichupté Lagoon system—are relevant here.

The NLS is a relatively isolated water body whose capacity to withstand direct modification and discharges depends to a great extent on their flushing processes. Rain is the only important flushing force in the NLS. As a result, the average flushing time for the whole system is about two years. This figure indicates that the NLS is a system that would be very sensitive to any changes in water quality or quantity. Many lagoon and estuarine systems exhibit flushing times in the order of days or weeks.

One of the most evident adverse impacts of Cancún's development is the eutrophic pollution of the Nichupté Lagoon system. The main actions that have over-fertilized and degraded the Nichupté Lagoon system have been identified (4). These are dredging of wide areas of the original bottom, continuous sewage discharge, increasing boat traffic, and the destruction and filling of surrounding mangroves.

The eutrophication is severe where there is restricted circulation and multiple inputs of nutrients. The large fill area created for the golf course particularly increased the constricted circulation and decreased flushing. The primary expression of degradation is the proliferation of a diverse macroalgae array, which forms large floating mats but also forms mats that cover both the seagrasses and the lagoon bottom. In summer months the floating mats are odiferous and ugly. FONATUR (the Mexican tourism agency) operates a number of algae-gathering barges to reduce the problem.

A further loss of amenity is decreased clarity of the water. Descriptions of the lagoon in the early 1970s note the transparency and the alluring turquoise color. The inputs of nutrients have increased the density of phytoplankton and decreased the transparency.

Dredging activities have eliminated seagrass meadows in the dredge areas and have released an excess of nutrients into the lagoon system. A soft layer of organic matter has covered the bottom of many dredged areas which, coupled with light reduction caused by increased depth, prevents seagrass recolonization of these areas. The reduction of the seagrass community has had at least two effects. It has reduced the habitat quality of the lagoon for fisheries and wildlife. Seagrasses also prevent bottom sediments from being re-suspended in the water column by boat wakes and wind waves—which in turn increases both the turbidity and the nutrient enrichment of the lagoon system. Also, the loss of the macrophytes—particularly the seagrass beds—removes a natural filtering process that removes particulate matter from the water column. In many parts of the lagoon system, the alluring turquoise colors have turned to muddy gray.

Nutrients, insecticides, pesticides, and oils are entering the NLS from many sources, such as street and gutter wash, fertilizers from the golf course, and all-around landscaping of hotels and residences. The amount of these inputs into the system and their effects are not known. However, the impacts of such inputs appear to be minor in comparison to the direct discharge of raw or partially treated sewage into the lagoon from the outfall pipe at the sewage plant next to the golf course. The direct discharge occurs during heavy rainfall and when the treatment plant malfunctions (it goes to a sub-lagoon named Bojórquez). When the system is overloaded, sewage flows out through the manhole covers in the boulevard and drains into the adjoining lagoon.

It is logical to conclude that the reduction in water clarity, the change of water color, the mats of stinking algae, the reduction in fish and wildlife species, and a proliferation of jellyfish have combined to have three significant adverse impacts: reduction of the recreational potential of the lagoon system, reduction in the value of properties and businesses that front on the lagoon, and a slump in biodiversity. Many parts of the lagoon system have turned from a recreational and visual attraction into a repulsive feature.

The ocean beaches and dunes have been greatly affected, too. A comparison of vertical and oblique aerial photographs taken before any development on Cancún Island and with 1991 development clearly shows a reduction in beach width. The total sand dune system has been reduced to 16 percent of the original coverage—from 18.5 hectares in 1969 to 3 hectares in 1991. The most dramatic—and probably the largest—reduction of beach sand (and beach width) occurred when Hurricane Gilbert directly hit Cancún Island in September 1988.

Solutions

The following environmental objectives were articulated in the 1982 FONATUR Report and Master Plan (2).

- The development of Cancún should be done within a nature-preserving framework, described as "environmental protection," "creative utilization of natural resources," "re-encounter of man with nature," and "aesthetic and functional design of tourist spaces."
- Design concepts were more precisely defined as "achieving a homogeneous, dynamic and ordered complex, in which architectural and urban components harmonize with, and integrate themselves to the natural environment."
- Other specifications were creative use of natural resources and design of tourist spaces that join beauty and function.

Whether the environmental qualities of the Nichupté Lagoon System (NLS) will continue to degrade is unknown. With few exceptions the government does not seem to have changed its way of managing the NLS despite the apparent degradation of the system. One new policy has been adopted to reduce pollution impacts. No dredging is allowed in the NLS to create landfills. Fill must be taken from mainland excavations, as was done for Lot 18-A and the Aoki Hotel lot. This is helpful but still leaves empty all the promises above.

Environmental impact assessment requirements, a law since 1984, have been used to a limited extent to minimize adverse impacts on the NLS. For example, proposals for fills of significant size, such as the Aoki Hotel lot, have plans that include an environmental impact assessment and that were available, to a limited degree, for public review. The assessment of impact was largely done to placate the Mexican environment agency (SEDASOL) and citizen groups (NGOs). In respect to the large-scale fill projects, partial benefits were achieved—notably the prohibition on dredging to create fill material. However, three recently proposed large-scale projects show a re-emergence of disregard for both the purpose and the principles of environmental impact assessment.

But there are some hopeful signs on the horizon for Cancún's environment, including the NLS. A new state government has just been elected in Quintana Roo. The new governor believes that environmental protection is required to achieve sustained economic growth.

Contributed by — Jens Sorensen, Harbor and Coastal Center, University of Masachusetts (Boston, USA); Martin Merino, Instituto de Ciencias del Mar y Limnologia, Universidad Nacional Autonoma de Mexico (Mexico City); and David Gutiérrez, BIOCENOSIS, Cancún (Mexico).

REFERENCES

1. Bosselman, F. 1978. Mexico reaches for the moon. In: *In the Wake of the Tourists: Managing Special Places in Eight Countries.* The Conservation Foundation.
2. FONATUR. 1982. *Un Dessarrollo Turistico en la Costa Turquesa: Cancun, Quintana Roo.* 103 pp.
3. FONATUR. 1988. Uso del Suelo y Zonificacíon (scale 1:20,000).
4. Merino, M., A. Gonzalez, E. Reyes, M. Gallegos, and S. Czitrom. 1992. Eutrophication in the lagoons of Cancún, Mexico. *Science of the Total Environment, Suppl.* Pp. 861–870.
5. Sorensen, J., M. Merino, and D. Gutiérrez. 1993. *Cancun: The Visions and the Realities in the Planning and Development of an International Coastal Tourism Center.* Intl Coastal Res Mgmt Prog, University of Rhode Island.

4.23 MEXICO, YUCATAN: PROVINCIAL COASTAL ZONE MANAGEMENT

Background

The Yucatan Peninsula is a flat, limestone platform. The rain dissolves the lime, creating a typical relief called *karst* (1), a stony surface pitted with sinkholes; hence, there is almost no surface drainage. The water rapidly enters the freatic mantle and a complex system of caverns and underground rivers. Eventually, the water reaches the coast and gets to the surface in springs and sinkholes; such freshwater springs can be found sometimes below the sea surface. Nevertheless, most of the freshwater is discharged in coastal lagoons and mudflats separated from the sea by long, narrow sand barrier islands. Politically, the peninsula is divided into three states; Campeche at the west, Yucatan at the north, and Quintana Roo in the east.

Coastal ecosystems around the peninsula include coastal lagoons, mangrove forests, sand dune vegetation, seasonally flooded tropical deciduous forest, freshwater and brackish water marshes, coral reefs, and extensive mudflats. This environmental diversity harbors a variety of flora and fauna—e.g., there are 348 bird species reported for the coast of Yucatan state (4).

In the peninsula there are several federal and state coastal protected areas from west to east: Ria Celestún Special Biosphere Reserve, El Palmar State Park, Bocas de Dzilam State Park, Rio Lagartos Special Biosphere Reserve, Isla Contoy Special Biosphere Reserve, Tulum National Park, and Sian Ka'an Biosphere Reserve.

Problem

The coast of Yucatan State has been developed, particularly for holiday-season housing and tourism. Salt production is a mayor concern as well because of the excessive extraction of sand from the dunes (2, 5). On the other hand, fishermen have had conflicts with the salt company over the rights to fish/

close productive interior bays. Obstruction of the natural water flow patterns is a major problem on the Yucatan coast, where 13 roads going to coastal towns lack almost completely appropriate culverts. Further development is on its way and major changes should be prevented.

Solutions

As was mentioned by Clark (3), there are situations when a state or province may consider the possibility of establishing its own integrated coastal zone management program. The reasons mentioned are that (a) nationwide programs may lose priority when benefits are diluted over a whole country's coastline or (b) they may encounter higher levels of political resistance than the programs generated for a specific state or region. This was the situation when in 1988 Yucatan State elected a new governor. A group of scientists and citizens approached the new governor and asked him to take special care of the protected areas and the coast. The government appointed a State Consulting Council for Ecology (CECE) with representatives of local citizen organizations, research and higher education institutions, as well as the state government.

One of the first assignments for CECE was to put together a series of guidelines to control coastal zone development in a document called "Yucatan State Integrated Coastal Zone Management Program" (PMIZCY). The PMIZCY first defined the coastal zone from the administrative point of view but with biological, hydrological, and economic basis as "a zone extending from the 20.84-meter (12 fathom) bathymetric line (where 70% of the coastal fishing takes place) to the the furthest inland extent of the tide's influence on the ground water table (a 5 km average band)."

The coast was divided into seven zones based on ecological criteria, and critical habitats were identified. Areas to be put under protection were selected. Administrative procedures and guidelines to control coastal development were recommended. In the case of major coastal development projects, environmental impact analysis modifications were recommended by CECE.

Lessons Learned

The nomination of a collegiate body such as CECE was a very important breakthrough in Mexico. At that time there was no equivalent in any other state in the country; The General Law on Ecology had just been issued, so Yucatan rapidly took advantage of the autonomy now given to the states. Direct communication between government officers and concerned citizens was the key to producing valuable documents. However, the implementation of plans was halted because the governor resigned prematurely. This is a typical situation in a country with centralized executive power.

Some of the problems curtailing the implementation of a national program in Mexico—lack of recognition of the problems, poor coordination between agencies, and lack of apropriate knowledge of coastal ecosystems—were overlooked at the state level. What it could never ignore was the lack of funds and the discontinuity in the government. These two factors should be carefully analyzed. The first probably could be solved as soon as society becomes conscious of the real value of coastal resources and asks the government to direct more funds to work on integrated coastal zone management programs. The second could be worked out by the establishment of a regional, flexible network of research institutions and NGOs that would have time scales different from the governmental periods. Such a network could help as well with overlapping political boundaries.

Contributed by — Jorge Correa-Sandoval, Centro de Investigaciones de Quintana Roo, Apartado Postal 424, Chetumal 77000, Quintana Roo, México.

REFERENCES

1. Batliori, E. 1991. Caracterización hidrológica del Refugio Faunístico Ría Celestún al Noreste de la Península de Yucatán. (unpublished). CINVESTAV-Mérida, Yucatán, México. 36 pp.
2. Clark, J. 1990. Coastal zone management: a state program for Yucatan, Mexico. In: *Coastal Zone '89: Proceedings of the Sixth Symposium on Coastal and Ocean Management*, Vol. 1. Pp. 1466–1478.

3. Clark, J. R. 1992. Integrated management of coastal zones. FAO Fisheries Technical Paper No. 327. United Nations/FAO, Rome. 167 pp.
4. Correa, J. and J. García. 1993. Avifauna de Ría Celestún y Ría Lagartos. In: *Biodiversidad marina y costera de México,* S. I. Salazar-Vallejo and N. E. Gonzalez (Eds.). Com. Nal. Biodiversidad y CIQRO, Mexico. pp. 641–649.
5. Murguía, R., J. Correa, E. Batliori, E. Boege, and G. de la Cruz. 1989. Estudio interdisciplinario de la Ría de Lagartos. *Avance y Perspectiva,* Vol. 8, No. 39. Pp. 13–24.

4.24 NETHERLANDS: ENVIRONMENTAL PRIORITY FOR DELTA PLAN

Background

The Eastern Scheldt (surface area 500 km^2) is a "riverless estuary" situated in the delta region in the southwest Netherlands. In 1953 disaster struck when the area became flooded, killing more than 1,800 and destroying homes and livestock. After ample consideration, it was decided to prevent such catastrophes from happening again by implementing the "Delta Plan." Of paramount importance in the decision-making process were considerations of safety, problems for agriculture caused by saltwater intrusion, the technology available, and financial factors.

Indeed, three estuaries were closed off by dikes: the Haringvliet basin, now one of Holland's most polluted freshwater areas; Lake Veere, a brackish, small lake; and Lake Grevelingen (Figure 4.24.1), which developed into a more or less healthy saltwater lake (after long discussions and decision-making processes, parliament decided to keep the saltwater lake instead of creating a freshwater lake).

Problems

When the diking of the Eastern Scheldt, the last and biggest project, had already started, various reasons led to a reconsideration of the whole scheme. People were concerned about environmental effects of large-scale hydraulic engineering works. The closure would finish the large mussel-culturing area and destroy some fishing grounds. Furthermore, the closure would create a freshwater lake. Much of the water in the lake would originate from the rivers Rhine and Meuse, which had worsening water quality at that time.

During a study, the following were taken into account:

1. Total closure with desalinization
2. Leaving the Eastern Scheldt open with dike reinforcements
3. Partial closure with a sluice dam or storm surge barrier

Solutions

For safety reasons, Alternative 2 was unacceptable at that time, whereas the environmental aspects were the major reasons to adopt Alternative 3, even if it was almost 5 times more expensive than the original plan! After considering the effects of the different alternatives on function (e.g., water quality, fishing, aquaculture, tourism, etc.) and the financial consequences, a report was issued recommending the construction of a storm-surge barrier across the mouth of the Eastern Scheldt and two compartment dams. As a result, a normal tidal regime would remain in the western part (100 km^2), a reduced tidal regime would be maintained in the central part (300 km^2), and there would be a freshwater lake (100 km^2) to the east of the compartment dams.

After some consideration, both government and parliament, in 1974, accepted the proposed solution under three conditions:

1. The project had to be technically feasible.
2. The total costs should not exceed Dfl 3.8 billion (the original plan was estimated at 0.8 billion),
3. The barrier was to be completed by 1985 at the latest.

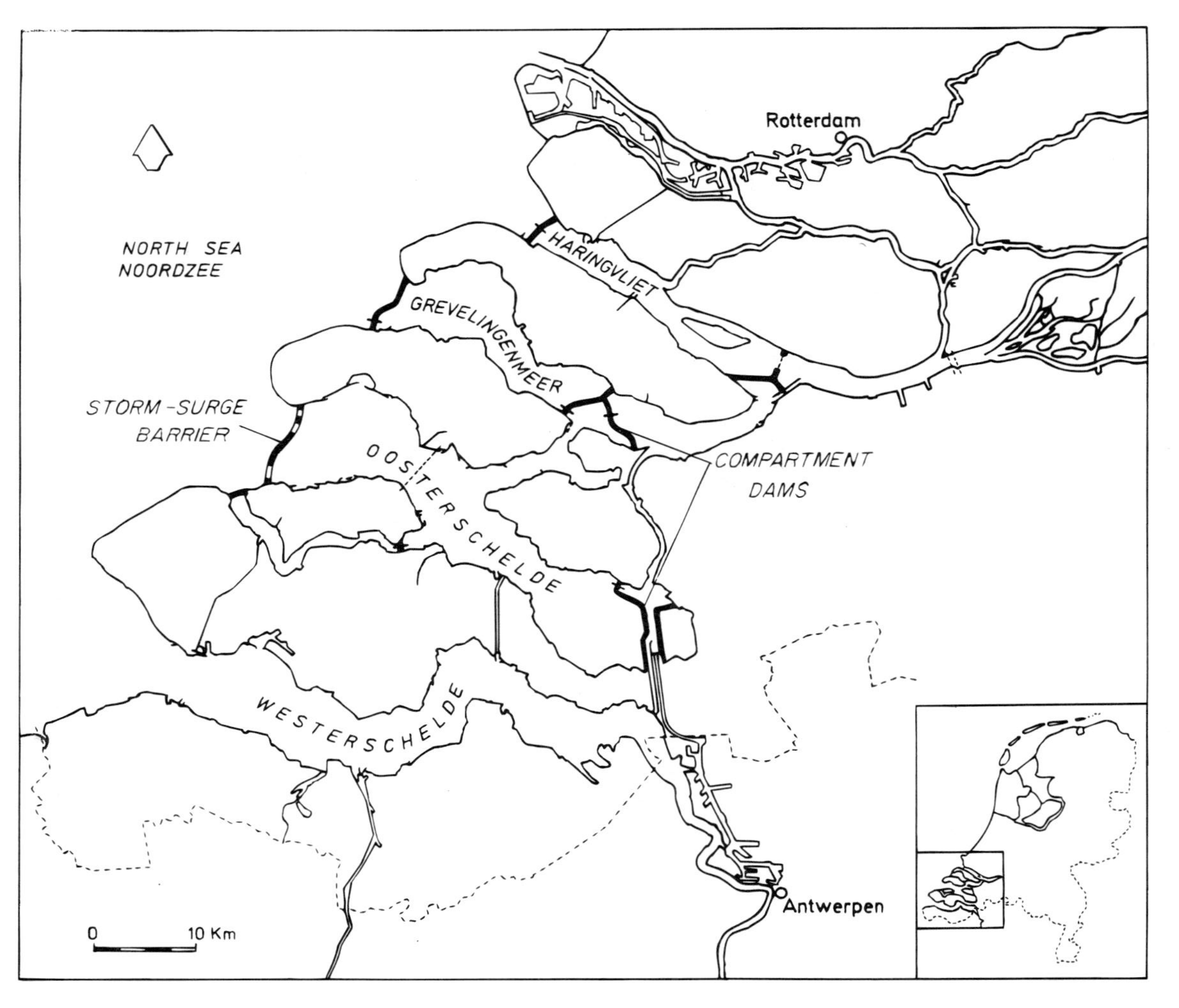

Figure 4.24.1. The Eastern Scheldt are where major alterations of the coastal environment were implemented.

In 1986 the storm-surge barrier in the Eastern Scheldt was taken into operation by the Queen of the Netherlands. The total costs came close to Dfl 8 billion ($US3.9 billion).

In 1994, the Eastern Scheldt is still a very productive estuary with good water quality. Compared to the predictions made in the 1980s, the salinity and the tidal amplitude are higher. In contrast, habitats—salt marsh and tidal flat—are decreasing much more rapidly. In the future, this might limit the value of the estuary as a bird sanctuary of international importance.

Lessons Learned

The lessons we have learned from the Eastern Scheldt project include the following:

- Actions should not only be judged on engineering technical merits alone, but also should include all aspects (multi-disciplinary and multi-functional).
- Large-scale interventions that take place over many decades require periodic assessment. A plan must be implemented in a phased manner so that major changes in design are possible.
- One should take into account the changing "spirit of the time." In the 1960s only a total closure was acceptable; in the 1970s the most expensive solution was possible; in the 1980s the decision would have been to raise the dikes around the Eastern Scheldt.
- The alternative (*no* project) should be given serious consideration at the decision-making stage.
- Research must play an integrated part in controlling the project and should continue throughout the whole period, as well as for some time afterward.
- There should be sufficient freedom to introduce technical modifications.
- The project does not end when the construction work is completed. For the Eastern Scheldt a special "Situation Management" ICZM commission was formed in which local governments, interest groups, and the decision-making bodies (provincial and national government) are represented and which discusses all development plans in the area (fishing, tourism, industry, etc.), recommending to local and national authorities what decisions should be made.
- It is vitally important that all authorities are actively involved during the course of the projects, along with appropriate non-government organizations.
- Natural variations in the ecosystem may be larger than human-induced changes. For proper long-term predictions, the collection of baseline data over a decade or more is necessary.
- The development of systems can, within certain limits, be controlled and guided.

Contributed by — H. J. Lindeboom, Head, Department of Applied Research, Netherlands Institute for Sea Research (Texel).

4.25 OMAN: COASTAL MANAGEMENT BY NETWORKING

Background

In 1984, the government of the Sultanate of Oman began to develop an ICZM plan for the 200-kilometer stretch of coast spanning Muscat, the capital. At the time, there was no government authority with a mandate for ICZM. However, the project, assisted by IUCN, was lodged in the Department of Tourism in anticipation of the coast providing a major focus of tourism development. At the time there was no schedule for opening the country for this activity. Following completion of the first ICZM plan, the project continued through a series of extensions over nearly eight years to cover the entire coast. Thus, it started with situation management and ended up with virtually a nationwide program.

Problem

There were clearly several authorities with interests in the coastal zone (CZ), some with active programs, thereby posing the first objective:

- To design the project to avoid duplicating effort and antagonizing other agencies and to collaborate with the many agencies having a stake in the CZ.

The absence of any authority with the mandate for ICZM posed the second and more important objective:

- To design a practical, implementable plan without a particular agency to lodge it and no one agency with the directive to execute it.

The challenge was to deal effectively with typical ICZM problems, such as the following examples:

- Establishment of conservation areas
- Management of recreational resources
- Management of fisheries resources
- Protection of cultural and archaeological resources
- Maintenance of coastal aesthetic resources
- Conservation of threatened wildlife resources, such as turtles and seabirds
- Management of special environments, such as mangroves, coral reefs, sand dunes, beaches, lagoons, dry river valleys, and flood plains
- Proper design and construction of coastal developments such as harbors, marinas, seawalls, grones, piers, breakwaters, coastal roads, and dams

Solutions

It was clear that Oman required a customized approach to resolve its particular problems in ICZM planning and implementation. The Oman formula proved a successful solution. It evolved over time through a flexible process of dialogue, trial and error, and adaptation. The essential elements of this process are summarized below.

The participatory element was initiated by official communication with all concerned authorities to inform them of the project (or new phase) and its goals and objectives and to request the identification of a focal person for liaison and technical collaboration. All correspondence was at undersecretary (deputy minister) level, while all ensuing technical contact was at director-general or director level. Technical experts were involved as needed on a case-by-case basis.

A series of meetings was initiated with the focal person from a very early stage of each project phase. The first meetings aimed to ascertain the interests and activities of each authority in the CZ, their concerns and constraints, and to explain the objectives and anticipated outputs of the ICZM plan. Later meetings focused on specific management issues and observations and sought to identify solutions for their resolution.

The early meetings helped the project focus its activities in a way that built on and complemented existing initiatives and avoided duplication of effort. For example, water resources, agriculture, and fisheries were being studied by teams of experts, so they were left to their respective agencies to plan and implement. However, aspects of relevance to ICZM that were overlooked, such as impact on coastal resources and conflicts with other activities, were incorporated into the project.

For example, specific attention was given to the identification of fish landing sites and their classification by numbers of standardized boat units to ensure that conflicting developments considered the impact on coastal communities and their boat anchorages or landing sites. Also, the impact of fisheries on environment (e.g., nets in corals) and wildlife (e.g., turtles, whales, and dolphins) was recorded and monitored in detail by the project.

During the course of data analysis and exploration of mechanisms to resolve management issues, the novel approach to ICZM was devised. As there was no single authority with the mandate to implement ICZM, the responsibility for execution of issue-specific actions was shared among all concerned agencies. Thus, a single plan was produced for each target stretch of coast through a participatory process. This required each concerned government authority to endorse the plan and its recommendations and accept its respective responsibilities for implementation of the agreed actions. For example, different activities were assigned as follows:

- Coordination of CZM to the Ministry of Regional Municipalities and Environment
- Coordination of coastal development policies to the Supreme Committee for Town Planning
- Land use policy (including setbacks), planning, and allocation to the Ministry of Housing
- Recreation and tourism policy, planning, and management to the Ministry of Commerce and Industry through the Department of Tourism
- Critical fisheries habitat management to the Ministry of Agriculture and Fisheries through the Directorate General of Fisheries
- Development and implementation of environmental policy to the Ministry of Regional Municipalities and Environment
- Protected areas establishment and management to the Ministry of Regional Municipalities and Environment
- Control of roads, harbors, and coastal works to the Ministry of Communications
- Site implementation of policies and clean-up to the respective municipalities
- Enforcement of environmental and species protection regulations to the Royal Oman Police

Endorsements were achieved through a process of regular contact, meetings, workshops, and opportunities to contribute through comments and revisions to a consensus in the content of the plan. In several instances this process led to actions being executed and issues resolved before the plan was printed, so that implementation overtook the planning process.

The Ministry of Regional Municipalities and Environment, which is well represented throughout the Sultanate, now has the firm mandate to implement ICZM. The ministry follows a policy of coordination and networking, following the philosophy of the original plans (i.e., issue-specific actions are implemented by the responsible agency).

Conflict resolution was never a simple process, inevitably involving several research, implementation, and planning authorities. The policy of coordination through regular meetings pursued by the Ministry of Regional Municipalities and Environment has proved to be an effective means to achieve cooperation in conflict resolution among different authorities.

Lessons Learned

Sharing of authority and responsibility for the execution of ICZM actions may be a practical mechanism for a country to achieve effective implementation, as it was for Oman. However, this is most difficult to accomplish unless the actions and responsibilities are identified and agreed through a fully participatory process. Identification of a lead agency with the primary responsibility for coordinating these two tasks will also facilitate implementation.

The success of any plan that implicates different ministries, as do the CZM plans for Oman, depends on three principal factors:

- Regular consultation, information exchange, and opportunity for plan review and input
- Consensus on responsibilities and actions for implementation
- Effective coordination of plan implementation

Contributed by — Rodney V. Salm, Coordinator, Marine and Coastal Conservation Programme, World Conservation Union (IUCN), Nairobi, Kenya (70, 71).

4.26 PHILIPPINES, PALAWAN: ECONOMIC ANALYSIS OF RESOURCE CONFLICT

Background

This case study illustrates how the combined use of ecological and economic analyses can provide useful information for government planners seeking to maximize net economic benefits while minimizing social and environmental costs. The case described here brought three major industries into conflict—fishing, tourism, and logging.

In such a resource use conflict, the market-based, property-right solution traditionally prevails: the logging concession owner harvests the trees and moves on, leaving those dependent on the bay's resources to adjust. Here it is shown that such an action would be extraordinary wasteful.

Problem

Bacuit Bay on Palawan Island, Philippines, supports major fishing and tourist economies. The surrounding hills contain valuable timber resources harvested under a government concession (Figure 4.26.1). The logging activity has led to substantial soil erosion and sedimentation of the bay, which has killed corals and disrupted the natural food chain. This decreased fish populations and made the bay less attractive to tourists.

The stage was set with a triad of common sources of conflict that beset coastal zones around the world (29.2):

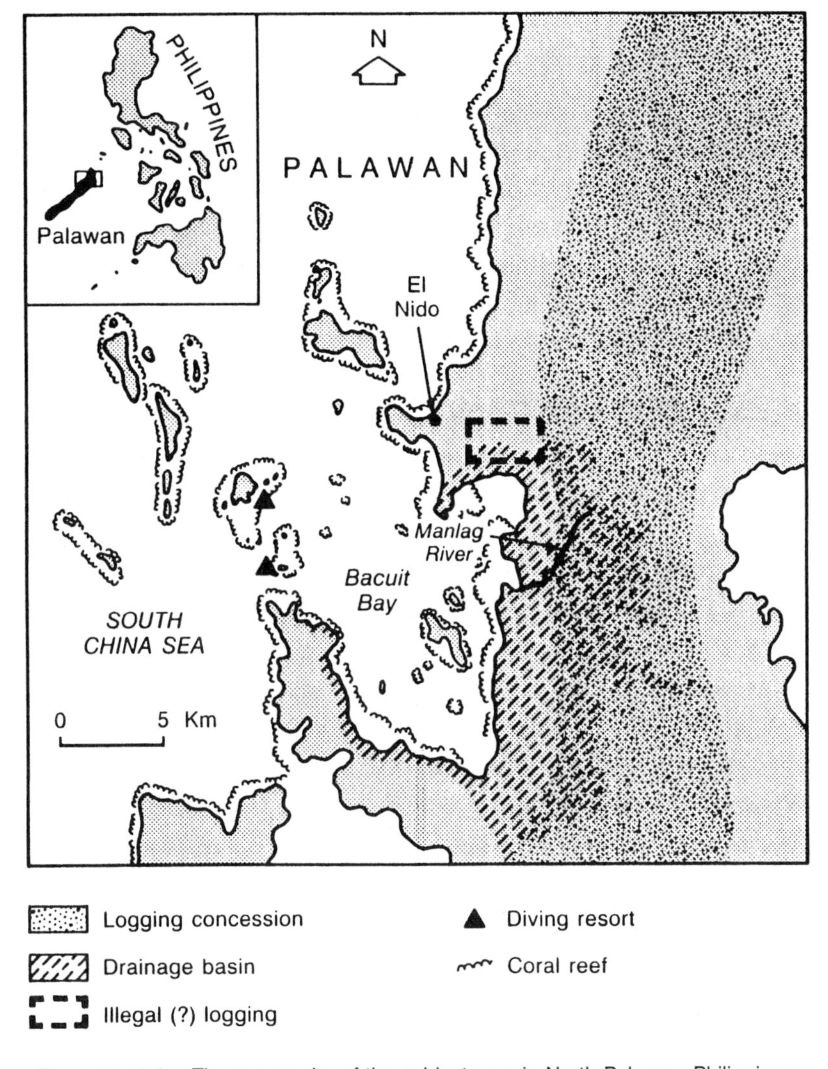

Figure 4.26.1. The geography of the subject area in North Palawan, Philippines.

1. ***Common property resource.*** The bay, its coral, fish population, and clean water are a common property resource. Traditionally fished by El Nido residents and other fishing communities, the newly developed resort industry was a complementary, nonconsumptive user of the same resource. Both fishermen and resort operators had a stake in maintaining a healthy, productive marine ecosystem. In fact, they had cooperated to protect the bay's fishery resource from over-exploitation by outside groups.
2. ***Property rights.*** In contrast to the bay's open access nature, the logging concession owner had a legal right to harvest timber from its concession area. It had the approval to harvest and export logs via a temporary earth pier built in the bay.
3. ***Environmental and economic externalities.*** Eroded soil from the logging concession enters the bay as suspended sediment. This in turn affects the coral reef, causes coral death in some cases, and disrupts the food chain upon which the bay and reef fish populations depend. There is also evidence of impact on offshore pelagic species. Such economic externalities include:

- Decreased fish catch by local, subsistence fishermen
- Decreased attractiveness of Bacuit Bay to local and foreign divers, with a resulting decline in tourism
- Possible adverse effects on pelagic fish caught in offshore waters

Solutions

To assess the situation and predict its outcome, a combined ecological economic analysis was conducted by G. Hodgson and J. A. Dixon (170) for two options—(a) cessation of logging and (b) continued logging. The first option would prevent further damage to the bay's ecosystem due to logging-induced sedimentation and thus the tourism and marine fisheries dependent on it. The second option would maximize logging revenue but reduce revenue from the other industries.

The results tabulated below show quite clearly that over a ten-year period the combined revenues of the three industries would be conservatively (at 10 percent discount rate) 50 percent greater if logging ceased rather than continued (values in $1,000s US).

	Option 1: Cessation	Option 2: Continuation
Gross Revenue		
Tourism	47,415	8,178
Fisheries	28,070	12,844
Logging	0	12,885
Total	75,485	33,907
	Present Value (10%)	
Tourism	25,481	6,280
Fisheries	17,248	9,108
Logging	0	9,769
Total	42,729	25,157
Present value (15%)		
Tourism	19,511	5,591
Fisheries	14,088	7,895
Logging	0	8,639
Total	33,599	22,125

The difference is due to projected losses in tourism and fisheries. Present value analysis was performed using both a 10 and a 15 percent discount rate. Even with the higher discount rate, the present value of lost revenue exceeds $11 million under Option 2—continued logging. Sensitivity analysis shows that significant deviation from predicted effects of sedimentation damage do not alter the conclusion. In addition to these quantitative results, consideration of qualitative factors reveals that the social, economic, and environmental benefits of fisheries and tourism outweigh those of logging in this location (Figure 4.26.2).

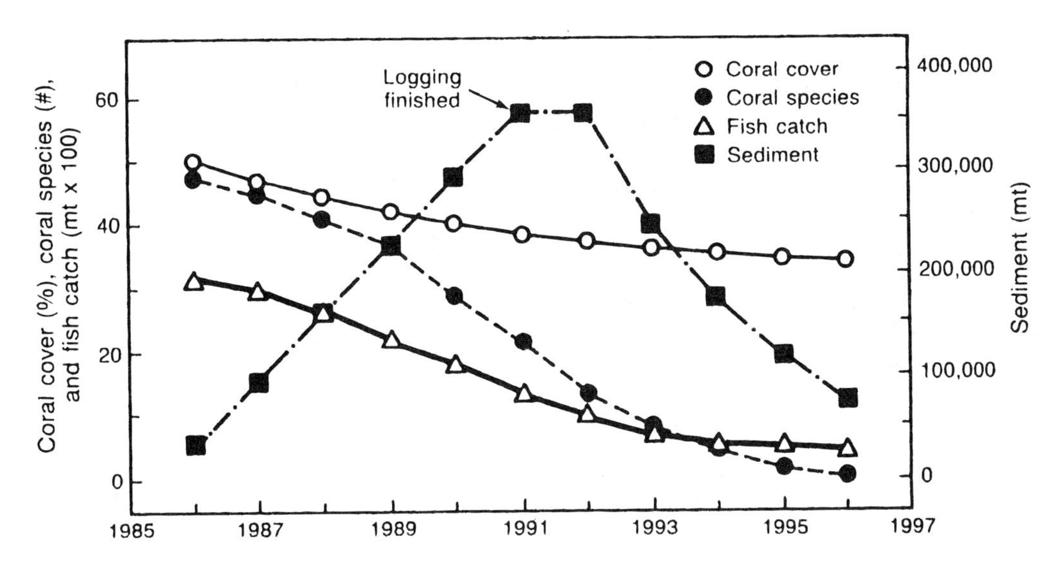

Figure 4.26.2. Predicted changes over ten years in coral cover, coral species, and fish catches associated with sediment runoff from logging.

Lessons Learned

An important lesson learned is that, with good methodology and dedicated effort, it is possible to demonstrate the economic logic for conservation.* However, incorporation of this logic into government or private sector action may be orders of magnitude more difficult because politics follows fashion, not logic.

In this analysis, the evidence is unarguably in favor of stopping soil erosion, but stopping the logging is only one solution to the erosion problem. And in spite of the evidence (Figure 4.26.2), the logging industry would not be expected voluntarily to cease operation. This would require strongly focused government action, which could be achieved by any of several multi-sectoral mechanisms, including an ICZM-type mechanism.

An ICZM program could attempt to balance socio-economic needs with conservation of coastal resources. However, the cease-or-continue outcome may be too simplistic, so an ICZM approach might encourage continuance of logging under strict controls on sediment runoff, logging method, and replanting. In this manner sustainability of the resource system might be accomplished.

Main Source — John A. Dixon, World Bank, Washington, D.C. USA (Reference 109 in Part 5) and G. Hodgson, Singapore (Reference 170 in Part 5).

4.27 PHILIPPINES: COMMUNITY MANAGEMENT OF CORAL REEF RESOURCES

Background

Fisheries are an important sector of the national economy of the Philippines, which is ranked 13th in world fish production. Coral reefs supply 10 to 15 percent of the total yield of fishes (4). About 27,000 km^2 of coral reef area exists, which is pristine in only a few remote areas and is seriously degraded or destroyed in many areas of intense fishing (6, 15).

* The "Lessons Learned" were prepared by the author of this book.

The country, with its extensive coastline and concentration of people in coastal areas, is experimenting with various forms of coastal zone situation management. This case describes the results of a two-year program on three islands that initiated community-based management programs in 1985, which are still functioning without significant outside support.

Problem

Destruction of coral reef habitats, over-fishing, and a consequent decline in fish catches plague small-scale fishermen throughout the Philippines (6). The Marine Conservation and Development Program (MCDP) of Silliman University was set up in 1985 to help correct this problem (12). The MCDP grew out of a self-help program initiated at Sumilon Island in 1974 (2, 7, 8) and at Apo Island in 1978, by Alcala (1).

Although each island had its own set of problems, exploitive fishing methods were common to each, usually done by "outsiders" using explosives, fine mesh nets, scare-in techniques, and spears. Fish catches were declining along with disposable income derived from sales of valuable fish. Increasing poverty was forcing people to find more efficient and, often, more destructive fishing methods (5).

Solutions

One approach that has proved effective for coral reefs surrounding small islands and along some large island shorelines is a *marine reserve and sanctuary* model, which encourages local communities to be responsible for their fishery and coral reef resources. This reserve model includes limited protection for the coral reef and fishery surrounding the entire island and strict protection from all extraction or damaging activities in a small "sanctuary" normally covering up to 20 percent of the coral reef area (8). This reserve and sanctuary approach has provided real benefits to local fishing communities through increased or stable fish yields from coral reefs that are maintained and protected (2, 10).

The MCDP operated on the premise that resource management must be rooted in local communities and thus was designed to enable local communities to conserve their own marine resources. At the heart of the project was initiation of local marine management programs in three Visayan fishing communities—Apo Island (Negros) and Pamilacan and Balicasag Islands (Bohol) (12,13). Local governments have authority to regulate marine resources through ordinance, but only in the first 5 km offshore. Also, they cannot provide exclusive rights to local citizens—outsiders have equal rights of access to fishing areas.

The general objectives of the MCDP included (12):

1. Institutional development by raising awareness of resource management technologies and community development skills among faculty, staff and students at Silliman University
2. Implementation of marine resource management programs at the three sites by setting up marine reserves with fish sanctuaries and buffer areas surrounding the islands to prevent destructive fishing, to increase the abundance and diversity of coral reef fish at the island, and to increase long-term fish yields
3. Community development programs to establish working groups of local people for accomplishing marine resource management, alternative livelihood projects, and a community education center/rest area adjacent to the sanctuary
4. A small agro-forestry and water development component
5. An outreach and replication component to extend programs to neighboring fishing communities and establish linkages with groups having similar interests

Implementation at the three project sites of the MCDP included five groups of activities. The activity groups provide a framework for the community development plan leading to marine resource management and aid in clarifying the progression of events (13). They were:

1. ***Integration into the community.*** In the three-month initial period, field workers introduced the project, met with leaders, attended community meetings, and generally became acculturated

to the island situation with its particular problem set. Baseline data for later project evaluation were collected, including socio-economic and demographic surveys; a pretest of environmental and resource knowledge and perceived problems of local people; and an environmental survey to document the status of the coral reefs by means of substrate cover, fish diversity/abundance, and other indicators.

2. ***Education.*** Education was continuous throughout the project but emphasized in the initial stages. Most forms of education were non-formal, in small groups and by one-on-one contact. Focus was on marine ecology and resource management rationale and methods.
3. ***Core group building.*** Identifying existing community groups and/or facilitating the formation of new groups was a central focus because management solutions could best be implemented through special island work groups with close ties to the island political structure. The first group to emerge was the one responsible for the education center construction. As the concept for the marine reserve took shape, those interested in marine conservation formed a "marine management committee" (MMC) for each island. Each MMC gained high respect from implementing the marine reserves.

4/5. ***Formalizing*** and ***strengthening organizations.*** These last two steps are difficult to separate. The idea of both was to provide continuing support, in real and symbolic terms, to the core group and its management efforts. This was accomplished by helping the group to identify new projects such as reforestation, planting giant clams, refining marine reserve guidelines, training MMC members as tourist guides, collecting fees for entrance to the sanctuary, and initiating alternative income schemes such as mat weaving. In addition, one island (Apo) has become a training site where MMC helps conduct workshops for sharing experiences of the Apo success with fishermen from other places. This activity has strengthened the core group and solidified support for the marine reserve within the community.

The results of the MCDP have been substantial. Three island-wide marine reserves with municipal legal support exist. Island-resident committees patrol for rule infractions. Municipal ordinances are posted. Enforcement varies—e.g., effectiveness has waned on Balicasag Island because of intrusions of the Philippine Tourism Authority. Some active (but mostly moral) support is received from the Philippine police. It is noteworthy that diving tourism to both Apo (Figure 4.27.1) and Balicasag Island has increased significantly in response to the sanctuaries, which are teeming with fish.

Fishermen members of the three island MMCs in 1992 all said that the marine reserve/sanctuary had caused no decrease in fish catch or personal income since establishment of the reserves, and most believed the sanctuary had significantly improved fishing. They all said that the sanctuary served as a *semilyahan* (breeding place) for fish (11). Fish yield studies in the fishing areas outside the sanctuaries indicate that yields have been at least stable and probably increased at Apo Island, which had a yield of more than 30 tons per km^2 in 1986, more than the yield measured by a similar study done in 1981 (1, 14).

Comparison of baseline data of 1985–86 with a survey made in 1992 shows increases in fish diversity (species richness) and abundances within the fish sanctuary at Apo Island, where 500 m^2 of reef were sampled (9, 11):

Increase	1985–86	1992	%
Species richness	52.4	56.0	6.8
Abundance			
Food fishes	1,286	2,352	83
All fishes	3,895	5,153	32

Note: Samples consisted of 19 families of fish, and totals represent the mean of five replicate samples in both years using the same method by the same observer.

It is encouraging that the coral reef bottom substrate cover in the sanctuary and non-sanctuary areas of the three islands has remained stable or improved slightly since 1984, which is generally not the case for other coral reef areas in the Philippines (Figure 4.27.2).

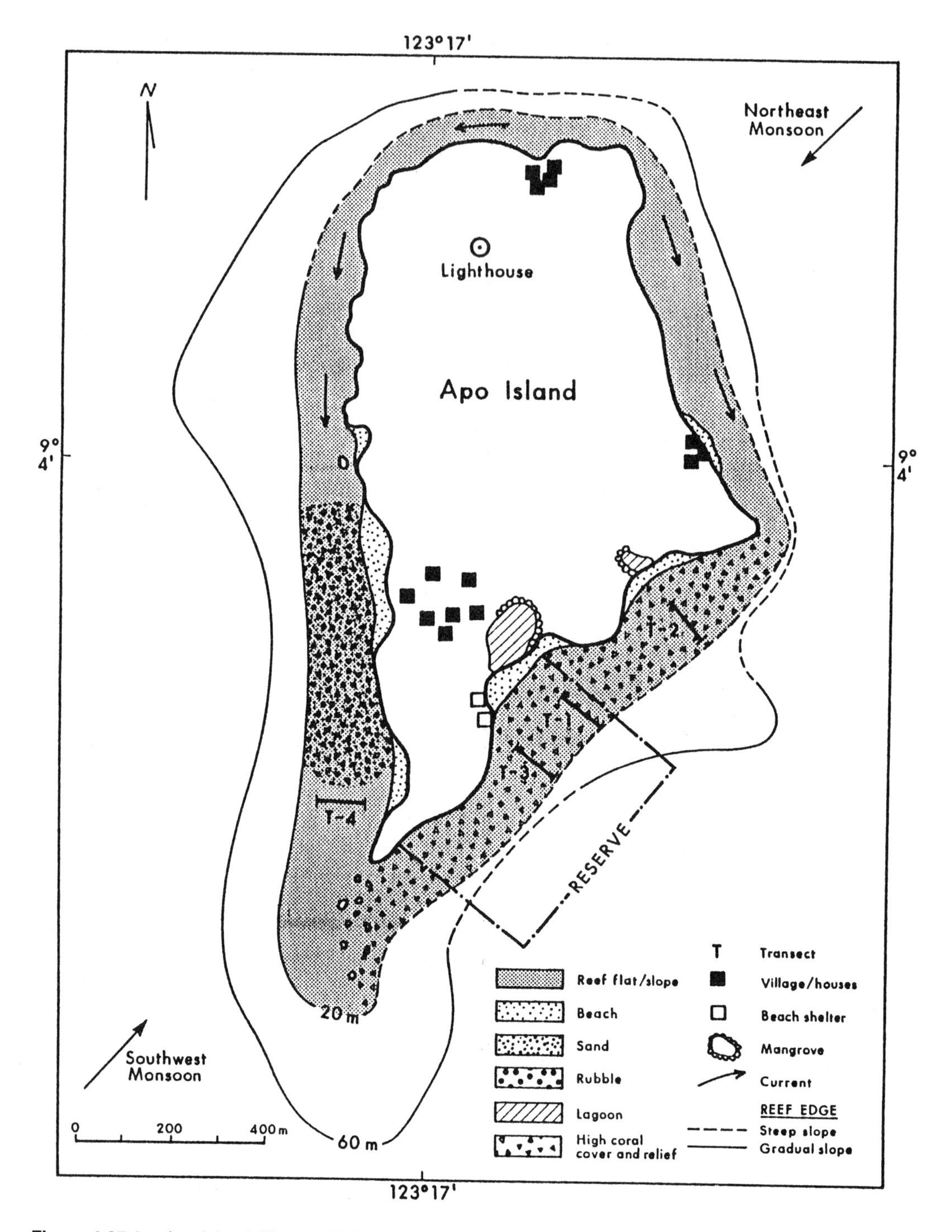

Figure 4.27.1. Apo Island (Negros, Philippines), where community-based conservation of coastal resources succeeded through integration, education, core group building, and institutional strengthening.

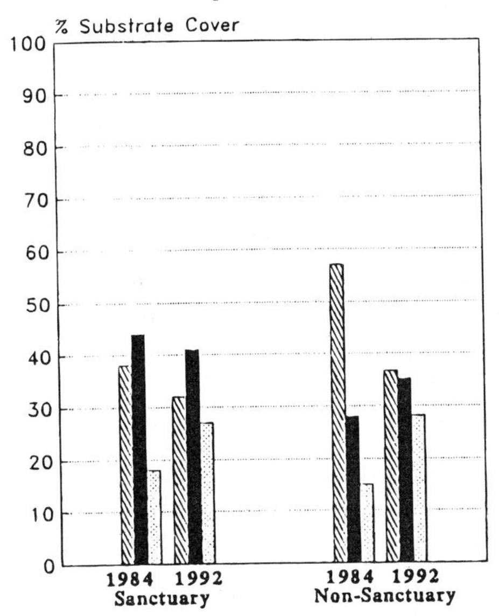

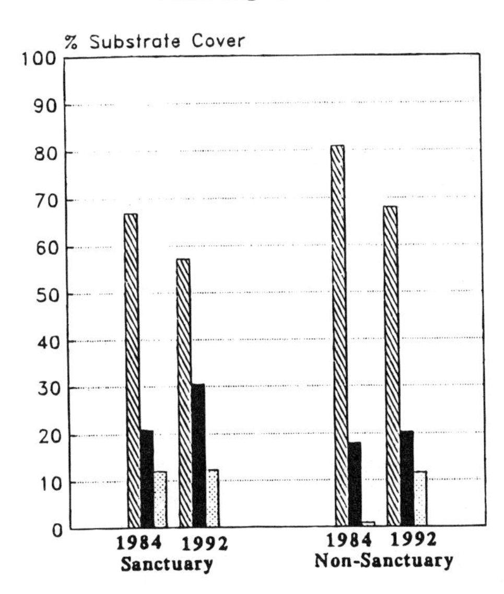

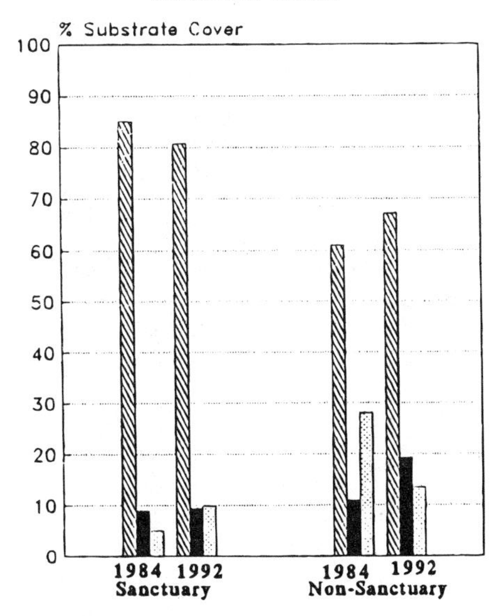

Figure 4.27.2. Comparison of mean percent living and dead substrate cover between sanctuary and non-sanctuary reefs at the three island sites from the start of management in 1984 until 1992. (*Sources:* References 9 and 14).

Lessons Learned

Many lessons have been learned from the MCDP experience in small-scale coastal management, but it is uncertain how broadly they can be applied. A small island and community setting may be critical to the success of such a program, but some results may be helpful for larger scale coastal management schemes:

1. It is possible to manage small island coral reef resources in a manner that benefits local users and those interested in sustainable resource use. Benefits measured in terms of fish catch and quality of the coral reef can be accrued with the installation of a regime that (a) prevents destructive uses of the resource and ensures that only ecologically sound fishing methods are permitted; (b) limits the fishing effort by establishing a marine reserve inclusive of a sanctuary where no fishing or collecting is allowed; and (c) monitors the impact of the management and feeds back the results to the resource users in the form of understandable information and real benefits (13).
2. Small islands provide a geographic advantage to marine resource management because of decreased access to "outsiders," and the community can more easily identify with its own marine resources. To gain this advantage on the mainland, each segment of the coast should be assigned to local residents (12).
3. People must see some immediate results of their management efforts if they are to continue a management program intended to improve their marine environment. Education provides the initial understanding, but only observable results sustain a program (10).
4. Baseline data and monitoring of coral reef resources are required to illustrate to fishermen the condition of their environment and to reinforce their management participation. Data on the increase in fish abundance and diversity have been used to convince policy makers and government officials, both local and national, about the effectiveness of the marine reserve management.
5. Local residents must learn how management will solve a problem they think is important. For example, if they see no link between a damaged coral reef and decreased fish catch, they will not be motivated to improve the reef. This is not a trivial point, since most islanders believed that coral was stone.
6. Formation of community working groups is essential. Successful implementation of resource management requires more than simply a mayor's approval; it requires groups working together on projects with real outputs.
7. Local fishermen can design a marine sanctuary with the assistance of community organizers and technical inputs from marine scientists.
8. Complete and realistic environmental and resource use surveys and analyses are a prerequisite to helping a community design a management plan (3).
9. Apparently successful management for small-scale settings such as the three islands' coral reef fisheries discussed is vulnerable to local changes in political leadership. Results are also sometimes dependent on moral and physical support from outside entities not obvious during the initial implementation phase (7, 11).
10. All coastal management projects need to consider linkages among potential participants— community leaders, town mayor and council, local law enforcement officers, private business with local interests, and national government organizations like tourist authorities or fishery groups.

In summary, effective coastal resource management in populated coastal areas is more than a problem of environmental considerations or law enforcement. Community-based approaches that mobilize those people who use the resources daily are necessary to ensure wide participation and potentially long-lasting results. Strictly legal approaches have had few successes in the Philippines. Equally, good environmental surveys and information have not been sufficient to bring about rational use of marine resources. Combining community, environmental, and legal approaches in a manner appropriate for a particular site offers some possibility of success, as it has for these three island communities in the southern Philippines.

Contributed by — Alan T. White, Coastal Resources Management Project, No.3 St. Kilda's Lane, Colombo 3, Sri Lanka.

REFERENCES

1. Alcala, A. C. and T. Luchavez. 1981. Fish yield of the coral reef surrounding Apo Island, Negros

Oriental, Central Visayas, Philippines. In: *Proceedings of the Fourth International Coral Reef Symposium,* May, Manila, Vol. 1. Pp. 69–73.
2. Alcala, A. C. and G. R. Russ 1990. A direct test of the effects of protective management on abundance and yield of tropical marine resources. *J. Cons. Intl. Explor. Mer,* Vol. 46. Pp. 40–47.
3. Calumpong, H. 1993. The role of academe in community-based coastal resource management: the case of Apo Island. Case Presentation 4. In: *Our Sea Our Life,* Proceedings of the Seminar Workshop on Community-based Coastal Resources Management, February 7–12, Dumaguete, Philippines. Pp. 49–57.
4. Murdy, E. and C. Ferraris. 1980. The contribution of coral reef fisheries to Philippine fisheries production. *ICLARM Newsletter,* Vol. 6, No. 1. Manila. Pp. 3–4.
5. Savina, G. C. and A. T. White. 1986. The tale of two islands: some lessons for marine resource management. *Environmental Conservation,* Vol. 13, No. 2. Pp. 107–113.
6. White, A. T. 1987. *Coral Reefs, Valuable Resources of Southeast Asia,* ICLARM Education Series 1, 36 Pp. International Center for Living Aquatic Resources Management, Manila.
7. White, A. T. 1987. Philippine Marine Park Pilot Site: benefits and management conflicts. *Environmental Conservation,* Vol. 14, No. 1. Pp. 355–359.
8. White, A. T. 1988. Marine parks and reserves: management for coastal environments in Southeast Asia. International Center for Living Aquatic Resources Management, Manila, Philippines.
9. White, A. T. 1988. The effect of community-managed marine reserves in the Philippines on their associated coral reef fish populations. *Asian Fisheries Science,* Vol. 1, No. 2. Pp. 27–42.
10. White, A. T. 1989. Two community-based marine reserves: lessons for coastal management. In: *Coastal area management in Southeast Asia: policies, management strategies and case studies,* T.-E. Chua and D. Pauly (Eds.). ICLARM Conference Proceedings 19, 254 Pp. Manila. Pp. 85–96.
11. White, A. T. and H. Calumpong. 1992. Saving Tubbataha Reefs and monitoring marine reserves in the Central Visayas. Summary Field Report, Earthwatch Expedition, Philippines, April–May 1992, unpublished.
12. White, A., E. Delfin, and F. Tiempo 1986. The Marine Conservation and Development Program of Silliman University, Philippines. *Tropical Coastal Area Management,* Vol. 1, No. 2. Pp. 1–4, Manila.
13. White, A. T. and G. C. Savina. 1987. Community-based marine reserves, a Philippine first. In: *Proceedings of Coastal Zone '87,* 26–30 May. Seattle.
14. White, A. T. and G. C. Savina 1987. Reef fish yield and nonreef catch of Apo Island, Negros, Philippines. *Asian Marine Biology,* Vol. 4. Pp. 67–76.
15. Yap, H. T. and E. D. Gomez. 1986. Coral reef degradation and pollution in the East Asian Seas region. *Environment and Resources in the Pacific.* UNEP Regional Seas Reports and Studies No. 69.

Acknowledgments

Results discussed in this case were made possible by the author's long-term association with Silliman University, which has encouraged this kind of research. Dr. Angel Alcala, Dr. Nida Calumpong, Ms. Felina Tiempo, Ms. Ester Delfin, and Ms. Louella Dolar of Silliman have all been very helpful in this regard. The East-West Center supported field research in 1983, the International Center for Living Aquatic Resources Management (ICLARM) encouraged analysis of these findings, Earthwatch International supported a 1992 research expedition and now the Coastal Resources Center, University of Rhode Island, has provided support to revisit and discuss these sites and experiences, the funds for which were provided by USAID.

4.28 ST. LUCIA: COMMUNITY PARTICIPATION IN RESOURCES MANAGEMENT

Background

For almost 50 years, the hard, heavy wood removed from the Man Kòtè mangroves of southern St. Lucia has been the lifeblood of a handful of families. The fruit of their toil, charcoal, remains the principal cooking fuel for most of the nearby town of Vieux Fort. As the town has grown (to 15,000 residents), however, so has the demand for charcoal, which remains the fuel of choice because of the favorable taste it imparts to foods and because it is still the cheapest alternative. It can be bought and sold in small quantities, a critical matter to cash-poor residents. The common unit of measure is the handful or canful, instead of the much more expensive tank of cooking gas.

Problem

By the 1970s, the 40 hectares of Man Kòtè's mangrove, once covered with well-developed trees, contained, only skeletal remainders, reflecting over-harvesting. In the 1940s, before heavy demand for charcoal developed, the average tree diameter was almost 8 inches. Continued heavy cutting reduced stems to no more than 2 inches in diameter; charcoal producers reported a marked decrease in trees of a size suitable for cutting.

Solutions

An innovative project to develop new approaches to managing the mangroves and address the needs of charcoal makers and users was developed cooperatively by the Eastern Caribbean Natural Area Management Program, the government of St. Lucia, and local community groups. It was guided by Yves Renard.

The program's objectives were to test, demonstrate, document, and disseminate community-based approaches to managing renewable natural resources. Started in 1981, the overall program was recognized as one of the most successful of its kind in the eastern Caribbean, with tangible successes in community participation, appropriate technology, small-scale aquaculture, community organization, and sustainable development activities.

The first step taken toward easing the pressures exerted on Man Kòtè was a survey of the area's charcoal producers, conducted by local secondary school students. Among the main conclusions and recommendations of the survey were:

- Producers were aware of the main management requirements for sustainability. However, several problems, including difficult access to other areas and economic difficulties after a devastating hurricane in 1980, forced them to continue over-exploiting the Man Kòtè area.
- A number of traditional practices were beneficial to the mangroves.
- Work in the mangroves was so difficult that producers preferred to work anywhere else and actively sought better employment opportunities.
- The only feasible solution identified was the provision of an alternative source for fuelwood (i.e., a plantation).

Based on the information provided by the survey, the project was designed to reduce drastically cutting pressure on Man Kòtè. With the active support and involvement of the Forestry Division of the Ministry of Agriculture, idle government land near the homes of the charcoal makers was converted to a fuelwood plantation based on *Leucaena, Gmelina,* and other fast-growing exotic tree species.

The project started in 1984, and to date nearly 7 hectares have been planted. Approximately 2 additional hectares are scheduled to be planted annually. It is hoped that the woodlot, established and maintained by the community, will be able to reduce the demand for mangrove charcoal. In support of this activity, there is ongoing research and monitoring of the mangroves and research on charcoal production.

After the initial success of the fuelwood plantation, the government provided each of the 12 charcoal-producing families with 1-hectare plots of unused agricultural land. The plots are now producing a combination of vegetables and fruits for home consumption, as well as cash crops to be sold at some of the tourist hotels. At this stage, the project is also planning to construct a nature trail, train community members as tour guides, and offer educational tours.

These efforts have made a noticeable difference in the lives of the small community and have reduced the burden on the over-exploited Man Kòtè mangroves. The problem was approached not as a mangrove problem, but rather as a people problem. By creating alternative sources of fuelwood and income, the project has allowed the people to take control of their future, while protecting the mangroves.

In doing so, the charcoal producers have become participants rather than bystanders in development. Great efforts have been made to establish a communal sense of responsibility for this publicly held resource and for its use and management. A contract is now under negotiation between the government and the group of producers to set the conditions for that responsibility.

Lessons Learned

The main lesson gained from this one small project is that, when appropriate economic alternatives can be created for the rural poor, environmental amenities and values can be maintained. In the case of Man Kòtè, socio-economic and conservation objectives have been met.

This project can serve as an example to neighboring islands faced with similar economic hardships and decisions.

Source — Richard Bossi and Gilberto Cintron, Department of Natural Resources, Puerto Rico (See Reference 28, Part 5).

4.29 SOLOMON ISLANDS: SOCIAL CHAOS FROM TOURISM

Background

Anuha is a small island of about 150 acres lying just half a mile off the north coast of central Ngela, about 45 sea miles from Honiaira, the capital of Solomon Islands. With four hills clothed in virgin tropical rain forests, 5 miles of white sandy beaches, a fringing lagoon with two small palm-crowned inlets on the fringing coral reefs, and a freshwater lake thrown in for good measure, it is the tropical island of tourists dreams.

Problem

An Australian resort company negotiated to lease the island from its customary owners, a small village of 60 people on the Ngela mainland led by a retired Anglican priest, Father Pule. Negotiations began in 1981, and a 75-year lease was signed with Father Pule and other landowners on April 8, 1983. Most of the activity during this two-year period was taken up by claims of ownership of the island by other Solomon Islands groups.

One claim, by a group of landowners from Isobel Island, some 50 miles away, went all the way to the High Court in appeal. This group lodged a claim as the alleged owners of the fringing reef and its resources. Their claim was based on the undisputed fact that about 150 years earlier they had been granted refuge on Anuha Island by the people of Ngela, as they sought escape from the marauding head hunters of Marovo to the west.

Such folks claimed they had rights to live on the island and use its land and its resources (but not own it) in accordance with Melanesian *kastom* because their ancestors had paid compensation for the surrounding reefs and marine resources. For a period this case threatened access to the island, the right of the developers to moor their boats or even swim in the lagoon, to snorkle over the coral, and their right to catch fish, lobsters, and so forth. Then the Supreme Court rejected the claim in favor of Father Pule's village.

The lease was registered with the Land's Department on April 29, 1983. Under the terms of the lease, yearly payments on a gradually increasing scale were agreed, so that up to three generations of customary owners would benefit monetarily over the 75-year period. A supplementary agreement to construct an airstrip on an old coconut plantation on Anuha was negotiated in 1983, with the customary landowners being paid compensation for the coconut trees at standard rates set by the government.

This was the first resort of any size in the Solomon Islands (there were only two other small resorts, Uepi Island in the Marovo Lagoon and Tambea Beach, West Guadalcanal). Therefore, the lease agreements were seen as a test case for foreign investments and a model for future tourism development.

After two years of operations, the original Australian shareholders were bought out. The initial management had been sympathetic to the customary owners, taking care to identify special items such as canoe trees and fruit trees and paying compensation when development needs necessitated the removal of a valuable tree. They understood that ownership in a social sense extended beyond the land to the jobs created by the resort, and made a point of employing local villagers and their kin.

But the new owners had little understanding of *kastom*. They considered that the lease gave them the right to manage the island and its resources as they saw fit. In their first few weeks they dismissed the original management team and sacked many of the local workforce, replacing them with tribespeople from other islands. They bulldozed a section of the rainforest to double the number of bungalows without consultation or compensation payments and in various ways insulted and demeaned the customary owners.

Father Pule was incensed. He demanded to know why the new expatriate management team had been granted work permits and resident permits when there had been no negotiations with the customary owners. Father Pule claimed that, since there had been no agreement with the new owners, the lease agreement was null and void. He began a campaign for a new lease, seeking the multiple payments as a single $2 million lump sum. Future generations were forgotten in pursuit of immediate wealth.

Relations between the new expatriate management and the customary owners deteriorated rapidly and a period of increasing conflict began. Overt hostility first gained national attention when villagers "invaded" their island and dug holes in the airstrip. Other incidents included a raid just before Christmas 1986 when Father Pule sent warriors in war paint to do the following: force guests off the island and close down operations; make threats against non-Ngela employees, coercing them to flee the island; damage equipment such as outboard motors; and so forth.

The ownership of the resort passed to developer Mike Gore in August 1987. The owner was unaware of the legacy of distrust he inherited from his predecessors. Tension escalated when an Australian chef, one John Smith, fell out with the Gore management and became influential with Father Pule. At this point the stance of the customary landowners changed; from demanding a new lease they now demanded compensation and repossesion of their island.

In December 1987 Father Pule's warriors again invaded Anuha, apparently at the exhortation of Smith, forced the expatriate management team off the island at spear and arrow point, and held guests and construction crew hostage for several days. Some parts of the resort were damaged. Finally, the Royal Soloman Islands Police field force went to the rescue and detained a number of Father Pule's men.

The chef, John Smith, gained access to government ministers, including the prime minister, and muddied the waters in respect to not only the Anuha resort owners, but also a significant number of other Australian businessmen and interests in Solomon Islands. Death threats against the Australian High Commissioner and his staff by Smith (who alleged that he had his own band of warriors) resulted in protection of diplomatic residences being undertaken by the Royal Solomon Islands Police. Australian Broadcasting Corporation journalists were banned from the Solomon Islands in April 1988 by the prime minister when a Radio Australia report alleged that he had allowed himself to be unwisely influenced by Smith.

In May 1988 the resort's central complex burned down: damage was estimated at about US$1 million. Arson at the instigation of Smith was suspected following the submission of an affidavit to the police by the resort's owner, which named Smith. Before the John Smith saga ended with a prohibition order against him in June 1988, media reports indicated that he had successfully assisted in the expulsion of 6 to 12 Australians. The resort remained closed.

Father Pule withdrew from all attempts to negotiate a way out of the situation, and the Australian investor resorted to the courts to force him to recognize and respect the validity of the lease agreement.

Solutions

A court order against Father Pule and his supporters for damages and wrongful entry when they had raided the resort in December 1987 was obtained. They were forbidden to visit the island unless gainfully employed by the resort.

Father Pule retaliated by contesting the lease on the grounds that he had reached agreement with the first lessees only. In a judgment at the end of August 1988, Chief Justice Gordon Ward rejected Father Pule's claims, ruling that the lease was a proper document. He stated that he found the clergyman to be "totally unconcerned about the truth. He lied easily, frequently and unhesitatingly when it suited his case."

Following the court ruling, Anuha's lessees said they would terminate the claims for damage and wrongful entry in the interests of resolving the deadlock, but Father Pule refused to re-enter negotiations. In mid-1989 the airstrip was replanted with coconuts by customary owners, but the plants were uprooted by the resort's caretaker management and legal action was again contemplated but not pursued by the investor. In May 1990 the resort and airstrip remained closed, although the present Solomon Islands government, concerned about damage to its reputation as a tourist destination, was undertaking moves at the ministerial level to defuse the situation and get the parties together.

As good an example as any of the different culturally determined attitudes to the legal system may be found in reactions to the Chief Justice's comment that Father Pule was an unmitigated liar. To the Australian owner's representative (and, one suspects, the majority of the expatriate populations of the Solomans), the Chief Justice's criticism was a vindication of their stance. In Father Pule's case there was at least understanding, if not complete acceptance, of his position: loss of face and "bigman" status was slight.

Lessons Learned

This case study reveals conflict at a number of levels. First, the desire of a foreign investor to develop Anuha Island as a tourist resort provoked conflict and dissention among the Soloman Islanders claiming customary ownership of the land. In the short term this was dysfunctional, since it prevented negotiations and development from proceding. But in the longer term it proved functional because it resulted in a clear and unambiguous identification of the customary owners and their title to the land in dispute.

Second, the change of company ownership resulted in conflict between the customary owners and the resort lessees. This conflict has been highly dysfunctional. It has extended well beyond the confines of the small island of Anuha. It has had far-reaching consequences on investor confidence in Soloman Islands, on the security of foreign businesses and foreign residents in Soloman Islands, and on tourism to the country.

It tested the legal system, an imported regime which, in the view of many Soloman Islanders, is inadequate to fairly and contructively manage disputes involving *kastom.* They consider that by its very nature it favors the expatriate.

In examination of the role of the legal system in the conflict between the parties to the Anuha (dis)agreement, the affair could in a broad sense be deemed functional because of the educative impact on the court proceedings. Reports were broadcast regularly throughout the Soloman Islands over the national radio news network and provoked considerable discussion in village communities (where 60 percent of the population has access to radio). The conflict could thus be held to have increased internal awareness about some of the implications of tourism-related development.

It could be interpreted as dysfunctional, however, in that a conclusion to be drawn by Soloman Islanders and investors alike was that a victory in the courts was not necessarily a victory out in the provinces. No matter what rulings were obtained through the courts, no matter what penalties were imposed, when it came to the final analysis the customary landowner was likely to achieve at least a "negative victory"—that is, he could prevent a venture proceeding because of his local control.

No tourist operator can stay in business for very long if he must ring his resort with barbed wire fences and employ armed guards to keep tourists in and local people out! In one other respect the conflict between foreign investor and customary owner can be judged as highly functional because it precipitated a review of tourism policy and foreign investment procedures by two successive Soloman Islands governments. New policies and regulations now in place, which are designed to provide better protection for both foreign investor and customary landowner, owe their genesis directly to the Anuha imbroglio.

Finally, there was conflict between two neighboring states, which soured relations between them for a considerable period. Questions were raised in both the Australian and Soloman Islands parliaments, and criticisms were voiced on both sides. The incident impeded normal working relations. The Solomon Islands government was deeply concerned about the intrusion of Australian diplomats and journalists into its internal affairs and the exploitation of Soloman Islanders by what it regarded as unscrupulous businessmen.

For its part Australia was highly concerned about the ban on the Australian Broadcasting Corporation and the expulsion of Australian citizens for seemingly inconsequential reasons. The strain was sufficient to cause Australia to reconsider its aid program and its delivery of that assistance to the Solomon Islands. A change of government in Solomon Islands in 1989 assisted in the recovery of the bilateral relationship, but some vexation over the Anuha affair lingers. At this level, the conflict has been dysfunctional.

Source — T.H.B. Sofield, James Cook University, Townsville, Queensland, Australia (Reference 74.1).

4.30 SOUTH PACIFIC: COASTAL CONSTRUCTION IMPACTS

Background

The series of graphic case studies that follows (Figures 4.30.1–4.30.9) are adapted from the work of James E. Maragos (231) in the South Pacific over a 20-year period for a major U.S. government agency.

Problem

Maragos notes that dredging, filling, and other construction in mangrove swamps, seagrass beds, estuaries, and beaches result in major ecological impacts.

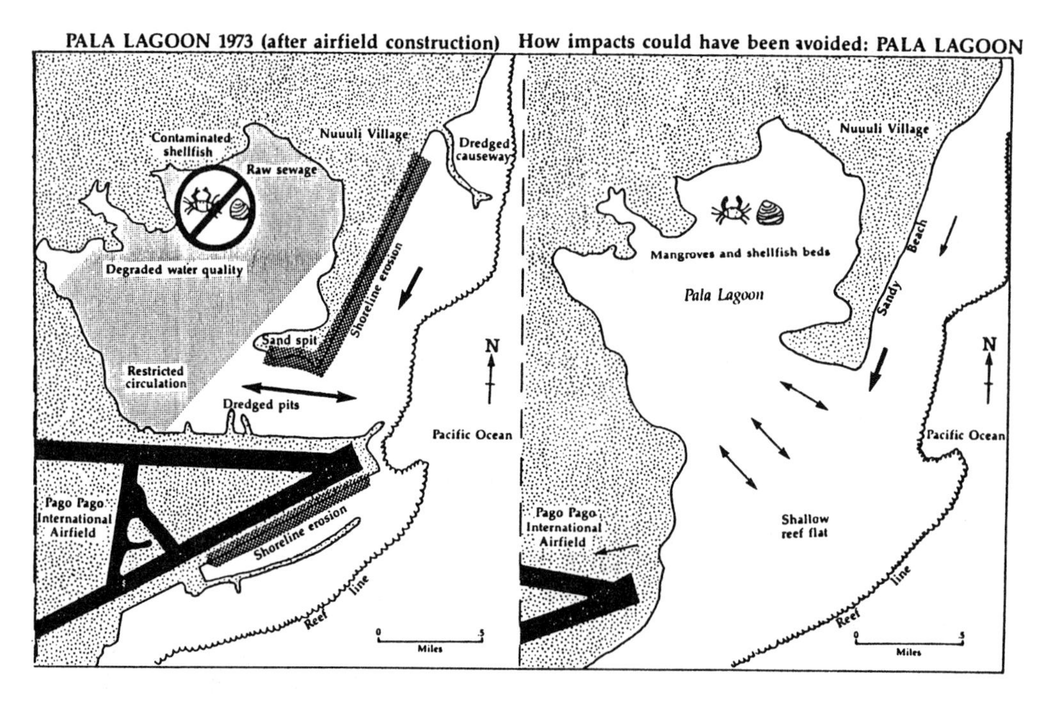

Figure 4.30.1. Adverse effects of coastal airfield construction, dredging, and filling at *Pala Lagoon*, American Samoa. Before airfield construction (about 1960), Pala Lagoon was home to American Samoa's most important shellfish grounds. Dredging and filling along the coast disrupted longshore drift, prevented sand replenishment along the coast and probably caused shoreline erosion at Coconut Point. The airfield partially blocked the entrance to the lagoon, restricting water exchange between the ocean and lagoon, degrading shellfish and water quality in the lagoon. To avoid the impacts, the airfield should have been located inland from the coast.

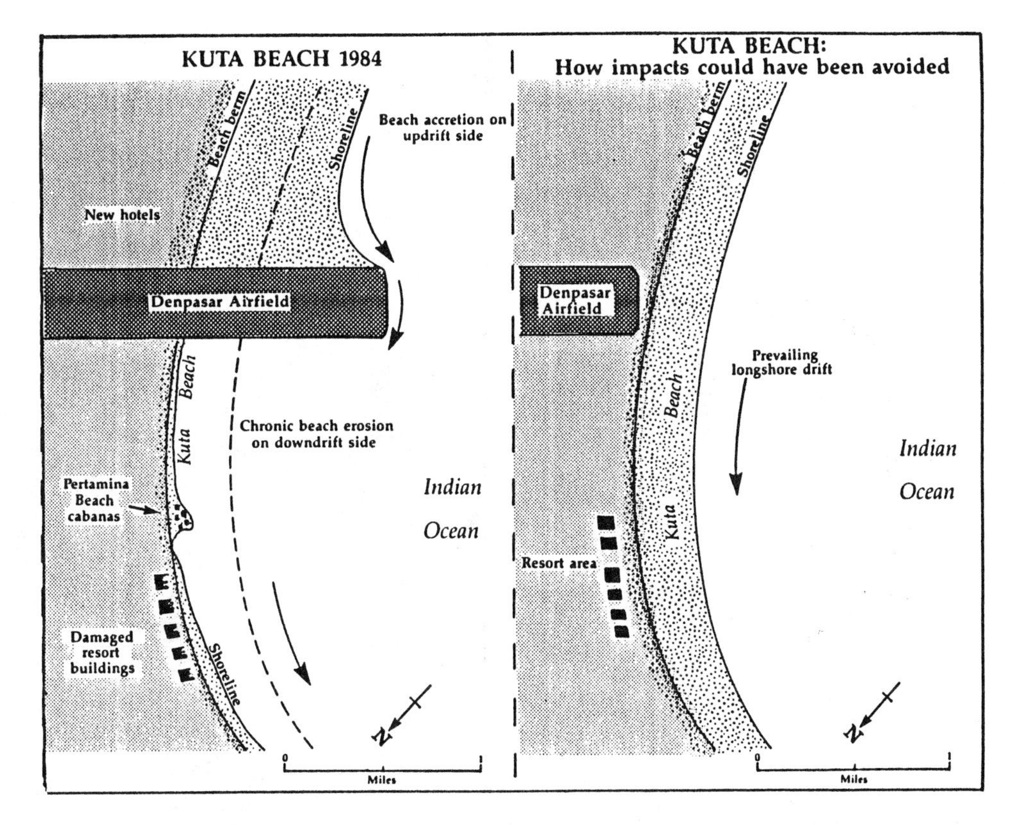

Figure 4.30.2. Adverse effects of coastal airfield construction near Kuta Beach, *Bali*, Indonesia. Before 1967, Kuta was Bali's largest resort and beach area. In 1967, the new Denpasar airfield was constructed, projecting 3,000 ft. offshore beyond the beach. Since that time the beach has eroded more than 1,000 ft. on the downstream side of the airfield. Uninformed officials blamed the erosion on traditional coral mining activities but, in fact, the airfield acted as a huge groin blocking beach replenishment and causing the erosion. While restaurants and hotels fall into the water on eroding side of the airfield, entrepreneurs construct new resorts on the accreting side. The airfield could have been moved landward of the beach dunes and berm to avoid beach erosion caused by the airfield projecting offshore.

Solutions

Many of the negative impacts of coastal construction could be avoided or reduced to minor levels if improvements in the design, location, and construction of such projects were implemented by including environmental objectives into the early planning phases of the projects.

Lessons Learned

Long-range planning and research can look to cumulative impacts of a series of projects on a regional basis. Regional management provides the best approach to acheive an appropriate balance of development and conservation of coastal resources in the framework of ICZM.

Source — James Maragos, East/West Center, Hawaii (See Reference 231, Part 5).

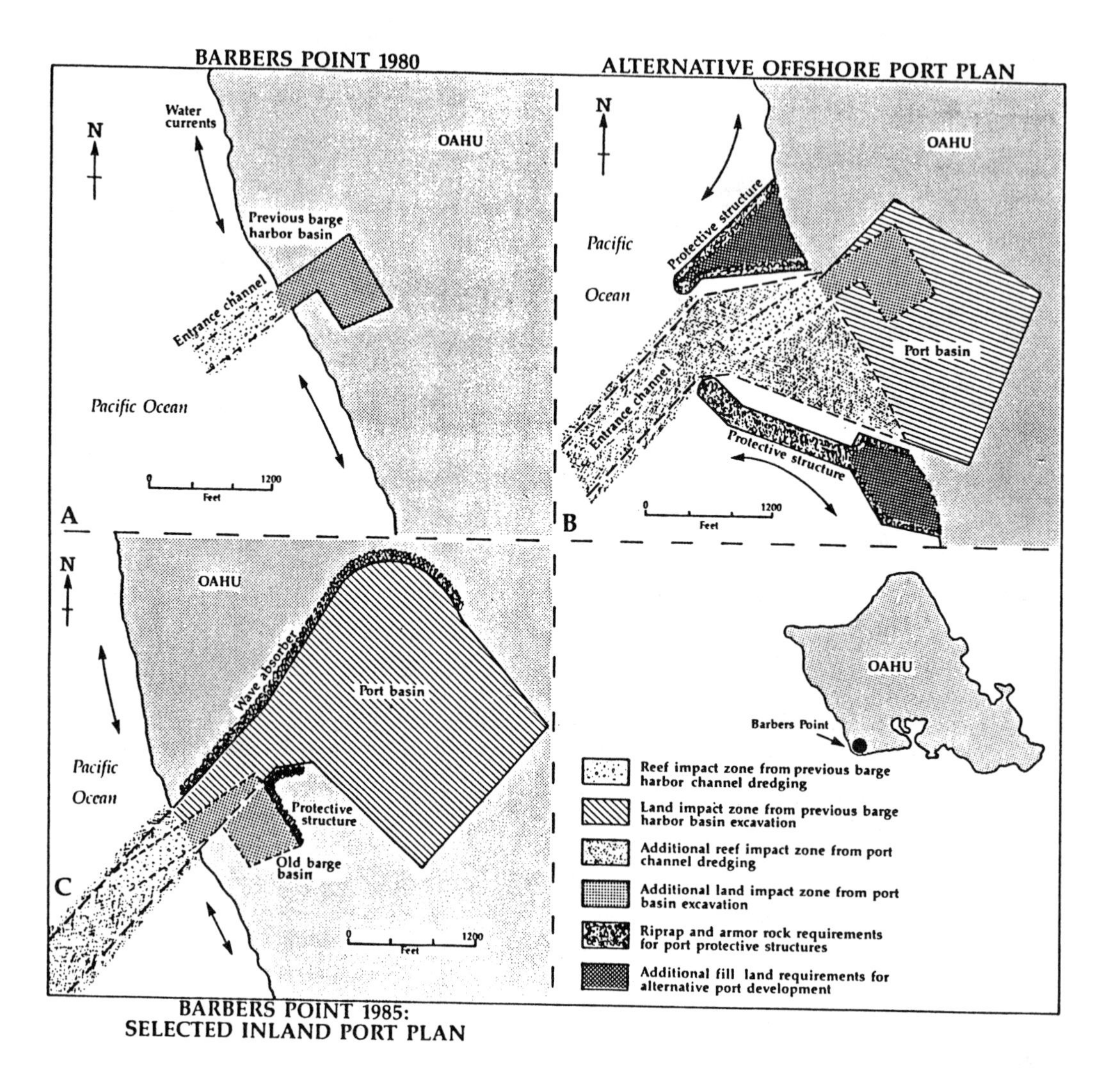

Figure 4.30.3. Port development off *Barbers Point, Oahu*, Hawaii. Deep draft port development invariably results in significant environmental and economic impacts. An existing, medium draft, barge harbor off the SW corner of the island of Oahu (A) was too shallow and small for commercial ships. A conventional offshore port plan (B) would have caused greater marine impacts from dredging and filling and moderate loss of land to port basin excavation. An inland port concept (C) was selected for development. This port design resulted in substantially less marine impacts but greater loss of land to basin development. No permanent loss of marine habitat to fill land resulted. Material dredged from the basin was stockpiled on adjacent land and is being sold for construction purposes to offset the cost of the harbor. Excavation of the inland basin was isolated from the sea by an earthen barrier, further limiting adverse environmental effects to coastal reefs from turbidity and sedimentation during construction. Overall development of the inland prot plan resulted in less economic cost and environmental impacts.

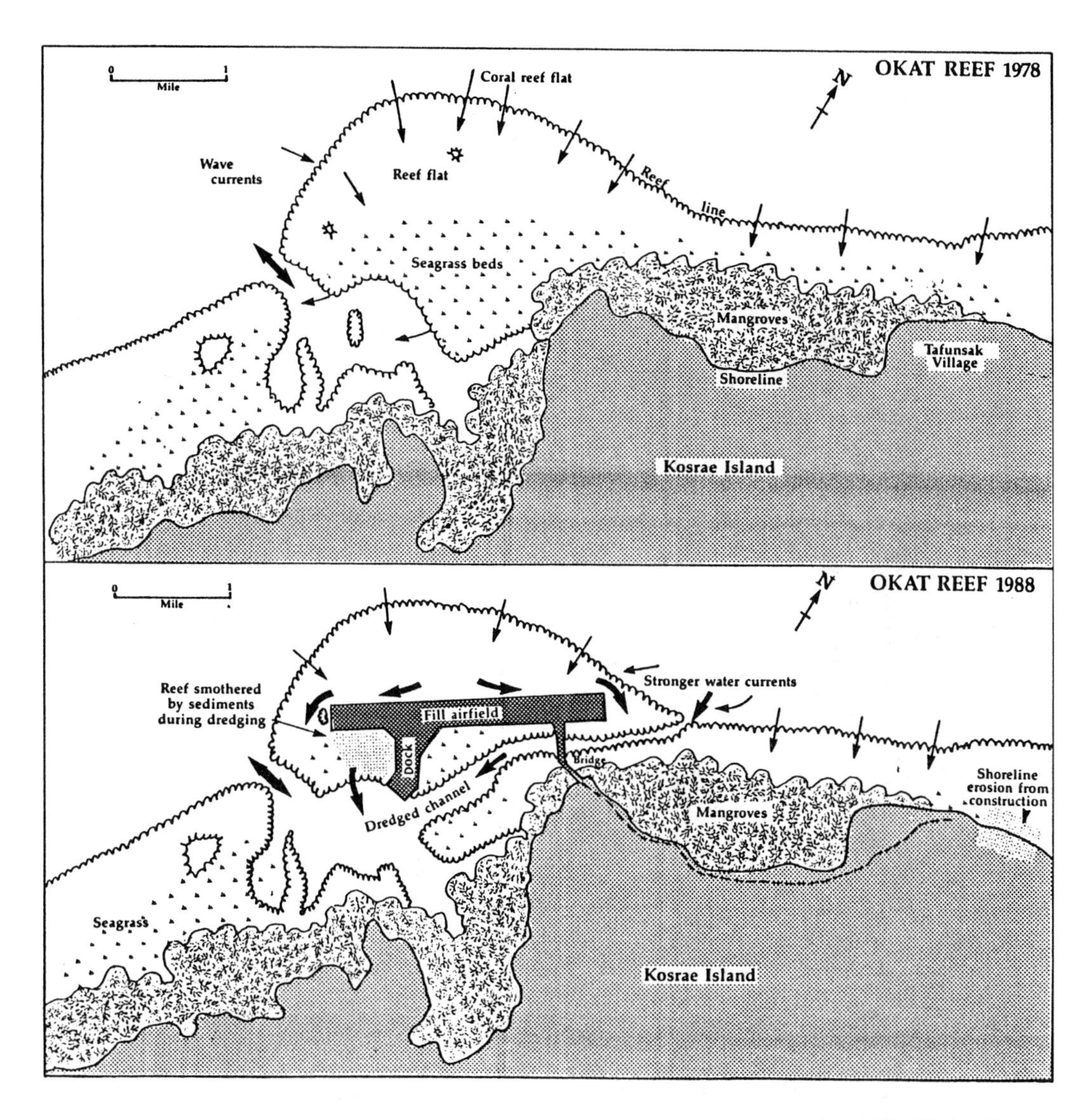

Figure 4.30.4. Adverse effects of dredge and fill for reef flat runway and dock construction at Okat Harbor, *Kosrae Island*, Federated States of Micronesia. Construction buried most of the offshore seagrass beds and much o reef flat under fill land. Dredging further destroyed nearshore reef and seagrass beds and greatly altered circulation in the harbor. The stronger water currents have been implicated as causing shoreline erosion near the airfield and Tafunsak Village. Once Kosrae's most important fishing ground, Okat reef's fish yields have declined to half of preconstruction level.

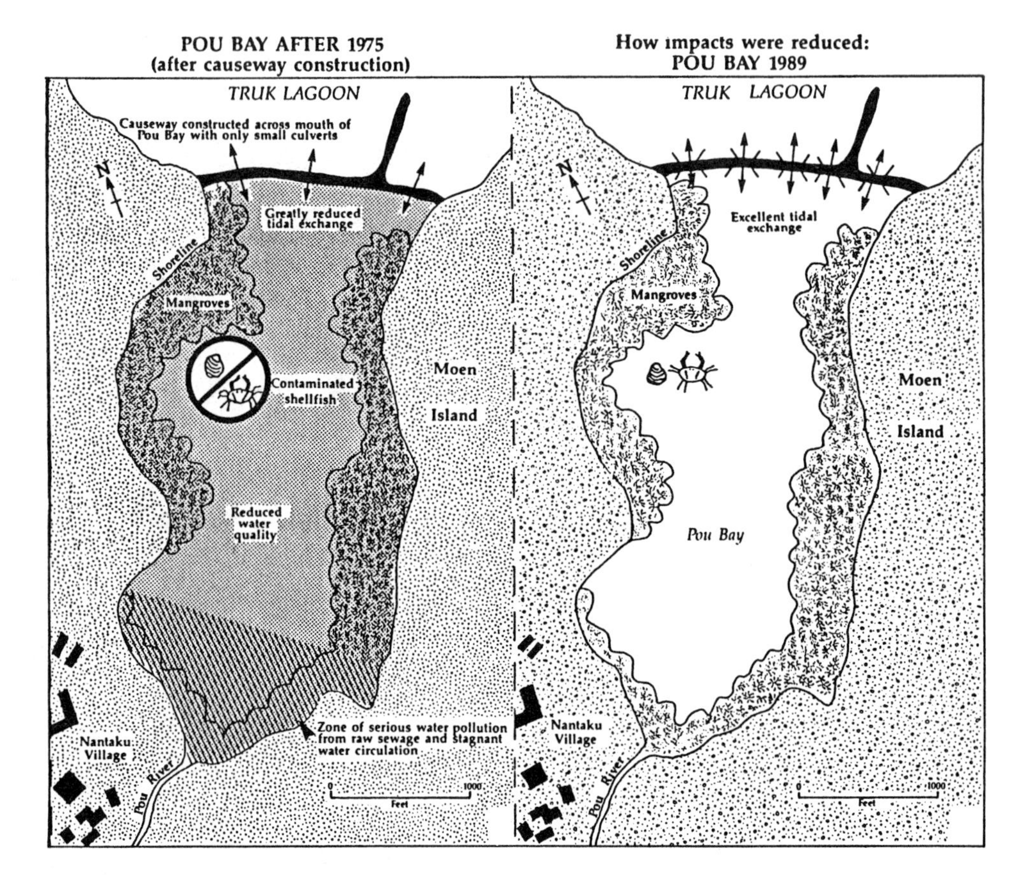

Figure 4.30.5. Adverse effects of road causeway construction across Pou Bay, *Moen Island* (Truk Lagoon, Federated States of Micronesia). Original causeway construction blocked circulation into Pou Bay, degrading water quality and contaminating Moen Island's most important shellfish resources. Causeway construction after 1975 required installation of larger culverts to increase circulation and reduce pollution. Additional culverts would result in greater improvement and water quality.

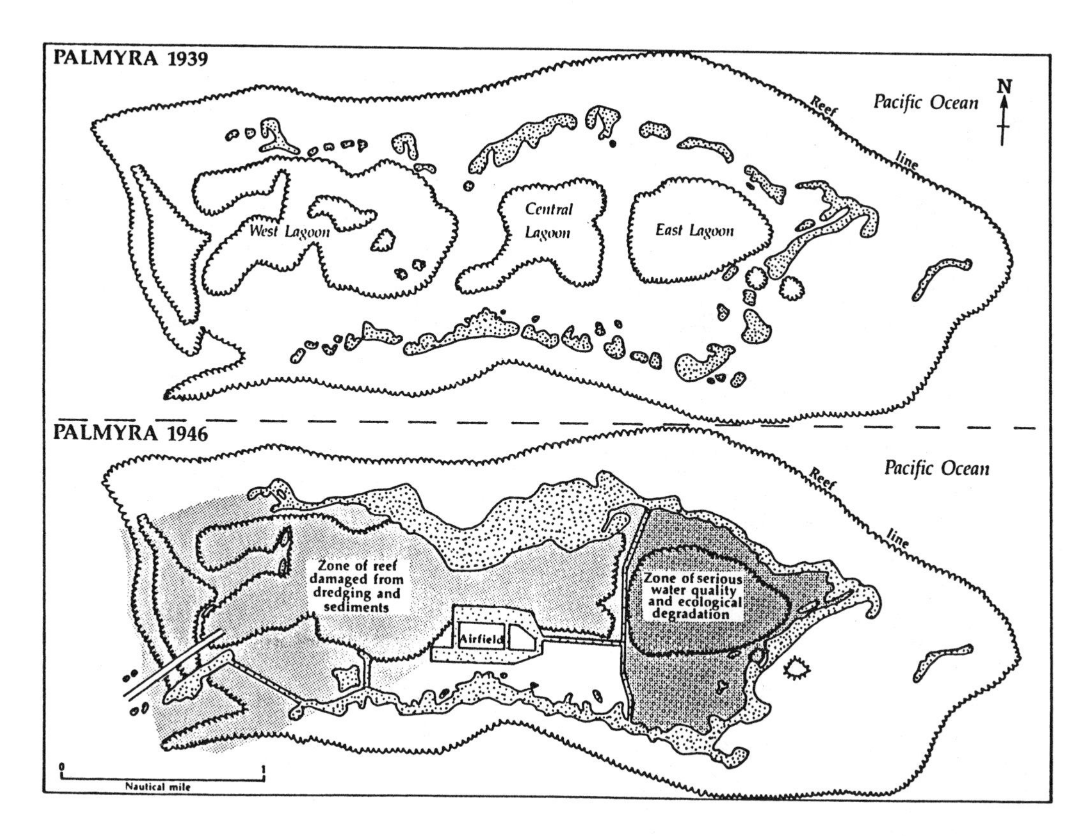

Figure 4.30.6. Adverse effects of dredge-and-fill operations at *Palmyra Atoll*, U.S. Line Islands. Pre-World War II construction of road causeways around East Lagoon by the U.S. Navy completely blocked circulation, causing collapse of coral reef communities. Dredging of a channel through the western reef and between the Central and East Lagoons destroyed reefs and altered water circulation. sediments drifting west from the dredge-and-fill areas probably damaged reef communities off the western end of the atoll. By 1979, some of the northern causeways had breached restoring some exchange between the East Lagoon and the ocean. Observations in 1987 reveal only partial recovery of the reefs from the military construction.

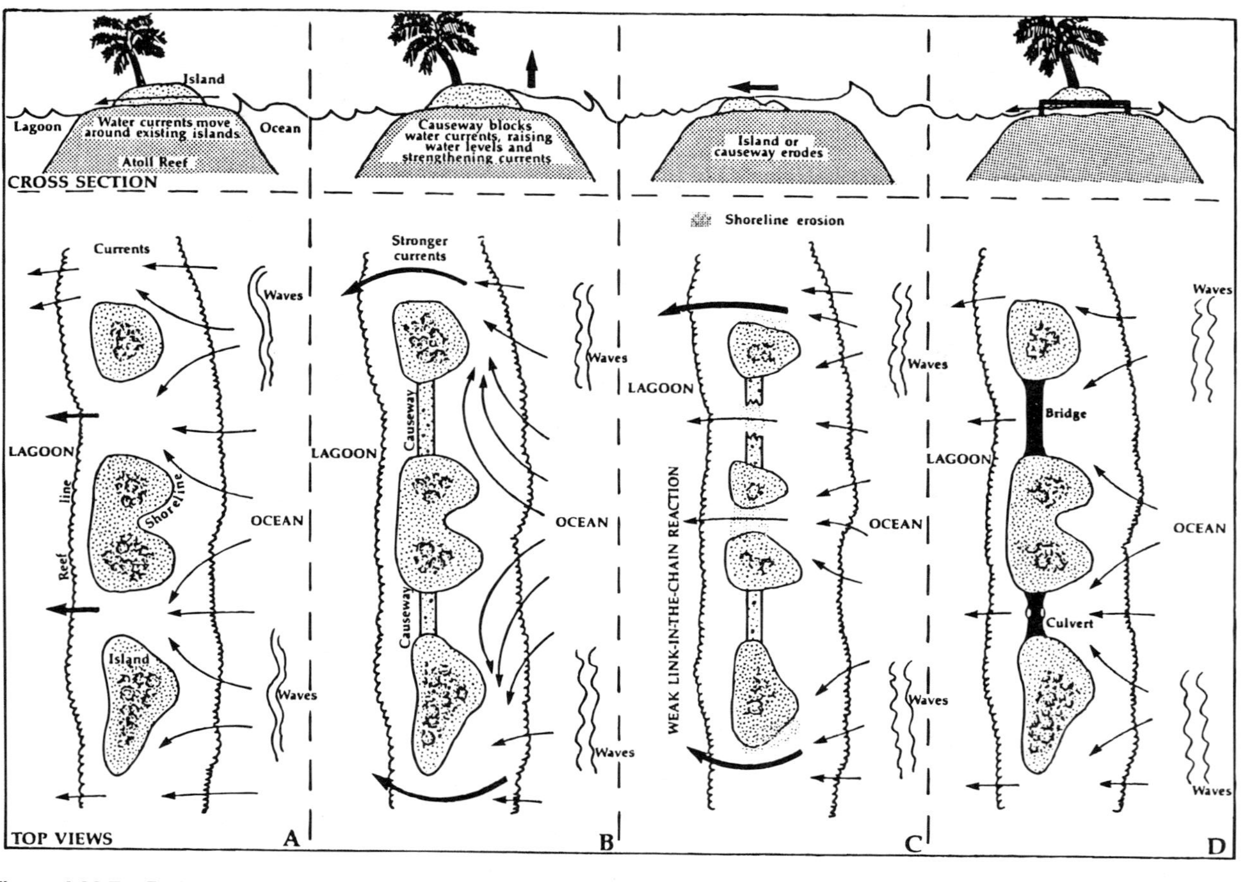

Figure 4.30.7. Problems from response of water currents and island configurations to causeway road fills to connect islands along a windward atoll reef, and solutions: (A) natural configurations before causeway construction, with waves pushing water currents over the reef into the lagoon; (B) after construction, causeways cause currents to deflect and strengthen, and water levels are pushed higher against islands and causeways by waves; (C) eventually currents and waves break through lowest lying portions of causeways or island, causing erosion; and (D) impacts can be eliminated by placing culverts or bridges through the causeway sections to maintain current flow between islands and reduce sea level rise from wave setup.

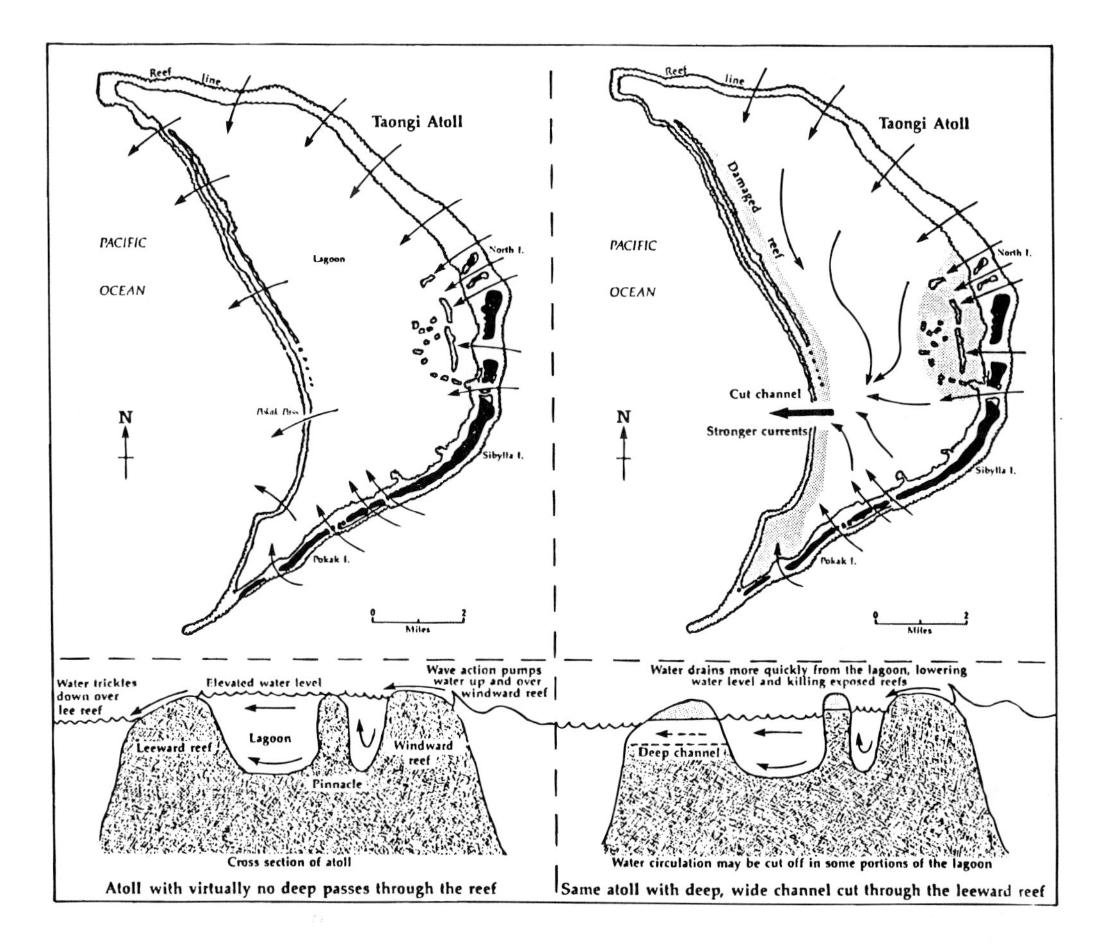

Figure 4.30.8. Possible adverse effects of cutting channels through semi-enclosed atoll lagoons. Many atolls (such as *Taongi*, which is pictured) have elevated lagoon water levels because of wave action pumping water over windward reefs and the lack of large, deep channels to drain the excess water. The reefs grow above normal ocean level because of constant water flow and in response to higher lagoon water level. Cutting a deep channel through such an atoll reef would cause waters to drain more quickly, lowering lagoon water level and killing emergent reefs.

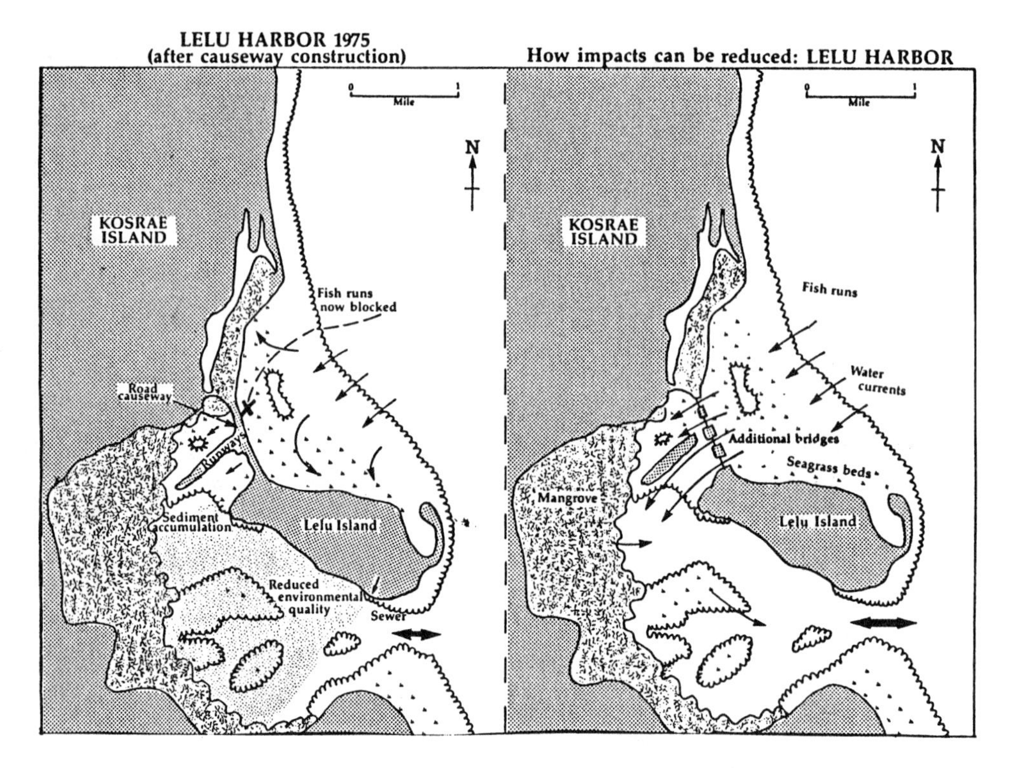

Figure 4.30.9. Adverse effects of causeway construction between *Lelu and Kosrae Islands*, Federated States of Micronesia. Original causeway blocked circulation and fish runs into inner Lelu Harbor, leading to a decline in seagrasses and fish catches. Main culvert in causeway was later blocked for runway fill expansion, further reducing circulation, fish yields, water quality, seagrasses, and coral reefs in Lelu Harbor—once Kosrae's most valuable fishery. Impacts could have been reduced or avoided by adding a continuous line of culverts through the causeway, replacing a section of the causeway with a bridge, and unblocking the main culvert.

4.31 SRI LANKA: ISSUE-DRIVEN COASTAL MANAGEMENT

Background

As the first tropical country with a centrally managed, full-scale ICZM program, Sri Lanka demonstrates how coastal zone program creation is often issue driven. The original issue was coral mining and the jeopardy to the shoreline caused by the mining of coral reefs, which had protected the shore for millenia.

The ICZM program was developed in close coordination with other Sri Lanka agencies concerned about and responsible for coastal resources. It provided both a policy framework and a practical strategy for dealing with problems. It was supported partially by USAID through the University of Rhode Island.

Considerable experience has been gained during the 12 years of the program. Subjects of particular interest are the permit program, setback requirements, locus of authority for permits (decentralization), the need for Situation Management at the local level, and the problems of job denial by program actions.

The success of the coastal zone erosion management program in the mid-1980s led to expansion of program goals to include two other coastal problem areas—reduction in biodiversity and natural resources in the coastal zone and inadequate preservation of coastal historical and religious sites.

Problem

The original issue—the one that propelled Sri Lanka into a centrally operated ICZM program—was severe shoreline erosion caused by the mining of coral deposits to produce construction materials and some other activities, as shown in the following Table (1):

Causal agent	Process	Effect	Areas
Beach sand mining	Reduction of sand in beach maintenance system posing possible threats to renewal	Increased erosion	Panadura, Lunawa, Angulana, Palliyawatta
River sand mining	Reduction of sand in beach maintenance system posing possible threats to renewal	Increased erosion of adjacent beaches, erosion of river banks	Kalu Ganga, Kelani Ganga, Maha Oya
Inland coral mining	Conversion of productive land into waterlogged area	Development of inland waste dumps & abandoned pits, reduction of coastal stability by creation of low-lying areas	Akurala, Kahawa, Ahangama, Midigama
Collection of coral from beaches	Reduction of beach nourishment material	Increased erosion	Ambalangoda to Hikkaduwa, Midigama, Ahangama and Polhena
Reef "breaking"	Reduction of reef size, creation of gaps in reef	Increased wave energy on beaches resulting in erosion	Ambalangoda to Hikkaduwa, Koggala, Midigama, Polhena, Rekawa Pasikudah, Kuchchaveli, Nilaveli
Reef dynamiting and "breaking" for fishery activity	Reduction of reef size, creation of gaps in reef	Increased wave energy on beaches resulting in erosion	Ambalangoda to Hikkaduwa, Koggala, Midigama, Polhena, Rekawa Pasikudah, Kuchchaveli, Nilaveli
Improperly sited groins, harbors, revetments, jetties	Interference with natural sand transport process	Erosion in some places, accretion in others	Beruwala Fishery Harbor, Kirinde Fishery Harbor
Improperly sited coastal buildings	Interference with dynamics of coastal processes	Loss of structures, other property due to retreat	Hikkaduwa, Bentota, Beruwela, Negombo
Improper removal of same	Exposed area subject to more rapid rates of wind erosion	Erosion, retreat	Palliyawatte, Koggala, Polhena Negombo

Coral mining—the removal of coral materials—stimulated by a surge of economic development in the 1960s and 1970s, was so extensive that it left the shoreline increasingly exposed to natural hazards. Serious erosion of beaches led to loss of shorelands and exposure of the coast to storm surges. Also, mangroves, small lagoons, and coconut groves were lost to shore erosion, and local wells became contaminated with saltwater. Coral mining contributed to the collapse of a local fishery as well (1, 2).

The coral is "mined" in the water (12 percent in 1984), along the beaches (30 percent), and in relic land deposits (58 percent) in and beyond the defined coastal zone (2). In the water, coral is rarely removed from the crest of the living reef; rather, it is removed from the lagoon or from dead reef patches, usually by women who must "free dive" quite deep to find it. Coral mining employed nearly 6,000 people, 1,200 people directly (3) with perhaps 4,700 involved in processing, sales, transport, etc. Much of the work was done by women.

Solutions

Sri Lanka reacted to its erosion problem by enacting a centrally operated coastal management program in 1982 (2). Even though the original problem was localized to one region and specific to one issue, a national format was set up as the best way to deal with it. Later, the same format was ideal for broadening the program to include other issues and to cover the entire coastline as it became a full-scale ICZM program (1, 7). The Coast Conservation Department (CCD) is the lead agency.

A major event in extending and consolidating Sri Lanka's ICZM program was the preparation of a two-volume report entitled "Coastal 2000: Recommendations for a Resource Management Strategy for Sri Lanka's Coastal Region," which was officially approved in 1990. Coastal 2000 makes a number of significant recommendations, which are being implemented and which expand the program to embrace four national priorities (6):

- Erosion management and land use
- Cessation of coral mining and control of sand mining
- Prevention of loss and degradation of coastal natural habitats
- Protection of scenic areas and cultural and religious sites

Regulatory measures are used to control the construction of shoreline protection works. A designated setback line is located to ensure that structures are not sited so close to the shoreline that they contribute to or are affected by erosion. There is a general setback 60 meters wide, but the demands of development do not always permit reserving a setback 60 meters wide (2), and so there are variations according to physical features of the coast and some exceptions. Setbacks for particular *water-dependent* activities, such as hatcheries for aquaculture and boatyards, are determined on a case-by-case basis in the Sri Lanka program.

Included in the erosion management strategy are a public education campaign targeted at coral and sand miners regarding the impacts of their activities, a program to identify alternative employment for displaced coral miners, and a research effort to identify alternate sources of lime for the building industry. Complementing these management efforts is a public investment program to build shoreline protection works in appropriate areas.

Permit System: The primary mechanism for implementing Sri Lanka's coastal management program has been a permit system for development activities in the 300-meter-wide coastal zone (Figure 4.31.1).

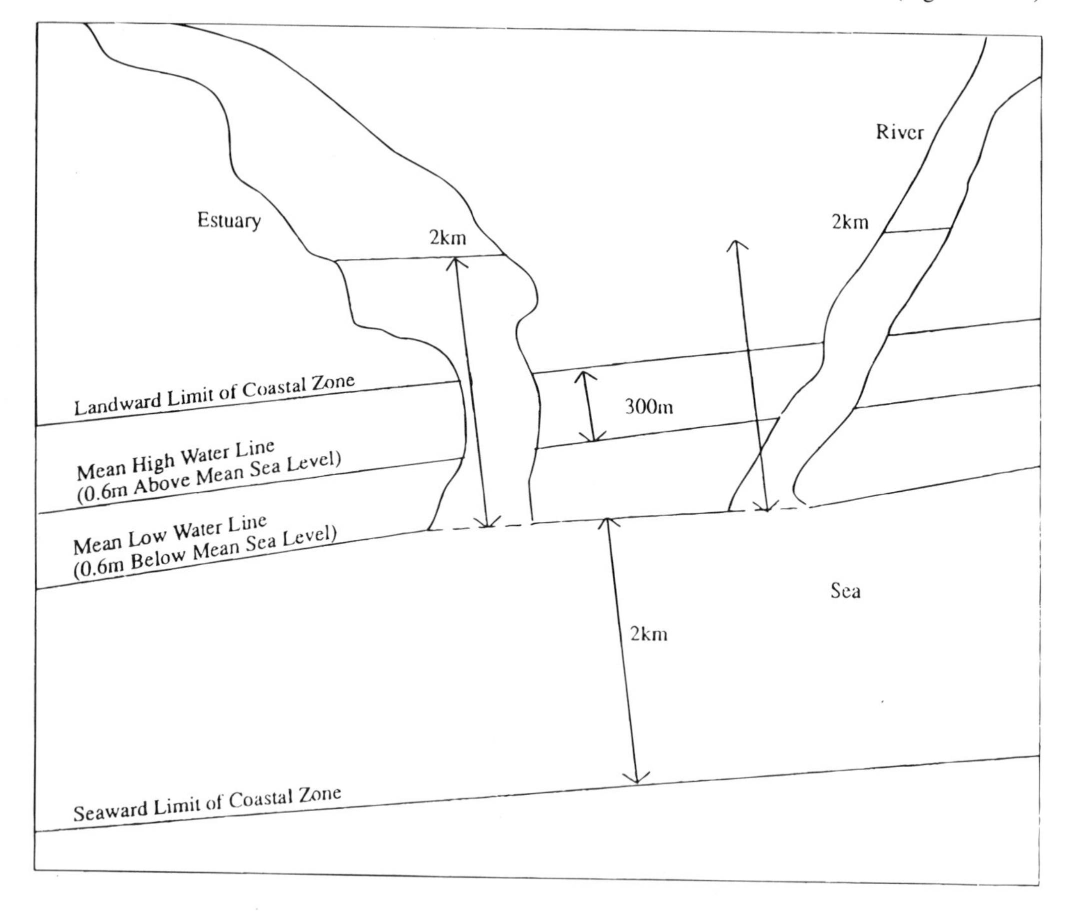

Figure 4.31.1. The Sri Lanka coastal management program official boundary limits are shown: 2 km seaward and 300 m landward from the coastline, except 2 km into littoral basins.

Any person desiring to engage in a Development Activity within the Sri Lanka Coastal Zone will be required to obtain a permit issued by the Director prior to commencing the activity. A permit may be issued if:

- The activity is consistent with management policies.
- The activity is not prohibited by this plan.
- The activity is outside designated setback lines.
- The National Standards for the relevant environmental parameters are met.

Examples of development types requiring permits include:

- Residential and other structures
- Commercial and industrial structures
- Recreational structures
- Public roads, bridges, and railroads
- Shoreline protection works
- Sewage treatment facilities
- Wastewater discharge facilities
- Harbor structures and navigation channels
- Waste disposal
- Dredging
- Filling
- Mining
- Removal of sand/shells
- Removal of vegetation
- Aquaculture facilities
- Grading
- Breaching of sand bars

Applicants for a coastal permit fill out a simple form indicating the name and address of the applicant, nature and location of the proposed activity, existing uses, and whether the area is subject to erosion or accretion. Site designs are required for houses, hotels, and other structures. Applicants for mining activities must estimate the amount of material to be removed and the duration of the mining activities. Typical applications are reviewed in three weeks (4).

Since 1983, when the permit system was initially implemented, 2,186 applications have been approved—about 95 percent of the total requested. Most of the permits have been for houses and for sand mining. Although CCD staff have approved 95 percent of permit applications, they have been able to exercise influence over coastal development by discouraging some applicants from submitting applications that do not comply with the law. Second, CCD staff frequently encourage applicants to alter their site plans by moving proposed structures further inland away from erosion-prone areas.

In addition, CCD staff frequently attach conditions to permits, such as setback requirements, which are intended to reduce potential adverse impacts (4). CCD now handles about 400 permits per year, each requiring assessment of its environmental impacts.

For permit applications that do not require an EIA, a decision on the application will usually be made within three weeks of receiving all required information. Consultation with CCD early in the project planning stage is encouraged to facilitate the permit process.

Activities within the Coastal Zone prohibited by the CCD (no permit to be issued) are:

- Removal of coral other than for research purposes
- Mining of sand except in areas identified by CCD
- Development within 200 m of designated archaeological sites
- Any development activity that will significantly degrade the quality of officially designated natural areas

Exceptions may be granted to engage in prohibited activities only if and when the applicant has demonstrated that the proposed activity serves a compelling public purpose that provides benefits to the public as a whole or when an extreme hardship on the proponent can be proved.

Activities that may be engaged in without a permit within the Sri Lanka Coastal Zone include:

- Fishing
- Cultivation of crops
- Planting of trees and other vegetation
- Construction and maintenance of coastal protection works by CCD in accordance with the Master Plan for Coast Erosion

The Coast Conservation Act specifies penalties for contravention of the provisions of the act. Engaging in any development activity prior to obtaining a permit issued by the director and/or noncompliance with conditions stipulated in the permit are contraventions. Penalties may include a fine of up to 25,000 rupees or imprisonment for one year with additional penalties for second offenses. The magistrate can order the confiscation of any vessel, craft, boat, vehicle, equipment, or machinery used in committing the offense. The CCD director is empowered to order the demolition of any unauthorized structures.

The CCD ensures compliance with permit conditions through a monitoring system that includes one or more of the following:

1. Monitoring key stages in a project in which the developer shall obtain clearance from the CCD prior to proceeding to the subsequent stages
2. Requiring the developer to furnish a certificate of conformity from a nominated authority that the permit conditions have been adhered to
3. Direct supervision by CCD personnel or by a nominated authority
4. Periodic site visits by CCD personnel
5. Requiring the developer to conform to a stipulated time schedule for the development activity or the stipulated quotas
6. Requiring the developer to submit reports, carry out surveys, tests, and so forth

Attached to the ICZM process is a policy group, the Coast Conservation Advisory Council (CCAC), which reviews all major permit applications along with the accompanying EIA and CCD staff recommendations. In virtually all cases the CCAC agree with the staff and support their decision in their advice to the director of CCD. This is a very important aspect of the Sri Lanka CCD and seems to operate with a minimum of political interference.

Permit applications requesting a variance are reviewed by the CCAC, who gives credence to staff-produced technical analysis in assuring that regulatory decisions are not arbitrary or subjective (4). Clear management goals, careful analysis, and public and political support for ICZM all contribute to its effectiveness. Trouble arises with "cumulative impacts," where there are many coastal users, such as sand miners or fishermen, and where the impacts of any single user's activities are minimal (4).

Also included in the erosion management strategy are a public education campaign targeted at coral and sand miners regarding the impacts of their activities, a program to identify alternative employment for displaced coral miners, and a research effort to identify alternate sources of lime for the building industry. Complementing these management efforts is a public investment program whereby CCD builds priority shoreline protection works in appropriate areas (6).

The Coral Mining Problem: An issue-driven program is in trouble when the program cannot resolve the main issue. As of 1991, while "the permit program for . . . sand mining and shorefront construction is proceeding with increased credibility . . . [but] enforcing a prohibition on coral mining has proved much more difficult and politically very troublesome," according to Hale and Kumin (3).

The situation has several facets (3). First, the coral mining prohibition directly affects the livelihood of thousands of individuals engaged in or dependent on that industry. The Sri Lankan government believes it has a responsibility and it is a political necessity to find alternative livelihoods for displaced coral workers. Second, local police departments who are in charge of stopping illicit coral mining are often either undermanned, and therefore unable to enforce the prohibition adequately, or reluctant to act in the face of opposing public sentiment. And, of course, Sri Lanka has been distracted by civil strife.

As part of the effort to discourage coral mining, CCD staff worked with other agencies to help develop alternative sources of employment for the miners as well as alternative sources of lime (4). But CCD's attempt encountered difficulties. The sea-faring miners are well paid and the alternative employment offered by CCD (e.g., farming) was not attractive.

Therefore, many continued under various ruses or in blatant violation. The kilns continued to produce quicklime more or less openly, even in the face of ministerial orders and persuasion campaigns. One CCD officer reported threats on his life (3). Finally, CCD destroyed all kilns in the 300-meter-wide jurisdictional zone. Still, those farther inland continued to operate, some legally, using land deposits of carbonate materials.

The difficulty of controlling coral mining threatens to undermine the credibility that the Coast Conservation Department has built for itself and its projects over nearly a decade (3). To assure long-term progress toward managing Sri Lanka's coastal zone, the CCD cannot ignore law breaking, of course, and so the agency is changing its approach. But, at present, there are only about 100 families still engaged in illegal mining of coral.

Other problems with the program are reported to be that (a) most mangrove forests are outside the designated coastal zone (300 m inland from upper tideline and 2 km seaward and up rivers) and (b) building regulations are difficult to enforce, particularly for poor people who cannot manage to get permits and build illegally. Yet CCD has not given up on these problems and continues to try various approaches, particularly decentralization and situation management.

Decentralization: The 1990 Coastal Plan refers to a "second-generation coastal resources management program (which) must be implemented simultaneously at national, provincial, district and local levels." CCD has begun the process of delegating authority for issuing "minor" permits to local authorities and has organized an extensive program of training local officials to implement a minor permit system. This devolution process also requires the development of a monitoring system for ensuring that permit procedures comply with the Coast Conservation Act (4). A major criterion of success for the decentralization effort is the success of sharing of authority between national and local levels of government.

The "minor permits" will be given for modest houses, small quantity sand removal from beaches, and other activities whose impacts are limited and predictable. This task is going well and is considered to be essential to the smooth functioning of the CZM program in hope that decentralization will earn local support for the program. Local support is the key to finally putting an end to coral mining.

Situation Management: The "bottom-up" approach of decentralization by CCD involves the use of situation management (SM); specifically, the designation of special area management planning areas (SAM), as they are called here, in which residents become actively involved in both design and implementation of the coastal management program. Small-scale SAM management projects rely primarily on reciprocation through mutual education, persuasion, and cooperation to minimize activities that deplete or degrade coastal habitats to the detriment of all residents. SAM projects were already under way in 1994 in two southwest coast sites, Hikkaduwa and Rekawa Lagoon.

Hikkaduwa is a resort and fishing community known for its beaches, which feaures a designated marine sanctuary; its problems are with maintaining clean water and beaches protecting a small coral reef, and soothing simmering conflict between hotel operators and fishermen. Rekawa Lagoon is rural and has valuable mangroves and fisheries but is threatened with hydrological problems (insufficient incoming waterflow to keep the lagoon open) and with major encroachment by aquaculture. Illegal coral miners are suspected to inhabit both areas.

It was a key consideration for CCD in choosing the present two SAM sites that there be a major, and obvious, environmental problem to be solved at the site. This is bit of a gamble because the key problem at each site is most difficult of solution. Both are beach sediment problems—one is blockage of seawater flow into a constricted lagoon and the other a boat harbor that is impacted with sand. Solving these problems will, it is hoped, gain local support for CCD's program.

In the SM areas the whole community will have to be fully aware of and integrated into the planning effort so that it is truly participatory. A key aspect is that the SAM activities have to be very closely coordinated with the ICZM regulatory program, including decentralization.

Lowry and Wickremeratne (5) concluded after a detailed examination of Sri Lanka's program that its strength and vigor are due in large part to the following: (a) strong coastal orientation of the country; (b) widely shared agreement about what the coastal problems are, what the causes of the problems are and, to a lesser extent, what the appropriate roles of government are in dealing with the problems; (c) a law that provides a strong legal basis for management; (d) strong program leadership; (e) adequate political support for planning and management; and (f) an adaptive, incremental approach to the development of the planning and management program.

Lessons Learned

It is quite possible to begin a coastal management program with a single issue affecting one region of a country and expand into a full-scale ICZM program.

Enforcement of coastal management rules can be very difficult if jobs are threatened and if practicable alternative employment is not available.

Full success in coastal management programs requires the active participation of the stakeholders, the people who will be affected. This implies some form of decentralization and reciprocation, either program-wide or for selected special situations (situation management).

The jurisdictional area in which a permit system is administered should not be too large; Sri Lanka's 300-meter-wide coastal zone is narrow enough for effective focusing of management efforts (4).

The goals of the coastal management program should be well understood and accepted as legitimate by public officials and by a vast majority of coastal residents. (4)

Regulatory staff should not deal only with yes or no answers; they should show applicants how their proposed development activities can be made compatible with ICZM in project scale or site design (4). Performance standards are helpful in encouraging project sponsors to innovate environmental protection approaches.

Technical backup is extremely important to management staff in resolving conflicts over different uses. The technical experts do not have to be staff members; they can be collaborators or consultants.

Contributed by — Author.

REFERENCES

1. Amarasinghe, S. R., H. J. M. Wickremeratne, and G. K. Lowry, Jr. 1987. Coastal zone management in Sri Lanka 1978–1986. In: *Coastal Zone '87,* Vol. III, Amer. Soc. Civ. Eng., New York. Pp. 2822–2841.
2. CCD. 1990. *Coastal Zone Management Plan.* Sri Lanka Coast Conservation Department, Colombo, Sri Lanka. 110 pp.
3. Guzman, H. M. 1991. Restoration of coral reefs in Pacific Costa Rica. *Conservation Biology,* Vol. 5, No. 2. Pp. 189–195.
4. Lowry, K. 1993. Coastal management in Sri Lanka. In: *Coastal Management in Tropical Asia,* No. 1, Coastal Resources Center, University of Rhode Island. Pp. 1–7.
5. Lowry, K. and H. Wickremeratne. 1989. Coastal management in Sri Lanka. In: *Ocean Yearbook 7.* University of Chicago Press. Pp. 263–293.
6. Olsen, S. 1987. Sri Lanka completes its coastal zone management plan. *CAMP Newsletter,* August 1987. P. 7.
7. URI. 1986. *The Management of Coastal Habitats in Sri Lanka.* Report of Workshop, May 12–15, 1986. University of Rhode Island, Coastal Resources Center. 36 pp.

4.32 TANZANIA, MAFIA ISLAND: PARTICIPATORY CONTROL OF CORAL MINING

Background

Mafia Island, Tanzania (Figure 4.32.1), like other island communities, has limited land-based resources, especially materials for construction. Demand for aggregate and cement substitutes has encouraged the use of live corals to satisfy this need. Coral mining has been identified as one of the greatest threats to the continued high levels of productivity of Mafia's waters (5). On Mafia Island, coral is widely used in traditional building both to pack floors and walls of dwellings and as a whitewash. Live coral has also been used for large projects such as hotel building and airport and road construction (3, 5).

Problem

Coral mining is carried out manually. Collectors remove corals from back reef areas at spring low tide using iron bars. Slow-growing, massive coral forms such as *Porites* are preferred. The mined coral is left to dry on land before being broken up for use as aggregate or for conversion to lime. Lime, a cement substitute, is produced by roasting fragmented coral on a mangrove wood pyre. Approximately 150–300, 3–6-m-high mangrove trees are required to produce up to 300 50-kg sacks of lime (Figure 4.32.2). The efficiency of open kilns is low, with as little as 20–50 percent of the coral rock converted to lime (3, 5).

On Mafia, 5,500 m^3 of reef are known to be removed annually and sold through a commercial dealer. With the inclusion of local sales throughout the rest of the islands, the total amount at least doubles this quantity. Driven by housing demands and combined with the building demands of a predicted fivefold increase in tourism, the scale of reef removal could be expected to increase.

A study undertaken to assess the effects of coral mining revealed obvious biological and physical impacts. Fish abundance and diversity were reduced over the mined site compared with the control

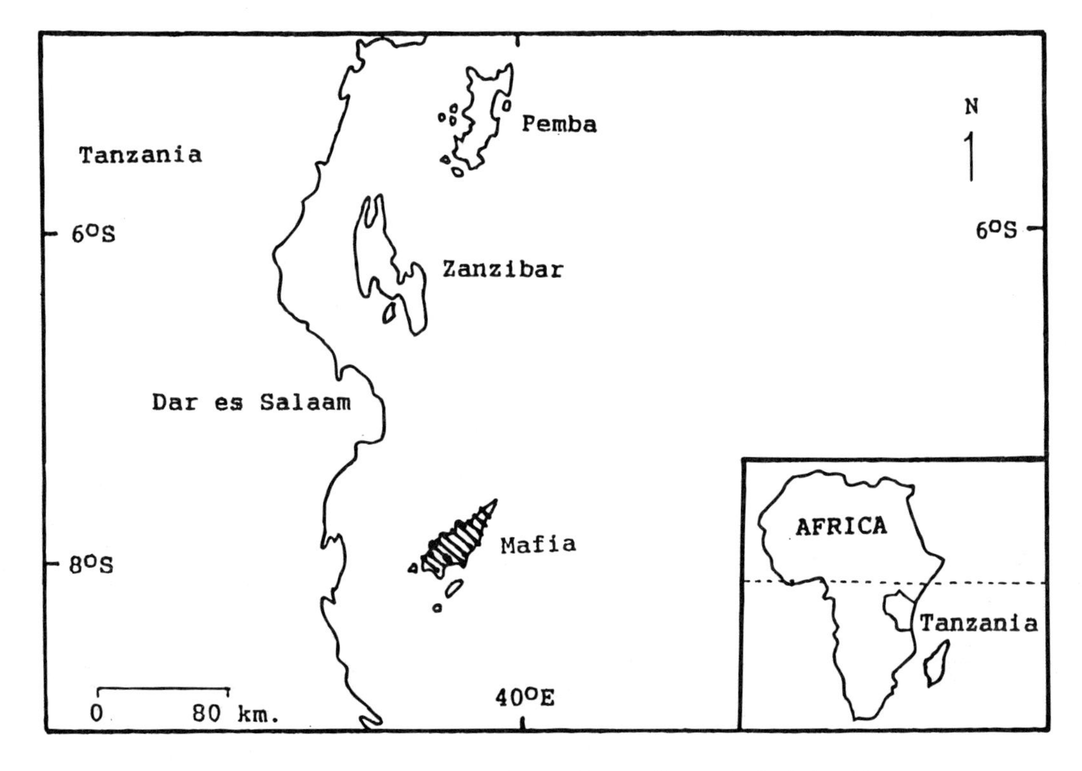

Figure 4.32.1. Mafia Island is southeast of Dar es Salaam in the Indian Ocean.

Figure 4.32.2. On Mafia Island, lime is produced by burning fragmented coral on a mangrove wood pyre.

site: 41 percent reduction in total number of fish and 24 percent reduction in number of species (Figure 4.32.3). This is an expected result; for example, a reduction in fish biomass and numbers was observed on mined sites in the Maldives (2).

On some sites in the Maldives, little recovery has been noted 16 years after mining activity (4). At Mafia, little sign of coral regeneration was observed 3 years after mining (3).

The removal of live coral has also led to a reduction in the efficiency of reefs as breakwaters for curbing erosion of shorelines. For example, on neighboring Bwejuu Island (a major center of mining activity) large numbers of coconut trees have been washed away.

Solutions

Substitutes to mined coral are available; however, the cost is nearly prohibitive. Imported portland cement retails at $US 6.50 for 50 kg, whereas the equivalent quantity of lime retails at $US 0.85. An immediate solution could be a ban on all live coral collection using existing legislation; however, alternative building materials and income sources would have to be identified.

To balance the conflict between conservation and the needs of the people, a two-stage strategy for management was developed following extensive consultation of local users, commercial interests, tourism developers, and construction contractors. The strategy involves alternative construction materials and income sources and an awareness program. Potential alternative materials and income sources include:

- Production of sun-dried mud bricks using Mafia's extensive clay deposits
- Utilization of land-based limestone sources
- Construction of closed kilns, aimed at reducing the consumption of live coral (closed kilns can have up to a 70 percent coral-to-lime conversion efficiency) using coconut wood instead of mangrove for fuel

Alternative income sources could be provided by both the expansion of the tourist industry and the operation of kilns and brick presses as cooperatives. In Stage 1, use of live coral would be restricted

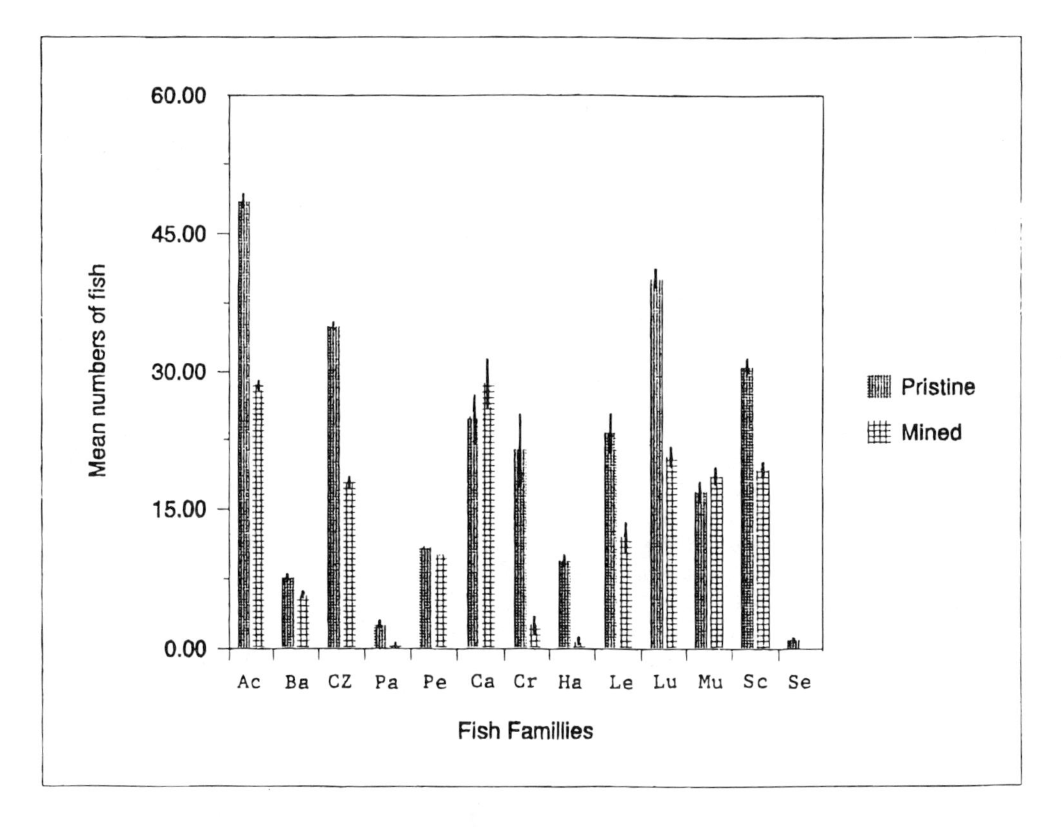

Figure 4.32.3. Coral mining is associated with a reduction of 41 percent in total number of fish and 24 percent in number of species.

to traditional users only. The development of alternative construction materials and income sources would be pursued during this period. Stage 2 starts after alternative materials and employment are identified, with the aim of gradually limiting collection of live coral by imposing quota systems. Mining activity would also be restricted to less sensitive areas with respect to tourist areas, important breakwater areas, and key fishing sites. Pilot projects for production of alternative materials have already been successful, including testing closed lime kilns and the suitability of Mafia clays for bricks.

Tanzania has no ICZM program that could be utilized for reef conservation. However, there is a proposal for a marine park for Mafia Island (operating as a resource management entity), but it is yet to be gazetted. Nevertheless, significant progress has already been made toward a reduction in mining activity.

Lessons Learned

1. Destruction of back reef areas by coral mining leads to biological and physical degradation of reefs and associated habitats. There are reductions in reef fish abundance and diversity-limiting fisheries production and increased rates of erosion-reducing terrestrial productivity.
2. Management must carefully balance people's dependence on the resource and the need to sustain the productive fisheries supported by reefs.
3. Alternative livelihoods must be part of any coastal conservation project. In the present case alternative income sources have been created through tourism and construction.
4. With sufficient consultation of concerned parties, workable alternatives can be provided instead of the extreme of protectionism. With prior consultation and planning, the resource users can play a key role in conserving their resources and protecting the environment.

Acknowledgment

Field support for this study was provided by the Frontier-Tanzania Marine Research Program, a joint initiative of the University of Dar es Salaam, Tanzania and the Society for Environmental Exploration, United Kingdom.

Contributed by — N. Dulvy and W.R.T. Darwell, Frontier-Tanzania, The Society for Environmental Exploration, London.

REFERENCES

1. Brown, B. E and R. P. Dunne. 1988. The impact of coral mining on coral reefs in the Maldives. *Env. Conserv.,* Vol. 15. Pp. 159–163.
2. Dawson Sheppard, A. R., R. M. Warwick, K. R. Clarke, and B. E. Brown. 1992. An analysis of fish community responses to coral mining in the Maldives. *Env. Biol. of Fishes,* Vol. 33. Pp. 367–380.
3. Dulvy, N. K., D. P. Stanwell-Smith, W. R. T. Darwall, and J. C. Horrill. 1994. Coral mining at Mafia island, Tanzania: management dilemma (manuscript).
4. Horrill, J. C. and C. J. Mayers. 1991. Census and economic survey of local coastal resource users in the proposed Mafia Island marine reserve. *World Wildlife Fund for Nature, Tanzania.* 29 Pp.
5. Horrill, J. C. and M. A. K. Ngoile. 1991. Mafia Island Report Number 2: Results of the physical, biological and resource use surveys: rationale for the development of a management strategy. *The Society for Environmental Exploration, London.*

4.33 THAILAND: SHIPPING PORTS AND CUMULATIVE IMPACTS

Background

Most large harbors have evolved from smaller harbors that have been situated in a particular position for hundreds of years. Mostly they are situated at the head of estuaries or inside rivers where relatively small bridges could cross the river and where industrial or merchandising towns or villages had developed. The natural depth of the waterways was sufficient for the shipping of that time (2–4-m draft). Many of these harbors have grown to such an extent that ships with a draft of 10 m or more have to be accommodated. Yearly dredging of more than 20 million cubic meters of sediment has become quite common (Rotterdam, Hamburg, London, Charleston). It is clear that harbors of this size are situated in the wrong place, but they cannot easily be moved because in the present place a special infrastructure exists.

Problem

Major harbors tend to have big problems with dredging. Dumpsites for dredge spoil tend to be filled up or to take up an unacceptable area of the estuaries. Tidal exchange in these parts is decreased and fisheries, mariculture, and receiving capacity for wastewater are negatively influenced. Dredged materials from harbors in industrial surroundings are always polluted, and legal requirements for quality standards of spoil are getting tighter.

As an example, the harbor of Surat Thani in Thailand is a typical harbor in a river delta with a natural depth of ± 2 m. It is connected to the deep water outside Ban Don Bay by a semi-natural gully 10 km long through a shallow bay. With limited dredging this gully can be kept at the required depth. The harbor is now being renovated, and new piers and wharfs are being constructed. The harbor and its approaches will be dredged to a depth of 4 m. In this region where a river with heavy sediment load meets the saline water of the bay, there will be much sedimentation. Natural currents will not be enough to keep the depth at 4 m.

It is clear that the harbor will require much maintenance, which will have a negative impact on other users, such as fisheries, shrimp and oyster culture, ornamental fish fisheries, agriculture, reed cutting and basket manufacture, rice cultivation, and tourism.

If each future small development is accompanied by only a small impact, the result may not seem severe and the activity allowed. However, in several decades we may end up with an accumulated large impact—this is the problem of cumulative impacts, whereby the addition of many small impacts equals a most adverse situation.

At Surat Thani, several activities might require room for expansion in the future. One might think of oyster culture, mussel culture, fish culture, shrimp culture, fishery, mangrove forestry, urbanization, infrastructure for roads and industry, navigation channels and harbors, recreation areas, and so forth. Each of these would have an impact on the environment or directly on another activity. Some maricultural activities only require space and a healthy environment; others cause eutrophication. If space is required in the estuary, this will have an impact on activities like fishery and water transport. If culture basins are constructed in the mangroves, this will impact all mangrove uses and the nursery function for the fishery in the bay.

Extensive dredging would cause increased turbidity, which destroys the possibilities for oyster culture. Pollution by industrial activities has an impact on most other activities. Also, the sewage of large, urbanized areas will make a region unsuitable for mariculture. If an activity is affected by another, it will try to compensate and expand elsewhere, thereby impacting others. Eventually, the effects of all activities might accumulate and destroy the natural values and sustainable uses of the whole area.

At present the population of Surat Thani is approximately 50,000. It is expected that the town will grow to 500,000 in the very near future. A town of that size, with a well-developed inland region, will require space for housing, agriculture, fishery, and recreation and a harbor with a depth of close to 10 m. It will be almost impossible to maintain that depth at reasonable cost in both monetary and environmental terms.

Solutions

The main solution is to plan for the distant future. For example, development of an alternative major harbor 50 km away and the development of Surat Thani as a "feeder" port, serving light draft ships only, will solve a lot of future problems. To prevent capital loss, very few investments would be put into the harbor of Surat Thani or the harbor-connected industrial development in that place. That will be difficult because the alternative harbor does not yet have any significant infrastructure. On the other hand, an evolution of Surat Thani into a major port would present the town with even bigger problems than those of Hamburg, Rotterdam, Emden, London, or Charleston.

When thinking about planning and environmental impact, one should first plan for future developments based upon an investigation of the present environment and of the social and economical functions it fulfills (1). Also, future functions of the natural environment should be taken into account. When an integrated, sometimes "utopian" Master Plan has been designed, the environmental impact of the whole development should be evaluated, and from several alternatives the best one should be chosen. Stepwise development can then proceed.

A method for environmental impact assessment (EIA) which has proven to be fruitful was described by Dankers (2). It is often helpful to set up a factor train (3); an example is given in Figure 4.33.1. Based on complete knowledge of the system, the physical, chemical, and biological effects of a large development can be identified. Specialists can be asked specific questions without involving them in the whole project. When all functions of the environment have been evaluated, it is also possible to predict their value when the development is carried out (1).

The evaluation does not necessarily have to be in monetary terms. On the basis of the total impact of different alternatives, decisions can be taken and the best scenario chosen. When the Master Plan has been laid down, development has a target. For each specific development, a detailed EIA can be made to guide the process leading to implementation of the final plan.

Lessons Learned

It is not a good idea to start planning in small steps based on project-by-project EAs. At Step 1 there should be an assessment of the present (and future) environment, and all environmental functions

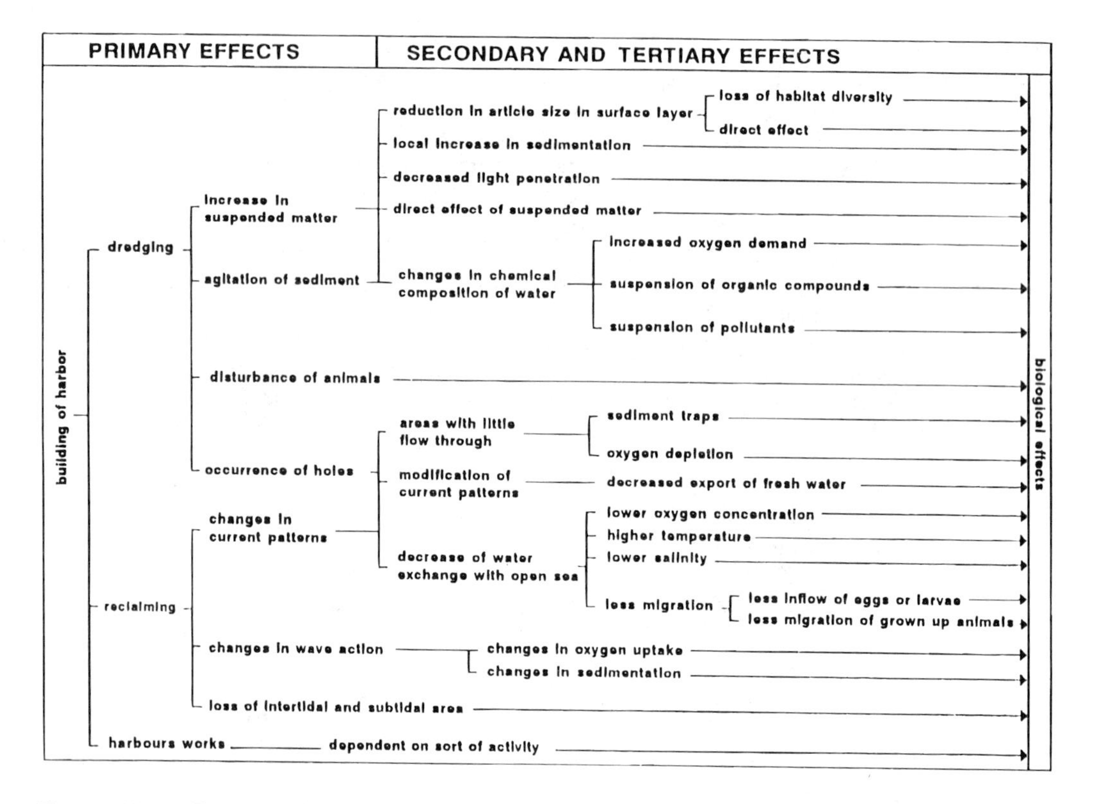

Figure 4.33.1. Factor train for the description of the biological effects of harbor construction.

(fisheries, aquaculture, tourism, etc.) should be investigated. Then a Master Plan with some alternatives should be set up.

As Step 2, the complete Master Plan should be subjected to a more or less detailed EA, taking into account the impact on each sector. On the basis of that study, the best alternative should be chosen. Only then can one speak of integrated coastal zone management (ICZM). With proper management some of the uses may be sustainable. Trying out this exercise in an area that is certainly going to be developed would be a great challenge.

Contributed by — Norbert Dankers, Instituutvoor Bosen Natuuronderzoek (IBN-DLO), Den Burg (Texel), Netherlands.

REFERENCES

1. Burbridge, P. R., N. Dankers, and J. R. Clark. 1989. Multiple use assessment for coastal management. In: *Coastal Zone '89.* Proceedings of the 6th Symposium on Coastal and Ocean Management, Charleston, SC, 11–14 July 1989. American Society of Civil Engineers, New York. Pp. 33–45.
2. Dankers, N. 1982. Quantifying the loss of functions of a natural area as a means of impact assessment for reclamation projects. In: *International Symposium Polders of the World, Lelystad, The Netherlands,* Vol. 2. ILRI, Wageningen. Pp. 639–652.
3. Darnell, R. M. 1976. *Impacts of Construction Activities in Wetlands of the United States.* E.P.A.-600/3-76-045 Washington, D.C.

4.34 THAILAND: TANTALUM RIOT AT PHUKET

Background

It would have been one of only five tantalum processing plants in the world and the only one in Asia. Designed to produce 600,000 pounds of tantalum pentoxide each year, or approximately a quarter of the current world demand of 2.5 million pounds, it would also have produced an equivalent quantity of niobium pentoxide. The entire production of tantalum would have been exported, earning an estimated 1,100 million *baht* a year. Claimed to incorporate the latest pollution control technology, the plant—owned by the Thailand Tantalum Industry Corporation (TTIC)—was shortly to open. But then, on June 23, 1986, a mob broke in and burned the expensive plant virtually to the ground. What had gone wrong?

Problem

The first problem was that the authorities had badly underestimated how far the public might go to stop a project with which it was not in sympathy. In retrospect, the planning of the project had not been sufficiently sensitive to public feelings at a time when the reverberations of the Bhopal disaster were still being felt and the Chernobyl disaster was headline news. Among the worst fears were that the hydrofluoric acid used by the plant might leak, causing another Bhopal, or that the sheer environmental impact of normal operations would undermine the booming local tourism industry.

The residents of Phuket first began to voice their concern in May 1986, culminating in mass demonstrations by about 50,000 people in June. Local community leaders were threatening to boycott the national election on July 27, but the real surprise was the mass break-in on June 23, the day that a public hearing was meant to begin, with the Minister of Industry present.

On the surface, things seemed to be going well. TTIC had received considerable support from the Board of Investment, with the bulk of the project's financing coming from a US$53.5 million package arranged by the International Finance Corporation (IFC), the World Bank's investment arm. The project's original proponent, S.A. Minerals, was the largest investor, holding 45 percent of the equity, with the IFC, three Thai banks, the quasi-governmental Industrial Finance Corporation of Thailand and local tin-mining families holding the rest.

Tantalum is very much a high technology metal, exploited for its high tensile strength and heat resistance. It is used to make capacitors, special alloys, aerospace components, and nuclear reactor equipment. It is normally derived either from tantalite ore or from tin slag, the major source in Thailand. The tin slags have tantalum pentoxcide varying from 2.5 percent in low grade ores to 14 percent in high grade ores. The original plan for the Phuket tantalum plant called for two components: an enrichment plant to produce slag with a 20–30 percent tantalum pentoxide content (later deferred because of falling tantalum prices) and an extraction plant using hydrofluoric acid to dissolve the tantalum in the ground ore or slag and methyl isobutyl ketone, a common solvent, to extract the tantalum-rich material.

The wastes from the plant would have included 380 cubic meters per day of acidic wastewater. The plan was that the acidic wastewater and ammoniacal digestion residues would be neutralized with lime and that the resulting 48 tons per day of waste would be filtered and the solids landfilled. The treated wastewater would be discharged to a canal in the rainy season and to an abandoned mine during the dry season. Air pollutants, it had been promised, would be collected and treated.

Solutions

Solutions came too late and were too narrow. An environmental assessment (EA) was completed only shortly before the plant was burned down and was not used as a means of promoting public participation. The TTIC plant was also approved before the World Bank issued its post-Bhopal guidelines for assessing the major hazards associated with industrial development projects in developing countries. The TTIC plant would now come within those guidelines, since they cover any plant handling more than 10 tons of hydrogen fluoride. Such assessments are not a legal requirement under Thai law, but the lack of such an assessment can leave the door open for rumor and speculation.

The plant was located on the northern outskirts of the city, just a few hundred meters from a teacher training college. The site was apparently chosen because of its proximity to the tin smelter that produces the tantalum-rich slags, but the siting of such plants in residential areas is questionable at the best of times. Even if the local people had read and understood the EA report, which concluded that the plant could be run safely (if sufficient effort was devoted to making sure that it was), the lack of effective enforcement of pollution control standards in Thailand would have raised question marks over the apparently rosy picture painted in the EA.

Lessons Learned

The tantalum riot, reprehensible though it may have been, underscored a number of points about the planning requirements for such projects. The local public remained grossly uninformed of what was likely to be involved throughout the project. But the facts that did percolate through—for example, that tantalum is used in nuclear warheads—simply fanned the flames. The fact that the public could be gradually swayed into staging a riot should be a lesson to us all.

Given that there were probably other, deeper-rooted political forces at work, it is not possible to predict outcomes of public hostility. While full participation of those affected will help greatly, it is still not possible to guarantee that developers who do their homework, basing it on a thorough environmental assessment of the major hazard potential involved, and who do their best to ensure that the local community is involved in the decision-making process will succeed. But TTIC's experience confirms that, if they do not do these things, they will certainly increase their chances of failure.

Clearly, the failure to effect a partnership with the surrounding community was the main cause of the disaster, a mistake repeated over and over by coastal zone developers.

Source — A. Abrhabhirama, National Environment Board, Thailand (See Reference 1, Part 5).

4.35 THAILAND: TROUBLE WITH MANAGING COASTAL AQUACULTURE

Background

Thailand has 2,614 km of coastline, which are richly endowed with natural resources such as fertile soil, minerals, beautiful scenery, and mangrove and hardwood forests. Coastal seas support coral reefs, seagrass beds, and diverse fish stocks. However, this same coastal zone happens to be where human activities tend to be concentrated, all over the world, including cities, harbors, beach resorts, and aquaculture.

Coastal aquaculture in Thailand has been conducted for more than 30 years. Shrimp is the most important cultured resource because of increase in national consumption and demand for frozen shrimp export. Shrimp farming has rapidly increased from 24,000 tons in 1987 to 162,000 tons in 1991. However, while an increase in shrimp farming improves local and national income, it also can cause serious adverse impacts on the environment, with economic losses that cannot be ignored.

Problems

Aquaculture interacts with the environment and may increasingly be subject to a range of environmental, resource, and market constraints. At present, shrimp cultivation in Thailand is expanding and trespasses on mangrove forest. It is questionable whether the environmental losses are more than the perceived gains in earnings; that is, the altered system may not be economically sustainable.

Aquaculture in Thailand has been developed through isolated and uncoordinated effort and has not been guided by national development plans. It has been driven by market forces toward immediate profitability while the costs of environmental losses have been ignored. But shrimp farming is now facing increased international price and demand fluctuations and marketability problems because of insufficient attention to long-range planning.

Aquaculture development plans have been prepared mainly based on the following questionable bases: (a) preoccupation with what is technically possible rather than with what is economically feasible and socially acceptable; (b) inadequate attention to the design of facilities needed for a sustainable,

Table 4.35.1 Changes of Mangrove Areas Utilized for Various Activities (One Rai = 6.25 ha)

Activities	Area (rai) Before 1980	Area (rai) During 1980–86	Total	Percentage
Aquaculture	162,725	526,395	689,120	64.3
Mining	5,787	28,279	34,066	3.2
Salt pan	66,000	—	66,000	6.2
Other activities	269,188	13,327	282,515	26.3
Agriculture	—	4,386	—	—
Urban expansion	—	3,125	—	—
Road	—	1,467	—	—
Industry and power plant	—	1,135	—	—
Jetty	—	2,684	—	—
Dredging	—	530	—	—
Total	503,700	568,001	1,071,701	100

profitable venture; and (c) insufficient linkage with a national aquaculture policy (UNDP/Norway/FAO, 1987, "Thematic Evaluation of Aquaculture").

Aquaculture operations have brought about significant adverse effects on mangrove ecosystems. In 1989 the total mangrove area in Thailand was only about 1,128,500 rais (180,560 ha), of which more than 120,000 rais were classified as deteriorated mangrove forest. The annual rate of loss (about 81,136 rai/year) was highest during 1979–86. About 64 percent of the mangrove area has been exploited for aquaculture (Table 4.35.1).

Fish and shrimp ponds in mangrove areas are often accompanied by salinization and sulfate acidification of soil and aquifers. Excessive cutting of mangroves for aquaculture (also fuel wood and timber) has caused serious losses in nurturing areas of commercially important finfish and shrimp.

Few operators reserved a strip of mangrove along the shoreline to serve as a buffer zone. Such factors as forest abundance and soil and water quality were seldom taken into their consideration during site selection. Disease control chemicals have affected the ecosystems in the vicinity of the ponds, including antibiotic resistance in microbial communities. Economic disasters have been attributed to self-pollution. Also, increased coastal water pollution (feces and waste food) has triggered disease outbreaks and phytoplankton blooms.

Solutions

In Thailand, there is little legislation with application to aquaculture. Additionally, there is no single national agency responsible for coastal management or which has jurisdiction over both marine areas and coastal lands. For example, the Department of Fisheries manages fisheries resources and aquaculture, while mangrove areas come under the authority of the Royal Forestry Department. Although intersectoral cooperation has been attempted, it is rarely achieved. Thus, legislative and administrative measures aiming at the environmental compatibility of the various aquaculture practices should be considered within the broader legislative context governing coastal aquaculture.

Several regional environmental management plans have been developed in Thailand. The experiences in coastal resources management in Thailand and other developing nations suggests the need for an integrated, interdisciplinary, and multisectoral approach in developing management plans. Integrated coastal zone management (ICZM) plans have been prepared over the past decade in various regions of Thailand, as described in the reports on situation management below:

1. The Eastern Seaboard Project managed the development of an emerging urban and industrial region along the east coast of the Gulf of Thailand. This included preparation of an environmental management plan, including fisheries and aquaculture management.
2. The Songkhla Lake Basin Project was initiated due to the need for guiding continuing economic development in the province by incorporating environmental parameters into the development planning process. This included a coastal zone management plan that recommended the use of a conservation zone.

3. The Upper South Coastal Resources Management Project was an integrated plan showing how environmental management actions can be implemented and coordinated. One of the plan's components was designed to increase awareness of trends in coastal resources utilization and impacts from development.

However, actual experience has shown that the planning noted above has not resulted in development of sufficient action to control mangrove degradation and that the degradation continues to increase. The ecological and socio-economic benefits and costs of aquaculture activities are potentially so significant that policies are necessary to ensure formulation of national aquaculture development plans to minimize developmental, economic, and environmental conflicts. Planned and properly managed aquaculture development should be undertaken within the broader framework of integrated coastal area management planning according to national economic objectives and national goals for sustainable development.

Lessons Learned

Planned and properly managed aquaculture development should be undertaken within the broader framework of ICZM according to national economic objectives and national goals for sustainable development.

Policies related to resources and environment need to be coordinated and a linkage among various policies should be established.

When formulating or modifying coastal aquaculture development plans, it may prove useful to address specifically limitations to development which should be imposed by (a) the present state of aquaculture technology and expertise, (b) prevailing socio-economic conditions, (c) institutional and regulatory weaknesses, (d) access to and availability of resources required, and (e) the current state of the coastal environment and its capacity for further assimilation of pollution loadings and of physical modifications.

Because the cost efficiency of aquaculture has decreased in the past few years, further expansion of shrimp culture should be carefully evaluated. Certainly, existing shrimp pond culture should be intensified to reduce encroachment into mangrove areas while achieving the same or more yield through intensification by improved aquaculture technologies.

Lack of cross-sectoral coordination leads to mangrove depletion, coastal water degradation, and resource use conflicts. Most problems occur because of weakness in enforcement due to lack of monitoring and enforcement officers, budgets, and capabilities.

Mismanagement of the ponds and unsuitable sites results in negative economic impacts on the aquaculture sector, with the operators leaving the ponds to move and encroach upon more mangrove areas for new ponds.

Prevention and control of aquatic pollution are required for sustainable development of aquaculture because this depends critically upon availability of good quality coastal water resources.

Appropriate environmental management of aquaculture in coastal areas can be achieved by combination of improved aquaculture technology, integration of aquaculture to be part of the overall coastal economic system, and appropriate attention to the need, for environmental protection.

Contributed by — P. Kiravanich, P. Chongprasith, and C. Rattikhasukha, Pollution Control Department, Ministry of Science, Technology and Environment, Bangkok 10400 Thailand.

4.36 TRINIDAD AND TOBAGO: A SMALL COUNTRY TESTS AND REJECTS INTEGRATED MANAGEMENT

Background

Trinidad and Tobago, like most small island, developing states, are faced with a variety of problems that make the achievement of sustainable development a challenge. In small island nations the whole island is often considered a coastal zone for land use control purposes. The problems are their size, limited natural resource base, fragile ecosystems, susceptibility to disasters, depletion of non-renewable resources, economic openness, and development pressures caused by increasing population demands, which contribute to environmental degradation.

Trinidad and Tobago lack the resources, institutional capacity, and economic resilience to deal with these problems and are therefore vulnerable. Management solutions for sustainable development must take into account the need to make a special case for development assistance from the international community. Technical and financial assistance are required in the areas of capacity building and institutional strengthening, waste disposal and management, integrated management of natural resources, management of the impacts of natural and anthropogenic disasters, and implementation of integrated development planning.

Trinidad and Tobago are the most southerly of the Caribbean island archipelago. Both islands are located between 10°–12° north latitude and 60°–62° west longitude. They lie on the continental shelf of South America, Trinidad being closer to Venezuela, approximately 12 km (Figure 4.36.1). Together they comprise an area of approximately 5,128 km^2. The total population in 1990 was 1.2 million.

Trinidad and Tobago were transformed from a plantation-type agricultural economy to an oil-based economy. In the 1970s the government of the Republic of Trinidad and Tobago attempted to diversify the economy by diverting revenues earned from oil into the establishment of heavy industries, which would use the recently discovered extensive reserves of natural gas for fuel and feedstock and earn foreign exchange when oil prices fell.

The development thrust in Trinidad (the larger island) has been concentrated in the western half of the island. This area is therefore experiencing tremendous development pressure. Here are located the two cities, Port of Spain and San Fernando, which are major ports and heavy industrial areas including the oil refinery and petrochemical industries. Over 81 percent of the population of the whole nation lives in this area (1).

Tobago (the smaller island) remained mostly in agriculture until oil prices fell drastically in 1986. The government then decided to make Tobago the center of its new development thrust and alternative foreign exchange earner, that is, tourism development.

In Tobago, the major infrastructural developments and proposed projects for the tourism sector are all to be concentrated in the coastal areas of Southwest Tobago. These include an international airport, a cruise ship harbor, and resorts with marinas and golf courses.

The high cost to the state of providing supporting infrastructure to large-scale development projects has led to clustering of such projects. While the clustering of large-scale development projects in Southwest Tobago—and in the west coastal area of Trinidad—benefits the state, it leads to an intensification of environmental impacts and the creation of environmentally stressed regions.

Development activities have already resulted in signs of environmental pollution and degradation, including:

- Pollution of rivers and seas by wastes from industries, oil exploration and production, agriculture, sewage treatment plants, and quarrying activities
- Degradation of watershed areas by subsistence farming, unplanned settlements, and quarrying
- Inadequate disposal of solid wastes, including hazardous wastes
- Air pollution from industries and motor vehicle exhausts
- Land use conflicts, particularly housing developments on prime agricultural lands

The islands are vulnerable to natural disasters such as hurricanes. The concentration and degree of industrialization in Trinidad poses the risk of industrial disasters. The country's vulnerability to disasters is exacerbated by deficiencies in the existing institutional framework for disaster preparedness.

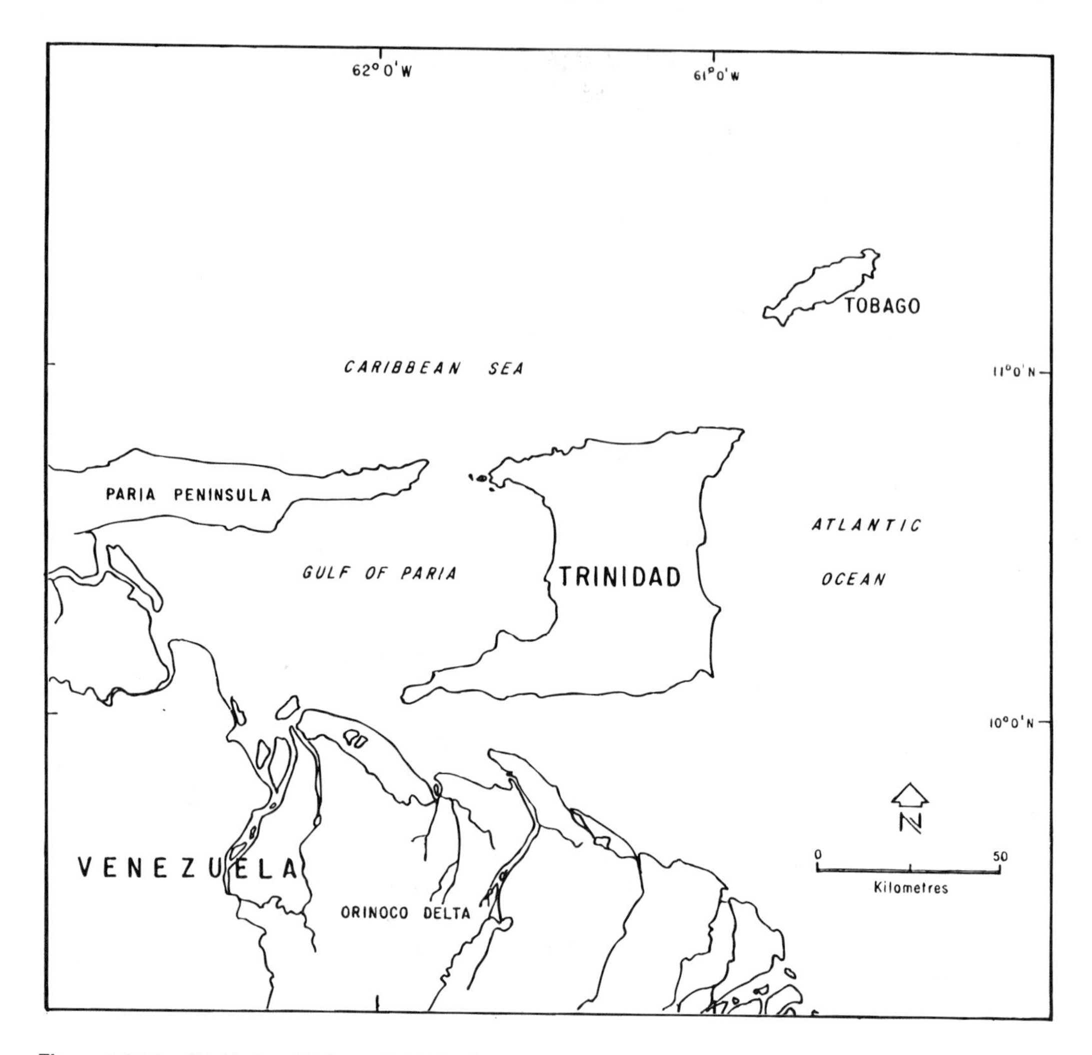

Figure 4.36.1. Trinidad and Tobago (5,128 km^2) is the most southerly country of the Caribbean island archipelago. Lying on the continental shelf of South America, Trinidad is as close as 12 km to Venezuela; population is near 1.2 million.

The adjustment of the economy since the mid 1980s to lower income levels, resulting from structural changes in the petroleum sector, has led to erosion in the gains from development. The per capital gross domestic product (GDP) at market prices was estimated to be US$6,851 in 1983 but declined to US$4,360 in 1992. Real growth in GDP fell from 5.3 percent in 1982 to 0.2 percent in 1992 (7).

There has been a deterioration in living standards of the population, especially among the more vulnerable social groups. This has been exacerbated by the substantial depreciation of the domestic currency, removal of government subsidies and expenditure on social benefits, and increases in utility rates. The level of unemployment rose to 20.2 percent in 1992.

Solutions

In small island nations the whole island is sometimes considered a coastal zone, at least for planning purposes. Solutions to coastal area or environmental problems must usually include the whole island. National planning in Trinidad and Tobago was initially that of planning for economic development. In 1969, the concept of regional planning was introduced into the national planning and development process.

The National Physical Development Plan was prepared by the Town and Country Planning Division (TCPD) of the Ministry of Finance and Planning and approved by Parliament in 1984. The plan sets out to formulate a coherent and comprehensive land use policy that will provide the criteria for the execution and enforcement of development control (6). It seeks to ensure consistency and co-ordination by providing guidance on matters of land use to government agencies as well as to private developers and investors. The plan provides a framework for the preparation of regional and local plans and for the integration of spatial planning with socio-economic sectoral policy making. The time perspective of the plan is to the year 2000. The plan articulated the need for an environmental management policy as a strategy for environmental planning and conservation.

The idea that there was a need to establish a separate agency to deal with the coastal zone management in Trinidad and Tobago was introduced as early as 1974. Ten years later TCPD provided opportunity for proper co-ordination and management of development within the land sector only of the coastal zone. But there were barriers to success of planning and environmental management functions in the coastal zone due to (a) the lack of knowledge and shortage of expertise, (b) the institutional problem, and (c) the jurisdiction problem.

The proposed solution to Problems a and b was the establishment of a special multi-disciplinary coastal zone agency. But such an agency would not, of itself, solve existing jurisdictional problems. It was concluded that a full-scale coastal area development and management study using the West Coastal Area of Trinidad as the pilot area should be undertaken.

The Institute of Marine Affairs (IMA) was established by Act 15 of 1976. It was decided that one of the research programs of the IMA should be a Coastal Zone Management Programme, which would carry out the multi-disciplinary coastal area planning and management study. The intent of this study was to suggest a pattern of growth that would permit and encourage development in a balanced manner. The West Coastal Area of Trinidad was to be used as the priority coastal area for the reasons stated in the preceding section. *The techniques developed in the study would be utilized for a nationwide Coastal Area Plan.*

At the end of 1984, it was decided that the IMA should begin to integrate its efforts with those of the TCPD in the interest of developing a unified approach to coastal zone management in Trinidad and Tobago. Based on discussions between the two organizations, it was felt that IMA could contribute to the collaborative effort by making input into three main areas of TCPD's work, that is, development plans, research projects, and review of applications from developers. The planning aspects would be covered by TCPD while IMA would deal with environmental aspects (3).

The IMA continued to carry out environmental assessments to establish permissible land and water uses for the coastal area development study as envisaged (2). The idea of a special ICZM agency was abandoned when IMA recognized the National Physical Development Plan of the TCPD as being the overall plan for the coastal zone for Trinidad and Tobago, which is in agreement with the premise that, in small island nations, the whole island is the coastal zone for the purposes of land use control.

Subsequently, a National Planning Commission was established and produced a planning framework for a medium-term development plan for Trinidad and Tobago for 1989–95 (4). The Ministry of Planning has had inputs for this exercise from other organizations, cognizance having been taken of the existing sectoral policy statements and the sectoral plans. The areas covered by these plans were Agriculture, Energy, Education, Housing, Health, Industry, Investment, Settlements, and Tourism. The document was concerned with planning for economic development. The strategy was directed toward fostering a climate conducive to investment activity in general and the growth of exports in particular. Environmental issues did not inform the plan. The formulation of sectoral policies, all of which impact on the environment and few of which have taken this into account, emphasized the need for a national environmental management policy.

In 1989, a Ministry of the Environment was created to address the various environmental concerns and solve the problems mentioned in the preceding paragraph. The creation of this Ministry was considered a landmark development in Trinidad and Tobago. However, it failed to gather enough support and in 1992 its functions were incorporated into the Ministry of Planning and Development.

Lessons Learned

The problems and vulnerabilities of small island developing states, such as Trinidad and Tobago, require specific kinds of action to be taken to bring about sustainable development. This should be done through integrated development planning, the rationalization of institutions to facilitate environmental management, the strengthening of environmental legislation and the development of public awareness and education programs.

The institutional framework for environmental management requires the establishment of an environmental management authority. Institutional strengthening will be required for such intersectoral issues as coastal zone management, climate change, and disaster preparedness and mitigation.

Public education programs are required to increase public awareness, including that of the politicians, of how certain policies, consumption patterns, and life-styles could affect the quality of life and what is required on their part to achieve sustainable development. To implement these actions at a time of shrinking financial resources, small countries will have to seek external financial and technical assistance. This requires convincing the international community that its ecological fragility and economic vulnerability merit special assistance.

Contributed by — Hazel McShine, Institute of Marine Affairs, Chaguaramas, Trinidad and Tobago.

REFERENCES

1. Central Statistical Office. 1992. 1990 Population and Housing Census Listing of Area Register. Ministry of Finance, Republic of Trinidad and Tobago. 97 pp.
2. Manwarring, G. and H. McShine. 1991. Economic aspects of the Point Lisas case study. In: *Caribbean Ecology and Economics,* N. P. Girvan and D. Simmons (Eds.). Caribbean Conservation Association, Barbados. Pp 233–258.
3. McShine-Mutunhu, H. 1987. The status of coastal zone management in Trinidad and Tobago. In: *Coastal Zone Management of the Caribbean Region: A Status Report. Environmental Planning Programme. Coastal Zone Management of Tropical Islands.* Commonwealth Science Council Technical Publication No. 227. Commonwealth Secretariat, London. Pp 9–21.
4. National Planning Commission. 1988. Restructuring for Economic Independence—Draft Medium Term Macro Planning Framework 1989–1995. Government of the Republic of Trinidad and Tobago. 283 pp.
5. Town and Country Planning Division. 1974. Planning and Environmental Management in the Gulf Coast of Trinidad. Prepared for Meeting of Group of Experts on Coastal Area Development. United Nations, New York. Ministry of Planning and Development, Trinidad and Tobago. 40 pp.
6. Town and Country Planning Division. 1982. Planning for Development—National Physical Development Plan Trinidad and Tobago. *Development Planning Series,* No. TT3. Ministry of Finance and Planning, Trinidad and Tobago. Vol. 1: Survey and Analysis. Pp 1–22. Vol. 2: Strategies and Proposals. Pp 123–272.
7. Trinidad and Tobago Delegation. 1993. Position Paper for Caribbean/Atlantic Regional Technical Meeting on the Sustainable Development of Small Island Developing States, Trinidad and Tobago, July 12–16, 1993. 48 pp.

4.37 TURKS AND CAICOS: RESTORING WETLANDS WITH MINING EFFLUENT

Background

A major problem facing coastal developments is determining acceptable methods of waste disposal. Severe environmental impact has resulted from discharge of effluents into the sea, especially around small islands, where this has been a common practice.

A proposal to extract aragonite sand and stockpile it on a Caribbean island would have resulted in large quantities of runoff wastewater with high sediment load contaminating productive coastal ecosystems. Of particular concern were fringing coral reefs and the clear inshore waters on which much of the island's dive-tourism industry depended. Ecological studies identified an alternative disposal pathway, which avoided reef damage while at the same time having beneficial impacts on wetlands adjacent to the project.

West Caicos is uninhabited, although historically land areas were cultivated in sisal and cotton and wetlands were modified extensively for salt production. Field survey in 1989 (1) found upland areas covered by low secondary scrub interspersed with shallow, hypersaline basins with a sparse, ephemeral aquatic biota and stunted mangrove trees on the margins. These salinas are isolated from the sea, whereas the western water body known as Lake Catherine (Figure 4.37.1) retains a subterranean marine connection and supports a varied flora and fauna.

Problem

Valuable aragonite sand reserves occur on the Caicos Bank, with readily accessible, high quality deposits near West Caicos (2). Proposed mining operations included (a) hydraulic dredging from a barge anchored over the seabed, with (b) slurry pumped via pipeline to the northern end of the island, where (c) sand would be stockpiled for (d) trans-shipment via a dock at the northwest corner, as shown in Figure 4.37.1. As the wet sand dried, wastewater would drain from the stockpile at a rate of several million gallons per day, carrying fine sediment in suspension.

A well-developed fringing reef along the western coast is one of the most important dive sites in the Caicos Islands (6). Coral reefs are known to be particularly sensitive to any increase in suspended sediment, which can lead to smothering and reduction of light needed for photosynthesis (4, 6). Damage to the reef and reduction in coastal water quality would have serious ecological and economic implications. The government approval process for an aragonite mining operation at West Caicos depended on weighing the benefits of mining against potential costs to tourism and the environment and on identifying an effluent discharge strategy that would avoid negative environmental impact. Each of several identified discharge options has a potential for significant adverse environmental impact:

- Effluent release to the west coast puts coral reefs at risk.
- Release to the east or north coasts might contaminate recreational beaches and/or nearshore seagrass beds important to the conch and lobster fishery.
- Discharge via Lake Catherine would be unacceptable, as this is a national park, and discharge might affect the diverse aquatic biota and rare roseate flamingo population.
- Routing the effluent through Eastside Salina would produce major changes in hydrological conditions and ecology.

Solutions

It was recognized that settlement ponds would be required at the stockpile area to reduce sediment load in the effluent seawater and for recycling freshwater used in washing the sand. Potential reef damage could be mitigated further by discharging effluent to the sea via a salina, but this might result in damage being displaced from the reef and seagrass beds to another ecosystem.

Discharge via Eastside Salina (Figure 4.37.2) overcomes two aspects of the problem. Adverse ecological impact is unlikely as flora and fauna are sparse, and passage through the salina would ensure maximum settlement of remaining suspended sediment before wastewater re-entered the sea.

Of importance to a management solution was the realization that land and wetland ecosystems on West Caicos were not "pristine" or as sensitive to perturbation as the coral reefs. In this instance, both the biodiversity and productivity of Eastside Salina were limited by extreme hypersalinity resulting from historical use of the site for salt production, compounded by seasonal drought. While flooding of the basin with seawater could do little damage, effluent use could enhance biodiversity and productivity by creating a wetland habitat similar to neighboring Lake Catherine.

Thus, while preventing potential damage to important coastal marine ecosystems, the project wastewater could be used to improve the quality of a degraded wetland ecosystem on West Caicos (2). More permanent flooding would enhance dwarf mangrove growth on the margins, increase aquatic invertebrate diversity, improve habitat for resident and migratory waterbirds, give access to marine fish and their juveniles, and produce a water feature that could increase landscape value for future resort development. Permanently raised levels of moving water would mitigate a serious mosquito problem, also.

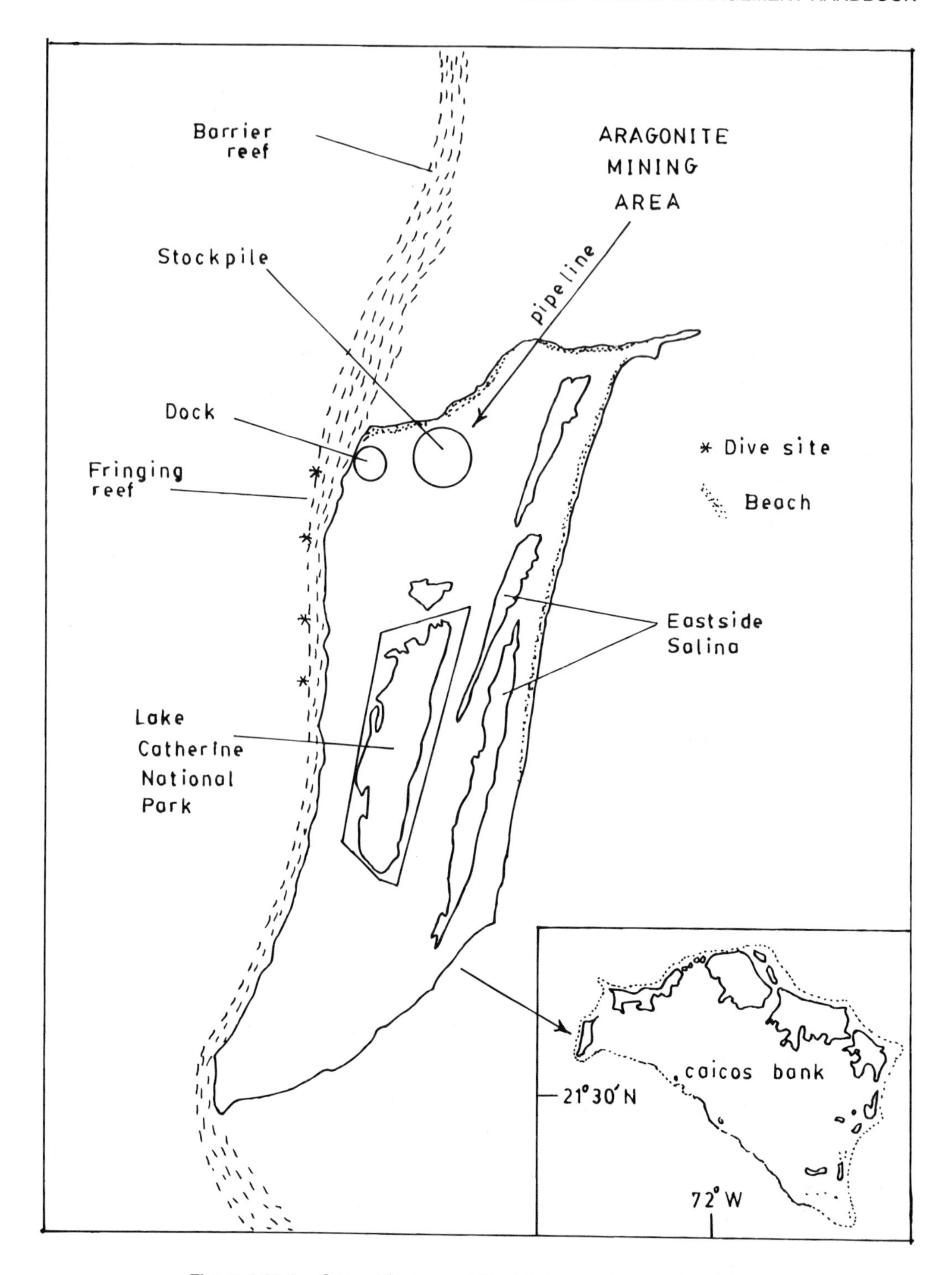

Figure 4.37.1. General features of West Caicos and aragonite mining.

Figure 4.37.2. Eastside Salina looking north.

Lessons Learned

1. The need to dispose of effluents from coastal development projects to sensitive nearshore environments can be reduced if the waste is considered as a resource of potential benefit through integration with other coastal activities.
2. A cost-effective solution can be found in re-use of wastewater to enhance degraded ecosystems, particularly coastal wetlands in low rainfall areas. This has been done elsewhere effectively for irrigation and aquaculture but needs to be considered for wildlife conservation and recreational or residential landscape enhancement in coastal areas.
3. In environmental impact analysis it must be remembered that few of the coastal areas proposed for development are pristine, particularly those of small tropical islands. Coastal planners should appreciate that impacts on already disturbed ecosystems will be different from and probably less severe than those on natural systems. Furthermore, if an area is degraded already, any development project has the potential not only to perturb but also to enhance ecosystems, both in the project area and adjacent to it.

REFERENCES

1. Bacon, P. R. 1990. *Assessment of the Environmental Impact of Resort and Residential Development on West Caicos.* Report to West Caicos Construction and Investment Co. 115 pp.
2. Bacon, P. R. 1992. *Proposal for Mitigation of Adverse Environmental Impacts of Aragonite Mining on West Caicos.* Report to the Government of the Turks & Caicos Islands and Aragonite Mining Co. Ltd. 55 pp.
3. Mitchell, A. J. B., A. R. Dawson, and H. L. Wakeling. 1985. *Environmental Appraisal of a Proposed Aragonite Mining Operation and Other Proposed Activities at West Caicos Island.* Land Resources

Development Division, Overseas Development Administration, London. 73 pp.
4. Rogers, C. S. 1983. Sublethal and lethal effects of sediments applied to common Caribbean reef corals in the field. *Marine Pollution Bulletin,* Vol. 14, No. 10. Pp. 378–382.
5. Rogers, C. S. 1990. Responses of coral reefs and reef organisms to sedimentation. *Marine Ecology Progress Series,* Vol. 62. Pp. 185–202.
6. Wells, S. 1988. Turks and Caicos. In: *A Directory of Coral Reefs of International Importance,* Vol. 1. *Atlantic and Eastern Caribbean.* IUCN Conservation Monitoring Centre, Cambridge. Pp. 534–549.

Contributed by — Peter R. Bacon, University of the West Indies, St. Augustine, Trinidad.

4.38 UNITED STATES: A NATIONAL COASTAL MANAGEMENT PROGRAM

Background

Operating for over two decades, the ICZM program of the United States has used a combination of grants and organizational incentives to enlist 29 states and territories covering 94 percent of the nation's shoreline in an effective intergovernmental coastal management network. At a relatively low cost to the federal treasury, the program has resulted in major improvements in land use regulation, natural resource protection, public access, urban waterfront re-development, hazard mitigation, resource development, and ports and marinas. It has created 18 National Estuarine Research Reserve sites protecting over 262,000 acres of biologically productive estuaries as teaching and research laboratories.

Core program values, stated in the Coastal Zone Management Act (CZMA) of 1972, are to manage growth so as to *balance* conservation and development under a *voluntary* national program with guidelines that allow individual states to design their coastal plans to meet their particular needs and require federal activities to be *consistent* with federally approved state plans. Unique aspects are (a) the use of incentives to create a coastal policy coalition spanning national, state, and local levels of government; (b) the need to deal with the varied geography and management problems of the U.S. coast; and (c) the dynamics of a participatory democratic approach, in which attitudes of a new national administration cannot dismantle the established ICZM program.

Problem

One challenge facing coastal management is how to define the scope and nature of the problem. The multiple-objective nature of the U.S. coastal program stems from its origin in a national debate during the 1960s and 1970s over the effectiveness of environmental protection and land use planning. Involved interests included advocates of outdoor recreation, marine resource development, estuarine protection, and land use policy, all of whom believed that a "coastal crisis" was arising due to local governments' inability to cope with problems of competing demands for coastal resources.

Marine resource users saw the problem as the lack of a national program to tap the wealth of the oceans through public-private development initiatives. Estuarine protection advocates saw the problem as the lack of a national program to prevent the ongoing destruction of ecologically fragile and valuable estuaries. And land use policy advocates saw the problem as the lack of a national policy for coordinating land use planning, both on the coast and in the rest of the nation (5). Over time, the problem definition has enlarged to deal with a national energy crisis, non-point source pollution, sea level rise, wetlands loss, marine debris, erosion, and storm hazards.

Another challenge was how to design a program that can cope with the tremendous physical, economic, political, and cultural diversity of the nation's coasts. The 95,000-mile coastline includes rocky headlands in Maine, barrier islands in the Carolinas, wetlands in Louisiana, harbors in Lake Michigan, glaciers in Alaska, and coral reefs in American Samoa. Some areas are heavily urbanized, others are resorts, and others are dependent on rural agriculture, forestry, and fisheries. Some areas face high development pressures, while others suffer economic decline and high unemployment. Governance responsibilities are divided among 35 states, over 400 coastal counties, and thousands of municipalities and special purpose authorities. This diversity makes the design of one simple, top-down program with uniform national coastal regulations nearly impossible. (1)

Another challenge was how to protect the coastal environment while accommodating ongoing development (i.e., to effect sustainable use that allows growth without "killing the golden goose"). The coastal area encompasses a high percentage of the nation's population and economic activity. Thirty-two percent of the 1985 U.S. gross national product—almost $1.3 trillion—originated in 413 coastal counties, providing 28.3 million jobs and $479.9 billion in payroll (3).

Solution

In forming its ICZM approach, the U.S. government considered two program designs: (a) a centralized national program with mandatory regulations and uniform standards oriented toward environmental protection (similar to the Clean Air and Clean Water programs, in which states must comply or face penalties) or (b) a decentralized, incentive-based program where states participate voluntarily under broad national guidelines to establish and implement ICZM programs to meet their diverse needs for conservation and development. The first of these is the "top-down" alternative; the second is the "bottom-up" alternative.

The solution chosen in 1972 was a mixed alternative, directed toward comprehensive coastal planning and management. It incorporated both bottom-up state plans and top-down oversight by the national coastal agency, Congress, and national interest groups. It required balance between development and conservation goals within a voluntary, decentralized program in which individual states and territories are strongly encouraged to participate through federal financial grants for planning and implementation. Each participating state must submit its plan to the national coastal agency for approval and in return is promised that future federal actions in coastal areas will be "consistent" with approved state plans. The consistency process covers the actions of all federal agencies operating within the coastal zone.

Over time, 29 of the eligible 35 states and territories applied for and received grants. Each participating state operates a coastal planning office and implements a management plan under federal oversight.

A successful element of the national ICZM program is the use of program funds to deal with a broad range of coastal management needs, tailored to individual state conditions. The bulk of grant funds have been spent by states on two types of activities: improving government decision making and protecting natural resources. Thirty-nine percent of the federal implementation grants between 1982 and 1987 were devoted to work on improving decision making through facilitated permit reviews, technical assistance to local governments, plan implementation, building management data systems, and public participation in coastal management. Twenty-eight percent were devoted to natural resource protection through issuing permits for development in sensitive areas, such as wetlands, beaches, and estuarine waters; controlling coastal pollution; managing special areas; and acquiring land for parks, natural areas, and open space preservation. Remaining grant funds have been spent on improving coastal access (11 percent), natural resource development, hazards mitigation and urban waterfront development (each about 6–7 percent), and port projects (2 percent). (4)

The technical basis of the program focuses on comprehensive land use planning and environmental controls as the means to protect coastal resources. It directs federal agency programs—e.g., Department of Interior leasing of continental shelf lands for oil and gas extraction—to be consistent with state-approved ICZM plans. It encourages states and localities to use a variety of innovations.

Increasingly, this technical basis is being linked to cumulative impact assessment in which watersheds are used as the natural planning units. Environmental impacts are considered in combination with social and economic impacts to achieve sustainable development goals. This requires not only assessment of impacts on water quality and ecological system health, but also impacts on residents' livelihoods and equitable participation by the public in planning and decision making.

Lessons Learned

Major lessons are that program survival can be ensured through the creation of policy and that the program needs to be adaptable to changing circumstances over time through social learning. (2)

That the US ICZM program was durable and effective is due to the intergovernmental collaboration that sprang from federal funding and oversight, voluntary state participation, and the promise of federal consistency with approved state plans. States enter the program because they want to, not because they are required to, with federal funding as a major motivation. They design plans to deal with their individual coastal needs as they see them. This recognition of the states as responsible co-producers of coastal policy and management programs builds their commitment and loyalty, which help the national program survive when it is attacked, as it was by the Reagan administration. Rather than strong federal control, the spirit of the U.S. coastal program is based on collaboration and capacity building.

To what extent could the U.S. program serve as model for coastal management in another country? Clearly, each country needs to build its coastal program on its own needs and circumstances. In a country without a similar structure of strong national, state, and local governments or in a smaller country with a less diverse coastal area or with different coastal problems, the U.S. model would not fit. However, the principles of treating coastal issues within a comprehensive management framework aimed at sustainable development, of building a supportive coastal advocacy coalition to co-produce coastal management, and of focusing on land use planning to implement coastal policy objectives would seem to be universal (e.g., see "USA, Alaska" case history below).

Contributed by — David R. Godschalk, Department of City and Regional Planning, University of North Carolina, Chapel Hill, North Carolina.

REFERENCES

1. Center for Urban and Regional Studies. 1991. Evaluation of the national coastal zone management program. Newport, OR: National Coastal Resources Research and Development Institute.
2. Godschalk, D. R. 1992. Implementing coastal zone management; 1972–1990. *Coastal Management,* Vol. 20. Pp. 93–116.
3. Luger, M. I. 1993. The economic value of the coastal zone. *Journal of Environmental Systems,* Vol. 21, No. 4—1991–92. Pp. 279–301.
4. Owens, D. W. 1992. National goals, state flexibility, and accountability in coastal zone management. *Coastal Management,* Vol. 22. Pp. 143–165.
5. Zile, Z. L. 1974. A legislative-political history of the Coastal Zone Management Act of 1972. *Coastal Zone Management Journal,* Vol. 1, No. 3. Pp. 235–274.

4.39 UNITED STATES, ALASKA: PARTICIPATORY COASTAL ZONE MANAGEMENT

Background

In the United States the first comprehensive coastal zone management (CZM) program was initiated in October 1972, under the Coastal Zone Management Act. Implementation of the act was delegated to the Department of Commerce. The purpose of the CZM Act was to provide incentive to each of the U.S. coastal states to conserve its coastal resources (see previous case history).

While the United States ICZM program is centrally funded, it is not centrally operated—it is operated by the individual states, each with its own unique program. At the national level, ICZM requires the participation of numerous federal agencies and active coordination among them. One reason for the focus at the state level is that strong federal participation in coastal environmental matters is provided for by many other national laws and programs. The political motivation is federal funding for states that agree to certain federal requirements. The federal office in Washington, D.C., was mandated to ensure federal consistency with the state programs and vice versa, to provide technical assistance, to set up and operate protected baseline "Estuarine Reserves" for research purposes, and to provide policy liaison with other federal agencies.

The State of Alaska's CZM program can be used as a model of participatory planning and management for small countries with strong central governments, for which Alaska is a reasonable simulation. Alaska's oceans and coastal areas are vital places, rich in natural resources that provide food and jobs for local residents, create economic opportunities for private industry, generate revenue for government, and supply energy and mineral resources that benefit the nation. The coasts are of cultural, recreational, and spiritual value to all Alaskans.

Alaska's coastal waters, wetlands, and watersheds also nurture productive fishing stocks and marine mammals, waterfowl, caribou, and other animals. Daily decisions made by the state and federal governments regarding coastal development projects determine the fate of Alaska's 33,000-mile coastline (Figure 4.39.1).

Problem

Three major questions regarding coastal development in Alaska are (a) Where should offshore oil and gas exploration and development occur, and what should be done to protect marine mammals and valuable fisheries from oil spills, noise disturbance, and other impacts? (b) Which wetland and waterfront area in Alaska's villages and cities should be developed for community growth and which protected from development? (c) How can conflicts be resolved between coastal residents who have traditionally used fish and wildlife for subsistence and non-residents whose use of coastal areas for recreational fishing and hunting is increasing?

Alaska viewed its main overall problem as How can we create opportunities for coastal residents, communities, and regions to participate in decisions about the use, development, and protection of coastal resources?

Solutions

In developing an ICZM program, Alaska has been successful in involving coastal residents, communities, and regions in decisions affecting coastal resources. Locally based groups write management plans for the coastal areas in which they live, setting their own priorities and working with the state and federal governments to implement the plans. Local residents and government agencies have learned to

Figure 4.39.1. Alaska is the northernmost state of the United States, has the longest coastline, and is rich in coastal resources.

balance their interests with those of private industry, special interest groups, and other members of the public. This is accomplished through discussion and negotiation during the planning process.

Thirty-four local and coastal areas known as coastal resource districts have responsibility to plan and implement local plans, which must be approved by the Coastal Policy Council and the federal government. In four large, rural coastal regions, not represented by local government, coastal resource service areas have been created to permit involvement of local residents. Local coastal management plans are created and implemented primarily through the consistency review process of development projects that require permits.

Coastal resource districts write local management plans that:

- Inventory resources in the region
- Consider issues of concern to local residents
- Define an appropriate coastal zone boundary
- Adopt policies to guide coastal development decisions
- Describe how the plan will be implemented
- Write more specific management plans for areas with unique coastal values or where there are particular conflicts over the use of the area

If a coastal district disagrees with the results of a state project review, it can appeal the decision to higher levels in state government, the governor, and the Coastal Policy Council.

Coastal districts closely involve the public during plan development and implementation. Coastal districts talk with local communities, native corporations, private industry, and other public interest groups. Districts use workshops, formal public hearings, questionares, newspaper and radio stories, and brochures or newsletters to educate the public and ask for public input.

But local views cannot solely control decisions on where and how coastal development will occur. There are often legitimate conflicts between the goals of local residents and those of state and federal agencies, private industry, native corporations, or environmental organizations. Achieving the correct "balance of power" between local interests and those of the state and federal governments and ensuring that private industry and other interest groups are also treated fairly is a challenge during both the development and the implementation of each local coastal management plan.

In Alaska each coastal district plan accomplishes something different, depending upon the needs and interests of people in the area. Plans for rural areas have often emphasized protection of fish and wildlife and subsistence activities. Coastal plans for Alaska's urban areas have often focused on streamlining government approvals for waterfront and wetland development projects to encourage community growth and economic opportunity. Coastal districts have also completed special management plans and projects related to specific local concerns, including floodplain management and drainage control, port and harbor development, protection of watersheds for city drinking water supplies, enhancement of coastal public access, and prevention of marine debris.

Lessons Learned

The U.S. program appears to be a legislative success and to have accomplished coastal conservation to the extent politically possible. While the United States CZM program serves as one model of a state-operated, permit-based national ICZM program, with its focus on control of the use of private land, a model that might be more relevant for many developing countries with authoritarian government—at least in the initial stage—is one with more direct control and influence by the central government. Other lessons include:

1. Local involvement in coastal resource decisions can be successfully achieved by preparing coastal management plans at a local rather than state level. Local management plans can be used to influence where and how coastal development projects occur and to ensure that projects are compatible with the views of local residents regarding how the coastal areas should be used and protected.
2. Involving local people in coastal management planning takes time and money. The amount of time and money that must be invested to produce local plans can be reduced by (a)

providing suitable examples of successful plans for local planners to use, (b) providing training for local planning staff, and (c) relying on private consultants and, to a lesser degree, state government agencies, that are experienced in coastal management planning.
3. Coastal management laws should establish structure that specifically provides for local involvement, through (a) local participation on the coastal policy-making body, (b) local responsibility for plan preparation, (c) a strong role for local people in plan implementation, and (d) strong public participation requirements.
4. Local people must have access to adequate funding during plan development and during implementation of the plan to ensure that they can fully affect coastal management decisions from the local point of view.
5. There may be legitimate conflicts between the goals of local residents, state and federal government agencies, and non-government interests such as private industry and native corporations. Local coastal districts have the responsibility to balance their goals with the views and needs of these other parties. Likewise, the degree to which local concerns are met depends on the willingness of the state and federal governments to work in good faith with local people to help them achieve their goals through negotiation with all affected parties.

Source — Division of Governmental Coordination, Office of the Governor, Juneau, Alaska.

4.40 UNITED STATES, FLORIDA: DISTANT INFLUENCE ON CORAL REEFS

Background

Lying parallel to the Florida coast is a barrier reef structure more than 200 miles long. Parts of the reef system were designated as marine protected areas some time ago: John Pennekamp Coral Reef Park combined with Key Largo National Marine Sanctuary (103 square nautical miles), designated in 1961 and 1975, and the Looe Key National Marine Sanctuary (5.32 square nautical miles), in 1981. The management of these protected areas was focused primarily on various sources of direct, *in situ* impact to the coral reef resources and depletion of reef populations. The idea was to prevent damage from boat anchors, boat groundings, harvesting, and diver impact—which proved to have some limited success in protecting the coral reefs. Other management strategies, such as gear restrictions and prohibitions on spearfishing and the harvest of invertebrates, have shown some success. However, something was missing from the protection scheme.

In 1990, the U.S. Congress expanded the coral reef program greatly by designating an area of 2,800 square nautical miles for conservation. This area—the Florida Keys National Marine Sanctuary—includes the previously protected areas and encompasses 220 miles of coral reef tract that parallels the island chain of the Florida Keys. Also contained within its boundary are vast seagrass communities, mangrove habitats, hardbottom communities, thousands of patch reefs, over two dozen shallow bank reefs, and a reef habitat (intermediate to deep) that runs almost continuously for the length of the reef tract.

Problem

The problem is that, regardless of all management arrangements, the coral reefs and coral reef resources of the Florida Keys are declining. Clearly, the management program for Key Largo and Looe Key was not sufficient to prevent the decline of the coral reef resources from impacts originating outside their boundaries. No matter how many fixed mooring buoys were installed in the sanctuaries or how many times the education efforts helped keep boats from running aground, it is our opinion that the health of the coral continued to decline.

Managing Key Largo and Looe Key has been like managing "islands" in the ecosystem. Regardless of the intensity of management effort, scientists have documented a decrease in the amount of living coral cover, as well as recruitment of new corals. Following studies that characterized the flow of water in and around the Looe Key, it became clear that the potential for water quality impacts originating in the "Zone of Influence" outside the boundaries of the Sanctuaries was very high.

Even after its great expansion, the Florida Keys marine protected area is dwarfed by the huge Central and South Florida mega-ecosystem that drains into it. This system begins at the headwaters of the Kissimmee River, includes Lake Okeechobee, the Everglades Agricultural Area, Everglades National Park, and Florida Bay. There is extensive agricultural and urban activity in this area, which is the source of directed pollution discharge (point source) and runoff pollution (non-point source), all of which runs south into the waters surrounding the Florida Keys (Figure 4.40.1).

Solutions

The establishment of the much larger Florida Keys protected area made it possible for sanctuary management (National Oceanic and Atmospheric Administration—NOAA) to look at holistic management of the coral reef community. In addition, the act that established the sanctuary also called for the Federal Environmental Protection Agency and the State of Florida, in coordination with NOAA, to develop a comprehensive water quality protection program for the sanctuary. The focus of this plan is mainly on land-based sources of water quality impacts, including pollution that originates outside the boundary of the enlarged sanctuary. Currently, a Federal Task Force has been established to develop a plan to restore the South Florida mega-ecosystem as it is described above. Although complete restoration may not be feasible, major hydrological linkages and ecological linkages can be restored to the point that a naturally functioning ecosystem can be maintained.

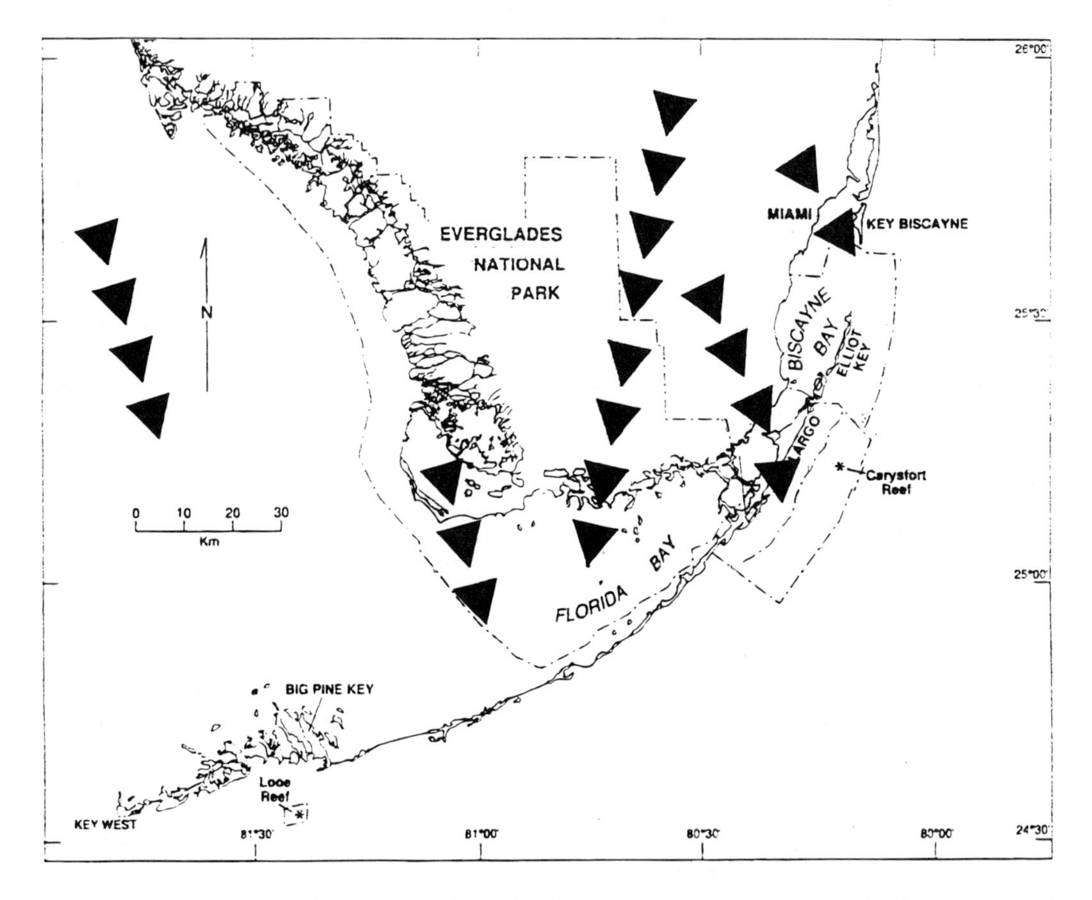

Figure 4.40.1. The Florida Keys Sanctuary from Miami to Key West is affected by a variety of agricultural and urban pollutants entering from the north (as shown by the arrowheads) and by major changes in the amount and periodicity of flow through the Everglades.

Lessons Learned

Today, it is obvious that successful management of the coral reef resources of the Florida Keys depends on the ability of sanctuary management to address impacts that come from The Zone of Influence outside the physical boundaries of the sanctuary (which go only up to mean high water). To be successful in protecting the resources of the Keys, sanctuary managers must be capable of influencing federal, state, and local agency activities that affect the quality of the water that flows through the Keys and out to the reefs.

To be fully successful in protecting the resources of the sanctuary, it is important to be able to address the impacts affecting the entire South Florida ecosystem through a type of "Situation Management" approach. Whereas eight years ago sanctuary managers felt they were managing the "coral reef ecosystem" of the Florida Keys, today it is recognized that the coral reef community of the Florida Keys is only a small portion of an enormous ecosystem.

Contributed by — Billy Causey, Sanctuary Superintendent, National Oceanic and Atmospheric Administration, Marathon, Florida.

4.41 UNITED STATES, FLORIDA: SUCCESS WITH OCEAN OUTFALLS

Background

The southeast coast of Florida from Miami north to Palm Beach, a distance of about 115 km, is one of the most densely populated areas of the state. Tourism is the largest industry; the subtropical climate and the ocean are the basis for attracting tourists. Thus, economic considerations as well as environmental concerns provide a strong incentive for local, state, and federal agencies to maintain the marine resources. As with other populous coastal areas, the adjacent ocean is the repository for local sewage. Southeast Florida illustrates the fate of sewage pollution in a setting where the marine ecosystem is well flushed and consequently the impact of sewage is low.

Problem

Ten or more ocean outfalls discharge enormous quantities of sewage along southern Florida's Atlantic coast (Figure 4.41.1). In the recent past, most of the sewage was untreated. In most cases, the effects of the sewage on the nearby marine ecosystems was minimal, mainly because the effluents were rapidly diluted and dispersed by the Florida Current. Depending on prevailing winds and local current, however, effluent plumes would occasionally be transported landward and pollute the bathing beaches. Originally, 1.77×10^8 gallons of sewage effluent per day were discharged 1400 m offshore at a depth of 5 m, between the offshore reefs and a popular Miami bathing beach.

Solutions

Concern for the health of bathers and consumers of seafood and aesthetic considerations resulted in extension of the Miami ocean outfall from depth of 5 m to a depth of 30 m at 5.6 km (3 1/2 miles) seaward of the beach. Extension of the Miami outfall resulted in true ocean disposal of the effluents and a marked reduction of coastal pollution.

Appreciable quantities of indicator bacteria (fecal coliforms) and enteroviruses are found in the surface waters within 200 m of outfalls. Pathogens are not found in the water column at distances greater than 2000 m from the outfall.

Lessons Learned

We learned that coral reefs and other coastal resources are best protected from sewage pollution by ensuring that effluents are discharged outside of coastal waters. This is most economically accom-

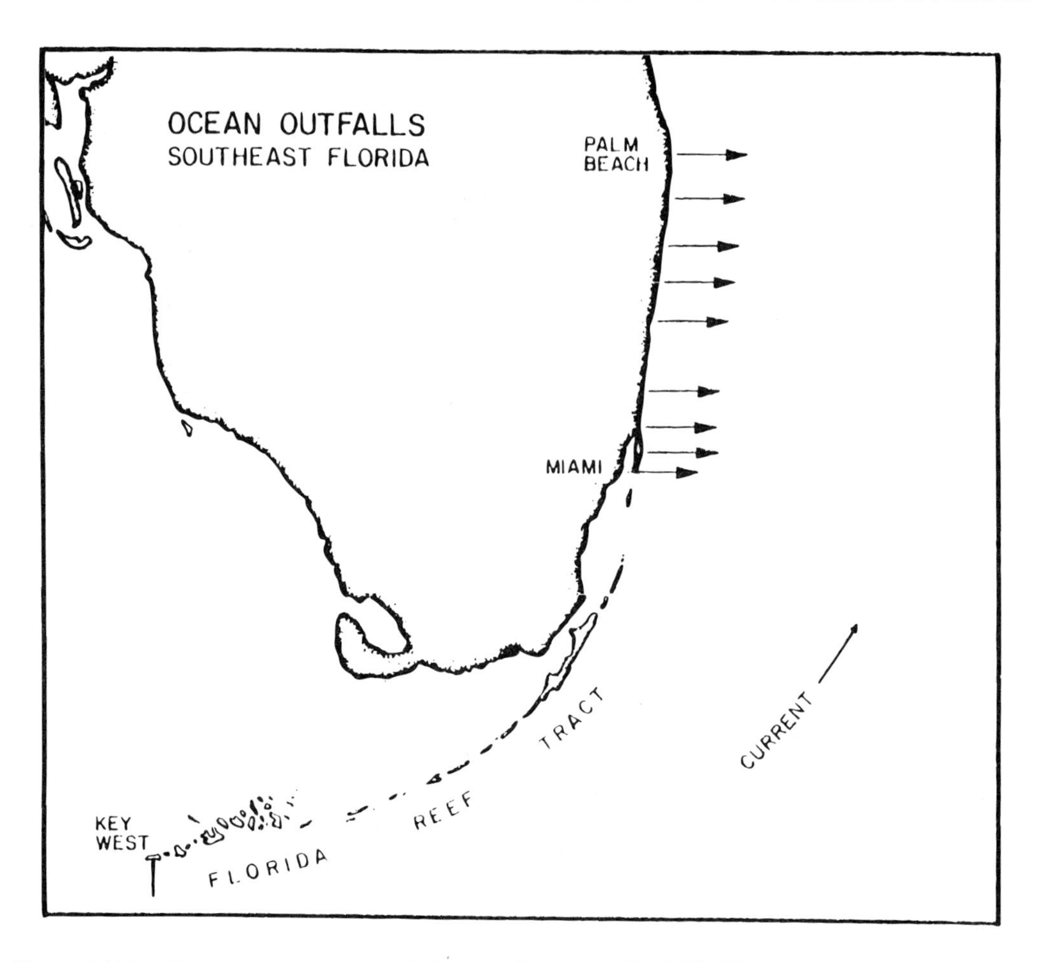

Figure 4.41.1. The numerous sewage outfalls located along the South Florida coast have worked effectively and had little environmental impact.

plished by the use of ocean outfalls that discharge into oceanic (not coastal) waters. Where ocean outfalls are not practical, more expensive tertiary treatment of effluents may be required.

Diversion to deeper water seaward and down-current from the coral reefs is especially desirable because dilution and dispersal by large volumes of seawater neutralizes the pollution potential of the sewage. In Florida and elsewhere, ocean dispersal has proved to be an effective and efficient way to dispose of sewage without polluting the coastal environment. Both examples also dramatically illustrate the need for adequate planning while sewage disposal systems are still in the design stage. The potential effects of sewage effluents on the local marine environment were obviously not adequately considered when the Kanehoe Bay and Miami outfalls were emplaced. Both systems later required modification at great additional expense. In the case of Kanehoe Bay, an entirely new ocean outfall was constructed and bay outfalls were abandoned (see "United States, Hawaii" case history below). At Miami, the existing outfall was extended to discharge into the Florida Current (Gulf Stream).

In the case of southeast Florida, abatement of sewage pollution of the coastal environment was achieved through the combined efforts of several interest groups. These included environmentalists concerned with preserving the reefs and other marine life, health officials concerned with the spread of disease, and the tourist industry concerned with maintaining a local resource and the income derived from it. Sewage pollution was eliminated, or at least significantly reduced, in two ways: the upgrading of sewage treatment facilitates resulting in the elimination of outfalls discharging raw (untreated) sewage and the extension ocean outfalls made the point of discharge remote from reefs and bathing beaches.

Before ocean outfalls are designed, a thorough pre-construction survey should be made to document the physical (currents, bathymetry, sediments, etc.) and biological (biotopes and their distribution) characteristics of the area to provide data for effective outfall design and function. Also, a marine monitoring program should be part of any waste management plan to recognize early signs of sewage pollution.

Source — D. S. Marszelak, Boca Raton, Florida (See also Reference 236 in Part 5).

4.42 UNITED STATES, HAWAII: MONITORING NEW HARBOR CORAL GROWTH

Background

Honokohau small boat harbor is situated 3 miles north of the town of Kailua-Kona off the west coast of Hawaii Island, the largest and farthest southeast of the Hawaiian Islands. The harbor was constructed in 1970 and expanded in 1978 to accommodate more recreational, tourist, and fishing vessels. Its maximum depth is about 6 m, and its total area is about 8 hectares. The harbor is now filled to capacity. The U.S. Army Corps of Engineers designed and constructed the harbor, and the State of Hawaii operates the harbor. The Corps also sponsored most of the ecological and water quality monitoring surveys conducted six times during 1971–91 (1). The duration and scope of the studies are among the most comprehensive of any monitoring project in coral reef environments.

Problem

Dredging, blasting, and filling have been documented to cause major impacts to coral reefs in the Pacific. Since 1970 and the passage of several U.S. environmental laws, the success rate for approving and constructing harbor projects has diminished, partly out of environmental concerns. Dredging degrades coral reefs in several ways, including re-suspension of sediments, which settle on corals, reducing light penetration, clogging pores, and blocking feeding and photosynthesis. Severe sediment accumulation from dredging and other coastal construction can bury and smother corals or lead to anoxic bottom conditions where water quality conditions are eutrophic. Blasting can fracture, shear, or pulverize corals and injure or kill fish, shellfish, sea turtles, or other animals through shock and concussion. Filling of coral reefs to create backup land areas for harbor is also among the most severe of impacts because coral reef habitat is permanently lost to make way for the new fill land.

Solutions

Honokohau was one of the first harbors in the United States to be designed and constructed with environmental protection as an objective. The harbor was quarried from basalt lava inland from the shoreline and then opened to the sea. As such, the harbor basin constituted a sterile and unpolluted new marine environment. Only some minor dredging of the harbor entrance channel impinged on pre-existing coral reef habitats. The inland configuration and small size of the harbor facilitated water quality and ecological monitoring, and surveys of fish, corals, other macro-invertebrates, water circulation, and water quality were accomplished in 1971, 1972, 1973, 1976, 1981, and 1991. The series of six surveys also allowed the documentation of the colonization and ecological development of the harbor over a 21-year period.

The surveys revealed that the new harbor environment was favorable for colonization by coral reef organisms. The results of these surveys revealed that the new harbor supported very favorable conditions for both coral reef and estuarine communities of plants and animals. The inland configuration of the harbor intercepted groundwater flow, resulting in the discharge of brackish water into the harbor at a rate of about 2 million gallons per day. This set up a stratified water column in the harbor, with the upper 1-m layer brackish and the lower 4–5-m-thick layer marine. Groundwater discharges through the walls and floor of the harbor set up an efficient water circulation system in the harbor basin. The brackish surface water layer flows quickly out of the entrance channel, entraining some of the water of the deeper marine layer with it. Tidal exchange also helps to flush the deeper layer. As a result the average residence time of the surface water layer is only a couple of hours, while the average residence time of the deeper layer is less than a day. This efficient flushing system improves the quality of the harbor's waters, especially within the outer half of the harbor. However, the water residence time in some stagnant interior pockets may approach several days.

Eleven species of corals had already colonized the outer half of the harbor by the time of the first survey in 1971, about 18 months after the harbor entrance had been opened to the sea. Coral species richness continued to increase through 1981 but declined slightly after that. Average coral colony size (measured as each colony's maximum diameter) has increased steadily for most of the common species over the 21-year period. Initially, branching corals of the genus *Pocillopora* colonized first and in greater numbers compared to other species. Eventually, however, massive and encrusting species of *Porites* caught up with and then surpassed *Pocillopora* in terms of abundance, average size, and maximum size. This shift in species follows the same trends as observed on recent dated lava flows in Hawaii, which have been colonized by corals during the past century. Average coral coverage in the harbor has also increased steadily to about 6 percent in 1991. However, outer harbor environments show coverage of 25 percent or more and colonies up to 70 cm in diameter, approaching what is observed outside the harbor in comparable undisturbed environments.

Harbor expansion in 1978 may have caused some adverse impact to corals in the harbor. For example, the 1981 survey revealed many dead colonies of *Pocillopora,* a species known to be sensitive to sediments, which may have been generated during the 1978 construction work. Furthermore, the newer (expanded) portion of the harbor is located furthest from the entrance channel and has been slow to attract coral colonizers. Water residence time is high in the innermost harbor, and phytoplankton populations are very heavy, reducing underwater visibility to only a few feet, compared to 20–30 in most outer harbor environments. The degree of coral colonization in the innermost harbor is only a fraction of that observed in the outer harbor over a similar time interval (12–13 years). Litter and fish waste are also discharged into the harbor, with the former gradually accumulating over time on the harbor floor. The debris, junk, and litter present during the 1991 survey were sufficient to discourage coral development in some areas of the harbor. Nevertheless, coral development remains healthy and high in Honokohau. In contrast, fishing pressure *within* the harbor is heavy, and fish species diversity has also declined since the 1981 survey.

Lessons Learned

The long-term monitoring studies at Honokohau harbor over two decades were invaluable in revealing conditions favorable for coral reef development in an active harbor. The studies further revealed that marine life may find the harbor a haven, protected from the heavy wave action that periodically strikes the west or "kona" coast. Corals achieved abundance, size, diversity, and density comparable to those observed on natural habitats outside the harbor. The monitoring program also documented that harbor expansion in 1978 may have degraded water quality in the inner harbor and discouraged further coral development. Long-term littering and fish waste disposal in the harbor may also discourage colonization and degrade the relatively healthy coral reef conditions inside the harbor. Monitoring programs such as those completed at Honohokau may help to influence procedures and designs for future harbor construction and optimize harbor development and operation that both minimize impacts to natural coral reefs but also help account for other impacts (such as littering) that may have been overlooked during pre-construction environmental impact assessments.

Contributed by — J. E. Maragos, Program on Environment, East-West Center, Honolulu, HI 96848.

REFERENCE

1. OI Consultants, Inc., 1991 studies of water quality, ecology, and mixing processes at Honokohau and Kawaihae harbors on the Island of Hawaii. Prepared for Mauna Lani Resort, Inc., P.O. Box 4959, Kohala Coast, HI 96743.

4.43 UNITED STATES, HAWAII: SUCCESS WITH KANEOHE BAY OCEAN OUTFALL

Background

Kaneohe Bay, along the northeast coast of Oahu, is the largest embayment in Hawaii, measuring 7 miles long and 2 miles wide. A broad barrier reef extends offshore across the mouth of the bay, maintaining a lagoon inshore with numerous pinnacle and patch reefs (Figure 4.43.1). It is one of only two barrier reef systems in Hawaii. Two passes, one in the north and one in the south, bisect the barrier reef and promote tidal exchange between the inner bay (lagoon) and ocean. Fringing reefs line the entire shoreline inside the bay. The southern third of the bay is a nearly enclosed basin surrounded on three sides by land, with restricted water exchange with the rest of the bay due to broad, shallow reef flats and islands. The waters of the middle bay are protected by the shallowest part of the barrier reef. The barrier reef deepens in the more open northern third of the bay, allowing more vigorous wave action and tidal exchange inside the bay. Surrounding land uses are predominantly residential and military along the southern shoreline of the bay and mostly rural and agricultural along the northern bay shoreline. Before 1940 the coral reefs in Kaneohe Bay were the best developed and most diverse of any in Hawaii.

Problem

Beginning in 1939, a U.S. naval air station was constructed on the southern peninsula (Mokapu) fronting the bay, resulting in the dredging and filling of coral reefs mostly within the southern lagoon. Over 15,000,000 cubic yards of reef materials were dredged, with most placed over reef flats to enlarge the land area of the naval base. Excess sediments from the dredging operations were also discharged back into the lagoon. A ship's channel was also dredged along the entire length of the inner bay, with

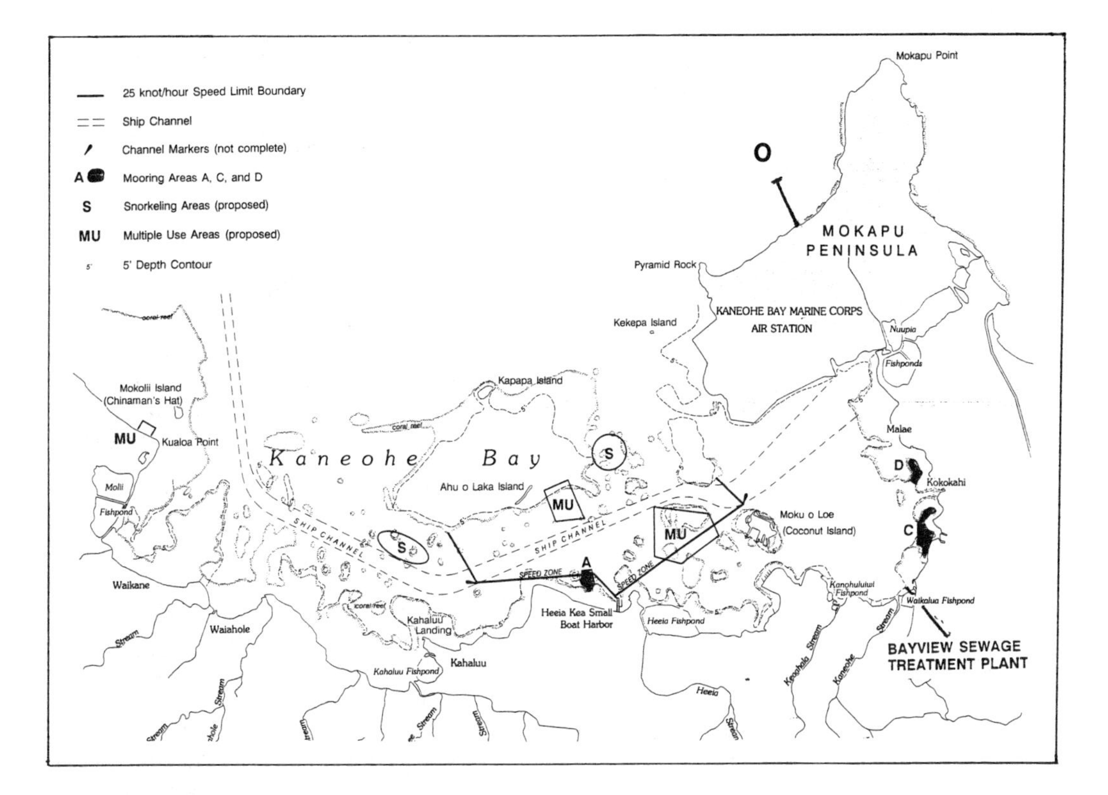

Figure 4.43.1. The Kaneohe Bay (Oahu, Hawaii) environment: approximate location of ocean outfall.

the northern pass deepened to accommodate ship navigation between the bay and ocean (8). Shopping centers and residential neighborhoods later sprang up, mostly in the southern end of the bay at Kaneohe town. By the early 1960s the need for improved sewage disposal led to the construction of a sewage treatment plant and outfalls. The navy outfall discharged near the southeast corner of the bay, and the new municipal outfall discharged into the southern basin near the town of Kaneohe. As the population grew, sewage discharges continued to increase. By 1977 the combined effluents of three sewage treatment plants and outfalls totaled more than 7.5 million gallons per day (20,000 cubic meters/day), with 95 percent of the sewage discharged into the southern lagoon.

Reef decline began with the military dredging operations during 1939–50, mostly in the southern lagoon (8). By the late 1960s, when the first extensive surveys of coral reefs were conducted in the bay, very little of the reef had recovered from the previous dredge-and-fill operations, and more south bay reefs had deteriorated because of poor water quality attributed to sewage discharges (1, 8). The dissolved nutrients in the discharged sewage were stimulating massive growths of phytoplankton and eutrophic conditions that in turn led to the proliferation of suspension and filter feeders (soft corals, tunicates, bryozoans, feather-duster worms, clams, etc.). Corals and other normal reef life were all but wiped out from the south lagoon. Periodic freshwater floods and discharges of eroded soils from land contributed to the further demise of the reefs in the lagoon (4, 6). The lack of oxygen and presence of toxic sulfides in the sediments may have caused some of the coral deaths in the south lagoon (8).

Furthermore, lagoon reefs in the middle bay also began to decline in the mid 1960s. Growths of a green bubble algae (*Dictyosphaeria*) formed massive mats, which overgrew and smothered most living corals on the reef slopes inside the bay (lagoon). Marine scientists believed that the algae were stimulated by higher nutrient levels contributed by sewage moving up from the south bay (1, 8). In contrast, bubble algae coverage was greatly reduced in the north bay, where greater water flushing would discourage high nutrient levels. The coverage of bubble algae was actually more abundant in the lagoon in 1971 than that of all species of live corals combined.

Solutions

Beginning about 1970, the continuing decline of Kaneohe Bay's coral reefs, recreational values, and other values generated considerable public outcry, convincing the local government to move the discharges from the two main outfalls in the south bay to a new deep ocean outfall outside the bay and off the Mokapu peninsula. Water conditions at the new discharge site were well flushed, with residence time measured in terms of hours compared to several weeks at the old lagoon outfall sites.

Within months of sewage diversion from the south lagoon in 1978, water quality conditions throughout the entire lagoon improved dramatically (10). Within six years, middle bay coral reefs showed remarkable recovery while south bay reefs showed dramatic recolonization—the initial phase of full recovery (3, 9). Survey results in 1983 revealed that both dominant and rare species of corals were much more abundant in the lagoon. In contrast, the bubble algae (*Dictyosphaeria*) declined to one fifth of its pre-diversion levels by 1983 (see 1983 sketch in Figure 4.43.2). Coral coverage also more than doubled in the bay from 1971 to 1983 and showed increases in all areas of the bay except in the north, where live coral cover was already high (3, 9).

In 1990, the coral reefs in the lagoon were re-surveyed at the same sites, revealing that coral recovery had slowed. The most dramatic increases in live coral cover occurred in the south bay (2, 6). In the middle and north bay, coral coverage remained about the same (at a high level). However, the green bubble algae (*Dictyosphaeria*) had begun to make a comeback in the bay, already achieving about half of its 1971 levels (2, 6). This increase had not yet led to the demise of corals, but the trend, if it continues, could lead to decline of corals from the smothering growths of the algae. This turn of events has puzzled marine biologists (2, 5, 7, 11). Many now believe that non-point sources of sewage pollution, especially from cesspools, improperly maintained septic tanks, raw sewage "bypass" discharges into the bay during power outages, and sewage leaking from the now aging sewerage system underneath Kaneohe town, have all contributed to higher levels of nutrients in the lagoon. These in turn may have now reached the threshold needed to re-stimulate bubble algal growths. It may take years of research to determine the true cause of the reversals and implement the remedial actions to sustain the high levels of coral development that now occur in the bay (6, 11).

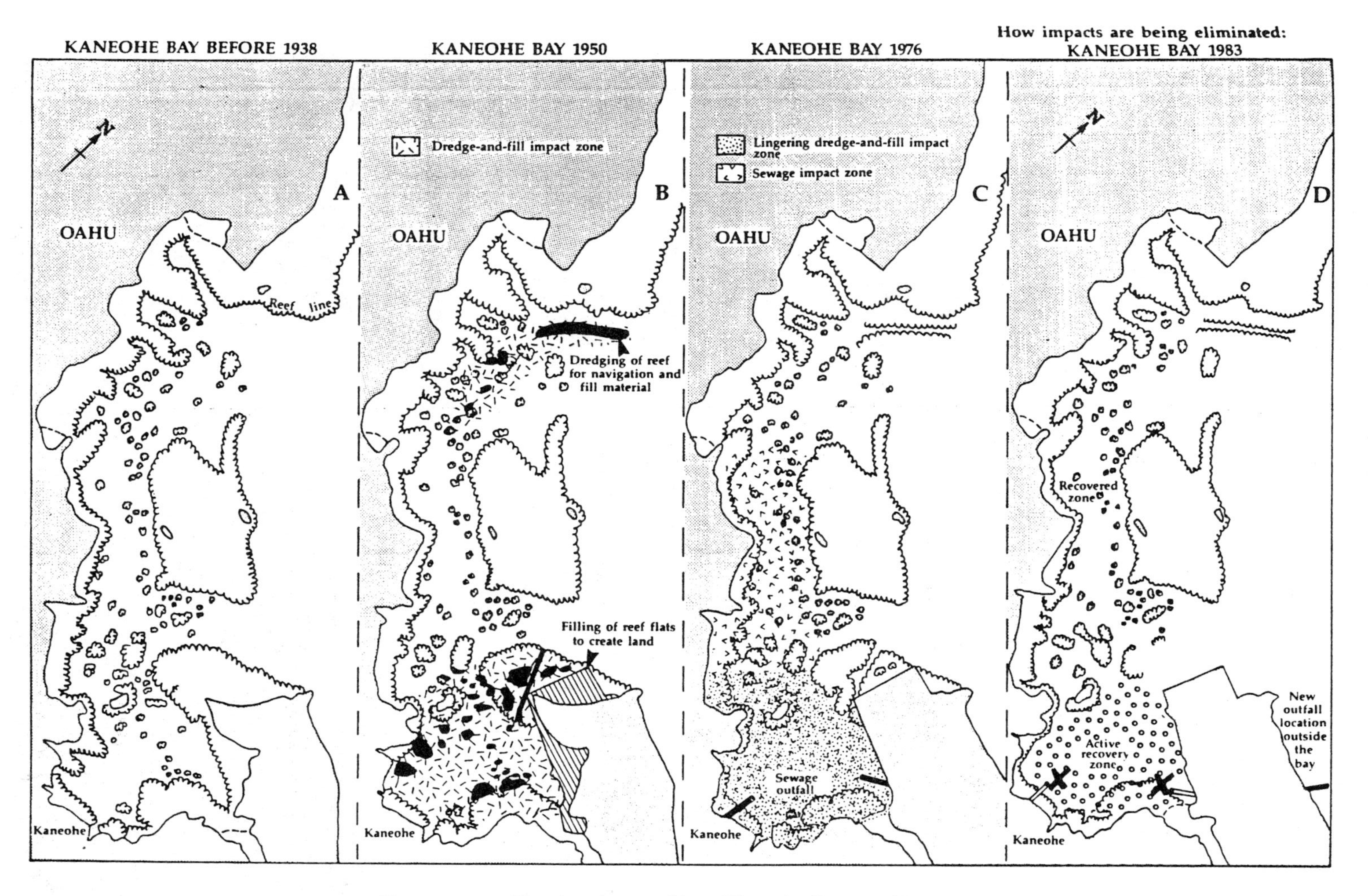

Figure 4.43.2. The changing condition of Kaneohe Bay over 45 years.

Storm water runoff into the bay has also periodically killed shallow water coral communities in the lagoon, especially near stream mouths, as occurred in 1965 and 1988. Coral recovery appears to be rapid on flood-damaged reefs, but a series of catastrophic rainstorms could cause coral kills and further complicate predictions of coral recovery in the lagoon (4, 6).

Lessons Learned

The discharge of secondary treated sewage into southern Kaneohe Bay was analogous to dumping sewage into a swimming pool (8) (Figure 4.43.3). The confined nature and slow turnover of the south bay waters favored an offshore location for sewage disposal. However, only point discharges of sewage have been eliminated from Kaneohe Bay.

Non-point discharges from cesspools, non-operational septic tanks, leaking sewerage (underground pipelines), and emergency discharges during power failures collectively constitute an increasing source of stress to lagoon coral reefs, which may lead to reversals of the earlier gains in coral reef recovery. Although a 1992 Kaneohe Bay master plan has addressed non-point source pollution (11), it may take many years before remedial actions are completed to reduce non-point source pollution into Kaneohe Bay.

Careful evaluation of ecological, recreational, commercial, and other values of proposed receiving waters should enter the equations along with standard engineering, sanitation, chemical, and mixing parameters. These studies indicate that the detrimental effects of sewage on corals are magnified in confined embayments with restricted circulation and that coral recovery from previous stresses (dredging, filling, flooding impacts) is postponed until sewage stresses are reduced.

Termination of point source sewage discharges reversed the earlier trend toward eutrophication and led to rapid declines of plankton, south bay suspension feeders, reef cryptofauna, and mid-bay *Dictyosphaeria* populations (10). Corals are now rapidly re-colonizing all previously degraded coral habitats. Although coral recovery appears to be nearing completion in the middle lagoon, a recent resurgence of *Dictyosphaeria* growth may prevent full recovery until populations of the bubble algae again decline to "normal" levels (2, 6). At least a decade or more may be required for recovery to peak in the southern lagoon, where coral reefs were initially more severely degraded. Some deteriorated south lagoon reefs covered with sediment will probably take longer to recover.

Figure 4.43.3. Kaneohe Bay on the north side of Oahu (photo by John Clark).

After 1977–78, sewage discharges outside the bay at the Mokapu ocean outfall have not caused problems because of rapid dilution and mixing of sewage with open waters. The outfall is also placed at more than a 100-ft depth to promote dilution and reduce sewage effects at the surface (3, 9). It may seem an enigma that the same amount of sewage that caused catastrophic damage to corals inside the bay would have only negligible impact on the reefs outside the bay. The shorter residence time of water outside the bay prevented high nutrient levels and a buildup of phytoplankton and eutrophic waters that occurred inside the bay, where water residence time is much longer.

In retrospect, the location of the original outfalls in the south lagoon was perhaps the worst of all alternatives from the standpoint of ecosystem health, public health, and recreational/commercial values. This case points to the need to site sewage outfalls in well-flushed waters to avoid adverse effects to coral reefs. Proper siting of outfalls is as important as establishing proper levels of sewage treatment.

Certainly, public expenditures for sewage disposal were higher than necessary because of the need to re-locate the sewage outfall to an open ocean environment, where construction and maintenance costs are higher. Even though the level of sewage treatment was high (secondary), sewage discharges in the south bay were still stressful to coral reefs. Adequate treatment was insufficient by itself to avoid impacts to coral reefs. Secondary treatment actually generates more labile forms of nutrients (nitrogen and phosphorus) that are more readily taken up by marine plants (phytoplankton and seaweeds), which in turn could enable them to compete better against corals.

Outfalls need to be sited where natural dilution forces (waves, tides, currents) can assist in avoiding major damage to coral reef ecosystems. The experience with Kaneohe Bay suggests that, in general, sewage outfalls should never be placed in protected coral lagoons and embayments with restricted water exchange, regardless of the level of sewage treatment.

Contributed by — J. E. Maragos, Program on Environment, East-West Center, Honolulu, HI 96848.

REFERENCES

1. Banner, A. H. 1974. Kanehoe Bay, Hawaii: urban pollution and a coral reef ecosystem. In: *Proceedings at the 2nd International Coral Reef Symposium,* Vol. 2. Great Barrier Reef Committee, Brisbane.
2. Evans, C. W. and C. L. Hunter. 1993. Kaneohe Bay, an update on recovery and trends to the contrary (Abstract). In: *7th International Coral Reef Symposium, Agana, Guam,* Vol. 10 P. 346.
3. Evans, C. W., J. E. Maragos, and P. F. Holthus. 1986. Reef corals in Kaneohe Bay six years before and after termination of sewage discharges. In: *Coral Reef Population Biology,* Hawaii Institute of Marine Biology, Tech. Rep. 37. Pp. 76–90.
4. Holthus, P. F., J. E. Maragos, and C. W. Evans. 1989. Coral reef recovery subsequent to the freshwater kill of 1965 in Kaneohe Bay, Oahu, Hawaii. *Pacific Science,* Vol. 43, No. 2. Pp. 122–134.
5. Hunter C. L. and C. W. Evans. 1993. Reefs in Kaneohe Bay, Hawaii: two centuries of western influence and two decades of data. In: *Proceedings of the Colloquium on Global Aspects of Coral Reefs: Health, Hazards and History,* R. N. Ginsburg (Compiler). University of Miami, Rosenstiel School of Marine and Atmospheric Science, Miami. Pp. 339–345.
6. Jokiel, P. L., C. L. Hunter, S. Taguchi, and L. Watarai. 1993. Ecological impact of a fresh-water "reef kill" in Kaneohe Bay, Oahu, Hawaii. *Coral Reefs,* Vol. 12, No. 3/4. Pp. 177–184.
7. Kinsey, D. W. 1988. Coral reef system response to some natural and anthropogenic stresses. *Galaxea,* Vol. 7. Pp. 113–128.
8. Maragos, J. E. 1972. A study of the ecology of Hawaiian reef corals. Ph.D. dissertation, University of Hawaii, Honolulu. 292 pp.
9. Maragos, J. E., C. Evans, and P. Holthus. 1985. Reef corals in Kaneohe Bay six years before and after termination of sewage discharges. *Proceedings of the Fifth International Coral Reef Congress, Tahiti,* Vol. 4. Pp. 189–194.
10. Smith, S. V., W. J. Kimmerer, E. A. Laws, R. E. Brock, and T. W. Walsh. 1981. Kaneohe Bay sewage diversion experiment: perspectives on ecosystem responses to nutritional perturbation. *Pacific Sciences,* Vol. 35, No. 4. Pp. 279–395.
11. State of Hawaii Office of State Planning. 1992. Kaneohe Bay Master Plan. Prepared by the Kaneohe Bay Master Plan Task Force. Office of the Governor, State of Hawaii, Honolulu. 115 pp.

4.44 UNITED STATES, HAWAII: TOURISM THREATENS HAUNAMA BAY

Background

The explosive growth in tourist use of tiny Haunama Bay, in Hawaii, has in the last 15 years degraded both the experience and the coral reef ecosystem (Figure 4.44.1). Hanauma, on the island of O'ahu, was until recently an isolated cove visited by a few fisherfolk and weekend campers. Today the bay is choked with beach people and shows signs of acute environmental overload.

Problems

First, an absence of guiding concepts of desirable and unacceptable states of the reef ecosystem precluded any real management of tourist use of the bay or even the collection of relevant data. Second, authorities have a contradictory mandate to conserve the bay while promoting greater public use. Third, the short-term viewpoint of commercial interests is not recognized as inherently conflicting with the long-term interests of residents or the sustainability of the reef itself.

Increased use of the bay by residents and visitors prompted government investments in infrastructure, which further improved access and encouraged use. Eventually, the city funded a study to determine optimal use levels for the park and to recommend other changes. In 1977, the study concluded that the park was too crowded and that maximum beach occupancy should be 1,000 people.

But the number of visitors continued to increase. In 1983, the city attempted to control one of the main commercial uses of the bay by instituting a permit system for dive and snorkel charters. The total failure of these controls allowed continued rise. The rapid change alienated local residents, who largely avoid the bay or leave when the tour buses arrive.

Unfortunately, the interactions of reef fish with other components of the ecosystem are very complex and mostly unknown. It is, therefore, difficult reliably to infer changes in the ecosystem from changes in the fish numbers and species. However, the total fish biomass inside the reef crest has steadily

Figure 4.44.1. Haunama Bay near Honolulu, Hawaii, is an example of tourism gone out of control, resulting in serious resource degradation (photo by John Clark).

declined, although it is still well above what the reef could be expected to support naturally. Concurrently, the number of fish in the area has more than doubled, but they are individually of much smaller size.

The high biomass inside the reef crest is almost certainly being supported by artificial feeding, but the steady decline is troubling. Biomass density measures ecological "carrying capacity," and a long-term reduction implies that the ability of this environment to support fish, and probably other organisms, is declining. Assuming that artificial food has become more, not less, available during this period, some unmeasured factor or interaction (perhaps chemically induced) must be invoked to explain the drop.

The fact that these changes are occurring only in the high use area inside the reef crest, where most swimming, wading, and fish feeding occurs, strongly implicates human impacts.

Solutions

Three fundamental economic concepts are egregiously at odds with preservation of a sensitive ecosystem such as Hanauma. First, it is assumed that the environment can sustain an indefinite expansion of commercial activity and that continuous growth is both feasible and desirable. Second, the idea of discounting makes a dollar earned today worth more than the same dollar earned next year, and high but destructive short-term profits may therefore be more desirable than lower but sustainable long-term yields. Third, degradative aspects of an activity which are difficult to trace are "externalities" rather than costs and, erroneously, the environment is assumed to absorb an infinite amount of this abuse at no cost to business.

Because the environmental consequences of these basic concepts are not generally appreciated, ecological damage is conventionally viewed as an aberration rather than a predictable feature of unrestrained commercial activity. Tourism, in particular, is seen as environmentally benign, which perhaps explains why the commercial capture and degradation of Hanauma came as a surprise, but this exploitation is almost a classic "tragedy of the commons." Individual tour operators expand their business without limit because the immediate profit accrues to them, while the future costs of environmental degradation are borne by the community and the ecosystem.

The plight of Hanauma is being addressed by managers of the city's park. New restrictions on commercial access and other changes should reduce the visitor load to 2,000 at any given time, and future rules may sharply reduce fish feeding. A monitoring program will document some biotic responses to these changes. While such measures are necessary and laudable, they treat only the symptoms, not the disease.

Why haven't the real problems been addressed? One often-cited impediment is that responsibility for Hanauma is divided between the State of Hawai'i and the City and County of Honolulu, which oversee the beach and the surrounding land. Eleven separate city and state agencies and divisions have some management authority at Hanauma. Both the state's Division of Aquatic Resources and its Parks Department are technically responsible for management of the underwater part of the bay, but enforcement of their regulations falls to yet a third agency. Effective management of Hanauma is blocked primarily by administrative and communication barriers, not by technical shortcomings.

Any management scheme would be crippled by the present fragmentation of authority over Hanauma, so creation of an over-arching, autonomous body would seem to be necessary. Perhaps the Situation Management approach of ICZM could be effective.

Lessons Learned

An ultimate goal of maintaining a balanced reef will be compatible with tourist use only if the quality, as well as the quantity, of use is examined and controlled. Use types and levels might be adjusted periodically based on the status of reef populations and processes. Education to promote benign activities would be fundamental and would be most effective if the predominant users were repeat visitors, rather than transients.

Too often commercial tourist pressure on shoreline ecosystems will continue to grow despite *ad hoc,* localized management actions. In this case, the city parks director is trying "to encourage tour operators to look at the other sites around the island." The hope appears to be that negative impacts can be "diluted away" if spread thinly enough. This may transfer impacts from Hanauma to alternative sites, but, in the long run, even linear growth of tourism will prove excessive.

Human effects on the coral reef are likely to be multi-faceted. For instance, the foods offered to fish are suspected to damage fish health. When not being fed, these fish may intensely graze reef algae and coral rock, reducing cover and food needed by smaller animals. Such grazing, plus freshwater from showers, parking lot runoff, silt stirred up by wading, suntan lotion, and direct trampling may all be involved in the near elimination of large algae, corals, and coral-feeding fishes.

Hawai'i clearly needs an alternative management vision, one that emphasizes sustainability over growth. Hanauma is an ideal incubator for an ecocentric management approach because of its controllability and degraded state. Ecocentric management would share many features with the traditional Hawaiian system.

Establishing Hanauma as an example of a controlled and sustainable interaction between people and nature would both ensure the bay's survival and set a crucial precedent in Hawai'i. Ultimately, however, we must deal with the unsustainability of unregulated growth and realize that these islands and their ecosystems are finite and fragile.

Source — J. Burgett, East-West Center, Honolulu (from Reference 37, Part 5).

4.45 UNITED STATES, SOUTH CAROLINA: SOFT ENGINEERING BEACH RESTORATION

Background

Shorelines near tidal inlets tend to be more dynamic than those along straight coastlines. Inlets and their associated deltas trap littoral sand moving along the coast; they sometimes migrate and usually alter wave direction and energy along adjacent beaches. Management of development and erosion around inlets is therefore quite complex.

A project at Seabrook Island, South Carolina (Figure 4.45.1), illustrates the effect of an unstabilized, migrating inlet on a new development and how the community responded to severe erosion. These case follows a sequence of poorly planned development setbacks in the early 1970s, the use of hard structures (seawalls) to control erosion in the late 1970s, and then an innovative "soft engineering" solution (inlet relocation) in the early 1980s.

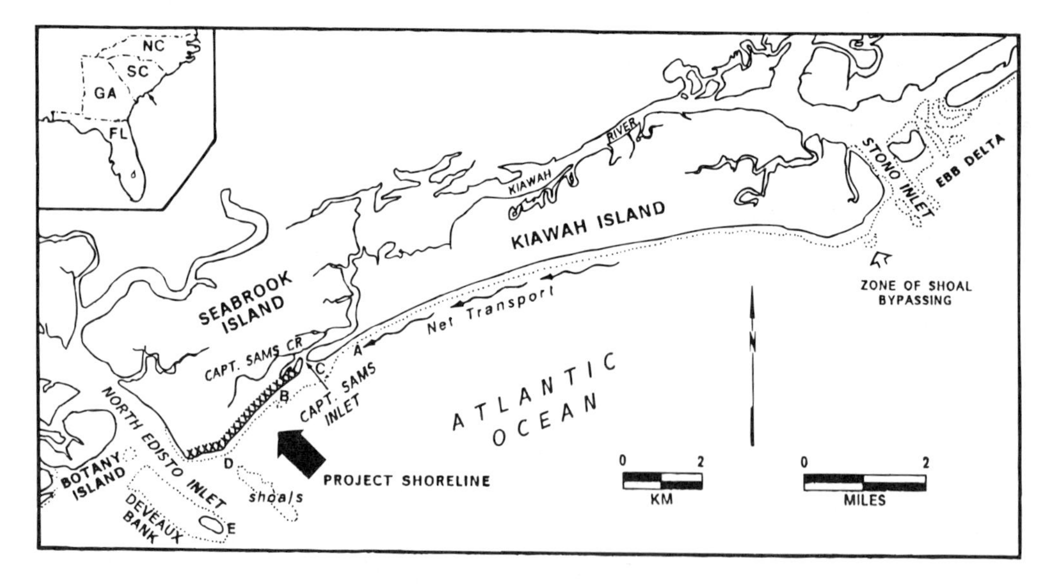

Figure 4.45.1. Seabrook Island and Captain Sams Inlet project area and regional setting 30 km southwest of Charleston, South Carolina, USA. A = 1948 inlet, B = 1939 and 1982 inlet, C = 1963 and 1983 inlet position.

Like many privately owned barrier islands, Seabrook Island was a prime candidate for development in the early 1970s. Located 30 km south of Charleston, South Carolina, it contains about 1,000 ha of high ground fronted by 5 km of ocean beach. Unlike classic barrier islands in low tide range settings, which tend to be long and narrow (4), Seabrook Island is short and broad. During the past century, in fact, it has built seaward by excess sand supplied from the updrift island, Kiawah (6). A tide range averaging about 2 m allows more inlets to persist along this section of coast.

Seabrook is bounded at the north by Captain Sams Inlet, a shallow, unstable channel that tends to migrate in response to spit growth at the south end of Kiawah Island (see Figure 4.45.1). The southern boundary of Seabrook is North Edisto Inlet, a drowned river inlet that is one of the largest on the U.S. coast. In contrast to Captain Sams Inlet, the North Edisto has a naturally stable channel, deeply incised in older sediments. Both inlets are backed by mature marshes and tidal creeks on their landward ends and fronted by sandy tidal deltas at their seaward ends.

Problem

Given its short length, Seabrook Island's entire shoreline is strongly influenced by the adjacent tidal inlets. As waves propagate toward shore, they change direction over the deltas, lose energy at varying rates, and cause random wave patterns at the beach. It is this variation in wave energy that produces an irregular shoreline response. Some sections of beach erode while others accrete. While not recognized in the early 1970s, much of the Seabrook shoreline has experienced decadal cycles of erosion and accretion. These cycles are superimposed on a long-term trend of accretion (2). Thus, Seabrook Island, while growing in a 100-year time frame, has eroded severely during certain 10–20-year periods.

The first development in the early 1970s followed a period of natural beach building along the oceanfront. At that time, lots were laid out based on the position of the high tide mark, allowing a 100-m width for seaward properties. Some reaches had been growing *seaward* by as much as 10 m per year, suggesting that the oceanfront was reasonably safe for development (9). But soon after the first lots were platted, a period of rapid erosion began, placing many properties at risk. As owners watched, the sea reclaimed half of their lots. Whether or not a building existed, the pressure was there to protect the remainder of each lot. In the course of ten years, from the mid-1970s to the mid-1980s, the philosophy of erosion control at Seabrook evolved from shoreline hardening (the structural approach) to soft-engineering solutions (the non-structural approach). This paralleled recognition among professionals that the *most cost-effective shore-protection schemes are those that work with nature.*

Seabrook's problem in the 1970s was typical of many oceanfront resorts—the need to protect property values and the rights of owners while preserving an aesthetic shoreline in the face of erosion. Initial shore-protection efforts were piecemeal and left to individuals. Cheap solutions, such as small sandbags, were tried; then large sandbags were placed at the backshore. This was followed by vertical concrete bulkheads at US$500/m (1975) and composite seawalls incorporating vertical sheeting and a facing of quarry stone at US$1,000/m (1978). By 1981, 80 percent of the shoreline was armored and lacked any beach at mid-tide. Along some reaches, low tide water depth exceeded 3 m at the base of the seawall. Continued loss of sand fronting the seawall placed the structure in jeopardy, culminating with the massive failure of a 500-m section during Hurricane David in September 1979 (6).

Solutions

Inlet relocation, costing US$300,000 (1983) restored the beach by freeing over 1 million cubic meters of sand trapped in the abandoned tidal delta. The natural energy in breaking waves did most of the beach rebuilding (Figure 4.45.2), making this an extremely cost-effective erosion control project. A brief history of the project follows.

Using comparative analysis of vertical aerial photographs, Hayes (4) confirmed the controlling effects of Captain Sams Inlet and the shoals of North Edisto Inlet. Three basic causes of erosion were identified (labeled A to E in Figure 4.45.1):

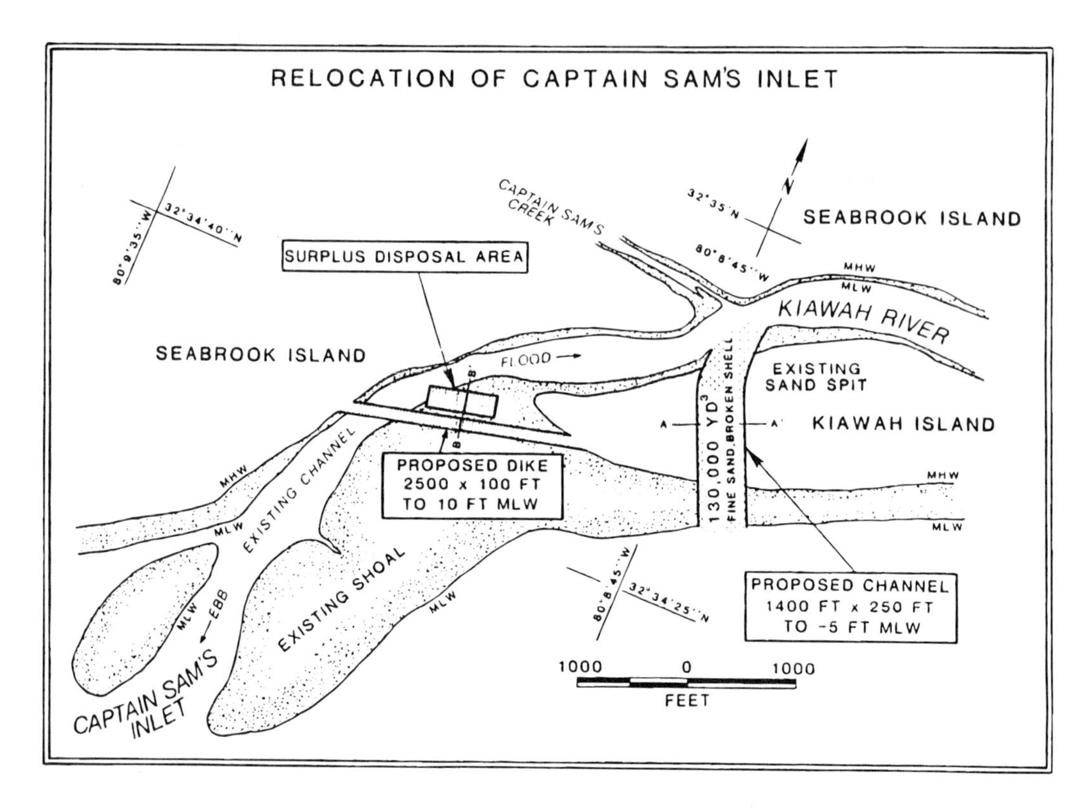

Figure 4.45.2. The major specifications for relocation of Captain Sam's Inlet—a "soft engineering" project.

1. Southerly migration of Captain Sams Inlet and associated sand trapping by its ebb-tidal delta (A to C)
2. Encroachment of a secondary channel to North Edisto Inlet along one segment (D)
3. Erosion of a large offshore island in North Edisto Inlet that had served as a natural breakwater

In 1983, the first solution was implemented—for re-locating Captain Sams inlet ~2.5 km to the north by breaching Kiawah spit and closing off the existing channel (8). In theory, this would free sand trapped in the ebb-tidal delta and allow it to move onshore by wave action. The natural southerly transport would gradually rebuild the eroded beach. Figure 4.45.3 shows regional coastal processes in 1983 after inlet relocation and the result in 1987. While relief was not immediate, four years after the inlet relocation a major section of beach had been restored. In fact, where no beach existed in 1983 at low tide, a 300-m-wide dune field had developed. Wind-blown sand eventually covered much of the seawall, restoring the aesthetic character of the shoreline.

Ten years after inlet relocation, the northern half of Seabrook Island retains one of the widest beaches in South Carolina. Natural nourishment by inlet relocation in this case added more than 1 million cubic meters to the beach. The project was relatively inexpensive to construct because less than 150,000 cubic meters of sand had to be moved to create the new inlet and close the old inlet. Earthmovers and trucks performed the work at a total cost of US$300,000.

Whereas a number of agencies and individuals opposed the project on environmental grounds before permits were issued, the inlet relocation largely shifted habitat from one place to another, just as natural breaches in earlier times had done (1). There was a reduction in intertidal sand bars as the abandoned delta attached to the shoreline and built up a dune line and an increase in supratidal nesting area favored by least terns. Accreting dunes enclosed a new lagoon (Figure 4.45.4) that eventually accumulated fine sediments and allowed the propagation of a 5-ha salt marsh.

The inlet relocation was not a permanent solution because, after completion, the new inlet was allowed to resume its migration. At the time of this writing (1994), the new inlet is situated 1,000 m from its relocation point. In fact, erosion from channel migration is cutting away newly formed marsh

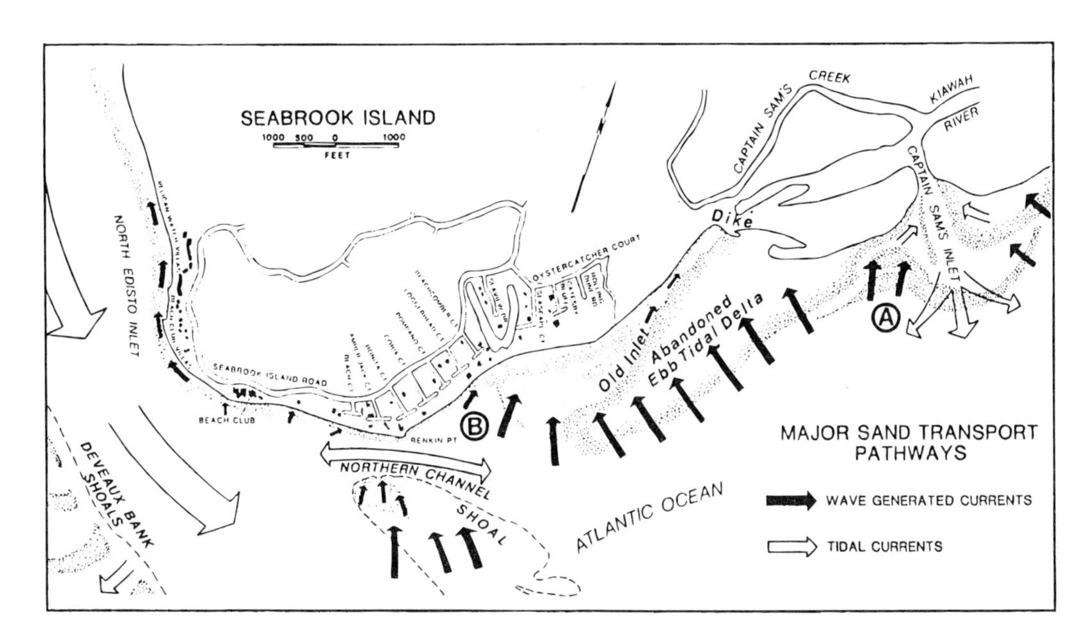

Figure 4.45.3. Inferred coastal sand transport in 1983 after inlet relocation. With the elimination of tidal flow, the abandoned tidal delta of the old inlet was expected to move onshore by waves and naturally restore the beach along the developed areas southwest of B.

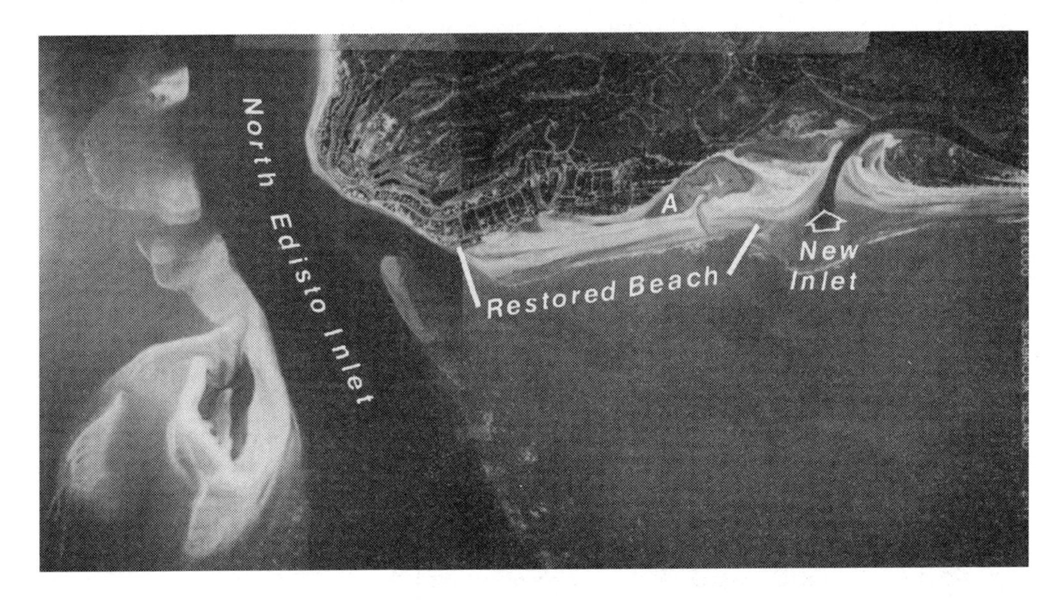

Figure 4.45.4. Vertical aerial photo in April 1987, showing the restored beach four years after inlet relocation. The new beach enclosed añ 10-ha lagoon (A), half of which filled with new salt marsh by 1990.

in a conservation zone that was established at the time of construction to prevent development in the historical floodway. The Seabrook community is making plans to repeat the project at intervals of perhaps 15 years.

Lessons Learned

Seabrook's erosion control experience confirms the importance of regional coastal and inlet processes in site-specific shoreline problems. Property owners first tried structures to protect their investments.

But, as erosion continued, even expensive seawalls failed. By addressing the root causes of erosion, including sand trapping by an inlet well removed from oceanfront properties, the Seabrook community took a chance that nature might help solve the problem.

Inlet relocation restored 70 percent of the shoreline and will last about 15 years before it must be repeated. Its cost was one-tenth the amount applied to seawalls. While such opportunities may not exist at many places, given Seabrook's unusual setting, the community's experience shows the general value of:

- Regional shoreline assessments and erosion analyses before development
- Wider development setbacks around inlets
- Community-wide response to erosion
- Maintenance of the littoral sand flow
- Soft engineering solutions like beach nourishment and inlet relocation

Contributed by -- Timothy W. Kana, Coastal Science and Engineering, Inc., 1316 Main St., P.O. Box 8056, Columbia, SC 29202 USA.

REFERENCES

1. Baca, B. J. and T. E. Lankford. 1987. Environmental report on the Captain Sams Inlet relocation project (March 1983 to July 1987). Prepared for Seabrook Island POA; Coastal Science & Engineering, Inc., Columbia, S.C. 32 pp.
2. CSE. 1989. Beach restoration and shore-protection alternatives along the south end of Seabrook Island: Appendix III—technical criteria and estimates of beach nourishment requirements (18 pp.). In: *Feasibility Study for Seabrook Island POA,* CSE, Columbia, S.C. 38 pp. + appendices.
3. CSE. 1993. Seabrook Island, South Carolina, beach nourishment project. Survey Report No. 4 to Seabrook Island POA; CSE, Columbia, S.C. 34 pp. + Appendix I.
4. Hayes, M. O. 1979. Barrier island morphology as a function of tidal and wave regime. In: *Barrier Islands,* S. Leatherman (Ed.). Academic Press, New York. Pp. 1–26.
5. Hayes, M. O., T. W. Kana, and J. H. Barwis. 1980. Soft designs for coastal protection at Seabrook Island, S.C. In: *Proceedings of the 17th Conference on Coastal Engineering, ASCE, New York.* Ch. 56, pp. 897–912.
6. Hayes, M. O., W. J. Sexton, D. D. Domeracki, T. W. Kana, J. Michel, J. H. Barwis, and T. M. Moslow. 1979. Assessment of shoreline changes, Seabrook Island, South Carolina. Summary Report for Seabrook Island Company; Research Planning Institute, Inc., Columbia, S.C. 82 pp.
7. Kana, T. W. 1989. Erosion and beach restoration at Seabrook Island, South Carolina. *Shore and Beach,* Vol. 57, No. 3. Pp. 3–18.
8. Kana, T. W., W. J. Sexton, L. C. Thebeau, and M. O. Hayes. 1981. Preliminary design and permit application for breaching Kiawah spit north of Captain Sams Inlet. Final Report for Seabrook Island Company; Research Planning Institute, Inc., Columbia, S.C. 43 pp.
9. Kana, T. W., S. J. Siah, and M. L. Williams. 1984. Alternatives for beach restoration and future shoreline management, Seabrook Island, S.C. In: *Feasibility Study for Seabrook Island POA.* RPI Coastal Science & Engineering Division, Columbia, S.C. 130 pp.

4.46 VIETNAM, MEKONG: DIFFICULTY OF REPAIRING DAMAGED WETLANDS

Background

Wetlands play an important role in the Vietnamese economy. The sustainable use of the wetlands of the Red River and Mekong Delta systems is critically important to the long-term economic and social welfare of the Vietnamese people. The two deltas form the foci for concentrations of population and generate a very high percentage of the food and raw materials produced for both domestic and export markets. However, the maximization of food production in the delta's wetland areas has been achieved almost entirely through conversion of the natural ecosystems into agricultural land and aquaculture ponds.

The natural *Melaleuca* wetland forest of the Mekong Delta contains some 77 plant species, many of which are of direct economic importance in terms of timber and secondary forest products, such as firewood and fodder for animals. In addition, the forest system provides between 5 and 6 liters of honey per hectare from wild bees, a species of insectivorous plant is used to form a decoction for treating diarrhea, and many forms of animal are harvested as a source of food.

Despite repeated failures in the conversion of wetland to alternative uses in the Mekong Delta, the conviction remains among some international donors, national agencies, and local people that, given adequate supplies of fresh water to leach out the acid sulphates from wetland soils, rice cultivation will eventually be possible throughout most of the remaining wetlands. However, there are very large reserves of potential acid sulphate in the soils of much of the Mekong Delta, and the feasibility of successfully leaching the acid is extremely questionable. There also remains the question of the environmental and economic impact of the acids on the estuarine and coastal fisheries and other activities downstream.

Problem

Few, if any, conversions of peat soils underlain by acid sulphate clays have created agricultural conditions that are sustainable under levels of investment and management that are economically and technically feasible at this stage in Vietnam's development. In some cases, these man-made changes may prove irreversible and the formerly very productive wetlands may become permanently unusable wastelands. In the Mekong Delta, conversion of floodplain wetlands has not proceeded to the same extent as in the Red River Delta because of the extensive areas of acid sulphate potential clays which, if exposed through drainage, severely limit agricultural production.

A large part of the area has had a complex system of dredged canals cut for agriculture, for water transport, and to form firebreaks to help protect re-forestation sites. This has led to great ecological resource damage: oxidation of peat, subsidence of the soil, compaction, acidification (acid sulphate and carbonic acids) of the drained soils, loss of organic and mineralized soil nutrients, and acidification of ponds and adjacent streams and canals.

Many of the *Melaleuca* forests were drained and sprayed with defoliants (e.g., Agent Orange) and herbicides, and then some where napalmed during the last war. The practice of draining the *Melaleuca* was continued after the war in an attempt to control flooding and to lower the water table to assist in the reclamation of the land for rice-based agriculture.

Fire is the major problem in conserving the remaining forest and replanted areas. The firebreaks are less than 6 meters wide because the foresters fear a loss of productive area. In one area, firebreaks had not been maintained and people using fire to smoke out wild animals for food had set fire to 1,000 ha of 2-year-old *Melaleuca.* The foresters are dredging more canals to act as firebreaks. This is counter-productive because it leads to drying of the peat, loss of soil moisture, prolonged drought conditions, and reduced micro/meso atmospheric humidity. This increases the hazard of fire—underground fires in the peat layer are now common—and increases the costs and difficulties of establishing and conserving replanted areas.

The natural hazard of flooding may also have been increased due to shrinkage of the soils and subsequent subsidence of the ground surface. A greater area of the delta between major river systems may now be subject to more frequent and severe flooding. Over the longer term this may be compounded by the predicted sea level rise for the region. The extent of possible flooding due to sea level rise is difficult to map because the spot heights shown on maps may be well out of date in respect to the changes in the land surface brought about by subsidence resulting from past attempts at reclamation.

Solutions

Under the Socialist government, *Melaleuca* forests are public lands. As such they form common property resources, and the task of conserving and rehabilitating the forests has proven beyond the abilities of the government. The local forestry officials are preparing to give local people a proprietary interest in the management of the forest and plan to allocate 10-ha plots of former forest to local farmers on long leases. The objective is to form a reserve area of production forest surrounded by a series of

10-hectare plots where a system of agro-forestry will be practiced by farmers. Each farmer will be required to replant 7.5 ha of *Melaleuca* and to practice permanent agriculture on the remaining 2.5 ha. Space for a dwelling is to be included in the 10 ha (see Figure 4.46.1).

Although the model is innovative in that it encourages the farmers to rehabilitate and protect the forest by granting them rights of use, it requires modification of the agricultural activities and rehabilitation of the hydrology of the units if it is to prove feasible. It is most likely that field crops would not be sustainable because of acidification and the peaty composition of the soil.

In the Tri Ton area of the northern Mekong Delta, district officials secured international help (via IUCN) for developing a strategy and improved practical techniques for rehabilitating 35,000 ha of former *Melaleuca leucodendron* wetland forest that were destroyed by war and non-sustainable agricultural conversions. A Pilot Project has thus been initiated by the district officials and local farmers to address the following basic problems:

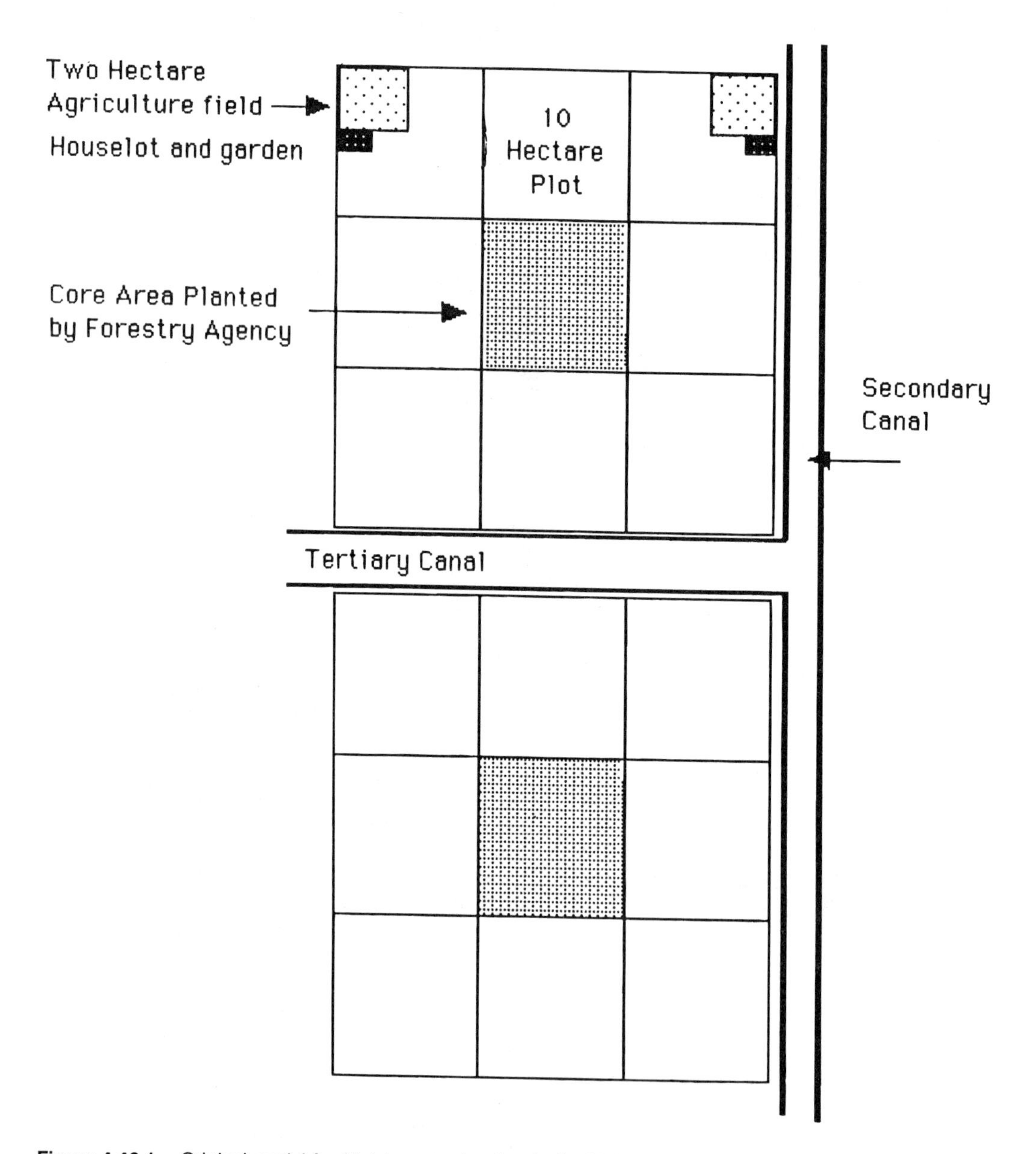

Figure 4.46.1. Original model for *Melaleuca* restoration in the Mekong Delta (Vietnam) project.

1. Restoration of the hydrological characteristics of the wetland as a basic pre-requisite for rehabilitation.
2. Reformation of land holdings and the design of a land use management system to promote sustainable forest utilization. Filed studies by IUCN experts have suggested the following modifications to the original model: (a) designing a more diversified farming system based upon an intensive raised bed mixed cropping system on 0.25–0.3 ha to provide food, fodder, fuelwood, medicinal plants, and cash crops like citrus (similar to the Pekarangan or home gardens in Indonesia), (b) increased re-forestation in place of field crops combined with diversification of use of the secondary resources such as honey production, and (c) diversification of the economic opportunities open to the farmers through the distillation of essential oils from the *Melaleuca* and the collection of medicinal plants.
3. Improvement in the forest management techniques, such as widening of firebreaks to a minimum of 15 meters, trial plantings of pineapple as a cash crop and edible grasses for feeding animals using a "cut and carry" system where the animals are penned and fed the cut grass, and improvement of the selection of seed used to replant areas to improve growth rates and to encourage early flowering to support the start up of honey production as a crop within the third year.

Lessons Learned

1. Local communities have grown to recognize that the wise use of the *Melaleuca* forests provides a greater measure of economic and social value than their reclamation for agriculture. The national and international significance of these wetland systems requires an improved community-based management plan for sustainable use of the *Melaleuca* forests of the Mekong. If international assistance here proves successful, it could be used as a model that can be transferred to other areas of Asia.
2. The agricultural potential of much of the area normally colonized by *Melaleuca* is limited. The reclamation of areas with acid sulphate potential soils and peat more than 1 meter thick leads to very poor agricultural returns and the subsequent abandonment of large areas.
3. Growing recognition of the damage to economic and social welfare resulting from non-sustainable wetland conversion is beginning to change the way some governmental officials approach the development of wetland ecosystems. However, generally it is only where wetland conversions have not proven successful for agriculture or aquaculture or where environmental hazards, such as increased lowland flooding, have been created that government officials and local people are beginning to promote the rehabilitation of wetlands.
4. There are profound problems that hinder national governments in the accomplishment of sustainable management of wetland resources and environmental functions generated by wetlands. These include rapid population growth, marginal agricultural quality of the lands that have not yet been intensively cultivated, degradation of major areas through inappropriate agricultural development, a paucity of information on the ability of land and water resources to respond to development pressures, a critical shortage of domestic capital and a heavy burden of debt associated with the repayment of moneys borrowed to finance the past wars, lack of planning and management skills, and a basic lack of coordination and integration of development initiatives, plans, and investment at all levels of government.
5. The real power to implement changes in wetlands management often lies with provincial and district authorities. In practice, national governments may have limited control. Unless there is popular local support for wetlands conservation, national policies may have little practical influence on the management of wetland ecosystems. Therefore, any national wetland conservation initiative will have to be able to illustrate the value to the economic and social well-being of local people, perhaps through an ICZM approach.

Contributed by — Peter Burbridge, Department of Environmental Science, University of Stirling, Stirling, Scotland, UK.

REFERENCES

1. Burbridge, P. R. 1990. Tidal wetland resources in the tropics. *Journal of Resource Management and Optimization,* Vol. 7, No. 1–4. Pp. 115–138.
2. Burbridge, P. R. and Koesoebiono, 1981. Coastal zone management in Southeast Asia. In: L. S. Chia and C. MacAndrews (Eds.), *Southeast Asian Seas; Frontiers for Development.* McGraw-Hill, Siongapore and London. Pp. 110–135.
3. Burbridge, P. R. and J. E. Maragos. 1985. Coastal Resources Management and Environmental Assessment Needs for Aquatic Resources Development in Indonesia, Jakarta. U.S. Agency for International Development and International Institute for Environment and Development, Washington, D.C.

4.47 WEST INDIES: MONITORING COASTAL EROSION

Background

Many of the Eastern Caribbean Islands are small in size and population, have few natural resources, and are classified as developing countries. The major industries are agriculture, fishing, tourism, and small-scale manufacturing; in addition, service industries such as offshore financing are growing enterprises. Tourism is the fastest growing sector in these economies, and in some countries it is the most important industry. Sandy beaches, clear waters, and coral reefs are vital parts of the overall tourism product.

As coastal development increased in the past 30 years in the smaller Eastern Caribbean Islands, coastal erosion became a major problem, creating environmental and economic problems. The beaches have high value to the tourist industry. A regional program of UNESCO to strengthen local expertise to monitor erosion and to develop solutions resulted in trained personnel in nine islands.

Problem

Before the advent of mass tourism in the 1960s, most of the coastline was relatively undeveloped, with the exception of the main towns, which tended to grow around the ports. However, in the past 30 years, the coastal area immediately behind the beach or cliff has undergone intense development in many islands, with resultant impacts on the beaches.

The development of the shoreline raised serious erosion problems. The erosion is a result of both natural causes, such as hurricanes, winter swells, and land subsidence/sea level rise, and man-made causes, such as beach sand mining, badly placed sea defense structures, and buildings placed too close to the active beach zone. Whether natural or induced, the cause is structures too close to the beach.

While coastal erosion and its resultant problems are clearly evident, there is often little specific information for particular beaches. Most of the islands have populations of less than 100,000 persons and do not have the local technical expertise to manage coastal erosion effectively.

Solutions

UNESCO established a regional program—COSALC I: Beach and Coastal Stability in the Lesser Antilles—within its COMAR (COastal MARine) program in 1985. The overall goal of this program is to strengthen local expertise within the small islands of the Eastern Caribbean to identify, measure, manage, and find solutions to their coastal erosion problems. Since 1993, COSALC I has been administered jointly by UNESCO and the University of Puerto Rico Sea Grant College Program.

The program commenced with assessments of the coastal erosion problems in six islands: Grenada, St. Vincent, St. Lucia, Dominica, Antigua, and St. Kitts Nevis (1). Following this, in-country workshops were held on each island in 1985. The aim of these workshops was to create awareness and to determine each island's specific needs. The major concerns expressed at these workshops were the need for audiovisual aids to prompt awareness and assistance in monitoring the nature and extent of the erosion. As a response to these needs, audio-visual packages were prepared for each island in 1986 to 87, consisting of slides, booklets, and video cassettes individual to each island.

In 1987, a start was made on establishing coastal monitoring programs, with Dominica as the target. A counterpart agency within the country was designated to do the monitoring—in Dominica this was the Forestry and Wildlife Division. Sites for measurement were selected at problem beaches, control sites, beaches affected by sea defense works, and beaches used for sand mining. A UNESCO consultant assisted in the site selection and in training personnel to do the monitoring.

The monitoring consists of measuring beach cross-sections from a fixed reference point behind the beach with simple surveying techniques that involve the use of Abney levels, tape measures, and ranging poles. The sites were measured at three-month intervals.

By 1993 beach monitoring programs were ongoing in nine islands: Anguilla, Antigua, British Virgin Islands, Dominica, Grenada, Montserrat, Nevis, St. Kitts, and St. Lucia, with help from the Organization of Eastern Caribbean States and the Organization of American States (OAS). Some of the longer running programs have shown interesting results, as in Dominica, where the effects of sand mining and the long-term impacts of two major hurricanes in 1989 have been quantified (2).

To date the beach monitoring data have been applied mainly to sea defense planning and port structures. The step from beach monitoring to application of the data to everyday coastal problems, such as coastal development setbacks, planning of coastal developments and infrastructure, and selection of beaches for sand mining, is a major one and will require considerable further training of personnel in the islands, particularly within the planning and environmental agencies.

Lessons Learned

Coastal monitoring programs are an important component of ICZM. They provide necessary information on how the coastal systems are changing due to natural and human-induced influences and thus provide the foundation upon which decisions can be based. The need for continuing quantitative monitoring is as relevant to coastal systems such as coral reefs, mangroves, and seagrass beds as it is to beaches.

For any monitoring program to be effective, clear goals must be defined. For instance, the COSALC beach monitoring program was designed to provide information on medium- to long-term (+5 years) beach changes. However, as the program evolved, some islanders wished to obtain different information from the data (e.g., how much sand can "safely" be mined from an accreting beach). The COSALC monitoring program could be modified to answer such questions.

Key issues in the design of the monitoring program include the following needs:

1. Keep measurement and data analysis techniques as simple as possible.
2. Train several persons from each place in field and analysis techniques, for there is rapid turnover in government staff.
3. Involve persons from different government agencies; this helps to promote and publicize the program and also begin the integration and coordination needed for ICZM.
4. Schedule visits each year by program coordinators to discuss the results and evaluate the work. Such visits also help to ensure that the program remains on schedule.

One of the major lessons learned from COLSAC I is the magnitude of the step from data collection to data application. It is relatively easy to set up monitoring systems to collect data. Actually to use the beach change data within the decision-making process requires considerable additional training of personnel at all levels of government.

Contributed by — Gillian Cambers, COSALC I, Sea Grant College Program, University of Puerto Rico, Mayaguez, P.R.

REFERENCES

1. Cambers, G. 1985. *An Overview of Coastal Zone Management in Six Eastern Caribbean Islands: Grenada, St. Vincent, St. Lucia, Dominica, St. Kitts, Antigua.* ROSTLAC, UNESCO Reg. Off. for Sci. & Tech. for Latin America and the Caribbean (Montevideo).
2. Cambers, G. and A. James. 1994. *Sandy Coast Monitoring: The Dominica Example (1987–1992).* UNESCO Reports in Marine Sciences 63. (In Press).

PART 5

References

1. Abrhabhirama, A. (Ed.). 1987. The Tantalum riot: some lessons learned. In: *Thailand Natural Resources Profile, 1987.* Thailand Development Research Institute, National Environment Board and USAID. Pp. 282–283.
2. ADB. 1993. *Environmental Assessment Requirements and Environmental Review Procedures of the Asian Development Bank.* Asian Development Bank, Office of the Environment (Tokyo). 44 pp.
3. Ahmad, Y. J. and G. R. Sammy. 1985. *Guidelines to Environmental Impact Assessment in Developing Countries.* Hodder and Stoughton, London. 52 pp.
4. Alcala, A. C. and E. D. Gomez. 1987. Dynamiting coral reefs for fish: a resource-destructive fishing method. In: *Human Impacts on Coral Reefs: Facts and Recommendations,* Salvat, B. (Ed.), Antenne Museum E.P.H.E., Fr. Polynesia. Pp. 51–60.
5. Allen, K. O. and J. W. Hardy. 1980. *Impacts of Navigational Dredging on Fish and Wildlife: A Literature Review.* FWS/OBS-80/07. Biological Service Program, U.S. Fish and Wildlife Service, Washington, D.C. 81 pp.
6. Amarasinghe, S. R. and H. J. M. Wickremeratne. 1983. The evolution and implementation of legislation for coastal zone management in a developing country—The Sri Lankan experience. In: *Coastal Zone '83,* Vol. III. American Society of Civil Engineers, New York. Pp. 2822–2841.
7. Amarasinghe, S. R., H. J. M. Wickremeratne, and G. K. Lowry, Jr. 1987. Coastal zone management in Sri Lanka 1978–1986. In: *Coastal Zone '87,* Vol. III, American Society of Civil Engineers, New York. Pp. 2822–2841.
8. Anderson, D. M. and A. W. White. 1989. Toxic dinoflagellates and marine mammal mortalities. In: *Proceedings of Expert Consultation.* Woods Hole Ocean. Inst. Tech. Report 89–36.
9. ASPO. 1972. *Regulations for Flood Plains.* Am. Soc. of Planning Officials, Advisory No. 277.
10. Baden, D. 1990. Toxic fish. *Sea Frontiers,* May–June, 1990. Pp. 11–14.
11. Bagnis, R. 1987. Ciguatera fish poisoning: an objective witness of the coral reef stress. In: *Human Impacts on Coral Reefs: Facts and Recommendations,* Salvat, B. (Ed.). Antenne Museum E.P.H.E., French Polynesia. Pp. 141–253.
12. Baines, G. 1986. Unpublished report on Morovo Lagoon Resource Management Project: Solomon Islands.
13. Baker, I. and P. Kaeonian (Eds.). 1986. *Manual of Coastal Development Planning and Management for Thailand.* U.N. Educational, Scientific and Cultural Organisation, Jakarta. 179 pp.
14. Baldwin, C. L. (Ed.), 1987. *Nutrients in the Great Barrier Reef Region,* Great Barrier Reef Marine Park Authority, Workshop Series. No. 10. 191 pp.
15. Barg, U. C. 1992. *Guidelines for the Promotion of Environmental Management of Coastal Aquaculture Development.* FAO Fisheries Technical Paper No. 328. U.N. Food and Agriculture Organisation, Rome. 122 pp.
16. Bascom, W. 1974. The disposal of waste in the ocean. *Scientific American,* Vol. 231, August 1974.
17. Bassey, G. 1990. Personal communication.
18. Beanlands, G. E. 1985. Basic approaches to EIA. In: *Proceedings of the Caribbean Seminar on Environmental Impact Assessment,* T. Goeghegan (Ed.). Caribbean Conservation Association, Barbados. pp. 64–81.
19. Bell, M. 1973. *Fisheries Handbook of Engineering Requirements and Biological Criteria.* U.S. Army Corps of Engineers, Fisheries-Engineering Research Program, Portland, OR.

20. Bell, P. R. 1991. Impact of wastewater discharges from tourist resorts on eutrophication in coral reef regions and recommended methods of treatment. In: *Proceedings of the 1990 Congress on Coastal and Marine Tourism,* Vol. II. National Coastal Resources Research & Development Institute, Newport, Oregon, USA. Pp. 317–322.
21. Bell, P. R. and P. F. Greenfield. 1988. Monitoring treatment and management of nutrients in wastewater discharges to the GBRMP. In: *Nutrients in the Great Barrier Reef Region,* Baldwin, C. L. (Ed.), Great Barrier Reef Marine Park Authority, Workshop Series No. 10. Pp. 55–65.
22. Bell, P. R., P. F. Greenfield, D. Hawker, and D. Connell. 1989. The impacts of waste discharges on coral reef regions. *Water Science Technology,* Vol. 21, No. 1. Pp. 121–130.
23. Bender, W. H. 1972. *Soils and Septic Tanks.* Soil Conservation Service, Ag. Info. Bull. 349. U.S. Department of Agriculture, Washington, D.C.
24. Bigford, T. E. 1991. Sea-level rise, nearshore fisheries, and the fishing industry. *Coastal Management,* Vol. 19. Pp. 417–437.
25. Birkeland, C. 1988. Second-order ecological effects of nutrient input into coral communities. *Galaxea* 7. Pp 91–100.
26. Blankenship, K. 1990. *Saving SAV.* Chesapeake Citizen Report/US Environmental Protection Administration. Pp. 6–7.
27. Blommestein, E. 1988. Environment in Caribbean development, a regional view. In: *Environmentally Sound Tourism in the Caribbean,* F. Edwards (Ed.). University of Calgary Press, Calgary, Alberta, Canada.
28. Bossi, R. and G. Cintron. 1990. *Mangroves of the Wider Caribbean.* U.N. Environment Program, Nairobi. 33 pp.
29. Brodie, J. E. and Ryan, P. A. 1987. *Environmental Monitoring of the Treasure Island Marine Sewage Outfall.* Environmental Studies Report No. 36. Institute of Natural Resources, University of the South Pacific. Suva. 23 pp.
30. Brown, L. R. 1985. *State of the World.* W.W. Norton & Co., New York. 301 pp.
31. Burbridge, P. R. 1986. The importance of multiple use management strategies for the sustainable development of coastal resource systems (unpublished manuscript). 14 pp.
32. Burbridge, P. R. 1987. Economic considerations in management of coastal resources. *CAMP Network Newsletter,* May 1987. Pp. 2–3.
33. Burbridge, P. R., N. Dankers, and J. R. Clark. 1989. Multiple-use assessment for coastal management. In: *Coastal Zone '89,* Vol. I, American Society of Civil Engineers, New York. Pp. 33–45.
34. Burbridge, P. R. and Koesoebiona. 1981. Coastal zone management in Southeast Asia. In: *Southeast Asia Seas: Frontiers for Development.* McGraw-Hill. Pp. 110–135.
35. Burbridge, P. R. and J. E. Maragos. 1985. Analysis of Environmental Assessment and Coastal Resources Management Needs (Indonesia). International Institute for Environmental Development, Washington, D.C. (Draft report)
36. Burbridge, P. R. and J. A. Stanturf. (in press). The Development and Management of Tidal Wetlands in Southeast Asia. U.N. Educational, Scientific and Cultural Organisation, Paris.
37. Burgett, J. 1991. Diving deeper into Hanauma: identifying conceptual barriers to effective management. In: *Proceedings of the 1990 Congress on Coastal and Marine Tourism,* Vol. II, National Coastal Resources Research & Development Institute, Newport, Oregon, USA. Pp. 100–105.
38. Caddy, J. 1990. Enclosed and semi-enclosed seas: principal issues and future actions. Informal paper. U.N. Food and Agriculture Organization, Fisheries/Environment Division, Rome. 5 pp.
39. Caddy, J. 1990. The economic exploitation of renewable resources of natural ecosystems. Unpublished manuscript. U.N. Food and Agriculture Organization, Rome.
40. Cambers, G. 1985. *An Overview of Coastal Zone Management in Six East Caribbean Islands.* U.N. Educational, Scientific and Cultural Organization Office for Science and Technology, Montevideo. 69 pp.
41. Carpenter, R. A. and J. E. Maragos. 1989. *How to Assess Environmental Impacts on Tropical Islands and Coastal Areas.* East-West Center, Honolulu, Hawaii. 345 pp.
42. Carroll, A. 1976. *Developer's Handbook,* Connecticut Department of Environmental Protection, Hartford, Conn. 60 pp.
43. CCD. 1990. *Coastal Zone Management Plan.* Sri Lanka Coast Conservation Department, Colombo, Sri Lanka. 110 pp.

44. CEQ. 1978. Definition of mitigation in U.S. Code 40 CFR Pt. 1508.30 (a–e). U.S. Council on Environmental Quality, Washington, D.C.
45. Chapman, C. 1983. Freshwater discharge. In: *Coastal Ecosystems Management: A Technical Manual for the Conservation of Coastal Zone Resources,* J. R. Clark (Ed.), Robt. E. Krieger Publishing Co., Melbourne, FL. Pp. 634–639.
46. Chaverri, R. 1989. Coastal management: the Costa Rica experience. In: *Coastal Zone '87,* Vol. V. American Society of Civil Engineers, New York. Pp. 5273–5285.
47. Chong, K. C. and I. Manwan. 1987. Incorporating integrated rural development into coastal resources use and management. In: *Studi Kasus Penelitian Agro-ekosistem.* KEPAS, Badan Penelitian dan Pemengambang Pertanian, Jakarta. Pp. 131–169.
48. Chou, L. M. 1984. A review of coral reef survey and management methods in Singapore. In: *Comparing Coral Reef Survey Methods.* U.N. Educational, Scientific and Cultural Organization, Paris. Pp. 36–46.
49. Chou, L. M. 1991. Some guidelines in the establishment of artificial reefs. *Tropical Coastal Area Management.* Vol. 6, Nos. 1/2, International Center for Living Aquatic Resources, Management, Manila. Pp. 4–5.
50. Choudhury, A. K. 1987. Climatology of the estuarine region of West Bengal. U.N. Educational, Scientific and Cultural Organization training course (Calcutta). Wildlife Institute of India, Dehradun. Pp. 63–73.
51. Chua, T. E. 1986. Managing ASEAN coastal resources. *Tropical Coastal Area Management,* Vol. 1, No. 1. International Center for Living Aquatic Resources Management, Manila. Pp. 8–10.
52. Chua, T. E. 1991. Personal communication.
53. Chua, T. E. and J. R. Charles. 1984. *Coastal Resources of East Coast Peninsular Malaysia.* Penerbit Universiti Sains Malaysia. 306 pp.
54. Chua, T. E. and L. F. Scura. 1992. *Integrative Framework and Methods for Coastal Area Management.* International Center for Living Aquatic Resources Management, Manila. 169 pp.
55. Chua, T. E. and A. T. White. 1988. Policy recommendations for coastal area management in the ASEAN region. *ICLARM Contributions,* No. 544. Pp. 5–7.
56. Claridge, G. 1988. Assessing development proposals. In: *Coral Reef Management Handbook,* Kenchington, R. A. and B. E. T. Hudson (Eds.). UNESCO Regional Office for Science and Technology for Southeast Asia, Jakarta. Pp. 131–138.
57. Clark, J. R. 1967. *Fish and Man: Conflict in the Atlantic Estuaries.* American Littoral Society, Special Publication No. 5. Highlands, N.J. 78 pp.
58. Clark, J. R. 1980. *Coastal Environmental Management, Guidelines for Conservation of Resources and Protection against Storm Hazards.* Federal Emergency Management Agency, Washington, D.C. 161 pp.
59. Clark, J. R. 1983. *Coastal Ecosystems Management: A Technical Manual for the Conservation of Coastal Zone Resources.* Robert E. Krieger Publishing Co., Melbourne, Fla. (First edition by John Wiley-Interscience, New York, 1977). 928 pp.
60. Clark, J. R. 1985. *Recommendations for a Combined Coastal Management and Protected Areas Program for the Saudi Arabia Red Sea Coast.* Summary Report submitted to the International Union for the Conservation of Nature, Gland (Switzerland) by the National Park Service (USA), Washington, D.C. 63 pp.
61. Clark, J. R. (Ed.). 1985. *Coastal Resources Management: Development Case Studies.* Coastal Management Publication No. 3, NPS/AID Series, Research Planning Institute, Columbia, S.C. 749 pp.
62. Clark, J. R. 1985. *Snorkeling, a Complete Guide to the Underwater Experience.* Prentice Hall, Englewood Cliffs, N.J. 191 pp.
63. Clark, J. R. 1990. *Evaluation of Khulna Embankment Rehabilitation Project.* Report submitted to UNDP/Asian Development Bank, Dhaka, Bangladesh. 41 pp. plus Figures and Annexes.
64. Clark, J. R. (Ed.). 1991. *Carrying Capacity, a Status Report on Marine and Coastal Parks and Reserves.* University of Miami/RSMAS, Miami, Fla. 73 pp.
65. Clark, J. R. (Ed.). 1991. *The Status of Integrated Coastal Zone Management: A Global Assessment.* CAMPNET, University of Miami/RSMAS, Miami, Fla. 118 pp. and appendices.
66. Clark, J. R. 1992. *Integrated Management of Coastal Zones.* FAO Fisheries Technical Paper No. 327. United Nations/FAO, Rome. 167 pp.
67. Clark, J. R. 1992. *Ujung Pandang Port Urgent Rehabilitation Project: Environmental Report.* Report to Nippon Koei, Tokyo, Japan, for Indonesia Department of Communications. 3 Sections plus Appendix.

68. Clark, J. R. 1992. *Urgent Bali Beach Conservation Project: Environmental Study Report,* Vol. 1: *Main Report.* Report to Nippon Koei, Tokyo, Japan. 88 pp.
69. Clark, J. R. 1992. *Urgent Bali Beach Conservation Project: Environmental Management Report,* Vol. 2. Report to Nippon Koei, Tokyo, Japan. 127 pp.
70. Clark, J. R. 1992. *Urgent Bali Beach Conservation Project: RPL Environmental Monitoring Plan and Data Collection Report,* Vol. 3. Report to Nippon Koei, Tokyo, Japan. 92 pp.
71. Clark, J. R. 1994. *Initial Environmental Examination Report for the Tree and Palm Plantation Projects in Cyclone Prone Areas.* Fountain Renewable Resources, London, and Asian Development Bank, Dhaka. 59 pp.
72. Clark, J. R. 1980. Progress in management of coastal ecosystems. *Helgolander Meeresunters* No. 33. Pp. 721–731.
73. Clark, J. R. 1983. Mitigation and grassroots conservation of wetlands—urban issues. In: *The Mitigation Symposium.* U.S. Forest Service, Ft. Collins, Colo., General Technical Report RM-65. Pp. 141–151.
74. Clark, J. R. 1983. Foreword. *The Future of Wetlands, Assessing Visual-Cultural Values,* Smardon, R. C. (Ed.). Allanhead, Osmun, Publishers, Totowa, N.J. 226 pp.
75. Clark, J. R. 1986. Assessment for wetlands restoration. In: *Proceedings: National Wetlands Assessment Symposium, Portland, Maine,* Kusler, J. and P. Riexinger (Eds.). Association of State Wetland Managers, Berne, N.Y. Pp. 250–253.
76. Clark, J. R. 1987. Combining hazards mitigation and resources management through coastal area planning. In: *Proceedings of the International Symposium on Post-Disaster Response and Mitigation of Future Losses.* American Bar Association Committee on Housing and Urban Development, Washington, D.C.
77. Clark, J. R. 1987. The role of protected areas in regional development. In: *Ecological Development in the Humid Tropics: Guidelines for Planners,* A. Lugo, J. R. Clark, and R. Child (Eds.). Winrock International (with USAID and the U.S. National Park Service), Arkansas. Pp. 139–168.
78. Clark, J. R. 1988. Rehabilitation of coral reef habitats. Report of a Science Workshop held at St. John, USVI, December 1987 (unpub.). University of Miami and U.S. National Park Service. 16 pp. and attachments.
79. Clark, J. R. 1990. Management of environment and natural disasters in coastal zones. Paper presented at Colloqium on the Environment and Natural Disaster Management, World Bank, Washington, D.C., June 1990. 20 pp.
80. Clark, J. R. 1990. Regional aspects of wetlands restoration and enhancement in the urban waterfront environment. In: *Wetland Creation and Restoration, the Status of the Science,* Kusler, J. and M. Kentula (Eds.). Island Press, Washington, D.C. Pp. 497–515.
81. Clark, J. R. 1991. Management of coastal barrier biosphere reserves. *Bioscience,* Vol. 41, No. 5. Pp. 331–336.
82. Clark, J. R. 1992. Carrying capacity and tourism in coastal and marine areas. *Parks Magazine,* Vol. 2, No. 3. Pp. 13–17. (Reprinted in *Marine Parks Journal,* 95, Mar. 1992, Tokyo).
83. Clark, J. R., Al Gain, A., and Chiffings, T. 1987. *A Coastal Management Program for the Saudi Arabian Red Sea Coast.* In: *Coastal Zone '87,* Vol. II. American Society of Civil Engineers, New York. Pp. 1673–1681.
84. Clark, J. R., J. S. Banta and J. A. Zinn. 1980. *Coastal Environmental Management, Guidelines for Conservation of Resources and Protection against Storm Hazards.* U.S. Government Printing Office, Washington, D.C. 161 pp.
85. Clark, J. R., B. Causey, and J. A. Bohnsack. 1989. Benefits from coral reef protection: Looe Key Reef, Florida. In: *Coastal Zone '89,* Vol. 4. American Society of Civil Engineers, New York. Pp. 3076–3086.
86. Clark, J. R., J. Sorensen, and G. Schultink. 1993. *Integrated Coastal Zone Management Strategy.* Tropical Resources and Development, Inc., Gainesville, Fla. Report to UNDP and Planning Commission of Bangladesh, Dhaka. 62 pp.
87. Clarke, W. 1989. Human ecology in the pre-commercial Pacific: An application of social science methods. In: *How to Assess Environmental Impacts on Tropical Islands and Coastal Areas,* Carpenter, R. A. and J. E. Maragos (Eds.). East-West Center, Honolulu, Hawaii. Pp. 70–72.
88. Classen, D. B. van R. 1989. Map analysis technique. In: *How to Assess Environmental Impacts on Tropical Islands and Coastal Areas,* Carpenter, R. A. and J. E. Maragos (Eds.). East-West Center, Honolulu, Hawaii. Pp. 105–108.

89. Classen, D. B. van R. and P. A. Pirazzoli. 1988. Remote sensing: a tool for management. In: *Coral Reef Management Handbook,* Kenchington, R. A. and B. E. T. Hudson (Eds.). United Nations, UNESCO Regional Office for Science and Technology for South-East Asia, Jakarta. Pp. 69–88.
90. C. M. F. 1993. Seaweed culture. Centre for Marine Fisheries Research, Cochin, India. 2 pp.
91. Coche, A. G. 1985. *Soil and Freshwater Fish Culture.* FAO Training Series No. 6. U.N. Food and Agriculture Organization, Rome. 174 pp.
92. Cocks, K. D. 1984. A systematic method of public use zoning of the Great Barrier Reef Marine Park, Australia. *Coastal Zone Management Journal,* Vol. 12, No. 4. Pp. 359–382.
93. Coetzee, M. D. 1991. The importance of public participation in coastal management. In: *The Status of Integrated Coastal Zone Management: A Global Assessment,* Clark, J. R. (Ed.). CAMPNET/ University of Miami/RSMAS, Miami, Fla. Pp. 29–30.
94. Cohen, F. G. 1985. Social impact assessment in the context of socio-cultural change. In: *Proceedings of the Caribbean Seminar on Environmental Impact Assessment.* Caribbean Conservation Association, Barbados. Pp. 82–89.
95. Conservation Foundation. 1979. *Apalachicola Symposium and Workshops.* Conservation Foundation, Washington, D.C. 70 pp.
96. Cotter, P. J. 1982. Barbados' new marine reserve. *Parks,* Vol. 7, No. 1. Pp. 8–11.
97. CRD. 1992. The planning process and public involvement. Information Paper No. 1. Capital Regional District, Victoria, B.C., Canada. 8 pp.
98. Cruikshank, K. 1991. Where's the beach? Managers report. *Journal for Community Associations,* Vol. 5, No. 4.
99. CSE. 1990. *Erosion Assessment and Beach Restoration Alternatives for Hunting Island, South Carolina.* Coastal Science and Engineering, Columbia, S.C. 66 pp. and appendices.
100. Dahl, A. L. 1984. *Coral Reef Monitoring Handbook.* Reference Methods for Marine Pollution Studies No. 25. U.N. Environment Programme, Nairobi. 25 pp.
101. Dahuri, R. 1991. The Need for Integrated Coastal Resource Management. In: *The Status of Integrated Coastal Zone Management: A Global Assessment,* Clark, J. R. (Ed.). CAMPNET/University of Miami/ RSMAS, Miami, Fla. Pp. 108–111.
102. Dames and Moore, Inc. 1981. *New Jersey Shore Protection Master Plan.* N.J. Dept of Environmental Protection, Trenton, N.J.
103. Davies-Colley, R. J. 1988. Secchi disc transparency and turbidity. *Journal of Environmental Engineering,* Vol. 116, No. 4. Pp. 796–800.
104. Day, J. W., C. A. S. Hall, W. M. Kemp, and A. Yáñez-Arancibia. 1989. *Estuarine Ecology.* John Wiley and Sons, New York. 558 pp.
105. Debrot, A. O. 1988. Speech to Plenary Session on coral reefs. Caribbean Conservation Association Annual Meeting. Pp. 30–33.
106. DeSilva, M. W. R. N. 1984. Coral reef assessment and management methodologies curently used in Malaysia. In: *Comparing Coral Reef Survey Methods,* UNESCO Reports in Marine Science 21. U.N. Educational, Scientific and Cultural Organization, Paris. Pp. 47–56.
107. Division of Coastal Management. 1986. *Protecting Coastal Waters through Local Planning,* North Carolina Department of Natural Resources, Raleigh, N.C. 92 pp.
108. Division of Planning. 1973. Bay and estuarine system management in the Texas coastal zone. In: *Project and Report Summaries,* Texas Division of Planning Coordination, Austin, Texas.
109. Dixon, J. A. 1989. Coastal resources: assessing alternatives. In: *Coastal Area Management in Southeast Asia: Policies, Management Strategies and Case Studies,* Chua, T. E. and D. Pauly (Eds.). ICLARM Conference Proceedings 19, International Center for Living Aquatic Resources Management, Manila. Pp. 153–162.
110. Dixon, J. A., R. A. Carpenter, L. A. Fallon, P. B. Sherman, and S. Manopimoke. 1988. *Economic Analysis of the Environmental Impacts of Development Projects.* Earthscan Publications, Ltd., London, in association with the Asian Development Bank, Manila, Philippines. 109 pp.
111. Dixon, J. A., L. F. Scura, and T. van't Hof. 1993. Meeting ecological and economic goals: marine parks in the Caribbean. *Ambio,* Vol. 22, No. 2–3. Pp. 117–125.
112. Dizon, C. P. 1986. Coral reef ecosystems of the Philippines: its problems, needs and management. In: *Proceedings of MAB-COMAR Regional Workshop on Coral Reef Ecosystems: Their Management*

Practises and Research/Training Needs, Soemodihardjo, S. (Ed.). United Nations/UNESCO. Pp. 129–132.

113. DOE. 1984. Environmental guidelines for siting of industry. India Department of Environment, New Delhi. 10 pp.
114. Dovel, W. L. 1971. *Fish Eggs and Larvae of the Upper Chesapeake Bay.* Special Report No. 4. National Research Institute, University of Maryland, College Park, Maryland.
115. Dubois, R. 1984. Personal communication.
116. Dugan, P. J. (Ed.). 1990. *Wetland Conservation: A Review of Current Issues and Required Action.* International Union for the Conservation of Nature, Gland, Switzerland. (Unpublished)
117. Dunkel, D. R. 1984. Tourism and the environment: a review of the literature and issues. *Environmental Sociology,* No. 37, Spring.
118. Dunne, T. and L. B. Leopold. 1978. *Water in Environmental Planning.* W. H. Freeman, San Francisco. 818 pp.
119. Eaton, P. 1985. Land tenure and conservation in protected areas in the South Pacific. *SPREP Topic Review,* Vol. 17.
120. ECLAC/UNEP. 1985. *Tourism and Environment in Caribbean Development with Emphasis on the Eastern Caribbean.* WP/ETCD/L85/2. Economic Commission for Latin America and the Caribbean, Subregional Headquarters for the Caribbean, Port of Spain, Trinidad and Tobago.
121. ECSOC. 1987. Development of marine areas under national jurisdiction: problems and approaches in policy-making, planning and management. U.N. Economic and Social Council, Rome. 22 pp.
122. EFL. 1991. *National Environmental Act.* Environmental Foundation, Ltd., Colombo, Sri Lanka. 62 pp.
123. EMDI. 1990. *Zoning in Marine and Coastal Areas.* Environmental Management Development in Indonesia, Jakarta.
124. ESCAP. 1985. *Environmental Impact Assessment: Guidelines for Planners and Decision Makers.* U.N. Publication ST/ESCAP/351. U.N. Economic and Social Commission for Asia and the Pacific, New York.
125. ESCAP. 1990. *Environmental Impact Assessment: Guidelines for Industrial Development.* U.N. Publication ST/ESCAP/784. U.N. Economic and Social Commission for Asia and the Pacific, New York. 62 pp.
126. FAO. 1982. *Management and Utilization of Mangroves in Asia and the Pacific.* FAO Environment Paper 3. U.N. Food and Agriculture Organization, Rome. 160 pp.
127. FAO. 1985. *Mangrove Management in Thailand, Malaysia and Indonesia.* FAO Environment Paper 4, U.N. Food and Agriculture Organization, Rome. 60 pp.
128. Fay, J. A. 1971. *Jamaica Bay and Kennedy Airport, a Multidisciplinary Environmental Study,* Vols. I and II. National Academy of Sciences-National Academy of Engineering, Washington, D.C.
129. Fleming, C. A. 1990. *Guide on the Uses of Groynes in Coastal Engineering.* Report No. 119, Construction Industry Research and Information Association, London.
130. Forqurean, J. W., J. C. Zieman, and G. V. N. Powell. 1992. Phosphorous limitation of primary production in Florida Bay: evidence from C:N:P ratios of the dominant seagrass *Thalassia testudinum. Limnology and Oceanography,* Vol. 37, No. 1. Pp. 162–171.
131. FPCO. 1992. *Guidelines for Environmental Impact Assessment (EIA).* Flood Plain Coordination Organization, Ministry of Irrigation, Water Development and Flood Control. Dhaka, Bangladesh. 75 pp.
132. Francis, C. H. 1990. Coastal beach erosion and a building setback policy at Grand Anse Bay, Grenada. *Proceedings of Workshop on Coastal Protected Areas in the Lesser Antilles.* University of the Virgin Islands, St. Thomas, V.I. Pp. 35–39.
133. Fraser, R. J., Jr. 1993. Removing contaminated sediments from the coastal environment: the New Bedford Harbor Project example. *Coastal Management,* Vol. 19, No. 2. Pp. 155–162.
134. Freestone, D. 1991. ICZM: The problem of boundaries. In: *The Status of Integrated Coastal Zone Management: A Global Assessment,* J. R. Clark (Ed.). CAMPNET, University Of Miami/RSMAS, Miami, Fla. Pp. 75–77.
135. FRRL. 1994. *Tree and Palm Plantation Project in Cyclone-prone Areas of Bangladesh (The Coastal Greenbelt Proposal).* Main Report. ADB TA No. 1816-BAN. Fountain Renewable Resources Ltd., Brackley, Northamptonshire, U.K., for Asian Development Bank. 109 pp.
136. Gammon, J. K. and S. T. McCreary. 1988. Suggestions for integrating EIA and economic development in the Caribbean region. In: *Environmental Impact Assessment Review,* Vol. 8. Elsevier, Amsterdam.
137. Gawel, M. 1988. Involvement of the users of coral reef resources in management plans. In: *Coral*

Reef Management Handbook, Kenchington, R. A. and B. E. T. Hudson (Eds.). United Nations, UNESCO Regional Office for Science and Technology for South-East Asia, Jakarta. Pp. 107–117.

138. GBRMPA. 1992. Spearfishing in the marine park. *Reflections,* Great Barrier Reef Marine Park Authority. Pp. 14–15.
139. GBRMPA. 1986. *Michaelmas Management Plan.* Great Barrier Reef Marine Park Authority and Queensland National Parks and Wildlife Service, Queensland, Australia.
140. Geoghegan, T. 1983. *Guidelines for Integrated Marine Resource Management in the Eastern Caribbean.* Caribbean Conservation Association, Caribbean Environment Technical Report No. 2. 56 pp.
141. Geoghegan, T., I. Jackson, A. Putney, and Y. Renard. 1984. *Environmental Guidelines for Development in the Lesser Antilles.* Eastern Caribbean Natural Areas Management Program, St. Croix, U.S.V.I. 44 pp.
142. Glineur, N. 1991. Conservation of biodiversity in the Mediterranean: protection of coastal ecosystems. Mediterranean Environment Program, World Bank, European Investment Bank, Working Paper No. 7. World Bank, Washington, D.C. 35 pp and Annexes.
143. Glynn, P. 1988. El Niño warming, coral mortality and reef framework destruction by echinoid bioerosion in the Eastern Pacific. In: *Proceedings of MAB/COMAR MICE IV Meeting: Asian and Pacific Regional Workshop and International Symposium on the Conservation and Management of Coral Reef and Mangrove Ecosystems,* September–October, 1987. University of the Ryukyus, Okinawa, Japan. Pp. 129–160.
144. Goldberg, E. D. 1993. Competitors for coastal ocean space. *Oceanus,* Vol. 36, No. 1. Pp. 12–18.
145. Gomez, E. D. and H. T. Yap. 1988. Monitoring reef conditions. In: *Coral Reef Management Handbook,* Kenchington, R. A. and B. E. T. Hudson (Eds.). United Nations, UNESCO Regional Office for Science, Jakarta. Pp. 187–195.
146. Goodwin, M. 1989. South Carolina Seagrant. Personal communication.
147. Gosselink, J. 1984. *The Ecology of Delta Marshes of Coastal Louisiana: a Community Profile.* FWS/OBS-84/09, U.S. Fish and Wildlife Service, Washington, D.C. 134 pp.
148. Gow, D. D. 1992. Personal communication.
149. Griffith, S. A. and E. Williams. 1985. *Corals and Coral Reefs in the Caribbean: A Manual for Students.* Caribbean Conservation Association, Barbados. 48 pp.
150. Gritzner, J. A. 1986. National Research Council, Washington, D.C. Personal communication (letter of 22 April).
151. Guzman, H. M. 1991. Restoration of coral reefs in Pacific Costa Rica. *Conservation Biology,* Vol. 5, No. 2. Pp. 189–195.
152. Haider, R., A. A. Rahman, and S. Huq. 1991. *Cyclone '91: An Environmental and Perceptional Study.* Bangladesh Center for Advanced Studies, Dhaka. 91 pp.
153. Hallock-Muller, P. 1990. Coastal pollution and coral communities. *Underwater Naturalist,* Vol. 19, No. 1. Pp. 15–18.
154. Hallock, P. and W. Schlager. 1986. Nutient excess and the demise of coral reefs and carbonate platforms. *PALAIOS,* Vol. 1, Pp. 389–398.
155. Hamilton, L. 1984. Targets and non-targets for action by World Bank in watershed management and rehabilitation. Fourth Agricultural Sector Symposium, World Bank, Washington, D.C.
156. Hamilton, L. and S. Snedaker. 1984. *Handbook for Mangrove Area Management.* East/West Center, Hawaii. 123 pp.
157. Hammill, S. M. 1974. *Environmental Profile and Guidelines.* Township of Middletown, N.J.
158. Hammond, W. G. 1876. *Introduction to the American Edition of Thomas Collett Sandars' Institutes of Justinian.* First American Edition, Callaghan & Co., Chicago. 683 pp.
159. Hanson, A. J. 1987. The concept of integrated planning and management. In: *Ecological Development in the Humid Tropics: Guidelines for Planners,* A. Lugo, J. R. Clark, and R. Child (Eds.). Winrock International, Arkansas, and USAID and U.S. National Park Service, Washington, D.C. Pp. 39–63.
160. Hanson, A. J. and Kosoebiona. 1977. *Settling Coastal Swamplands in Sumatra, a Case Study for Integrated Resource Management.* Center for Natural Resources Management and Environmental Studies, Bogor Agricultural University, Indonesia. 46 pp.
161. Hansen, J. 1993. Goddard Institute for Space Studies. Personal communication.
162. Hawker, D. W. and D. W. Connell. 1992. Standards and criteria for pollution control in coral reef areas. In: *Pollution in Tropical Aquatic Systems.* CRC Press, Boca Raton, Ann Arbor, London, Pp. 169–191.
163. Hayes, M. O. 1985. Beach erosion. In: *Coastal Resources Management: Development Case Studies,* J. R. Clark (Ed.). Coastal Publication No. 3, Research Planning Institute, Columbia, S.C. Pp. 67–200.

164. Healy, R. G. and J. A. Zinn. 1985. Environment and development conflict in coastal zone management. *Journal of the American Planning Association.* Vol. 51, No. 3. Pp. 299–311.
165. Heinen, E. T. (Ed.). 1985. *Coastal Marinas Assessment Handbook.* U.S. Environmental Protection Agency, Atlanta, Ga.
166. Heyman, A. M. 1986. *Inventory of Caribbean Marine and Coastal Protected Areas.* Organization of American States, Washington, D.C. 146 pp.
167. Hildebrand, L. P. 1989. *Canada's Experience with Coastal Zone Management.* Oceans Institute of Canada, Halifax. 118 pp.
168. Hinsley, T. D. 1973. Sludge recycling: the most reasonable choice? *Water Spectrum,* Vol. 5, No. 1. U.S. Army Corps of Engineers.
169. Ho, K. C. and I. J. Hodgkiss. 1993. Macronutrients in Tolo Harbor and their relations to red tide. *EMECS Newsletter,* No. 4. Hyogo Prefecture Government, Japan. Pp. 10–11.
170. Hodgson, G. and J. A. Dixon. 1988. *Logging versus Fisheries and Tourism in Palawan.* Occasional Paper No. 7, Environment and Policy Institute, East-West Center, Hawaii. 95 pp.
171. Holder, J. S. 1988. The pattern and impact of tourism on the environment of the Caribbean. In: *Environmentally Sound Tourism in the Caribbean,* F. Edwards (Ed.). University of Calgary Press, Calgary, Alberta, Canada.
172. Holl, K., G. Daily, and P. R. Erlich. 1990. Integrated pest management in Latin America. *Environmental Conservation,* Vol. 17. Pp. 341–350.
173. Holzer, T. L. 1985. Land subsidence. *Ground Failure,* No. 2. National Research Council, Committee on Ground Failures. Pp. 10–12.
174. Hopley, D. 1978. Aerial photography and other remote sensing techniques. In: *Coral Reefs: Research Methods,* D. R. Stoddard and R. E. Johannes (Eds.). U.N. Educational, Scientific and Cultural Organization, Paris. Pp. 23–44.
175. House of Representatives. 1991. *The Injured Coastline: Protection of the Coastal Environment.* Report of House of Representatives, Parliament of the Commonwealth of Australia. 126 pp.
176. Howard, L. S. and B. E. Brown. 1984. Heavy metals and reef corals. *Oceanography and Marine Biology Annual Review.* Pp. 195–210.
177. Hudson, B. E. T. 1988. User and public information. In: *Coral Reef Management Handbook,* R. A. Kenchington and B. E. T. Hudson (Eds.). UNESCO Regional Office for Science and Technology for South-East Asia, Jakarta. Pp. 163–176.
178. Hudson, H. 1992. Personal communication.
179. Hughes, P. J. 1989. Archaeological surveys and cultural resource management. In: *How to Assess Environmental Impacts on Tropical Islands and Coastal Areas,* R. A. Carpenter and J. E. Maragos (Eds.). East-West Center, Hawaii. Pp. 73–76.
180. Hutchins, P. and P. Saenger. 1987. *Ecology of Mangroves.* University of Queensland Press, St. Lucia, Australia. 388 pp.
181. Ibé, A. C. and R. E. Quélennec. 1989. *Methodology for Assessment and Control of Coastal Erosion in West and Central Africa.* Regional Seas Reports and Studies 107, U.N. Environmental Program, Nairobi. 100 pp.
182. ICES. 1985. *Revised Code of Practise to Reduce the Risks for Adverse Effects Arising from Introduction and Transfer of Marine Species.* International Council for Exploration of the Sea, Charlottenlund, Denmark.
183. ICLARM. 1986. *Tropical Coastal Area Management Newsletter.* Vol. 1, No. 1. International Center for Living Aquatic Resources Management, Manila.
184. ICLARM. 1988. Blast fishing: a Phillipine case study. In: *Tropical Coastal Area Management,* Vol. 3, No. 1. International Center for Living Aquatic Resources Management, Manila. Pp. 11–13.
185. ICPRB. 1974. *Potomac River Basin Water Quality Status and Trend.* Interstate Commission on the Potomac River Basin, Washington, D.C.
186. IRF. 1977. *Water, Sediments and Ecology of the Mangrove Lagoon and Benner Bay, St. Thomas.* Island Resources Foundation, St. Thomas, USVI.
187. IUCN. 1990. *Caring for the World: A Strategy for Sustainability.* World Conservation Union (IUCN), Gland, Switzerland. Second Draft, June 1990. 135 pp.
188. Jackson, J. 1993. Quoted in the "Island Navigator," Big Pine Key, Fla.
189. Jhingran, A. G. and P. K. Chakrabarti. 1990. *An Approach to Coastal Zone Management and Planning*

in West Bengal. Inland Fisheries Society of India, Cochin, India. 53 pp.

190. Johannes, R. E. 1982. Traditional conservation methods and protected marine areas in Oceania. *Ambio,* Vol. 5, No. 2. Pp. 258–261.
191. Johannes, R. E. 1988. Traditional use of reef and lagoon resources. In: *Coral Reef Management Handbook,* R. A. Kenchington and B. E. T. Hudson (Eds.). UNESCO Regional Office for Science and Technology for South-East Asia, Jakarta. Pp. 119–121.
192. Johannes, R. E. 1988. The role of marine resource tenure systems (TURFs) in sustainable nearshore marine resource development and management in U.S.-affiliated tropical Pacific islands. In: *Topic Reviews in Insular Resource Development and Management in the Pacific U.S.-affiliated Islands,* B. D. Smith (Ed.). University of Guam Marine Laboratory Technical Report 88.
193. Johnson, L. and W. V. McGuinness, Jr. 1975. *Guidelines for Material Placement in Marsh Creation.* Dredged Material Research Program, U.S. Army Engineer Waterways Experiment Station, Vicksburg, Miss.
194. Joliffe, I. P. and C. R. Patman. 1985. The coastal zone: the challenge. *Journal of Shoreline Management,* Vol. 1, No. 1. Pp. 3–36.
195. Jothy, A. A. 1984. Capture fisheries and the mangrove ecosystem. In: *Proceedings of the Workshop on Productivity of the Mangrove Ecosystem: Management Implications,* J. E. Ong and W. K. Gong (Eds.). Universiti Sains Malaysia, Penang, Malaysia. Pp. 129–141.
196. Kaliapermul, N. 1993. Seaweeds. Paper presented at INDAQUA Conference, Madras, India (20–23 March 1993). 4 pp.
197. Kana, T. W. 1988. *Beach Erosion in South Carolina.* S.C. Sea Grant Consortium, Charleston, S.C. 56 pp.
198. Kana, T. W. 1991. Treating the coast as a dynamic system. In: *The Status of Integrated Coastal Zone Management: A Global Assessment,* J. R. Clark (Ed.). CAMPNET. University of Miami/RSMAS, Miami, Fla. Pp. 60–61.
199. Kana, T. W., J. E. Mason, and M. L. Williams. 1987. A sediment budget for a relocated tidal inlet. In: *Proc. Coastal Seds.* 1987. Amer. Soc. of Civ. Eng., N.Y. Pp. 2094–2109.
200. Kapetsky, J. M. 1987. Development of the mangrove ecosystem for forestry, fisheries and aquaculture. In: *Final Report, Symposium on Ecosystem Redevelopment: Ecological, Economic and Social Aspects,* 5–10 April 1987, Budapest, Hungary. Pp. 5–36.
201. Kapetsky, J. M., L. McGregor, and H. Nanne. 1987. *A Geographical Information System and Satellite Remote Sensing to Plan for Aquaculture Development.* FAO Fisheries Technical Paper 287. U.N. Food and Agriculture Organization, Rome. 51 pp.
202. Kaplan, O. B. 1991. *Septic Systems Handbook.* Lewis Publishers, Chelsea, Mich. 434 pp.
203. Kelleher, G. and R. A. Kenchington. 1992. Guidelines for Establishing Marine Protected Areas (draft). Fourth World Congress on National Parks and Protected Areas, World Conservation Union (IUCN), Gland, Switzerland. 90 pp.
204. Kenchington, R. A. 1990. *Managing Marine Environments.* Taylor and Francis, New York.
205. Kenchington, R. A. and D. Crawford. 1993. On the meaning of integration in coastal zone management. *Ocean and Coastal Management,* Vol. 21, Nos. 1–3. Pp. 109–127.
206. Kenchington, R. A. and B. E. T. Hudson (Eds.). 1988. *Coral Reef Management Handbook,* R. A. Kenchington and B. E. T. Hudson (Eds.). UNESCO Regional Office for Science and Technology for South-East Asia, Jakarta. 321 pp.
207. Ketelle, M. 1972. Physical Considerations for the Location of Septic Disposal Systems. Inland Lake Renewal and Shoreland Management Demonstration Project. University of Wisconsin, Madison, Wisc.
208. Kinsey, D. W. 1988. Responses of coral reef systems to elevated levels of nutrients. In: *Nutrients in the Great Barrier Reef Region,* C. L. Baldwin (Ed.). Workshop Series No. 10. Great Barrier Reef Marine Park Authority, Townsville, Queensland, Australia. Pp. 55–65.
209. Kiravanich, P. and S. Bumpapong. 1991. Coastal area management planning: Thailand's experience. In: *The Status of Integrated Coastal Zone Management: A Global Assessment,* J. R. Clark (Ed.). CAMPNET. University of Miami/RSMAS, Miami, Fla. Pp. 97–104.
210. Kittel, C. 1993. Personal communication.
211. KLH. 1988. *Guideline for Marine Water Quality.* Indonesia Ministry for Population and Environment, KEP-02/MENKLH/1/1988. Pp. 4–6.
212. Knecht, R. W. 1979. Policy issues for coastal area development/management: legislation, agencies,

programs. In: *Proceedings of the Workshop on Coastal Area Development and Management in Asia and the Pacific,* N. J. Valencia (Ed.). East-West Center, Hawaii. Pp. 119–127.

213. Kuhlman, D. H. H. 1985. *Living Coral Reefs of the World.* Arco Publishing, New York. 185 pp.
214. Lal, P. N. 1984. Environmental implications of coastal development in Fiji. *Ambio,* Vol. 13, No. 5–6. Pp. 316–321.
215. Larson, J., P. Adamus, and E. Clairain. 1989. *Functional Assessment of Freshwater Wetlands: A Manual and Training Guide.* Environmental Institute, University of Massachusetts and World Wildlife Foundation, Washington, D.C. 62 pp.
216. Leksakundilok, A. 1988. Carrying capacity for tourism development of Ko Samui. Paper presented to International Symposium on Nature Conservation and Tourism Development (Suratthani, Thailand). 17 pp.
217. Leopold, L. B. 1968. Hydrology for Urban Land Planning—A Guidebook on the Hydrologic Effects of Urban Land Use. Circular 554. U.S. Geological Survey, Washington, D.C. 18 pp.
218. LGED. 1992. *Guidelines on Environmental Issues Related to Physical Planning.* Local Government Engineering Department, Ministry of Local Government, Rural Development and Cooperatives, Government of the People's Republic of Bangladesh, Dhaka, Bangladesh. 45 pp.
219. Lincer, J. L. 1983. Toxic substances. In: *Coastal Ecosystems Management: A Technical Manual for the Conservation of Coastal Zone Resources,* J. R. Clark (Ed.). Robert E. Krieger Publishing Co., Melbourne, Fla. Pp. 740–749.
220. Lindeboom, H. J. 1983. From treating symptoms towards a controlled ecosystem management in the Eastern Scheldt. In: *Coastal Ecosystems Management: A Technical Manual for the Conservation of Coastal Zone Resources,* J. R. Clark (Ed.). Robert E. Krieger Publishing Co., Melbourne, Fla. Pp. 37–39.
221. Lindsay, J. J. 1986. Carrying capacity for tourism development in national parks of the United States. *UNEP Industry and Environment,* January/February/March/1986. Pp. 17–20.
222. Lowry, K. 1993. Coastal management in Sri Lanka. *Coastal Management in Tropical Asia,* No. 1. Coastal Resources Center, University of Rhode Island. Pp. 1–7.
223. Lowry, K. and H. Wickremeratne. 1989. Coastal management in Sri Lanka. In: *Ocean Yearbook 7.* University of Chicago Press. Pp. 263–293.
224. Loya, Y. 1978. Plotless and transect methods. In: *Coral Reefs: Research Methods,* D. R. Stoddard and R. E. Johannes (Eds.). U.N. Educational, Scientific and Cultural Organization, Paris. Pp. 53–66.
225. Lugo, A. E. 1984. Principles of sound watershed management. In: *Watershed Management in the Caribbean, Proceedings of Second Workshop of Caribbean Foresters,* A. E. Lugo and S. Brown (Eds.). Institute of Tropical Foresters, Puerto Rico. Pp. 11–15.
226. Maclean, J. L. 1989. *Pyrodinium* red tides cause paralytic shellfish poisoning in tropical Pacific. *Tropical Coastal Area Management,* Vol. 4, No. 2. Pp. 14–15.
227. Maclean, J. L. and L. B. Dixon. 1984. *ICLARM Report 1983.* International Center for Living Aquatic Resources Management, Manila. 115 pp.
228. Magos, L. 1990. Marine health hazards of anthropogenic and natural origin, Annex X. In: *Technical Annexes to the Report on the State of the Marine Environment.* UNEP Regional Seas Reports and Studies No. 114/2., U.N. Environmental Program, Nairobi. Pp. 448–507.
229. Mahtab, F. U. 1989. *Effects of Climate Change and Sea-level Rise on Bangladesh.* Sponsored by Commonwealth Secretariat for Expert Group on Climate Change and Sea-level Rise. Prokaushaki Sangsad Ltd., Dhaka. 302 pp.
230. Mapstone, G. M. 1990. *Reef Corals and Sponges of Indonesia,* MARINF/75b, U.N. Educational, Scientific and Cultural Organization, Paris. 65 pp.
231. Maragos, J. E. 1989. Impact of Coastal Construction on Nearshore Ecosystems in Oceania: A Review (unpub. manuscript). 50 pp.
232. Maragos, J. E., A. Soegiarto, E. D. Gomez, and M. A. Dow. 1983. Development planning for tropical coastal ecosystems. In: *Natural Systems for Development,* R. A. Carpenter (Ed.). MacMillan, New York. Pp. 229–298.
233. Marchand, M. 1991. Introduction to the ecology of the coastal zone. Coastal Zone Management Seminar, Delft, Netherlands.
234. Marcos, I. 1983. *National Environmental Enhancement Program.* National Environmental Protection Council, Imelda Marcos, Chairperson, Manila. 137 pp.

235. Marks, R. H. (Ed.). 1967. *Waste-water Treatment.* Power, Special Report, New York.
236. Marszalek, D. S. 1987. Sewage and eutrophication. In: *Human Impacts on Coral D Reefs: Facts and Recommendations,* B. Salvat (Ed.). Antenne Museum E.P.H.E., French Polynesia. Pp. 77–90.
237. Matuszeski, W. E., E. Perez, and S. Olsen. 1988. *Structure and Objectives of a Coastal Resources Management Program for Ecuador.* URI/AID Technical Report, University of Rhode Island. 38 pp.
238. Maul, G. A. (Ed.). UNEP. 1989. Sea level variability in the intra-American Sea with concentration on Key West as a regional example. In: *Implications of Climatic Changes in the Wider Caribbean Region,* G. A. Maul (Ed.). U.N. Environment Program, Nairobi. Pp. 125–138.
239. McConchie, D. M. 1990. A Short Course in Delta Morphology and Sedimentology with Particular Reference to Coastal Land Stability Problems in Bangladesh. Southern Cross University, Center for Coastal Management, E. Lismore, Australia.
240. McGuinness, W. V., Jr. and R. Patchai. 1972. *Integrated Water Supply and Waste Water Disposal on Long Island.* CEM-4103-456. Center for the Environment and Man, Inc., Hartford, Conn.
241. McHarg, I. 1969. *Design with Nature.* Natural History Press, Garden City, N.Y. 197 pp.
242. McManus, J. W., C. L. Nanola, R. B. Reyes, and K. N. Kesner. 1992. *Resource Ecology of the Balinao Coral Reef System.* ICLARM Contributions 844, International Center for Living Aquatic Resources Management, Manila. 115 pp.
243. McNulty, J. K. 1983. Discharge of sewage. In: *Coastal Ecosystems Management: A Technical Manual for the Conservation of Coastal Zone Resources,* J. R. Clark (Ed.). Robert E. Krieger Publishing Co., Melbourne, Fla. Pp. 604–610.
244. McShine, H. 1985. The status of coastal zone management in Trinidad and Tobago. Unpublished paper prepared for the Workshop and Project Planning Meeting on Coastal Zone Management, 22–26 July, 1985, Castries, St. Lucia. 25 pp.
245. MFA. 1993. *COT Newsletter.* Marine Research Section, Ministry of Fisheries and Agriculture, Republic of Maldives.
246. Miller, K. R. 1981. A strategy for the conservation of living marine resources and processes in the Caribbean region. *Alumni Newsletter,* Dept. of Natural Resources, University of Michigan. Pp. 2–7.
247. Miller, L. G. 1988. Sectoral plans for tourism development. In: *Environmentally Sound Tourism in the Caribbean,* F. Edwards (Ed.). Banff Centre for Continuing Education, University of Calgary Press, Calgary, Alberta, Canada. 143 pp.
248. Mitchell, M. K. and W. B. Stapp. 1986. *Field Manual for Water Quality Monitoring.* Thompson-Shore, Dexter, Mich. 143 pp.
249. MMAG. 1993. Scientific consensus on the state of the receiving environment around the Clover and Macaulay Point sewage outfalls: knowledge and concerns. Marine Monitoring Advisory Group, Victoria Capital Regional District, Canada. 20 pp.
250. Morton, J. 1990. *The Shore Ecology of the Tropical Pacific.* U.N. Regional Office for Science and Technology for South-East Asia, Jakarta. 290 pp.
251. Murphy, P. 1983. Tourism in Canada: selected issues and options. *Western Geographical Series,* Vol. 21. Pp. 3–23.
252. NACO. n.d. Community Action Program for Water Pollution Control. National Association of Counties Research Foundation, Washington, D.C.
253. National Audubon Society. 1983. Quoted in *National Audubon Society v. Superior Court* (1983), 33 Cal. 3rd 419, 433–34.
254. NCDNR. 1986. *Protecting Coastal Waters through Local Planning.* Division of Coastal Management, North Carolina Department of Natural Resources and Community Development, Raleigh, N.C.
255. NCWQ. 1975. Staff draft report. National Committee on Water Quality, Washington, D.C.
256. Neelakantan, K. S. 1994. *Management Plan for the Gulf of Mannar Marine Biosphere Reserve.* Forest Department, Tamil Nadu, India. 118 pp.
257. New Jersey (USA). 1988. Basis and background for the proposed surface water quality standards. Department of Environmental Protection, Division of Water Resources. 10 pp.
258. New York (USA). 1985. *Water Quality Regulations.* New York State Codes, Rules and Regulations, Title 6, Chapter X, Parts 700–705. New York State Department of Environmental Conservation, New York.
259. Niering, W. A. 1983. Salt marshes. In: *Coastal Ecosystems Management: A Technical Manual for the Conservation of Coastal Zone Resources,* J. R. Clark (Ed.). Robert E. Krieger Publishing Co., Melbourne, Fla. Pp. 697–702.

260. NOAA/CZM. 1980. *Coastal Management Program for the Commonwealth of Puerto Rico.* National Oceanic and Atmospheric Administration, Office of CZM, Washington, D.C. 168 pp.
261. North. K. 1991. *Environmental Business Management.* Organization for Economic Cooperation and Development/International Labour Office, Geneva. 170 pp.
262. NPS. 1986. Position paper on coastal and marine parks and protected areas. Prepared by participants of the International Seminar on Marine and Coastal Parks and Protected Areas (April 1986). U.S. National Park Service, Office of International Affairs, Washington, D.C. 12 pp.
263. OAS. 1984. *Integrated Regional Development Planning: Guidelines and Case Studies from OAS Experience.* Organization of American States, Washington, D.C. 230 pp.
264. O'Brien, R. J. 1988. Western Australia's nonstatutory approach to coastal zone management: an evaluation. *Coastal Zone Management,* Vol. 16, No. 3. Pp. 201–214.
265. OECD. 1990. Integrated coastal zone management and marine living resources. AGR/FI/ENV, 90 (1). Committee on Fisheries, Organization for Economic Cooperation and Development, Paris. 13 pp.
266. Ogden, J. C. and E. H. Gladfelter (Eds.). 1983. *Coral Reefs, Seagrass Beds and Mangroves: Their Interaction in the Coastal Zones of the Caribbean.* Reports in Marine Science No. 23., U.N. Educational, Scientific and Cultural Organization, Paris. 133 pp.
267. Ohta, P. 1991. Development in Hawaii. In: *Case Studies of Coastal Management: Experience from the United States,* B. Needham (Ed.). Coastal Resources Center, University of Rhode Island, Narragansett, R.I. Pp. 97–106.
268. Olsen, S. 1986. The Ecuador CAMP test. *CAMP Newsletter,* April 1986. National Park Service, Office of International Affairs, Washington, D.C. Pp. 5–7.
269. Olsen, S. 1987. Sri Lanka completes its coastal zone management plan. *CAMP Newsletter,* August 1987. P. 7.
270. Olsen, S., L. Z. Hale, R. DuBois, D. Robadue Jr., and G. Foer. 1989. *Integrated Resources Management for Coastal Environments in the Asia Near East Region.* Coastal Resources Center, University of Rhode Island, Narragansett, R.I. 77 pp.
271. Ormond, A., A. D. Shepherd, A. Price, and R. Pitts. 1985. *Management of Red Sea Coastal Resources: Recommendations for Protected Areas.* Report No. 5. Saudi Arabia Marine Conservation Programme, International Union for the Conservation of Nature, Gland, Switzerland. 113 pp.
272. OTA. 1987. *Integrated Renewable Resource Management for U.S. Insular Areas.* U.S. Congress, Office of Technology Assessment, Washington, D.C. 443 pp.
273. Owens, D. 1991. The coastal management program in North Carolina. In: *Case Studies of Coastal Management: Experience from the United States,* B. Needham (Ed.). Coastal Resources Center, University of Rhode Island, Narragansett, R.I. Pp. 23–34.
274. Pabla, H. S., S. Pandey, and R. Badola. 1993. *Guidelines for Ecodevelopment Planning.* UNDP/FAO Project, Wildlife Institute of India, Dehradun, India. 40 pp.
275. van Pagee, J. A. 1991. Case Study: "Jakarta Bay." Coastal Zone Management Seminar 1991. Delft, Netherlands.
276. van Pagee, J. A. and J. C. Winterwerp. 1991. Integrated modeling as a tool for assessment of environmental impacts from dumping of polluted dredging spoil. Coastal Zone Management Seminar 1991. Delft, Netherlands.
277. Parish, D. 1991. Moving boundaries: erosion cycles and coastal protected areas. In: *The Status of Integrated Coastal Zone Management: A Global Assessment,* J. R. Clark (Ed.). CAMPNET, University of Miami/RSMAS. Pp. 58–59.
278. Patterson, R. G., R. J. Burby, and A. C. Nelson. 1991. Sewering the coast: bane or blessing to marine water quality. *Coastal Management,* Vol. 19. Pp. 239–252.
279. Pauley, D. 1988. Some definitions of overfishing relevant to coastal zone management in Southeast Asia. *Tropical Coastal Area Management,* Vol. 3, No. 1. Pp. 14–15.
280. Paw, J. N. and T. E. Chua. 1991. An assessment of the ecological and economic impact of mangrove conversion in Southeast Asia. *ICLARM Conference Proceedings 22.* International Center for Living Aquatic Resources Management, Manila. Pp. 201–212.
281. PCI. 1992. *The Development Study on Wastewater Disposal for Denpasar.* Progress Report II. Pacific Consultants International for Japan International Cooperation Agency, Tokyo.
282. Pheng, K. S. and W. P. Kam. 1989. Geographic information systems in resource assessment and planning. *Tropical Coastal Area Management,* Vol. 4, No. 2.

283. Pheng, K. S., J. N. Paw, and M. Loo. 1992. The use of remote sensing and geographic information systems in coastal zone management. In: *Integrative Framework and Methods for Coastal Area Management,* T.-E. Chua and L.-F. Scura (Eds.). International Center for Living Aquatic Resources Management, Manila.
284. PIANC. 1986. Disposal of Dredged Material at Sea. Bulletin No. 52, Permanent International Association of Navigation Congresses. 48 pp.
285. Plogg, S. C. 1973. Why destination areas rise and fall in popularity. *Cornell Hotel and Restaurant Quarterly XIV,* Vol. 4. Pp. 55–58.
286. Polovina, J. J. 1991. Ecological considerations on the application of artificial reefs in the management of artisanal fishery. *Tropical Coastal Area Management,* Vol. 6, Nos. 1/2. Pp. 1–4.
287. Porter, M. 1987. The state of strategic thinking. *Economist,* May 23, 1987. Pp. 17–24.
288. Pound, C. E. and R. W. Crites. 1973. *Wastewater Treatment and Reuse by Land Application.* U.S. Environmental Protection Agency, Washington, D.C.
289. Qureshi, T. 1994. *Experimental Plantation for Rehabilitation on Mangrove Forests in Pakistan.* UNDP/ UNESCO Regional Mangrove Project and Sind Forest Department, Karachi. 35 pp.
290. Ramey, E. H. 1973. National Ocean Survey. National Oceanographic and Atmospheric Administration, Washington, D.C. Unpublished data.
291. Ray, G. C. 1991. Personal communication. University of Virginia, 1991.
292. Ray, G. C. and M. G. McCormick-Ray. 1992. *Marine and Estuarine Protected Areas.* Australian National Parks and Wildlife Service, Canberra. 52 pp.
293. Renard, Y. 1986. Citizen Participation in Coastal Area Planning and Management. *CAMP Newsletter,* October 1986. National Park Service, Office of International Affairs, Washington, D.C. Pp. 1–3.
294. Revelle, R. 1983. Probable future changes in sea level resulting from increased atmospheric carbon dioxide. In: *Changing Climate.* National Academy Press, Washington, D.C. Pp. 433–448.
295. Riley, C. W. 1990. Bermuda's cruise ship industry: headed for the rocks? In: *Proceedings of the 1990 Congress on Coastal and Marine Tourism,* M. L. Miller and J. Auyong (Eds.). National Coastal Resources Research and Development Institute, Newport, Oregon, USA. Pp. 159–165.
296. Robinson, A. H. 1988. Staff training for coral reef and other marine area management. In: *Coral Reef Management Handbook,* R. A. Kenchington and B. E. T. Hudson (Eds.). UNESCO Regional Office for Science and Technology for South-East Asia, Jakarta. Pp. 147–162.
297. Rubec, P. J. 1988. Cyanide fishing and the international marine life alliance net-training program. *Tropical Coastal Area Management,* Vol. 3, No. 1. Pp. 11–13.
298. Ruddle, K. 1982. Environmental pollution and fishery resources in Southeast Asian coastal waters. In: *Man, Land and Sea,* H. S. Soysa, C. L. Sien, and W. L. Collier (Eds.). Agricultural Development Council, Bangkok. Pp. 15–35.
299. Ruddle, K. and R. E. Johannes (Eds.). 1985. *The Traditional Knowledge and Management of Coastal Systems in Asia and the Pacific.* UNESCO Regional Office for Science and Technology for South-East Asia, Jakarta. 313 pp.
300. Sadler, B. 1988. Sustaining tomorrow and endless summer: on linking tourism and environment in the Caribbean. In: *Environmentally Sound Tourism in the Caribbean,* F. Edwards (Ed.). Banff Center for Continuing Education, University of Calgary Press, Calgary, Alberta, Canada. Pp. ix–xxiii.
301. Saenger, P. 1989. Use allocation via zoning, Great Barrier Reef. Unpublished paper presented at International Seminar on Coastal and Marine Parks and Protected Areas, Miami, June 1989.
302. Saenger, P. 1989. Environmental impacts of coastal tourism: an overview and guide to relevant literature. Informal paper, Centre for Coastal Management, University of New England, NSW, Australia. 17 pp.
303. Saenger, P. 1993. Land from the sea: the mangrove afforestation program of Bangladesh. *Ocean and Coastal Management,* No. 20. Pp. 23–29.
304. Saenger, P., E. J. Hegerld, and J. D. S. Davie. 1983. Global status of mangrove ecosystems. *Environmentalist* 3, Supplement No. 3. Pp. 1–88.
305. Saenger, P. and N. Holmes. 1991. Physiological, temperature tolerance, and behavioral differences between tropical and temperate organisms. In: *Pollution in Tropical Systems,* D. W. Connell and D. W. Harker, (Eds.). CRC Press, Melbourne, Fla. Pp. 69–95.
306. Salm, R. V. 1986. *Oman Coastal Zone Management Plan: Greater Capital Area.* International Union for the Conservation of Nature, Gland, Switzerland. 78 pp.

307. Salm, R. V. 1987. Coastal zone management in the Sultanate of Oman. *CAMP Newsletter,* Aug. 1987. National Park Service, Office of International Affairs, Washington, D.C. Pp. 4–5.
308. Salm, R. V. and J. R. Clark. 1989. *Marine and Coastal Protected Areas: A Guide for Planners and Managers,* Second Edition. International Union for the Conservation of Nature, Gland, Switzerland. 302 pp.
309. Salvat, B. 1987. Dredging in coral reefs. In: *Human Impacts on Coral Reefs: Facts and Recommendations,* B. Salvat (Ed.). Antenne Museum E.P.H.E., French Polynesia. Pp. 165–184.
310. Singh, H. S. 1994. Manager, Gulf of Kutch Marine Park, Gudjarat, India. Personal communication.
311. Smardon, R. C. (Ed.). 1983. *The Future of Wetlands. Assessing Visual-Cultural Values.* Allanheld, Osmun Publishers, Totowa, N.J. 219 pp.
312. Smith, I. 1984. Social feasibility of coastal aquaculture: packaged technology from above or participatory rural development? In: *Consultation on Social Feasibility of Coastal Aquaculture,* App. 3. U.N. Food and Agriculture Organization, Bay of Bengal Programme. Pp. 12–34.
313. Smith, R. A. 1994. Coastal tourism in the Asian Pacific: environmental planning needs. *Coastal Management in Tropical Asia,* No. 2, Columbo, Sri Lanka. Pp. 9–11.
314. Smith, S. H. 1988. Cruise ships: a serious threat to coral reefs and associated organisms. In: *Ocean and Shoreline Management,* Vol. 11. Pp. 231–248.
315. Snedaker, S. C., J. C. Dickinson III, M. S. Brown, and E. J. Lahmann. 1986. *Shrimp Pond Siting and Management Alternatives in Mangrove Ecosystems in Ecuador.* Report for USAID, Office of Science Advisor, Washington, D.C. 67 pp.
316. Snedaker, S. C. and C. D. Getter. 1985. *Coastal Resources Management Guidelines.* Coastal Management Publication No. 2, NPS/AID Series, Research Planning Institute, Columbia, S.C. 205 pp.
317. Snyder, R. M. and J. R. Clark. 1985. Coastal wetlands restoration—developer's challenge. In: *Coastal Zone '85.* American Society of Civil Engineers, New York. Pp. 339–350.
318. Sobers, H. A. S. 1988. Some thoughts on tourism and its impact on the environment. Ministry of Foreign Affairs, Antigua and Barbuda.
319. Sofield, T. H. B. 1990. The impact of tourism development on traditional socio-cultural values in the South Pacific: conflict, coexistence, and symbiosis. In: *Proceedings of the 1990 Congress on Coastal and Marine Tourism,* Vol. 2, M. L. Miller and J. Auyong, (Eds.). National Coastal Resources Research and Development Institute, Newport, OR. Pp. 49–66.
320. Solbrig, O. B. (Ed.). 1991. From genes to ecosystems: a research agenda for biodiversity (excerpt). *U.S. MAB Bulletin,* Vol. 15, No. 3A. P. 6.
321. Sorensen, J. 1982. Towards an overall strategy in designing wetland restoration. In: *Wetland Restoration and Enhancement in California,* J. Zedler (Ed.). California Sea Grant, University of California, La Jolla. Pp. 85–96.
322. Sorensen, J. 1990. An assessment of Costa Rica's coastal management program. *Coastal Management,* Vol. 18. Pp. 37–63.
323. Sorensen, J. 1990. The lure of ecotourism and carrying capacity. In: *CAMP Network Newsletter,* No. 10, September 1990, University of Rhode Island, Narragansett, R.I. P. 6.
324. Sorensen, J. 1993. A conceptual framework for integrated coastal zone management. OECD, World Coast Conference 1993.
325. Sorensen, J. 1994. Manual of Operating Procedures Territorial Arrangement and Management of the Black Sea Coastal Region. (Draft).
326. Sorensen, J. C. and S. T. McCreary. 1990. *Institutional Arrangements for Managing Coastal Resources and Environments,* Coastal Management Publication No. 1 [Rev.], NPS/US AID Series, National Park Service, Office of International Affairs, Washington, D.C. 194 pp.
327. Sorensen, J. and N. West. 1993. *A Guide to Impact Assessment in Coastal Environments.* Coastal Resources Center, University of Rhode Island. 100 pp.
328. SPREP. 1985. South Pacific coastal zone management program. Commonwealth Science Council and South Pacific Regional Environmental Progam. 17 pp.
329. Spurgeon, J. 1992. The economic valuation of coral reefs. *Marine Pollution Bulletin,* Vol. 24, No. 11. Pp. 529–536.
330. Squibb, K. S. 1992. Overview of toxics in the harbor estuary. *Tidal Exchange,* Vol. 3, No. 1. Pp. 1–2.
331. Stankey, G. H. Conservation, recreation and tourism in marine settings: the good, the bad, and the ugly? In: *Proceedings of the 1990 Congress on Coastal and Marine Tourism,* Vol. 2, M. L. Miller

and J. Auyong (Eds.). National Coastal Resources Research and Development Institute, Newport, Oreg. Pp. 159–164.

332. Stoddard, D. R. 1978. Mechanical analysis of reef sediments. In: *Coral Reefs: Research Methods,* D. R. Stoddard and R. E. Johannes (Eds.). U.N. Educational, Scientific and Cultural Organization, Paris. Pp. 53–66.
333. Stoddard, D. R. and R. E. Johannes, 1978. *Coral Reefs: Research Methods.* U.N. Educational, Scientific and Cultural Organization, Paris. 581 pp.
334. Storm, D. R. 1972. Hydrologic and water quality aspects of the Tomales Bay environmental study. Storm Engineering Co., Davis, Calif.
335. Sudirman, S. 1987. Multiple-use practises for establishing eco-development policies in Indonesia. In: *Report of the Workshop for Mangrove Managers* (Phuket, Thailand, Sept. 9–10, 1986). UNESCO/UNDP, New Delhi. Pp. 36–40.
336. Sukarno. 1984. A review of coral reef survey and assessment methods currently in use in Indonesia. In: *Comparing Coral Reef Survey Methods.* Report of a regional Unesco/UNEP Workshop, Phuket Marine Biological Centre, Thailand, 13–17 December 1982. UNESCO Reports in Marine Science. U.N. Educational, Scientific and Cultural Organisation, Paris. Pp. 74–82.
337. Sukarno, N. N. and M. Hutomo. 1986. The status of coral reef in Indonesia. In: *Proceedings of MAB-COMAR Regional Workshop on Coral Reef Ecosystems: Their Management Practises and Research/Training Needs.* S. Soemodihardjo (Ed.). U.N. Educational, Scientific and Cultural Organization, Paris. Pp. 24–33.
338. Sumardja, E. 1992. Personal communication.
339. Sun, J. T. (Ed.). 1985. *USGS Coastal Research Studies and Maps—A Source of Information for Coastal Decision-making.* U.S. Geological Survey Circular 883, U.S. Geological Survey, Washington, D.C. 80 pp.
340. Talge, H. 1990. Reported in *Miami Herald,* October 23, 1990.
341. Templet, P. H. 1987. The policy roots of Louisiana's land loss crisis. In: *Coastal Zone '87,* Vol. 1. American Society of Civil Engineers, New York. Pp. 714–725.
342. Thompson, T. P. 1985. Appropriate methodologies for the identification of environmental impact and the evaluation of alternatives. Occasional Paper No. 42. Island Resources Foundation, Washington, D.C. 24 pp.
343. Tobin, R. J. 1992. Legal and organizational considerations in the management of coastal areas. In: *Integrative Framework and Methods for Coastal Area Management,* T. E. Chua and L. F. Scura (Eds.). International Center for Living Aquatic Resources Management, Manila. Pp. 93–105.
344. Tolentino, A. S. 1984. How to protect coastal and marine ecosystems: lessons from the Philippines. In: *National Parks, Conservation and Development,* J. A. McNeeley and K. R. Miller (Eds.). International Union for the Conservation of Nature, Gland, Switzerland. Pp. 160–164.
345. Tomascik, T. 1992. *Environmental Management Guidelines for Coral Reef Ecosystems.* Prepared for State Ministry for Population and Environment (Kependudukan dan Lingkungan Hidup). Jakarta.
346. Tomascik, T. 1991. Settlement patterns of Caribbean scleractinian corals on artificial substrata along a eutrophication gradient, Barbados, West Indies. *Marine Ecology Progress Series,* Vol. 77, Nos. 2/3. Pp. 261–269.
347. Tomascik, T. and F. Sander. 1985. Effects of eutrophication on reef building corals: Part I. Growth rates of reef building coral *Montastrea annlaris. Marine Biology,* Vol. 87, No. 2. Pp. 143–155.
348. Tourbier, J. and R. Westmacott. 1974. *Water Resources Protection Measures in Land Development—A Handbook.* Water Resources Center, University of Delaware, Newark, Del.
349. Towle, E. L. 1985. The Virgin Islands Mangrove Lagoon (1967–1984). In: *Proceedings of the Caribbean Seminar on Environmental Impact Assessment,* T. Goeghegan (Ed.). Caribbean Conservation Association, Barbados. Pp. 244–266.
350. UNEP. 1988. A common methodological framework for integrated planning and management in Mediterranean coastal areas. PAP-4/EM.5/2. U.N. Environmental Program, Nairobi. 20 pp. and App.
351. UNEP. 1990. *GESAMP: The State of the Marine Environment.* UNEP Regional Seas Reports No. 115. U.N. Environmental Program, Nairobi. 111 pp.
352. UNEP. 1990. *Environmental Impact Assessment: Sea-outfall for the Larnaca Sewerage System.* Regional Seas Reports, No. 131. U.N. Environmental Program, Nairobi. 46 pp.
353. UNEP. 1991. *Environmental Impact Assessment: Sewage Treatment Plant for Port Said.* Regional Seas Reports, No. 123. U.N. Environmental Program. 32 pp.

354. UNESC. 1987. Development of marine areas under national jurisdiction: problems and approaches in policy-making, planning, and management. United Nations Economic and Social Council, New York. 22 pp.
355. UNESCO. 1984. *Comparing Coral Reef Survey Methods.* Report of a regional UNESCO/UNEP Workshop, Phuket Marine Biological Centre, Thailand, 13–17 December 1982. UNESCO Reports in Marine Science. U.N. Educational, Scientific and Cultural Organisation, Paris. 170 pp.
356. UNESCO/WHO. 1978. *Water Quality Surveys.* Studies and Reports in Hydrology, No. 23. U.N. Educational, Scientific and Cultural Organization and World Health Organization, Paris. 350 pp.
357. UNISYSTEM. 1993. *Development of a Procedure for Evaluation of Nature Based Tourism.* Unisystem Utama Consults, Jakarta. Report to Natural Resources Management Project, USAID/ARD, Jakarta. 73 pp.
358. URI. 1986. *The Management of Coastal Habitats in Sri Lanka.* Report of Workshop, May 12–15, 1986. Coastal Resources Center, University of Rhode Island, Naragansett, R.I. 36 pp.
359. USAID. 1984. *Project Identification Document: Coastal Resources Management Project.* USAID Bureau for Science and Technology, Washington, D.C.
360. U.S. Army Corps of Engineers. 1971. *National Shoreline Study.* U.S. Army Corps of Engineers, Washington, D.C.
361. U.S. Army Corps of Engineers. 1973. *Ocean Dumping in the New York Bight.* Technical Mememorandum No. 39, Coastal Engineering Research Center, U.S. Army Corps of Engineers, Washington, D.C.
362. U.S. Army Corps of Engineers. 1984. *Shore Protection Manual.* Vols. 1, 2, and 3. Coastal Engineering Research Center, U.S. Army Corps of Engineers, Washington, D.C.
363. U.S. Atomic Energy Commission. 1972. *Final Environmental Statement—Shoreham Nuclear Power Station.* Washington, D.C.
364. U.S. Department of Interior. 1970. *The National Estuarine Pollution Study.* U.S. Congress, Doc. No. 91-58. U.S. Government Printing Office, Washington, D.C. 633 pp.
365. USEPA. 1971. Noise from construction equipment. United States Environmental Protection Agency, Washington, D.C.
366. USEPA. 1976 and 1986. *Quality Criteria for Water.* EPA 440/5-86-001. U.S. Environmental Protection Agency, Washington, D.C.
367. Vannucci, M. 1991. Saving a crucial ecosystem. In: *The Hindu Survey of the Environment, 1991.* Madras, India. Pp. 157–161.
368. Velsink, H. 1992. Personal communication.
369. Veri, A., W. W. Jenna, and D. E. Bermaschi. 1975. *Environmental Quality by Design: South Florida.* University of Miami Press, Miami, Fla.
370. Veri, A. and L. Warner. 1975. *Interior Wetlands Water Quality Management.* Conservation Foundation, Washington, D.C.
371. Vesilind, P. A., J. J. Peirce, and R. F. Weiner. 1990. *Environmental Pollution and Control.* Ann Arbor Science Publishers, Ann Arbor, Mich. (3d edition). 389 pp.
372. van de Vusse, F. J. 1991. The use and abuse of artificial reefs: the Philippine experience. *Tropical Coastal Area Management,* Vol. 6, Nos. 1/2. Pp. 8–11.
373. Wafar, M. 1992. Management and conservation options for Indian coral reefs. In: *Tropical Ecosystems: Ecology and Management,* K. Singh and J. Singh (Eds.). Wiley Eastern Ltd., New Delhi. Pp. 173–183.
374. Wager, J., R. Bisset, P. Bacon, and J. McLoughlin. 1988. *Turks and Caicos Islands: Ecological Survey and Environmental Policies.* Report to U.N. Center for Human Settlements. Cobham Resource Consultants, Manchester, U.K. Pp. 185–226.
375. Weinberger, L. W., D. G. Stephan, and F. M. Middleton. 1968. Solving our water problems—water renovation and reuse. *Annals of the N.Y. Academy of Sciences,* Vol. 136.
376. Wells, S. M. (Ed.). 1988. *Coral Reefs of the World,* Vols. 1–3. United Nations Environment Program, Nairobi.
377. Wells, S. M. 1992. "Successful" coral reef management programmes. *Intercoast Network,* No. 17, Coastal Resources Center, University of Rhode Island, Narragansett, R.I. Pp. 1–3.
378. Wells, S. M. and J. G. Barzdo. 1991. International trade in marine species: is CITES a useful control mechanism? *Coastal Management,* Vol. 19, No. 1. Pp. 135–154.
379. Wenzel, L. 1989. Social science techniques: overview, 1989. In: *How to Assess Environmental Impacts on Tropical Islands and Coastal Areas,* R. A. Carpenter and J. E. Maragos (Eds.). East-West Center, Hawaii. Pp. 65–69.

380. White, A. T. 1988. *Marine Parks and Reserves: Management for Coastal Environments in Southeast Asia.* ICLARM Education Series 2. International Center for Living Aquatic Resources Management, Manila. 36 pp.
381. White, A. T. 1987. Why public participation is important for marine protected areas. *CAMPNET Newsletter,* August 1987. National Park Service. Office of International Affairs, Washington, D.C. Pp. 5–6.
382. White, A. T. 1989. The marine conservation and development program of Silliman University as an example for the Lingayen Gulf. *ICLARM Conference Proceedings 17.* International Center for Living Aquatic Resources Management, Manila. Pp. 119–123.
383. White, A. T. and A. A. Alcala. 1988. Options for management. In: *Coral Reef Management Handbook,* 2nd Edition, R. A. Kenchington, and B. E. T. Hudson (Eds.). UNESCO Regional Office for Science and Technology for South-East Asia, Jakarta. Pp. 37–46.
384. White, A. T., L. Z. Hale, Y. Renard, and L. Cortesi (Eds.). 1994. *Collaborative and Community-Based Management of Coral Reefs: Lessons from Experience.* Kumarian Press.
385. White, A. T. and J. I. Samarakoon. 1994. Special area management for coastal resources: a first for Sri Lanka. *Coastal Management in Tropical Asia,* No. 2, Columbo, Sri Lanka. Pp. 9–11.
386. Wickremeratne, H. J. M. and A. White. 1992. Concept paper on special area management for Sri Lankan coasts. Coastal Resources Center, University of Rhode Island. 8 pp.
387. Wiese, P. 1992. (Vulcan Materials Co., Birmingham, Ala.) Personal communication.
388. van Woesik, R. 1992. *Urgent Bali Beach Conservation Project: Environmental Monitoring Report.* Nippon Koei, Tokyo. 56 pp.
389. Wood, M. W. and S. M. Wells. 1988. *The Marine Curio Trade: Conservation Issues.* A Report for the Marine Conservation Society, Herefordshire, U.K. 120 pp. and appendices.
390. Woodley, J. D. and J. R. Clark. 1989. Rehabilitation of degraded coral reefs. *Coastal Zone '89,* Vol. 4. American Society of Civil Engineers, New York. Pp. 3059–3075.
391. WRI. 1986. *World Resources 1986.* World Resources Institute, Washington, D.C. 353 pp.
392. Wycoff, R. L. and R. D. G. Pyne. 1975. Urban water management and coastal wetland protection in Collier County, Florida. *Water Resources Bulletin,* Vol. 2, No. 3. Pp. 455–468.
393. Yamazato, K. 1986. The effects of suspended particles on reef building corals. In: *Proceedings of MAB-COMAR Regional Workshop on Coral Reef Ecosystems: Their Management Practises and Research/Training Needs,* S. Soemodihardjo (Ed.). U.N. Educational, Scientific and Cultural Organization, Paris. Pp. 86–90.

PART 6

Glossary

abiotic the non-living component of the environment; not pertaining to life or living organisms.
activated sludge a process that removes organic matter from sewage by saturating it with air and biologically active sludge.
aerobic living or active only in the presence of oxygen; taking place in the presence of oxygen.
algae bloom an overgrowth of algae in water that can shade out other aquatic plants and use up the water's oxygen supply as the plants decompose; blooms are often caused by pollution from excessive nutrient input.
alternative livelihoods jobs offered to people who are displaced from their current jobs because of resource conservation programs.
anoxic devoid of free oxygen.
anthropogenic effects from the influence of human beings on natural systems.
aquaculture the cultivation of fish, shellfish, and/or other aquatic animals or plants, including the processing of these products for human use.
aquifer a geologic stratum that contains water than can be economically removed and used for water supply.
archaeology the scientific study of material remains of past human life and activities.
armor a marine engineering term that means providing structural protection for shorelines; e.g., bulkheads, seawalls, etc.
artificial reef any marine habitat constructed for the purpose of attracting marine species or enhancing marine resources to improve fisheries; usually made of terriginous substances such as used auto tires, concrete rubble, old ship hulls, automobile bodies, etc.
artisanal activities based on personal skills and manual dexterity; generally "low tech" work, often resource oriented.
backshore the accretion or erosion zone, located landward of the line of ordinary high tide, which is normally wetted only by storm tides; a narrow storm berm (ridge of wave-heaped sand and/or gravel) or a complex of berms, marshes, or dunes landward of the line of ordinary high tide.
barrier islands elongate seafront islands of sand formed by the action of the sea and having an elongate lagunal or estuarine embayment behind them.
baseline study an inventory of a natural community or environment to provide a measure of its condition at a point of time—often done to describe the status of diversity and abundance against which future change can be gauged (usally development driven); baseline data may be used in models for planning or for establishing goals for allowable environmental impacts (also, "reference study" sites).
beach a zone, or strip, of unstable unconsolidated material (e.g., sand, gravel) along the shoreline that is moved by waves, wind and tidal currents.
beach nourishment a process of replenishing a beach; it may be done naturally, by manipulating longshore drift, or artificially by the deposition of imported materials.
beach profile the intersection of the surface plane of the beach with vertical planes; may extend from the top of the dune line to the seaward limit of sand movement.
bench mark a fixed physical object or mark used as reference for a vertical datum; a tidal bench mark is often near a tide station to which the tide staff and tidal datums are referred.
benthic of, pertaining to, or living on or in the bottom of the sea; upon or attached to the sea bottom (as opposed to pelagic).

berm a ridge of sand or gravel deposited by wave action on the shore just above the normal high water mark.
bioaccumulation the uptake of substances—e.g., heavy metals or chlorinated hydrocarbons—leading to elevated concentrations of those substances within marine organisms.
biochemical oxygen demand (BOD) a measure of the amount of dissolved oxygen required by biochemical processes to oxidize organic wastes in water.
bioerosion the process by which biota erode environmental structure; e.g., the effects of drilling, grazing, and burrowing animals that can lead to slow disintegration of coral reefs.
biogenic produced by the action of living organisms.
biological diversity (biodiversity) the variety of faunal and floral species living within a certain habitat; also the social advocacy of protecting species and saving them from extinction.
biomass the total mass of living matter, usually within a given area or volume of environment.
biosphere reserve a designated resource area featuring multiple use management systems whereby nature protection and uses for farming, forestry, fisheries, etc are accomodated (note: an international system of such reserves is endorsed and guided by UNESCO).
brackish water a dilution of fresh water by the sea; brackish water may be defined as containing between 5 and 30 parts per thousand (ppt) of dissolved solids.
breakwater an artificial offshore structure aligned parallel to shore usually to provide protection of the shore from large waves.
buffer area a protective, often transitional, area of controlled use—in coastal management, a peripheral zone separating a developed area from a protected natural area.
bulkhead a wall erected parallel to and near the high water mark for the purpose of protecting adjacent uplands from waves and current action.
carrying capacity the limit to the amount of life, or economic activity, that can be supported by the environment; often the specific number of individuals of a species that can be supported by a particular habitat or, in the sense of resource management, the reasonable limits of human occupancy and/or resource use.
chemical oxygen demand (COD) a measure of the amount of oxygen required to oxidize (with a chemical oxidant) the amount of organic and oxidizable inorganic compounds in water. Note: both COD and BOD (see above) test biological demands on oxygen resources.
ciguatera a toxic reaction to consumption of certain tropical fishes, usually the high level predators, caused by a toxic dinoflagellate microorganism; the toxin is passed up the food chain from vegetarian to predaceous fishes and causes a variety of cardiovascular and nervous disorders (rarely fatal).
CITES the Convention on International Trade in Endangered Species (1973), ratified by 109 nations, which prohibits commercial trade in species whose existence is seriously threatened.
coastal baseline a constructed, geo-specific, line from which the distance to the edge of the Territorial Sea of a country is plotted.
coastal waters littoral waters that contain a measurable quantity or percentage of seawater (e.g., more than 0.5 parts per thousand).
coastal zone (U.S.legal definition) the coastal waters (including the lands therein and thereunder) and the adjacent shorelands (including the waters therein and thereunder), strongly influenced by each and in proximity to the shorelines of the several coastal states, and includes islands, transitional and intertidal areas, salt marshes, wetlands and beaches.
coliform bacteria widely distributed micro-orgamisms found in the intestinal tract of humans and other animals and in soils; their presence in water indicates fecal pollution and potentially dangerous contamination by disease-causing micro-organisms.
commons publicly owned areas of land or water, often managed by government as a public trust for the people; common property.
compensation plan is the portion of an environmental management plan that describes compensation measures to be undertaken (usually by the development entity) to offset or compensate for negative environmental impacts caused by a development project.
conservation the political/social/economic process by which the wise use of resources is excercised and environments are protected.
containment (retention) basin a containment constructed of dikes for the purpose of retaining dredge spoil until much of the suspended material has settled out when the water itself is released.

contaminant a substance that is not naturally present in the environment in damaging amounts but when present in sufficient concentrations adversely affects the environment and becomes a "pollutant" (see pollutant below).
contingency plan a set of countermeasures planned in advance to mitigate damage from an accident (oil spill, cyclone, or etc).
Coriolis effect the deflection relative to the earth's surface of any object moving across or above the earth, caused by the earth's interial force—an object moving horizontally along the Earth's orbit is deflected to the right in the northern hemisphere, to the left in the southern.
critical habitat see "ecologically critical area" below.
cumulative impacts environmental impacts caused by multiple human activities; that is, the combined environmental impacts that accrue from a number of individual actions, contaminants, or projects, whereby actions which may each be acceptable individually have a significant impact in combination.
datum (vertical) for marine applications, a base elevation used as a reference from which to reckon heights or depths. It is called a tidal datum when defined in terms of tidal phenomena and is based on a 19-year tide cycle (in the USA)—the datum is referenced to a fixed point typically known as a bench mark.
detritus floatable particles produced by erosion/decomposition of larger pieces of material; generally refers to fine organic matter suspended in the water or settled on the sea bed, such as particles of plant matter (e.g., from marsh grass or mangrove leaves) in varying stages of decomposition.
dissolved oxygen (DO) the quantity of oxygen dissolved in a unit volume of water; expressed as milligrams per liter (mg/L) or parts per million (ppm).
diversity index any one of several methods of calculation created to provide a numerical measure of the diversity of species in particular habitats.
dredging excavation, digging, scraping, draglining, suction dredging, or other means of removing sand, silt, rubble, rock, coral or other underwater bottom materials.
dunes accumulations of sand in ridges or mounds landward of the beach berm formed by natural processes and usually parallel to the shoreline.
dryside that part of the coastal zone that has dry soils and that is not regularly flooded, including lowlands and hinterlands.
ecodevelopment a process of socio-economic development in which the sustainable use of environmental resources has priority.
ecologically critical area (ECA) an area of highly concentrated biological activity of a type that is especially valuable for maintaining biodiversity and/or resource productivity; an ecologically sensitive area (ESA).
ecosystem the complete ecological system operating in a given geographic unit, including the biological community and the physical environment, functioning as an ecological unit in nature.
ecotone the transition or border area lying between two different ecological communities, as between a marsh system and a forest system.
ecotourism tourist activity attracted to environmental resources and based, usually, on a conservation theme.
effluent the outflow of a sewer, industry pipe, or other waste discharge.
endemic a species of limited geographic extent, confined to or indigenous to a region; that is, restricted to or peculiar to a particular locality or habitat.
environmental audit a management tool employed to evaluate the environmental performance of a private sector operation; usually an ongoing enterprise.
environmental impact assessment (EIA) detailed prediction of the impact of a development project on environment and natural resources with recommendations as to acceptability of the project, needs for minimizing/eliminating/offsetting adverse effects, a management plan to accomplish these countermeasures, and a project effects monitoring program; a generic term for all types of impact assessment is "environmental assessment" (EA).
environmental management plan a plan that describes specific conservation actions that will be undertaken during project planning, construction, operation, and maintenance to lessen the effects of the project on the environment and to ensure that sustainable development is achieved; it includes real time and retroactive monitoring of project effects.

estuary a semi-enclosed littoral basin (embayment) of the coast in which fresh river water entering at its head mixes with saline water entering from the ocean; usually associated with a river's intersection of the coast.

eutrophication the process of enrichment of water which leads to excessive growth of algae and other aquatic plants from the introduction of an over supply of nutrients such as nitrates or phosphates.

Exclusive Economic Zone the maritime zone adjacent to and extending 200 nautical miles beyond the baseline from which the territorial sea is measured—internationally authorized by the Third United Nations Conference on the Law of the Sea; the coastal state has sovereign rights to explore, exploit, conserve and manage the natural resources in this zone.

exotic species alien species; species not endemic to a geographic area.

extensification an increase of production (e.g. of shrimps) in an aquacultural or agricultural system which is the rersult of expanding the size of facilities; e.g, adding new pond area in a shrimp farming facility.

externality a term from economics meaning the effects that one resource user's activities have on other users or stakeholders; costs external to a development activity and imposed upon persons other than the developer.

feeder beach an offshore stockpile of artificially placed sand or gravel with a prupose of supplying replenishment material to an eroding beach.

filling elevating the land surface with artificial deposits using excavated, dredged, or waste materials.

finfish a term used to distinguish vertebrate fishes from invertebrate animals such as molluscs and crustacea.

floodplain the area of shorelands that is subject to frequent storm flooding and is often defined by the statistical probability of flooding; e.g., 1% ("100-year flood") or 5% ("20-year flood").

flushing time the time required to replace the water in a basin and therefore to remove or reduce (to a permissible concentration) any dissolved or suspended contaminant in an estuary or harbor.

food chain the step-by-step transfer of food energy and materials, by consumption, from the primary source (plantlife) through to the highest level carnivorous predators.

food web interactive system of transfer of food energy throughout an ecosystem whereby primary producers (such as plankton) are eaten by herbivores (such as zooplankton) which are eaten by predators (such as herring) which are eaten by top level carnivores (such as larger fish); the network of feeding relationships or the complex of interlocking food chains in a biological community.

foreshore the intertidal part of a beach or the part of the shorefront lying between the beach head (or upper limit of wave wash at high tide) and the ordinary low water mark that is ordinarily traversed by the uprush and backrush of the waves as the tides rise and fall.

geographic information system (GIS) computer-assisted systems that can input, store, retrieve, analyze and display geographically referenced information and enhance the analysis and display of interpreted geographic data, often by digitizing mapped or photographed data so that the GIS database is typically composed of a large number of map-like spatial representations.

goals what a project or program hopes to achieve in the long-term expressed as broad generalities, such as, "to improve the production from coastal fisheries".

greenbelt a strip of vegetation, usually along a transition zone boundary, which separates one type of resource area from another.

greenhouse effect heating of the Earth from the increase in the gases such as CO_2, methane, CFCs, etc that make up the atmospheric envelope that surrounds the globe; the term was coined by the scientist Svante Arrhenius in the late 1800s.

groin elongate structure of large rock, concrete, or wood piles and planks, built perpendicular to the shoreline in order to intercept longshore drift of sand and reduce localized erosion.

ground truth ground level direct observations made to verify interpretations from remotely sensed data.

growout facility in aquaculture, a containment where the target species, having gone through their critical early life stages, are encouraged to grow at an optimum (usually rapid) rate.

halocline an abrupt vertical change of salinity with depth, usually from fresher to saltier water occurring within a rather narrow horizontal layer; it shows on sonar as a sharp discontinuity and has important effects on distribution of life in the ocean ("pycnocline").

heavy industry manufacturing and processing industries which are characterized by disturbance (to the local ambience) and tend to produce polluting effluents and have other negative impacts on the

environment; examples include pulp mills, cement plants, copra mills, canneries, power plants, desalination plants, oil storage and transhipment facilities, refineries, etc.
heavy metals metallic elements (e.g., mercury, lead, nickel, zinc, copper and cadmium) that are of environmental concern because they do not degrade over time—many are essential to life but can be toxic in high concentrations.
hermatypic corals stony, reef building, coral species.
high water the maximum elevation reached by the rising tide; mean high water is the average of such tidal elevations (see "Tides" in Part 3).
hinterlands the land away from the coast which influences the coast including plains, hills, watersheds, water courses, etc (the "uplands").
hydrology the science dealing with the properties, distribution, and circulation of water on earth.
impact assessment the evaluation of ecological effects to determine their impact on human needs, environmental, social, and economic (see "environmental impact assessment" above).
incremental management program step-by-step ICZM whereby the program starts with a limited number of areas, issues, resources, or etc and then methodically expands.
indicator species one or more species elected to represent the condition of the environment in which they live.
initial environmental evaluation (IEE) the initial environmental assessment of a development activity at the project feasibility phase in order to provide early identification of potential environmental impacts and to determine whether a full EIA will be necessary.
integrated coastal zone management (ICZM) a program designed to manage the resources of the coastal zone whereby participation by all affected economic sectors, government agencies, and non-government organizations is involved.
infrastructure usually the publicly constructed support system for a community including roads, electricity, water, sewage, etc.
integrated regional development planning large-scale development planning for a region which incorporates all salient planning parameters including economic, socioeconomic, environmental and others.
intensification the increase of production in an aquacultural or agricultural system through increasing the stock or planting density (and expected production) in the existing water or wetland area.
intertidal zone the transition zone between the sea and the land, often defined as the zone that lies between mean higher high water and mean lower low water lines.
issues analysis the exploration, definition, and evaluation of the basic resource management issues to be faced in an ICZM program.
Jackson Turbidity Unit (JTU) the standard unit used in measuring the turbidity of a water sample; originally defined in terms of the depth of water beyond which a candle flame cannot be clearly distinguished.
jetty a structure projecting out into the sea, usually at the mouth of a harbor or river intersecting the coast, for the purpose of protecting a navigation channel or harbor, or to influence water currents; often built in pairs along both sides of an entrance channel.
lagoon a semi-enclosed littoral basin with limited fresh water input, high salinity, and restricted circulation; lagoons often lie behind sand dunes, barrier islands, or other protective features.
LANDSAT an unmanned earth-orbiting NASA (National Aeronautical and Space Administration) satellite that transmits multispectral images (0.4-1.1 micrometer range) of the electromagnetic spectrum to earth-receiving stations; the digital data and/or images produced are used to identify earth features and resources.
land use planning planning for allocations for use of the regional (or national) land resources to achieve a strategic objective, often for sustainable use of a particular resource (water resources, fisheries, wildlife) or to meet certain social equity or economic objectives.
LD_{50} the dose of a toxicant that is lethal (fatal) to 50 percent of the organisms tested under specified conditions during a standard time period.
liquifaction the hydrogeological process by which moist soil becomes liquified; e.g., earthquakes may cause liquification of wetland soils that underlie real estate fills.
littoral of or pertaining to the shore, especially of the sea; coastal.

littoral drift the movement of sand and other material by littoral (longshore) currents in a direction parallel to the beach along the shore; usually wind driven.
littoral zone in coastal engineering, the area from the shoreline to just beyond the breaker zone; in ecology the littoral system extends farther and is divided into eulittoral and sublittoral zones, separated at a depth of about 50 meters.
longshore current a current, created by waves, which moves parallel to the shore, particularly in shallow water, and which is most noticeable in the surf or breaker zone; littoral drift current.
mangroves any of the many genera of trees that are capable of living and growing in salt water or salty soils; often includes the rich biological community that is supported by the mangrove forests or fringing strips of mangrove.
master plan the operational plan which defines rules, resources, conservation issues, performance standards, authorities, objectives, use rights (permitted uses), development restrictions, participation, coordination mechanisms, permit/EIA conditions, protected areas, setbacks, staff, training, etc. under an ICZM plan.
mean sea level the average height of the sea; a datum, or "plane of zero elevation", established by averaging hourly tidal elevations over a 19-year tidal cycle or "epoch" and corrected for curvature of the earth which is the standard reference for elevations on the earth's surface.
micrograms per liter (μg/l) the concentration at which 1 millionth of a gram (10^{-6}g) is contained in a volume of 1 liter of water (parts per billion).
milligrams per liter (mg/l) the concentration at which 1 milligram (10^{-3}g) is contained in a volume of 1 liter of water (parts per million); note that there are 453.59 grams in a pound.
mitigation the elimination, reduction or control of the adverse environmental impacts of a project, including countermeasures against negative environmental impacts of development which may include: (a) avoiding the impact altogether, (b) minimizing impacts by limiting the type or magnitude of an action, (c) rectifying an impact by repairing, rehabilitating, or restoring the affected environment, (d) reducing or eliminating the impact over time by preservation and maintenance operations, or (e) offsetting, or compensating for the impact by replacing or providing substitute resources or environments.
mixing zone a limited water area surrounding a point of pollution discharge where the restriction on amount of contaminants is waived to allow dilution to take place before the contaminant reaches the water body at large;
monitoring systematic recording and periodic analysis of information over time, often to determine the environmental effects of a development project.
multiple use the concept of providing for multiple activities in particular areas or for particular resources by managing them for sustainable resource use.
natural hazards natural events that can damage property or jeopardize lives, such as cyclonic seastorms, floods, erosion, landslides, or soil liquifaction.
nature synchronous development that is in synchrony with natural forces; engineering design that coincides with natural processes, rather than resisting or confronting them.
Nephalometric Turbidity Unit (NTU) the measure of light penetration in seawater or another liquid used in electronic turbidity meters which corresponds closely to the Jackson Candle, or Jackson Turbidity Unit, because all instruments are calibrated to the equivalent of: 1 mg/l of SiO_2 = 1 NTU.
nurture (or nurturing) area any place in the coastal zone where larval, juvenile, or young stages of aquatic life concentrate for feeding or refuge; a nursery area.
nutrient any substance assimilated by living things for bodily maintenance or to promote growth. The term is often applied to nitrogen and phosphorus, but may also be applied to other essential and trace elements such as carbon and silica.
oligotrophic an aquatic environment typified by a low amount of nutrient; the opposite of "eutrophic".
opportunity cost a term from multiple use economics that means the value of options that are lost (excluded) because of choosing one particular mode of use.
organic detritus suspensible small organic particles, usually of vegetative origin.
overlay mapping superimposition of several theme maps to spatially analyze environmental resources and development modes, particularly useful in studying the interactions between various components of land use.
oxidation in sewage treatment, the consuming or breaking down of organic wastes or chemicals in sewage by bacteria and chemical oxidants.

oxidation pond a man-made lake or body of water in which wastes are treated, mostly by bacterial consumption; a sewage treament lagoon.
PAHs (polynuclear or polycyclic aromatic hydrocarbons) a class of complex organic compounds, having more than one benzene ring, some of which are persistent and cancer-causing—PAH's are often produced by oil refineries and may be formed by the combustion of petroleum.
participation the active involvement of resource stakeholders in decisions relating to development activities and impacts in order to encourage community self-determination and to foster sustainable development.
pelagic capable of living any place from top to bottom in the oceanic water column; not restricted to living at the bottom.
percolation the downward movement of water into the soils.
performance standards specific standards for control of an activity whereby the end result is the measure of effect rather than the process or proceedure.
pesticide any chemical substance used to kill plant and animal (and insect) pests; some pesticides contaminate water, air, or soil and can accumulate in humans, plants, animals and the environment with negative effects.
plankton small suspended aquatic organisms (plants or animals) that drift passively or swim weakly.
point source pollution pollution that is discharged from a fixed location such as the end of a pipe.
policy a prescribed course based on guiding principles adopted and followed by a government, institution, or individual.
pollutant a contaminant that in a certain concentration or amount will adversely alter the physical, chemical, or biological properties of the environment—includes pathogens, heavy metals, carcinogens, oxygen-demanding materials, and all other harmful substances, including dredged spoil, solid waste, incinerator residue, sewage, garbage, sewage sludge, munitions, chemical wastes, biological materials, radioactive materials, and industrial, municipal, and agricultural wastes discharged into coastal waters.
primary production the amount of plant life produced in a given area or environment.
primary waste treatment a process that removes material that floats or will settle in sewage, accomplished by using screens to catch the floating objects and tanks for the heavy matter to settle in and often including chlorination; results in the removal of about 30 percent of BOD from domestic sewage and something less than half of the metals and toxic organic substances.
proprietary control the right of a property owner to dictate, exclusively, the entrance to and the uses of a parcel of property.
protected natural area an area of land or water set aside by governmental action to protect its resources and biodiversity from degradation and managed through legal institutional means.
quadrat a square area of specified size selected for enumeration or evaluation of biota (land or sea bottom).
rapid rural assessment (RRA) a procedure for gathering and analyzing information about community socio-economic conditions preparatory to making development decisions; where community participation is a priority, the process may be termed " participatory rural assessment" (PRA).
real time the situation whereby reporting or recording of events is simultaneous with the event.
recruitment addition to a population of organisms by reproduction or immigration of new individuals.
red tide a massive "bloom" of dinoflagellate microscopic organisms that may produce neurotoxins such as paralytic shellfish poisoning (PSP) that infest marine organisms and humans that eat them, that kill fish, and that may pollute the air with an irritating substance—the name "red tide" comes from the red or reddish brown discoloration of the sea caused by the blooms.
remote sensing the acquisition and processing of information about a distant object or phenomenon without any physical contact; often done from satellites orbiting the earth (e.g., see "LANDSAT" above).
replenishment zone an area within a coastal reserve designated and managed as a non-exploitation sanctuary to enhance replenishment of fishery stocks.
reserve (nature reserve) an area designated for protection (and restoration) of environmental resources, including biodiversity, which usually requires suspension of all exploitive use.
restoration repair or rehabilitation of an environmental resource to restore its structure and ecological processes.
retreat a coastal land use strategy whereby structural development is withdrawn from the coast to a designated setback line farther inland.

revetment a structure built to protect the shore from erosion, usually constructed from stones laid with a sloping face.

riprap a layer, facing, or protective mound of stones placed laterally to prevent erosion, scour, or sloughing of a structure or embankment; also, the stone so used.

runoff that part of precipitation, snow melt, or irrigation water that runs off the land into streams or other water bodies, including coastal waters; it often carries pollutants from the land into the receiving waters.

salt marshes low, wet, muddy areas periodically or continuously flooded by brackish or salt water to a shallow depth, usually characterized by grasses and other low plants (but not trees); lands transitional between terrestrial and aquatic systems where saturation with water is the dominant factor controlling plant and animal communities and soils.

satellite imagery visual representation of energy recorded by remote sensing instruments. These imageries are taken by satellites using various sensors that record electromagnetic energy associated with an environmental phenomenon or feature.

scoping the process involved in environmental impact assessment, by which the important environmental issues, project alternatives and important environmental components are identified by the interested parties in advance of the assessment.

sea level rise the increase in elevation of the sea caused by the Global Warming Phenomenon which results from heat expansion of the ocean waters and melting of the Antarctic Icecap; first recognized by the writer Jules Verne nearly a century ago.

seawall a structure built along the shore to protect the shore by preventing erosion and other damage from wave action; generally more massive and capable of resisting greater wave forces than a bulkhead.

Secchi Disk a disk about 20 cm (8 in) in diameter with a four-part propellor design of alternating black and white triangles painted on its surface; it is lowered with a rope fastened at its center to measure vertical transparency of the water (via the depth of its "disappearance").

secondary treatment waste treatment that follows primary treatment primarily to consume the organic part of the wastes, It accomplished by bringing sewage and bacteria together in "trickling filters" or in "activated sludge processes". Secondary treatment may remove up to 90 percent of BOD by converting organic materials to inorganic materials such as phosphate and nitrate which are then released to the environment.

septic tank an individual household sewage treatment system using underground tanks to receive the waste water; the bacteria in the sewage decomposes the organic waste, the sludge settles on the bottom of the tank, and the effluent flows out of the tank below ground through drainage pipes (sludge must be pumped out of the tanks at intervals.

setback a prescriptive buffer strip or space which is often specified in shoreline management programs, to separate development sites from natural areas or to remove structures inland away from the danger of sea storms ot erosion.

shorelands the dryside of the coastal zone; lowlying areas that are affected by coastal waters through flooding, air borne salt, or other coastal processes.

silt curtain (silt screen) a curtain suspended in the water to prevent silt from escaping from an aquatic construction site.

situation management coastal management programs that focus on particular problems or areas, rather than the whole coastal zone of a country; geo-specific or issue-specic coastal managment programs.

sludge is the solid matter that settles to the bottom of sedimentation tanks in sewage treatment facilities and must be disposed of by digestion or other methods to complete waste treatment.

social impact assessment (SIA) prediction of social effects on from environmental changes caused by any of a variety of economic development types.

spoil bottom material excavated during dredging.

stakeholder a person (or entity) having a vested interest in decisions affecting the use and conservation of coastal resources.

storm sewer a system of drains and culverts or pipes that carries rainwater runoff from urbanized areas to receiving water bodies.

storm surge a rise of sea elevation caused by water piling up against a coast under the force of strong onshore winds such as those accompanying a hurricane or other intense storm; reduced atmospheric pressure may contribute to the rise.

strategy plan the first stage in coastal planning whereby the basic national strategy for ICZM is decided, including analysis of issues, needs, goals, objectives, and equities.
subsidence sinking of the earth surface (downward local mass movement) often caused by excessive groundwater removal or by settling/compacting of fill.
suspended load amount of particulate matter moving in suspension in water.
suspended solids particles suspended in water by hydraulic motion forces—such as upward components of turbulent currents and colloidal suspension—including, e.g., sediment and organic detritus.
sustainable development development that ensures the continuance of natural resource productivity and a high level of environmental quality, thereby providing for economic growth to meet the needs of the present without compromising the needs of future generations; also the continuance of community benefits after external inputs (subsidies) have ceased.
swamp a wetland community characterized by woody vegetation—usually trees and shrubs that in combination rise higher than six meters from grade level.
tambak a brackish water coastal aquaculture pond.
terms of reference (TOR) a statement of the specific work to be done under a consulting agreement or similar contract.
thermocline a sharp vertical temperature gradient in the water whereby the temperature changes rapidly with depth (usually decreases)—occurring within a narrow horizontal layer, it shows on sonar as a sharp discontinuity and has important effects on distribution of life in the ocean.
tideflat an unvegetated intertidal area (usually mud or sand).
tidelands the area of land covered by the ebb and flow of the tide; the area that lies between the higher high water mark and lower low water mark (see "intertidal zone" above).
traditional use the traditional, localized, mechanisms for allocation of marine resources at the tribal or village levelin the absence of central government intervention; the process of sharing and conserving resources ("customary rights").
tsunami a shallow water progressive wave, potentially catastrophic, caused by an underwater earthquake or volcano, that can rise to great heights and catostrophically inundate shorelands.
turbidity reduced water clarity resulting from the presence of suspended matter; also a measure of the amount of material suspended in the water.
turbidity meter A photoelectronic device for measuring turbidity against a standard (usually referenced to "JTUs" or Jackson candle turbidimeter units); a standardized measuring instrument which compares the amount of light penetrating a given water sample with that penetrating a known sample.
upland land areas sufficiently inland from the shoreline so as to have limited interaction with the sea.
watershed the geographically defined region within which all water drains through a particular system of rivers, streams, or other water bodies; watershed are defined by "watershed divides" (high points or ridges on the land) and includes hills, slopes, lowlands, floodplains and receiving body of water.
water table the upper surface of groundwater; that level below which the soil is saturated with water.
wave period time period of the passage of two successive crests (or troughs) of a wave past a specific point.
wave refraction the process by which a wave moving in shallow water at an angle to the bed contours is changed in direction.
wave set-up the elevation of the water surface over normal surge storm elevation due to onshore mass transport of water by wave action alone.
wetlands lowlying areas that are flooded at a sufficient frequency to support vegetation adapted for life in saturated soils, including mangrove swamps, salt marshes, and other wet vegetated areas (often between low water and the yearly normal maximum flood water level); lands transitional between terrestrial and aquatic sytems that have a water table near or above the surface (shallow covering of water), hydric soils, and a prevalence of hydrophytic vegetation.
wetside that part of the coastal zone that is usually, or frequently, flooded with salt water, including wetlands, lagoons, estuaries, beaches, coastal bays, channels, and the near ocean.
zoning a system of designating areas of land or water to be allocated to specific (often exclusive) uses; the division of a particular area into several zones, each of which is scheduled for a particular use or set of uses.

Appendices

APPENDIX 1: COUNTRIES WITH COASTAL MANAGEMENT PROGRAMS

This material, generously provided by Dr. Jens C. Sorensen, University of Rhode Island, appeared previously in the "Intercoast Network: International Newsletter of Coastal Management" (Coastal Resources Center, University of Rhode Island, Narragansett, R.I. 02882) as "The Roster of Integrated Coastal Area Management Programs, Pilot Programs and Feasibility Studies" in nearly this same form—the main change is the addition of Brunei Darussalam. Programs marked with an asterisk (*) are being implemented as of the current year of this list (1993).

A. EXISTING NATIONAL PROGRAMS

Barbados. Coastal Conservation Plan. A 3-year study, 1991–1994. Contact: Leonard Nurse, Coast Conservation Unit, Oistens Government Complex, Oistens Christ Church. Telephone: 1-809-428-5945. Fax: 6023.

Belize. National Coastal Management Program. 1991. Contact: Vincent Gillett, Fisheries Administration, Princess Margret Drive, P.O. Box 148, Belize City. Telephone: 501 2 44-552. Fax: 32-983.

Brazil. National Coastal Zone Management Plan.* Created by national law in 1988. Prepared by the Inter-ministerial Commission for the Resources of the Sea (CIRM). Contact: Renato Herz, Instit. de Oceanographico, Univ. de Sao Paulo. Telephone: 55 11 813-3222. Fax: 210-3092.

British Virgin Islands. Coast Conservation and Development Program. Almost enacted into law in 1991. Contact: Gillian Cambers, Conservation and Fisheries Department, Government of the British Virgin Islands, Tortola. Telephone: 1-809-494-5682. Fax: 495-2100.

Brunei Darussalam. Integrated Coastal Management Plans. 1991. Contact: Haji Matdanan bin Haji Jaafar, Dept. of Fisheries, P.O. Box 2161, Bandar Seri Begawan 1921. Telephone: 673 242-067. Fax: 069.

Bulgaria. Black Sea Coastal Management Program. Interim rules drafted in 1992. The program will be supported by the World Bank as one component of the national water and sanitation program. Contact: Tzveta Kamenova, Regional Development Division, Ministry of Regional Development, 17-19 Cyril and Methodius Stra. 1202, Sophia. Fax: 35 92 87-25-17.

China, People's Republic of. National Project of Comprehensive Investigation in the Mainland Coastal Zone. Initiated in 1980 with an inventory of the coastal zone. A national law is currently being drafted. Currently, regulations are also being drafted for the provinces of Liaoning, Shangdong, Guangdong, and Hainan, as well as for the city of Tianjin. Contact: Chen Degong, Institute for Marine Development Strategy, No. 8. Dahuisi St., Haidian District, Beijing 100081. Telephone: 86 1 831-7542. Fax: 3612.

Costa Rica. The Planning and Management of the Marine and Terrestrial Zone.* Enacted by national law in 1977. Contact: Robert Chaverri, Sinengia 69, Apartado 672-2010 San Pedro, San Jose. Telephone: 506 532-766. Fax: same.

Ecuador. National Coastal Management Program.* Enacted by executive decree in 1989. Contact: Miguel Fierro, Coastal Management Programs Office, Av. Quito y Padre Solano, Edificio del MAG, Piso 20, Casilla 5850, Guayaquil. Telephone: 593 4 284-453. Fax: 285-038.

France. Planning, Protection, and Development of Coastal Space.* Enacted by national !aw in 1986. Contact: Dominique Ligraim, Conservatoire Littoral, 72 Rue Regnault, 75013 Paris. Telephone: 33 1 457-07475.

Guinea Bissau. Coastal Planning Project. Initiated in 1989 by the Forestry Department of the Ministry of Rural Development (MDRA) with technical assistance and support from IUCN and the Swiss Directorate of Development Cooperation and Humanitarian Aid. Contact: Rui Miranda, IUCN-MDRA Coastal Plan Director or Abilio Rachid Said, Director of the Natural Research (INEP) Environmental Department, Bissau, Guinea Bissau.

Israel. Master Plan for the Mediterranean Coast.* First phase enacted in 1981 by government regulation. Contact: Valyrie Brachya, Environmental Planning, Ministry of the Environment. Jerusalem, Israel. Telephone: 972 2 701-606. Fax: 385-038.

Japan. Tentative Guidelines for the Planning concerning the Comprehensive Utilization of the Coastal Zone. Government declaration in 1990. Contact: Yoshimi Nagao, Nihon University, 7-24-1, Narashinodai, Funabashi, 274 Japan. Telephone: 81 3 547-01919.

New Zealand. New Zealand Coastal Policy Statement. Enacted by the Resources Assessment Act of 1991. Contact: Hamish Rennie, Department of Conservation. P.O. Box 10420, Wellington. Telephone: 64 4 471-3066. Fax: 1082.

Oman. Coastal Zone Management Plan.* The first phase of planning was initiated in 1986. Contact: Ali Amer Al-Kiyumi, National Conservation Strategy, Ministry of Environment, P.O. Box 332, Muscat. Telephone: 968 696-458. Fax: 602-320.

Spain. A Program to Define Protect and Regulate the Use of the Shores.* Created by the Shores Act of 1988. Contact: Francisco Montoya, Ministry of Public Works and Urban Planning, Plaza Imperial, Tarraco #4, Tarragona 43071. Telephone: 34 77 216-613. Fax: 230-536.

Sri Lanka. National Coastal Zone Management Plan.* Adopted 1990 by Cabinet. Contact: B. Kahawita, Coast Conservation Department, Maligawatte Secretariat, Colombo, 10. Telephone: 94 1 588-311. Fax: 575-032.

St. Kitts (St. Christopher)—Nevis. South-East Peninsula Development and Land Use Management Plan.* Created by legislation in 1986. Contact: Patrick Williams, South-East Peninsula Development Authority, Government House, St. Kitts. Telephone: 1 809 465-2277. Fax: 1106.

Thailand. National Coral Management Strategy.* Phuket Island Action Plan.* Action Plan for the Upper South Sub-region's Coastal Zone. 1991. Contact: A. Suphapodok, Office of the National Environment Board, 60/1 Soi Phibun Watthana 7, Rama VI Road, Bangkok, 10400. Telephone: 66 2 279-7180. Fax: 0672.

Turkey. Definition, Use and Planning of the Coastal Zones. Enacted by national law in 1990. Also pilot programs in Antalya Province. Contact: Semra Ongan, Resik Belender Sok 91-5, Cankaya, Ankara. Telephone: 90 4 141-0249.

United States. The National Coastal Zone Management Program.* Enacted by national law in 1972. The program currently includes the states of Alabama, Alaska, California, Connecticut, Delaware, Florida, Hawaii, Louisiana, Maine, Maryland, Michigan, Mississippi, New Hampshire, New Jersey, New York, North Carolina, Oregon, Rhode Island, South Carolina, Virginia, Washington, and Wisconsin. Contact: Trudy Coxe, Office of Ocean and Coastal Resource Management, NOAA, 1825 Connecticut Ave. NW, Washington, D.C. 20235. Telephone: 1-202-606-4111. Fax: 4329. Also, the National Estuary Program. The following estuary systems are each in the process of preparing or implementing a Comprehensive Conservation and Management Plan: Albermarle/Pamlico Sounds (North Carolina), Barataria and Terrebonne Bays (Louisiana), Buzzards Bay (Massachusetts), Casco Bay (Maine), Delaware Bay (Delaware and New Jersey), Delaware Inland Bays (Delaware), Galveston Bay (Texas), Indian River Lagoon (Florida), Long Island Sound (New York and Connecticut), Massachusetts Bay, Narragansett Bay (Rhode Island), New York/New Jersey Harbor, Puget Sound (Washington State), San Francisco Bay (California), Santa Monica Bay (California), Sarasota Bay (Florida), and Tampa Bay (Florida). Contact: Robert Wayland, Wetlands, Oceans and Watersheds Office, U.S. Environmental Protection Agency, 401 M Street NW, Washington 20460. Telephone: 1 202 260-7166. Fax: 6294.

B. EXISTING STATE/PROVINCIAL OR REGIONAL PROGRAMS

American Samoa. Coastal Zone Management Program.* Approved in 1980. Contact: Lelei Peau, Development Planning Office, Government of American Samoa, Pago Pago, 96799. Telephone: 684 633-5155. Fax: 4195.

Australia (Great Barrier Reef). The Great Barrier Reef Marine Park.* Enacted by national legislation in 1975. Contact: Graeme Kelleher, Great Barrier Reef Marine Park Authority, GPO Box 791 Canberra, ACT. Telephone: 61 62 47-0211. Fax: 5761.

Australia (New South Wales, State). New South Wales Coastal Policy. 1990. The Coastal Committee of New South Wales. Contact: Bruce Thom, Research, Main Quadrangle, Sydney University, NSW 2006. Telephone: 61 2 692 2273. Fax: 3636.

Australia (Western Australian State). Coastal Planning and Management in Western Australia. A Government Position Paper. 1983. Coastal Management Coordination Committee was formed in 1982. Contact: Ian Eliot, Coastal Management and Co-ordinating Committee, Albert Facey House, 469 Wellington St., Perth 6000 WA. Telephone: 61 9 264-7777. Fax: 321-1617.

Canada (Fraser River Estuary). Fraser River Estuary Management Program.* Created by agreement between the Province of British Columbia and the federal government in 1985. Contact: Mike McPhee, F.R.E.M.P, 708 Clarkson St., New Westminster, B.C. V3M 1E2. Telephone: 604 525-1047. Fax: 3005.

China, People's Republic of (Jiangsu Province). Regulation on Coastal Zone Management in Jiangsu Province. Provincial law enacted in 1991. Contact: Chen Degong (see listing for China above).

Cote d'Ivoire or Ivory Coast (Abidjan Lagoon). Abidjan Environment Project. Contact: Mr. Kakadie, Director of Sanitation, Ministry of Environment. Fax: 225 21-22-71.

Guam. Coastal Zone Management Program. Approved in 1979. Contact: Peter Leon Guerrero, Bureau of Planning, Government of Guam, P.O. Box 2950, Agana 96910. Telephone: 671 472-4201. Fax: 477-1812.

Italy (Venice Lagoon). The Comprehensive Management of the Venice Lagoon Ecosystem.* Contact: Francesco Bandarin, Consorzio Venezia Nuova, San Marco 2803, Venice. Telephone: 39 41 52-93-511. Fax: 89-252.

Kosrae State (Federated States of Micronesia). Kosrae Island Natural Resources Management Plan. 1989. Lou Brooks, Department of Conservation, P.O. Box CD, Kosrae State 96944. Telephone: 691 370-3044. Fax: 2066.

Mexico (Campeche State, Laguna De Terminos). Management Plan for the Laguna de Terminos System. Planning program initiated in 1990. Contact: Alejandro Yáñez-Arancibia, Programa EPOMEX, Universidad de Campeche, Apartado Postal, 520, Campeche 24030. Telephone: 52 981 11600. Fax: 65954.

Mexico (Sian Káan, State of Quintana Roo). Management Program for the Biosphere Reserve.* Created by presidential decree in 1986. Contact: Sebastian Estrella Pool, General Dirección de Parques, Reservas y Areas Protegidas, SEDUE, Rio Elbe #20, District Federal 06500. Telephone: 52 5 286-6231.

Netherlands (Wadden Sea, the Netherlands' component). The Dutch Wadden Sea Management Plan.* Enacted by national legislation in 1981. Contact: Norbert Dankers, Research Institute for Nature, Postbus 167, 1790 AD Den Burg. Telephone: 31 2220-69-700. Fax: 19-235.

Nigeria. Management Program for Lagos Lagoon. Contact: Dr. Aina, Federal Environmental Protection Agency, Phase 2, PMB 12620, Lagos. Telephone 234 1 687-600.

Pohnpei State (Federated States of Micronesia). Pohnpei Island Natural Resources: Proposed Management Plan. 1987. Contact: Director, Conservation and Natural Resources Surveillance, Kolonia 96941. Telephone: 691 320-2735. Fax: 2331.

Saint Kitts (Christopher)—Nevis. South-East Peninsula Development and Land Use Management Plan.* Created by legislation in 1986. Contact: Patrick Willians, South-East Peninsula Development Authority, Government House, St. Kitts. Telephone: 1 809 465-2277. Fax: 1106.

United States of America (Northern Marianas). Coastal Resources Management Program.* Approved in 1980. Contact: Joaquin Villagomez, CRM Office, Nauru Building, Saipan 96950. Telephone: 670-234-6623. Fax: 0007.

United States of America (Puerto Rico). Coastal Zone Management Program.* Approved 1978. Contact: Ines Monefeldt, Coastal Management Office, P.O. Box 5887, Puerta de Tierra 00906. Telephone: 1-809-724-5516. Fax: 722-2785.

United States of America (San Francisco Bay, California). San Francisco Bay Plan.* Enacted by California state legislation in 1965. Contact: Alan Pendleton, S.F.B.C.D.C., 30 Van Ness Ave., Room 2011, San Francisco, Calif. 94102. Telephone: 1-415-557-3686. Fax: 3767.

United States Virgin Islands. Coastal Zone Management Program.* Approved in 1979. Contact: Joan Harrigan-Farrelly, Department of Planning and Natural Resources, Nisky Center, Suite 231, No. 45A Estate Nisky, Charlotte Amalie, St. Thomas 00802. Telephone: 1-809-774-3320. Fax: 5706. Also: Biosphere Reserve Program for St. John Island. Contact: Mark Koenigs, Virgin Islands National Park, St. John, U.S.V.I. 00831. Telephone: 1-809-776-6201. Fax: 4714.

Venezuela (Isla de Margarita, State of Nueva Esparta). Coastal Planning Arrangement for the Region.* Created by presidential decree in 1987. Contact: Ernesto Diaz, Ministry of Natural and Renewable Resources, Torre Sur, Piso 18, Centro Simón Bolívar A.A. 6623, Caracas 1010-A. Telephone: 58 2 408-1230. Fax: 483-2445.

Yap State (Federated States of Micronesia). Yap Island Marine and Coastal Resources Management Plan, Draft. 1991. Marine Resources Division. Contact: John Iou, Marine Resources Division, P.O. 251, Colonia. Telephone: 691 350 2294. Fax: 4113.

C. NATIONAL STATE/PROVINCIAL OR REGIONAL PILOT PROGRAMS OR FEASIBILITY STUDIES

Albania. Coastal Area Management Program for the Durres-Vlore Region. Established in 1993 in conjunction with the Priority Action Program of UNEP's Mediterranean Action Plan. Contact: Lirim Selfo, Chairman, Committee for Environmental Protection, Tirana. Telephone and Fax: 355 42 27907.

Argentina (Buenos Aires Province). Surveys of the Interdisciplinary Commission for the Management of the Coastal Zone. 1985–88. Contact: Aldo Brandani, 4834 El Salvador, 1414 Buenos Aires. Telephone: 54 1 727-061. Fax: 313-7267.

Australia. Resource Assessment Commission: "Coastal Zone Inquiry." Created by national law in 1989. The final report(s) must be submitted by November 25, 1993. Also: Commonwealth of Australia, Department of the Arts, Sport, the Environment and Territories: "A Draft Policy for Commonwealth Responsibilities in the Coastal Zone". In the past 25 years there have been 25 reports in respect to coastal zone management. Contact: Richard Kenchington, Resource Assessment Commission, Canberra. Telephone: 61 6 271-5280. Fax: 5073.

Australia (Northern Territory). Northern Territory Coastal Management, 1982. Contact: Janice Warren, Coastal Management Committee, Conservation Commission, PO Box 496, Palmerston 0831, NT. Telephone: 61 89 894-469. Fax: 657.

Australia (Tasmania, State). Footprints in the Sand; A Proposal and Discussion Paper on Coastal Zone Management Strategies. 1991. Contact: Geoff Tregenza, Coastal Project Team, Department of Environment and Planning, Box 510E, GPO Hobart 7001, Tasmania. Telephone: 61 2 306-737. Fax: 348-730.

Australia (Queensland, State). Coastal Protection Strategy: Green Paper. Proposal for Managing Queenslands Coast. Promulgated in 1991 for review and comment. Contact: Chris Pattearson, Division of Coastal Protection, PO Box 155, North Quay, 4002 Queensland. Telephone: 61 7 227-7111.

Australia (Victoria, State). Coastal Policy for Victoria. 1988. Contact: Chris Lester, Coastal Management and Coordination Committee, Department of Conservation and Environment, 7th Floor, P.O. Box 41, East Melbourne 3002, Victoria. Telephone: 61 3 412-4987. Fax: 4160.

Bangladesh. The feasibility study on the preparation of an Integrated Coastal Area Management Program. 1991. Contact: Zaman Mozumder, Planning Commission, Shere Bangla, Nagar, Dhaka. Telephone: 880 2 314-724.

Brunei Darussalam. Integrated Coastal Management Plan. 1991. This project emanated from the memorandum of understanding signed in 1986 between the Association of South East Asian Nations and USAID. The project for Brunei Darussalam was initiated in 1987. Contact: Haji Matdanan bin Haji Jaafar, Department of Fisheries, PO Box 2161, Bandar Seri Begawan 1921. Telephone: 673 242-067. Fax: 242-069.

Canada. Atlantic Coastal Action Program. Thirteen demonstration projects in the four maritime provinces (New Brunswick, New foundland and Labrador, Nova Scotia, and Prince Edward Island). Created as a component of the National Green Plan in 1991. Contact: Larry Hildebrand, Environment Canada, 4th floor, 45 Aiderney Drive, Dartmouth, Nova Scotia. Telephone: 1-902-426-9632. Fax: 2690.

Canada (Prince Edward Island, Province). Coastal Zone Planning Initiative. Initiated by a Memorandum of Understanding on Conservation and Development between the Province and the federal government in 1987. Contact: Christopher Leach, Department of Community and Cultural Affairs, P.O. Box 2000, Charlottetown, P.E.I. CIA 7N8. Telephone: 902-368-4882. Fax: 5544.

Colombia. The Development and Administration of the Colombian Coastal Zone in the Caribbean Sea. 1984. Produced by the Centro de Investigaciones Oceangraficas e Hidrograficas. Contact: Rafael Steer, Comision de Oceanografia, Apartado Aero 28466, Bogota. Telephone: 57 1 222-0449. Fax: 0416.

Cote d'Ivoire (Ivory Coast). Interministry Working Group on Coastal Protection Plan. 1984. Contact: Jacques Abe, Center for Oceanographic Research, 29 Fisheries Street, BP, v 18, Abidjan. Telephone: 225 214-235. Fax: 351-155.

Croatia. Coastal Area Management Program for Kastela Bay. Established in 1988 in conjunction with the Priority Actions Program of UNEP's Mediterranean Action Plan. Contact: Ivica Trumbic, Priority Actions Program Regional Activity Center, UNEP, Kraj. sv. Ivana 11, 58000 Split, Croatia. Telephone: 38 58 43-499. Fax: 361-677.

Dominica. Demonstration project for the Environment and Coastal Resources (ENCOR) Project. Cooperative agreement signed by USAID and the Organization of Eastern Caribbean States in 1992. Contact: Ophelia Marie, ENCORE National Coordinator, Ministry of Agriculture, Roseau. Telephone: 809 448-4577. Fax: 448-7999.

Egypt. Coastal Area Management Program for the Fuka Matrouh region. Established in 1993 in conjunction with the Priority Actions Program of UNEP's Mediterranean Action Plan. Contact: M. A. Fawzi, Project Coordinator, Egyptian Environmental Affairs Agency, 11-A Hassan Sabry Street, Zamalek, Cairo. Telephone: 20 2 341-1323. Fax: 342-0768. Also: Ras Mohammed National Park. Created by national law in 1983. Contact: Omar Hassan, Egyptian Environmental Affairs Agency.

El Salvador. The Environmental Protection Project of El Salvador; South Central Coast Component (PROMESA). Initiated by presidential decree in 1991. Contact: The Executive Secretary, SEMA of CONAMA, Ministry of Agriculture, 83 Avenida Norte y 11 Calle Poniente #704, Escalaon, El Salvador. Telephone: 503 791-579. Fax: 242-944.

Greece. Island of Rhodes. Coastal Area Management Program for the Island of Rhodes. Established in 1988 in conjunction with the Priority Actions Program of the UNEP's Mediterranean Action Plan. Contact: Dimitrios Tsotsos, PERPA, Ministry of the Environment, Physical Planning and Public Works, Patission 147, Athens. Telephone: 30 1 865-0106. Fax: 864-7420.

Honduras. Environmental Plan for the Tourism Development of Islas de Bahias. Established in 1991 by presidential decree. Contact: Nico Agurcia, Comisión pro Desarrollo Turistico de Las Islas Bahias, Tegucigalpa. Telephone: 504 32-7376. Fax: 7362.

Indonesia. Integrated Management Plan for Segara Anakan-Cilacap, Central Java, 1991. This project emanated from the memorandum of understanding signed in 1986 between the Association of South East Asian Nations and USAID. Also, Action Plan for the Sustainable Development of Indonesia's Marine and Coastal Resources. 1988. Contact: Sukotjo Adisukresno, Directorate General of Fisheries, 6th Floor B Building, Ministry of Agriculture, Jalan Harsono RM No. 3, Ragunan-Pasar Mingu, Jakarta Selatan. Telephone: 62 21 790-4026. Fax: 80-3196.

Ireland. Bord Failte Eireann and An Faras Forbartha, National Coastline Study. 1972. Contact: Dorothy Corr, Librarian, Environmental Research Unit, Department of the Environment, St. Martins House, Dublin. Telephone: 353 1 660-2511. Fax: 668-0009.

Malaysia. South Johore Coastal Resources Plan. 1991. Contact: Ching Kim Looi, ASEAN Secretariat for COST, Ministry of Science, Technology and the Environment, 14th Floor Wisma Sims Darby, Jalan Raja Laut, 50662, Kuala Lumpur. Telephone: 60 3 293-8955. Fax: 6006.

Maldives. Report on the Mission to the Republic of Maldives in Respect to Coastal Management. 1985. Contact: Alex Dawson Shepard, Ministry of Planning, Ghaz Building 2005, Male. Telephone: 960 324-861. Fax: 327-351.

Philippines. Integrated Coastal Management Plan for Lingayen Gulf. Contact: Leonardo Quitos, NEDA-Region 1, Don Pedro Building, Pagdaraoan, San Fernando, La Union. Telephone: 63 2 412-679. Fax: 413-066.

Portugal (Ria de Aveiro). Natural Resource Management Plan for the Estuary of the Aveiro River. Interdisciplinary and intergovernmental study group created in 1988. Contact: Carlos Borrego, Gabinete da Ria de Aveiro, University de Aveiro, Rua San Jose, Rabumba, No. 3, 3800 Aveiro. Telephone: 351 34 25085.

Saint Lucia. Demonstration project for the Environment and Coastal Resources (ENCOR) Project. Cooperative agreement signed by USAID and the Organization of Eastern Caribbean States in 1992. Contact: Crispin D'Auvergne, ENCORE National Coordinator, Ministry of Planning, New Government Buildings, Castries. Telephone: 809 451-6957. Fax: 451-6958.

Saudi Arabia. An Assessment of Management for the Saudi Arabian Red Sea Coastal Zone. 1987. Contact: Khaled Ibraheem Al Mubarek, Ministry of Agriculture and Water, Riyadh 11195. Telephone: 966 1 402-0268. Fax: same.

Singapore. Strategies for Urban Coastal Area Management. 1991. Contact: Chia Lin Sien, Department of Geography, National University of Singapore, 10 Kent Ridge Crescent, Singapore 05111. Telephone: 65 775-6666. Fax: 777-3091.

South Africa. A Policy for Coastal Zone Management in the Republic of South Africa. 1989. Also: Coastal Action Strategy (COAST). 1991. Contact: Bruce Glavovic, Department of Environment Affairs, Bag X2, Roggebay 0012. Telephone: 27 21 403-3035. Fax: 252-920.

Syria. Coastal Area Management Programme for the Syrian Coastal Region. Established in 1988 in conjunction with the Priority Actions Programme of UNEP's Mediterranean Action Plan. Contact: Yahia Owidah, General Commission for Environmental Affairs, Tolyani St., Damascus. Telephone: 963 11 215-426. Telex: 412-686 ENV SY.

Taiwan. Coastal Protection Plan. 1984. Adopted by executive authority. Also, the Ministry of Interior in co-operation with other units of government is drafting a coastal zone management law. Contact: Dah-Win Sheih, Department of Fisheries, 37 Nan Hai Road, Taipei 100. Telephone: 886 2 312 4601. Fax: 331 6408.

Thailand. Phuket Island Action Plan.* The preparation of the Phuket Island Plan followed the signing of a cooperative agreement in 1986 by USAID, the Government of Thailand, and the University of Rhode Island. Action Plan for the Upper South Sub-region's Coastal Zone. 1991. This project emanated from the memorandum of understanding signed in 1986 between the Association of South East Asian Nations and USAID. Contact: A. Suphapodok, Office of the National Environment Board, 60/1 Soi Phibun Watthana 7, Rama VI Road, Bangkok, 10400. Telephone: 66 2 279-7180. Fax: 0672.

Tunisia. Coastal Area Management Program for the City of Sfax. Established in 1993 in conjunction with the Priority Actions Program of UNEP's Mediterranean Action Plan. Contact: Mohammed Ennabli, Director General, Agence Nationale de Protection de l'Environment, 12 Avenue Kheireddine Pacha, 1002 Tunis, BP 52. Telephone: 216 1 799-201. Fax: 789-844.

Turkey. Coastal Area Management Program for the Bay of Izmir. Established in 1988 in conjunction with the Priority Actions Program of UNEP's Mediterranean Action Plan. Contact: Muran Talu, Undersecretariat of the Environment, Ataturk Bulvari 143, Ankara. Telephone: 90 41 174-455. Fax: 117-7971.

Turks and Caicos (British West Indies). Ecological Survey and Environmental Policies for the National Physical Development Plan. 1990. Contact: Oswald Williams, Department of Planning, Heritage and Parks, Grand Turk Island. Telephone: 1 809 946-2220. Fax: 2448.

United Kingdom. A Future for the Coast? Proposals for a U.K. Coastal Zone Management Plan. 1990. Contact: Susan Gubbay, Marine Conservation Society, 9 Gloucester Road, Ross-on-Wye, HR9 5BU. Telephone: 44 989 66-017. Fax: 68-715.

APPENDIX 2: CONVERSION DATA

Table A.2.1 Conversion of Basic Measures Used in Hydrology

Multiply	By	To obtain
Acre feet	43.560	Cubic feet
	325.851	Gallons
	1,233.49	Cubic meters
Meter/second	0.3281	Feet/second
	0.36	Kilometers/hour
	0.2237	Miles/hours
Cubic feet	1,728	Cubic inches
	0.02832	Cubic meters
	0.03704	Cubic yards
	7.48052	Gallons
	28.32	Liters
Cubic inches	16.39	Cubic centimeters
	5.787×10^{-4}	Cubic feet
	4.329×10^{-3}	Gallons
Cubic meters	35.31	Cubic feet
	264.2	Gallons
	10^3	Liters
Cubic yards	0.7646	Cubic meters
Gallons U.S.	0.1337	Cubic feet
	231	Cubic inches
	3.785	Liters
	0.83267	Imperial gallons
Gallons, water	8.3453	Pounds of water
Liters	0.03531	Cubic feet
	61.02	Cubic inches
	0.2642	Gallons
Ounces (fluid)	1.805	Cubic inches
	0.02957	Liters

From Reference 18.

Appendix A.2.2

CONVERSION: BRITISH AND METRIC UNITS OF MEASUREMENT

Conversion factors for basic British and metric units.
Note: Figures in bold type are exact measurements.

	Multiply number of	by	to obtain equivalent number of	Multiply number of	by	to obtain equivalent number of
Length	Inches (in)	**25·4**	millimetres (mm)	Millimetres	0·03937	inches
		2·54	centimetres (cm)	Centimetres	0·3937	inches
	Feet (ft)	**30·48**	centimetres	Metres	39·3701	inches
		0·3048	metres (m)		3·2808	feet
	Yards (yd)	**0·9144**	metres		1·0936	yards
	Fathoms (6ft)	**1·8288**	metres		0·54681	fathoms
	Miles (land: 5,280 ft)	**1·609344**	kilometres (km)	Kilometres	0·62137	miles (land)
	Miles (UK sea: 6,080 ft)	**1·853184**	kilometres (km)		0·53961	miles (UK sea)
	Miles, international nautical	**1·852**	kilometres (km)		0·53996	miles, international nautical
Area	Sq. inches (in²)	**645·16**	sq. millimetres (mm²)	Sq. millimetres	0·00155	sq. inches
		6·4516	sq. centimetres (cm²)	Sq. centimetres	0·1550	sq. inches
	Sq. feet (ft²)	**929·0304**	sq. centimetres	Sq. metres	10·7639	sq. feet
		0·092903	sq. metres (m²)		1·19599	sq. yards
	Sq. yards (yd²)	0·836127	sq. metres	Hectares	2·47105	acres
	Acres	4,046·86	sq. metres	Sq. kilometres	247·105	acres
		0·404686	hectares (ha)		0·3861	sq. miles
		0·004047	sq. kilometres (km²)			
	Sq. miles	2·58999	sq. kilometres			
Volume and capacity	Cu. inches (in³)	**16·387064**	cu. centimetres (cm³)	Cu. centimetres	0·06102	cu. inches
	UK pints	34·6774	cu. inches	Litres	61·024	cu. inches
	UK pints	0·568	litres (l)		0·0353	cu. feet
	UK gallons	4·546	litres (l)		0·2642	US gallons
	US gallons	3·785	litres (l)		0·2200	UK gallons
	Cu. feet (ft³)	28·317	litres (l)	Hectolitres	26·417	US gallons
	Cu. feet	0·028317	cu. metres (m³)		21·997	UK gallons
	UK bushels	0·3637	hectolitres (hl)		2·838	US bushels
	US bushels	0·3524	hectolitres (hl)		2·750	UK bushels
	UK gallons	1·20095	US gallons	Cu. metres	35·3147	cu. feet
	US gallons	0·832674	UK gallons		1·30795	cu. yards
	UK bulk barrels	36	UK gallons		264·172	US gallons
		43·2342	US gallons		219·969	UK gallons
		0·1637	cu. metres		6·11025	UK bulk barrels
Weight (mass)	Grains (gr)	**64·79891**	milligrams (mg)	Milligrams	0·01543	grains
	Ounces, avoirdupois (oz)	28·3495	grams (g)	Grams	0·03527	ounces, avoirdupois
	Ounces, troy (oz tr)	31·1035	grams (g)		0·03215	ounces, troy
	Ounces, avoirdupois	0·9115	ounces, troy	Kilograms	2·20462	pounds, avoirdupois
	Pounds, avoirdupois (lb)	**453·59237**	grams	Metric quintals	220·462	pounds, avoirdupois
		0·45359	kilograms (kg)	Tonnes	2,204·62	pounds, avoirdupois
	Hundredweights (cwt) (112 lb)	0·05	long tons		1·10231	short tons
		0·508023	metric quintals (q)		0·984207	long tons
	Short tons (2,000 lb)	0·892857	long tons			
		0·907185	tonnes (t)			
	Long tons (2,240 lb)	1·12	short tons			
		1·01605	tonnes			

Appendix A.2.3

CONVERSION: SALINITY UNITS

Conversion of several methods of expressing salinity. Read across. For example, a specific conductance of 8.6 millimhos per centimeter is equivalent to 16 percent sea strength.

SOURCE. Soil Conservation Service, *Louisiana Gulf Coast Marsh Handbook*, U.S. Department of Agriculture, Alexandria, Louisiana, November 1966.

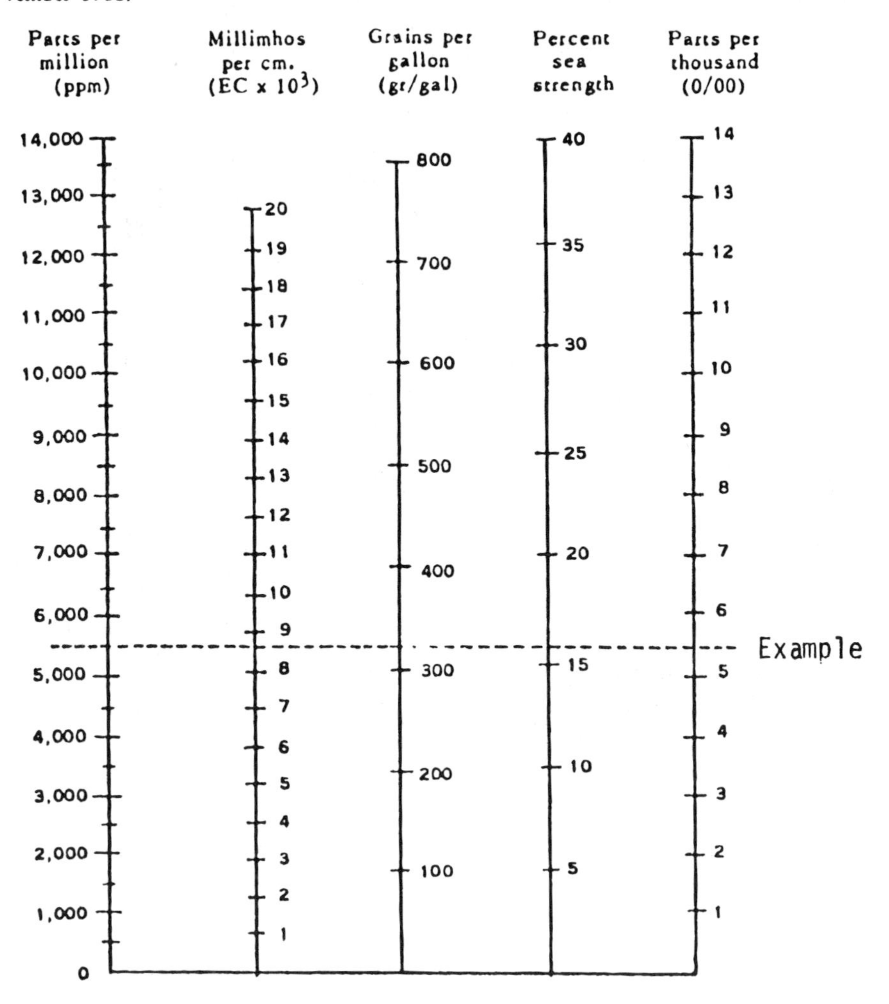

Appendix A.2.4

AERIAL PHOTOGRAPH INTERPRETATION

Scale conversions for vertical aerial photographs.

SOURCE. T. Eugene Avery, *Forester's Guide to Aerial Photo Interpretation*, U.S. Department of Agriculture, Forest Service, Washington, D.C., October 1966.

Representative fraction (scale)	Feet per inch	Chains per inch	Inches per mile	Acres per square inch	Square miles per square inch
(1)	(2)	(3)	(4)	(5)	(6)
1:7,920	660.00	10.00	8.00	10.00	0.0156
1:8,000	666.67	10.10	7.92	10.20	.0159
1:8,400	700.00	10.61	7.54	11.25	.0176
1:9,000	750.00	11.36	7.04	12.91	.0202
1:9,600	800.00	12.12	6.60	14.69	.0230
1:10,000	833.33	12.63	6.34	15.94	.0249
1:10,800	900.00	13.64	5.87	18.60	.0291
1:12,000	1,000.00	15.15	5.28	22.96	.0359
1:13,200	1,100.00	16.67	4.80	27.78	.0434
1:14,400	1,200.00	18.18	4.40	33.06	.0517
1:15,000	1,250.00	18.94	4.22	35.87	.0560
1:15,600	1,300.00	19.70	4.06	38.80	.0606
1:15,840	1,320.00	20.00	4.00	40.00	.0625
1:16,000	1,333.33	20.20	3.96	40.81	.0638
1:16,800	1,400.00	21.21	3.77	45.00	.0703
1:18,000	1,500.00	22.73	3.52	51.65	.0807
1:19,200	1,600.00	24.24	3.30	58.77	.0918
1:20,000	1,666.67	25.25	3.17	63.77	.0996
1:20,400	1,700.00	25.76	3.11	66.34	.1037
1:21,120	1,760.00	26.67	3.00	71.11	.1111
1:21,600	1,800.00	27.27	2.93	74.38	.1162
1:22,800	1,900.00	28.79	2.78	82.87	.1295
1:24,000	2,000.00	30.30	2.64	91.83	.1435
1:25,000	2,083.33	31.57	2.53	99.64	.1557
1:31,680	2,640.00	40.00	2.00	160.00	.2500
Method of calculation	$\frac{RFD}{12}$	$\frac{RFD}{792}$	$\frac{63,360}{RFD}$	$\frac{(RFD)^2}{6,272,640}$	$\frac{\text{Acres/sq. in.}}{640}$

[1] Conversions for scales not shown can be made from the relationships listed at the bottom of each column. With the scale of 1:7,920 as an example (column 1, line 1), the number of feet per inch is computed by dividing the representative fraction denominator (RFD) by 12 (number of inches per foot). Thus, 7,920 ÷ 12 = 660 feet per inch (column 2). By dividing the RFD by 792 (inches per chain), the number of chains per inch is derived (column 3). Other calculations can be made similarly. Under column 4, the figure 63,360 represents the number of inches in one mile; in column 5, the figure 6,272,640 is the number of square inches in one acre; and in column 6, the number 640 is acres per square mile.

Index